PRINCIPLES OF
FOUNDATION ENGINEERING

PRINCIPLES OF
FOUNDATION ENGINEERING

Fourth Edition

BRAJA M. DAS
California State University, Sacramento

PWS PUBLISHING
An Imprint of Brooks/Cole Publishing Company
I(T)P® An International Thomson Publishing Company

Pacific Grove · Albany · Belmont · Bonn · Boston · Cincinnati · Detroit
Johannesburg · London · Madrid · Melbourne · Mexico City · New York
Paris · Singapore · Tokyo · Toronto · Washington

Sponsoring Editor: *Suzanne Jeans*
Marketing Representative: *Nathan Wilbur*
Marketing Manager: *Nathan Wilbur*
Production Editor: *Janet Hill*
Book Production/Interior Design: *Rockwell Production Services*
Manufacturing Buyer: *Vena Dyer*

Cover Design: *Roger Knox*
Cover Photo: *Hayward Baker, Inc.*
Composition: *The PRD Group*
Cover Printing: *Phoenix Color Corporation*
Printing and Binding: *R. R. Donnelley & Sons Company*

For more information, contact PWS Publishing at Brooks/Cole Publishing Company

BROOKS/COLE PUBLISHING COMPANY
511 Forest Lodge Road
Pacific Grove, CA 93950
USA

International Thomson Publishing Europe
Berkshire House 168-173
High Holborn
London WC1V 7AA
England

Thomas Nelson Australia
102 Dodds Street
South Melbourne, 3205
Victoria, Australia

Nelson Canada
1120 Birchmont Road
Scarborough, Ontario
Canada M1K 5G4

International Thomson Editores
Seneca 53
Col. Polanco
11560 Mexico D.F., México

International Thomson Publishing GmbH
Königswinterer Strasse 418
53227 Bonn
Germany

International Thomson Publishing Asia
60 Albert Street #15-01
Albert Complex
Singapore 189969

International Thomson Publishing Japan
Hirakawacho Kyowa Building, 3F
2-2-1 Hirakawacho
Chiyoda-ku, Tokyo 102
Japan

Printed in the United States of America

10 9 8 7 6 5 4 3 2

Library of Congress Cataloging-in-Publication Data

Das, Braja M., 1941-
    Principles of foundation engineering / Braja M. Das, — 4th ed.
        p. cm.
    Includes bibliographical references and index.
    ISBN 0-534-95403-0 (alk. paper)
    1. Foundations. I. Title.
TA775.D227 1998
624.1'5—dc21   98-36450

In the memory of my father,
and to Janice and Valerie

CONTENTS

FOUR

Shallow Foundations:
Allowable Bearing Capacity and Settlement   219

FIVE   # Mat Foundations   293

SIX   # Lateral Earth Pressure   334

NINE   Pile Foundations   564

TEN Drilled-Shaft and Caisson Foundations **674**

ELEVEN Foundations on Difficult Soils **728**

PREFACE

The first edition of *Principles of Foundation Engineering,* published in 1984, was intended for use as a text by undergraduate civil engineering students. The text was well received by students and practicing geotechnical engineers. Their encouragement and comments helped to develop the second edition in 1990, the third edition in 1995, and finally this fourth edition of the text.

The core of the original text has not changed greatly. There has been some rearrangement of the contents in this edition compared to the third edition. It now contains 12 chapters and 5 appendices. A brief overview of the changes follows:

- Chapter 1 has been renamed "Geotechnical Properties of Soil and Soil Reinforcement." A brief overview of the soil reinforcement materials such as galvanized metallic strips, geotextile, and geogrid has been added to this chapter.
- Details of the pressuremeter test and flat plate dilatometer test have been added to Chapter 2.
- Chapter 3 now contains only the theories and applications of the ultimate bearing capacity of shallow foundations. Relationships for the ultimate bearing capacity of foundations on the top of a slope and recent developments on the ultimate bearing capacity of shallow foundations in soils reinforced with geotextile and geogrid are now included in this chapter.
- The allowable bearing capacity of shallow foundations including those on reinforced earth, which are based on allowable settlement criteria, are now presented in Chapter 4.
- The material presented in Chapter 5 of the third edition is now given in Chapters 6 and 7. Chapter 6 contains lateral earth pressure theories, and Chapter 7 contains the design principles of retaining walls including mechanically stabilized earth walls.
- The materials on sheet pile walls and braced cuts presented in Chapters 6 and 7, respectively, of the third edition are now combined into one chapter (Chapter 8) under the theme of sheet pile structures.

Because the text introduces civil engineering students to the fundamental concepts and application of foundation analysis and design, the mathematical derivations of some equations are not presented; instead, just the final forms of those equations are given. However, a list of references for further information and study is included in every chapter.

In the preparation of the text, where applicable, multiple theories and empirical correlations have been presented. The reason for presenting different theories and correlations is to acquaint the readers with the chronological development of the subject and the answers obtained from theories based on different assumptions. It also convinces the readers that the soil parameters obtained from different empirical correlations will not always be the same, with some being better than others for a given soil condition. This is primarily because soil profiles are seldom homogeneous, elastic, and isotropic. The judgment needed to properly apply the theories, equations, and graphs to the evaluation of soils and foundation design cannot be overemphasized or completely taught by any textbook. Field experience must supplement classroom work.

Acknowledgments

Thanks are due to my wife, Janice, who was the main driving force in the completion of this edition. She typed the revisions and completed the original graphs and figures.

I wish to acknowledge the late Paul C. Hassler, formerly of the University of Texas at El Paso, Gerald R. Seeley of Valparaiso University, Ronald B. McPherson of New Mexico State University, and Said Larbi-Cherif, formerly of Navarro and Associates, El Paso, Texas, for their help, support, and encouragement during the preparation of the first edition of the manuscript. In addition, I want to thank Ronald P. Anderson of Tensar Earth Technologies, Inc., and Henry Ng, consulting engineer, El Paso, Texas, for their help in the development of the fourth edition.

I also wish to acknowledge the following reviewers whose helpful comments have helped immensely:

- For the second edition:

M. Sherif Aggour
University of Maryland

Yong S. Chae
Rutgers-The State University

Samuel Clemence
Syracuse University

George Gezetas
SUNY at Buffalo

- For the third edition:

A. G. Altschaeffl
Purdue University

Jeffrey C. Evans
Bucknell University

Thomas F. Zimmie
Rensselaer Polytechnic Institute

Norman D. Dennis, Jr.
United States Military Academy

Steven Perkins
Montana State University

- For this edition:

Yong S. Chae
Rutgers—The State University

Manjriker Gunaratne
University of South Florida

William F. Kane
University of the Pacific

Anil Misra
University of Missouri

Rodrigo Salgado
Purdue University

Colby C. Swan
University of Iowa

Kenneth McManis
University of New Orleans

John P. Turner
University of Wyoming

I extend thanks also the staff of PWS Publishing Company for their interest and patience during the preparation and production of this text.

As a final note, I welcome suggestions from students and instructors for use in further editions.

Braja M. Das
Sacramento, California

CHAPTER ONE

GEOTECHNICAL PROPERTIES OF SOIL AND OF REINFORCED SOIL

1.1 INTRODUCTION

The design of foundations of structures such as buildings, bridges, and dams generally requires a knowledge of such factors as (a) the load that will be transmitted by the superstructure to the foundation system, (b) the requirements of the local building code, (c) the behavior and stress-related deformability of soils that will support the foundation system, and (d) the geological conditions of the soil under consideration. To a foundation engineer, the last two factors are extremely important because they concern soil mechanics.

The geotechnical properties of a soil — such as the grain-size distribution, plasticity, compressibility, and shear strength — can be assessed by proper laboratory testing. And, recently, emphasis has been placed on *in situ* determination of strength and deformation properties of soil, because this process avoids the sample disturbances that occur during field exploration. However, under certain circumstances, all of the needed parameters cannot be determined or are not determined because of economic or other reasons. In such cases, the engineer must make certain assumptions regarding the properties of the soil. To assess the accuracy of soil parameters — whether they were determined in the laboratory and the field or were assumed — the engineer must have a good grasp of the basic principles of soil mechanics. At the same time, he or she must realize that the natural soil deposits on which foundations are constructed are not homogeneous in most cases. Thus the engineer must have a thorough understanding of the geology of the area — that is, the origin and nature of soil stratification and also the groundwater conditions. Foundation engineering is a clever combination of soil mechanics, engineering geology, and proper judgment derived from past experience. To a certain extent, it may be called an "art."

When determining which foundation is the most economical, the engineer must consider the superstructure load, the subsoil conditions, and the desired tolerable settlement. In general, foundations of buildings and bridges may be divided into two major categories: (1) *shallow foundations* and (2) *deep foundations. Spread foot-*

ings, wall footings, and *mat foundations* are all shallow foundations. In most shallow foundations, the *depth of embedment can be equal to or less than three to four times the width of the foundation. Pile* and *drilled shaft* foundations are deep foundations. They are used when top layers have poor load-bearing capacity and when use of shallow foundations will cause considerable structural damage and/or instability. The problems relating to shallow foundations and mat foundations are considered in Chapters 3, 4, and 5. Chapter 9 discusses pile foundations, and Chapter 10 examines drilled shafts.

More recently, soil reinforcement has been used in the construction and design of foundations, retaining walls, embankment slopes, and other structures. Depending on the type of construction, the reinforcements may be galvanized metal strips, geotextiles, geogrids, and geocomposites. Use of soil reinforcement in the design of shallow foundations is introduced in Chapters 3 and 4. Chapter 7 gives an outline of the principles of soil reinforcement in the design of retaining walls. Some of the physical properties of soil reinforcement are discussed in the latter part of this chapter.

This chapter serves primarily as a review of the basic geotechnical properties of soils. It includes topics such as grain-size distribution, plasticity, soil classification, effective stress, consolidation, and shear strength parameters. It is based on the assumption that you have already been exposed to these concepts in a basic soil mechanics course.

1.2   GRAIN-SIZE DISTRIBUTION

In any soil mass, the sizes of various soil grains vary greatly. To classify a soil properly, you must know its *grain-size distribution.* The grain-size distribution of *coarse-grained* soil is generally determined by means of *sieve analysis.* For a *fine-grained* soil, the grain-size distribution can be obtained by means of *hydrometer analysis.* The fundamental features of these analyses are presented in this section. For detailed descriptions, see any soil mechanics laboratory manual (for example, Das, 1997).

Sieve Analysis

A sieve analysis is conducted by taking a measured amount of dry, well-pulverized soil and passing it through a stack of progressively finer sieves with a pan at the bottom. The amount of soil retained on each sieve is measured, and the cumulative percentage of soil passing through each sieve is determined. This percentage is generally referred to as *percent finer.* Table 1.1 contains a list of U.S. sieve numbers and the corresponding size of their hole openings. These sieves are commonly used for the analysis of soil for classification purposes.

The percent finer for each sieve determined by a sieve analysis is plotted on *semilogarithmic graph paper,* as shown in Figure 1.1. Note that the grain diameter, *D,* is plotted on the *logarithmic scale,* and the percent finer is plotted on the *arithmetic scale.*

Two parameters can be determined from the grain-size distribution curves of

▼ **TABLE 1.1** U.S. Standard Sieve Sizes

Sieve no.	Opening (mm)
4	4.750
6	3.350
8	2.360
10	2.000
16	1.180
20	0.850
30	0.600
40	0.425
50	0.300
60	0.250
80	0.180
100	0.150
140	0.106
170	0.088
200	0.075
270	0.053

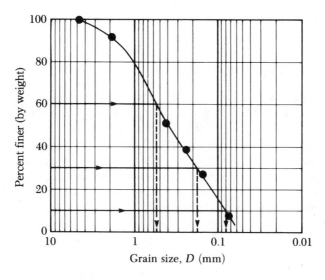

▼ **FIGURE 1.1**   Grain-size distribution curve of a coarse-
grained soil obtained from sieve analysis

coarse-grained soils: (1) the *uniformity coefficient* (C_u) and (2) the *coefficient of gradation,* or *coefficient of curvature* (C_z). These coefficients are

$$C_u = \frac{D_{60}}{D_{10}}$$

(1.1)

$$C_z = \frac{D_{30}^2}{(D_{60})(D_{10})}$$

(1.2)

where D_{10}, D_{30}, and D_{60} are the diameters corresponding to percents finer than 10, 30, and 60%, respectively

For the grain-size distribution curve shown in Figure 1.1, $D_{10} = 0.08$ mm, $D_{30} = 0.17$ mm, and $D_{60} = 0.57$ mm. Thus the values of C_u and C_z are

$$C_u = \frac{0.57}{0.08} = 7.13$$

$$C_z = \frac{0.17^2}{(0.57)(0.08)} = 0.63$$

Parameters C_u and C_z are used in the *Unified Soil Classification System,* which is described later in this chapter.

Hydrometer Analysis

Hydrometer analysis is based on the principle of sedimentation of soil particles in water. This test involves the use of 50 grams of dry, pulverized soil. A *deflocculating* agent is always added to the soil. The most common deflocculating agent used for hydrometer analysis is 125 cc of 4% solution of sodium hexametaphosphate. The soil is allowed to soak for at least 16 hours in the deflocculating agent. After the soaking period, distilled water is added, and the soil-deflocculating agent mixture is thoroughly agitated. The sample is then transferred to a 1000-ml glass cylinder. More distilled water is added to the cylinder to fill it to the 1000-ml mark, and then the mixture is again thoroughly agitated. A hydrometer is placed in the cylinder to measure — usually over a 24-hour period — the specific gravity of the soil–water suspension in the vicinity of its bulb (Figure 1.2). Hydrometers are calibrated to show the amount of soil that is still in suspension at any given time, t. The largest diameter of the soil particles still in suspension at time t can be determined by Stokes' law:

$$D = \sqrt{\frac{18\eta}{(G_s - 1)\gamma_w}} \sqrt{\frac{L}{t}}$$

(1.3)

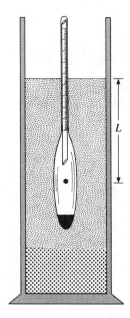

▼ **FIGURE 1.2** Hydrometer analysis

where D = diameter of the soil particle
G_s = specific gravity of soil solids
η = viscosity of water
γ_w = unit weight of water
L = effective length (that is, length measured from the water surface in the cylinder to the center of gravity of the hydrometer; see Figure 1.2)
t = time

Soil particles having diameters larger than those calculated by Eq. (1.3) would have settled beyond the zone of measurement. In this manner, with hydrometer readings taken at various times, the soil *percent finer* than a given diameter D can be calculated, and a grain-size distribution plot can be prepared. The sieve and hydrometer techniques may be combined for a soil having both coarse-grained and fine-grained soil constituents.

1.3. SIZE LIMITS FOR SOILS

Several organizations have attempted to develop the size limits for *gravel, sand, silt,* and *clay* based on the grain sizes present in soils. Table 1.2 presents the size limits recommended by the American Association of State Highway and Transportation Officials (AASHTO) and the Unified Soil Classification systems (Corps of Engineers, Department of the Army, and Bureau of Reclamation). Table 1.2 shows that soil particles smaller than 0.002 mm have been classified as *clay*. However, clays by nature are cohesive and can be rolled into a thread when moist. This property is

▼   **TABLE 1.2** Soil-Separate Size Limits

Classification system	Grain size (mm)
Unified	Gravel: 75 mm to 4.75 mm Sand: 4.75 mm to 0.075 mm Silt and clay (fines): <0.075 mm
AASHTO	Gravel: 75 mm to 2 mm Sand: 2 mm to 0.05 mm Silt: 0.05 mm to 0.002 mm Clay: <0.002 mm

caused by the presence of *clay minerals* such as *kaolinite, illite,* and *montmorillonite.* In contrast, some minerals such as *quartz* and *feldspar* may be present in a soil in particle sizes as small as clay minerals. But these particles will not have the cohesive property of clay minerals. Hence they are called *clay-size particles,* not *clay particles.*

1.4   WEIGHT–VOLUME RELATIONSHIPS

In nature, soils are three-phase systems consisting of solid soil particles, water, and air (or gas). To develop the *weight–volume relationships* for a soil, the three phases can be separated as shown in Figure 1.3a. Based on this separation, the volume relationships can be defined in the following manner.

Void ratio, e, is the ratio of the volume of voids to the volume of soil solids in a given soil mass, or

$$e = \frac{V_v}{V_s} \tag{1.4}$$

where V_v = volume of voids
   V_s = volume of soil solids

Porosity, n, is the ratio of the volume of voids to the volume of the soil specimen, or

$$n = \frac{V_v}{V} \tag{1.5}$$

where V = total volume of soil

Moreover,

$$n = \frac{V_v}{V} = \frac{V_v}{V_s + V_v} = \frac{\dfrac{V_v}{V_s}}{\dfrac{V_s}{V_s} + \dfrac{V_v}{V_s}} = \frac{e}{1 + e} \tag{1.6}$$

Degree of saturation, S, is the ratio of the volume of water in the void spaces to the volume of voids, generally expressed as a percentage, or

Note: $V_a + V_w + V_s = V$
$W_w + W_s = W$

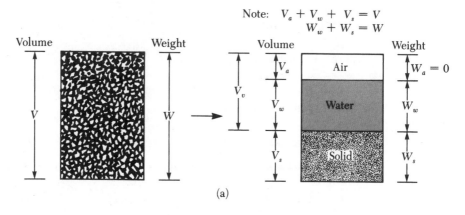

(a)

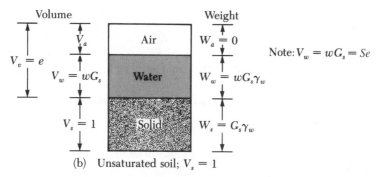

(b) Unsaturated soil; $V_s = 1$

Note: $V_w = wG_s = Se$

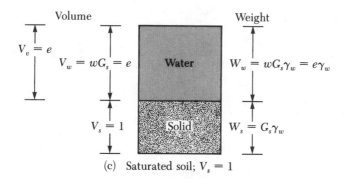

(c) Saturated soil; $V_s = 1$

▼ **FIGURE 1.3** Weight–volume relationships

$$S(\%) = \frac{V_w}{V_v} \times 100 \qquad\qquad (1.7)$$

where V_w = volume of water

Note that, for saturated soils, the degree of saturation is 100%.

The weight relationships are *moisture content, moist unit weight, dry unit weight,* and *saturated unit weight.* They can be defined as follows:

$$\text{Moisture content} = w(\%) = \frac{W_w}{W_s} \times 100 \tag{1.8}$$

where W_s = weight of the soil solids
           W_w = weight of water

$$\text{Moist unit weight} = \gamma = \frac{W}{V} \tag{1.9}$$

where W = total weight of the soil specimen = $W_s + W_w$

The weight of air, W_a, in the soil mass is assumed to be negligible.

$$\text{Dry unit weight} = \gamma_d = \frac{W_s}{V} \tag{1.10}$$

When a soil mass is completely saturated (that is, all the void volume is occupied by water), the moist unit weight of a soil [Eq. (1.9)] becomes equal to the *saturated unit weight* (γ_{sat}). So $\gamma = \gamma_{\text{sat}}$ if $V_v = V_w$.

More useful relations can now be developed by considering a representative soil specimen in which the volume of soil solids is equal to *unity*, as shown in Figure 1.3b. Note that if $V_s = 1$, from Eq. (1.4), $V_v = e$ and the weight of the soil solids is

$$W_w = G_s\gamma_w$$

where G_s = specific gravity of soil solids
           γ_w = unit weight of water (9.81 kN/m³, or 62.4 lb/ft³)

Also, from Eq. (1.8), the weight of water $W_w = wW_s$. Thus, for the soil specimen under consideration, $W_w = wW_s = wG_s\gamma_w$. Now, for the general relation for moist unit weight given in Eq. (1.9),

$$\gamma = \frac{W}{V} = \frac{W_s + W_w}{V_s + V_v} = \frac{G_s\gamma_w(1 + w)}{1 + e} \tag{1.11}$$

Similarly, the dry unit weight [Eq. (1.10)] is

$$\gamma_d = \frac{W_s}{V} = \frac{W_s}{V_s + V_v} = \frac{G_s\gamma_w}{1 + e} \tag{1.12}$$

From Eqs. (1.11) and (1.12), note that

$$\gamma_d = \frac{\gamma}{1 + w} \tag{1.13}$$

If a soil specimen is completely saturated as shown in Figure 1.3c,

$$V_v = e$$

Also, for this case

$$V_v = \frac{W_w}{\gamma_w} = \frac{wG_s\gamma_w}{\gamma_w} = wG_s$$

Thus

$$e = wG_s \qquad \text{(for saturated soil } only) \tag{1.14}$$

The saturated unit weight of soil becomes

$$\gamma_{\text{sat}} = \frac{W_s + W_w}{V_s + V_v} = \frac{G_s\gamma_w + e\gamma_w}{1 + e} \tag{1.15}$$

Relationships similar to Eqs. (1.11), (1.12), and (1.15) in terms of porosity can also be obtained by considering a representative soil specimen with a unit volume. These relationships are

$$\gamma = G_s\gamma_w(1 - n)(1 + w) \tag{1.16}$$

$$\gamma_d = (1 - n)G_s\gamma_w \tag{1.17}$$

$$\gamma_{\text{sat}} = [(1 - n)G_s + n]\gamma_w \tag{1.18}$$

Table 1.3 provides several useful relationships for γ, γ_d, and γ_{sat}.

Except for peat and highly organic soils, the general range of the values of specific gravity of soil solids (G_s) found in nature is rather small. Table 1.4 gives some representative values. For practical purposes, a reasonable value can be assumed in lieu of running a test.

Table 1.5 presents some representative values for the void ratio, dry unit weight, and moisture content (in a saturated state) of some naturally occurring soils. Note that in most cohesionless soils the void ratio varies from about 0.4 to 0.8. The dry unit weights in these soils generally fall within a range of about 90–120 lb/ft^3 (14–19 kN/m^3).

1.5 RELATIVE DENSITY

In *granular soils,* the degree of compaction in the field can be measured according to *relative density, D_r,* which is defined as

$$D_r(\%) = \frac{e_{\max} - e}{e_{\max} - e_{\min}} \times 100 \tag{1.19}$$

where $e_{\max}$ = void ratio of the soil in the loosest state
           $e_{\min}$ = void ratio in the densest state
           e = *in situ* void ratio

Moist unit weight, γ	
Given	Relationship
w, G_s, e	$\dfrac{(1 + w) G_s \gamma_w}{1 + e}$
S, G_s, e	$\dfrac{(G_s + Se) \gamma_w}{1 + e}$
w, G_s, S	$\dfrac{(1 + w) G_s \gamma_w}{1 + \dfrac{wG_s}{S}}$
w, G_s, n	$G_s \gamma_w (1 - n)(1 + w)$
S, G_s, n	$G_s \gamma_w (1 - n) + nS\gamma_w$

Dry unit weight (γ_d)	
γ, w	$\dfrac{\gamma}{1 + w}$
G_s, e	$\dfrac{G_s \gamma_w}{1 + e}$
G_s, n	$G_s \gamma_w (1 - n)$
G_s, w, S	$\dfrac{G_s \gamma_w}{1 + \left(\dfrac{wG_s}{S}\right)}$
e, w, S	$\dfrac{eS\gamma_w}{(1 + e) w}$
γ_{sat}, e	$\gamma_{sat} - \dfrac{e\gamma_w}{1 + e}$
γ_{sat}, n	$\gamma_{sat} - n\gamma_w$
γ_{sat}, G_s	$\dfrac{(\gamma_{sat} - \gamma_w) G_s}{(G_s - 1)}$

Saturated unit weight (γ_{sat})	
G_s, e	$\dfrac{(G_s + e) \gamma_w}{1 + e}$
G_s, n	$[(1 - n) G_s + n]\gamma_w$
G_s, w_{sat}	$\left(\dfrac{1 + w_{sat}}{1 + w_{sat} G_s}\right) G_s \gamma_w$
e, w_{sat}	$\left(\dfrac{e}{w_{sat}}\right)\left(\dfrac{1 + w_{sat}}{1 + e}\right) \gamma_w$
n, w_{sat}	$n\left(\dfrac{1 + w_{sat}}{w_{sat}}\right) \gamma_w$
γ_d, e	$\gamma_d + \left(\dfrac{e}{1 + e}\right) \gamma_w$
γ_d, n	$\gamma_d + n\gamma_w$
γ_d, S	$\left(1 - \dfrac{1}{G_s}\right)\gamma_d + \gamma_w$
γ_d, w_{sat}	$\gamma_d(1 + w_{sat})$

▼ **TABLE 1.4** Specific Gravities of Some Soils

Soil type	G_s
Quartz sand	2.64–2.66
Silt	2.67–2.73
Clay	2.70–2.9
Chalk	2.60–2.75
Loess	2.65–2.73
Peat	1.30–1.9

▼ **TABLE 1.5** Typical Void Ratio, Moisture Content, and Dry Unit Weight for Some Soils

Type of soil	Void ratio e	Natural moisture content in saturated condition (%)	Dry unit weight, γ_d (lb/ft³)	(kN/m³)
Loose uniform sand	0.8	30	92	14.5
Dense uniform sand	0.45	16	115	18
Loose angular-grained silty sand	0.65	25	102	16
Dense angular-grained silty sand	0.4	15	120	19
Stiff clay	0.6	21	108	17
Soft clay	0.9–1.4	30–50	73–92	11.5–14.5
Loess	0.9	25	86	13.5
Soft organic clay	2.5–3.2	90–120	38–51	6–8
Glacial till	0.3	10	134	21

The values of e_{max} are determined in the laboratory in accordance with the test procedures outlined in the American Society for Testing and Materials, *ASTM Standards* (1997, Test Designation D-4254).

The relative density can also be expressed in terms of dry unit weight, or

$$D_r(\%) = \left\{ \frac{\gamma_d - \gamma_{d(min)}}{\gamma_{d(max)} - \gamma_{d(min)}} \right\} \frac{\gamma_{d(max)}}{\gamma_d} \times 100 \tag{1.20}$$

where γ_d = *in situ* dry unit weight

$\gamma_{d(max)}$ = dry unit weight in the *densest* state; that is, when the void ratio is e_{min}

$\gamma_{d(min)}$ = dry unit weight in the *loosest* state, that is, when the void ratio is e_{max}

The denseness of a granular soil is sometimes related to its relative density. Table 1.6 gives a general correlation of the denseness and D_r. For naturally occurring sands, the magnitudes of e_{max} and e_{min} [Eq. (1.19)] may vary widely. The main reasons for such wide variations are the uniformity coefficient, C_u, and the roundness of the particles, R. The uniformity coefficient is defined in Eq. (1.1). *Roundness* is defined as

▼ **TABLE 1.6**   Denseness of a Granular Soil

Relative density, D_r(%)	Description
0–20	Very loose
20–40	Loose
40–60	Medium
60–80	Dense
80–100	Very dense

$$R = \frac{\text{minimum radius of the particle edges}}{\text{inscribed radius of the entire particle}} \qquad (1.21)$$

Measuring R is difficult, but it can be estimated. Figure 1.4 shows the general range of the magnitude of R with particle roundness. Figure 1.5 shows the variation of e_{max} and e_{min} with the uniformity coefficient for various values of particle roundness (Youd, 1973). This range is applicable to clean sand with normal to moderately skewed particle-size distribution.

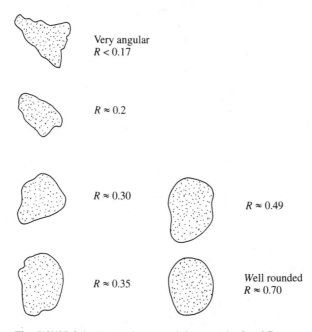

Very angular
$R < 0.17$

$R \approx 0.2$

$R \approx 0.30$

$R \approx 0.49$

$R \approx 0.35$

Well rounded
$R \approx 0.70$

▼ **FIGURE 1.4**   General range of the magnitude of R

▼ **EXAMPLE 1.1** _____

A 0.25 ft³ moist soil weighs 30.8 lb. When dried in an oven, this soil weighs 28.2 lb. Given $G_s = 2.7$. Determine the

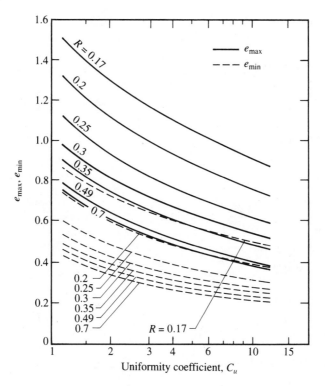

▼ FIGURE 1.5 Approximate variation of e_{max} and e_{min} with uniformity coefficient (based on Youd, 1973)

a. Moist unit weight, γ
b. Moisture content, w
c. Dry unit weight, γ_d
d. Void ratio, e
e. Porosity, n
f. Degree of saturation, S

Solution

Part a. Moist Unit Weight
From Eq. (1.9),

$$\gamma = \frac{W}{V} = \frac{30.8}{0.25} = \mathbf{123.2\ lb/ft^3}$$

Part b. Moisture Content
From Eq. (1.8),

$$w = \frac{W_w}{W_s} = \frac{30.8 - 28.2}{28.2} \times 100 = \mathbf{9.2\%}$$

Part c. Dry Unit Weight

From Eq. (1.10),

$$\gamma_d = \frac{W_s}{V} = \frac{28.2}{0.25} = \textbf{112.8 lb/ft}^3$$

Part d. Void Ratio

From Eq. (1.4),

$$e = \frac{V_v}{V_s}$$

$$V_s = \frac{W_s}{G_s \gamma_w} = \frac{28.2}{(2.67)(62.4)} = 0.169 \text{ ft}^3$$

$$e = \frac{0.25 - 0.169}{0.169} = \textbf{0.479}$$

Part e. Porosity

From Eq. (1.6),

$$n = \frac{e}{1 + e} = \frac{0.479}{1 + 0.479} = \textbf{0.324}$$

Part f. Degree of Saturation

From Eq. (1.7),

$$S(\%) = \frac{V_w}{V_v} \times 100$$

$$V_v = V - V_s = 0.25 - 0.169 = 0.081 \text{ ft}^3$$

$$V_w = \frac{W_w}{\gamma_w} = \frac{30.8 - 28.2}{62.4} = 0.042 \text{ ft}^3$$

$$S = \frac{0.042}{0.081} \times 100 = \textbf{51.9\%}$$

▲

▼ **EXAMPLE 1.2**_____

A soil has a void ratio of 0.72, moisture content = 12%, and G_s = 2.72. Determine the

   a. Dry unit weight (kN/m³)
   b. Moist unit weight (kN/m³)
   c. Weight of water in kN/m³ to be added to make the soil saturated

Solution

Part a. Dry Unit Weight

From Eq. (1.12),

$$\gamma_d = \frac{G_s\gamma_w}{1+e} = \frac{(2.72)(9.81)}{1+0.72} = \textbf{15.51 kN/m}^3$$

Part b. Moist Unit Weight

From Eq. (1.11),

$$\gamma = \frac{G_s\gamma_w(1+w)}{1+e} = \frac{(2.72)(9.81)(1+0.12)}{1+0.72} = \textbf{17.38 kN/m}^3$$

Part c. Mass of Water to be Added

From Eq. (1.15),

$$\gamma_{sat} = \frac{(G_s+e)\gamma_w}{1+e} = \frac{(2.72+0.72)(9.81)}{1+0.72} = 19.62\ \text{kN/m}^3$$

Water to be added $= \gamma_{sat} - \gamma = 19.62 - 17.38 = \textbf{2.24 kN/m}^3$ ▲

▼ **EXAMPLE 1.3**

The maximum and minimum dry unit weights of a sand are 17.1 kN/m³ and 14.2 kN/m³, respectively. The sand in the field has a relative density of 70% with a moisture content of 8%. Determine the moist unit weight of the sand in the field.

Solution   From Eq. (1.20),

$$D_r = \left[\frac{\gamma_d - \gamma_{d(min)}}{\gamma_{d(max)} - \gamma_{d(min)}}\right]\left[\frac{\gamma_{d(max)}}{\gamma_d}\right]$$

$$0.7 = \left[\frac{\gamma_d - 14.2}{17.1 - 14.2}\right]\left[\frac{17.1}{\gamma_d}\right]$$

$$\gamma_d = 16.11\ \text{kN/m}^3$$

$$\gamma = \gamma_d(1+w) = 16.11\left(1 + \frac{8}{100}\right) = \textbf{17.4 kN/m}^3$$ ▲

1.6 ATTERBERG LIMITS

When a clayey soil is mixed with an excessive amount of water, it may flow like a *semiliquid.* If the soil is gradually dried, it will behave like a *plastic, semisolid,* or *solid* material depending on its moisture content. The moisture content, in percent, at which the soil changes from a liquid to a plastic state is defined as the *liquid*

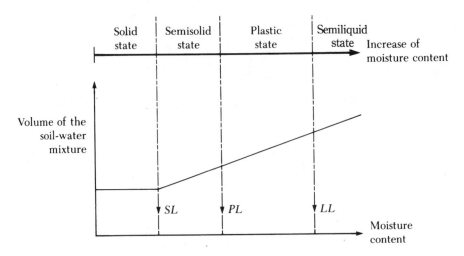

▼ **FIGURE 1.6**   Definition of Atterberg limits

limit (LL). Similarly, the moisture contents, in percent, at which the soil changes from a plastic to a semisolid state and from a semisolid to a solid state are defined as the *plastic limit (PL)* and the *shrinkage limit (SL)*, respectively. These limits are referred to as *Atterberg limits* (Figure 1.6).

▶ The liquid limit of a soil is determined by Casagrande's liquid device (ASTM Test Designation D-4318) and is defined as the moisture content at which a groove closure of $\frac{1}{2}$ in. (12.7 mm) occurs at 25 blows.

▶ The *plastic limit* is defined as the moisture content at which the soil crumbles when rolled into a thread of $\frac{1}{8}$ in. (3.18 mm) in diameter (ASTM Test Designation D-4318).

▶ The *shrinkage limit* is defined as the moisture content at which the soil does not undergo further volume change with loss of moisture (ASTM Test Designation D-427). Figure 1.6 shows this limit.

The difference between the liquid limit and the plastic limit of a soil is defined as the *plasticity index (PI)*, or

$$PI = LL - PL \qquad (1.22)$$

Table 1.7 gives some representative values of liquid limit and plastic limit for several clay minerals and soils. However, Atterberg limits for various soils will vary considerably, depending on the soil's origin and the nature and amount of clay minerals in it.

▼ **TABLE 1.7** Typical Liquid and Plastic Limits for Some Clay Minerals and Soils

Description	Liquid limit	Plastic limit
Kaolinite	35–100	25–35
Illite	50–100	30–60
Montmorillonite	100–800	50–100
Boston Blue clay	40	20
Chicago clay	60	20
Louisiana clay	75	25
London clay	66	27
Cambridge clay	39	21
Montana clay	52	18
Mississippi gumbo	95	32
Loessial soils in north and northwest China	25–35	15–20

1.7 SOIL CLASSIFICATION SYSTEMS

Soil classification systems divide soils into groups and subgroups based on common engineering properties such as *grain-size distribution, liquid limit,* and *plastic limit.* The two major classification systems presently in use are (1) the *AASHTO* (*American Association of State Highway and Transportation Officials*) *System* and (2) the *Unified Soil Classification System* (also *ASTM*). The AASHTO classification system is used mainly for classification of highway subgrades. It is not used in foundation construction.

AASHTO System

The AASHTO Soil Classification System was originally proposed by the Highway Research Board's Committee on Classification of Materials for Subgrades and Granular Type Roads (1945). According to the present form of this system, soils can be classified according to eight major groups, A-1 through A-8, based on their grain-size distribution, liquid limit, and plasticity indices. Soils listed in groups A-1, A-2, and A-3 are coarse-grained materials, and those in groups A-4, A-5, A-6, and A-7 are fine-grained materials. Peat, muck, and other highly organic soils are classified under A-8. They are identified by visual inspection.

The AASHTO classification system (for soils A-1 through A-7) is presented in Table 1.8. Note that group A-7 includes two types of soil. For the A-7-5 type, the plasticity index of the soil is less than or equal to the liquid limit minus 30. For the A-7-6 type, the plasticity index is greater than the liquid limit minus 30.

For qualitative evaluation of the desirability of a soil as a highway subgrade material, a number referred to as the *group index* has also been developed. The higher the value of the group index for a given soil, the weaker will be the soil's performance as a subgrade. A group index of 20 or more indicates a very poor subgrade material. The formula for group index, *GI*, is

▼   **TABLE 1.8**   AASHTO Soil Classification System

General classification	Granular materials (35% or less of total sample passing no. 200 sieve)						
	A-1		A-3	*A-2*			
Group classification	A-1-a	A-1-b	A-3	A-2-4	A-2-5	A-2-6	A-2-7
Sieve analysis (% passing)							
No. 10 sieve	50 max						
No. 40 sieve	30 max	50 max	51 min				
No. 200 sieve	15 max	25 max	10 max	35 max	35 max	35 max	35 max
For fraction passing No. 40 sieve							
Liquid limit (*LL*)				40 max	41 min	40 max	41 min
Plasticity index (*PI*)		6 max	Nonplastic	10 max	10 max	11 min	11 min
Usual type of material	Stone fragments, gravel, and sand		Fine sand	Silty or clayey gravel and sand			
Subgrade rating	Excellent to good						

General classification	Silt-clay materials (More than 35% of total sample passing no. 200 sieve)			
Group classification	A-4	A-5	A-6	A-7 A-7-5[a] A-7-6[b]
Sieve analysis (% passing)				
No. 10 sieve				
No. 40 sieve				
No. 200 sieve	36 min	36 min	36 min	36 min
For fraction passing No. 40 sieve				
Liquid limit (*LL*)	40 max	41 min	40 max	41 min
Plasticity index (*PI*)	10 max	10 max	11 min	11 min
Usual types of material	Mostly silty soils		Mostly clayey soils	
Subgrade rating	Fair to poor			

[a] If $PI \le LL - 30$, it is A-7-5.
[b] If $PI > LL - 30$, it is A-7-6.

$$GI = (F_{200} - 35)\,[0.2 + 0.005\,(LL - 40)] + 0.01\,(F_{200} - 15)\,(PI - 10) \qquad (1.23)$$

where F_{200} = percent passing no. 200 sieve, expressed as a whole number
       LL = liquid limit
       PI = plasticity index

When calculating the group index for a soil belonging to groups A-2-6 or A-2-7, use only the partial group-index equation relating to the plasticity index:

$$GI = 0.01\,(F_{200} - 15)\,(PI - 10) \qquad (1.24)$$

The group index is rounded to the nearest whole number and written next to the soil group in parentheses; for example,

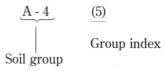

A - 4   (5)

Soil group   Group index

Unified System

The Unified Soil Classification System was originally proposed by A. Casagrande in 1942 and was later revised and adopted by the United States Bureau of Reclamation and the Corps of Engineers. This system is presently used in practically all geotechnical work.

In the Unified System, the following symbols are used for identification.

Symbol	G	S	M	C	O	Pt	H	L	W	P
Description	Gravel	Sand	Silt	Clay	Organic silts and clay	Peat and highly organic soils	High plasticity	Low plasticity	Well graded	Poorly graded

The plasticity chart (Figure 1.7) and Table 1.9 show the procedure for determining the group symbols for various types of soil. When classifying a soil be sure to provide the group name that generally describes the soil, along with the group

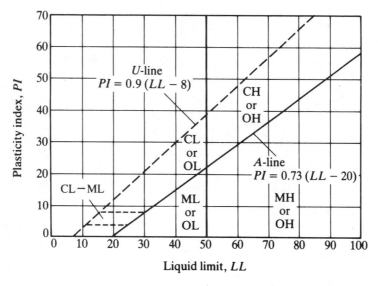

▼ **FIGURE 1.7** Plasticity chart

▼ **TABLE 1.9** Group Symbols for Soil According to the Unified Classification System [Based on Material Passing 3-in. (75-mm) Sieve]

Major division	Criteria	Group symbol
Coarse-grained soil $R_{200} > 50$	$F_{200} < 5$, $C_u \geq 4$, $1 \leq C_z \leq 3$	GW
	$F_{200} < 5$, $C_u < 4$, and/or C_z not between 1 and 3	GP
Gravelly soil $R_4 > 0.5R_{200}$	$F_{200} > 12$, $PI < 4$, or Atterberg limits plot below A line (Figure 1.7)	GM
	$F_{200} > 12$, $PI > 7$, and Atterberg limits plot on or above A line (Figure 1.7)	GC
	$F_{200} > 12$, $LL < 50$, $4 \leq PI \leq 7$, and Atterberg limits plot on or above A line	GC-GM[a]
	$5 \leq F_{200} \leq 12$; meets the gradation criteria of GW and the plasticity criteria of GM	GW-GM[a]
	$5 \leq F_{200} \leq 12$; meets the gradation criteria of GW and the plasticity criteria of GC	GW-GC[a]
	$5 \leq F_{200} \leq 12$; meets the gradation criteria of GP and the plasticity criteria of GM	GP-GM[a]
	$5 \leq F_{200} \leq 12$; meets the gradation criteria of GP and the plasticity criteria of GC	GP-GC[a]
Sandy soil $R_4 \leq 0.5R_{200}$	$F_{200} < 5$, $C_u \geq 6$, $1 \leq C_z \leq 3$	SW
	$F_{200} < 5$, $C_u < 6$, and/or C_z not between 1 and 3	SP
	$F_{200} > 12$, $PI < 4$, or Atterberg limits plot below A line (Figure 1.7)	SM
	$F_{200} > 12$, $PI > 7$, and Atterberg limits plot on or above A line (Figure 1.7)	SC
	$F_{200} > 12$, $LL > 50$, $4 \leq PI \leq 7$, and Atterberg limits plot on or above A line (Figure 1.7)	SC-SM[a]
	$5 \leq F_{200} \leq 12$; meets the gradation criteria of SW and the plasticity criteria of SM	SW-SM[a]
	$5 \leq F_{200} \leq 12$; meets the gradation criteria of SW and the plasticity criteria of SC	SW-SC[a]
	$5 \leq F_{200} \leq 12$; meets the gradation criteria of SP and the plasticity criteria of SM	SP-SM[a]
	$5 \leq F_{200} \leq 12$; meets the gradation criteria of SP and the plasticity criteria of SC	SP-SC[a]
Fine-grained soil (inorganic), $R_{200} \leq 50$ Silty and clayey soil $LL < 50$	$PI < 4$, or Atterberg limits plot below A line (Figure 1.7)	ML
	$PI > 7$, and Atterberg limits plot on or above A line (Figure 1.7)	CL
	$4 \leq PI \leq 7$, and Atterberg limits plot above A line (Figure 1.7)	CL-ML[a]
Silty and clayey soil $LL \geq 50$	Atterberg limits plot below A line (Figure 1.7)	MH
	Atterberg limits plot on or above A line (Figure 1.7)	CH
Fine-grained soil (organic) Organic silt and clay $LL < 50$	$\dfrac{LL_{\text{not oven dry}}}{LL_{\text{oven dry}}} < 0.75$	OL
Organic silt and clay $LL \geq 50$	$\dfrac{LL_{\text{not oven dry}}}{LL_{\text{oven dry}}} < 0.75$	OH

Note: F_{200} = percent finer than no. 200 sieve; R_{200} = percent retained on no. 200 sieve; R_4 = percent retained on no. 4 sieve; C_u = uniformity coefficient; C_z = coefficient of gradation; LL = liquid limit; PI = plasticity index; Atterberg limits based on minus no. 40 fraction
[a] Borderline case; dual classification.

symbol. Tables 1.10, 1.11, and 1.12, respectively, give the criteria for obtaining the group names for coarse-grained soil, inorganic fine-grained soil, and organic fine-grained soil. These tables are based on ASTM Designation D-2487.

▼ EXAMPLE 1.4

Classify the following soil by the AASHTO classification system:

> Percent passing no. 4 sieve = 82
> Percent passing no. 10 sieve = 71
> Percent passing no. 40 sieve = 64
> Percent passing no. 200 sieve = 41
> Liquid limit = 31
> Plasticity index = 12

Solution Refer to Table 1.8. More than 35% passes through a no. 200 sieve, so it is a silt-clay material. It could be A-4, A-5, A-6, or A-7. Because $LL = 31$ (that is, less than 40) and $PI = 12$ (that is, greater than 11), this soil falls in group A-6. From Eq. (1.23),

$$GI = (F_{200} - 35)[0.02 + 0.005(LL - 40)] + 0.01(F_{200} - 15)(PI - 10)$$

So

$$GI = (41 - 35)[0.02 + 0.005(31 - 40)] + 0.01(41 - 15)(12 - 10)$$
$$= 0.37 \approx 0$$

Thus the soil is **A-6(0).** ▲

▼ EXAMPLE 1.5

Classify the following soil by the AASHTO classification system.

> Percent passing no. 4 sieve = 92
> Percent passing no. 10 sieve = 87
> Percent passing no. 40 sieve = 65
> Percent passing no. 200 sieve = 30
> Liquid limit = 22
> Plasticity index = 8

Solution Table 1.8 shows that it is a granular material because less than 35% is passing a no. 200 sieve. With $LL = 22$ (that is, less than 40) and $PI = 8$ (that is, less than 10), the soil falls in group A-2-4. From Eq. (1.24),

$$GI = 0.01(F_{200} - 15)(PI - 10) = 0.01(30 - 15)(8 - 10)$$
$$= -0.3 \approx 0$$

The soil is **A-2-4(0).** ▲

▼ **TABLE 1.10** Group Names for Coarse-Grained Soils (Based on ASTM D-2487)

Group symbol	Criteria		Group name
	Gravel fraction (%)	Sand fraction (%)	
GW		<15	Well-graded gravel
		≥15	Well-graded gravel with sand
GP		<15	Poorly graded gravel
		≥15	Poorly graded gravel with sand
GM		<15	Silty gravel
		≥15	Silty gravel with sand
GC		<15	Clayey gravel
		≥15	Clayey gravel with sand
GC-GM		<15	Silty clayey gravel
		≥15	Silty clayey gravel with sand
GW-GM		<15	Well-graded gravel with silt
		≥15	Well-graded gravel with silt and sand
GW-GC		<15	Well-graded gravel with clay
		≥15	Well-graded gravel with clay and sand
GP-GM		<15	Poorly graded gravel with silt
		≥15	Poorly graded gravel with silt and sand
GP-GC		<15	Poorly graded gravel with clay
		≥15	Poorly graded gravel with clay and sand
SW	<15		Well-graded sand
	≥15		Well-graded sand with gravel
SP	<15		Poorly graded sand
	≥15		Poorly graded sand with gravel
SM	<15		Silty sand
	≥15		Silty sand with gravel
SC	<15		Clayey sand
	≥15		Clayey sand with gravel
SM-SC	<15		Silty clayey sand
	≥15		Silty clayey sand with gravel
SW-SM	<15		Well-graded sand with silt
	≥15		Well-graded sand with silt and gravel
SW-SC	<15		Well-graded sand with clay
	≥15		Well-graded sand with clay and gravel
SP-SM	<15		Poorly graded sand with silt
	≥15		Poorly graded sand with silt and gravel
SP-SC	<15		Poorly graded sand with clay
	≥15		Poorly graded sand with clay and gravel

Note: Sand fraction = percent of soil passing no. 4 sieve but retained on no. 200 sieve = $R_{200} - R_4$; gravel fraction = percent of soil passing 3-in. sieve but retained on no. 4 sieve = R_4.

▼ **TABLE 1.11** Group Names for Inorganic Fine-Grained Soils (Based on ASTM D-2487)

Group symbol	R_{200}	Criteria — Sand fraction — Gravel fraction	Gravel fraction	Sand fraction	Group name
CL	<15				Lean clay
	15 to 29	≥1			Lean clay with sand
		<1			Lean clay with gravel
	≥30	≥1	<15		Sandy lean clay
		≥1	≥15		Sandy lean clay with gravel
		<1		<15	Gravelly lean clay
		<1		≥15	Gravelly lean clay with sand
ML	<15				Silt
	15 to 29	≥1			Silt with sand
		<1			Silt with gravel
	≥30	≥1	<15		Sandy silt
		≥1	≥15		Sandy silt with gravel
		<1		<15	Gravelly silt
		<1		≥15	Gravelly silt with sand
CL-ML	<15				Silty clay
	15 to 29	≥1			Silty clay with sand
		<1			Silty clay with gravel
	≥30	≥1	<15		Sandy silty clay
		≥1	≥15		Sandy silty clay with gravel
		<1		<15	Gravelly silty clay
		<1		≥15	Gravelly silty clay with sand
CH	<15				Fat clay
	15 to 29	≥1			Fat clay with sand
		<1			Fat clay with gravel
	≥30	≥1	<15		Sandy fat clay
		≥1	≥15		Sandy fat clay with gravel
		<1		<15	Gravelly fat clay
		<1		≥15	Gravelly fat clay with sand
MH	<15				Elastic silt
	15 to 29	≥1			Elastic silt with sand
		<1			Elastic silt with gravel
	≥30	≥1	<15		Sandy elastic silt
		≥1	≥15		Sandy elastic silt with gravel
		<1		<15	Gravelly elastic silt
		<1		≥15	Gravelly elastic silt with sand

Note: R_{200} = percent of soil retained on no. 200 sieve; sand fraction = percent of soil passing no. 4 sieve but retained on no. 200 sieve = $R_{200} - R_4$; gravel fraction = percent of soil passing 3-in. sieve but retained on no. 4 sieve = R_4

▼ **TABLE 1.12** Group Names for Organic Fine-Grained Soils (Based on ASTM D-2487)

Group symbol	Plasticity	R_{200}	Sand fraction / Gravel fraction	Gravel fraction	Sand fraction	Group name
OL	$PI \geq 4$, and	<15				Organic clay
	Atterberg	15 to 29	≥1			Organic clay with sand
	limits on		<1			Organic clay with gravel
	or above	≥30	≥1	<15		Sandy organic clay
	A line		≥1	≥15		Sandy organic clay with gravel
			<1		<15	Gravelly organic clay
			<1		≥15	Gravelly organic clay with sand
	$PI < 4$, and	<15				Organic silt
	Atterberg	15 to 29	≥1			Organic silt with sand
	limits plot		<1			Organic silt with gravel
	below A line	≥30	≥1	<15		Sandy organic silt
			≥1	≥15		Sandy organic silt with gravel
			<1		<15	Gravelly organic silt
			<1		≥15	Gravelly organic silt with sand
OH	Atterberg	<15				Organic clay
	limits plot	15 to 29	≥1			Organic clay with sand
	on or above		<1			Organic clay with gravel
	A line	≥30	≥1	<15		Sandy organic clay
			≥1	≥15		Sandy organic clay with gravel
			<1		<15	Gravelly organic clay
			<1		≥15	Gravelly organic clay with sand
	Atterberg	<15				Organic silt
	limits plot	15 to 29	≥1			Organic silt with sand
	below A line		<1			Organic silt with gravel
		≥30	≥1	<15		Sandy organic silt
			≥1	≥15		Sandy organic silt with gravel
			<1		<15	Gravelly organic silt
			<1		≥15	Gravelly organic silt with sand

Note: R_{200} = percent of soil retained on no. 200 sieve; sand fraction = percent of soil passing no. 4 sieve but retained on no. 200 sieve = $R_{200} - R_4$; gravel fraction = percent of soil passing 3-in. sieve but retained on no. 4 sieve = R_4

▼ **EXAMPLE 1.6**

Classify the soil described in Example 1.5 according to the Unified Soil Classification System.

Solution For $F_{200} = 30$,

$$R_{200} = 100 - F_{200} = 100 - 30 = 70$$

As $R_{200} > 50$, it is a coarse-grained soil.

$$R_4 = 100 - \text{percent passing no. 4 sieve}$$
$$= 100 - 92 = 8$$

As $R_4 = 8 < 0.5R_{200} = 35$, it is a sandy soil. Now, refer to Table 1.9. Because F_{200} is greater than 12, the group symbol would be SM or SC. As the *PI* is greater than 7 and the Atterberg limits plot above the *A* line in Figure 1.7, it is *SC*.

For the group name, refer to Table 1.10. The gravel fraction is less than 15%, so the group name is **clayey sand.** ▲

1.8 HYDRAULIC CONDUCTIVITY OF SOIL

The void spaces or pores between soil grains allow water to flow through them. In soil mechanics and foundation engineering, you must know how much water is flowing through a soil in unit time. This knowledge is required to design earth dams, determine the quantity of seepage under hydraulic structures, and dewater before and during the construction of foundations. Darcy (1856) proposed the following equation (Figure 1.8) for calculating the velocity of flow of water through a soil.

$$v = ki \qquad (1.25)$$

where v = Darcy velocity (unit: cm/sec)
k = hydraulic conductivity of soil (unit: cm/sec)
i = hydraulic gradient

The hydraulic gradient, i, is defined as

$$i = \frac{\Delta h}{L} \qquad (1.26)$$

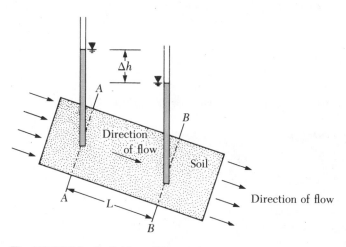

▼ **FIGURE 1.8** Definition of Darcy's law

where Δh = piezometric head difference between the sections at AA and BB
    L = distance between the sections at AA and BB

(*Note:* Sections AA and BB are perpendicular to the direction of flow.)

Darcy's law [Eq. (1.25)] is valid for a wide range of soil types. However, with materials like clean gravel and open-graded rockfills, Darcy's law breaks down because of the turbulent nature of flow through them.

The value of the hydraulic conductivity of soils varies greatly. In the laboratory, it can be determined by means of *constant head* or *falling head* permeability tests. The constant head test is more suitable for granular soils. Table 1.13 provides the general range for the values of k for various soils. In granular soils, the value primarily depends on the void ratio. In the past, several equations have been proposed to relate the value of k with the void ratio in the granular soil:

$$\frac{k_1}{k_2} = \frac{e_1^2}{e_2^2} \tag{1.27}$$

$$\frac{k_1}{k_2} = \frac{\left(\dfrac{e_1^2}{1 + e_1}\right)}{\left(\dfrac{e_2^2}{1 + e_2}\right)} \tag{1.28}$$

$$\frac{k_1}{k_2} = \frac{\left(\dfrac{e_1^3}{1 + e_1}\right)}{\left(\dfrac{e_2^3}{1 + e_2}\right)} \tag{1.29}$$

where k_1 and k_2 are the hydraulic conductivities of a given soil at void ratios e_1 and e_2, respectively

Hazen (1930) proposed an equation for the hydraulic conductivity of fairly uniform sand as

$$k = AD_{10}^2 \tag{1.30}$$

▼ **TABLE 1.13** Range of the Hydraulic Conductivity for Various Soils

Type of soil	Hydraulic conductivity, k (cm/sec)
Medium to coarse gravel	Greater than 10^{-1}
Coarse to fine sand	10^{-1} to 10^{-3}
Fine sand, silty sand	10^{-3} to 10^{-5}
Silt, clayey silt, silty clay	10^{-4} to 10^{-6}
Clays	10^{-7} or less

where k is in mm/sec
A = a constant that varies between 10 and 15
D_{10} = effective soil size, in mm

For clayey soils in the field, a practical relationship for estimating the hydraulic conductivity (Tavenas et al., 1983) is

$$\log k = \log k_0 - \frac{e_0 - e}{C_k} \tag{1.31}$$

where k = hydraulic conductivity at a void ratio e
k_0 = *in situ* hydraulic conductivity at a void ratio e_0
C_k = conductivity change index $\approx 0.5 e_0$

For clayey soils, the hydraulic conductivity for flow in the vertical and horizontal directions may vary substantially. The hydraulic conductivity for flow in the vertical direction (k_v) for *in situ* soils can be estimated from Figure 1.9. For marine and other massive clay deposits

$$\frac{k_h}{k_v} < 1.5 \tag{1.32}$$

where k_h = hydraulic conductivity for flow in the horizontal direction

For varved clays, the ratio of k_h/k_v may exceed 10.

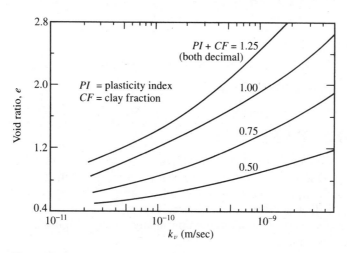

▼ **FIGURE 1.9**   Variation of *in situ* k_v for clay soils (after Tavenas et al., 1983)

▼ **EXAMPLE 1.7**

For a fine sand, the following are given:
Dry unit weight = 15.1 kN/m³
G_s = 2.67
Hydraulic conductivity = 0.14 cm/sec

If the sand is compacted to a dry unit weight of 16.3 kN/m³, estimate its hydraulic conductivity. Use Eq. (1.29).

Solution

$$\gamma_d = \frac{G_s \gamma_w}{1 + e}$$

For $\gamma_d = 15.1$ kN/m³,

$$e = \frac{G_s \gamma_w}{\gamma_d} - 1 = \frac{(2.67)(9.81)}{15.1} - 1 = 0.735$$

For $\gamma_d = 16.3$ kN/m³,

$$e = \frac{(2.67)(9.81)}{16.3} - 1 = 0.607$$

From Eq. (1.29),

$$\frac{k_1}{k_2} = \frac{\dfrac{e_1^3}{1 + e_1}}{\dfrac{e_2^3}{1 + e_2}}$$

$$\frac{0.14}{k_2} = \left[\frac{(0.735)^3}{1 + 0.735} \right] \left[\frac{1 + 0.607}{(0.607)^3} \right]$$

$$k_2 = \mathbf{0.085 \ cm/sec} \qquad \blacktriangle$$

1.9   STEADY-STATE SEEPAGE

For most cases of seepage under hydraulic structures, the flow path changes direction and is not uniform over the entire area. In such cases, one of the ways of determining the rate of seepage is by a graphical construction referred to as *flow net*. The flow net is based on Laplace's theory of continuity. According to this theory, for a steady flow condition, the flow at any point A (Figure 1.10) can be represented by the equation

$$k_x \frac{\partial^2 h}{\partial x^2} + k_y \frac{\partial^2 h}{\partial y^2} + k_z \frac{\partial^2 h}{\partial z^2} = 0 \qquad (1.33)$$

where k_x, k_y, k_z = hydraulic conductivity of the soil in the x, y, and z directions, respectively

h = hydraulic head at point A (that is, the head of water that a piezometer placed at A would show with the *downstream water level* as *datum* as shown in Figure 1.10)

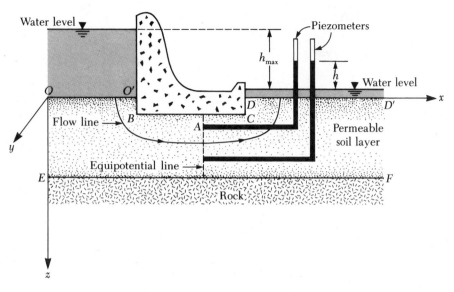

▼ **FIGURE 1.10**   Steady-state seepage

For a two-dimensional flow condition as shown in Figure 1.10,

$$\frac{\partial^2 h}{\partial^2 y} = 0$$

So Eq. (1.33) takes the form

$$k_x \frac{\partial^2 h}{\partial x^2} + k_z \frac{\partial^2 h}{\partial z^2} = 0 \tag{1.34}$$

If the soil is isotropic with respect to hydraulic conductivity, $k_x = k_z = k$, and

$$\frac{\partial^2 h}{\partial x^2} + \frac{\partial^2 h}{\partial z^2} = 0 \tag{1.35}$$

Equation (1.35), which is referred to as Laplace's equation and is valid for confined flow, represents two orthogonal sets of curves that are known as *flow lines* and *equipotential lines*. A flow net is a combination of numerous equipotential lines and flow lines. A flow line is a path that a water particle would follow in traveling from the upstream side to the downstream side. An equipotential line is a line along which water in piezometers would rise to the same elevation (see Figure 1.10).

   In drawing a flow net, you need to establish the *boundary conditions*. For example, in Figure 1.10 the ground surfaces on the upstream (OO') and downstream (DD') sides are equipotential lines. The base of the dam below the ground surface, $O'BCD$, is a flow line. The top of the rock surface, EF, is also a flow line. Once the boundary conditions are established, a number of flow lines and equipotential lines are drawn

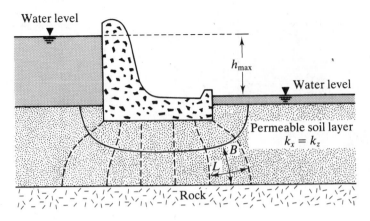

by trial and error so that all the flow elements in the net have the same length-to-width ratio (L/B). In most cases, the L/B ratio is kept as 1 — that is, the flow elements are drawn as curvilinear "squares." This method is illustrated by the flow net shown in Figure 1.11. Note that all flow lines must intersect all equipotential lines at *right angles*.

Once the flow net is drawn, the seepage in unit time per unit length of the structure can be calculated as

$$q = k h_{max} \frac{N_f}{N_d} n \tag{1.36}$$

where N_f = number of flow channels
$\quad\quad\quad N_d$ = number of drops
$\quad\quad\quad n$ = width-to-length ratio of the flow elements in the flow net (B/L)
$\quad\quad\quad h_{max}$ = difference in water level between the upstream and downstream sides

The space between two consecutive flow lines is defined as a *flow channel,* and the space between two consecutive equipotential lines is called a *drop*. In Figure 1.11, $N_f = 2$, $N_d = 7$, and $n = 1$. When square elements are drawn in a flow net,

$$\boxed{q = k h_{max} \frac{N_f}{N_d}} \tag{1.37}$$

1.10   FILTER DESIGN CRITERIA

In the design of earth structures the engineer often encounters problems caused by the flow of water, such as soil erosion, which may result in structural instability.

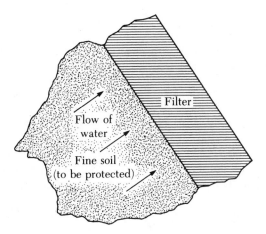

▼ **FIGURE 1.12** Filter design

Erosion is generally prevented by building soil zones that are referred to as *filters* (see Figure 1.12). Two main factors influence the choice of filter material: The grain-size distribution of the filter materials should be such that (a) the soil to be protected is not washed into the filter and (b) excessive hydrostatic pressure head is not created in the soil that has a lower coefficient of permeability.

The preceding conditions can be satisfied if the following requirements are met (Terzaghi and Peck, 1967):

$$\frac{D_{15(F)}}{D_{85(B)}} < 5 \qquad \text{[to satisfy condition (a)]} \tag{1.38}$$

$$\frac{D_{15(F)}}{D_{15(B)}} > 4 \qquad \text{[to satisfy condition (b)]} \tag{1.39}$$

In these relations, the subscripts F and B refer to the *filter* and the *base* material (that is, the soil to be protected). Also D_{15} and D_{85} refer to the diameters through which 15% and 85% of the soil (filter or base, as the case may be) will pass.

The U.S. Department of the Navy (1971) provides some additional requirements for filter design to satisfy condition (a):

$$\frac{D_{50(F)}}{D_{50(B)}} < 25 \tag{1.40}$$

$$\frac{D_{15(F)}}{D_{15(B)}} < 20 \tag{1.41}$$

Currently, geotextiles are also used as filter materials.

1.11 EFFECTIVE STRESS CONCEPT

Consider the vertical stress at a point A located at a depth $h_1 + h_2$ below the ground surface, as shown in Figure 1.13a. The total vertical stress, σ, at A is

$$\sigma = h_1\gamma + h_2\gamma_{sat} \tag{1.42}$$

where γ and γ_{sat} are unit weights of soil above and below the water table, respectively.

The total stress is carried partially by the *pore water* in the void spaces and partially by the *soil solids* at their points of contact. For example, consider a wavy plane AB drawn through point A (see Figure 1.13a) that passes through the points of contact of soil grains. The plan of this section is shown in Figure 1.13b. The small

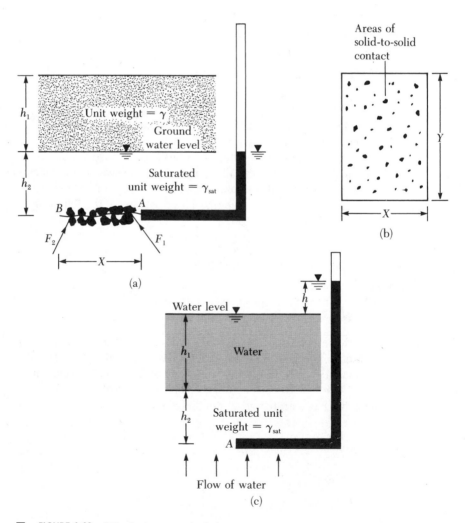

▼ **FIGURE 1.13**   Effective stress calculation

dots in Figure 1.13b represent the areas in which there is solid-to-solid contact. If the sum of these areas equals A', the area filled by water equals $XY - A$. The force carried by the pore water over the area shown in Figure 1.13b then is

$$F_w = (XY - A')u \qquad (1.43)$$

$$\text{where} \quad u = \text{pore water pressure} = \gamma_w h_2 \qquad (1.44)$$

Now let $F_1, F_2, \ldots$ be the forces at the contact points of the soil solids as shown in Figure 1.13a. The sum of the vertical components of these forces over a horizontal area XY is

$$F_s = \Sigma F_{1(v)} + F_{2(v)} + \cdots \qquad (1.45)$$

where $F_{1(v)}, F_{2(v)}, \ldots$ are vertical components of forces $F_1, F_2, \ldots$, respectively

Based on the principles of statics,

$$(\sigma)XY = F_w + F_s$$

or

$$(\sigma)XY = (XY - A')u + F_s$$

so

$$\sigma = (1 - a)u + \sigma' \qquad (1.46)$$

where $a = A'/XY$ = fraction of the unit cross-sectional area occupied by solid-to-solid contact
$\sigma' = F_s/(XY)$ = vertical component of forces at solid-to-solid contact points over a unit cross-sectional area

The term σ' in Eq. (1.46) is generally referred to as the *vertical effective stress.* Also, the quantity a in Eq. (1.46) is very small. Thus

$$\boxed{\sigma = u + \sigma'} \qquad (1.47)$$

Note that the effective stress is a *derived* quantity. Also, because the effective stress σ' is related to the contact between the soil solids, changes in effective stress will induce volume changes. It is also responsible for producing *frictional resistance* in soils and rocks. For dry soils, $u = 0$; hence $\sigma = \sigma'$.

For the problem under consideration in Figure 1.13a, $u = h_2 \gamma_w$ (γ_w = unit weight of water). Thus the effective stress at point A is

$$\sigma' = \sigma - u = (h_1 \gamma + h_2 \gamma_{sat}) - h_2 \gamma_w$$

$$= h_1 \gamma + h_2(\gamma_{sat} - \gamma_w) = h_1 \gamma + h_2 \gamma' \qquad (1.48)$$

where γ' = effective or the submerged unit weight of soil
$= \gamma_{sat} - \gamma_w$

From Eq. (1.15),

$$\gamma_{sat} = \frac{G_s\gamma_w + e\gamma_w}{1 + e}$$

so

$$\gamma' = \gamma_{sat} - \gamma_w = \frac{G_s\gamma_w + e\gamma_w}{1 + e} - \gamma_w = \frac{\gamma_w(G_s - 1)}{1 + e} \qquad (1.49)$$

For the problem in Figure 1.13a and 1.13b, there was *no seepage of water* in the soil. Figure 1.13c shows a simple condition in a soil profile where there is upward seepage. For this case, at point A

$$\sigma = h_1\gamma_w + h_2\gamma_{sat}$$

$$u = (h_1 + h_2 + h)\gamma_w$$

Thus from Eq. (1.47),

$$\sigma' = \sigma - u = (h_1\gamma_w + h_2\gamma_{sat}) - (h_1 + h_2 + h)\gamma_w$$

$$= h_2(\gamma_{sat} - \gamma_w) - h\gamma_w = h_2\gamma' - h\gamma_w$$

or

$$\sigma' = h_2\left(\gamma' - \frac{h}{h_2}\gamma_w\right) = h_2(\gamma' - i\gamma_w) \qquad (1.50)$$

Note in Eq. (1.50) that h/h_2 is the hydraulic gradient, i. If the hydraulic gradient is very high, so that $\gamma' - i\gamma_w$ becomes zero, *the effective stress will become zero*. In other words, there is no contact stress between the soil particles, and the soil structure will break up. This situation is referred to as the *quick condition*, or *failure by heave*. So, for heave,

$$\boxed{i = i_{cr} = \frac{\gamma'}{\gamma_w} = \frac{G_s - 1}{1 + e}} \qquad (1.51)$$

where i_{cr} = critical hydraulic gradient

For most sandy soils, i_{cr} ranges from 0.9 to 1.1, with an average of about 1.

▼ **EXAMPLE 1.8**_____

For the soil profile shown in Figure 1.14, determine the total vertical stress, pore water pressure, and effective vertical stress at A, B, and C.

Solution   The following table can now be prepared.

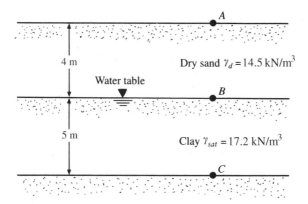

A

4 m

Dry sand $\gamma_d = 14.5\ \text{kN/m}^3$

Water table

B

5 m

Clay $\gamma_{sat} = 17.2\ \text{kN/m}^3$

C

▼ FIGURE 1.14

Point	σ (kN/m²)	u (kN/m²)	$\sigma' = \sigma - u$ (kN/m²)
A	0	0	0
B	$(4)(\gamma_d) = (4)(14.5) = \mathbf{58}$	0	58
C	$58 + (\gamma_{sat})(5) = 58 + (17.2)(5) = \mathbf{144}$	$(5)(\gamma_w) = (5)(9.81) = 49.05$	94.95

▲

1.12 CAPILLARY RISE IN SOIL

When a capillary tube is placed in water, the water level in the tube rises (Figure 1.15a). This rise is caused by the *surface tension* effect. According to Figure 1.15a, the pressure at any point A in the capillary tube (with respect to the atmospheric

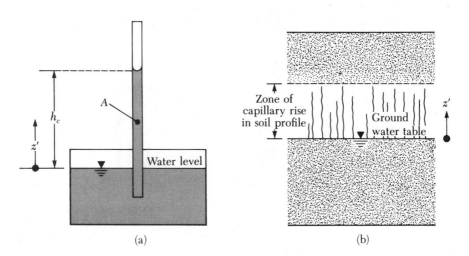

(a)

(b)

▼ FIGURE 1.15   Capillary rise

pressure) can be expressed as

$$u = -\gamma_w z' \qquad \text{(for } z' = 0 \text{ to } h_c)$$

and

$$u = 0 \qquad \text{(for } z' \geq h_c)$$

In a given soil mass, the interconnected void spaces can behave like a number of capillary tubes with varying diameters. The surface tension force may cause water in the soil to rise above the water table, as shown in Figure 1.15b. The height of the capillary rise will depend on the diameter of the capillary tubes. The capillary rise will *decrease* with the increase of the tube diameter. Because the capillary tubes in soil have variable diameters, the height of capillary rise will be nonuniform. The pore water pressure at any point in the zone of capillary rise in soil can be approximated as

$$u = -S\gamma_w z' \tag{1.52}$$

where S = degree of saturation of soil [Eq. (1.7)]
$\quad\quad z'$ = distance measured above the water table

1.13 CONSOLIDATION—GENERAL

In the field, when the stress on a saturated clay layer is increased — for example, by the construction of a foundation — the pore water pressure in the clay will increase. Because the hydraulic conductivity of clays is very small, some time will be required for the excess pore water pressure to dissipate and the stress increase to be transferred to the soil skeleton gradually. According to Figure 1.16, if Δp is a surcharge at the ground surface over a very large area, the increase of total stress, $\Delta\sigma$, at any depth of the clay layer will be equal to Δp, or

$$\Delta\sigma = \Delta p$$

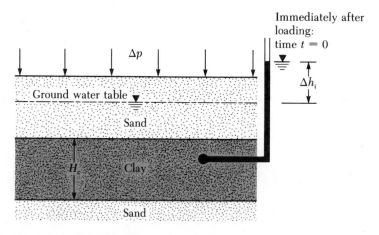

▼ **FIGURE 1.16** Principles of consolidation

However, at time $t = 0$ (that is, immediately after the stress application), the excess pore water pressure at any depth, Δu, will equal Δp, or

$$\Delta u = \Delta h_i \gamma_w = \Delta p \qquad \text{(at time } t = 0)$$

Hence the increase of effective stress at time $t = 0$ will be

$$\Delta \sigma' = \Delta \sigma - \Delta u = 0$$

Theoretically, at time $t = \infty$, when all the excess pore water pressure in the clay layer has dissipated as a result of drainage into the sand layers,

$$\Delta u = 0 \qquad \text{(at time } t = \infty)$$

Then the increase of effective stress in the clay layer is

$$\Delta \sigma' = \Delta \sigma - \Delta u = \Delta p - 0 = \Delta p$$

This gradual increase in the effective stress in the clay layer will cause settlement over a period of time and is referred to as *consolidation*.

Laboratory tests on undisturbed saturated clay specimens can be conducted (ASTM Test Designation D-2435) to determine the consolidation settlement caused by various incremental loadings. The test specimens are usually 2.5 in. (63.5 mm) in diameter and 1 in. (25.4 mm) in height. Specimens are placed inside a ring, with one porous stone at the top and one at the bottom of the specimen (Figure 1.17a). Load on the specimen is then applied so that the total vertical stress is equal to p. Settlement readings for the specimen are taken for 24 hours. After that, the load on the specimen is doubled and settlement readings are taken. At all times during the test the specimen is kept under water. This procedure is continued until the desired limit of stress on the clay specimen is reached.

Based on the laboratory tests, a graph can be plotted showing the variation of the void ratio e at the *end* of consolidation against the corresponding vertical stress p (semilogarithmic graph: e on the arithmetic scale and p on the log scale). The

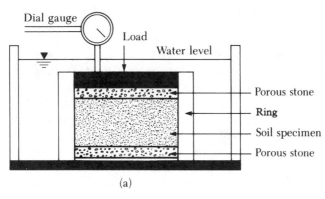

(a)

▼ **FIGURE 1.17** (a) Schematic diagram of consolidation test arrangement; (b) e–log p curve for a soft clay from East St. Louis, Illinois

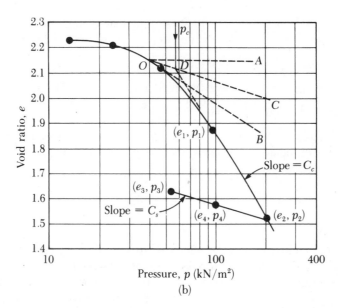

▼ FIGURE 1.17   (*Continued*)

nature of variation of e against $\log p$ for a clay specimen is shown in Figure 1.17b. After the desired consolidation pressure has been reached, the specimen can be gradually unloaded, which will result in the swelling of the specimen. Figure 1.17b also shows the variation of the void ratio during the unloading period.

From the e–$\log p$ curve shown in Figure 1.17b, three parameters necessary for calculating settlement in the field can be determined.

1. The *preconsolidation pressure, p_c,* is the *maximum past effective overburden pressure* to which the soil specimen has been subjected. It can be determined by using a simple graphical procedure as proposed by Casagrande (1936). This procedure for determining the preconsolidation pressure, with reference to Figure 1.17b, involves five steps:
   a. Determine the point O on the e–$\log p$ curve that has the sharpest curvature (that is, the smallest radius of curvature).
   b. Draw a horizontal line OA.
   c. Draw a line OB that is tangent to the e–$\log p$ curve at O.
   d. Draw a line OC that bisects the angle AOB.
   e. Produce the straight-line portion of the e–$\log p$ curve backward to intersect OC. This is point D. The pressure that corresponds to point p is the preconsolidation pressure, p_c.

   Natural soil deposits can be *normally consolidated* or *overconsolidated* (or *preconsolidated*). If the present effective overburden pressure $p = p_o$ is equal to the preconsolidated pressure p_c, the soil is *normally consolidated*. However, if $p_o < p_c$, the soil is *overconsolidated*.

Preconsolidation pressure (p_c) has been correlated with the index parameters by several investigators. Stas and Kulhawy (1984) suggested that

$$\frac{p_c}{\sigma_a} = 10^{(1.11-1.62LI)} \tag{1.53a}$$

where σ_a = atmospheric stress in derived unit
$\quad\quad\quad LI$ = liquidity index

The liquidity index of a soil is defined as

$$LI = \frac{w - PL}{LL - PL} \tag{1.53b}$$

where w = *in situ* moisture content
$\quad\quad\quad LL$ = liquid limit
$\quad\quad\quad PL$ = plastic limit

Nagaraj and Murthy (1985) provided an empirical relation to calculate p_c, which is as follows:

$$\overset{\overset{\textstyle kN/m^2}{\downarrow}}{\log p_c = \frac{1.122 - \left(\dfrac{e_o}{e_L}\right) - 0.0463 \log p_o}{0.188}} \tag{1.54}$$

$$\underset{\underset{\textstyle kN/m^2}{\uparrow}}{}$$

where e_o = *in situ* void ratio
$\quad\quad\quad p_o$ = *in situ* effective overburden pressure
$\quad\quad\quad e_L$ = void ratio of the soil at liquid limit

$$e_L = \left[\frac{LL(\%)}{100}\right] G_s \tag{1.55}$$

The U.S. Department of the Navy (1982) also provided generalized relationships between p_c, LI, and the sensitivity of clayey soils (S_t). This relationship was also recommended by Kulhawy and Mayne (1990). The definition of sensitivity is given in Section 1.19. Figure 1.18 shows the relationship.

2. The *compression index*, C_c, is the slope of the straight-line portion (latter part of the loading curve), or

$$C_c = \frac{e_1 - e_2}{\log p_2 - \log p_1} = \frac{e_1 - e_2}{\log\left(\dfrac{p_2}{p_1}\right)} \tag{1.56}$$

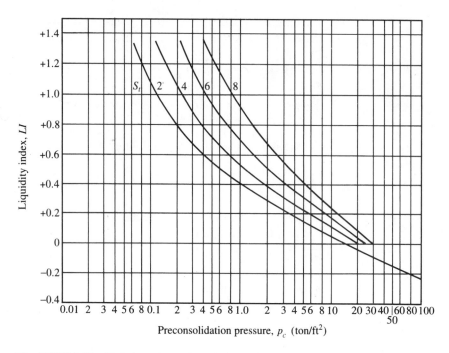

▼ **FIGURE 1.18**   Variation of p_c with *LI* (after U.S. Department of the Navy, 1982)

where e_1 and e_2 are the void ratios at the end of consolidation under stresses p_1 and p_2, respectively

The *compression index,* as determined from the laboratory e–log p curve, will be somewhat different from that encountered in the field. The primary reason is that the soil remolds to some degree during the field exploration. The nature of variation of the e–log p curve in the field for a normally consolidated clay is shown in Figure 1.19. It is generally referred to as the *virgin compression curve*. The virgin curve approximately intersects the laboratory curve at a void ratio of $0.42e_o$ (Terzaghi and Peck, 1967). Note that e_o is the void ratio of the clay in the field. Knowing the values of e_o and p_c, you can easily construct the virgin curve and calculate the compression index of the virgin curve by using Eq. (1.56).

The value of C_c can vary widely depending on the soil. Skempton (1944) has given an empirical correlation for the compression index in which

$$C_c = 0.009(LL - 10)$$

(1.57)

where LL = liquid limit

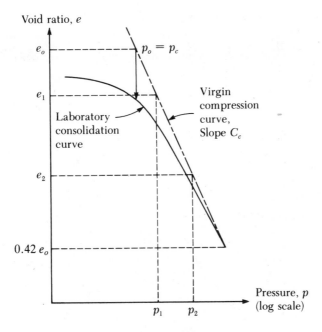

▼ **FIGURE 1.19** Construction of virgin compression curve
for normally consolidated clay

Besides Skempton, other investigators have proposed correlations for the
compression index. Some of these correlations are summarized in
Table 1.14.

3. The *swelling index, C_s,* is the slope of the unloading portion of the e–log p
curve. In Figure 1.17b, it can be defined as

$$C_s = \frac{e_3 - e_4}{\log\left(\dfrac{p_4}{p_3}\right)}$$

(1.58)

In most cases the value of the swelling index (C_s) is $\frac{1}{4}$ to $\frac{1}{5}$ of the compression
index. Following are some representative values of C_s/C_c for natural soil
deposits. The swelling index is also referred to as the *recompression index.*

Description of soil	C_s/C_c
Boston Blue clay	0.24–0.33
Chicago clay	0.15–0.3
New Orleans clay	0.15–0.28
St. Lawrence clay	0.05–0.1

▼ **TABLE 1.14** Correlations for Compression Index

Reference	Correlation
Azzouz, Krizek, and Corotis (1976)	$C_c = 0.01w_n$ (Chicago clay) $C_c = 0.208e_o + 0.0083$ (Chicago clay) $C_c = 0.0115w_n$ (organic soils, peat) $C_c = 0.0046(LL - 9)$ (Brazilian clay)
Rendon-Herrero (1980)	$C_c = 0.141G_s^{1.2} \left(\dfrac{1 + e_o}{G_s} \right)^{2.38}$
Nagaraj and Murthy (1985)	$C_c = 0.2343 \left(\dfrac{LL}{100} \right) G_s$
Wroth and Wood (1978)	$C_c = 0.5G_s \left(\dfrac{PI}{100} \right)$
Leroueil, Tavenas, and LeBihan (1983)	

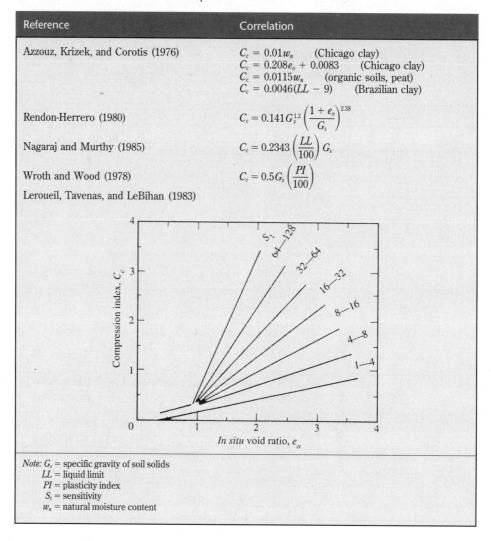

Note: G_s = specific gravity of soil solids
 LL = liquid limit
 PI = plasticity index
 S_t = sensitivity
 w_n = natural moisture content

The swelling index determination is important in the estimation of consolidation settlement of *overconsolidated clays*. In the field, depending on the pressure increase, an overconsolidated clay will follow an e–log p path *abc*, as shown in Figure 1.20. Note that point *a*, with coordinates of p_o and e_o, corresponds to the field conditions before any pressure increase. Point *b* corresponds to the preconsolidation pressure (p_c) of the clay. Line *ab* is approximately parallel to the laboratory unloading curve *cd* (Schmertmann,

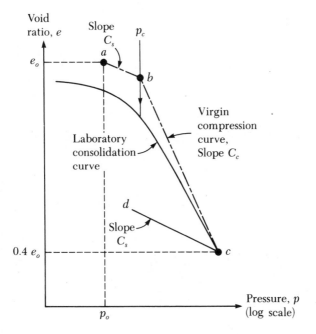

▼ **FIGURE 1.20**   Construction of field consolidation curve
for overconsolidated clay

1953). Hence, if you know e_o, p_o, p_c, C_c, and C_s, you can easily construct the field consolidation curve.

Nagaraj and Murthy (1985) expressed the swelling index as

$$C_s = 0.0463 \left(\frac{LL}{100} \right) G_s \tag{1.59}$$

It is essential to point out that any of the empirical correlations for C_c and C_s given in this section are only approximate. It may be valid for a given soil for which the relationship was developed but may not hold good for other soils. As an example, Figure 1.21 shows the plots of C_c and C_s with liquid limit for soils from Richmond, Virginia (Martin et al., 1995). For these soils,

$$C_c = 0.0326(LL - 43.4) \tag{1.60}$$

and

$$C_s = 0.00045(LL + 11.9) \tag{1.61}$$

The C_s/C_c ratio is about $\frac{1}{25}$; whereas, the typical range is about $\frac{1}{5}$ to $\frac{1}{10}$.

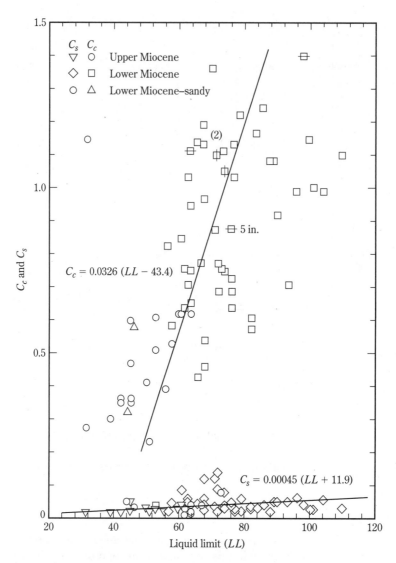

▼ **FIGURE 1.21**   Variation of C_c and C_s with liquid limit for soils from
Richmond, Virginia (after Martin et al., 1995)

1.14 CONSOLIDATION SETTLEMENT CALCULATION

The one-dimensional consolidation settlement (caused by an additional load) of a
clay layer (Figure 1.22a) having a thickness H_c may be calculated as

$$S = \frac{\Delta e}{1 + e_o} H_c$$

(1.62)

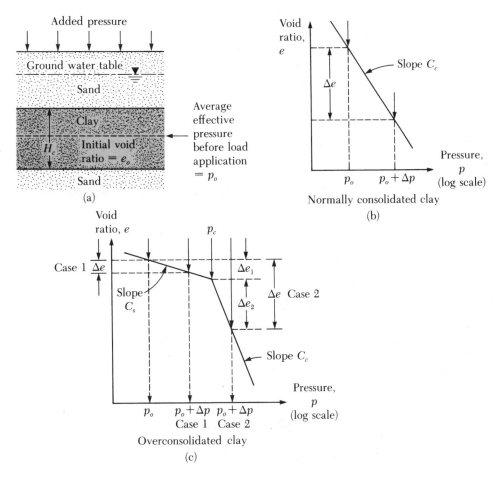

▼ FIGURE 1.22 One-dimensional settlement calculation: (b) is for Eq. (1.64); (c) is for Eqs. (1.66) and (1.68)

where S = settlement

Δe = total change of void ratio caused by the additional load application

e_o = the void ratio of the clay before the application of load

Note that

$$\frac{\Delta e}{1 + e_o} = \varepsilon_v = \text{vertical strain}$$

For normally consolidated clay, the field e–log p curve will be like the one shown in Figure 1.22b. If p_o = initial average effective overburden pressure on the clay layer and Δp = average pressure increase on the clay layer caused by the added load, the change of void ratio caused by the load increase is

$$\Delta e = C_c \log \frac{p_o + \Delta p}{p_o}$$

(1.63)

Now, combining Eqs. (1.62) and (1.63) yields

$$S = \frac{C_c H_c}{1 + e_o} \log \frac{p_o + \Delta p}{p_o}$$

(1.64)

For overconsolidated clay, the field e–log p curve will be like the one shown in Figure 1.22c. In this case, depending on the value of Δp, two conditions may arise. First, if $p_o + \Delta p < p_c$,

$$\Delta e = C_s \log \frac{p_o + \Delta p}{p_o}$$

(1.65)

Combining Eqs. (1.62) and (1.65) gives

$$S = \frac{H_c C_s}{1 + e_o} \log \frac{p_o + \Delta p}{p_o}$$

(1.66)

Second, if $p_o < p_c < p_o + \Delta p$,

$$\Delta e = \Delta e_1 + \Delta e_2 = C_s \log \frac{p_c}{p_o} + C_c \log \frac{p_o + \Delta p}{p_o}$$

(1.67)

Now, combining Eqs. (1.62) and (1.67) yields

$$S = \frac{C_s H_c}{1 + e_o} \log \frac{p_c}{p_o} + \frac{C_c H_c}{1 + e_o} \log \frac{p_o + \Delta p}{p_c}$$

(1.68)

1.15   TIME RATE OF CONSOLIDATION

In Section 1.13 (see Figure 1.16) we showed that consolidation is the result of gradual dissipation of the excess pore water pressure from a clay layer. Pore water pressure dissipation, in turn, increases the effective stress, which induces settlement. Hence, to estimate the degree of consolidation of a clay layer at some time t after the load application, you need to know the rate of dissipation of the excess pore water pressure.

Figure 1.23 shows a clay layer of thickness H_c that has highly permeable sand layers at its top and bottom. Here, the excess pore water pressure at any point A

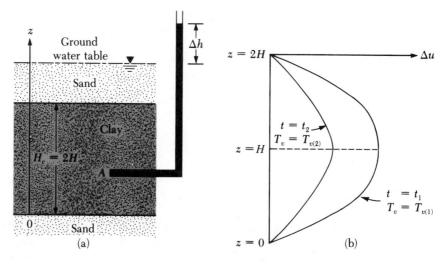

▼ **FIGURE 1.23**   (a) Derivation of Eq. (1.71); (b) nature of variation of Δu with time

at any time t after the load application is $\Delta u = (\Delta h)\gamma_w$. For a vertical drainage condition (that is, in the direction of z only) from the clay layer, Terzaghi derived the following differential equation:

$$\frac{\partial(\Delta u)}{\partial t} = C_v \frac{\partial^2(\Delta u)}{\partial z^2} \tag{1.69}$$

where C_v = coefficient of consolidation

$$C_v = \frac{k}{m_v \gamma_w} = \frac{k}{\dfrac{\Delta e}{\Delta p(1 + e_{av})}\gamma_w} \tag{1.70}$$

where k = hydraulic conductivity of the clay
   Δe = total change of void ratio caused by a stress increase of Δp
   e_{av} = average void ratio during consolidation
   m_v = volume coefficient of compressibility = $\Delta e / [\Delta p(1 + e_{av})]$

Equation (1.69) can be solved to obtain Δu as a function of time t with the following boundary conditions:

1.   Because highly permeable sand layers are located at $z = 0$ and $z = H_c$, the excess pore water pressure developed in the clay at those points will be immediately dissipated. Hence

$\Delta u = 0$ at $z = 0$

$\Delta u = 0$ at $z = H_c = 2H$

where H = length of maximum drainage path (due to two-way drainage condition — that is, at the top and bottom of the clay)

2.   At time $t = 0$,

$\Delta u = \Delta u_o$ = initial excess pore water pressure after the load application

With the preceding boundary conditions, Eq. (1.69) yields

$$\Delta u = \sum_{m=0}^{m=\infty} \left[\frac{2(\Delta u_o)}{M} \sin \left(\frac{Mz}{H} \right) \right] e^{-M^2 T_v} \qquad (1.71)$$

where $M = [(2m + 1)\pi]/2$
        m = an integer = 1, 2, ...
        T_v = nondimensional time factor = $(C_v t)/H^2$ $\qquad$ (1.72)

Determining the field value of C_v is difficult. Figure 1.24 provides a first-order

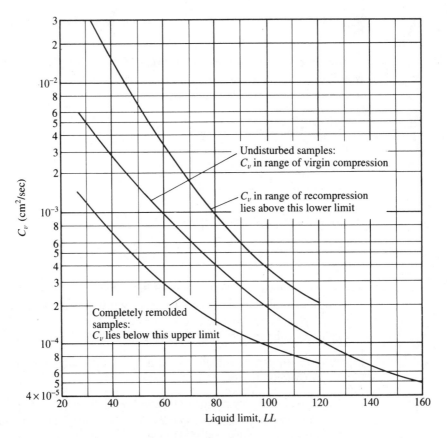

▼ **FIGURE 1.24**   Range of C_v (after U.S. Department of the Navy, 1971)

determination of C_v using the liquid limit (U.S. Department of the Navy, 1971). The value of Δu for various depths (that is, $z = 0$ to $z = 2H$) at any given time t (thus T_v) can be calculated from Eq. (1.71). The nature of this variation of Δu is shown in Figure 1.23b.

The *average degree of consolidation* of the clay layer can be defined as

$$U = \frac{S_t}{S_{max}} \tag{1.73}$$

where U = average degree of consolidation
   S_t = settlement of a clay layer at time t after the load application
   S_{max} = maximum consolidation settlement that the clay will undergo under a given loading

If the initial pore water pressure (Δu_o) distribution is constant with depth as shown in Figure 1.25a, the average degree of consolidation can also be expressed as

$$U = \frac{S_t}{S_{max}} = \frac{\int_0^{2H} (\Delta u_0)\,dz - \int_0^{2H} (\Delta u)\,dz}{\int_0^{2H} (\Delta u_0)\,dz} \tag{1.74}$$

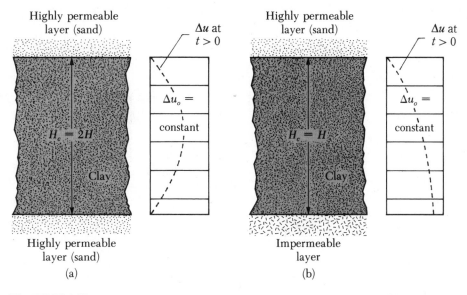

▼ **FIGURE 1.25**   Drainage condition for consolidation: (a) two-way drainage; (b) one-way drainage

or

$$U = \frac{(\Delta u_0)2H - \int_0^{2H} (\Delta u)\,dz}{(\Delta u_o)2H} = 1 - \frac{\int_0^{2H} (\Delta u)\,dz}{2H(\Delta u_o)} \tag{1.75}$$

Now, combining Eqs. (1.71) and (1.75), we obtain.

$$U = \frac{S_t}{S_{\max}} = 1 - \sum_{m=0}^{m=\infty} \left(\frac{2}{M^2}\right) e^{-M^2 T_v} \tag{1.76}$$

The variation of U with T_v can be calculated from Eq. (1.76) and is plotted in Figure 1.26. Note that Eq. (1.76) and thus Figure 1.26 are also valid when an impermeable layer is located at the bottom of the clay layer (Figure 1.25b). In that case, excess pore water pressure dissipation can take place in one direction only. The length of the *maximum drainage path* then is equal to $H = H_c$.

The variation of T_v with U shown in Figure 1.26 can also be approximated by

$$T_v = \frac{\pi}{4}\left(\frac{U\%}{100}\right)^2 \qquad (\text{for } U = 0\text{–}60\%) \tag{1.77}$$

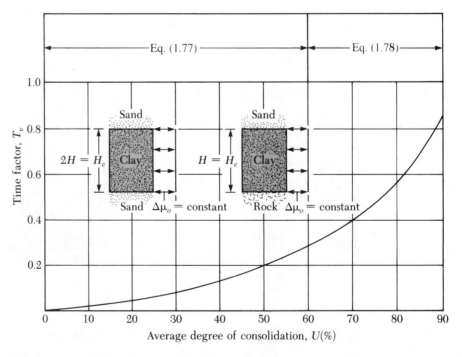

▼ **FIGURE 1.26** Plot of time factor against average degree of consolidation ($\Delta u_o =$ constant)

and

$$T_v = 1.781 - 0.933 \log (100 - U\%) \qquad \text{(for } U > 60\%) \qquad (1.78)$$

Sivaram and Swamee (1977) have also developed an empirical relationship between T_v and U that is valid for U varying from 0 to 100%. It is of the form

$$T_v = \frac{\left(\dfrac{\pi}{4}\right)\left(\dfrac{U\%}{100}\right)^2}{\left[1 - \left(\dfrac{U\%}{100}\right)^{5.6}\right]^{0.357}} \qquad (1.79)$$

In some cases, initial excess pore water pressure may not be constant with depth as shown in Figure 1.25. Following are a few cases of those and the solutions for the average degree of consolidation.

Trapezoidal Variation   Figure 1.27 shows a trapezoidal variation of initial excess pore water pressure with *two-way drainage*. For this case the variation of T_v with U will be the same as shown in Figure 1.26.

Sinusoidal Variation   This variation is shown in Figures 1.28a and 1.28b. For the initial excess pore water pressure variation shown in Figure 1.28a,

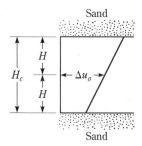

▼ **FIGURE 1.27**   Trapizoidal initial excess pore water pressure distribution

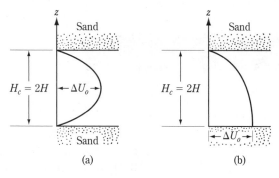

▼ **FIGURE 1.28**   Sinusoidal initial excess pore
water pressure distribution

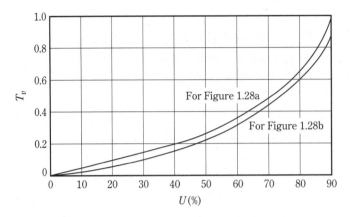

▼   **FIGURE 1.29**   Variation of U with T_v — sinusoidal variation of
initial excess pore water pressure distribution

$$\Delta u = \Delta u_o \sin \frac{\pi z}{2H} \qquad\qquad (1.80)$$

Similarly, for the case shown in Figure 1.28b,

$$\Delta u = \Delta u_o \cos \frac{\pi z}{4H} \qquad\qquad (1.81)$$

The variations of T_v with U for these two cases are shown in Figure 1.29.

Triangular Variation   Figures 1.30 and 1.31 show several types of initial pore
water pressure variation and the variations of T_v with the average degree of consolidation.

▼   **EXAMPLE 1.9** _____

A laboratory consolidation test on a normally consolidated clay showed the following:

Load, p (kN/m²)	Void ratio at the end of consolidation, e
140	0.92
212	0.86

The specimen tested was 25.4 mm in thickness and drained on both sides. The time
required for the specimen to reach 50% consolidation was 4.5 min.

A similar clay layer in the field, 2.8 m thick and drained on both sides, is
subjected to similar average presure increase (that is, $p_o = 140$ kN/m² and $p_o +
\Delta p = 212$ kN/m²). Determine the

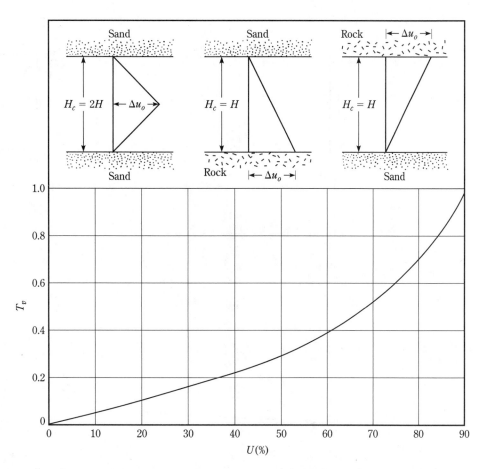

▼ **FIGURE 1.30**   Variation of U with T_v — triangular initial excess pore water pressure distribution

a. Expected maximum consolidation settlement in the field

b. Length of time required for the total settlement in the field to reach 40 mm (assume uniform initial excess pore water pressure increase with depth)

Solution

Part a

For normally consolidated clay [Eq. (1.56)]

$$C_c = \frac{e_1 - e_2}{\log\left(\dfrac{p_2}{p_1}\right)} = \frac{0.92 - 0.86}{\log\left(\dfrac{212}{140}\right)} = 0.333$$

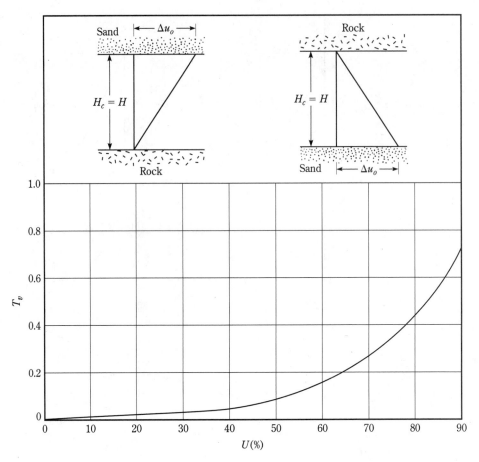

▼ **FIGURE 1.31**   Triangular initial excess pore water pressure distribution — variation of U with T_v

From Eq. (1.64)

$$S = \frac{C_c H_c}{1 + e_o} \log \frac{p_o + \Delta p}{p_o} = \frac{(0.333)(2.8)}{1 + 0.92} \log \frac{212}{140} = 0.0875 \text{ m} = \mathbf{87.5 \text{ mm}}$$

Part b

From Eq. (1.73) the average degree of consolidation is

$$U = \frac{S_t}{S_{\max}} = \frac{40}{87.5} (100) = 45.7\%$$

The coefficient of consolidation, C_v, can be calculated from the laboratory test. From Eq. (1.72)

$$T_v = \frac{C_v t}{H^2}$$

For 50% consolidation (Figure 1.26), $T_v = 0.197$, $t = 4.5$ min, and $H = H_c/2 = 12.7$ mm, so

$$C_v = T_{50} \frac{H^2}{t} = \frac{(0.197)(12.7)^2}{4.5} = 7.061 \text{ mm}^2/\text{min}$$

Again, for field consolidation, $U = 45.7\%$. From Eq. (1.77)

$$T_v = \frac{\pi}{4} \left(\frac{U\%}{100} \right)^2 = \frac{\pi}{4} \left(\frac{45.7}{100} \right)^2 = 0.164$$

But

$$T_v = \frac{C_v t}{H^2}$$

or

$$t = \frac{T_v H^2}{C_v} = \frac{0.164 \left(\dfrac{2.8 \times 1000}{2} \right)^2}{7.061} = 45{,}523 \text{ min} = \mathbf{31.6 \ days}$$

▲

1.16 SHEAR STRENGTH

The shear strength, s, of a soil, in terms of effective stress, is

$$\boxed{s = c + \sigma' \tan \phi} \tag{1.82}$$

where $\sigma' =$ effective normal stress on plane of shearing
$c =$ cohesion, or apparent cohesion
$\phi =$ angle of friction

Equation (1.82) is referred to as the *Mohr-Coulomb failure criteria*. The value of c for sands and normally consolidated clays is equal to zero. For overconsolidated clays, $c > 0$.

For most day-to-day work, the shear strength parameters of a soil (that is, c and ϕ) are determined by two standard laboratory tests. They are (a) the *direct shear test* and (b) the *triaxial test*.

Direct Shear Test

Dry sand can be conveniently tested by direct shear tests. The sand is placed in a shear box that is split into two halves (Figure 1.32a). A normal load is first applied to the specimen. Then a shear force is applied to the top half of the shear box to cause failure in the sand. The normal and shear stresses at failure are

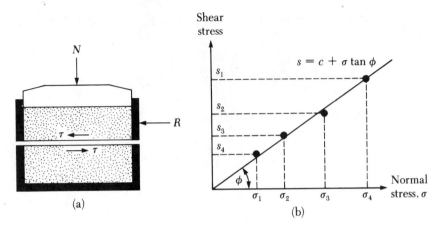

▼ **FIGURE 1.32**   Direct shear test in sand: (a) schematic diagram of test equipment; (b) plot of test results to obtain the friction angle, ϕ

$$\sigma' = \frac{N}{A}$$

$$s = \frac{R}{A}$$

where A = area of the failure plane in soil — that is, the area of cross section of the shear box

Several tests of this type can be conducted by varying the normal load. The angle of friction of the sand can be determined by plotting a graph of s against σ' ($= \sigma$ for dry sand), as shown in Figure 1.32b, or

$$\phi = \tan^{-1}\left(\frac{s}{\sigma'}\right) \tag{1.83}$$

For sands, the angle of friction usually ranges from 26° to 45°, increasing with the relative density of compaction. The approximate range of the relative density of compaction and the corresponding range of the angle of friction for various coarse-grained soils is shown in Figure 1.33.

Triaxial Tests

Triaxial compression tests can be conducted on sands and clays. Figure 1.34a shows a schematic diagram of the triaxial test arrangement. Essentially, it consists of placing a soil specimen confined by a rubber membrane in a lucite chamber. An all-around confining pressure (σ_3) is applied to the specimen by means of the

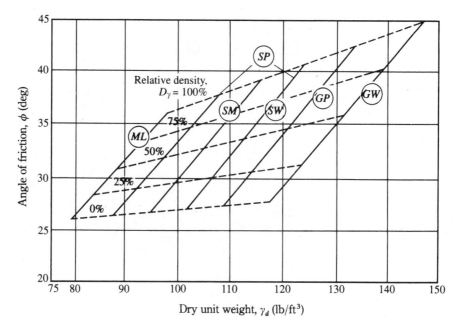

▼ **FIGURE 1.33** Range of relative density and corresponding range of angle of friction for coarse-grained soil (after U.S. Department of the Navy, 1971)

chamber fluid (generally water or glycerin). An added stress ($\Delta\sigma$) can also be applied to the specimen in the axial direction to cause failure ($\Delta\sigma = \Delta\sigma_f$ at failure). Drainage from the specimen can be allowed or stopped, depending on the test condition. For clays, three main types of tests can be conducted with triaxial equipment:

1. Consolidated-drained test (CD test)
2. Consolidated-undrained test (CU test)
3. Unconsolidated-undrained test (UU test)

Table 1.15 summarizes these three tests. For *consolidated-drained tests,* at failure,

Major principal effective stress $= \sigma_3 + \Delta\sigma_f = \sigma_1 = \sigma_1'$
Minor principal effective stress $= \sigma_3 = \sigma_3'$

Changing σ_3 allows several tests of this type to be conducted on various clay specimens. The shear strength parameters (c and ϕ) can now be determined by plotting Mohr's circle at failure, as shown in Figure 1.34b, and drawing a common tangent to the Mohr's circles. This is the *Mohr–Coulomb failure envelope.* (*Note:* For normally consolidated clay, $c \approx 0$.) At failure

$$\sigma_1' = \sigma_3' \tan^2\left(45 + \frac{\phi}{2}\right) + 2c \tan\left(45 + \frac{\phi}{2}\right)$$

(1.84)

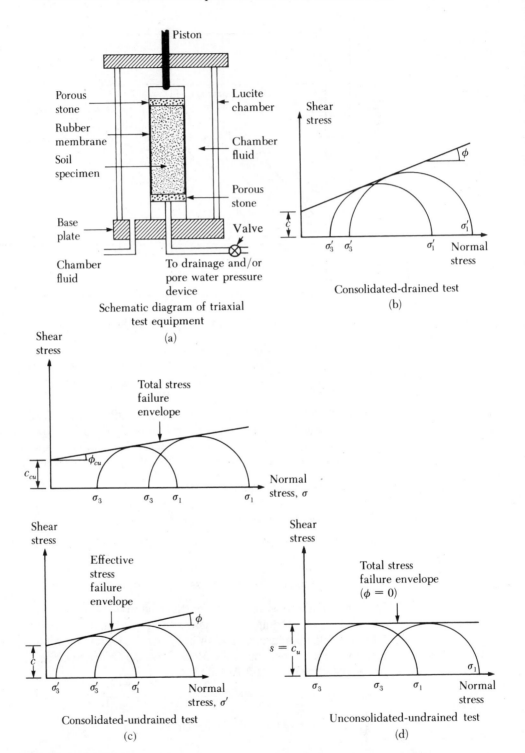

FIGURE 1.34   Triaxial test

▼ **TABLE 1.15** Summary of Triaxial Tests on Saturated Clays

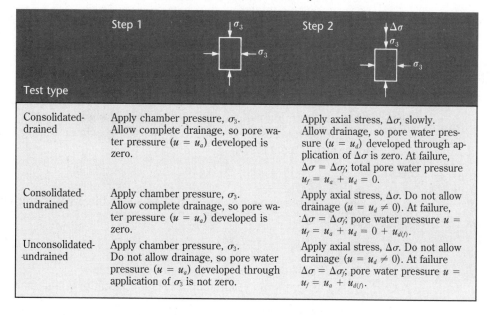

Test type	Step 1 σ_3	Step 2 $\Delta\sigma$
Consolidated-drained	Apply chamber pressure, σ_3. Allow complete drainage, so pore water pressure ($u = u_a$) developed is zero.	Apply axial stress, $\Delta\sigma$, slowly. Allow drainage, so pore water pressure ($u = u_d$) developed through application of $\Delta\sigma$ is zero. At failure, $\Delta\sigma = \Delta\sigma_f$; total pore water pressure $u_f = u_a + u_d = 0$.
Consolidated-undrained	Apply chamber pressure, σ_3. Allow complete drainage, so pore water pressure ($u = u_a$) developed is zero.	Apply axial stress, $\Delta\sigma$. Do not allow drainage ($u = u_d \neq 0$). At failure, $\Delta\sigma = \Delta\sigma_f$; pore water pressure $u = u_f = u_a + u_d = 0 + u_{d(f)}$.
Unconsolidated-undrained	Apply chamber pressure, σ_3. Do not allow drainage, so pore water pressure ($u = u_a$) developed through application of σ_3 is not zero.	Apply axial stress, $\Delta\sigma$. Do not allow drainage ($u = u_d \neq 0$). At failure $\Delta\sigma = \Delta\sigma_f$; pore water pressure $u = u_f = u_a + u_{d(f)}$.

For *consolidated-undrained* tests, at failure,

Major principal total stress = $\sigma_3 + \Delta\sigma_f = \sigma_1$

Minor principal total stress = σ_3

Major principal effective stress = $(\sigma_3 + \Delta\sigma_f) - u_f = \sigma_1'$

Minor principal effective stress = $\sigma_3 - u_f = \sigma_3'$

Changing σ_3 permits multiple tests of this type to be conducted on several soil specimens. The total stress Mohr's circles at failure can now be plotted, as shown in Figure 1.34c, and then a common tangent can be drawn to define the *failure envelope*. This *total stress failure envelope* is defined by the equation

$$s = c_{cu} + \sigma \tan \phi_{cu} \tag{1.85}$$

where c_{cu} and ϕ_{cu} are the *consolidated-undrained cohesion* and *angle of friction*, respectively (*Note: $c_{cu} \approx 0$ for normally consolidated clays*)

Similarly, effective stress Mohr's circles at failure can be drawn to determine the *effective stress failure envelopes* (Figure 1.34c). They follow the relation expressed in Eq. (1.82).

For *unconsolidated-undrained* triaxial tests

Major principal total stress = $\sigma_3 + \Delta\sigma_f = \sigma_1$

Minor principal total stress = σ_3

The total stress Mohr's circle at failure can now be drawn, as shown in Figure 1.34d. For saturated clays, the value of $\sigma_1 - \sigma_3 = \Delta\sigma_f$ is a constant, irrespective of the chamber confining pressure, σ_3 (also shown in Figure 1.34d). The tangent to these Mohr's circles will be a horizontal line, called the $\phi = 0$ condition. The shear stress for this condition is

$$s = c_u = \frac{\Delta \sigma_f}{2} \tag{1.86}$$

where c_u = undrained cohesion (or undrained shear strength)

The pore pressure developed in the soil specimen during the unconsolidated-undrained triaxial test is

$$u = u_a + u_d \tag{1.87}$$

The pore pressure u_a is the contribution of the hydrostatic chamber pressure, σ_3. Hence

$$u_a = B\sigma_3 \tag{1.88}$$

where B = Skempton's pore pressure parameter

Similarly, the pore parameter u_d is the result of added axial stress, $\Delta\sigma$, so

$$u_d = A\,\Delta\sigma \tag{1.89}$$

where A = Skempton's pore pressure parameter

However,

$$\Delta\sigma = \sigma_1 - \sigma_3 \tag{1.90}$$

Combining Eqs. (1.87), (1.88), (1.89), and (1.90) gives

$$u = u_a + u_d = B\sigma_3 + A(\sigma_1 - \sigma_3) \tag{1.91}$$

The pore water pressure parameter B in soft saturated soils is 1, so

$$u = \sigma_3 + A(\sigma_1 - \sigma_3) \tag{1.92}$$

The value of the pore water pressure parameter A at failure will vary with the type of soil. Following is a general range of the values of A at failure for various types of clayey soil encountered in nature.

Type of soil	A at failure
Sandy clays	0.5–0.7
Normally consolidated clays	0.5–1
Overconsolidated clays	−0.5–0

Figure 1.35 shows a photograph of laboratory triaxial equipment.

▼ **FIGURE 1.35** Triaxial test equipment

1.17 UNCONFINED COMPRESSION TEST

The *unconfined compression test* (Figure 1.36a) is a special type of unconsolidated-undrained triaxial test in which the confining pressure $\sigma_3 = 0$, as shown in Figure 1.36b. In this test an axial stress, $\Delta\sigma$, is applied to the specimen to cause failure (that is, $\Delta\sigma = \Delta\sigma_f$). The corresponding Mohr's circle is shown in Figure 1.36b. Note that, for this case,

Major principal total stress $= \Delta\sigma_f = q_u$
Minor principal total stress $= 0$

The axial stress at failure, $\Delta\sigma_f = q_u$, is generally referred to as the *unconfined compression strength*. The shear strength of saturated clays under this condition ($\phi = 0$), from Eq. (1.82), is

$$s = c_u = \frac{q_u}{2} \tag{1.93}$$

The unconfined compression strength can be used as an indicator for the consistency of clays.

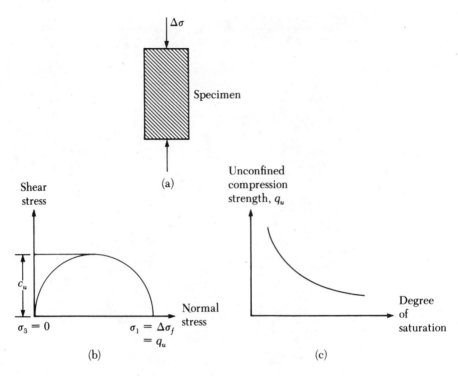

▼ **FIGURE 1.36**   Unconfined compression test: (a) soil specimen; (b) Mohr's circle for
the test; (c) variation of q_u with the degree of saturation

Unconfined compression tests are sometimes conducted on unsaturated soils. With the void ratio of a soil specimen remaining constant, the unconfined compression strength rapidly decreases with the degree of saturation (Figure 1.36c). Figure 1.37 shows an unconfined compression test in progress.

1.18 COMMENTS ON SHEAR STRENGTH PARAMETERS

Drained Friction Angle of Granular Soils

In general, the direct shear test yields a higher angle of friction compared to that obtained by the triaxial test. It also needs to be pointed out that the failure envelope for a given soil is actually curved. The Mohr-Coulomb failure criteria defined by Eq. (1.82) is only an approximation. Because of the curved nature of the failure envelope, a soil tested at higher normal stress will yield a lower value of ϕ. An example of that is shown in Figure 1.38, which is a plot of ϕ versus the void ratio, e, for Chattahoochee River sand near Atlanta, Georgia (Vesic, 1963). These friction angles were obtained from triaxial tests. Note that, for a given value of e, the

▼ **FIGURE 1.37** Unconfined compression test in progress (courtesy of Soiltest, Inc., Lake Bluff, Illinois)

magnitude of ϕ is about 4 to 5 degrees smaller when the confining pressure σ_3' is greater than 10 lb/in.^{2} (69 kN/m^2) compared to that when $\sigma_3' < 10$ lb/in.^{2}.

Drained Friction Angle of Cohesive Soils

Figure 1.39 shows the drained friction angle, ϕ, for several normally consolidated clays obtained by conducting triaxial tests (Bjerrum and Simons, 1960). It can be seen from this figure that, in general, the friction angle ϕ decreases with the increase in plasticity index. The value of ϕ generally decreases from about 37°–38° with a plasticity index of about 10, to about 25° or less with a plasticity index of about 100. Similar results were also provided by Kenney (1959). The consolidated undrained friction angle (ϕ_{cu}) of normally consolidated saturated clays generally ranges from 5°–20°.

The consolidated drained triaxial test is described in Section 1.16. Figure 1.40 shows the schematic diagram of the plot of $\Delta\sigma$ versus axial strain of a drained triaxial test for a clay. At failure, for this test, $\Delta\sigma = \Delta\sigma_f$. However, at large axial strain (i.e., ultimate strength condition),

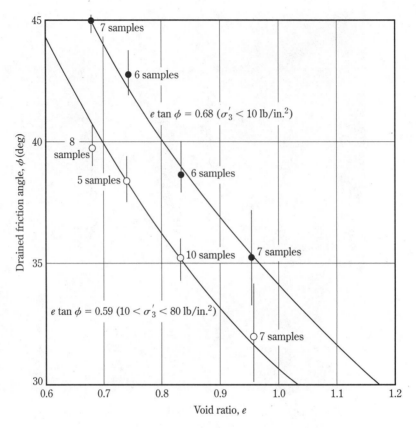

▼ **FIGURE 1.38**   Variation of friction angle ϕ with void ratio for Chatta-choochee River sand (after Vesic, 1963)

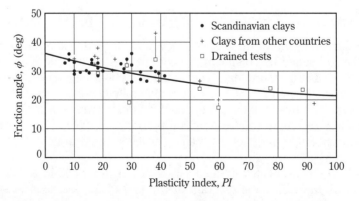

▼ **FIGURE 1.39**   Variation of friction angle ϕ with plasticity index for several clays (after Bjerrum and Simons, 1960)

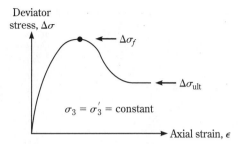

▼ **FIGURE 1.40** Plot of deviator stress vs. axial strain — drained triaxial test

Major principal stress: $\sigma'_{1\text{(ult)}} = \sigma_3 + \Delta\sigma_{\text{ult}}$
Minor principal stress: $\sigma'_{3\text{(ult)}} = \sigma_3$

At failure (that is, peak strength), the relationship between σ'_1 and σ'_3 was given by Eq. (1.84). However, for ultimate strength, it can be shown that

$$\sigma'_{1\text{(ult)}} = \sigma'_3 \tan^2\left(45 + \frac{\phi_r}{2}\right) \tag{1.94}$$

where ϕ_r = residual drained friction angle

Figure 1.41 shows the general nature of the failure envelopes at peak strength and ultimate strength (or *residual strength*). The residual shear strength of clays is important in the evaluation of long-term stability of new and existing slopes and the design of remedial measures. The drained residual friction angles (ϕ_r) of clays may

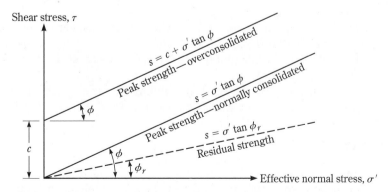

▼ **FIGURE 1.41** Peak and residual strength envelopes for clay

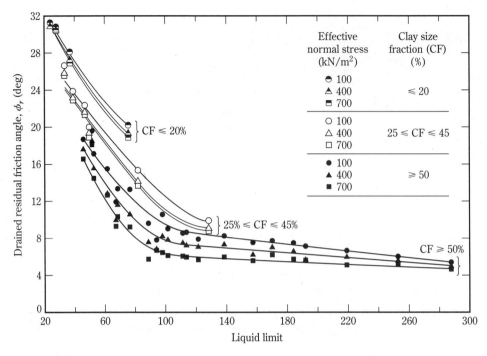

▼ **FIGURE 1.42**   Variation of ϕ_r with liquid limit for some clays (after Stark, 1995)

be substantially smaller than the drained peak friction angles. Figure 1.42 shows the variation of ϕ_r with liquid limit for some clays (Stark, 1995). It is important to note that

1. For a given clay, ϕ_r decreases with the increase in liquid limit.
2. For a given liquid limit and clay-size fractions present in the soil, the magnitude of ϕ_r decreases with the increase in the normal effective stress. This is due to the curvilinear nature of the failure envelope.

Undrained Shear Strength, c_u   The undrained shear strength, c_u, is an important parameter in the design of foundations. For normally consolidated clay deposits (Figure 1.43), the magnitude of c_u increases almost linearly with the increase of effective overburden pressure.

There are several empirical relations between c_u and the effective overburden pressure p in the field. Some of these relationships are summarized in Table 1.16.

1.19   SENSITIVITY

For many naturally deposited clay soils, the unconfined compression strength is much less when the soils are tested after remolding without any change in the moisture content. This property of clay soil is called *sensitivity*. The degree of

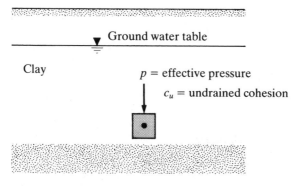

Ground water table

Clay

p = effective pressure

c_u = undrained cohesion

▼ **FIGURE 1.43** Clay deposit

▼ **TABLE 1.16** Empirical Equations Related to c_u and Effective Overburden Pressure

Reference	Relationship	Remarks
Skempton (1957)	$\dfrac{c_{u(VST)}}{p} = 0.11 + 0.0037 PI$ PI = plasticity index (%) $c_{u(VST)}$ = undrained shear strength from vane shear test (See Chapter 3 for details of vane shear test)	For normally consolidated clay
Chandler (1988)	$\dfrac{c_{u(VST)}}{p_c} = 0.11 + 0.0037 PI$ p_c = preconsolidation pressure	Can be used for over consolidated soil Accuracy ±25% Not valid for sensitive and fissured clays
Jamiolkowski et al. (1985)	$\dfrac{c_u}{p_c} = 0.23 \pm 0.04$	For lightly overconsolidated clays
Mesri (1989)	$\dfrac{c_u}{p} = 0.22$	
Bjerrum and Simons (1960)	$\dfrac{c_u}{p} = f(LI)$ LI = liquidity index [See Eq. (1.53b) for definition]	See Figure 1.44 for normally consolidated clays
Ladd et al. (1977)	$\dfrac{\left(\dfrac{c_u}{p}\right)_{overconsolidated}}{\left(\dfrac{c_u}{p}\right)_{normally\ consolidated}} = (OCR)^{0.8}$ OCR = overconsolidation ratio = $\dfrac{p_c}{p}$	

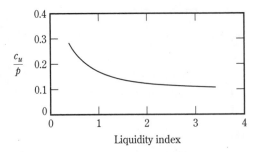

▼ **FIGURE 1.44**   Variation of c_u/p with liquidity index [based on Bjerrum and Simons (1960)]

sensitivity is the ratio of the unconfined compression strength in an undisturbed state to that in a remolded state, or

$$S_t = \frac{q_{u(\text{undisturbed})}}{q_{u(\text{remolded})}} \qquad\qquad (1.95)$$

The sensitivity ratio of most clays ranges from about 1 to 8; however, highly flocculent marine clay deposits may have sensitivity ratios ranging from about 10 to 80. Some clays turn to viscous liquids upon remolding, and these clays are referred to as "quick" clays. The loss of strength of clay soils from remolding is caused primarily by the destruction of the clay particle structure that was developed during the original process of sedimentation.

1.20 SOIL REINFORCEMENT—GENERAL

The use of reinforced earth is a recent development in the design and construction of foundations and earth-retaining structures. *Reinforced earth* is a construction material comprising soil that has been strengthened by tensile elements such as metal rods and/or strips, nonbiodegradable fabrics (geotextiles), geogrids, and the like. The fundamental idea of reinforcing soil is not new; in fact, it goes back several centuries. However, the present concept of systematic analysis and design was developed by a French engineer, H. Vidal (1966). The French Road Research Laboratory has done extensive research on the applicability and the beneficial effects of the use of reinforced earth as a construction material. This research has been documented in detail by Darbin (1970), Schlosser and Long (1974), and Schlosser and Vidal (1969). The tests conducted involved the use of metallic strips as reinforcing material.

Retaining walls with reinforced earth have been constructed around the world since Vidal began his work. The first reinforced earth retaining wall with metal strips as reinforcement in the United States was constructed in 1972 in southern California.

The beneficial effects of soil reinforcement derive from (a) the soil's increased tensile strength and (b) the shear resistance developed from the friction at the soil-reinforcement interfaces. Such reinforcement is comparable to that of concrete structures. Currently, most reinforced earth design is done with *free-draining granular soil only*. Thus the effect of pore water development in cohesive soils, which, in turn, reduces the shear strength of the soil, is avoided.

1.21 CONSIDERATIONS FOR SOIL REINFORCEMENT

Metal Strips

In most instances, galvanized steel strips are used as reinforcement in soil. However, galvanized steel is subject to corrosion. The rate of corrosion depends on several environmental factors. Binquet and Lee (1975) suggested that the average rate of corrosion of galvanized steel strips varies between 0.025 and 0.050 mm/yr. So, in the actual design of reinforcement, allowance must be made for the rate of corrosion. Thus

$$t_c = t_{design} + r \text{ (life span of structure)}$$

where t_c = actual thickness of reinforcing strips to be used in construction
  t_{design} = thickness of strips determined from design calculations
  r = rate of corrosion

Further research needs to be done on corrosion-resistant materials such as fiberglass before they can be used as reinforcing strips.

Nonbiodegradable Fabrics

Nonbiodegradable fabrics are generally referred to as *geotextiles*. Since 1970, the use of geotextiles in construction has increased tremendously around the world. The fabrics are usually made from petroleum products — polyester, polyethylene, and polypropylene. They may also be made from fiberglass. Geotextiles are not prepared from natural fabrics because they decay too quickly. Geotextiles may be woven, knitted, or nonwoven.

Woven geotextiles are made of two sets of parallel filaments or strands of yarn systematically interlaced to form a planar structure. *Knitted geotextiles* are formed by interlocking a series of loops of one or more filaments or strands of yarn to form a planar structure. *Nonwoven geotextiles* are formed from filaments or short fibers arranged in an oriented or random pattern in a planar structure. These filaments or short fibers are, in the beginning, arranged into a loose web. They are then bonded by one or a combination of the following processes:

1. *Chemical bonding* — by glue, rubber, latex, cellulose derivative, and the like
2. *Thermal bonding* — by heat for partial melting of filaments
3. *Mechanical bonding* — by needle punching

Needle-punched nonwoven geotextiles are thick and have high in-plane permeability.

Geotextiles have four primary uses in foundation engineering.

1. *Drainage:* The fabrics can rapidly channel water from soil to various outlets, thereby providing a higher soil shear strength and hence stability.

2. *Filtration:* When placed between two soil layers, one coarse grained and the other fine grained, the fabric allows free seepage of water from one layer to the other. However, it protects the fine-grained soil from being washed into the coarse-grained soil.

3. *Separation:* Geotextiles help keep various soil layers separate after construction and during the projected service period of the structure. For example, in the construction of highways, a clayey subgrade can be kept separate from a granular base course.

4. *Reinforcement:* The tensile strength of geofabrics increases the load-bearing capacity of the soil.

Geogrids

Geogrids are high-modulus polymer materials, such as polypropylene and polyethylene, and are prepared by tensile drawing. Netlon Ltd. of the United Kingdom was the first producer of geogrids. In 1982, the Tensar Corporation, presently Tensar Earth Technologies, Inc., introduced geogrids in the United States.

The major function of geogrids is *reinforcement.* Geogrids are relatively stiff netlike materials with large openings called *apertures.* These apertures are large enough to allow interlocking with the surrounding soil and/or rock to perform the function(s) of reinforcement and/or segregation.

Geogrids generally are of two types: (a) biaxial geogrids and (b) uniaxial geogrids. Figure 1.45a and 1.45b show the two types of geogrids just described, which are produced by Tensar Earth Technologies, Inc. Uniaxial TENSAR grids are manufactured by stretching a punched sheet of extruded high-density polyethylene in one direction under carefully controlled conditions. This process aligns the polymer's long-chain molecules in the direction of draw and results in a product with high one-directional tensile strength and modulus. Biaxial TENSAR grids are manufactured by stretching the punched sheet of polypropylene in two orthogonal directions. This process results in a product with high tensile strength and modulus in two perpendicular directions. The resulting grid apertures are either square or rectangular.

The commercial geogrids currently available for soil reinforcement have nominal rib thicknesses of about 0.02–0.06 in. (0.5–1.5 mm) and junctions of about 0.1–0.2 in. (2.5–5 mm). The grids used for soil reinforcement usually have apertures that are rectangular or elliptical in shape. The dimensions of the apertures vary from about 1–6 in. (25–150 mm). Geogrids are manufactured so that the open areas of the grids are greater than 50% of the total area. They develop reinforcing strength at low strain levels, such as 2% (Carroll, 1988). Table 1.17 gives some properties of the TENSAR biaxial geogrids currently available commercially.

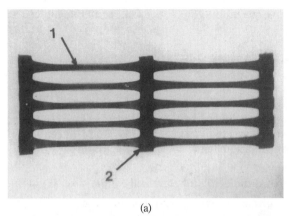

(a)

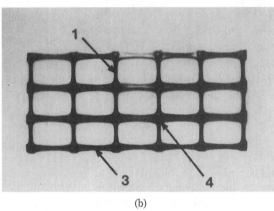

(b)

▼ **FIGURE 1.45** Geogrids: (a) uniaxial; (b) biaxial (*note:* 1–longitudinal rib; 2 — transverse bar; 3 — transverse rib; 4 — junction)

▼ **TABLE 1.17** Properties of TENSAR Biaxial Geogrids

Property	Geogrid		
	BX1000	BX1100	BX1200
Aperture size			
Machine direction	1 in. (nominal)	1 in. (nominal)	1 in. (nominal)
Cross-machine direction	1.3 in. (nominal)	1.3 in. (nominal)	1.3 in. (nominal)
Open area	70% (minimum)	74% (nominal)	77% (nominal)
Junction			
Thickness	0.09 in. (nominal)	0.11 in. (nominal)	0.16 in. (nominal)
Tensile modulus			
Machine direction	12,500 lb/ft (minimum)	14,000 lb/ft (minimum)	18,500 lb/ft (minimum)
Cross-machine direction	12,500 lb/ft (minimum)	20,000 lb/ft (minimum)	30,000 lb/ft (minimum)
Material			
Polypropylene	97% (minimum)	99% (nominal)	99% (nominal)
Carbon black	2% (minimum)	1% (nominal)	1% (nominal)

PROBLEMS

1.1   A soil specimen has a volume of 0.05 m^3 and a mass of 87.5 kg. Given: $w = 15\%$, $G_s = 2.68$. Determine
   a.   Void ratio
   b.   Porosity
   c.   Dry unit weight
   d.   Moist unit weight
   e.   Degree of saturation

1.2   The saturated unit weight of a soil is 20.1 kN/m^3 at a moisture content of 22%. Determine (a) the dry unit weight and (b) the specific gravity of soil solids, G_s.

1.3   For a soil, given: void ratio = 0.81, moisture content = 21%, and $G_s = 2.68$. Calculate the following:
   a.   Porosity
   b.   Degree of saturation
   c.   Moist unit weight in kN/m^3
   d.   Dry unit weight in kN/m^3

1.4   For a given soil, the following are given: moist unit weight = 122 lb/ft^3, moisture content = 14.7%, and $G_s = 2.68$. Calculate the following:
   a.   Void ratio
   b.   Porosity
   c.   Degree of saturation
   d.   Dry unit weight

1.5   For the soil described in Problem 1.4:
   a.   What would be the saturated unit weight in lb/ft^3?
   b.   How much water, in lb/ft^3, needs to be added to the soil for complete saturation?
   c.   What would be the moist unit weight in lb/ft^3 when the degree of saturation is 80%?

1.6   For a granular soil, given: $\gamma = 108$ lb/ft^3, $D_r = 82\%$, $w = 8\%$, and $G_s = 2.65$. For this soil, if $e_{min} = 0.44$, what would be e_{max}? What would be the dry unit weight in the loosest state?

1.7   The laboratory test results of six soils are given in the following table. Classify the soils by the AASHTO Soil Classification System and give the group indices.

▼   **Sieve Analysis — Percent Passing**

	Soil					
Sieve no.	A	B	C	D	E	F
4	100	100	95	95	100	100
10	95	80	80	90	94	94
40	82	61	54	79	76	86
200	65	55	8	64	33	76
Liquid limit	42	38	NP*	35	38	52
Plastic limit	26	25	NP	26	25	28

*NP = nonplastic

1.8   Classify the soils given in Problem 1.7 by the Unified Soil Classification System and determine the group symbols and group names.

1.9 For a sandy soil, given: void ratio, $e = 0.63$; hydraulic conductivity, $k = 0.22$ cm/sec; specific gravity of soil solids, $G_s = 2.68$. Estimate the hydraulic conductivity of the sand (cm/sec) when the dry unit weight of compaction is 117 lb/ft³. Use Eq. (1.29).

1.10 A sand has a hydraulic conductivity of 0.25 cm/sec at a void ratio of 0.7. Estimate the void ratio at which its hydraulic conductivity would be 0.115 cm/sec. Use Eq. (1.28).

1.11 The *in situ* hydraulic conductivity of a clay is 5.4×10^{-6} cm/sec at a void ratio of 0.92. What would be its hydraulic conductivity at a void ratio of 0.72? Use Eq. (1.31).

1.12 Refer to the soil profile shown in Figure P1.12. Determine the total stress, pore water pressure, and effective stress at *A, B, C,* and *D.*

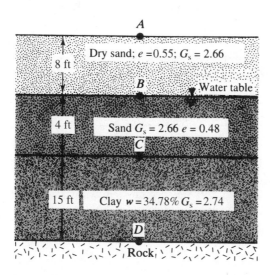

A

Dry sand; $e = 0.55$; $G_s = 2.66$

8 ft

B   Water table

Sand $G_s = 2.66$ $e = 0.48$

4 ft

C

15 ft   Clay $w = 34.78\%$ $G_s = 2.74$

D

Rock

▼ **FIGURE P1.12**

1.13 Refer to Problem 1.12. If the ground water table rises to 4 ft below the ground surface, what will be the change in the effective stress at *D*?

1.14 A sandy soil ($G_s = 2.65$), in its densest and loosest states, has void ratios of 0.42 and 0.85, respectively. Estimate the range of the critical hydraulic gradient in this soil at which a quicksand condition might occur.

1.15 A saturated clay deposit in the field has
Liquid limit = 52%
Plastic limit = 18%
Moisture content = 27%
Specific gravity of soil solids, $G_s = 2.69$
In situ effective overburden pressure = 79 kN/m²
Estimate the preconsolidation pressure, p_c
  a. By using Eqs. (1.53a) and (1.53b)
  b. By using Eqs. (1.54) and (1.55)

1.16 For a normally consolidated clay layer, the following are given:
Thickness = 3.7 m
Void ratio = 0.82
Liquid limit = 42
Average effective stress on the clay layer = 110 kN/m^2
How much consolidation settlement would the clay undergo if the average effective stress on the clay layer is increased to 155 kN/m^2 as the result of the construction of a foundation?

1.17 Refer to Problem 1.16. Assume that the clay layer is preconsolidated, p_c = 128 kN/m^2 and $C_s = \frac{1}{5}C_c$. Estimate the consolidation settlement.

1.18 Refer to the soil profile shown in Figure P1.12. The clay is normally consolidated. A laboratory consolidation test on the clay gave the following results:

Pressure (lb/in.^{2})	Void ratio
21	0.91
42	0.792

If the average effecive stress on the clay layer increases by 1000 lb/ft^2,
a. What would be the total consolidation settlement?
b. If $C_v = 1.45 \times 10^{-4}$ in^2/sec, how long will it take for half the consolidation settlement to take place?

1.19 For a normally consolidated soil, the following is given:

Pressure (kN/m^2)	Void ratio
120	0.82
360	0.64

Determine the following:
a. The compression index, C_c.
b. The void ratio corresponding to pressure of 200 kN/m^2.

1.20 A clay soil specimen, 1.5 in. thick (drained on top only) was tested in the laboratory. For a given load increment, the time for 60% consolidation was 8 min 10 sec. How long will it take for 50% consolidation for a similar clay layer in the field that is 10-ft thick and drained on both sides?

1.21 Refer to Figure P1.21. A total of 60 mm consolidation settlement is expected in the two clay layers due to a surcharge of Δp. Find the duration of surcharge application at which 30 mm of total settlement would take place.

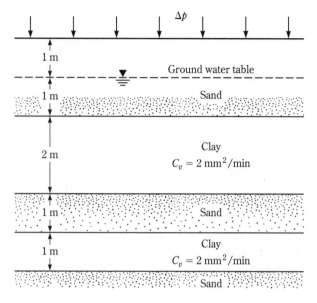

Δp

1 m

Ground water table

1 m   Sand

Clay
2 m
$C_v = 2 \ \text{mm}^2/\text{min}$

1 m   Sand

Clay
1 m
$C_v = 2 \ \text{mm}^2/\text{min}$

Sand

▼   **FIGURE P1.21**

1.22   A direct shear test was conducted on dry sand. The results were as follows:

Area of the specimen = 2 in. × 2 in.

Normal force (lb)	Shear force at failure (lb)
50	43.5
110	95.5
150	132.0

Graph the shear stress at failure against normal stress and determine the soil friction angle.

1.23   A consolidated-drained triaxial test on a sand yielded the following results:
All-around confining pressure = σ_3 = 30 lb/in^2
Added axial stress at failure = $\Delta\sigma$ = 96 lb/in^2
Determine the shear stress parameters.

1.24   Repeat Problem 1.23 with the following:
All-around confining pressure = σ_3 = 20 lb/in^2
Added axial stress at failure = $\Delta\sigma$ = 40 lb/in^2

1.25   A consolidated-drained triaxial test on a normally consolidated clay yielded a friction angle, ϕ, of 28°. If the all-around confining pressure during the test was 140 kN/m^2, what was the major principal stress at failure?

1.26 Following are the results of two consolidated-drained triaxial tests on a clay:

Test I: $\sigma_3 = 140$ kN/m^2; $\sigma_{1(\text{failure})} = 368$ kN/m^2

Test II: $\sigma_3 = 280$ kN/m^2; $\sigma_{1(\text{failure})} = 701$ kN/m^2

Determine the shear strength parameters; that is, c and ϕ.

1.27 A consolidated-undrained triaxial test was conducted on a saturated normally consolidated clay. Following are the test results:

$\sigma_3 = 13$ lb/in^2

$\sigma_{1(\text{failure})} = 32$ lb/in^2

Pore pressure at failure $= u_f = 5.5$ lb/in^2

Determine c_{cu}, ϕ_{cu}, c, and ϕ.

1.28 For a normally consolidated clay, given $\phi = 28°$ and $\phi_{cu} = 20°$. If a consolidated-undrained triaxial test is conducted on the same clay with $\sigma_3 = 150$ kN/m^2, what would be the pore water pressure at failure?

1.29 A saturated clay layer has

Saturated unit weight, $\gamma_{\text{sat}} = 19.6$ kN/m^3

Plasticity index = 21

The water table coincides with the ground surface. If the clay is normally consolidated, estimate the magnitude of c_u (kN/m^2) that can be obtained from a vane shear test at a depth of 8 m from the ground surface. Use the Skempton relationship given in Table 1.16.

REFERENCES

American Society for Testing and Materials (1997). *Annual Book of ASTM Standards,* Vol. 04.08, Conshohocken, Pa.

Azzouz, A. S., Krizek, R.J., and Corotis, R. B. (1976). "Regression Analysis of Soil Compressibility," *Soils and Foundations,* Vol. 16, No. 2, pp. 19–29.

Binquet, J., and Lee, K. L. (1975). "Bearing Capacity Analysis of Reinforced Earth Slabs," *Journal of the Geotechnical Engineering Division,* American Society of Civil Engineers, Vol. 101, No. GT12, pp. 1257–1276.

Bjerrum, L., and Simons, N. E. (1960). "Comparison of Shear Strength Characteristics of Normally Consolidated Clay," *Proceedings,* Research Conference on Shear Strength of Cohesive Soils, ASCE, 711–726.

Carroll, R., Jr. (1988). "Specifying Geogrids," *Geotechnical Fabric Report,* Industrial Fabric Association International, St. Paul, March/April.

Casagrande, A. (1936). "Determination of the Preconsolidation Load and Its Practical Significance," *Proceedings,* First International Conference on Soil Mechanics and Foundation Engineering, Cambridge, Mass., Vol. 3, pp. 60–64.

Chandler, R. J. (1988). "The *In Situ* Measurement of the Undrained Shear Strength of Clays Using the Field Vane," *STP 1014, Vane Shear Strength Testing in Soils: Field and Laboratory Studies,* ASTM, pp. 13–44.

Darbin, M. (1970). "Reinforced Earth for Construction of Freeways" (in French), *Revue Générale des Routes et Aerodromes,* No. 457, September.

Darcy, H. (1856). *Les Fontaines Publiques de la Ville de Dijon,* Paris.

Das, B. M. (1997). *Soil Mechanics Laboratory Manual,* 5th ed., Engineering Press, Austin, Tx.

Hazen, A. (1930). "Water Supply," *American Civil Engineers Handbook,* Wiley, New York.

Highway Research Board (1945). *Report of the Committee on Classification of Materials for Subgrades and Granular Type Roads,* Vol. 25, pp. 375–388.

Jamiolkowski, M., Ladd, C. C., Germaine, J. T., and Lancellotta, R. (1985). "New Developments in Field and Laboratory Testing of Soils," *Proceedings, XI International Conference on Soil Mechanics and Foundation Engineering,* San Francisco, Vol. 1, pp. 57–153.

Kenney, T. C. (1959). "Discussion," *Journal of the Soil Mechanics and Foundations Division,* American Society of Civil Engineers, Vol. 85, No. SM3, pp. 67–69.

Kulhawy, F. H., and Mayne, P. W. (1990). *Manual on Estimating Soil Properties for Foundation Design,* Report EL-6800, EPRI.

Ladd, C. C., and Foot, R. (1974). "New Design Procedure for Stability of Soft Clays," *Journal of the Geotechnical Engineering Division,* American Society of Civil Engineers, Vol. 92, No. GT2, pp. 79–103.

Ladd, C. C., Foote, R., Ishihara, K., Schlosser, F., and Poulos, H. G. (1977). "Stress Deformation and Strength Characteristics," *Proceedings,* 9th International Conference on Soil Mechanics and Foundation Engineering, Tokyo, Vol. 2, 421–494.

Leroueil, S., Tavenas, F., and LeBihan, J. P. (1983). "Propriétés Caractéristiques des Argiles de l'est due Canada," *Canadian Geotechnical Journal,* Vol. 20, No. 4, pp. 681–705.

Martin, R. E., Drahos, E. G., and Pappas, J. L. (1995). "Characterization of Preconsolidated Soils in Richmond, Virginia," *Transportation Research Record No. 1479,* National Research Council, Washington, D.C., pp. 89–98.

Mesri, G. (1989). "A Re-evaluation of $s_{u(\text{mob})} \approx 0.22\sigma_p$ Using Laboratory Shear Tests," *Canadian Geotechnical Journal,* Vol. 26, No. 1, pp. 162–164.

Nagaraj, T. S., and Murthy, B. R. S. (1985). "Prediction of the Preconsolidation Pressure and Recompression Index of Soils," *Geotechnical Testing Journal,* American Society for Testing and Materials, Vol. 8, No. 4, pp. 199–202.

Rendon-Herrero, O. (1980). "Universal Compression Index Equation," *Journal of the Geotechnical Engineering Division,* American Society of Civil Engineers, Vol. 106, No. GT11, pp. 1178–1200.

Schlosser, F., and Long, N. (1974). "Recent Results in French Research on Reinforced Earth," *Journal of the Construction Division,* American Society of Civil Engineers, Vol. 100, No. CO3, pp. 113–237.

Schlosser, F., and Vidal, H. (1969). "Reinforced Earth" (in French), *Bulletin de Liaison des Laboratoires Routier,* Ponts et Chassées, Paris, France, November, pp. 101–144.

Schmertmann, J. H. (1953). "Undisturbed Consolidation Behavior of Clay," *Transactions,* American Society of Civil Engineers, Vol. 120, p. 1201.

Sivaram, B., and Swamee, P. (1977). "A Computational Method for Consolidation Coefficient," *Soils and Foundations,* Tokyo, Japan, Vol. 17, No. 2, pp. 48–52.

Skempton, A. W. (1944). "Notes on the Compressibility of Clays," *Quarterly Journal of Geological Society,* London, Vol. C, pp. 119–135.

Skempton, A. W. (1957). "The Planning and Design of New Hong Kong Airport," *Proceedings,* The Institute of Civil Engineers, London, Vol. 7, pp. 305–307.

Skempton, A. W. (1957). "Discussion: The Planning and Design of New Hong Kong Airport," *Proceedings,* Institute of Civil Engineers, London, Vol. 7, 305–307.

Stark, T. D. (1995). "Measurement of Drained Residual Strength of Overconsolidated Clays," *Transportation Research Record No. 1479,* National Research Council, Washington, D.C., pp. 26–34.

Stas, C. V., and Kulhawy, F. H. (1984). *Critical Evaluation of Design Methods for Foundations Under Axial Uplift and Compression Loading,* Report EL-3771, EPRI.

Tavenas, F., Jean, P., Leblond, P., and Leroueil, S. (1983). "The Permeability of Natural Soft Clays. Part II: Permeability Characteristics," *Canadian Geotechnical Journal,* Vol. 20, No. 4, pp. 645–660.

Terzaghi, K., and Peck, R. B. (1967). *Soil Mechanics in Engineering Practice,* Wiley, New York.

U.S. Department of the Navy (1971). "Design Manual — Soil Mechanics, Foundations and Earth Structures," *NAVFAC DM-7,* U.S. Government Printing Office, Washington, D.C.

U.S. Department of the Navy (1982). "Soil Mechanics," *NAVFAC DM7.1,* U.S. Government Printing Office, Washington, D.C.

Vesic, A. S. (1963). "Bearing Capacity of Deep Foundations in Sand," *Highway Research Record No. 39,* National Academy of Sciences, Washington, D.C., pp. 112–154.

Vidal, H. (1966). "La terre Armee," *Anales de l'Institut Technique du Bâtiment et des Travaux Publiques,* France, July–August, pp. 888–938.

Wood, D. M. (1983). "Index Properties and Critical State Soil Mechanics," *Proceedings,* Symposium on Recent Developments in Laboratory and Field Tests and Analysis of Geotechnical Problems, Bangkok, pp. 301–309.

Wroth, C. P., and Wood, D. M. (1978). "The Correlation of Index Properties with Some Basic Engineering Properties of Soils," *Canadian Geotechnical Journal,* Vol. 15, No. 2, pp. 137–145.

Youd, T. L. (1973). "Factors Controlling Maximum and Minimum Densities of Sands," *Special Technical Publication No. 523,* American Society for Testing and Materials, pp. 98–122.

CHAPTER TWO

NATURAL SOIL DEPOSITS AND SUBSOIL EXPLORATION

2.1 INTRODUCTION

To design a foundation that will adequately support a structure, an engineer must understand the type of soil deposits that will support the foundation. Moreover, foundation engineers must remember that soil at any site frequently is non-homogeneous — that is, the soil profile may vary. Soil mechanics theories involve idealized conditions, so the application of these theories to foundation engineering problems involves judicious evaluation of site conditions and soil parameters. To do so requires some knowledge of the geological process by which the soil deposit at the site was formed, supplemented by subsurface exploration. Good professional judgment constitutes an essential part of geotechnical engineering — and it comes only with practice.

This chapter is divided into two parts. The first is a general overview of natural soil deposits generally encountered, and the second describes the general principles of subsoil exploration.

NATURAL SOIL DEPOSITS

2.2 SOIL ORIGIN

Most of the soils that cover the earth are formed by the weathering of various rocks. There are two general types of weathering: (1) mechanical weathering and (2) chemical weathering.

Mechanical weathering is the process by which rocks are broken into smaller and smaller pieces by physical forces. These physical forces may be running water, wind, ocean waves, glacier ice, frost action, and expansion and contraction caused by gain and loss of heat.

Chemical weathering is the process of chemical decomposition of the original rock. In the case of mechanical weathering, the rock breaks into smaller pieces without a change of chemical composition. However, in chemical weathering, the original material may be changed to something entirely different. For example, the chemical weathering of feldspar can produce clay minerals.

Soil produced by the weathering of rocks can be transported by physical processes to other places. These soil deposits are called *transported soils*. In contrast, some soils stay where they were formed and cover the rock surface from which they derive. These soils are referred to as *residual soils*.

Based on the *transporting agent,* transported soils can be subdivided into three major categories:

1. *Alluvial,* or *fluvial*: deposited by running water
2. *Glacial*: deposited by glacier action
3. *Aeolian*: deposited by wind action

In addition to transported and residual soils, there are *peats* and *organic soils,* which derive from the decomposition of organic materials.

2.3   RESIDUAL SOIL

Residual soil deposits are common in the tropics, Hawaii, and the southeastern United States. The nature of a residual soil deposit will generally depend on the parent rock. When hard rocks such as granite and gneiss undergo weathering, most of the materials are likely to remain in place. These soil deposits generally have a top layer of clayey or silty clay material below which are silty and/or sandy soil layers. They are generally underlain by a partially weathered rock and then sound bedrock. The depth of the sound bedrock may vary widely, even within a distance of a few meters. Figure 2.1 shows the log of a boring in a residual soil deposit derived from the weathering of granite.

In contrast to hard rocks, some chemical rocks, such as limestone, are made up chiefly of calcite ($CaCO_3$) mineral. Chalk and dolomite have large concentrations of dolomite minerals [$CaMg(CO_3)_2$]. These rocks have large amounts of soluble materials, some of which are removed by ground water, leaving behind the insoluble fraction of the rock. Residual soils that derive from chemical rocks possess a gradual transition zone to the bedrock, as shown in Figure 2.1. The residual soils derived from the weathering of limestonelike rocks are mostly gray in color. Although uniform in kind, the depth of weathering may vary greatly. The residual soils immediately above the bedrock may be normally consolidated. Large foundations with heavy loads may be susceptible to large consolidation settlements on these soils.

2.4   ALLUVIAL DEPOSITS

Alluvial soil deposits derive from the action of streams and rivers. They can be divided into two major categories: (1) *braided-stream deposits* and (2) deposits caused by the *meandering belt of streams.*

Deposits from Braided Streams

Braided streams are high-gradient, rapidly flowing streams. They are highly erosive and carry large amounts of sediment. Because of the high bed load, a minor change

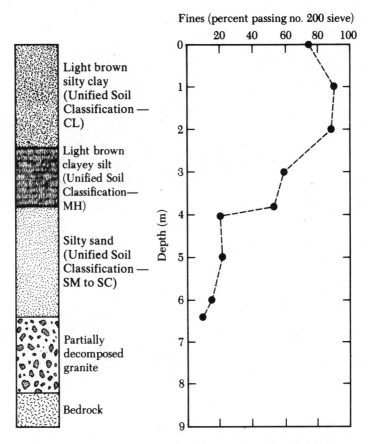

▼ **FIGURE 2.1**   Boring log for a residual soil derived from granite

in the velocity of flow will cause deposit of sediments. By this process, these streams may build up a complex tangle of converging and diverging channels separated by sandbars and islands.

The deposits formed from braided streams are very irregular in stratification and have a wide range of grain sizes. Figure 2.2 shows a cross section of such a deposit. These deposits share several characteristics.

1. The grain sizes usually range from gravel to silt. Clay-size particles are generally *not* found in these deposits.

2. Although grain size varies widely, the soil in a given pocket or lens is rather uniform.

3. At any given depth, the void ratio and unit weight may vary over a wide range within a lateral distance of only a few meters. This variation can be observed during soil exploration for construction of a foundation for a structure. The standard penetration resistance (*N* value) at a given depth obtained from various bore holes will be highly irregular and variable.

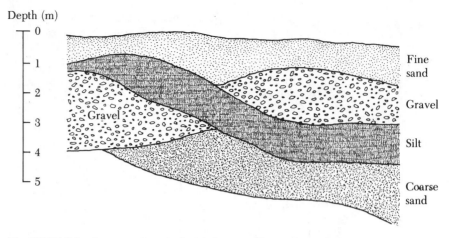

Depth (m)

Fine sand

Gravel

Gravel

Silt

Coarse sand

▼   **FIGURE 2.2**   Cross section of a braided-stream deposit

Alluvial deposits are present in several parts of the western United States, such as Southern California, Utah, and the basin and range sections of Nevada. Also, a large amount of sediment originally derived from the Rocky Mountain range was carried eastward to form the alluvial deposits of the Great Plains. On a smaller scale, this type of natural soil deposit, left by braided streams, can be encountered locally.

Meander Belt Deposits

The term *meander* is derived from the Greek word *maiandros,* which means "bends." Mature streams in a valley curve back and forth. The valley floor in which a river meanders is referred to as the *meander belt.* In a meandering river, the soil from the bank is continually eroded from the points where it is concave in shape and deposited at points where the bank is convex in shape, as shown in Figure 2.3. These deposits are called *point bar deposits,* and they usually consist of sand and silt-size particles. Sometimes, during the process of erosion and deposition, the river abandons a meander and cuts a shorter path. The abandoned meander, when filled with water, is called an *oxbow lake* (see Figure 2.3).

During floods, rivers overflow low-lying areas. The sand and silt-size particles carried by the river are deposited along the banks to form ridges known as *natural levees* (Figure 2.4). Finer soil particles consisting of silts and clays are carried by the water farther onto the flood plains. These particles settle at different rates to form what is referred to as *backswamp deposits* (Figure 2.4). These clays may be highly plastic. Table 2.1 gives the properties of soil deposits found in natural levees, point bars, abandoned channels, backswamps, and swamps in the Mississippi alluvial valley.

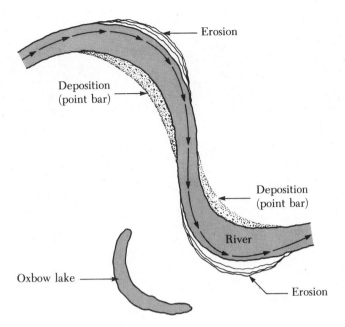

▼ FIGURE 2.3 Formation of point bar deposits and oxbow lake in a meandering stream

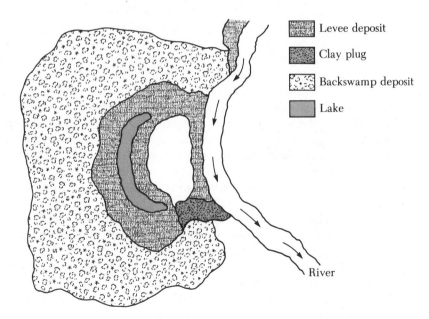

▼ FIGURE 2.4 Levee and backswamp deposit

▼ **TABLE 2.1** Properties of Deposits Within the Mississippi Alluvial Valley[a]

Environment	Soil texture	Natural water content (%)	Liquid limit	Plasticity index	Cohesion[b] (kN/m²)	Angle of friction (deg)
					Shear strength	
Natural levees	Clay (CL)	25–35	35–45	15–25	17–57	0
	Silt (ML)	15–35	NP[c]–35	NP–5	9–33	10–35
Point bar	Silt (ML) and silty sand (SM)	25–45	30–55	10–25	0–41	25–35
Abandoned channel	Clay (CL, CH)	30–95	30–100	10–65	14–57	0
Backswamps	Clay (CH)	25–70	40–115	25–100	19–120	0
Swamp	Organic clay (OH)	100–265	135–300	100–165	—	—

[a] After Kolb and Shockley (1959)
[b] Rounded off
[c] NP = nonplastic

2.5 GLACIAL DEPOSITS

During the Pleistocene Ice Age, glaciers covered large areas of the earth. The glaciers advanced and retreated with time. During their advance, the glaciers carried large amounts of sand, silt, clay, gravel, and boulders. *Drift* is a general term usually applied to the deposits laid down by glaciers. Unstratified deposits laid down by glaciers when they melt are referred to as *till*. The physical characteristics of till may vary from glacier to glacier.

The land forms that developed from the deposits of till are called *moraines*. A *terminal moraine* (Figure 2.5) is a ridge of till that marks the maximum limit of a glacier's advance. *Recessional moraines* are ridges of till developed behind the terminal moraine at varying distances apart. They are the result of temporary stabilization of the glacier during the recessional period. The till deposited by the glacier between the moraines is referred to as *ground moraine* (Figure 2.5). Ground moraine constitute large areas of the central United States and are called *till plains*.

The sand, silt, and gravel that are carried by the melting water from the front of a glacier are called *outwash*. In a pattern similar to the braided-stream deposits, the melted water deposits the outwash, forming *outwash plains* (Figure 2.5). They are usually called *glaciofluvial deposits*.

The range of grain sizes present in a till varies greatly. Figure 2.6 compares the grain-size distribution of *glacial till* and *dune sand* (see Section 2.6). The amount of clay-size fractions present and the plasticity indices of tills also vary widely. Field exploration may also reveal erratic values of standard penetration resistances.

Glacial water also carries with it silts and clays. The water finds its way to many basins and forms lakes. The silt particles initially tend to settle to the bottom of the lake when the water is still. During the winter, when the top of the lake freezes, the suspended clay particles gradually settle to the bottom. During the summer,

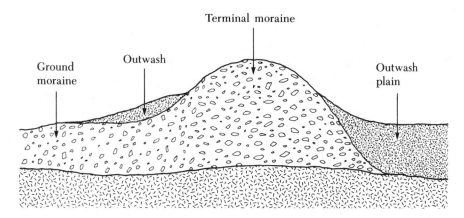

▼ **FIGURE 2.5**   Terminal moraine, ground moraine, and outwash plain

the snow on the lake melts. The supply of fresh water, loaded with sediments, repeats the process. As a result, the lacustrine soil formed from such a deposit has alternate layers of silt and clay. This soil is called *varved clay*. The varves are usually a few millimeters thick; however, in some instances they can be 50–100 mm (2–4 in.) thick. Varved clays can be found in the Northeast and the Pacific Northwest of the United States. They are mostly normally consolidated and may be sensitive. The hydraulic conductivity in the vertical direction is usually several times smaller than that in the horizontal direction. The load-bearing capacity of these deposits is quite low, and significant settlement of structures with shallow foundations may be anticipated.

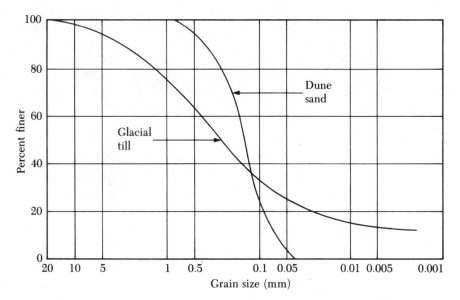

▼ **FIGURE 2.6**   Comparison of the grain-size distribution between glacial till and dune sand

2.6 AEOLIAN SOIL DEPOSITS

Wind is also a major transporting agent leading to formation of soil deposits. When large areas of sand lie exposed, wind can blow it away and redeposit it somewhere else. Deposits of windblown sand generally take the shape of *dunes* (Figure 2.7). As dunes are formed, the sand is blown over the crest by the wind. Beyond the crest, the sand particles roll down the slope. This process tends to form a *compact sand deposit* on the *windward side* and a rather *loose deposit* on the *leeward side*.

Dunes exist along the southern and eastern shores of Lake Michigan, the Atlantic Coast, the southern coast of California, and at various places along the coasts of Oregon and Washington. Sand dunes can also be found in the alluvial and rocky plains of the western United States. Following are some of the typical properties of *dune sand*:

1. The grain-size distribution of the sand at any particular location is surprisingly uniform. This uniformity can be attributed to the sorting action of the wind.

2. The general grain size decreases with the distance from the source because the wind carries the small particles farther than the large ones.

3. The relative density of sand deposited on the windward side of dunes may be as high as 50–65% and may decrease to about 0–15% on the leeward side.

Loess is an aeolian deposit consisting of silt and silt-size particles. The grain-size distribution of loess is rather uniform. The cohesion of loess is generally derived from a clay coating over the silt-size particles, which contributes to a stable soil structure in an unsaturated state. The cohesion may also be the result of the precipitation of chemicals leached by rain water. Loess is a *collapsing* soil, because when the soil becomes saturated, it loses its binding strength between the soil particles. Special precautions need to be taken for construction of foundations over loessial deposits. There are extensive deposits of loess in the United States — mostly in the midwestern states of Iowa, Missouri, Illinois, and Nebraska and for some distance along the Mississippi River in Tennessee and Mississippi.

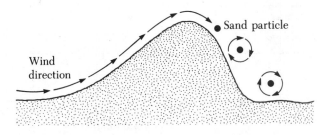

▼ FIGURE 2.7 Sand dune

2.7   ORGANIC SOIL

Organic soils are usually found in low-lying areas where the water table is near or above the ground surface. The presence of a high water table helps in the growth of aquatic plants that, when decomposed, form organic soil. This type of soil deposit is usually encountered in coastal areas and in glaciated regions. Organic soils show the following characteristics:

1. The natural moisture content may range from 200% to 300%.
2. They are highly compressible.
3. Laboratory tests have shown that, under loads, a large amount of settlement is derived from secondary consolidation.

2.8   SOME LOCAL TERMS FOR SOILS

Soils are sometimes referred to by local terms. Following are a few of these terms with a brief description of each:

1. *Caliche:* a Spanish word derived from the Latin word *calix,* meaning *lime.* It is found mostly in the desert southwestern United States. It is a mixture of sand, silt, and gravel bonded together by *calcareous deposits.* The calcareous deposits are brought to the surface by a net upward migration of water. The water evaporates in the high local temperature. Because of the sparse rainfall, the carbonates are not washed out of the top layer of soil.

2. *Gumbo:* a highly plastic, clayey soil.

3. *Adobe:* a highly plastic, clayey soil found in the southwestern United States.

4. *Terra Rossa:* residual soil deposits that are red in color and derive from limestone and dolomite.

5. *Muck:* organic soil with a very high moisture content.

6. *Muskeg:* organic soil deposit.

7. *Saprolite:* residual soil deposit derived from mostly insoluble rock.

8. *Loam:* a mixture of soil grains of various sizes, such as sand, silt, and clay.

9. *Laterite:* characterized by the accumulation of iron oxide (Fe_2O_3) and aluminum oxide (Al_2O_3) near the surface and the leaching of silica. Lateritic soils in Central America contain about 80–90% of clay and silt-size particles. In the United States, lateritic soils are present in the southeastern states of Alabama, Georgia, and the Carolinas.

10. *Peat:* partly decayed organic matter.

Figure 2.8 shows the general nature of the various soil deposits encountered in the United States.

Glacial Soils

Young and old drift, including associated sands and gravel

Lacustrine deposits, predominantly silts and clays

Loessial Soils

Silts and very fine sands

Soils of the Coastal Plain

Sand-clay; interbedded and mixed sands, gravels, clays, and silts; gravel and sand; or sand

Clay

Soils of the Filled Valleys and Great Plains Outwash Mantle

Predominantly sands and gravels with silts; sandy clays and clays

Residual Soils

All types

Recent Alluvium

Predominantly silts and clays

Nonsoil areas

▶ **FIGURE 2.8**   Soil deposits of the United States (adapted from *Foundation Engineering*, Second Edition, by R. B. Peck, W. E. Hanson, and T. H. Thornburn. Copyright 1974 by John Wiley and Sons. Reprinted by permission.)

SUBSURFACE EXPLORATION

2.9   PURPOSE OF SOIL EXPLORATION

The process of identifying the layers of deposits that underlie a proposed structure and their physical characteristics is generally referred to as *subsurface exploration*. The purpose of subsurface exploration is to obtain information that will aid the geotechnical engineer in

1. Selecting the type and depth of foundation suitable for a given structure.
2. Evaluating the load-bearing capacity of the foundation.
3. Estimating the probable settlement of a structure.
4. Determining potential foundation problems (for example, expansive soil, collapsible soil, sanitary landfill, and so on).
5. Determining the location of the water table.
6. Predicting lateral earth pressure for structures such as retaining walls, sheet pile bulkheads, and braced cuts.
7. Establishing construction methods for changing subsoil conditions.

Subsurface exploration may also be necessary when additions and alterations to existing structures are contemplated.

2.10   SUBSURFACE EXPLORATION PROGRAM

Subsurface exploration comprises several steps, including collection of preliminary information, reconnaissance, and site investigation.

Collection of Preliminary Information

This step includes obtaining information regarding the type of structure to be built and its general use. For the construction of buildings, the approximate column loads and their spacing and the local building-code and basement requirements should be known. The construction of bridges requires determining span length and the loading on piers and abutments.

A general idea of the topography and the type of soil to be encountered near and around the proposed site can be obtained from the following sources.

1. United States Geological Survey maps.
2. State government geological survey maps.
3. United States Department of Agriculture's Soil Conservation Service county soil reports.

4. Agronomy maps published by the agriculture departments of various states.
5. The hydrological information published by the United States Corps of Engineers. These include the records of stream flow, high flood levels, tidal records, and so on.
6. Highway department soils manuals published by several states.

The information collected from these sources can be extremely helpful in planning a site investigation. In some cases, substantial savings may be realized by anticipating problems that may be encountered later in the exploration program.

Reconnaissance

The engineer should always make a visual inspection of the site to obtain information about

1. The general topography of the site, possible existence of drainage ditches, abandoned dumps of debris, or other materials. Also, evidence of creep of slopes and deep, wide shrinkage cracks at regularly spaced intervals may be indicative of expansive soils.
2. Soil stratification from deep cuts, such as those made for construction of nearby highways and railroads.
3. Type of vegetation at the site, which may indicate the nature of the soil. For example, a mesquite cover in central Texas may indicate the existence of expansive clays that can cause possible foundation problems.
4. High-water marks on nearby buildings and bridge abutments.
5. Ground water levels, which can be determined by checking nearby wells.
6. Types of construction nearby and existence of any cracks in walls or other problems.

The nature of stratification and physical properties of the soil nearby can also be obtained from any available soil-exploration reports for existing structures.

Site Investigation

The site investigation phase of the exploration program consists of planning, making test boreholes, and collecting soil samples at desired intervals for subsequent observation and laboratory tests. The approximate required minimum depth of the borings should be predetermined. The depth can be changed during the drilling operation, depending on the subsoil encountered. To determine the approximate minimum depth of boring, engineers may use the rules established by the American Society of Civil Engineers (1972):

1. Determine the net increase of stress, $\Delta\sigma$, under a foundation with a depth as shown in Figure 2.9. (The general equations for estimating stress increase are given in Chapter 4.)

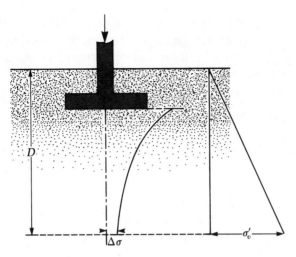

▼ **FIGURE 2.9**   Determination of the minimum depth
of boring

2. Estimate the variation of the vertical effective stress, σ_v', with depth.
3. Determine the depth, $D = D_1$, at which the stress increase $\Delta\sigma$ is equal to $(\frac{1}{10})q$ (q = estimated net stress on the foundation).
4. Determine the depth, $D = D_2$, at which $\Delta\sigma/\sigma_v' = 0.05$.
5. Unless bedrock is encountered, the smaller of the two depths, D_1 and D_2, just determined is the approximate minimum depth of boring required.

If the preceding rules are used, the depths of boring for a building with a width of 30.5 m (100 ft) will be approximately the following, according to Sowers and Sowers (1970):

No. of stories	Boring depth	
1	3.5 m	(11 ft)
2	6 m	(20 ft)
3	10 m	(33 ft)
4	16 m	(53 ft)
5	24 m	(79 ft)

For hospitals and office buildings, they also use the following rule to determine boring depth.

$$D_b = 3S^{0.7} \qquad \text{(for light steel or narrow concrete buildings)} \qquad (2.1a)$$

and

$$D_b = 6S^{0.7} \qquad \text{(for heavy steel or wide concrete buildings)} \qquad (2.1b)$$

where D_b = depth of boring, in meters
         S = number of stories

In English units, the preceding equations take the form

$$D_b \text{ (ft)} = 10S^{0.7} \qquad \text{(for light steel or narrow concrete buildings)} \qquad (2.2a)$$

and

$$D_b \text{ (ft)} = 20S^{0.7} \qquad \text{(for heavy steel or wide concrete buildings)} \qquad (2.2b)$$

When deep excavations are anticipated, the depth of boring should be at least 1.5 times the depth of excavation.

Sometimes subsoil conditions require that the foundation load be transmitted to bedrock. The minimum depth of core boring into the bedrock is about 3 m (10 ft). If the bedrock is irregular or weathered, the core borings may have to be deeper.

There are no hard and fast rules for borehole spacing. Table 2.2 gives some general guidelines. Spacing can be increased or decreased, depending on the subsoil condition. If various soil strata are more or less uniform and predictable, fewer boreholes are needed than in nonhomogeneous soil strata.

The engineer should also take into account the ultimate cost of the structure when making decisions regarding the extent of field exploration. The exploration cost generally should be 0.1–0.5% of the cost of the structure. Soil borings can be made by several methods, including auger boring, wash boring, percussion drilling, and rotary drilling.

▼ **TABLE 2.2** Approximate Spacing of Boreholes

Type of project	Spacing (m)	Spacing (ft)
Multistory building	10–30	30–100
One-store industrial plants	20–60	60–200
Highways	250–500	800–1600
Residential subdivision	250–500	800–1600
Dams and dikes	40–80	130–260

2.11 EXPLORATORY BORINGS IN THE FIELD

Auger boring is the simplest method of making exploratory boreholes. Figure 2.10 shows two types of hand auger — the *post hole auger* and the *helical auger*. Hand augers cannot be used for advancing holes to depths exceeding 3–5 m (10–16 ft). However, they can be used for soil exploration work for some highways and small structures. *Portable power-driven helical augers* (76.2 mm to 304.8 mm in diameter) are available for making deeper boreholes. The soil samples obtained from such borings are highly disturbed. In some noncohesive soils or soils having low cohesion, the walls of the boreholes will not stand unsupported. In such circumstances, a metal pipe is used as a *casing* to prevent the soil from caving in.

When power is available, *continuous-flight augers* are probably the most common method used for advancing a borehole. The power for drilling is delivered by truck- or tractor-mounted drilling rigs. Boreholes up to about 60–70 m (200–230 ft) can be easily made by this method. Continuous-flight augers are available in sections of about 1–2 m (3–6 ft) with either a solid or hollow stem. Some of the commonly used solid stem augers have outside diameters of 66.68 mm ($2\frac{5}{8}$ in.), 82.55 mm ($3\frac{1}{4}$ in.), 101.6 mm (4 in.), and 114.3 mm ($4\frac{1}{2}$ in.). Common hollow stem augers commercially available have dimensions of 63.5 mm ID and 158.75 mm OD (2.5 in. × 6.25 in.), 69.85 mm ID and 177.8 OD (2.75 in. × 7 in.), 76.2 mm ID and 203.2 OD (3 in. × 8 in.), and 82.55 mm ID and 228.6 mm OD (3.25 in. × 9 in.).

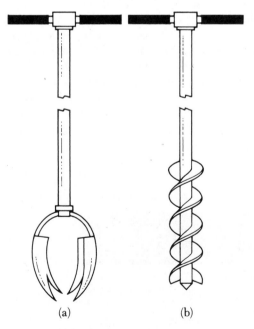

(a)   (b)

▼   **FIGURE 2.10**   Hand tools: (a) post hole auger; (b) helical auger

▼ **FIGURE 2.11**   Carbide-tipped cutting head on auger flight attached with bolt
(courtesy of William B. Ellis, El Paso Engineering and Testing,
Inc., El Paso, Texas)

The tip of the auger is attached to a cutter head (Figure 2.11). During the
drilling operation (Figure 2.12), section after section of auger can be added and the
hole extended downward. The flights of the augers bring the loose soil from the
bottom of the hole to the surface. The driller can detect changes in soil type by
noting changes in the speed and sound of drilling. When solid stem augers are
used, the auger must be withdrawn at regular intervals to obtain soil samples
and also to conduct other operations such as standard penetration tests. Hollow
stem augers have a distinct advantage over solid stem augers in that they do not
have to be removed frequently for sampling or other tests. As shown schematically in Figure 2.13, the outside of the hollow stem auger acts as a casing. A
removable plug is attached to the bottom of the auger by means of a center rod.
During the drilling, the plug can be pulled out with the auger in place, and soil

▼ **FIGURE 2.12**   Drilling with continuous-flight augers (courtesy of Danny R. Anderson, Danny R. Anderson Consultants, El Paso, Texas)

sampling and standard penetration tests can be performed. When hollow stem augers are used in sandy soils below the water table, the sand may be pushed several feet into the stem of the auger by excess hydrostatic pressure immediately after removal of the plug. Under such conditions, the plug should not be used. Instead, water inside the hollow stem should be maintained at a higher level than the water table.

Wash boring is another method of advancing boreholes. In this method, a casing about 2–3 m (6–10 ft) long is driven into the ground. The soil inside the casing is

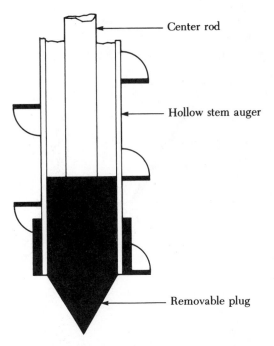

Center rod

Hollow stem auger

Removable plug

▼ **FIGURE 2.13** Schematic diagram of the hollow stem auger with removable plug

then removed by means of a chopping bit attached to a drilling rod. Water is forced through the drilling rod and exits at a very high velocity through the holes at the bottom of the chopping bit (Figure 2.14). The water and the chopped soil particles rise in the drill hole and overflow at the top of the casing through a T connection. The washwater is collected in a container. The casing can be extended with additional pieces as the borehole progresses; however, that is not required if the borehole will stay open and not cave in.

Rotary drilling is a procedure by which rapidly rotating drilling bits attached to the bottom of drilling rods cut and grind the soil and advance the borehole. There are several types of drilling bit. Rotary drilling can be used in sand, clay, and rocks (unless badly fissured). Water, or *drilling mud*, is forced down the drilling rods to the bits, and the return flow forces the cuttings to the surface. Boreholes with diameters of 50.8–203.2 mm (2–8 in.) can be easily made by this technique. The drilling mud is a slurry of water and bentonite. Generally, it is used when the soil encountered is likely to cave in. When soil samples are needed, the drilling rod is raised and the drilling bit is replaced by a sampler.

Percussion drilling is an alternative method of advancing a borehole, particularly through hard soil and rock. A heavy drilling bit is raised and lowered to chop the hard soil. The chopped soil particles are brought up by circulation of water. Percussion drilling may require casing.

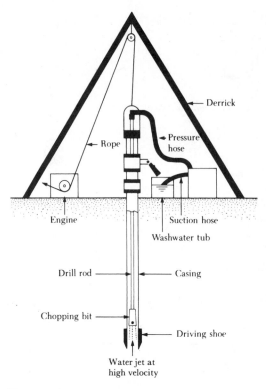

▼   **FIGURE 2.14**   Wash boring

2.12 PROCEDURES FOR SAMPLING SOIL

Two types of soil samples can be obtained during subsurface exploration: *disturbed* and *undisturbed*. Disturbed but representative samples can generally be used for the following types of laboratory test:

1. Grain-size analysis
2. Determination of liquid and plastic limits
3. Specific gravity of soil solids
4. Organic content determination
5. Classification of soil

Disturbed soil samples, however, cannot be used for consolidation, hydraulic conductivity, or shear strength tests. Undisturbed soil samples must be obtained for these types of laboratory tests.

Split-Spoon Sampling

Split-spoon samplers can be used in the field to obtain soil samples that are generally disturbed but still representative. A section of a *standard split-spoon sampler* is

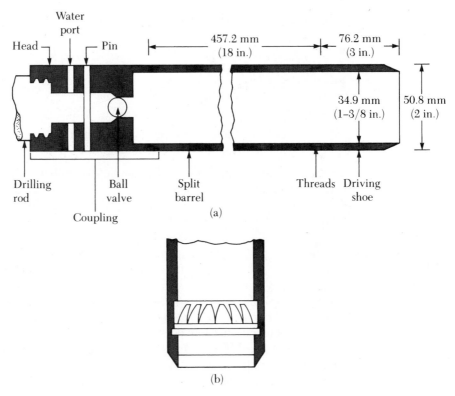

▼ **FIGURE 2.15** (a) Standard split-spoon sampler; (b) spring core catcher

shown in Figure 2.15a. It consists of a tool-steel driving shoe, a steel tube that is split longitudinally in half, and a coupling at the top. The coupling connects the sampler to the drill rod. The standard split tube has an inside diameter of 34.93 mm ($1\frac{3}{8}$ in.) and an outside diameter of 50.8 mm (2 in.); however, samplers having inside and outside diameters up to 63.5 mm ($2\frac{1}{2}$ in.) and 76.2 mm (3 in.), respectively, are also available. When a borehole is extended to a predetermined depth, the drill tools are removed and the sampler is lowered to the bottom of the borehole. The sampler is driven into the soil by hammer blows to the top of the drill rod. The standard weight of the hammer is 622.72 N (140 lb), and for each blow the hammer drops a distance of 0.762 m (30 in.). The number of blows required for spoon penetration of three 152.4-mm (6-in.) intervals are recorded. The number of blows required for the last two intervals are added to give the *standard penetration number* at that depth. This number is generally referred to as the *N value* (American Society for Testing and Materials, 1992, Designation D-1586-84). The sampler is then withdrawn, and the shoe and coupling are removed. The soil sample recovered from the tube is then placed in a glass bottle and transported to the laboratory.

The degree of disturbance for a soil sample is usually expressed as

$$A_R(\%) = \frac{D_o^2 - D_i^2}{D_i^2}(100)$$

(2.3)

where A_R = area ratio
D_o = outside diameter of the sampling tube
D_i = inside diameter of the sampling tube

When the area ratio is 10% or less, the sample generally is considered to be undisturbed. For a standard split-spoon sampler

$$\dot{A}_R(\%) = \frac{(50.8)^2 - (34.93)^2}{(34.93)^2}(100) = 111.5\%$$

Hence these samples are highly disturbed. Split-spoon samples generally are taken at intervals of about 1.53 m (5 ft). When the material encountered in the field is sand (particularly fine sand below the water table), sample recovery by a split-spoon sampler may be difficult. In that case, a device such as a *spring core catcher* may have to be placed inside the split spoon (Figure 2.15b).

Besides obtaining soil samples, standard penetration tests provide several useful correlations. For example, the consistency of clayey soils can often be estimated from the standard penetration number, N, as shown in Table 2.3. However, correlations for clays require tests to verify that the relationships are valid for the clay deposit being examined.

The literature contains many correlations between the standard penetration number and the undrained shear strength of clay, c_u. Based on the results of undrained triaxial tests conducted on insensitive clays, Stroud (1974) suggested that

$$c_u = KN$$

(2.4)

where K = constant = 3.5–6.5 kN/m² (0.507–0.942 lb/in²)
N = standard penetration number obtained from the field

The average value of K is about 4.4 kN/m² (0.638 lb/in²).

▼ **TABLE 2.3** Consistency of Clays and Approximate
Correlation to the Standard Penetration Number, N

Standard penetration number, N	Consistency	Unconfined compression strength, q_u (kN/m²)
0–2	Very soft	0–25
2–5	Soft	25–50
5–10	Medium stiff	50–100
10–20	Stiff	100–200
20–30	Very stiff	200–400
>30	Hard	>400

Hara et al. (1971) also suggested that

$$c_u \, (\mathrm{kN/m^2}) = 29N^{0.72}$$

(2.5)

The overconsolidation ratio, *OCR*, of a natural clay deposit can also be correlated with the standard penetration number. Based on the regression analysis of 110 data points, Mayne and Kemper (1988) obtained the relationship

$$OCR = 0.193 \left(\frac{N}{\sigma_v'} \right)^{0.689}$$

(2.6)

where σ_v' = effective vertical stress in MN/m²

It is important to point out that any correlation between c_u and N is only approximate. The sensitivity, S_t, of clay soils also plays an important role in the actual N value obtained from the field. Figure 2.16 shows a plot of $N_{(\text{measured})}/N_{(\text{at } S_t=1)}$ versus S_t as predicted by Schmertmann (1975).

In granular soils, the N value is affected by the effective overburden pressure, σ_v'. For that reason, the N value obtained from field exploration under different effective overburden pressures should be changed to correspond to a standard value of σ_v'. That is,

$$N_{\text{cor}} = C_N N_F$$

(2.7)

where N_{cor} = corrected N value to a standard value of σ_v' [95.6 kN/m² (1 ton/ft²)]
   C_N = correction factor
   N_F = N value obtained from the field

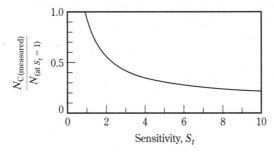

▼ **FIGURE 2.16**   Variation of $N_{(\text{measured})}/N_{(\text{at } S_t=1)}$ with S_t of clays (based on Schmertmann, 1975)

In the past, a number of empirical relations have been proposed for C_N. Some of the relationships are given in Table 2.4. The most commonly cited relationships are those given by Liao and Whitman (1986) and Skempton (1986). Figure 2.17 plots a comparison of C_N versus σ'_v obtained from those relationships.

An approximate relationship between the corrected standard penetration number and the relative density of sand is given in Table 2.5. However, these values are approximate, primarily because the effective overburden pressure and the stress history of the soil significantly influence the N_F values of sand. An extensive study conducted by Marcuson and Bieganousky (1977) produced the empirical relationship

$$D_r(\%) = 11.7 + 0.76(222N_F + 1600 - 53\sigma'_v - 50C_u^2)^{0.5} \tag{2.8}$$

where D_r = relative density
$\quad\quad\quad N_F$ = standard penetration number in the field
$\quad\quad\quad \sigma'_v$ = effective overburden pressure (lb/in^2)
$\quad\quad\quad C_u$ = uniformity coefficient of the sand

The *peak* angle of friction of granular soils, ϕ, has been correlated to the corrected standard penetration number. Peck, Hanson, and Thornburn (1974) give a correlation between N_{cor} and ϕ in a graphical form, which can be approximated as (Wolff, 1989)

$$\phi(\text{deg}) = 27.1 + 0.3N_{\text{cor}} - 0.00054N_{\text{cor}}^2 \tag{2.9}$$

▼ **TABLE 2.4** Empirical Relationships for C_N (*Note: σ'_v* is in U.S. ton/ft^2)

Source	C_N
Liao and Whitman (1986)	$\sqrt{\dfrac{1}{\sigma'_v}}$
Skempton (1986)	$\dfrac{2}{1 + \sigma'_v}$
Seed et al. (1975)	$1 - 1.25 \log\left(\dfrac{\sigma'_v}{\sigma'_1}\right)$
	where $\sigma'_1 = 1$ U.S. ton/ft^2
Peck et al. (1974)	$0.77 \log\left(\dfrac{20}{\sigma'_v}\right)$
	for $\sigma'_v \geq 0.25$ U.S. ton/ft^2

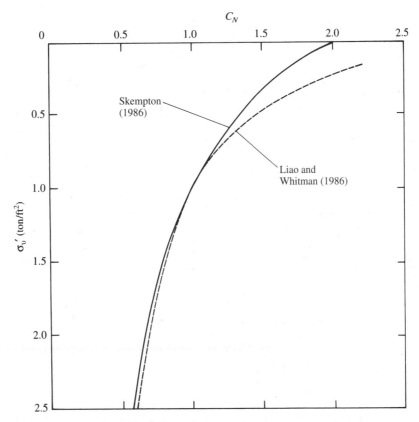

▼ **FIGURE 2.17** Comparison of C_N versus σ'_v plots obtained from the relationships given by Liao and Whitman (1986) and Skempton (1986)

▼ **TABLE 2.5** Relation between the Corrected N Values and the Relative Density in Sands

Standard penetration number, N_{cor}	Approximate relative density, D_r (%)
0–5	0–5
5–10	5–30
10–30	30–60
30–50	60–95

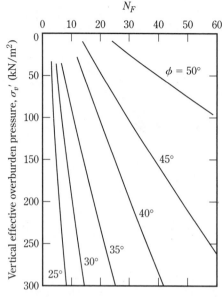

FIGURE 2.18 Schmertmann's (1975) correlation between N_F, σ_v', and ϕ for granular soils

Schmertmann (1975) provided a correlation between N_F, σ_v', and ϕ, which is shown in Figure 2.18. The correlation can be approximated as (Kulhawy and Mayne, 1990)

$$\phi = \tan^{-1}\left[\frac{N_F}{12.2 + 20.3\left(\dfrac{\sigma_v'}{p_a}\right)}\right]^{0.34}$$

(2.10)

where N_F = field standard penetration number
σ_v' = effective overburden pressure
p_a = atmospheric pressure in the same unit as σ_v'
ϕ = soil friction angle

More recently, Hatanaka and Uchida (1996) provided a simple correlation between ϕ and N_{cor} (Figure 2.19), which can be expressed as

$$\phi = \sqrt{20N_{\text{cor}}} + 20$$

(2.11)

When the standard penetration resistance values are used in the preceding correlations to estimate soil parameters, the following qualifications should be noted:

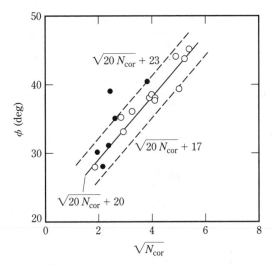

FIGURE 2.19   Laboratory test result of Hatanaka and Uchida (1996) for correlation between ϕ and $\sqrt{N_{cor}}$

1. The equations are approximate.
2. Because the soil is not homogeneous, the N_F values obtained from a given borehole vary widely.
3. In soil deposits that contain large boulders and gravel, standard penetration numbers may be erratic and unreliable.

Although approximate, with correct interpretation the standard penetration test provides a good evaluation of soil properties. The primary sources of errors in standard penetration tests are inadequate cleaning of the borehole, careless measurement of the blow count, eccentric hammer strikes on the drill rod, and inadequate maintenance of water head in the borehole.

Scraper Bucket

When soil deposits are sand mixed with pebbles, obtaining samples by split spoon with a spring core catcher may not be possible because the pebbles may prevent the springs from closing. In such cases, a scraper bucket may be used to obtain disturbed representative samples (Figure 2.20a). The scraper bucket has a driving point and can be attached to a drilling rod. The sampler is driven down into the soil and rotated, and the scrapings from the side fall into the bucket.

Thin Wall Tube

Thin wall tubes are sometimes referred to as *Shelby tubes*. They are made of seamless steel and are commonly used to obtain undisturbed clayey soils. The commonly used thin wall tube samplers have outside diameters of 50.8 mm (2 in.) and

76.2 mm (3 in.). The bottom end of the tube is sharpened. The tubes can be attached to drilling rods (Figure 2.20b). The drilling rod with the sampler attached is lowered to the bottom of the borehole and the sampler is pushed into the soil. The soil sample inside the tube is then pulled out. The two ends of the sampler are sealed, and it is sent to the laboratory for testing.

Samples obtained in this manner may be used for consolidation or shear tests. A thin wall tube with a 50.8-mm (2-in.) outside diameter has an inside diameter of about 47.63 mm ($1\frac{7}{8}$ in.). The area ratio is

$$A_R(\%) = \frac{D_o^2 - D_i^2}{D_i^2}\,(100) = \frac{(50.8)^2 - (47.63)^2}{(47.63)^2}\,(100) = 13.75\%$$

Increasing the diameters of samples increases the cost of obtaining them.

Piston Sampler

When undisturbed soil samples are very soft or larger than 76.2 mm (3 in.) in diameter, they tend to fall out of the sampler. Piston samplers are particularly useful under such conditions. There are several types of piston sampler; however, the sampler proposed by Osterberg (1952) is the most useful (see Figure 2.20c and 2.20 d). It consists of a thin wall tube with a piston. Initially, the piston closes the end of the thin wall tube. The sampler is lowered to the bottom of the borehole (Figure 2.20c), and the thin wall tube is pushed into the soil hydraulically, past the piston. Then the pressure is released through a hole in the piston rod (Figure 2.20d). To a large extent, the presence of the piston prevents distortion in the sample by not letting the soil squeeze into the sampling tube very fast and by not admitting excess soil. Consequently, samples obtained in this manner are less disturbed than those obtained by Shelby tubes.

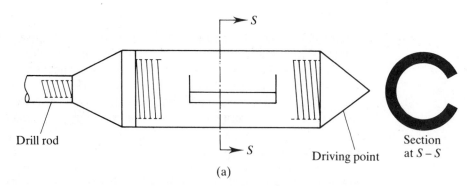

Drill rod   Driving point   Section
                                                      at $S-S$
(a)

▼ **FIGURE 2.20**   Sampling devices: (a) scraper bucket; (b) thin wall tube; (c) and (d) piston sampler

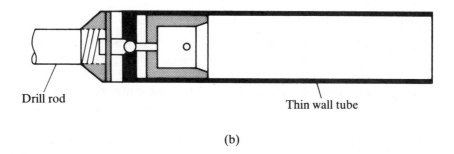

Drill rod

Thin wall tube

(b)

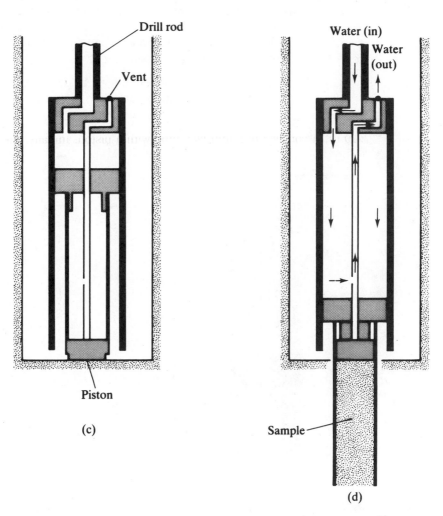

Drill rod

Vent

Piston

(c)

Water (in)

Water (out)

Sample

(d)

▼ **FIGURE 2.20** Continued

2.13   OBSERVATION OF WATER TABLES

The presence of a water table near a foundation significantly affects a foundation's load-bearing capacity and settlement, among other things. The water level will change seasonally. In many cases, establishing the highest and lowest possible levels of water during the life of a project may become necessary.

If water is encountered in a borehole during a field exploration, that fact should be recorded. In soils with high hydraulic conductivity, the level of water in a borehole will stabilize about 24 hours after completion of the boring. The depth of the water table can then be recorded by lowering a chain or tape into the borehole.

In highly impermeable layers, the water level in a borehole may not stabilize for several weeks. In such cases, if accurate water level measurements are required, a *piezometer* can be used. A piezometer basically consists of a porous stone or a perforated pipe with a plastic standpipe attached to it. Figure 2.21 shows the general placement of a piezometer in a borehole.

For silty soils, Hvorslev (1949) proposed a technique to determine the water level (see Figure 2.22) involving the following steps:

1.   Bail water out of the borehole to a level below the estimated water table.
2.   Observe the water levels in the borehole at times

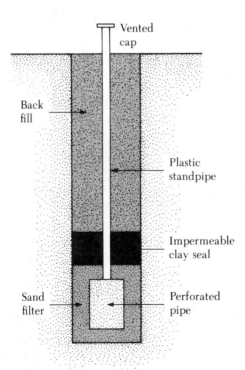

▼   **FIGURE 2.21**   Casagrande-type porous stone piezometer

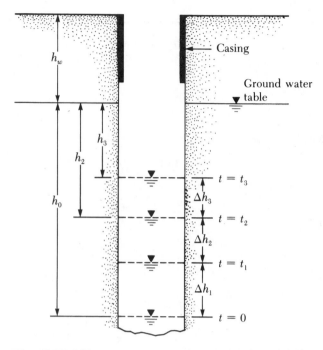

$$t = 0$$

$$t = t_1$$

$$t = t_2$$

$$t = t_3$$

Note that $t_1 - 0 = t_1 - t_2 = t_2 - t_3 = \Delta t$.

3.   Calculate Δh_1, Δh_2, and Δh_3 (see Figure 2.22).
4.   Calculate

$$h_0 = \frac{\Delta h_1^2}{\Delta h_1 - \Delta h_2} \qquad (2.12a)$$

$$h_2 = \frac{\Delta h_2^2}{\Delta h_1 - \Delta h_2} \qquad (2.12b)$$

$$h_3 = \frac{\Delta h_3^2}{\Delta h_2 - \Delta h_3} \qquad (2.12c)$$

5.   Plot h_0, h_2, and h_3 above the water levels observed at times $t = 0$, t_2, and t_3, respectively, to determine the final water level in the borehole.

▼   **EXAMPLE 2.1** _____

Refer to Figure 2.22. For a borehole, $h_w + h_0 = 9.5$ m

$\Delta t = 24$ hr

$\Delta h_1 = 0.9$ m

$\Delta h_2 = 0.70$ m

$\Delta h_3 = 0.54$ m

Make the necessary calculations and locate the water level.

Solution   Using Eq. (2.12),

$$h_0 = \frac{\Delta h_1^2}{\Delta h_1 - \Delta h_2} = \frac{0.9^2}{0.9 - 0.70} = 4.05 \text{ m}$$

$$h_2 = \frac{\Delta h_2^2}{\Delta h_1 - \Delta h_2} = \frac{0.7^2}{0.9 - 0.7} = 2.45 \text{ m}$$

$$h_3 = \frac{\Delta h_3^2}{\Delta h_2 - \Delta h_3} = \frac{0.54^2}{0.7 - 0.54} = 1.82 \text{ m} \qquad \blacktriangle$$

Figure 2.23 shows a plot of the preceding calculations and the estimated water levels. Note that $h_w = 5.5$ m.

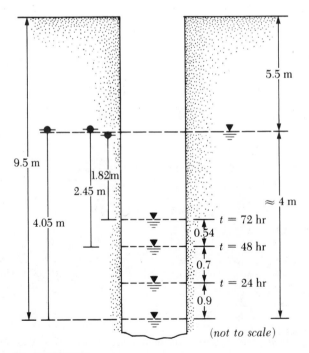

▼   **FIGURE 2.23**

2.14 VANE SHEAR TEST

The *vane shear test* (ASTM D-2573) may be used during the drilling operation to determine the *in situ* undrained shear strength (c_u) of clay soils — particularly soft clays. The vane shear apparatus consists of four blades on the end of a rod, as shown in Figure 2.24. The height, H, of the vane is twice the diameter, D. The vane can be either rectangular or tapered (see Figure 2.24). The dimensions of vanes used in the field are given in Table 2.6. The vanes of the apparatus are pushed into the soil at the bottom of a borehole without disturbing the soil appreciably. Torque is applied at the top of the rod to rotate the vanes at a standard rate of 0.1°/sec. This rotation will induce failure in a soil of cylindrical shape surrounding

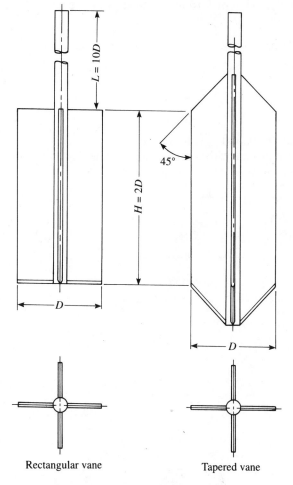

▼ FIGURE 2.24 Geometry of field vane (after ASTM, 1992)

▼ **TABLE 2.6** Recommended Dimensions of Field Vanes[a] (after ASTM, 1992)

Casing size	Diameter, D mm (in.)	Height, H mm (in.)	Thickness of blade mm (in.)	Diameter of rod mm (in.)
AX	38.1 ($1\frac{1}{2}$)	76.2 (3)	1.6 ($\frac{1}{16}$)	12.7 ($\frac{1}{2}$)
BX	50.8 (2)	101.6 (4)	1.6 ($\frac{1}{16}$)	12.7 ($\frac{1}{2}$)
NX	63.5 ($2\frac{1}{2}$)	127.0 (5)	3.2 ($\frac{1}{8}$)	12.7 ($\frac{1}{2}$)
4 in. (101.6 mm)[b]	92.1 ($3\frac{5}{8}$)	184.1 ($7\frac{1}{4}$)	3.2 ($\frac{1}{8}$)	12.7 ($\frac{1}{2}$)

[a] Selection of vane size is directly related to the consistency of the soil being tested; that is, the softer the soil, the larger the vane diameter should be.
[b] Inside diameter.

the vanes. The maximum torque, T, applied to cause failure is measured. Note that

$$T = f(c_u, H, \text{and } D) \tag{2.13}$$

or

$$c_u = \frac{T}{K} \tag{2.14}$$

where T is in N·m, and c_u is in kN/m^2
K = a constant with a magnitude depending on the dimension and shape of the vane

$$K = \left(\frac{\pi}{10^6}\right)\left(\frac{D^2 H}{2}\right)\left(1 + \frac{D}{3H}\right) \tag{2.15}$$

where D = diameter of vane in cm
H = measured height of vane in cm

If $H/D = 2$, Eq. (2.15) yields

$$K = 366 \times 10^{-8} D^3$$
$$\uparrow \tag{2.16}$$
$$(\text{cm})$$

In English units, if c_u and T in Eq. (2.14) are expressed in lb/ft^2 and lb-ft, respectively,

$$K = \left(\frac{\pi}{1728}\right)\left(\frac{D^2 H}{2}\right)\left(1 + \frac{D}{3H}\right) \tag{2.17}$$

If $H/D = 2$, Eq. (2.17) yields

$$K = 0.0021D^3$$
$$\uparrow$$
$$\text{(in.)}$$

(2.18)

Field vane shear tests are moderately rapid and economical and are used extensively in field soil-exploration programs. The test gives good results in soft and medium-stiff clays, and it is also an excellent test to determine the properties of sensitive clays.

Sources of significant error in the field vane shear test are poor calibration of torque measurement and damaged vanes. Other errors may be introduced if the rate of vane rotation is not properly controlled.

For actual design purposes, the undrained shear strength values obtained from field vane shear tests $[c_{u(\text{VST})}]$ are too high and it is recommended that they be corrected, or

$$c_{u(\text{corrected})} = \lambda c_{u(\text{VST})}$$

(2.19)

where λ = correction factor

Several correlations have been previously given for the correction factor, λ, and some are given in Table 2.7.

▼ **TABLE 2.7** Correlations for λ

Source	Correlation
Bjerrum (1972)	$\lambda = 1.7 - 0.54 \log(PI)$ PI = plasticity index (%)
Morris and Williams (1994)	$\lambda = 1.18e^{-0.08(PI)} + 0.57$ for $PI > 5$
Morris and Williams (1994)	$\lambda = 7.01e^{-0.08(LL)} + 0.57$ LL = liquid limit (%)
Aas et al. (1986)	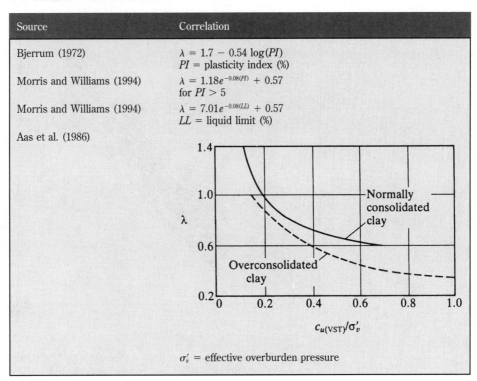

σ'_v = effective overburden pressure

The field vane shear strength can also be correlated to preconsolidation pressure and the overconsolidation ratio of the clay. Using 343 data points, Mayne and Mitchell (1988) derived the following empirical relationship for estimation of the preconsolidation pressure of a natural clay deposit.

$$p_c = 7.04 [c_{u(field)}]^{0.83} \tag{2.20}$$

where p_c = preconsolidation pressure (kN/m^2)

$c_{u(field)}$ = field vane shear strength (kN/m^2)

Figure 2.25 shows the plot of the data points from which they derived the relationship. They also showed that the overconsolidation ratio (OCR) can be correlated to $c_{u(field)}$ as

$$OCR = \beta \frac{c_{u(field)}}{\sigma'_v} \tag{2.21a}$$

where σ'_v = effective overburden pressure

$$\beta = 22(PI)^{-0.48} \tag{2.21b}$$

where PI = plasticity index

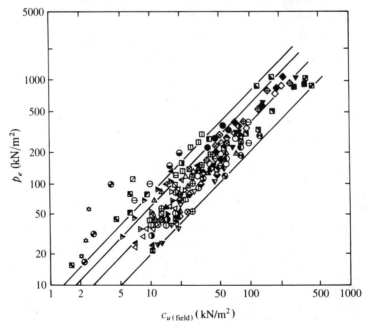

▼ FIGURE 2.25 Variation of preconsolidation pressure with field vane shear strength (after Mayne and Mitchell, 1988)

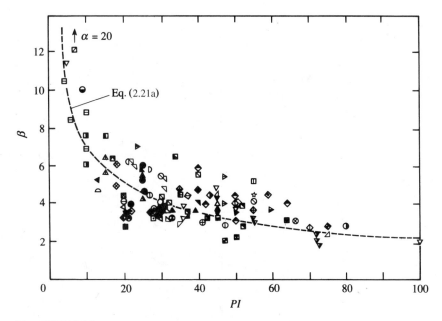

▼ **FIGURE 2.26** Variation of β with plasticity index (after Mayne and Mitchell, 1988)

Figure 2.26 shows the variation of β with plasticity index.

Other correlations for β presented in the literature are

Hansbo (1957)

$$\beta = \frac{222}{w\,(\%)} \tag{2.22}$$

Larsson (1980)

$$\beta = \frac{1}{0.08 + 0.0055\,(PI)} \tag{2.23}$$

Figure 2.27 compares the actual and predicted values of OCR obtained from Eq. (2.21) for six different sites.

2.15 CONE PENETRATION TEST

The cone penetration test (CPT), originally known as the Dutch cone penetration test, is a versatile sounding method that can be used to determine the materials in a soil profile and estimate their engineering properties. This test is also called the *static penetration test,* and no boreholes are necessary to perform it. In the original version, a 60° cone with a base area of 10 cm² was pushed into the ground at a steady rate of about 20 mm/sec, and the resistance to penetration (called the point resistance) was measured.

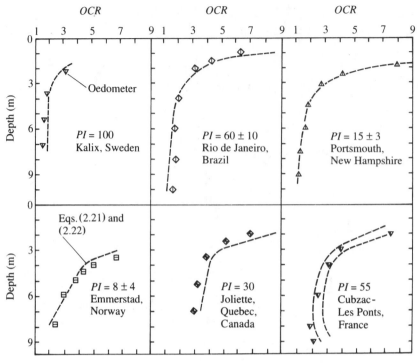

▼ **FIGURE 2.27** Measured *OCR* profiles at six sites from oedometer tests (individual points) and estimated profiles from field vane test data (dashed lines) (after Mayne and Mitchell, 1988)

The cone penetrometers in use at present measure (a) the *cone resistance* (q_c) to penetration developed by the cone, which is equal to the vertical force applied to the cone divided by its horizontally projected area, and (b) the *frictional resistance* (f_c), which is the resistance measured by a sleeve located above the cone with the local soil surrounding it. The frictional resistance is equal to the vertical force applied to the sleeve divided by its surface area — actually, the sum of friction and adhesion.

Generally, two types of penetrometers are used to measure q_c and f_c:

a. *Mechanical friction-cone penetrometer* (Figure 2.28). In this case the penetrometer tip is connected to an inner set of rods. The tip is first advanced about 40 mm giving the cone resistance. With further thrusting, the tip engages the friction sleeve. As the inner rod advances, the rod force is equal to the sum of the vertical force on the cone and sleeve. Subtracting the force on the cone gives the side resistance.

b. *Electric friction-cone penetrometer* (Figure 2.29). In this case the tip is attached to a string of steel rods. The tip is pushed into the ground at the rate of 20 mm/sec. Wires from the transducers are threaded through the center of the rods and continuously give the cone and side resistances.

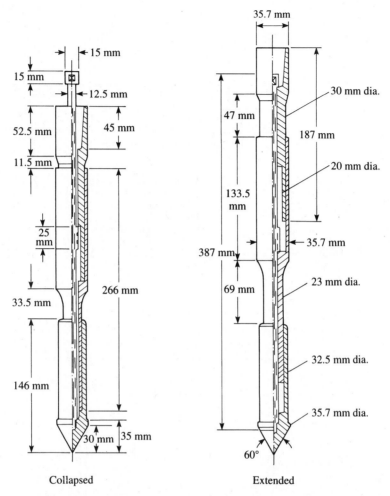

Collapsed   Extended

▼ **FIGURE 2.28** Mechanical friction-cone penetrometer (after ASTM, 1992)

Figure 2.30 shows the results of penetrometer tests in a soil profile with friction measurement by a mechanical friction-cone penetrometer and an electric friction-cone penetrometer.

Several correlations that are useful in estimating the properties of soils encountered during an exploration program have been developed for the point resistance (q_c) and the friction ratio (F_r) obtained from the cone penetration tests. The friction ratio, F_r, is defined as

$$F_r = \frac{\text{frictional resistance}}{\text{cone resistance}} = \frac{f_c}{q_c} \tag{2.24}$$

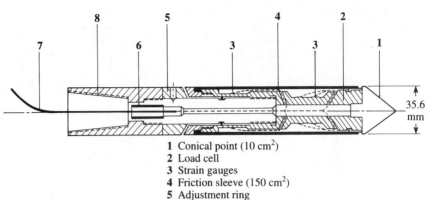

1 Conical point (10 cm^2)
2 Load cell
3 Strain gauges
4 Friction sleeve (150 cm^2)
5 Adjustment ring
6 Waterproof bushing
7 Cable
8 Connection with rods

▼ **FIGURE 2.29**   Electric friction-cone penetrometer (after ASTM, 1992)

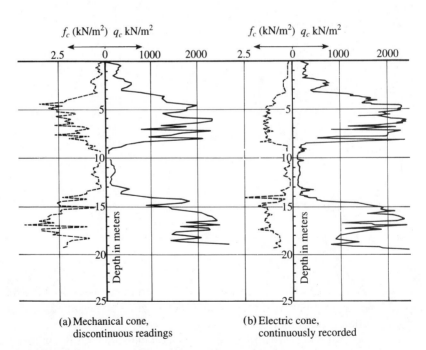

(a) Mechanical cone, discontinuous readings

(b) Electric cone, continuously recorded

▼ **FIGURE 2.30**   Penetrometer tests with friction measurement (after Ruiter, 1971)

Lancellotta (1983) and Jamiolkowski et al. (1985) showed that the relative density of normally consolidated sand, D_r, and q_c can be correlated as

$$D_r(\%) = A + B \log_{10}\left(\frac{q_c}{\sqrt{\sigma_v'}}\right)$$

(2.25)

where A, B = constants
$\qquad \sigma_v'$ = vertical effective stress

The values of A and B are

A	B	Unit of q_c and σ_v'
−98	66	metric ton/m²

Figure 2.31 shows the correlations obtained for several sands. Baldi et al. (1982), and Robertson and Campanella (1983) also recommended an empirical relationship

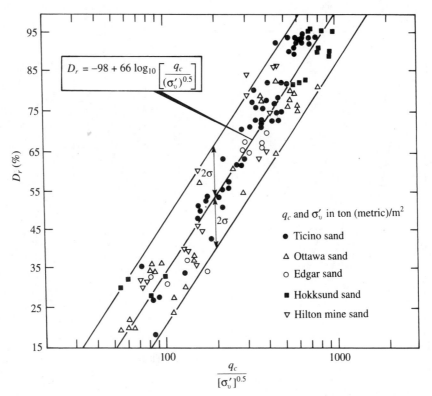

▼ **FIGURE 2.31**   Relationship between D_r and q_c (based on Lancellotta, 1983, and Jamiolkowski et al., 1985)

between vertical effective stress (σ_v'), relative density (D_r), and q_c for *normally consolidated sand*. This is shown in Figure 2.32.

Figure 2.33 shows a correlation between σ_v', q_c, and the peak friction angle ϕ for normally consolidated quartz sand. This correlation can be expressed as (Kulhawy and Mayne, 1990)

$$\phi = \tan^{-1}\left[0.1 + 0.38\log\left(\frac{q_c}{\sigma_v'}\right)\right]$$ (2.26)

Robertson and Campanella (1983) also provided a general correlation between q_c, friction ratio F_r, and the type of soil encountered in the field (Figure 2.34). Figure 2.35 shows the general range of q_c/N_F for various types of soil.

The correlations for q_c as proposed in Figures 2.31, 2.32, and 2.33, and Eqs. (2.25) and (2.26) are for normally consolidated sands. In fact, for a general condition, q_c is a function of σ_v', relative density, and vertical and lateral initial effective stress. A more rational theory for that correlation has been provided by Salgado, Mitchell, and Jamiolkowski (1997), and readers may refer to that paper for further information.

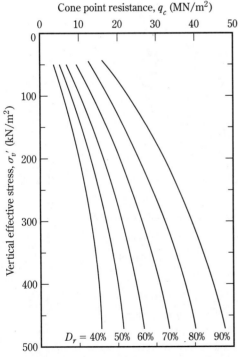

▼ **FIGURE 2.32**   Variation of q_c, σ_v', and D_r for normally consolidated quartz sand (based on Baldi et al., 1982, and Robertson and Campanella, 1983)

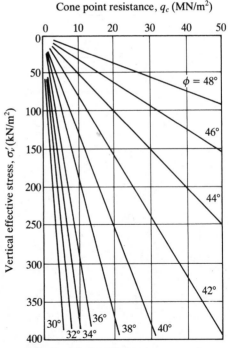

▼ **FIGURE 2.33**   Variation of q_c with σ_v' and ϕ in normally consolidated quartz sand (after Robertson and Campanella, 1983)

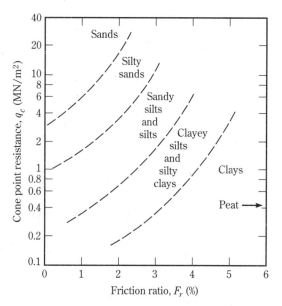

▼   **FIGURE 2.34**   Robertson and Campanellas' correlation (1983) between q_c, F_r, and the soil type

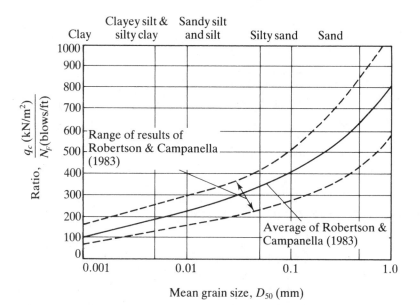

▼   **FIGURE 2.35**   General range of variation of q_c/N_F for various types of soil (after Robertson and Campanella, 1983)

According to Mayne and Kemper (1988), in clayey soil the undrained cohesion c_u, preconsolidation pressure p_c, and the overconsolidation ratio can be correlated as

$$\left(\frac{c_u}{\sigma_v'}\right) = \left(\frac{q_c - \sigma_v}{\sigma_v'}\right)\frac{1}{N_K} \tag{2.27}$$

or

$$c_u = \frac{q_c - \sigma_v}{N_K} \tag{2.27a}$$

where N_K = bearing capacity factor (N_K = 15 for electric cone and N_K = 20 for mechanical cone)
σ_v = *total* vertical stress
σ_v' = effective vertical stress

Consistent units of c_u, σ_v, σ_v', and q_c should be used with Eq. (2.27):

$$p_c = 0.243\,(q_c)^{0.96} \tag{2.28}$$

$$\uparrow \qquad\quad \uparrow$$

$$\text{MN/m}^2 \quad \text{MN/m}^2$$

and

$$OCR = 0.37\left(\frac{q_c - \sigma_v}{\sigma_v'}\right)^{1.01} \tag{2.29}$$

where σ_v and σ_v' = total and effective stress, respectively

2.16 PRESSUREMETER TEST (PMT)

The pressuremeter test is an *in situ* test conducted in a borehole. It was originally developed by Menard (1956) to measure the strength and deformability of soil. It has also been adopted by ASTM as Test Designation 4719. The Menard-type PMT essentially consists of a probe with three cells. The top and bottom ones are *guard cells* and the middle one is the *measuring cell,* as shown schematically in Figure 2.36a. The test is conducted in a pre-bored hole. The pre-bored hole should have a diameter that is between 1.03 to 1.2 times the nominal diameter of the probe. The probe that is most commonly used has a diameter of 58 mm and a length of

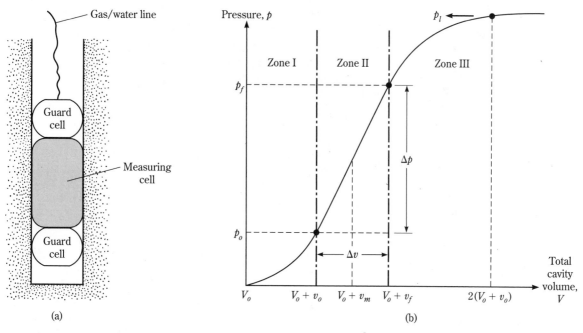

▼ **FIGURE 2.36** (a) Pressuremeter; (b) plot of pressure versus total cavity volume

420 mm. The probe cells can be expanded either by liquid or gas. The guard cells are expanded to reduce the end-condition effect on the measuring cell. The measuring cell has a volume (V_o) of 535 cm³. Following are the dimensions for the probe diameter and the diameter of the borehole as recommended by ASTM:

Probe diameter (mm)	Borehole diameter	
	Nominal (mm)	Maximum (mm)
44	45	53
58	60	70
74	76	89

In order to conduct a test, the measuring cell volume, V_o, is measured and the probe is inserted into the borehole. Pressure is applied in increments and the volumatic expansion of the cell is measured. This is continued until the soil fails or until the pressure limit of the device is reached. The soil is considered to have failed when the total volume of the expanded cavity (V) is about twice the volume of the original cavity. After the completion of the test, the probe is deflated and advanced for test at another depth.

The results of the pressuremeter test is expressed in a graphical form of pressure versus volume as shown in Figure 2.36b. In this figure, Zone I represents the reloading portion during which the soil around the borehole is pushed back into the initial state (that is, the state it was in before drilling). The pressure, p_o, represents the *in situ* total horizontal stress. Zone II represents a pseudo-elastic zone in which the cell volume versus cell pressure is practically linear. The pressure, p_f, represents the creep, or yield, pressure. The zone marked III is the plastic zone. The pressure, p_l, represents the limit pressure.

The pressuremeter modulus, E_p, of the soil is determined using the theory of expansion of an infinitely thick cylinder. Thus

$$E_p = 2(1 + \mu)(V_o + v_m)\left(\frac{\Delta p}{\Delta v}\right) \tag{2.30}$$

where $v_m = \dfrac{v_o + v_f}{2}$

$\Delta p = p_f - p_o$

$\Delta v = v_f - v_o$

μ = Poisson's ratio (which may be assumed to be 0.33)

The limit pressure, p_l, is usually obtained by extrapolation and not by direct measurement.

In order to overcome the difficulty of preparing the borehole to the proper size, self-boring pressuremeters (SBPMT) have also been developed. The details concerning SBPMTs can be found in the work of Baguelin et al. (1978).

Correlations between various soil parameters and the results obtained from the pressuremeter tests have been developed by various investigators. Kulhawy and Mayne (1990) proposed that

$$p_c = 0.45 p_l \tag{2.31}$$

where p_c = preconsolidation pressure

Based on the cavity expansion theory, Baguelin et al. (1978) proposed that

$$c_u = \frac{(p_l - p_o)}{N_p} \tag{2.32}$$

where c_u = undrained shear strength of a clay

$$N_p = 1 + \ln\left(\frac{E_p}{3c_u}\right)$$

Typical values of N_p vary between 5 to 12, with an average of about 8.5. Ohya et al. (1982) (see also Kulhawy and Mayne, 1990) correlated E_p with field standard penetration numbers (N_F) for sand and clay as follows:

Clay: $E_p (\text{kN}/\text{m}^2) = 1930 N_F^{0.63}$ \hfill (2.33)

Sand: $E_p (\text{kN}/\text{m}^2) = 908 N_F^{0.66}$ \hfill (2.34)

2.17 DILATOMETER TEST

The use of the flat-plate dilatometer test (DMT) is relatively recent (Marchetti, 1980; Schmertmann, 1986). The equipment essentially consists of a flat plate measuring 220 mm (length) × 95 mm (width) × 14 mm (thickness) (8.66 in. × 3.74 in. × 0.55 in.). A thin, flat circular expandable steel membrane having a diameter of 60 mm (2.36 in.) is located flush at the center on one side of the plate (Figure 2.37a). The dilatometer probe is inserted into the ground using a cone penetrometer testing rig (Figure 2.37b). Gas and electric lines extend from the surface control box through the penetrometer rod into the blade. At the required depth, high-pressure nitrogen gas is used to inflate the membrane. Two pressure readings are taken. They are

1. The pressure A to "lift off" the membrane, and
2. The pressure B at which the membrane expands 1.1 mm (0.4 in.) into the surrounding soil

The A and B readings are corrected as follows (Schmertmann, 1986)

$$\text{Contact stress, } p_o = 1.05(A + \Delta A - Z_m) - 0.05(B - \Delta B - Z_m) \tag{2.35}$$

$$\text{Expansion stress, } p_1 = B - Z_m - \Delta B \tag{2.36}$$

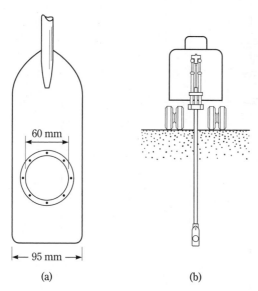

(a)   (b)

▼ **FIGURE 2.37**   (a) Schematic diagram of a flat-plate dilatometer; (b) dilatometer probe inserted into ground

where ΔA = vacuum pressure required to keep the membrane in contact
with its seating

ΔB = air pressure required inside the membrane to deflect it outward
to a center expansion of 1.1 mm

Z_m = gauge pressure deviation from zero when vented to atmospheric
pressure

The test is normally conducted at depths 200 mm to 300 mm apart. The result of a
given test is used to determine three parameters:

1. Material index, $I_D = \dfrac{p_1 - p_o}{p_o - u_o}$

2. Horizontal stress index, $K_D = \dfrac{p_o - u_o}{\sigma'_v}$

3. Dilatometer modulus, $E_D (\text{kN/m}^2) = 34.7(p_1 \, \text{kN/m}^2 - p_o \, \text{kN/m}^2)$

where u_o = pore water pressure

σ'_v = *in situ* vertical effective stress

Figure 2.38 shows the results of a dilatometer test conducted in Porto Tolle,
Italy (Marchetti, 1980). The subsoil consisted of recent, normally consolidated delta

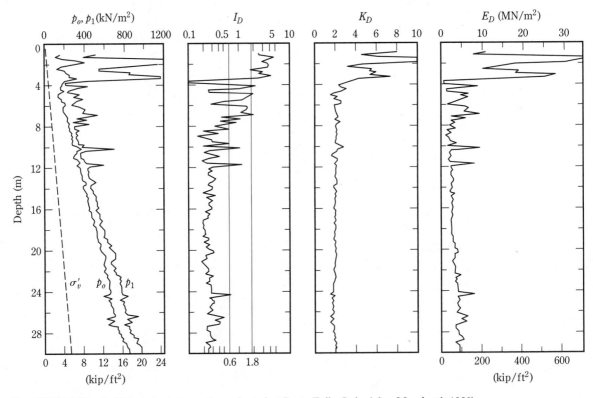

▼ **FIGURE 2.38**   A dilatometer test result conducted at Porto Tolle, Italy (after Marchetti, 1980)

deposits of the Po River. A thick layer of silty clay was found below a depth of 10 ft ($c = 0$; $\phi \approx 28°$). The results obtained from the dilatometer tests have been correlated to several soil properties (Marchetti, 1980). Some of these correlations are given below.

a. $K_o = \left(\dfrac{K_D}{1.5}\right)^{0.47} - 0.6$ (2.37)

b. $OCR = (0.5K_D)^{1.6}$ (2.38)

c. $\dfrac{c_u}{\sigma'_v} = 0.22$ (for normally consolidated clay) (2.39)

d. $\left(\dfrac{c_u}{\sigma'_v}\right)_{OC} = \left(\dfrac{c_u}{\sigma'_v}\right)_{NC} (0.5K_D)^{1.25}$ (2.40)

e. $E = (1 - \mu^2)E_D$ (2.41)

where K_o = coefficient of at-rest earth pressure (Chapter 5)
   OCR = overconsolidation ratio
   OC = overconsolidated soil
   NC = normally consolidated soil
   E = modulus of elasticity

Schmertmann (1986) also provided a correlation between the material index (I_D) and the dilatometer modulus (E_D) for determination of soil description and unit weight (γ). This is shown in Figure 2.39.

2.18 CORING OF ROCKS

When a rock layer is encountered during a drilling operation, rock coring may be necessary. For coring of rocks, a *core barrel* is attached to a drilling rod. A *coring bit* is attached to the bottom of the core barrel (Fig. 2.40). The cutting elements may be diamond, tungsten, carbide, and so on. Table 2.8 summarizes the various types of core barrel and their sizes, as well as the compatible drill rods commonly used for foundation exploration. The coring is advanced by rotary drilling. Water is circulated through the drilling rod during coring, and the cutting is washed out.

Two types of core barrel are available: the *single-tube core barrel* (Figure 2.40a) and the *double-tube core barrel* (Figure 2.40b). Rock cores obtained by single-tube core barrels can be highly disturbed and fractured because of torsion. Rock cores smaller than the BX size tend to fracture during the coring process.

When the core samples are recovered, the depth of recovery should be properly recorded for further evaluation in the laboratory. Based on the length of the rock core recovered from each run, the following quantities may be calculated for a general evaluation of the rock quality encountered.

$$\text{Recovery ratio} = \frac{\text{length of core recovered}}{\text{theoretical length of rock cored}}$$ (2.42)

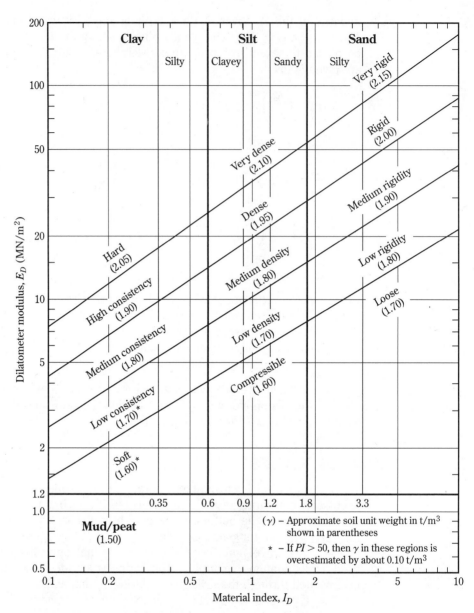

▼ **FIGURE 2.39**   Chart for determination of soil description and unit weight (after Schmertmann, 1986); *Note:* 1 t/m³ = 9.81 kN/m³

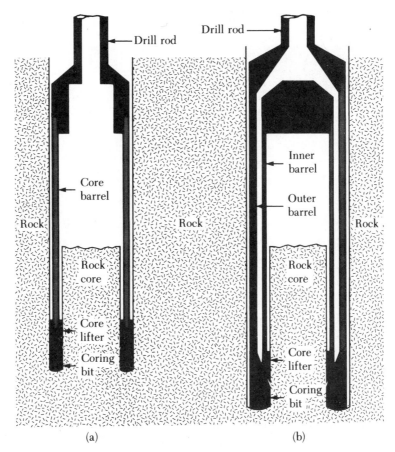

▼ FIGURE 2.40 Rock coring: (a) single-tube core barrel; (b) double-tube core barrel

▼ TABLE 2.8 Standard Size and Designation of Casing, Core Barrel, and Compatible Drill Rod

Casing and core barrel designation	Outside diameter of core barrel bit		Drill rod designation	Outside diameter of drill rod		Diameter of borehole		Diameter of core sample	
	(mm)	(in.)		(mm)	(in.)	(mm)	(in.)	(mm)	(in.)
EX	36.51	$1\frac{7}{16}$	E	33.34	$1\frac{5}{16}$	38.1	$1\frac{1}{2}$	22.23	$\frac{7}{8}$
AX	47.63	$1\frac{7}{8}$	A	41.28	$1\frac{5}{8}$	50.8	2	28.58	$1\frac{1}{8}$
BX	58.74	$2\frac{5}{16}$	B	47.63	$1\frac{7}{8}$	63.5	$2\frac{1}{2}$	41.28	$1\frac{5}{8}$
NX	74.61	$2\frac{15}{16}$	N	60.33	$2\frac{3}{8}$	76.2	3	53.98	$2\frac{1}{8}$

▼ **TABLE 2.9** Relation between *in situ* Rock Quality and *RQD*

RQD	Rock quality
0–0.25	Very poor
0.25–0.5	Poor
0.5–0.75	Fair
0.75–0.9	Good
0.9–1	Excellent

Rock quality designation (RQD) =

$$\frac{\Sigma \text{ length of recovered pieces equal to or larger than } 101.6 \text{ m (4 in.)}}{\text{theoretical length of rock cored}} \quad (2.43)$$

A recovery ratio of 1 will indicate the presence of intact rock; for highly fractured rocks, the recovery ratio may be 0.5 or smaller. Table 2.9 presents the general relationship (Deere, 1963) between the *RQD* and the *in situ* rock quality.

2.19 PREPARATION OF BORING LOGS

The detailed information gathered from each borehole is presented in a graphical form called the *boring log*. As a borehole is advanced downward, the driller generally should record the following information in a standard log:

1. Name and address of the drilling company
2. Driller's name
3. Job description and number
4. Number and type of boring and boring location
5. Date of boring
6. Subsurface stratification, which can be obtained by visual observation of the soil brought out by auger, split-spoon sampler, and thin wall Shelby tube sampler
7. Elevation of water table and date observed, use of casing and mud losses, and so on
8. Standard penetration resistance and the depth of *SPT*
9. Number, type, and depth of soil sample collected
10. In case of rock coring, type of core barrel used, and for each run, the actual length of coring, length of core recovery, and the *RQD*

This information should never be left to memory, because that often results in erroneous boring logs.

Boring Log

Name of the Project   Two-story apartment building

Location   Johnson & Olive St.   Date of Boring   March 2, 1982

Boring No.   3   Type of   Hollow stem auger   Ground Elevation   60.8 m
Boring

Soil description	Depth (m)	Soil sample type and number	N	w_n (%)	Comments
Light brown clay (fill)					
	1				
Silty sand (SM)	2	SS-1	9	8.2	
	3	SS-2	12	17.6	$LL = 38$ $PI = 11$
°G.W.T. 3.5 m	4				
Light gray clayey silt (ML)	5	ST-1		20.4	$LL = 36$ $q_u = 112 \, kN/m^2$
	6	SS-3	11	20.6	
Sand with some gravel (SP)	7				
End of boring @ 8 m	8	SS-4	27	9	

N = standard penetration number (below/304.8 mm)   °Ground water table
w_n = natural moisture content   observed after one
LL = liquid limit; PI = plasticity index   week of drilling
q_u = unconfined compression strength
SS = split-spoon sample; ST = Shelby tube sample

▼   FIGURE 2.41   A typical boring log

After completion of the necessary laboratory tests, the geotechnical engineer prepares a finished log that includes notes from the driller's field log and the results of tests conducted in the laboratory. Figure 2.41 shows a typical boring log. These logs have to be attached to the final soil-exploration report submitted to the client. Note that Figure 2.41 also lists the classifications of the soils in the left-hand column, along with the description of each soil (based on the Unified Soil Classification System).

2.20 DETERMINATION OF HYDRAULIC CONDUCTIVITY IN THE FIELD

Several types of field test are now available to determine the hydraulic conductivity of soil. Two fairly easy test procedures described by the U.S. Bureau of Reclamation (1974) are the *open end test* and the *packer test*.

Open End Test

The first step in the open end test (Figure 2.42) is to advance a borehole to the desired depth. A casing is then driven to extend to the bottom of the borehole. Water is supplied at a constant rate from the top of the casing, and it escapes at the bottom of the borehole. The water level in the casing must remain constant. Once the steady state of water supply is established, the hydraulic conductivity can be determined as

$$k = \frac{Q}{5.5rH} \tag{2.44}$$

where k = hydraulic conductivity
   Q = constant rate of supply of water to the borehole
   r = inside radius of the casing
   H = differential head of water

Any system of consistent units may be used in Eq. (2.44).

The head, H, has been defined in Figure 2.42. Note that for pressure tests (Figure 2.42c and 2.42d) the value of H is given as

$$H = H_{(gravity)} + H_{(pressure)} \tag{2.45}$$

The pressure head, $H_{(pressure)}$, given in Eq. (2.45) is expressed in meters (or feet) of water (1 kN/m^2 = 0.102 m; 1 lb/in^2 = 2.308 ft).

Packer Test

The packer test (Figure 2.43) can be conducted in a portion of the borehole during drilling or after drilling has been completed. Water is supplied to the portion of the borehole under test under constant pressure. The hydraulic conductivity can be determined from

$$k = \frac{Q}{2\pi LH} \log_e \left(\frac{L}{r}\right) \qquad \text{(for } L \geq 10r) \tag{2.46}$$

$$k = \frac{Q}{2\pi LH} \sinh^{-1} \frac{L}{2r} \qquad \text{(for } 10r > L \geq r) \tag{2.47}$$

where k = hydraulic conductivity
   Q = constant rate of flow into the hole
   L = length of portion of the hole under test

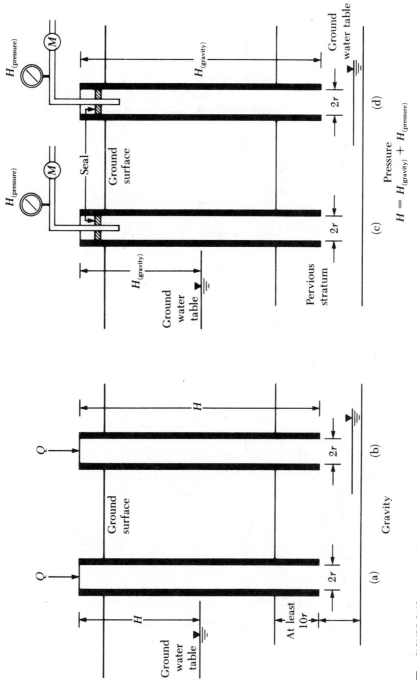

▶ **FIGURE 2.42** Hydraulic conductivity — open end test (redrawn after U.S. Bureau of Reclamation, 1974)

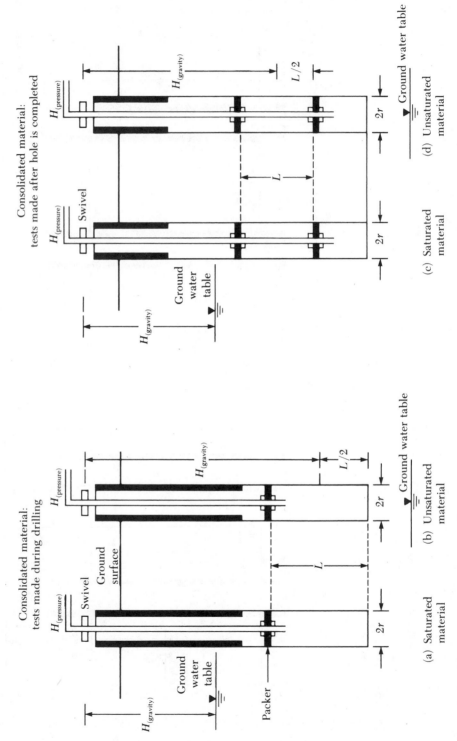

▼ **FIGURE 2.43** Hydraulic conductivity determination — packer test (redrawn after U.S. Bureau of Reclamation, 1974)

r = radius of the hole

H = differential pressure head

Note that the differential pressure head is the sum of the gravity head [$H_{(gravity)}$] and the pressure head [$H_{(pressure)}$]. The packer test is used primarily to determine the hydraulic conductivity of rock. However, as mentioned previously, it can also be used for soils.

2.21   GEOPHYSICAL EXPLORATION

Several types of geophysical exploration techniques permit rapid evaluation of subsoil characteristics. They allow rapid coverage of large areas and are less expensive than conventional exploration by drilling. However, in many cases, definitive interpretation of the results is difficult. For that reason, these techniques should be used for preliminary work only. Here, we discuss three types of geophysical exploration technique: seismic refraction survey, cross-hole seismic survey, and resistivity survey.

Seismic Refraction Survey

Seismic refraction surveys are useful in obtaining preliminary information about the thickness of the layering of various soils and the depth to rock or hard soil at a site. Refraction surveys are conducted by impacting the surface, as at point A in Figure 2.44a, and observing the first arrival of the disturbance (stress waves) at several other points (e.g., B, C, D, . . .). The impact can be created by a hammer blow or by a small explosive charge. The first arrival of disturbance waves at various points can be recorded by geophones.

The impact on the ground surface creates two types of *stress wave*: P waves (or *plane waves*) and S waves (or *shear waves*). The P waves travel faster than S waves; hence the first arrival of disturbance waves will be related to the velocities of the P waves in various layers. The velocity of P waves in a medium is

$$v = \sqrt{\frac{E}{\left(\dfrac{\gamma}{g}\right)} \frac{(1 - \mu)}{(1 - 2\mu)(1 + \mu)}} \tag{2.48}$$

where E = modulus of elasticity of the medium

γ = unit weight of the medium

g = acceleration due to gravity

μ = Poisson's ratio

To determine the velocity, v, of P waves in various layers and the thicknesses of those layers, use the following procedure.

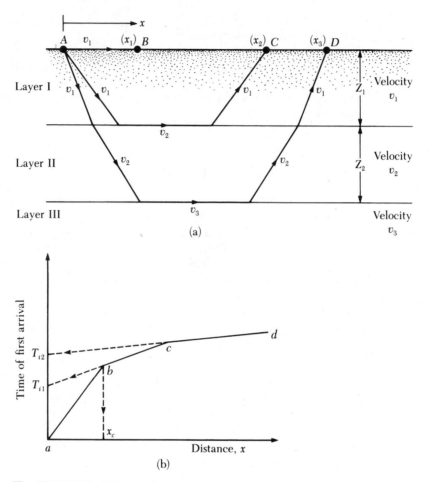

▼ **FIGURE 2.44**   Seismic refraction survey

1. Obtain the times of first arrival, t_1, t_2, t_3, ... , at various distances, x_1, x_2, x_3, ... , from the point of impact.
2. Plot a graph of time, t, against distance, x. The graph will look like the one shown in Figure 2.44b.
3. Determine the slopes of the lines *ab, bc, cd,* ...

$$\text{Slope of } ab = \frac{1}{v_1}$$

$$\text{Slope of } bc = \frac{1}{v_2}$$

$$\text{Slope of } cd = \frac{1}{v_3}$$

where $v_1, v_2, v_3, \ldots$ are the P-wave velocities in layers I, II, III, $\ldots$, respectively (Figure 2.44a)

4. Determine the thickness of the top layer as

$$Z_1 = \frac{1}{2} \sqrt{\frac{v_2 - v_1}{v_2 + v_1}} \, x_c \tag{2.49}$$

The value of x_c can be obtained from the plot, as shown in Figure 2.44b.

5. Determine the thickness of the second layer, Z_2, shown in Figure 2.44a, as

$$Z_2 = \frac{1}{2} \left[T_{i2} - 2Z_1 \frac{\sqrt{v_3^2 - v_1^2}}{v_3 v_1} \right] \frac{v_3 v_2}{\sqrt{v_3^2 - v_2^2}} \tag{2.50}$$

where T_{i2} is the time intercept of the line cd in Figure 2.44b extended backward

For detailed derivatives of these equations and other related information, refer to Dobrin (1960) and Das (1992).

Knowing the velocities of P waves in various layers indicates the types of soil or rock that are present below the ground surface. The range of the P-wave velocity that is generally encountered in various types of soil and rock at shallow depths is given in Table 2.10.

In analyzing the results of a refraction survey, two limitations need to be kept in mind:

1. The basic equations for the refraction survey — that is, Eqs. (2.49) and (2.50) — are based on the assumption that the P-wave velocity $v_1 < v_2 < v_3 < \cdots$.

2. When a soil is saturated below the water table, the P-wave velocity may be deceptive. P waves can travel with a velocity of about 1500 m/sec (5000 ft/sec) through water. For dry, loose soils, the velocity may be well

▼ **TABLE 2.10** Range of P-Wave Velocity in Various Soils and Rocks

	P-wave velocity	
Type of soil or rock	m/sec	ft/sec
Soil		
Sand, dry silt, and fine-grained top soil	200–1,000	650–3,300
Alluvium	500–2,000	1,650–6,600
Compacted clays, clayey gravel, and dense clayey sand	1,000–2,500	3,300–8,200
Loess	250–750	800–2,450
Rock		
Slate and shale	2,500–5,000	8,200–16,400
Sandstone	1,500–5,000	4,900–16,400
Granite	4,000–6,000	13,100–19,700
Sound limestone	5,000–10,000	16,400–32,800

below 1500 m/sec. However, in a saturated condition, the waves will travel through water present in the void spaces with a velocity of about 1500 m/sec (5000 ft/sec). If the presence of groundwater has not been detected, the *P*-wave velocity may be erroneously interpreted to indicate a stronger material (e.g., sandstone) than actually present *in situ.* In general, geophysical interpretations should always be verified by the results obtained from borings.

▼ EXAMPLE 2.2

The results of a refraction survey at a site are given in the following table. Determine the *P*-wave velocities and the thickness of the material encountered.

Distance from the source of disturbance (m)	Time of first arrival (sec × 10³)
2.5	11.2
5	23.3
7.5	33.5
10	42.4
15	50.9
20	57.2
25	64.4
30	68.6
35	71.1
40	72.1
50	75.5

Solution

Velocity   In Figure 2.45, the times of first arrival are plotted against the distance from the source of disturbance. The plot has three straight-line segments. The velocity of the top three layers can now be calculated as follows:

$$\text{Slope of segment } 0a = \frac{1}{v_1} = \frac{\text{time}}{\text{distance}} = \frac{23 \times 10^{-3}}{5.25}$$

or

$$v_1 = \frac{5.25 \times 10^3}{23} = \textbf{228 m/sec (top layer)}$$

$$\text{Slope of segment } ab = \frac{1}{v_2} = \frac{13.5 \times 10^{-3}}{11}$$

or

$$v_2 = \frac{11 \times 10^3}{13.5} = \textbf{814.8 m/sec (middle layer)}$$

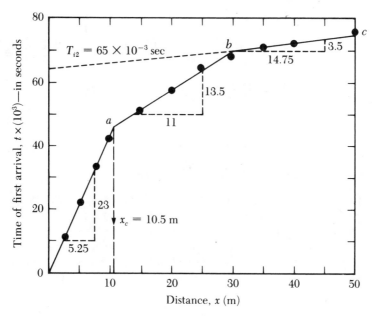

▼ FIGURE 2.45

$$\text{Slope of segment } bc = \frac{1}{v_3} = \frac{3.5 \times 10^{-3}}{14.75}$$

or

$$v_3 = \textbf{4214 m/sec (third layer)}$$

Comparing the velocities obtained here with those given in Table 2.10 indicates that the third layer is a *rock layer*.

Thickness of Layers

From Figure 2.45, $x_c = 10.5$ m, so

$$Z_1 = \frac{1}{2}\sqrt{\frac{v_2 - v_1}{v_2 + v_1}}\, x_c \qquad \text{[Eq. (2.49)]}$$

Thus

$$Z_1 = \frac{1}{2}\sqrt{\frac{814.8 - 228}{814.8 + 228}} \times 10.5 = \textbf{3.94 m}$$

Again, from Eq. (2.50)

$$Z_2 = \frac{1}{2}\left[T_{i2} - \frac{2Z_1\sqrt{v_3^2 - v_1^2}}{(v_3 v_1)} \right] \frac{(v_3)(v_2)}{\sqrt{v_3^2 - v_2^2}}$$

The value of T_{i2} (from Figure 2.45) is 65×10^{-3} sec. Hence

$$Z_2 = \frac{1}{2} \left[65 \times 10^{-3} - \frac{2(3.94)\sqrt{(4214)^2 - (228)^2}}{(4214)(228)} \right] \frac{(4214)(814.8)}{\sqrt{(4214)^2 - (814.8)^2}}$$

$$= \frac{1}{2}(0.065 - 0.0345)830.47 = \mathbf{12.66\,m}$$

The rock layer lies at a depth of $Z_1 + Z_2 = 3.94 + 12.66 = \mathbf{16.60\,m\ measured}$ **from the ground surface.**   ▲

Cross-Hole Seismic Survey

The velocity of shear waves created as the result of an impact to a given soil layer can be effectively determined by *cross-hole seismic survey* (Stokoe and Woods, 1972). The principle of this technique is illustrated in Figure 2.46, which shows two holes drilled into the ground at distance L apart. A vertical impulse is created at the bottom of one borehole by means of an impulse rod. The shear waves thus generated are recorded by a vertically sensitive transducer. The velocity of shear waves, v_s, can be calculated as

$$v_s = \frac{L}{t} \tag{2.51}$$

where t = travel time of shear waves

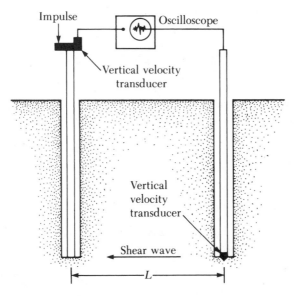

▼ **FIGURE 2.46**   Cross-hole method of seismic survey

The shear modulus of the soil at the depth of the test can be determined from v_s as

$$v_s = \sqrt{\frac{G}{(\gamma/g)}}$$

$$G = \frac{v_s^2 \gamma}{g} \tag{2.52}$$

where G = shear modulus of soil
   γ = soil unit weight
   g = acceleration due to gravity

The values of shear modulus are useful in the design of foundations to support vibrating machinery and the like.

Resistivity Survey

Another geophysical method for subsoil exploration is the *electrical resistivity survey*. The electrical resistivity, ρ, of any conducting material having a length L and an area of cross section A can be defined as

$$\rho = \frac{RA}{L} \tag{2.53}$$

where R = electrical resistance

The unit of resistivity is generally expressed as *ohm · centimeter* or *ohm · meter*. The resistivity of various soils depends primarily on the moisture content and also on the concentration of dissolved ions. Saturated clays have a very low resistivity; in contrast, dry soils and rocks have a high resistivity. The range of resistivity generally encountered in various soils and rocks is given in Table 2.11.

▼ TABLE 2.11 Representative Values of Resistivity

Material	Resistivity (ohm · m)
Sand	500–1500
Clays, saturated silt	0–100
Clayey sand	200–500
Gravel	1500–4000
Weathered rock	1500–2500
Sound rock	>5000

The most common procedure for measuring electrical resistivity of a soil profile makes use of four electrodes that are driven into the ground and spaced equally along a straight line. It is generally referred to as the *Wenner method* (Figure 2.47a). The two outside electrodes are used to send an electrical current, I, (usually a dc current with nonpolarizing potential electrodes) into the ground. The electrical current is typically in the range of 50–100 milliamperes. The voltage drop, V, is measured between the two inside electrodes. If the soil profile is homogeneous, its

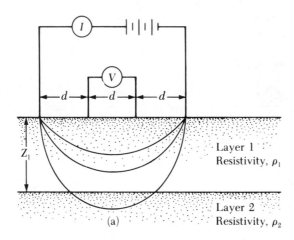

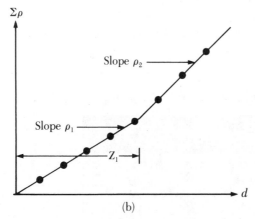

▼ **FIGURE 2.47** Electrical resistivity survey: (a) Wenner method; (b) empirical method for determination of resistivity and thickness of each layer

electrical resistivity is

$$\rho = \frac{2\pi dV}{I} \tag{2.54}$$

In most cases, the soil profile may consist of various layers with different resistivities and Eq. (2.54) will yield the *apparent resistivity*. To obtain the *actual resistivity* of various layers and their thicknesses, an empirical method may be used. It involves conducting tests at various electrode spacings (that is, d is changed). The sum of the apparent resistivities, $\Sigma\rho$, is plotted against the spacing d, as shown in Figure 2.47b. The plot thus obtained has relatively straight segments. The slopes of these straight segments give the resistivity of individual layers. The thicknesses of various layers can be estimated as shown in Figure 2.47b.

The resistivity survey is particularly useful in locating gravel deposits within a fine-grained soil.

2.22 SUBSOIL EXPLORATION REPORT

At the end of all soil exploration programs, the soil and/or rock specimens collected in the field are subject to visual observation and appropriate laboratory testing (the basic soil tests were described in Chapter 1). After all the required information has been compiled, a soil exploration report is prepared for the use of the design office and for reference during future construction work. Although the details and sequence of information in the report may vary to some degree, depending on the structure under consideration and the person compiling the report, each report should include the following items:

1. The scope of the investigation
2. A description of the proposed structure for which the subsoil exploration has been conducted
3. A description of the location of the site, including structure(s) nearby, drainage conditions of the site, nature of vegetation on the site and surrounding it, and any other feature(s) unique to the site
4. Geological setting of the site
5. Details of the field exploration — that is, number of borings, depths of borings, type of boring, and so on
6. General description of the subsoil conditions as determined from soil specimens and from related laboratory tests, standard penetration resistance and cone penetration resistance, and so on
7. Water-table conditions
8. Foundation recommendations, including the type of foundation recommended, allowable bearing pressure, and any special construction procedure that may be needed; alternative foundation design procedures should also be discussed in this portion of the report
9. Conclusions and limitations of the investigations

The following graphical presentations should be attached to the report:

1. Site location map
2. A plan view of the location of the borings with respect to the proposed structures and those existing nearby
3. Boring logs
4. Laboratory test results
5. Other special graphical presentations

The exploration reports should be well planned and documented. They will help in answering questions and solving foundation problems that may arise later during design and construction.

PROBLEMS

2.1 For a Shelby tube, given: outside diameter = 2 in.; inside diameter = $1\frac{7}{8}$ in.
  a. What is the area ratio of the tube?
  b. The outside diameter remaining the same, what should be the inside diameter of the tube to give an area ratio of 10%?

2.2 A soil profile is shown in Figure P2.2 along with the standard penetration numbers

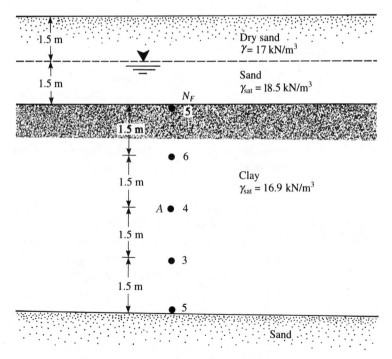

▼ **FIGURE P2.2**

in the clay layers. Use Eqs. (2.5) and (2.6) to determine and plot the variation of c_u and OCR with depth.

2.3 The average value of the field standard penetration number in a saturated clay layer is 6. Estimate the unconfined compression strength of the clay. Use Eq. (2.4) ($K \approx$ 4.2 kN/m²).

2.4 Following is the variation of the field standard penetration number (N_F) in a sand deposit:

Depth (m)	N_F
1.5	5
3	7
4.5	9
6	8
7.5	13
9	12

The ground water table is located at a depth of 5.5 m. Given: the dry unit weight of sand from 0 to a depth of 5.5 m is 18.08 kN/m³, and the saturated unit weight of sand for depth 5.5 m to 10.5 m is 19.34 kN/m³. Use the relationship of Liao and Whitman given in Table 2.4 to calculate the corrected penetration numbers.

2.5 For the soil profile described in Problem 2.4, estimate an average peak soil friction angle. Use Eq. (2.9).

2.6 Following are the standard penetration numbers determined from a sandy soil deposit in the field:

Depth (ft)	Unit weight of soil (lb/ft³)	N_F
10	106	7
15	106	9
20	106	11
25	118	16
30	118	18
35	118	20
40	118	22

Using Eq. (2.10), determine the variation of the peak soil friction angle, ϕ. Estimate an average value of ϕ for the design of a shallow foundation. *Note*: for depth greater than 20 ft, the unit weight of soil is 118 lb/ft³.

2.7   Redo Problem 2.6 using Skempton's relationship given in Table 2.4 and Eq. (2.11).

2.8   Following are the details for a soil deposit in sand:

Depth (ft)	Effective overburden pressure (lb/ft²)	Field standard penetration number, N_F
10	1150	9
15	1725	11
20	2030	12

Assume the uniformity coefficient (C_u) of the sand to be 3.2. Estimate the average relative density of the sand between the depth of 10 ft to 20 ft. Use Eq. (2.8).

2.9   Refer to Figure 2.22. For a borehole in a silty clay soil, the following are given:

$h_w + h_o = 25$ ft

$t_1 = 24$ hr $\Delta h_1 = 2.4$ ft

$t_2 = 48$ hr $\Delta h_2 = 1.7$ ft

$t_3 = 72$ hr $\Delta h_3 = 1.2$ ft

Determine the depth of the water table measured from the ground surface.

2.10   Repeat Problem 2.9 for

$h_w + h_o = 42$ ft

$t_1 = 24$ hr $\Delta h_1 = 6$ ft

$t_2 = 48$ hr $\Delta h_2 = 4.8$ ft

$t_3 = 72$ hr $\Delta h_3 = 3.8$ ft

2.11   Refer to Figure P2.2. Vane shear tests were conducted in the clay layer. The vane dimensions were 63.5 mm (D) × 127 mm (H). For the test at A, the torque required to cause failure was 0.051 N · m. For the clay, given: liquid limit = 46 and plastic limit = 21. Estimate the undrained cohesion of the clay for use in the design by using
  a.   Bjerrum's λ relationship (Table 2.7)
  b.   Morris and Williams' λ and PI relationship (Table 2.7)
  c.   Morris and Williams' λ and LL relationship (Table 2.7)

2.12   a.   A vane shear test was conducted in a saturated clay. The height and diameter of the vane were 4 in. and 2 in., respectively. During the test the maximum torque applied was 12.4 lb-ft. Determine the undrained shear strength of the clay.
  b.   The clay soil described in part (a) has a liquid limit of 64 and a plastic limit of 29. What would be the corrected undrained shear strength of the clay for design purposes? Use Bjerrum's relationship for λ.

2.13 Refer to Problem 2.11. Determine the overconsolidation ratio for the clay. Use Eqs. (2.21a) and (2.21b).

2.14 In a deposit of normally consolidated dry sand, a cone penetration test was conducted. Following are the results:

Depth (m)	Point resistance of cone, q_c (MN/m²)
1.5	2.05
3.0	4.23
4.5	6.01
6.0	8.18
7.5	9.97
9.0	12.42

Assuming the dry unit weight of sand to be 15.5 kN/m³,
a. Estimate the average peak friction angle, ϕ, of the sand. Use Eq. (2.26).
b. Estimate the average relative density of the sand. Use Figure 2.32.

2.15 Refer to the soil profile shown in Figure P2.15. If the cone penetration resistance (q_c) at A as determined by an electric friction-cone penetrometer is 0.6 MN/m², determine:
a. The undrained cohesion, c_u
b. The overconsolidation ratio, *OCR*

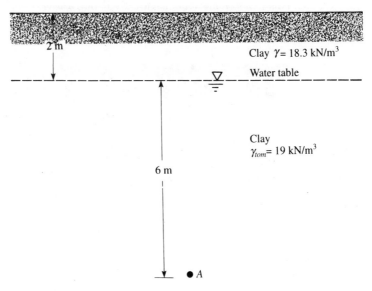

2 m

Clay $\gamma = 18.3$ kN/m³

Water table

Clay $\gamma_{tom} = 19$ kN/m³

6 m

● A

▼   **FIGURE P2.15**

2.16   Consider a pressuremeter test in a soft saturated clay. Given:

Measure cell volume, $V_o = 535 \text{ cm}^3$

$p_o = 42.4 \text{ kN/m}^2$ $v_o = 46 \text{ cm}^3$

$p_f = 326.5 \text{ kN/m}^2$ $v_f = 180 \text{ cm}^3$

Assuming Poisson's ratio (μ) to be 0.5, and referring to Figure 2.36, calculate the pressuremeter modulus, E_p.

2.17   A dilatometer test was conducted in a clay deposit. The ground water table was located at a depth of 3 m below the ground surface. At a depth of 8 m below the ground surface, the contact pressure (p_o) was 280 kN/m^2 and the expansion stress (p_1) was 350 kN/m^2. Determine the following:
   a.   Coefficient of at-rest earth pressure, K_o
   b.   Overconsolidation ratio, *OCR*
   c.   Modulus of elasticity, *E*
   Assume σ'_v at a depth of 8 m to be 95 kN/m^2, and $\mu = 0.35$.

2.18   During a field exploration, coring of rock was required. The core barrel was advanced 5 ft during the coring. The length of the core recovered was 3.2 ft. What was the recovery ratio?

2.19   An open-end permeability test was conducted in a borehole (refer to Figure 2.42a). The inside diameter of the casing was 2 in. The differential head of water was 23.4 ft. To maintain a constant head of 23.4 ft, a constant water supply rate of 4.8×10^{-2} ft^3/min was required. Calculate the hydraulic conductivity of the soil.

2.20   The *P*-wave velocity in a soil is 1900 m/sec. Assuming Poisson's ratio to be 0.32, calculate the modulus of elasticity of the soil. Assume that the unit weight of soil is 18 kN/m^3.

2.21   The results of a refraction survey (Figure 2.44a) at a site are given in the following table. Determine the thickness and the *P*-wave velocity of the materials encountered.

Distance from the source of disturbance (m)	Time of first arrival of *P* waves (sec $\times 10^3$)
2.5	5.08
5.0	10.16
7.5	15.24
10.0	17.01
15.0	20.02
20.0	24.2
25.0	27.1
30.0	28.0
40.0	31.1
50.0	33.9

2.22   Repeat Problem 2.21 for the following data.

Distance from the source of disturbance (ft)	Time of first arrival of P waves (sec $\times 10^3$)
25	49.06
50	81.96
75	122.8
100	148.2
150	174.2
200	202.8
250	228.6
300	256.7

REFERENCES

Aas, G., Lacasse, S., Lunne, I., and Høeg, K. (1986). "Use of In Situ Tests for Foundation Design in Clay," *Proceedings, In Situ '86,* American Society of Civil Engineers, pp. 1–30.

American Society for Testing and Materials (1992). *Annual Book of ASTM Standards,* Vol. 04.08, Philadelphia.

American Society of Civil Engineers (1972). "Subsurface Investigation for Design and Construction of Foundations of Buildings," *Journal of the Soil Mechanics and Foundations Division,* American Society of Civil Engineers, Vol. 98, No. SM5, pp. 481–490.

Baguelin, F., Jézéquel, J. F., and Shields, D. H. (1978). *The Pressuremeter and Foundation Engineering,* Trans Tech Publications, Clausthal.

Baldi, G., Bellotti, R., Ghionna, V., and Jamiolkowski, M. (1982). "Design Parameters for Sands from CPT," *Proceedings,* Second European Symposium on Penetration Testing, Amsterdam, Vol. 2, pp. 425–438.

Bjerrum, L. (1972). "Embankments on Soft Ground," *Proceedings of the Specialty Conference,* American Society of Civil Engineers, Vol. 2, pp. 1–54.

Das, B. M. (1992). *Principles of Soil Dynamics,* PWS Publishing Company, Boston.

Deere, D. U. (1963). "Technical Description of Rock Cores for Engineering Purposes," *Felsmechanik und Ingenieurgeologie,* Vol. 1, No. 1, pp. 16–22.

Dobrin, M. B. (1960). *Introduction to Geophysical Prospecting,* McGraw-Hill, New York.

Hansbo, S. (1957). *A New Approach to the Determination of the Shear Strength of Clay by the Fall Cone Test,* Swedish Geotechnical Institute, Report No. 114.

Hara, A., Ohata, T., and Niwa, M. (1971). "Shear Modulus and Shear Strength of Cohesive Soils," *Soils and Foundations,* Vol. 14, No. 3, pp. 1–12.

Hatanaka, M., and Uchida, A. (1996). "Empirical Correlation Between Penetration Resistance and Internal Friction Angle of Sandy Soils," *Soils and Foundations,* Vol. 36, No. 4, pp. 1–10.

Hvorslev, M. J. (1949). *Subsurface Exploration and Sampling of Soils for Civil Engineering Purposes,* Waterways Experiment Station, Vicksburg, Miss.

Jamiolkowski, M., Ladd, C. C., Germaine, J. T., and Lancellotta, R. (1985). "New Developments in Field and Laboratory Testing of Soils," *Proceedings,* 11th International Conference on Soil Mechanics and Foundation Engineering, Vol. 1, pp. 57–153.

Kolb, C. R., and Shockley, W. B. (1959). "Mississippi Valley Geology: Its Engineering Significance," *Proceedings,* American Society of Civil Engineers, Vol. 124, pp. 633–656.

Kulhawy, F. H., and Mayne, P. W. (1990). *Manual on Estimating Soil Properties for Foundation Design,* Electric Power Research Institute, Palo Alto, California.

Lancellotta, R. (1983). *Analisi di Affidabilità in Ingegneria Geotecnia,* Atti Istituto Sciencza Costruzioni, No. 625, Politecnico di Torino.

Larsson, R. (1980). "Undrained Shear Strength in Stability Calculation of Embankments and Foundations on Clay," *Canadian Geotechnical Journal,* Vol. 17, pp. 591–602.

Liao, S. S. C., and Whitman, R. V. (1986). "Overburden Correction Factors for SPT in Sand," *Journal of Geotechnical Engineering,* American Society of Civil Engineers, Vol. 112, No. 3, pp. 373–377.

Marchetti, S. (1980). "*In Situ* Test by Flat Dilatometer," *Journal of Geotechnical Engineering Division,* ASCE, Vol. 106, GT3, pp. 299–321.

Marcuson, W. F. III, and Bieganousky, W. A. (1977). "SPT and Relative Density in Coarse Sands," *Journal of Geotechnical Engineering Division,* American Society of Civil Engineers, Vol. 103, No. 11, pp. 1295–1309.

Mayne, P. W., and Kemper, J. B. (1988). "Profiling OCR in Stiff Clays by CPT and SPT," *Geotechnical Testing Journal,* ASTM, Vol. 11, No. 2, pp. 139–147.

Mayne, P. W., and Mitchell, J. K. (1988). "Profiling of Overconsolidation Ratio in Clays by Field Vane," *Canadian Geotechnical Journal,* Vol. 25, No. 1, pp. 150–158.

Menard, L. (1956). *An Apparatus for Measuring the Strength of Soils in Place,* M.S. Thesis, University of Illinois, Urbana, Illinois.

Morris, P. M., and Williams, D. T. (1994). "Effective Stress Vane Shear Strength Correction Factor Correlations," *Canadian Geotechnical Journal,* Vol. 31, No. 3, pp. 335–342.

Ohya, S., Imai, T., and Matsubara, M. (1982). "Relationships Between N Value by SPT and LLT Pressuremeter Results," *Proceedings,* 2nd European Symposium on Penetration Testing, Vol. 1, Amsterdam, pp. 125–130.

Osterberg, J. O. (1952). "New Piston-Type Soil Sampler," *Engineering News-Record,* April 24.

Peck, R. B., Hanson, W. E., and Thornburn, T. H. (1974). *Foundation Engineering,* 2nd ed., Wiley, New York.

Robertson, P. K., and Campanella, R. G. (1983). "Interpretation of Cone Penetration Tests. Part I: Sand," *Canadian Geotechnical Journal,* Vol. 20, No. 4, pp. 718–733.

Ruiter, J. (1971). "Electric Penetrometer for Site Investigations," *Journal of the Soil Mechanics and Foundations Division,* American Society of Civil Engineers, Vol. 97, No. 2, pp. 457–472.

Salgado, R., Mitchell, J. K., and Jamiolkowski, M. (1997). "Cavity Expansion and Penetration Resistance in Sand," *Journal of Geotechnical and Geoenvironmental Engineering,* American Society of Civil Engineers, Vol. 123, No. 4, pp. 344–354.

Schmertmann, J. H. (1975). "Measurement of *In Situ* Shear Strength," *Proceedings,* Specialty Conference on *In Situ* Measurement of Soil Properties, ASCE, Vol. 2, pp. 57–138.

Schmertmann, J. H. (1986). "Suggested Method for Performing the Flat Dilatometer Test," *Geotechnical Testing Journal,* ASTM, Vol. 9, No. 2, pp. 93–101.

Seed, H. B., Arango, I., and Chan, C. K. (1975). "Evaluation of Soil Liquefaction Potential During Earthquakes," *Report No. EERC 75-28,* Earthquake Engineering Research Center, University of California, Berkeley.

Skempton, A. W. (1986). "Standard Penetration Test Procedures and the Effect in Sands of Overburden Pressure, Relative Density, Particle Size, Aging and Overconsolidation," *Geotechnique,* Vol. 36, No. 3, pp. 425–447.

Sowers, G. B., and Sowers, G. F. (1970). *Introductory Soil Mechanics and Foundations,* 3rd ed., Macmillan, New York.

Stokoe, K. H., and Woods, R. D. (1972). "*In Situ* Shear Wave Velocity by Cross-Hole Method," *Journal of Soil Mechanics and Foundations Division,* American Society of Civil Engineers, Vol. 98, No. SM5, pp. 443–460.

Stroud, M. (1974). "SPT in Insensitive Clays," *Proceedings,* European Symposium on Penetration Testing, Vol. 2.2, pp. 367–375.

U.S. Bureau of Reclamation (1974). *Design of Small Dams,* 2nd. ed., U.S. Government Printing Office, Washington, D.C.

Wolff, T. F. (1989). "Pile Capacity Prediction Using Parameter Functions," in *Predicted and Observed Axial Behavior of Piles, Results of a Pile Prediction Symposium,* sponsored by the Geotechnical Engineering Division, ASCE, Evanston, IL, June, 1989, ASCE Geotechnical Special Publication No. 23, pp. 96–106.

CHAPTER THREE

SHALLOW FOUNDATIONS: ULTIMATE BEARING CAPACITY

3.1 INTRODUCTION

To perform satisfactorily, shallow foundations must have two main characteristics:

1. The foundation has to be safe against overall shear failure in the soil that supports it.
2. The foundation cannot undergo excessive displacement, that is, settlement. (The term *excessive* is relative, because the degree of settlement allowable for a structure depends on several considerations.)

The load per unit area of the foundation at which the shear failure in soil occurs is called the *ultimate bearing capacity,* which is the subject of this chapter.

3.2 GENERAL CONCEPT

Consider a strip foundation resting on the surface of a dense sand or stiff cohesive soil, as shown in Figure 3.1a, with a width of B. Now, if load is gradually applied to the foundation, settlement will increase. The variation of the load per unit area on the foundation, q, with the foundation settlement is also shown in Figure 3.1a. At a certain point — when the load per unit area equals q_u — a sudden failure in the soil supporting the foundation will take place, and the failure surface in the soil will extend to the ground surface. This load per unit area, q_u, is usually referred to as the *ultimate bearing capacity of the foundation*. When this type of sudden failure in soil takes place, it is called the *general shear failure*.

  If the foundation under consideration rests on sand or clayey soil of medium compaction (Figure 3.1b), an increase of load on the foundation will also be accompanied by an increase of settlement. However, in this case the failure surface in the soil will gradually extend outward from the foundation, as shown

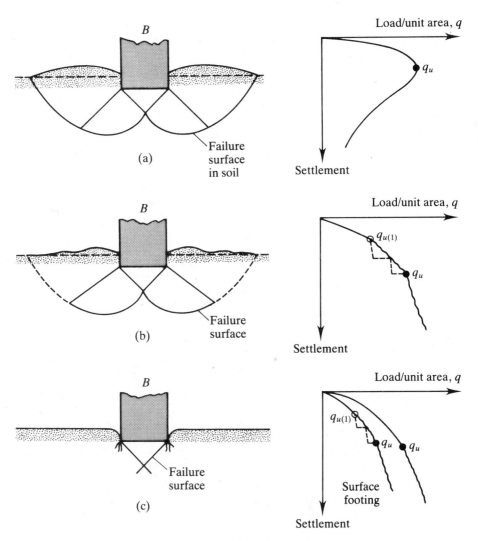

▼ FIGURE 3.1 Nature of bearing capacity failure in soil: (a) general shear failure; (b) local shear failure; (c) punching shear failure (redrawn after Vesic, 1973)

by the solid lines in Figure 3.1b. When the load per unit area on the foundation equals $q_{u(1)}$, the foundation movement will be accompanied by sudden jerks. A considerable movement of the foundation is then required for the failure surface in soil to extend to the ground surface (as shown by the broken lines in Figure 3.1b). The load per unit area at which this happens is the *ultimate bearing capacity, q_u.* Beyond this point, an increase of load will be accompanied by a large increase of foundation settlement. The load per unit area of the foundation, $q_{u(1)}$, is referred to as the *first failure load* (Vesic, 1963). Note that a peak value of q is not realized in this type of failure, which is called the *local shear failure* in soil.

If the foundation is supported by a fairly loose soil, the load–settlement plot will be like the one in Figure 3.1c. In this case, the failure surface in soil will not extend to the ground surface. Beyond the ultimate failure load, q_u, the load–settlement plot will be steep and practically linear. This type of failure in soil is called the *punching shear failure*.

Vesic (1963) conducted several laboratory load-bearing tests on circular and rectangular plates supported by a sand at various relative densities of compaction, D_r. The variations of $q_{u(1)}/\frac{1}{2}\gamma B$ and $q_u/\frac{1}{2}\gamma B$ obtained from those tests are shown in Figure 3.2 (B = diameter of circular plate or width of rectangular plate, and γ =

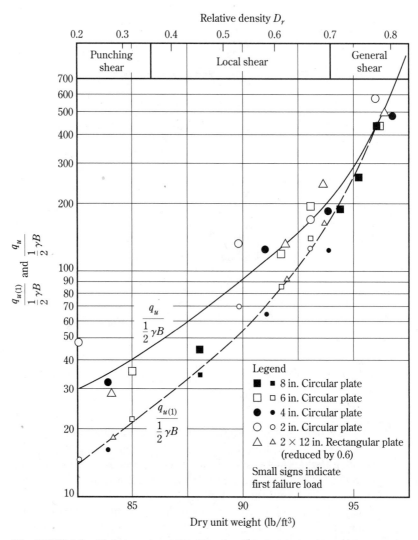

▼ **FIGURE 3.2**   Variation of $q_{u(1)}/0.5\gamma B$ and $q_u/0.5\gamma B$ for circular and rectangular plates on the surface of a sand (after Vesic, 1963)

dry unit weight of sand). It is important to note from this figure that, for $D_r \geq$ about 70%, the general shear type of failure in soil occurs.

Based on experimental results, Vesic (1973) proposed a relationship for the mode of bearing capacity failure of foundations resting on sands. Figure 3.3 shows this relationship, which involves the notation

D_r = relative density of sand

D_f = depth of foundation measured from the ground surface

$$B^\star = \frac{2BL}{B + L} \tag{3.1}$$

where B = width of foundation
        L = length of foundation

(*Note: L* is always greater than *B.*)

For square foundations, $B = L$; for circular foundations, $B = L$ = diameter, so

$$B^\star = B \tag{3.2}$$

Figure 3.4 shows the settlement, *S*, of the circular and rectangular plates on the surface of a sand at *ultimate load* as described in Figure 3.2. It shows a general range of *S/B* with the relative density of compaction of sand. So, in general, we can say that for foundations at a shallow depth (that is, small $D_f/B^\star$), the ultimate load may occur at a foundation settlement of 4–10% of *B*. This condition occurs when

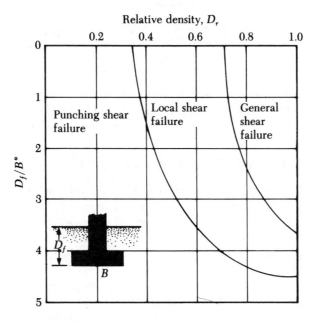

▼ FIGURE 3.3   Modes of foundation failure in sand (after Vesic, 1973)

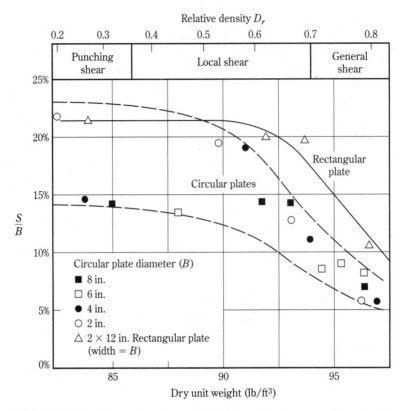

▼ **FIGURE 3.4**   Range of settlement of circular and rectangular plates at ultimate load ($D_f/B = 0$) in sand (after Vesic, 1963)

general shear failure in soil occurs; however, in the case of local or punching shear
failure, the ultimate load may occur at settlements of 15–25% of the width of the
foundation (B).

3.3 TERZAGHI'S BEARING CAPACITY THEORY

Terzaghi (1943) was the first to present a comprehensive theory for the evaluation
of the ultimate bearing capacity of rough shallow foundations. According to this
theory, a foundation is *shallow* if the depth, D_f (Figure 3.5), of the foundation is less
than or equal to the width of the foundation. Later investigators, however, have
suggested that foundations with D_f equal to 3–4 times the width of the foundation
may be defined as *shallow foundations*.

Terzaghi suggested that for a *continuous*, or *strip, foundation* (that is, the width-
to-length ratio of the foundation approaches zero), the failure surface in soil at
ultimate load may be assumed to be similar to that shown in Figure 3.5. (Note that
this is the case of general shear failure as defined in Figure 3.1a.) The effect of soil
above the bottom of the foundation may also be assumed to be replaced by an

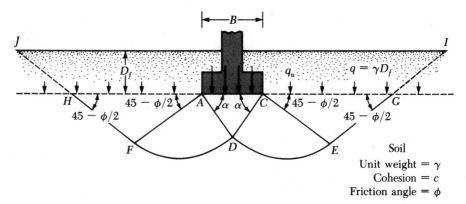

▼ **FIGURE 3.5**   Bearing capacity failure in soil under a rough rigid continuous foundation

equivalent surcharge, $q = \gamma D_f$ (where γ = unit weight of soil). The failure zone under the foundation can be separated into three parts (see Figure 3.5):

1. The *triangular zone ACD* immediately under the foundation
2. The *radial shear zones ADF* and *CDE*, with the curves *DE* and *DF* being arcs of a logarithmic spiral
3. Two triangular *Rankine passive zones AFH* and *CEG*

The angles *CAD* and *ACD* are assumed to be equal to the soil friction angle, ϕ. Note that, with the replacement of the soil above the bottom of the foundation by an equivalent surcharge q, the shear resistance of the soil along the failure surfaces *GI* and *HJ* was neglected.

Using the equilibrium analysis, Terzaghi expressed the ultimate bearing capacity in the form

$$q_u = cN_c + qN_q + \tfrac{1}{2}\gamma BN_\gamma \qquad \text{(strip foundation)} \tag{3.3}$$

where
$$c = \text{cohesion of soil}$$
$$\gamma = \text{unit weight of soil}$$
$$q = \gamma D_f$$
$$N_c, N_q, N_\gamma = \text{bearing capacity factors that are nondimensional and are only functions of the soil friction angle, } \phi$$

The bearing capacity factors, N_c, N_q, and N_γ are defined by

$$N_c = \cot\phi \left[\frac{e^{2(3\pi/4-\phi/2)\tan\phi}}{2\cos^2\left(\dfrac{\pi}{4}+\dfrac{\phi}{2}\right)} - 1 \right] = \cot\phi(N_q - 1) \tag{3.4}$$

$$N_q = \frac{e^{2(3\pi/4 - \phi/2)\tan\phi}}{2\cos^2\left(45 + \dfrac{\phi}{2}\right)} \tag{3.5}$$

$$N_\gamma = \frac{1}{2}\left(\frac{K_{p\gamma}}{\cos^2\phi} - 1\right)\tan\phi \tag{3.6}$$

where $K_{p\gamma}$ = passive pressure coefficient

The variations of the bearing capacity factors defined by Eqs. (3.4), (3.5), and (3.6) are given in Table 3.1.

▼ **TABLE 3.1** Terzaghi's Bearing Capacity Factors — Eqs. (3.4), (3.5), and (3.6)

ϕ	N_c	N_q	$N_\gamma{}^a$	ϕ	N_c	N_q	$N_\gamma{}^a$
0	5.70	1.00	0.00	26	27.09	14.21	9.84
1	6.00	1.1	0.01	27	29.24	15.90	11.60
2	6.30	1.22	0.04	28	31.61	17.81	13.70
3	6.62	1.35	0.06	29	34.24	19.98	16.18
4	6.97	1.49	0.10	30	37.16	22.46	19.13
5	7.34	1.64	0.14	31	40.41	25.28	22.65
6	7.73	1.81	0.20	32	44.04	28.52	26.87
7	8.15	2.00	0.27	33	48.09	32.23	31.94
8	8.60	2.21	0.35	34	52.64	36.50	38.04
9	9.09	2.44	0.44	35	57.75	41.44	45.41
10	9.61	2.69	0.56	36	63.53	47.16	54.36
11	10.16	2.98	0.69	37	70.01	53.80	65.27
12	10.76	3.29	0.85	38	77.50	61.55	78.61
13	11.41	3.63	1.04	39	85.97	70.61	95.03
14	12.11	4.02	1.26	40	95.66	81.27	115.31
15	12.86	4.45	1.52	41	106.81	93.85	140.51
16	13.68	4.92	1.82	42	119.67	108.75	171.99
17	14.60	5.45	2.18	43	134.58	126.50	211.56
18	15.12	6.04	2.59	44	151.95	147.74	261.60
19	16.56	6.70	3.07	45	172.28	173.28	325.34
20	17.69	7.44	3.64	46	196.22	204.19	407.11
21	18.92	8.26	4.31	47	224.55	241.80	512.84
22	20.27	9.19	5.09	48	258.28	287.85	650.67
23	21.75	10.23	6.00	49	298.71	344.63	831.99
24	23.36	11.40	7.08	50	347.50	415.14	1072.80
25	25.13	12.72	8.34				

[a] From Kumbhojkar (1993)

For estimating the ultimate bearing capacity of *square* or *circular foundations*, Eq. (3.1) may be modified to

$$q_u = 1.3cN_c + qN_q + 0.4\gamma BN_\gamma \qquad \text{(square foundation)} \tag{3.7}$$

and

$$q_u = 1.3cN_c + qN_q + 0.3\gamma BN_\gamma \qquad \text{(circular foundation)} \tag{3.8}$$

In Eq. (3.7), B equals the dimension of each side of the foundation; in Eq. (3.8), B equals the diameter of the foundation.

For foundations that exhibit the local shear failure mode in soils, Terzaghi suggested modifications to Eqs. (3.3), (3.7), and (3.8) as follows:

$$q_u = \frac{2}{3}cN_c' + qN_q' + \tfrac{1}{2}\gamma BN_\gamma' \qquad \text{(strip foundation)} \tag{3.9}$$

$$q_u = 0.867cN_c' + qN_q' + 0.4\gamma BN_\gamma' \qquad \text{(square foundation)} \tag{3.10}$$

$$q_u = 0.867cN_c' + qN_q' + 0.3\gamma BN_\gamma' \qquad \text{(circular foundation)} \tag{3.11}$$

N_c', N_q', and N_γ' are the *modified bearing capacity factors*. They can be calculated by using the bearing capacity factor equations (for N_c, N_q, and N_γ) by replacing ϕ by $\phi' = \tan^{-1}(\tfrac{2}{3}\tan\phi)$. The variation of N_c', N_q', and N_γ' with the soil friction angle, ϕ, is given in Table 3.2.

Terzaghi's bearing capacity equations have now been modified to take into account the effects of the foundation shape (B/L), depth of embedment (D_f), and the load inclination. This is given in Section 3.7. Many design engineers, however, still use Terzaghi's equation, which provides fairly good results considering the uncertainty of the soil conditions at various sites.

3.4 MODIFICATION OF BEARING CAPACITY EQUATIONS FOR WATER TABLE

Equations (3.3) and (3.7) to (3.11) have been developed for determining the ultimate bearing capacity based on the assumption that the water table is located well below the foundation. However, if the water table is close to the foundation, some modifications of the bearing capacity equations will be necessary, depending on the location of the water table (see Figure 3.6).

Case I

If the water table is located so that $0 \le D_1 \le D_f$, the factor q in the bearing capacity equations takes the form

▼ **TABLE 3.2** Terzaghi's Modified Bearing Capacity Factors, N'_c, N'_q, and N'_γ

ϕ	N'_c	N'_q	N'_γ	ϕ	N'_c	N'_q	N'_γ
0	5.70	1.00	0.00	26	15.53	6.05	2.59
1	5.90	1.07	0.005	27	16.30	6.54	2.88
2	6.10	1.14	0.02	28	17.13	7.07	3.29
3	6.30	1.22	0.04	29	18.03	7.66	3.76
4	6.51	1.30	0.055	30	18.99	8.31	4.39
5	6.74	1.39	0.074	31	20.03	9.03	4.83
6	6.97	1.49	0.10	32	21.16	9.82	5.51
7	7.22	1.59	0.128	33	22.39	10.69	6.32
8	7.47	1.70	0.16	34	23.72	11.67	7.22
9	7.74	1.82	0.20	35	25.18	12.75	8.35
10	8.02	1.94	0.24	36	26.77	13.97	9.41
11	8.32	2.08	0.30	37	28.51	15.32	10.90
12	8.63	2.22	0.35	38	30.43	16.85	12.75
13	8.96	2.38	0.42	39	32.53	18.56	14.71
14	9.31	2.55	0.48	40	34.87	20.50	17.22
15	9.67	2.73	0.57	41	37.45	22.70	19.75
16	10.06	2.92	0.67	42	40.33	25.21	22.50
17	10.47	3.13	0.76	43	43.54	28.06	26.25
18	10.90	3.36	0.88	44	47.13	31.34	30.40
19	11.36	3.61	1.03	45	51.17	35.11	36.00
20	11.85	3.88	1.12	46	55.73	39.48	41.70
21	12.37	4.17	1.35	47	60.91	44.45	49.30
22	12.92	4.48	1.55	48	66.80	50.46	59.25
23	13.51	4.82	1.74	49	73.55	57.41	71.45
24	14.14	5.20	1.97	50	81.31	65.60	85.75
25	14.80	5.60	2.25				

$$q = \text{effective surcharge} = D_1\gamma + D_2(\gamma_{sat} - \gamma_w) \qquad (3.12)$$

where γ_{sat} = saturated unit weight of soil
γ_w = unit weight of water

Also, the value of γ in the last term of the equations has to be replaced by $\gamma' = \gamma_{sat} - \gamma_w$.

Case II

For a water table located so that $0 \leq d \leq B$,

$$q = \gamma D_f \qquad (3.13)$$

The factor γ in the last term of the bearing capacity equations must be replaced by the factor

$$\boxed{\bar{\gamma} = \gamma' + \frac{d}{B}(\gamma - \gamma')} \qquad (3.14)$$

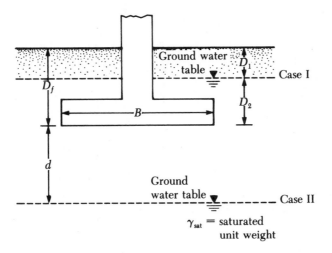

▼ **FIGURE 3.6**   Modification of bearing capacity equations
for water table

The preceding modifications are based on the assumption that there is no seepage
force in the soil.

Case III

When the water table is located so that $d \geq B$, the water will have no effect on
the ultimate bearing capacity.

3.5   CASE HISTORY: ULTIMATE BEARING CAPACITY IN SATURATED CLAY

Brand et al. (1972) reported field test results for small foundations on soft Bangkok
clay (a deposit of marine clay) in Rangsit, Thailand. The results of the soil
exploration are shown in Figure 3.7. Because of the sensitivity of the clay, the
laboratory test results for c_u (unconfined compression and unconsolidated undrained triaxial) were rather scattered; however, they obtained better results for
the variation of c_u with depth from field vane shear tests. The vane shear test
results showed that the average variations of the undrained cohesion were

Depth (m)	c_u (kN/m²)
0–1.5	≈35
1.5–2	Decreasing linearly from 35 to 24
2–8	≈24

Five small square foundations were tested for ultimate bearing capacity.
The sizes of the foundations were 0.6 m × 0.6 m, 0.675 m × 0.675 m, 0.75 m ×

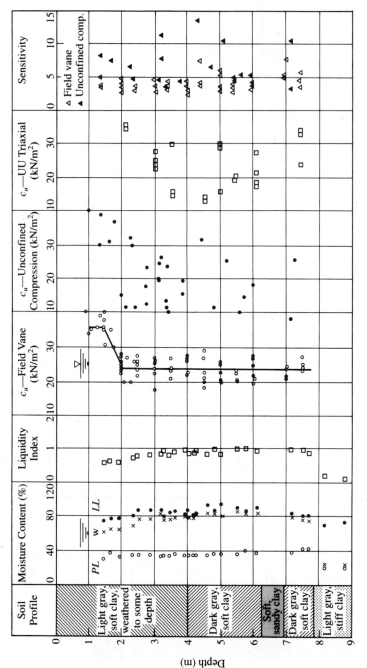

▶ **FIGURE 3.7** Results of soil exploration in soft Bangkok clay at Rangsit, Thailand (redrawn after Brand et al., 1972)

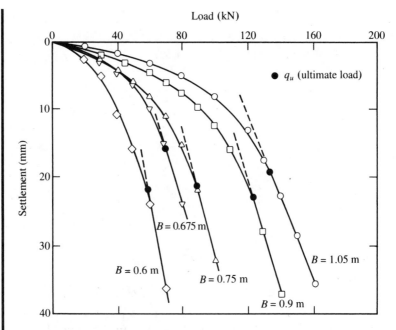

▼ **FIGURE 3.8** Loan–settlement plots obtained from bearing capacity tests

0.75 m, 0.9 m × 0.9 m, and 1.05 m × 1.05 m. The depth of the bottom of the foundations was 1.5 m measured from the ground surface. The load–settlement plots obtained from the bearing capacity tests are shown in Figure 3.8.

Analysis of the Field Test Results

The ultimate loads, Q_u, obtained from each test are also shown in Figure 3.8. The ultimate load is defined as the point where the load displacement becomes practically linear. The failure in soil below the foundation is of local shear type. Hence, from Eq. (3.10)

$$q_u = 0.867c_u N_c' + qN_q' + 0.4\gamma BN_\gamma'$$

For $\phi = 0$, $c = c_u$ and, from Table 3.2, $N_c' = 5.7$, $N_q' = 1$ and $N_\gamma' = 0$. Thus for $\phi = 0$

$$q_u = 4.94c_u + q \tag{3.15}$$

If we assume that the unit weight of soil is about 18.5 kN/m³, $q \approx D_f\gamma = (1.5)(18.5) = 27.75$ kN/m². We can then assume average values of c_u: for depths of 1.5 m to 2.0 m, $c_u \approx (35 + 24)/2 = 29.5$ kN/m²; for depths greater than 2.0 m, $c_u \approx 24$ kN/m². If we assume that the undrained cohesion of clay at depth $\leq B$ below the foundation controls the ultimate bearing capacity,

▼ **TABLE 3.3** Comparison of Theoretical and Field Ultimate Bearing Capacities

B (m)	$c_{u(average)}$[a] (kN/m²)	Plasticity index[b]	Correction factor, λ[c]	$c_{u(corrected)}$[d] (kN/m²)	$q_{u(theory)}$[e] (kN/m²)	$Q_{u(field)}$[f] (kN)	$q_{u(field)}$[g] (kN/m²)
0.6	28.58	40	0.84	24.01	146.4	60	166.6
0.675	28.07	40	0.84	23.58	144.2	71	155.8
0.75	27.67	40	0.84	23.24	142.6	90	160
0.9	27.06	40	0.84	22.73	140.0	124	153
1.05	26.62	40	0.84	22.36	138.2	140	127

[a] Eq. (3.16)
[b] From Figure 3.7
[c] From Table 2.7 [$\lambda = 1.7 - 0.54 \log(PI)$; Bjerrum (1972)]
[d] Eq. (2.19)
[e] Eq. (3.15)
[f] Figure 3.8
[g] $Q_{u(field)}/B^2$

$$c_{u(average)} \approx \frac{(29.5)(2.0 - 1.5) + (24)[B - (2.0 - 1.5)]}{B} \tag{3.16}$$

The $c_{u(average)}$ value obtained for each foundation needs to be corrected in view of Eq. (2.19). Table 3.3 presents the details of other calculations and a comparison of the theoretical and field ultimate bearing capacities.

Note that the ultimate bearing capacities obtained from the field are about 10% higher than those obtained from theory. One reason for such a difference is that the ratio D_f/B for the field tests varies from 1.5 to 2.5. The increase of the bearing capacity due to the depth of embedment has not been accounted for in Eq. (3.16).

3.6 FACTOR OF SAFETY

Calculating the gross *allowable load-bearing capacity* of shallow foundations requires application of a factor of safety (FS) to the gross ultimate bearing capacity, or

$$q_{all} = \frac{q_u}{FS} \tag{3.17}$$

However, some practicing engineers prefer to use a factor of safety of

$$\text{Net stress increase on soil} = \frac{\text{net ultimate bearing capacity}}{FS} \tag{3.18}$$

The net ultimate bearing capacity is defined as the ultimate pressure per unit area of the foundation that can be supported by the soil in excess of the pressure caused by the surrounding soil at the foundation level. If the difference between the unit weight of concrete used in the foundation and the unit weight of soil surrounding is assumed to be negligible,

$$q_{net(u)} = q_u - q \tag{3.19}$$

where $q_{net(u)}$ = net ultimate bearing capacity

$q = \gamma D_f$

So,

$$q_{all(net)} = \frac{q_u - q}{FS} \tag{3.20}$$

The factor of safety as defined by Eq. (3.20) may be at least 3 in all cases.

Another type of factor of safety for the bearing capacity of shallow foundations is often used. It is the factor of safety with respect to shear failure (FS_{shear}). In most cases, a value of $FS_{shear} = 1.4$–1.6 is desirable along with a *minimum* factor of safety of 3–4 against gross or net ultimate bearing capacity. The following procedure should be used to calculate the net allowable load for a given FS_{shear}.

1. Let c and ϕ be the cohesion and the angle of friction, respectively, of soil and let FS_{shear} be the required factor of safety with respect to shear failure. So the developed cohesion and the angle of friction are

$$c_d = \frac{c}{FS_{shear}} \tag{3.21}$$

$$\phi_d = \tan^{-1}\left(\frac{\tan \phi}{FS_{shear}}\right) \tag{3.22}$$

2. The gross allowable bearing capacity can now be calculated according to Eqs. (3.3), (3.7), (3.8), with c_d and ϕ_d as the shear strength parameters of the soil. For example, the gross allowable bearing capacity of a continuous foundation according to Terzaghi's equation is

$$q_{all} = c_d N_c + q N_q + \tfrac{1}{2} \gamma B N_\gamma \tag{3.23}$$

where N_c, N_q, and N_γ = bearing capacity factors for the friction angle, ϕ_d

3. The net allowable bearing capacity is thus

$$q_{all(net)} = q_{all} - q = c_d N_c + q(N_q - 1) + \tfrac{1}{2} \gamma B N_\gamma \tag{3.24}$$

Irrespective of the procedure by which the factor of safety is applied, the magnitude of *FS* should depend on the uncertainties and risks involved for the conditions encountered.

▼ **EXAMPLE 3.1** _____

A square foundation is 5 ft × 5 ft in plan. The soil supporting the foundation has a friction angle of $\phi = 20°$ and $c = 320$ lb/ft^2. The unit weight of soil, γ, is 115 lb/ft^3. Determine the allowable gross load on the foundation with a factor of safety (*FS*) of 4. Assume that the depth of the foundation (*D_f*) is 3 ft and that general shear failure occurs in the soil.

Solution   From Eq. (3.7)

$$q_u = 1.3cN_c + qN_q + 0.4\gamma BN_\gamma$$

From Table 3.1, for $\phi = 20°$,

$$N_c = 17.69$$

$$N_q = 7.44$$

$$N_\gamma = 3.64$$

Thus

$$q_u = (1.3)(320)(17.69) + (3 \times 115)(7.44) + (0.4)(115)(5)(3.64)$$

$$= 7359 + 2567 + 837 = 10,736 \, \text{lb}/\text{ft}^2$$

So, the allowable load per unit area of the foundation is

$$q_{\text{all}} = \frac{q_u}{FS} = \frac{10,736}{4} \approx 2691 \, \text{lb}/\text{ft}^2$$

Thus the total allowable gross load is

$$Q = (2691)B^2 = (2691)(5 \times 5) = 67,275 \, \text{lb}$$ ▲

3.7 THE GENERAL BEARING CAPACITY EQUATION

The ultimate bearing capacity equations presented in Eqs. (3.3), (3.7), and (3.8) are for continuous, square, and circular foundations only. They do not address the case of rectangular foundations ($0 < B/L < 1$). Also, the equations do not take into account the shearing resistance along the failure surface in soil above the bottom of the foundation (the portion of the failure surface marked as *GI* and *HJ* in Figure 3.5). In addition, the load on the foundation may be inclined. To account for all these shortcomings, Meyerhof (1963) suggested the following form of the general bearing capacity equation:

$$q_u = cN_cF_{cs}F_{cd}F_{ci} + qN_qF_{qs}F_{qd}F_{qi} + \tfrac{1}{2}\gamma BN_\gamma F_{\gamma s}F_{\gamma d}F_{\gamma i} \tag{3.25}$$

where
c = cohesion
q = effective stress at the level of the bottom of the foundation
γ = unit weight of soil
B = width of foundation (= diameter for a circular foundation)

$$F_{cs}, F_{qs}, F_{\gamma s} = \text{shape factors}$$
$$F_{cd}, F_{qd}, F_{\gamma d} = \text{depth factors}$$
$$F_{ci}, F_{qi}, F_{\gamma i} = \text{load inclination factors}$$
$$N_c, N_q, N_{\gamma} = \text{bearing capacity factors}$$

The equations for determining the various factors given in Eq. (3.25) are described briefly in the following sections. Note that the original equation for ultimate bearing capacity is derived only for the plane-strain case (that is, for continuous foundations). The shape, depth, and load inclination factors are empirical factors based on experimental data.

Bearing Capacity Factors

Based on laboratory and field studies of bearing capacity, the basic nature of the failure surface in soil suggested by Terzaghi now appears to be correct (Vesic, 1973). However, the angle α as shown in Figure 3.5 is closer to $45 + \phi/2$ than to ϕ. If this change is accepted, the values of N_c, N_q, and N_{γ} for a given soil friction angle will also change from those given in Table 3.1. With $\alpha = 45 + \phi/2$, the relations for N_c and N_q can be derived as

$$N_q = \tan^2\left(45 + \frac{\phi}{2}\right) e^{\pi \tan \phi} \tag{3.26}$$

$$N_c = (N_q - 1)\cot \phi \tag{3.27}$$

The equation for N_c given by Eq. (3.27) was originally derived by Prandtl (1921), and the relation for N_q [Eq. (3.26)] was presented by Reissner (1924). Caquot and Kerisel (1953) and Vesic (1973) gave the relation for N_{γ} as

$$N_{\gamma} = 2(N_q + 1)\tan \phi \tag{3.28}$$

Table 3.4 shows the variation of the preceding bearing capacity factors with soil friction angles.

In many texts and reference books, the relationship for N_{γ} may be different from that in Eq. (3.28). The reason is that there is still some controversy about the variation of N_{γ} with the soil friction angle, ϕ. *In this text, Eq. (3.28) is used.*

Other relationships for N_{γ} generally cited are those given by Meyerhof (1963), Hansen (1970), and Lundgren and Mortensen (1953). These N_{γ} values for various soil friction angles are given in Appendix B (Tables B-1, B-2, and B-3).

▼ **TABLE 3.4** Bearing Capacity Factors[a]

ϕ	N_c	N_q	N_γ	N_q/N_c	$\tan \phi$	ϕ	N_c	N_q	N_γ	N_q/N_c	$\tan \phi$
0	5.14	1.00	0.00	0.20	0.00	26	22.25	11.85	12.54	0.53	0.49
1	5.38	1.09	0.07	0.20	0.02	27	23.94	13.20	14.47	0.55	0.51
2	5.63	1.20	0.15	0.21	0.03	28	25.80	14.72	16.72	0.57	0.53
3	5.90	1.31	0.24	0.22	0.05	29	27.86	16.44	19.34	0.59	0.55
4	6.19	1.43	0.34	0.23	0.07	30	30.14	18.40	22.40	0.61	0.58
5	6.49	1.57	0.45	0.24	0.09	31	32.67	20.63	25.99	0.63	0.60
6	6.81	1.72	0.57	0.25	0.11	32	35.49	23.18	30.22	0.65	0.62
7	7.16	1.88	0.71	0.26	0.12	33	38.64	26.09	35.19	0.68	0.65
8	7.53	2.06	0.86	0.27	0.14	34	42.16	29.44	41.06	0.70	0.67
9	7.92	2.25	1.03	0.28	0.16	35	46.12	33.30	48.03	0.72	0.70
10	8.35	2.47	1.22	0.30	0.18	36	50.59	37.75	56.31	0.75	0.73
11	8.80	2.71	1.44	0.31	0.19	37	55.63	42.92	66.19	0.77	0.75
12	9.28	2.97	1.69	0.32	0.21	38	61.35	48.93	78.03	0.80	0.78
13	9.81	3.26	1.97	0.33	0.23	39	67.87	55.96	92.25	0.82	0.81
14	10.37	3.59	2.29	0.35	0.25	40	75.31	64.20	109.41	0.85	0.84
15	10.98	3.94	2.65	0.36	0.27	41	83.86	73.90	130.22	0.88	0.87
16	11.63	4.34	3.06	0.37	0.29	42	93.71	85.38	155.55	0.91	0.90
17	12.34	4.77	3.53	0.39	0.31	43	105.11	99.02	186.54	0.94	0.93
18	13.10	5.26	4.07	0.40	0.32	44	118.37	115.31	224.64	0.97	0.97
19	13.93	5.80	4.68	0.42	0.34	45	133.88	134.88	271.76	1.01	1.00
20	14.83	6.40	5.39	0.43	0.36	46	152.10	158.51	330.35	1.04	1.04
21	15.82	7.07	6.20	0.45	0.38	47	173.64	187.21	403.67	1.08	1.07
22	16.88	7.82	7.13	0.46	0.40	48	199.26	222.31	496.01	1.12	1.11
23	18.05	8.66	8.20	0.48	0.42	49	229.93	265.51	613.16	1.15	1.15
24	19.32	9.60	9.44	0.50	0.45	50	266.89	319.07	762.89	1.20	1.19
25	20.72	10.66	10.88	0.51	0.47						

[a] After Vesic (1973)

Shape, Depth, and Inclination Factors

The relationships for the shape factors, depth factors, and inclination factors *recommended for use* are shown in Table 3.5. Other relationships generally found in many texts and references are shown in Table B-4 (Appendix B).

General Comments

When the water table is present at or near the foundation, the factors q and γ given in the general bearing capacity equation, Eq. (3.25), will need modifications. The procedure for modifying them is the same as that described in Section 3.4.

For undrained loading conditions ($\phi = 0$ concept) in clayey soils, the general load-bearing capacity equation [Eq. (3.25)] takes the form (vertical load)

$$q_u = cN_cF_{cs}F_{cd} + q \tag{3.29}$$

▼ **TABLE 3.5** Shape, Depth, and Inclination Factors Recommended for Use

Factor	Relationship	Source
Shape[a]	$F_{cs} = 1 + \dfrac{B}{L}\dfrac{N_q}{N_c}$ $F_{qs} = 1 + \dfrac{B}{L}\tan\phi$ $F_{\gamma s} = 1 - 0.4\dfrac{B}{L}$ where L = length of the foundation $(L > B)$	De Beer (1970) Hansen (1970)
Depth[b]	*Condition (a):* $D_f/B \leq 1$ $F_{cd} = 1 + 0.4\dfrac{D_f}{B}$ $F_{qd} = 1 + 2\tan\phi(1 - \sin\phi)^2\dfrac{D_f}{B}$ $F_{\gamma d} = 1$ *Condition (b):* $D_f/B > 1$ $F_{cd} = 1 + (0.4)\tan^{-1}\left(\dfrac{D_f}{B}\right)$ $F_{qd} = 1 + 2\tan\phi(1 - \sin\phi)^2\tan^{-1}\left(\dfrac{D_f}{B}\right)$ $F_{\gamma d} = 1$	Hansen (1970)
Inclination	$F_{ci} = F_{qi} = \left(1 - \dfrac{\beta^\circ}{90^\circ}\right)^2$ $F_{\gamma i} = \left(1 - \dfrac{\beta}{\phi}\right)^2$ where β = inclination of the load on the foundation with respect to the vertical	Meyerhof (1963); Hanna and Meyerhof (1981)

[a] These shape factors are empirical relations based on extensive laboratory tests.
[b] The factor $\tan^{-1}(D_f/B)$ is in radians.

Hence the ultimate bearing capacity (vertical load) is

$$q_{net(u)} = q_u - q = cN_cF_{cs}F_{cd} \tag{3.30}$$

Skempton (1951) proposed an equation for the net ultimate bearing capacity for clayey soils ($\phi = 0$ condition), which is similar to Eq. (3.30):

$$q_{net(u)} = 5c\left(1 + 0.2\frac{D_f}{B}\right)\left(1 + 0.2\frac{B}{L}\right) \tag{3.31}$$

▼   **EXAMPLE 3.2**

A square foundation ($B \times B$) has to be constructed as shown in Figure 3.9. Assume that $\gamma = 105$ lb/ft³, $\gamma_{sat} = 118$ lb/ft³, $D_f = 4$ ft, and $D_1 = 2$ ft. The gross allowable load, Q_{all}, with $FS = 3$ is 150,000 lb. The field standard penetration resistance, N_F, values are as follows:

Depth (ft)	N_F (blow/ft)
5	4
10	6
15	6
20	10
25	5

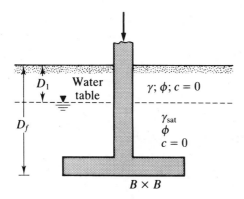

▼   **FIGURE 3.9**

Determine the size of the footing. Use Eq. (3.25).

Solution   Using Eq. (2.7) and the Liao and Whitman relationship (Table 2.4), the correct standard penetration number can be determined.

Depth (ft)	N_F	σ' (ton/ft²)	$N_{cor} = N_F \sqrt{\dfrac{1}{\sigma_v'}}$
5	4	$\dfrac{1}{2000}[2 \times 105 + 3 \times (118 - 62.4)] = 0.188$	12
10	6	$0.188 + \dfrac{1}{2000}(5)(118 - 62.4) = 0.327$	11
15	6	$0.327 + \dfrac{1}{2000}(5)(118 - 62.4) = 0.466$	9
20	10	$0.466 + \dfrac{1}{2000}(5)(118 - 62.4) = 0.605$	13
25	5	$0.605 + \dfrac{1}{2000}(5)(118 - 62.4) = 0.744$	6

The average N_{cor} can be taken to be about 11.
   From Eq. 2.11, $\phi \approx 35°$. Given

$$q_{all} = \frac{Q_{all}}{B^2} = \frac{150{,}000}{B^2} \text{ lb/ft}^2 \qquad \text{(a)}$$

From Eq. (3.25) (*note: c = 0*),

$$q_{all} = \frac{q_u}{FS} = \frac{1}{3}\left(qN_qF_{qs}F_{qd} + \frac{1}{2}\gamma'BN_\gamma F_{\gamma s}F_{\gamma d}\right)$$

For $\phi = 35°$, from Table 3.4, $N_q = 33.3$, $N_\gamma = 48.03$. From Table 3.5,

$$F_{qs} = 1 + \frac{B}{L}\tan\phi = 1 + \tan 35 = 1.7$$

$$F_{\gamma s} = 1 - 0.4\left(\frac{B}{L}\right) = 1 - 0.4 = 0.6$$

$$F_{qd} = 1 + 2\tan\phi(1 - \sin\phi)^2 = 1 + 2\tan 35(1 - \sin 35)^2\frac{4}{B} = 1 + \frac{1}{B}$$

$$F_{\gamma d} = 1$$

$$q = (2)(105) + 2(118 - 62.4) = 321.2 \text{ lb/ft}^2$$

So

$$q_{all} = \frac{1}{3}\left[(321.2)(33.3)(1.7)\left(1 + \frac{1}{B}\right) + \left(\frac{1}{2}\right)(118 - 62.4)(B)(48.03)(0.6)(1)\right]$$

$$= 6061.04 + \frac{6061.04}{B} + 267.05B \qquad \text{(b)}$$

Combining Eqs. (a) and (b)

$$\frac{150{,}000}{B^2} = 6061.04 + \frac{6061.04}{B} + 267.05B$$

By trial and error, $B \approx 4.2$ ft.   ▲

▼ **EXAMPLE 3.3**

Refer to Example 3.1. Use the definition of factor of safety given by Eq. (3.20) and
$FS = 5$ to determine the net allowable load for the foundation.

Solution   From Example 3.1,

$$q_u = 10{,}736 \text{ lb/ft}^2$$

$$q = (3)(115) = 345 \text{ lb/ft}^2$$

$$q_{all(net)} = \frac{10{,}736 - 345}{5} \approx 2078 \text{ lb/ft}^2$$

Hence

$$Q_{all(net)} = (2078)(5)(5) = 51{,}950 \text{ lb}$$ ▲

▼ **EXAMPLE 3.4** _____

Refer to Example 3.1. Use Eq. (3.7) and $FS_{shear} = 1.5$ to determine the net allowable load for the foundation.

Solution   For $c = 320 \ \text{lb/ft}^2$ and $\phi = 20°$,

$$c_d = \frac{c}{FS_{shear}} = \frac{320}{1.5} \approx 213 \ \text{lb/ft}^2$$

$$\phi_d = \tan^{-1}\left[\frac{\tan \phi}{FS_{shear}}\right] = \tan^{-1}\left[\frac{\tan 20}{1.5}\right] = 13.64°$$

From Eq. (3.7),

$$q_{all(net)} = 1.3 c_d N_c + q(N_q - 1) + 0.4\gamma B N_\gamma$$

For $\phi = 13.64°$, the values of the bearing capacity factors from Table 3.1 are

$$N_\gamma \approx 1.2, \qquad N_q \approx 3.8, \qquad \text{and} \qquad N_c \approx 12$$

Hence

$$q_{all(net)} = 1.3\,(213)\,(12) + (345)\,(3.8 - 1) + (0.4)\,(115)\,(5)\,(1.2) = 4565 \ \text{lb/ft}^2$$

and

$$Q_{all(net)} = (4565)\,(5)\,(5) = 114{,}125 \ \text{lb} \approx \textbf{57 ton}$$

Note: There appears to be a large discrepancy between the results of Examples 3.3 (or 3.1) and 3.4. The use of trial and error shows that, when FS_{shear} is about 1.2, the results are approximately equal.   ▲

3.8   EFFECT OF SOIL COMPRESSIBILITY

In Section 3.3, Eqs. (3.3), (3.7), and (3.8), which were for the case of general shear failure, were modified to Eqs. (3.9), (3.10), and (3.11) to take into account the change of failure mode in soil (that is, local shear failure). The change in failure mode is due to soil compressibility. In order to account for soil compressibility, Vesic (1973) proposed the following modification to Eq. (3.25),

$$q_u = cN_c F_{cs} F_{cd} F_{cc} + qN_q F_{qs} F_{qd} F_{qc} + \tfrac{1}{2}\gamma B N_\gamma F_{\gamma s} F_{\gamma d} F_{\gamma c} \tag{3.32}$$

where F_{cc}, F_{qc}, and $F_{\gamma c}$ = soil compressibility factors

The soil compressibility factors were derived by Vesic (1973) from the analogy of the expansion of cavities. According to that theory, in order to calculate F_{cc}, F_{qc}, and $F_{\gamma c}$, the following steps should be taken:

1. Calculate the *rigidity index, I_r,* of the soil at a depth approximately $B/2$ below the bottom of the foundation, or

$$I_r = \frac{G}{c + q' \tan \phi}$$
(3.33)

where G = shear modulus of the soil
q' = effective overburden pressure at a depth of $D_f + B/2$

2. The critical rigidity index, $I_{r(cr)}$, can be expressed as

$$I_{r(cr)} = \frac{1}{2} \left\{ \exp \left[\left(3.30 - 0.45 \frac{B}{L} \right) \cot \left(45 - \frac{\phi}{2} \right) \right] \right\}$$
(3.34)

The variations of $I_{r(cr)}$ for $B/L = 0$ and $B/L = 1$ are given in Table 3.6.

3. If $I_r \geq I_{r(cr)}$, then

$$F_{cc} = F_{qc} = F_{\gamma c} = 1$$

However, if $I_r < I_{r(cr)}$

$$F_{\gamma c} = F_{qc} = \exp \left\{ \left(-4.4 + 0.6 \frac{B}{L} \right) \tan \phi + \left[\frac{(3.07 \sin \phi)(\log 2I_r)}{1 + \sin \phi} \right] \right\}$$

(3.35)

▼ **TABLE 3.6** Variation of $I_{r(cr)}$ with ϕ and B/L[1]

ϕ (deg)	$I_{r(cr)}$ $\frac{B}{L} = 0$	$\frac{B}{L} = 1$
0	13	8
5	18	11
10	25	15
15	37	20
20	55	30
25	89	44
30	152	70
35	283	120
40	592	225
45	1442	482
50	4330	1258

[1] After Vesic (1973)

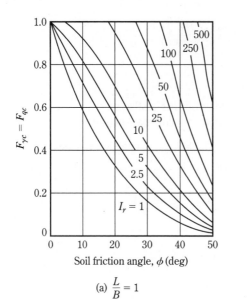

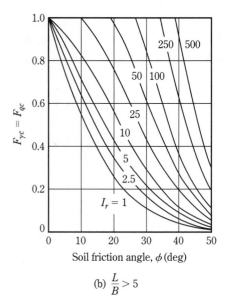

(a) $\dfrac{L}{B} = 1$ (b) $\dfrac{L}{B} > 5$

▼ **FIGURE 3.10**   Variation of $F_{\gamma c} = F_{qc}$ with I_r and ϕ

Figure 3.10 shows the variation of $F_{\gamma c} = F_{qc}$ [Eq. (3.35)] with ϕ and I_r. For $\phi = 0$,

$$F_{cc} = 0.32 + 0.12 \frac{B}{L} + 0.60 \log I_r \qquad (3.36)$$

For $\phi > 0$,

$$F_{cc} = F_{qc} - \frac{1 - F_{qc}}{N_q \tan \phi} \qquad (3.37)$$

▼ **EXAMPLE 3.5**

For a shallow foundation, the following are given: $B = 0.6$ m, $L = 1.2$ m, $D_f = 0.6$ m.

Soil: $\phi = 25°$

$c = 48 \text{ kN/m}^2$

$\gamma = 18 \text{ kN/m}^3$

Modulus of elasticity, $E = 620 \text{ kN/m}^2$

Poisson's ratio, $\mu = 0.3$

Calculate the ultimate bearing capacity.

Solution From Eq. (3.33)

$$I_r = \frac{G}{c + q' \tan \phi}$$

However,

$$G = \frac{E}{2(1 + \mu)}$$

So

$$I_r = \frac{E}{2(1 + \mu)[c + q' \tan \phi]}$$

$$q' = \gamma \left(D_f + \frac{B}{2} \right) = 18 \left(0.6 + \frac{0.6}{2} \right) = 162 \text{ kN/m}^2$$

$$I_r = \frac{620}{2(1 + 0.3)[48 + 16.2 \tan 25]} = 4.29$$

From Eq. (3.34)

$$I_{r(\text{cr})} = \frac{1}{2} \left\{ \exp \left[\left(3.3 - 0.45 \frac{B}{L} \right) \cot \left(45 - \frac{\phi}{2} \right) \right] \right\}$$

$$= \frac{1}{2} \left\{ \exp \left[\left(3.3 - 0.45 \frac{0.6}{1.2} \right) \cot \left(45 - \frac{25}{2} \right) \right] \right\} = 62.46$$

Since $I_{r(\text{cr})} > I_r$, use Eqs. (3.35) and (3.37).

$$F_{\gamma c} = F_{qc} = \exp \left\{ \left(-4.4 + 0.6 \frac{B}{L} \right) \tan \phi + \left[\frac{(3.07 \sin \phi) \log (2I_r)}{1 + \sin \phi} \right] \right\}$$

$$= \exp \left\{ \left(-4.4 + 0.6 \frac{0.6}{1.2} \right) \tan 25 \right.$$

$$\left. + \left[\frac{(3.07 \sin 25) \log (2 \times 4.29)}{1 + \sin 25} \right] \right\} = 0.347$$

$$F_{cc} = F_{qc} - \frac{1 - F_{qc}}{N_q \tan \phi}$$

For $\phi = 25°$, $N_q = 10.66$ (Table 3.4),

$$F_{cc} = 0.347 - \frac{1 - 0.347}{10.66 \tan 25} = 0.216$$

Now, from Eq. (3.32),

$$q_u = c N_c F_{cs} F_{cd} F_{cc} + q N_q F_{qs} F_{qd} F_{qc} + \tfrac{1}{2} \gamma B N_\gamma F_{\gamma s} F_{\gamma d} F_{\gamma c}$$

From Table 3.4, for $\phi = 25°$, $N_c = 20.72$, $N_q = 10.66$, $N_\gamma = 10.88$. From Table 3.5,

$$F_{cs} = 1 + \left(\frac{N_q}{N_c}\right)\left(\frac{B}{L}\right) = 1 + \left(\frac{10.66}{20.72}\right)\left(\frac{0.6}{1.2}\right) = 1.257$$

$$F_{qs} = 1 + \frac{B}{L}\tan\phi = 1 + \frac{0.6}{1.2}\tan 25 = 1.233$$

$$F_{\gamma s} = 1 - 0.4\frac{B}{L} = 1 - 0.4\frac{0.6}{1.2} = 0.8$$

$$F_{cd} = 1 + 0.4\left(\frac{D_f}{B}\right) = 1 + 0.4\left(\frac{0.6}{0.6}\right) = 1.4$$

$$F_{qd} = 1 + 2\tan\phi(1 - \sin\phi)^2\left(\frac{D_f}{B}\right) = 1 + 2\tan 25(1 - \sin 25)^2\left(\frac{0.6}{0.6}\right) = 1.311$$

$$F_{\gamma d} = 1$$

Thus

$$q_u = (48)(20.72)(1.257)(1.4)(0.216) + (0.6 \times 18)(10.66)(1.233)(1.311)(0.347)$$
$$+ (\tfrac{1}{2})(18)(0.6)(10.88)(0.8)(1)(0.347) = \mathbf{459\ kN/m^2} \qquad \blacktriangle$$

3.9 ECCENTRICALLY LOADED FOUNDATIONS

In several instances, as with the base of a retaining wall, foundations are subjected to moments in addition to the vertical load, as shown in Figure 3.11a. In such cases the distribution of pressure by the foundation on the soil is not uniform. The distribution of nominal pressure is

$$q_{max} = \frac{Q}{BL} + \frac{6M}{B^2 L} \tag{3.38}$$

and

$$q_{min} = \frac{Q}{BL} - \frac{6M}{B^2 L} \tag{3.39}$$

where Q = total vertical load
M = moment on the foundation

Figure 3.11b shows a force system equivalent to that shown in Figure 3.11a. The distance e is the eccentricity, or

$$e = \frac{M}{Q} \tag{3.40}$$

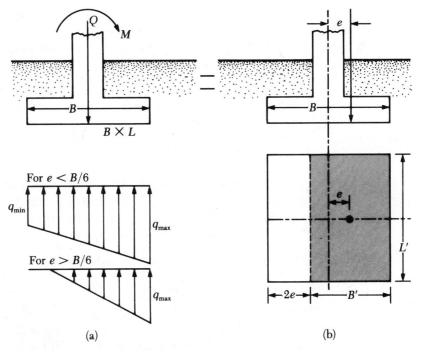

▼ FIGURE 3.11 Eccentrically loaded foundations

Substituting Eq. (3.40) in Eqs. (3.38) and (3.39) gives

$$q_{max} = \frac{Q}{BL}\left(1 + \frac{6e}{B}\right)$$

(3.41a)

and

$$q_{min} = \frac{Q}{BL}\left(1 - \frac{6e}{B}\right)$$

(3.41b)

Note that, in these equations, when the eccentricity, e, becomes $B/6$, q_{min} is zero. For $e > B/6$, q_{min} will be negative, which means that tension will develop. Because soil cannot take any tension, there will be a separation between the foundation and the soil underlying it. The nature of the pressure distribution on the soil will be as shown in Figure 3.11a. The value of q_{max} then is

$$q_{max} = \frac{4Q}{3L(B - 2e)}$$

(3.42)

The exact distribution of pressure is difficult to estimate.

The factor of safety for such types of loading against bearing capacity failure can be evaluated by using the procedure suggested by Meyerhof (1953), which is generally referred to as the *effective area* method. The following is Meyerhof's step-by-step procedure for determination of the ultimate load that the soil can support and the factor of safety against bearing capacity failure.

1. Determine the effective dimensions of the foundation as

   B' = effective width = $B - 2e$

   L' = effective length = L

   Note that, if the eccentricity were in the direction of the length of the foundation, the value of L' would be equal to $L - 2e$. The value of B' would equal B. The smaller of the two dimensions (that is, L' and B') is the effective width of the foundation.

2. Use Eq. (3.25) for the ultimate bearing capacity as

$$q'_u = cN_cF_{cs}F_{cd}F_{ci} + qN_qF_{qs}F_{qd}F_{qi} + \tfrac{1}{2}\gamma B'N_\gamma F_{\gamma s}F_{\gamma d}F_{\gamma i} \tag{3.43}$$

   To evaluate F_{cs}, F_{qs}, and $F_{\gamma s}$, use Table 3.5 with *effective length* and *effective width* dimensions instead of L and B, respectively. To determine F_{cd}, F_{qd}, and $F_{\gamma d}$, use Table 3.5 (*do not* replace B with B').

3. The total ultimate load that the foundation can sustain is

$$Q_{\text{ult}} = q'_u \overbrace{(B')(L')}^{A'} \tag{3.44}$$

   where A' = effective area

4. The factor of safety against bearing capacity failure is

$$FS = \frac{Q_{\text{ult}}}{Q} \tag{3.45}$$

5. Check the factor of safety against $q_{\max}$, or, $FS = q'_u/q_{\max}$.

Note that eccentricity tends to decrease the load-bearing capacity of a foundation. In such cases, placing foundation columns off center, as shown in Figure 3.12, probably is advantageous. Doing so, in effect, produces a centrally loaded foundation with uniformly distributed pressure.

Foundations with Two-Way Eccentricity

Consider a situation in which a foundation is subjected to a vertical ultimate load Q_{ult} and a moment M as shown in Figure 3.13a and b. For this case, the components of the moment, M, about the x and y axes can be determined as M_x and M_y, respectively

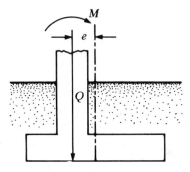

▼ **FIGURE 3.12** Foundation of columns with off-center loading

(Figure 3.13). This condition is equivalent to a load Q_{ult} placed eccentrically on the foundation with $x = e_B$ and $y = e_L$ (Figure 3.13d). Note that

$$e_B = \frac{M_y}{Q_{ult}}$$

(3.46)

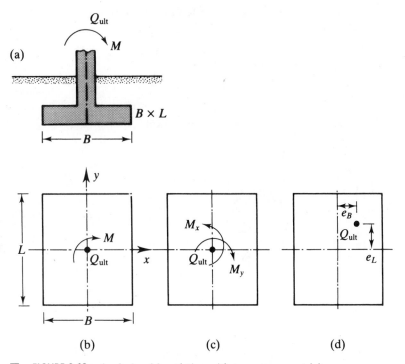

▼ **FIGURE 3.13** Analysis of foundation with two-way eccentricity

and

$$e_L = \frac{M_x}{Q_{\text{ult}}} \qquad (3.47)$$

If Q_{ult} is needed, it can be obtained as follows [Eq. (3.44)]:

$$Q_{\text{ult}} = q'_u A'$$

where, from Eq. (3.43)

$$q'_u = cN_c F_{cs} F_{cd} F_{ci} + qN_q F_{qs} F_{qd} F_{qi} + \tfrac{1}{2}\gamma B' N_\gamma F_{\gamma s} F_{\gamma d} F_{\gamma i}$$

and

$$A' = \text{effective area} = B'L'$$

As before, to evaluate F_{cs}, F_{qs}, and $F_{\gamma s}$ (Table 3.5), use the effective length (L') and effective width (B') dimensions instead of L and B, respectively. To calculate F_{cd}, F_{qd}, and $F_{\gamma d}$, use Table 3.5; however, do not replace B with B'. In determining the effective area (A'), effective width (B'), and effective length (L'), four possible cases may arise (Highter and Anders, 1985).

Case I

$e_L/L \ge \frac{1}{6}$ and $e_B/B \ge \frac{1}{6}$. The effective area for this condition is shown in Figure 3.14, or

$$A' = \tfrac{1}{2}B_1 L_1 \qquad (3.48)$$

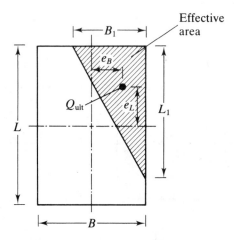

▼ **FIGURE 3.14**   Effective area for the case of $e_L/L \ge \frac{1}{6}$ and $e_B/B \ge \frac{1}{6}$

where

$$B_1 = B\left(1.5 - \frac{3e_B}{B}\right) \tag{3.49a}$$

$$L_1 = L\left(1.5 - \frac{3e_L}{L}\right) \tag{3.49b}$$

The effective length, L', is the larger of the two dimensions, that is, B_1 or L_1. So, the effective width is

$$B' = \frac{A'}{L'} \tag{3.50}$$

Case II

$e_L/L < 0.5$ and $0 < e_B/B < \frac{1}{6}$. The effective area for this case is shown in Figure 3.15a.

$$A' = \tfrac{1}{2}(L_1 + L_2)B \tag{3.51}$$

The magnitudes of L_1 and L_2 can be determined from Figure 3.15b. The effective width is

$$B' = \frac{A'}{L_1 \text{ or } L_2 \quad \text{(whichever is larger)}} \tag{3.52}$$

The effective length is

$$L' = L_1 \text{ or } L_2 \quad \text{(whichever is larger)} \tag{3.53}$$

Case III

$e_L/L < \frac{1}{6}$ and $0 < e_B/B < 0.5$. The effective area is shown in Figure 3.16a:

$$A' = \tfrac{1}{2}(B_1 + B_2)L \tag{3.54}$$

The effective width is

$$B' = \frac{A'}{L} \tag{3.55}$$

The effective length is equal to

$$L' = L \tag{3.56}$$

The magnitudes of B_1 and B_2 can be determined from Figure 3.16b.

Case IV

$e_L/L < \frac{1}{6}$ and $e_B/B < \frac{1}{6}$. Figure 3.17a shows the effective area for this case. The ratio B_2/B and thus B_2 can be determined by using the e_L/L curves that slope upward. Similarly, the ratio L_2/L and thus L_2 can be determined by using the e_L/L curves that slope downward. The effective area is then

$$A' = L_2 B + \tfrac{1}{2}(B + B_2)(L - L_2) \tag{3.57}$$

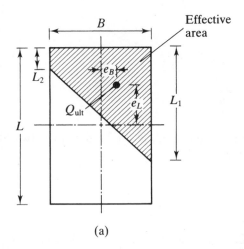

(a)

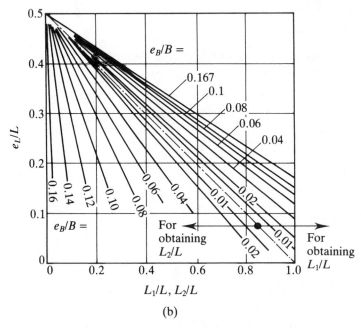

(b)

▼ **FIGURE 3.15** Effective area for the case of $e_L/L < 0.5$ and $0 < e_B/B < \frac{1}{6}$ (after Highter and Anders, 1985)

The effective width is

$$B' = \frac{A'}{L} \tag{3.58}$$

The effective length is

$$L' = L \tag{3.59}$$

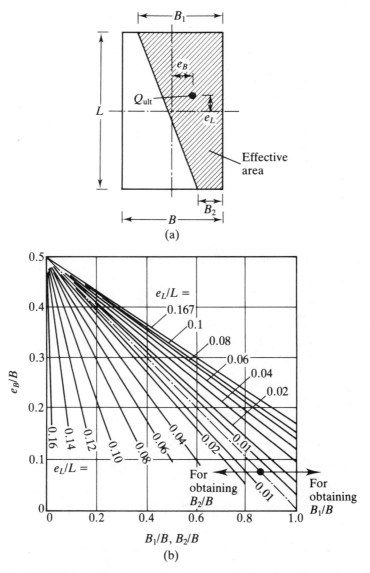

▼ **FIGURE 3.16** Effective area for the case of $e_L/L < \frac{1}{6}$ and $0 < e_B/B < 0.5$ (after Highter and Anders, 1985)

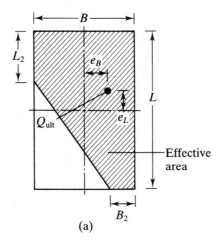

(a)

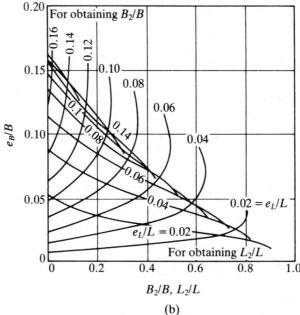

(b)

▼ **FIGURE 3.17**   Effective area for the case of $e_L/ < \frac{1}{6}$ and $e_B/B < \frac{1}{6}$ (after Highter and Anders, 1985)

▼ **EXAMPLE 3.6**

A square foundation is shown in Figure 3.18. Assume that the one-way load eccentricity $e = 0.15$ m. Determine the ultimate load, Q_{ult}.

Solution   With $c = 0$, Eq. (3.43) becomes

$$q'_u = qN_qF_{qs}F_{qd}F_{qi} + \tfrac{1}{2}\gamma B'N_\gamma F_{\gamma s}F_{\gamma d}F_{\gamma i}$$

$$q = (0.7)(18) = 12.6 \text{ kN/m}^2$$

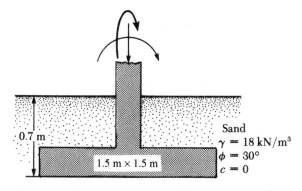

▼ FIGURE 3.18

For $\phi = 30°$, from Table 3.4, $N_q = 18.4$ and $N_\gamma = 22.4$.

$$B' = 1.5 - 2(0.15) = 1.2 \text{ m}$$
$$L' = 1.5 \text{ m}$$

From Table 3.5

$$F_{qs} = 1 + \frac{B'}{L'} \tan \phi = 1 + \left(\frac{1.2}{1.5}\right) \tan 30° = 1.462$$

$$F_{qd} = 1 + 2 \tan \phi (1 - \sin \phi)^2 \frac{D_f}{B} = 1 + \frac{(0.289)(0.7)}{1.5} = 1.135$$

$$F_{\gamma s} = 1 - 0.4 \left(\frac{B'}{L'}\right) = 1 - 0.4 \left(\frac{1.2}{1.5}\right) = 0.68$$

$$F_{\gamma d} = 1$$

So

$$q'_u = (12.6)(18.4)(1.462)(1.135) + \tfrac{1}{2}(18)(1.2)(22.4)(0.68)(1)$$
$$= 384.7 + 164.50 = 549.2 \text{ kN/m}^2$$

Hence

$$Q_{\text{ult}} = B'L'(q'_u) = (1.2)(1.5)(549.2) \approx \mathbf{988 \text{ kN}}$$

▲

▼ EXAMPLE 3.7

Refer to Example 3.6. Other quantities remaining the same, assume that the load has a two-way eccentricity. Given: $e_L = 0.3$ m and $e_B = 0.15$ m (Figure 3.19). Determine the ultimate load, Q_{ult}.

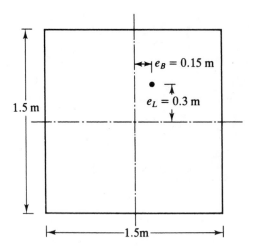

$eB = 0.15$ m

$eL = 0.3$ m

1.5 m

1.5m

▼　FIGURE 3.19

Solution

$$\frac{e_L}{L} = \frac{0.3}{1.5} = 0.2$$

$$\frac{e_B}{B} = \frac{0.15}{1.5} = 0.1$$

This case is similar to that shown in Figure 3.15a. From Figure 3.15b, for $e_L/L = 0.2$ and $e_B/B = 0.1$

$$\frac{L_1}{L} \approx 0.85; \qquad L_1 = (0.85)(1.5) = 1.275 \text{ m}$$

and

$$\frac{L_2}{L} \approx 0.21; \qquad L_2 = (0.21)(1.5) = 0.315 \text{ m}$$

From Eq. (3.51)

$$A' = \tfrac{1}{2}(L_1 + L_2)B = \tfrac{1}{2}(1.275 + 0.315)(1.5) = 1.193 \text{ m}^2$$

From Eq. (3.53)

$$L' = L_1 = 1.275 \text{ m}$$

From Eq. (3.52)

$$B' = \frac{A'}{L'} = \frac{1.193}{1.275} = 0.936 \text{ m}$$

Note, from Eq. (3.43), for $c = 0$

$$q_u' = qN_qF_{qs}F_{qd}F_{qi} + \tfrac{1}{2}\gamma B'N_\gamma F_{\gamma s}F_{\gamma d}F_{\gamma i}$$

$$q = (0.7)(18) = 12.6 \text{ kN/m}^2$$

For $\phi = 30°$, from Table 3.4, $N_q = 18.4$ and $N_\gamma = 22.4$. Thus

$$F_{qs} = 1 + \left(\frac{B'}{L'}\right)\tan\phi = 1 + \left(\frac{0.936}{1.275}\right)\tan 30° = 1.424$$

$$F_{\gamma s} = 1 - 0.4\left(\frac{B'}{L'}\right) = 1 - 0.4\left(\frac{0.936}{1.275}\right) = 0.706$$

$$F_{qd} = 1 + 2\tan\phi(1 - \sin\phi)^2\frac{D_f}{B} = 1 + \frac{(0.289)(0.7)}{1.5} = 1.135$$

$$F_{\gamma d} = 1$$

So

$$\begin{aligned}
Q_{\text{ult}} &= A'q_u' = A'(qN_qF_{qs}F_{qd} + \tfrac{1}{2}\gamma B'N_\gamma F_{\gamma s}F_{\gamma d})\\
&= (1.193)[(12.6)(18.4)(1.424)(1.135)\\
&\quad + (0.5)(18)(0.936)(22.4)(0.706)(1)]\\
&= \mathbf{605.95\ kN}
\end{aligned}$$

▲

3.10 BEARING CAPACITY OF LAYERED SOILS— STRONGER SOIL UNDERLAIN BY WEAKER SOIL

The bearing capacity equations presented in the preceding sections involve cases in which the soil supporting the foundation is homogeneous and extends to a considerable depth. Cohesion, angle of friction, and unit weight of soil were assumed to remain constant for the bearing capacity analysis. However, in practice, layered soil profiles are often encountered. In such instances, the failure surface at ultimate load may extend through two or more soil layers. Determination of ultimate bearing capacity in layered soils can be made in only a limited number of cases. This section features the procedure for estimating bearing capacity for layered soils proposed by Meyerhof and Hanna (1978) and Meyerhof (1974).

Figure 3.20 shows a shallow continuous foundation supported by a *stronger soil layer* underlain by a weaker soil, which extends to a great depth. For the two soil layers, the physical parameters are as follows:

Layer	Unit weight	Soil friction angle	Cohesion
Top	γ_1	ϕ_1	c_1
Bottom	γ_2	ϕ_2	c_2

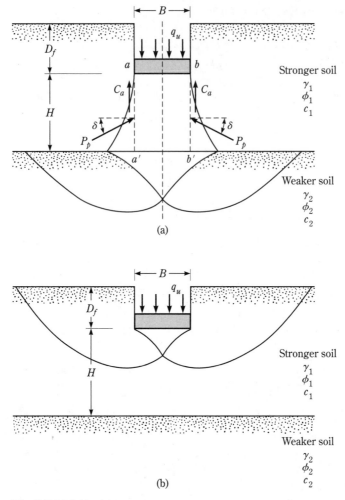

▼ **FIGURE 3.20**   Bearing capacity of a continuous foundation on
              layered soil

At ultimate load per unit area (q_u), the failure surface in soil will be as shown in Figure 3.20. If the depth H is relatively small compared to the foundation width B, a punching shear failure will occur in the top soil layer followed by a general shear failure in the bottom soil layer. This is shown in Figure 3.20a. However, if the depth H is relatively large, then the failure surface will be completely located in the top soil layer, which is the upper limit for the ultimate bearing capacity. This is shown in Figure 3.20b.

The ultimate bearing capacity, q_u, for this problem as shown in Figure 3.20a can be given as

$$q_u = q_b + \frac{2(C_a + P_p \sin \delta)}{B} - \gamma_1 H \qquad (3.60)$$

where B = width of the foundation
C_a = adhesive force
P_p = passive force per unit length of the faces aa' and bb'
q_b = bearing capacity of the bottom soil layer
δ = inclination of the passive force P_p with the horizontal

Note that, in Eq. (3.60),

$$C_a = c_a H \tag{3.61}$$

where c_a = adhesion

Equation (3.60) can be simplified to the form

$$q_u = q_b + \frac{2c_a H}{B} + \gamma_1 H^2 \left(1 + \frac{2D_f}{H}\right) \frac{K_{pH} \tan \delta}{B} - \gamma_1 H \tag{3.62}$$

where K_{pH} = horizontal component of passive earth pressure coefficient

However, let

$$K_{pH} \tan \delta = K_s \tan \phi_1 \tag{3.63}$$

where K_s = punching shear coefficient

So

$$q_u = q_b + \frac{2c_a H}{B} + \gamma_1 H^2 \left(1 + \frac{2D_f}{H}\right) \frac{K_s \tan \phi_1}{B} - \gamma_1 H \tag{3.64}$$

The punching shear coefficient, K_s, is a function of q_2/q_1 and ϕ_1, or

$$K_s = f\left(\frac{q_2}{q_1}, \phi_1\right)$$

Note that q_1 and q_2 are the ultimate bearing capacities of a continuous foundation of width B under vertical load on the surfaces of homogeneous thick beds of upper and lower soil, or

$$q_1 = c_1 N_{c(1)} + \tfrac{1}{2}\gamma_1 B N_{\gamma(1)} \tag{3.65}$$

and

$$q_2 = c_2 N_{c(2)} + \tfrac{1}{2}\gamma_2 B N_{\gamma(2)} \tag{3.66}$$

where $N_{c(1)}, N_{\gamma(1)}$ = bearing capacity factors for friction angle ϕ_1 (Table 3.4)
$N_{c(2)}, N_{\gamma(2)}$ = bearing capacity factors for friction angle ϕ_2 (Table 3.4)

It is important to note that, for the top layer to be a stronger soil, q_2/q_1 should be less than one.

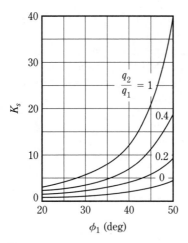

▼ **FIGURE 3.21** Meyerhof and Hanna's punching shear coefficient, K_s

The variation of K_s with q_2/q_1 and ϕ_1 is shown in Figure 3.21. The variation of c_a/c_1 with q_2/q_1 is shown in Figure 3.22. If the height H is relatively large, then the failure surface in soil will be completely located in the stronger upper-soil layer (Figure 3.20b). For this case,

$$q_u = q_t = c_1 N_{c(1)} + q N_{q(1)} + \tfrac{1}{2}\gamma_1 B N_{\gamma(1)} \tag{3.67}$$

where $N_{q(1)}$ = bearing capacity factor for $\phi = \phi_1$ (Table 3.4) and $q = \gamma_1 D_f$

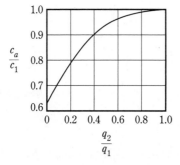

▼ **FIGURE 3.22** Variation of c_a/c_1 vs. q_2/q_1 based on the theory of Meyerhof and Hanna (1978)

Now, combining Eqs. (3.64) and (3.67)

$$q_u = q_b + \frac{2c_aH}{B} + \gamma_1H^2 \left(1 + \frac{2D_f}{H}\right) \frac{K_s \tan \phi_1}{B} - \gamma_1H \leq q_t \tag{3.68}$$

For rectangular foundations, the preceding equation can be extended to the form

$$q_u = q_b + \left(1 + \frac{B}{L}\right)\left(\frac{2c_aH}{B}\right)$$
$$+ \gamma_1H^2 \left(1 + \frac{B}{L}\right)\left(1 + \frac{2D_f}{H}\right)\left(\frac{K_s \tan \phi_1}{B}\right) - \gamma_1H \leq q_t \tag{3.68}$$

where

$$q_b = c_2N_{c(2)}F_{cs(2)} + \gamma_1(D_f + H)N_{q(2)}F_{qs(2)} + \frac{1}{2}\gamma_2BN_{\gamma(2)}F_{\gamma s(2)} \tag{3.69}$$

$$q_t = c_1N_{c(1)}F_{cs(1)} + \gamma_1D_fN_{q(1)}F_{qs(1)} + \frac{1}{2}\gamma_1BN_{\gamma(1)}F_{\gamma s(1)} \tag{3.70}$$

where $F_{cs(1)}, F_{qs(1)}, F_{\gamma s(1)}$ = shape factors with respect to top soil layer (Table 3.5)
$F_{cs(2)}, F_{qs(2)}, F_{\gamma s(2)}$ = shape factors with respect to bottom soil layer (Table 3.5)

Special Cases

1. *Top layer is strong sand and bottom layer is saturated soft clay* ($\phi_2 = 0$). From Eqs. (3.68), (3.69), and (3.70),

$$q_b = \left(1 + 0.2\frac{B}{L}\right)5.14c_2 + \gamma_1(D_f + H) \tag{3.71}$$

$$q_t = \gamma_1D_fN_{q(1)}F_{qs(1)} + \tfrac{1}{2}\gamma_1BN_{\gamma(1)}F_{\gamma s(1)} \tag{3.72}$$

Hence

$$q_u = \left(1 + 0.2\frac{B}{L}\right)5.14c_2 + \gamma_1H^2\left(1 + \frac{B}{L}\right)\left(1 + \frac{2D_f}{H}\right)\frac{K_s \tan \phi_1}{B}$$
$$+ \gamma_1D_f \leq \gamma_1D_fN_{q(1)}F_{qs(1)} + \frac{1}{2}\gamma_1BN_{\gamma(1)}F_{\gamma s(1)} \tag{3.73}$$

For determination of K_s from Figure 3.21,

$$\frac{q_2}{q_1} = \frac{c_2 N_{c(2)}}{\frac{1}{2}\gamma_1 B N_{\gamma(1)}} = \frac{5.14c_2}{0.5\gamma_1 B N_{\gamma(1)}} \tag{3.74}$$

2. *Top layer is stronger sand and bottom layer is weaker sand* ($c_1 = 0$, $c_2 = 0$). The ultimate bearing capacity can be given as

$$q_u = \left[\gamma_1(D_f + H)N_{q(2)}F_{qs(2)} + \frac{1}{2}\gamma_2 B N_{\gamma(2)}F_{\gamma s(2)}\right]$$
$$+ \gamma_1 H^2\left(1 + \frac{B}{L}\right)\left(1 + \frac{2D_f}{H}\right)\frac{K_s \tan\phi_1}{B} - \gamma_1 H \leq q_t \tag{3.75}$$

where

$$q_t = \gamma_1 D_f N_{q(1)}F_{qs(1)} + \frac{1}{2}\gamma_1 B N_{\gamma(1)}F_{\gamma s(1)} \tag{3.76}$$

$$\frac{q_2}{q_1} = \frac{\frac{1}{2}\gamma_2 B N_{\gamma(2)}}{\frac{1}{2}\gamma_1 B N_{\gamma(1)}} = \frac{\gamma_2 N_{\gamma(2)}}{\gamma_1 N_{\gamma(1)}} \tag{3.77}$$

3. *Top layer is stronger saturated clay* ($\phi_1 = 0$) *and bottom layer is weaker saturated clay* ($\phi_2 = 0$). The ultimate bearing capacity can be given as

$$q_u = \left(1 + 0.2\frac{B}{L}\right)5.14c_2 + \left(1 + \frac{B}{L}\right)\left(\frac{2c_a H}{B}\right) + \gamma_1 D_f \leq q_t \tag{3.78}$$

$$q_t = \left(1 + 0.2\frac{B}{L}\right)5.14c_1 + \gamma_1 D_f \tag{3.79}$$

For this case

$$\frac{q_2}{q_1} = \frac{5.14c_2}{5.14c_1} = \frac{c_2}{c_1} \tag{3.80}$$

▼ **EXAMPLE 3.8**

A foundation 1.5 m × 1 m is located at a depth, D_f, of 1 m in a stronger clay. A softer clay layer is located at a depth, H, of 1 m measured from the bottom of the foundation. For the top clay layer,

$$\text{Undrained shear strength} = 120 \text{ kN/m}^2$$

$$\text{Unit weight} = 16.8 \text{ kN/m}^3$$

and for the bottom clay layer,

$$\text{Undrained shear strength} = 48 \text{ kN/m}^2$$

$$\text{Unit weight} = 16.2 \text{ kN/m}^3$$

Determine the gross allowable load for the foundation with an *FS* of 4.

Solution For this problem, Eqs. (3.78), (3.79), and (3.80) will apply, or

$$q_u = \left(1 + 0.2\frac{B}{L}\right)5.14c_2 + \left(1 + \frac{B}{L}\right)\left(\frac{2c_aH}{B}\right) + \gamma_1 D_f$$

$$\leq \left(1 + 0.2\frac{B}{L}\right)5.14c_1 + \gamma_1 D_f$$

Given:

$$B = 1 \text{ m} \qquad H = 1 \text{ m} \qquad D_f = 1 \text{ m}$$

$$L = 1.5 \text{ m} \qquad \gamma_1 = 16.8 \text{ kN/m}^3$$

From Figure 3.22, $c_2/c_1 = 48/120 = 0.4$, the value of $c_a/c_1 \approx 0.9$, so

$$c_a = (0.9)(120) = 108 \text{ kN/m}^2$$

$$q_u = \left[1 + (0.2)\left(\frac{1}{1.5}\right)\right](5.14)(48) + \left(1 + \frac{1}{1.5}\right)\left[\frac{(2)(108)(1)}{1}\right] + (16.8)(1)$$

$$= 279.6 + 360 + 16.8 = 656.4 \text{ kN/m}^2$$

Check: From Eq. (3.79),

$$q_t = \left[1 + (0.2)\left(\frac{1}{1.5}\right)\right](5.14)(120) + (16.8)(1)$$

$$= 699 + 16.8 = 715.8 \text{ kN/m}^2$$

Thus $q_u = 656.4 \text{ kN/m}^2$ (that is, the smaller of the two values calculated above) and

$$q_{all} = \frac{q_u}{FS} = \frac{656.4}{4} = 164.1 \text{ kN/m}^2$$

The total allowable load is

$$(q_{all})(1 \times 1.5) = \textbf{246.15 kN} \qquad\qquad ▲$$

▼ **EXAMPLE 3.9**_____

Refer to Figure 3.20. Assume that the top layer is sand and the bottom layer is soft saturated clay. Given:

For the sand: $\gamma_1 = 117$ lb/ft³; $\phi_1 = 40°$

For the soft clay (bottom layer): $c_2 = 400$ lb/ft²; $\phi_2 = 0$

For the foundation: $B = 3$ ft; $D_f = 3$ ft; $L = 4.5$ ft; $H = 4$ ft

Determine the gross ultimate bearing capacity of the foundation.

Solution   For this case Eqs. (3.73) and (3.74) apply. For $\phi_1 = 40°$, from Table 3.4, $N_\gamma = 109.41$ and

$$\frac{q_2}{q_1} = \frac{c_2 N_{c(2)}}{0.5\gamma_1 BN_{\gamma(1)}} = \frac{(400)(5.14)}{(0.5)(117)(3)(109.41)} = 0.107$$

From Figure 3.21, for $c_2 N_{c(2)}/0.5\gamma_1 BN_{\gamma(1)} = 0.107$ and $\phi_1 = 40°$, the value of $K_s \approx 2.5$. Equation (3.73) gives

$$q_u = \left[1 + (0.2)\left(\frac{B}{L}\right)\right]5.14c_2 + \left(1 + \frac{B}{L}\right)\gamma_1 H^2 \left(1 + \frac{2D_f}{H}\right)K_s\frac{\tan\phi_1}{B} + \gamma_1 D_f$$

$$= \left[1 + (0.2)\left(\frac{3}{4.5}\right)\right](5.14)(400) + \left(1 + \frac{3}{4.5}\right)(117)(4)^2$$

$$\times \left[1 + \frac{(2)(3)}{4}\right](2.5)\frac{\tan 40}{3} + (117)(3)$$

$$= 2330 + 5454 + 351 = 8135 \text{ lb/ft}^2$$

Again, from Eq. (3.73)

$$q_t = \gamma_1 D_f N_{q(1)}F_{qs(1)} + \tfrac{1}{2}\gamma_1 BN_{\gamma(1)}F_{\gamma s(1)}$$

From Table 3.4, for $\phi_1 = 40°$, $N_\gamma = 109.4$, $N_q = 64.20$
From Table 3.5,

$$F_{qs(1)} = \left(1 + \frac{B}{L}\right)\tan\phi_1 = \left(1 + \frac{3}{4.5}\right)\tan 40 = 1.4$$

$$F_{\gamma s(1)} = 1 - 0.4\frac{B}{L} = 1 - (0.4)\left(\frac{3}{4.5}\right) = 0.733$$

$$q_t = (117)(3)(64.20)(1.4) + (\tfrac{1}{2})(117)(3)(109.4)(0.733) = 45,622 \text{ lb/ft}^2$$

Hence

$$q_u = \mathbf{8135 \text{ lb/ft}^2}$$

▲

3.11   BEARING CAPACITY OF FOUNDATIONS ON TOP OF A SLOPE

In some instances, shallow foundations need to be constructed on top of a slope (Figure 3.23). In Figure 3.23, the height of the slope is H, and the slope makes an angle β with the horizontal. The edge of the foundation is located at a distance b from the top of the slope. At ultimate load, q_u, the failure surface will be as shown in the figure.

Meyerhof developed the theoretical relation for the ultimate bearing capacity for *continuous foundations* in the form

$$q_u = cN_{cq} + \tfrac{1}{2}\gamma BN_{\gamma q} \tag{3.81}$$

For purely granular soil, $c = 0$. Thus

$$q_u = \tfrac{1}{2}\gamma BN_{\gamma q} \tag{3.82}$$

Again, for purely cohesive soil, $\phi = 0$. Hence

$$q_u = cN_{cq} \tag{3.83}$$

The variations of $N_{\gamma q}$ and N_{cq} defined by Eqs. (3.82) and (3.83) are shown in Figures 3.24 and 3.25. In using N_{cq} in Eq. (3.83) as given in Figure 3.25, the following points need to be kept in mind:

1. The term N_s is defined as the stability number:

$$N_s = \frac{\gamma H}{c} \tag{3.84}$$

2. If $B < H$,   use the curves for $N_s = 0$.
3. If $B \geq H$,   use the curves for the calculated stability number N_s.

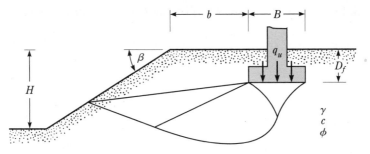

▼   **FIGURE 3.23**   Shallow foundation on top of a slope

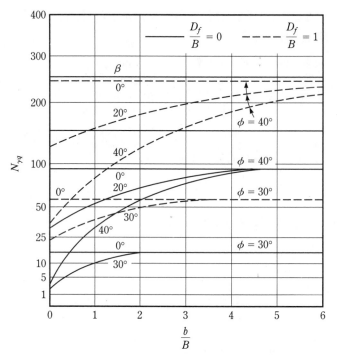

▼   **FIGURE 3.24**   Meyerhof's bearing capacity factor, $N_{\gamma q}$, for granular soil ($c = 0$)

▼ **EXAMPLE 3.10**————————————————————————————————

Refer to Figure 3.23. For a shallow continuous foundation in a clay, the following are given: $B = 1.2$ m; $D_f = 1.2$ m; $b = 0.8$ m; $H = 6.2$ m; $\beta = 30°$; unit weight of soil $= 17.5$ kN/m³; $\phi = 0$; $c = 50$ kN/m². Determine the gross allowable bearing capacity with a factor of safety $FS = 4$.

Solution   Since $B < H$, we will assume the stability number $N_s = 0$. From Eq. (3.83),

$$q_u = cN_{cq}$$

Given

$$\frac{D_f}{B} = \frac{1.2}{1.2} = 1$$

$$\frac{b}{B} = \frac{0.8}{1.2} = 0.75$$

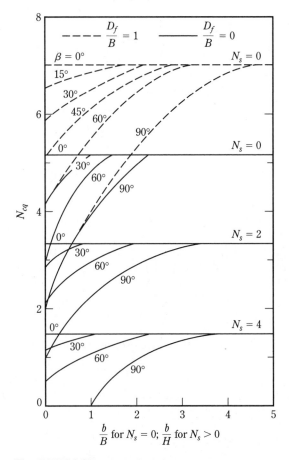

▼ **FIGURE 3.25** Meyerhof's bearing capacity factor, N_{cq}, for purely cohesive soil

For $\beta = 30°$, $D_f/B = 1$ and $b/B = 0.75$, Figure 3.25 gives $N_{cq} = 6.3$. Hence

$$q_u = (50)(6.3) = 315 \text{ kN/m}^2$$

$$q_{all} = \frac{q_u}{FS} = \frac{315}{4} = \mathbf{78.8 \text{ kN/m}^2}$$

▲

3.12 SEISMIC BEARING CAPACITY AND SETTLEMENT IN GRANULAR SOIL

In some instances shallow foundations may fail during seismic events. Published studies relating to the bearing capacity of shallow foundations in such instances are rare. Recently, however, Richards et al. (1993) developed a seismic bearing capacity

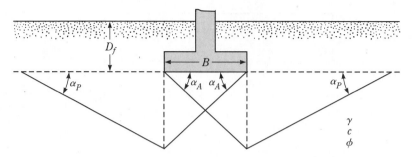

▼ **FIGURE 3.26**   Failure surface in soil for static bearing capacity analysis; *note:*
$\alpha_A = 45 + \phi/2$ and $\alpha_p = 45 - \phi/2$

theory that is presented in this section. It needs to be pointed out that this theory has not yet been supported by field data.

Figure 3.26 shows the nature of failure in soil assumed for this analysis for static conditions. Similarly, Figure 3.27 shows the failure surface under earthquake conditions. Note that, in Figures 3.26 and 3.27

α_A, α_{AE} = inclination angles for active pressure conditions

α_P, α_{PE} = inclination angles for passive pressure conditions

According to this theory, the ultimate bearing capacities for *continuous foundations* in granular soil are:

Static conditions: $q_u = qN_q + \frac{1}{2}\gamma BN_\gamma$ (3.85a)

Earthquake conditions: $q_{uE} = qN_{qE} + \frac{1}{2}\gamma BN_{\gamma E}$ (3.85b)

where $N_q, N_\gamma, N_{qE}, N_{\gamma E}$ = bearing capacity factors
$q = \gamma D_f$

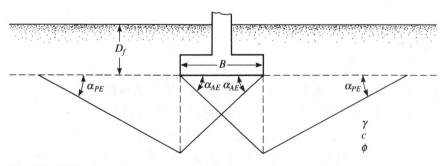

▼ **FIGURE 3.27**   Failure surface in soil for seismic bearing capacity analysis

Note that

$$N_q \text{ and } N_\gamma = f(\phi)$$

and

$$N_{qE} \text{ and } N_{\gamma E} = f(\phi, \tan \theta)$$

where $\quad \tan \theta = \dfrac{k_h}{1 - k_v}$

k_h = horizontal coefficient of acceleration due to an earthquake
k_v = vertical coefficient of acceleration due to an earthquake

The variations of N_q and N_γ with ϕ are shown in Figure 3.28. Figure 3.29 shows the variations of $N_{\gamma E}/N_\gamma$ and N_{qE}/N_q with $\tan \theta$ and the soil friction angle ϕ.

For static conditions, bearing capacity failure can lead to substantial sudden downward movement of the foundation. However, bearing capacity–related settlement in an earthquake takes place when the ratio $k_h/(1 - k_v)$ reaches a critical value $(k_h/1 - k_v)^*$. If $k_v = 0$, then $(k_h/1 - k_v)^*$ becomes equal to k_h^*. Figure 3.30 shows the variation of k_h^* (for $k_v = 0$ and $c = 0$; granular soil) with the factor of safety (*FS*) applied to the ultimate static bearing capacity [Eq. (3.84)], ϕ, and D_f/B.

The settlement of a strip foundation due to an earthquake (S_{Eq}) can be estimated (Richards et al., 1993) as

$$S_{Eq} \text{ (m)} = 0.174 \frac{V^2}{Ag} \left| \frac{k_h^*}{A} \right|^{-4} \tan \alpha_{AE} \tag{3.86}$$

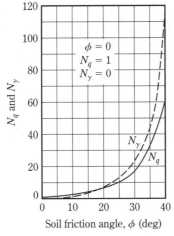

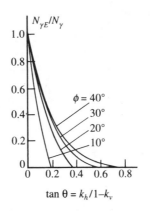

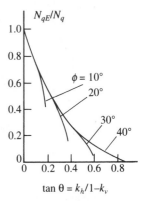

▼ **FIGURE 3.28** Variation of N_q and N_γ based on failure surface assumed in Figure 3.26

▼ **FIGURE 3.29** Variation of $N_{\gamma E}/N_\gamma$ and N_{qE}/N_q (after Richards et al., 1993)

where V = peak velocity for the design earthquake (m/sec)
   A = acceleration coefficient for the design earthquake
   g = acceleration due to gravity (9.18 m/sec²)

The values of k_h^* and α_{AE} can be obtained from Figures 3.30 and 3.31, respectively.

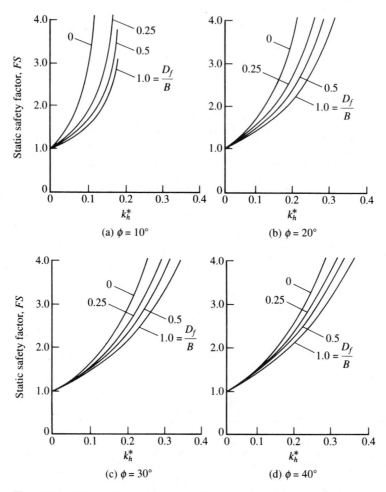

▼ FIGURE 3.30   Critical acceleration k_h^* for $c = 0$ (after Richards et al., 1993)

▼ EXAMPLE 3.11_____

A strip foundation is to be constructed on a sandy soil with $B = 2$ m, $D_f = 1.5$ m, $\gamma = 18$ kN/m³, $\phi = 30°$. Determine the gross ultimate bearing capacity q_{uE}. Assume $k_v = 0$ and $k_h = 0.176$.

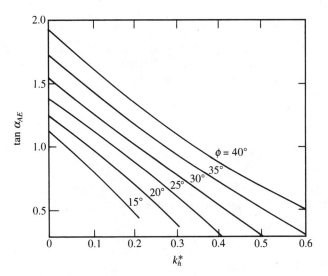

▼ **FIGURE 3.31**   Variation of tan α_{AE} with k_h^* and soil friction angle, ϕ (after Richards et al., 1993)

Solution   From Figure 3.28, for $\phi = 30°$, $N_q = 16.51$ and $N_\gamma = 23.76$.

$$\tan \theta = \frac{k_h}{1 - k_v} = 0.176$$

For $\tan \theta = 0.176$, Figure 3.29 gives

$$\frac{N_{\gamma E}}{N_\gamma} = 0.4 \qquad \text{and} \qquad \frac{N_{qE}}{N_q} = 0.6$$

Thus

$$N_{\gamma E} = (0.4)(23.76) = 9.5$$

$$N_{qE} = (0.6)(16.51) = 9.91$$

$$q_{uE} = qN_{qE} + \tfrac{1}{2}\gamma BN_{\gamma E}$$

$$= (1.5 \times 18)(9.91) + (\tfrac{1}{2})(18)(2)(9.5) = \mathbf{438.6\ kN/m^2}$$

▲

Refer to Example 3.11. If the design earthquake parameters are $V = 0.4$ m/sec and $A = 0.32$, determine the seismic settlement of the foundation. Use $FS = 3$ for obtaining static allowable bearing capacity.

Solution   For the foundation

$$\frac{D_f}{B} = \frac{1.5}{2} = 0.75$$

From Figure 3.30c, for $\phi = 30°$, $FS = 3$, and $D_f/B = 0.75$, the value of $k_h^* = 0.26$. Also from Figure 3.31, for $k_h^* = 0.26$ and $\phi = 30°$, the value of $\tan \alpha_{AE} = 0.88$. From Eq. (3.86)

$$S_{Eq} = 0.174 \left| \frac{k_h^*}{A} \right|^{-4} \tan \alpha_{AE} \left(\frac{V^2}{Ag} \right)$$

$$= 0.174 \frac{(0.4)^2}{(0.32)(9.81)} \left| \frac{0.26}{0.32} \right|^{-4} (0.88) = 0.0179 \text{ m} = \mathbf{17.9\ mm}$$

▲

RECENT ADVANCES IN BEARING CAPACITY OF FOUNDATIONS ON REINFORCED SOIL

During the last fifteen years, several studies have been conducted to evaluate the beneficial effects of reinforcing the soil as related to the bearing capacity of shallow foundations. The soil reinforcements that have been used are *metallic strips, geotextiles,* and *geogrids.* The design of shallow foundations with metallic strips as reinforcement is discussed in Chapter 4. The following sections describe some recent advances that have been made in evaluating the ultimate bearing capacity of foundations on soils reinforced with geotextiles and geomembranes.

3.13 FOUNDATIONS ON SAND WITH GEOTEXTILE REINFORCEMENT

Laboratory model tests for determining the bearing capacity of a *square* foundation supported by loose sand (relative density = 50%) and reinforced by layers of nonwoven heat-bonded geotextiles have been reported by Guido et al. (1985). Some of their test results are shown in Figure 3.32. For these tests, several parameters were varied: d, ΔH, and L_o (Figure 3.32); number of layers of geotextile, N; and tensile strength of geotextile, σ_G. In general, results show that, when the geotextile layers are placed within a depth equal to the width of the foundation, they increase the

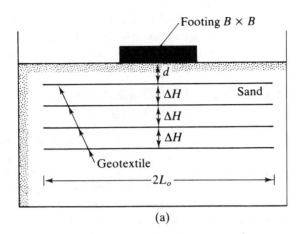

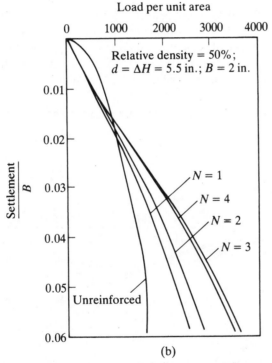

▼ **FIGURE 3.32** Bearing capacity test of square foundation on loose sand with geotextile reinforcement; N = number of layers of reinforcement (based on the model test results of Guido et al., 1985)

load-bearing capacity of the foundation — but only after a measurable settlement has occurred. This result is logical because the geotextile layers have to deform before their reinforcing benefits can be realized.

3.14   FOUNDATIONS ON SATURATED CLAY ($\phi = 0$) WITH GEOTEXTILE REINFORCEMENT

Studies relating to determination of the bearing capacity of a shallow foundation supported by a saturated clay layer reinforced by geotextile, similar to that described in Section 3.13, are rather limited. Sakti and Das (1987) reported some model test results on the bearing capacity of a *strip* foundation on saturated clay. They used a heat-bonded nonwoven geotextile for reinforcement (grab tensile strength = 534 N). Some of the load–settlement curves thus derived are shown in Figure 3.33.

From those tests, the following general conclusions may be drawn:

1. Beneficial effects of geotextile reinforcement are realized when reinforcement is placed within a distance equal to the width of the foundation.
2. The first layer of geotextile reinforcement should be placed at a distance $d = 0.35B$ (B = foundation width) for maximum benefit.
3. The most economical value of L_o/B is about 2. (See Figure 3.32a for a definition of L_o.)

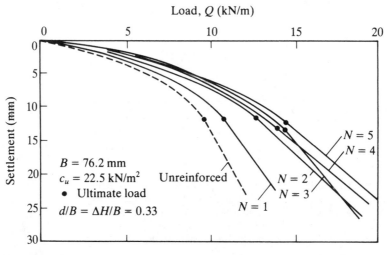

▼ **FIGURE 3.33**   Bearing capacity test of strip footing on saturated clay with geotextile reinforcement; N = number of layers of reinforcement (based on the results of Sakti and Das, 1987)

3.15 FOUNDATIONS ON SAND WITH GEOGRID REINFORCEMENT

As pointed out in Sections 3.13 and 3.14, the ultimate bearing capacity of shallow foundations increases when geotextiles are used for soil reinforcement. However, when the width of a shallow foundation is greater than about 3 ft (1 m), the design is primarily controlled by settlement rather than the ultimate bearing capacity. Figure 3.32 indicates that the flexibility of geotextiles does not improve load-bearing capacity at limited levels of settlement. For that reason, several studies of the possible use of geogrid layers as reinforcement in sand to support shallow foundations have been made (e.g., Guido et al., 1986; Guido et al., 1987; Khing et al., 1993; Omar et al., 1993a and 1993b). All were conducted in the laboratory on small-scale models. The results are summarized in this section.

Figure 3.34 shows a rectangular foundation of width B and length L being supported on a sand layer with N layers of geogrid as reinforcement. Each layer of reinforcement had dimensions of $2L_o \times 2L_l$. The first layer of reinforcement is located at a depth d from the bottom of the foundation. The total depth of geogrid reinforcement from the bottom of the foundation may be given as

$$u = d + (N - 1)(\Delta H) \tag{3.87}$$

In general, for any d, N, ΔH, L_o, and L_l, the load–settlement curve for a foundation with and without geogrid reinforcement will be as shown in Figure 3.35. Based on this concept, the increase in the bearing capacity due to reinforcement may be expressed in nondimensional form as

$$BCR_u = \frac{q_{u(R)}}{q_u} \tag{3.88}$$

and

$$BCR_s = \frac{q_R}{q_o} \tag{3.89}$$

where BCR_u = bearing capacity ratio with respect to the ultimate bearing capacity
            BCR_s = bearing capacity ratio at given settlement level, S, for the foundation
            q_R, q_o = load per unit area of the foundation (at a settlement level $S \leq S_u$) with and without geogrid reinforcement, respectively
            $q_{u(R)}$, q_u = ultimate bearing capacity with and without geogrid reinforcement, respectively

For a foundation on sand, the magnitude of BCR_u generally varies with d/B as shown in Figure 3.36. Beyond a critical value of d/B $[d/B \geq (d/B)_{cr}]$, the magnitude of BCR_u will decrease. With other parameters remaining constant, if the number of

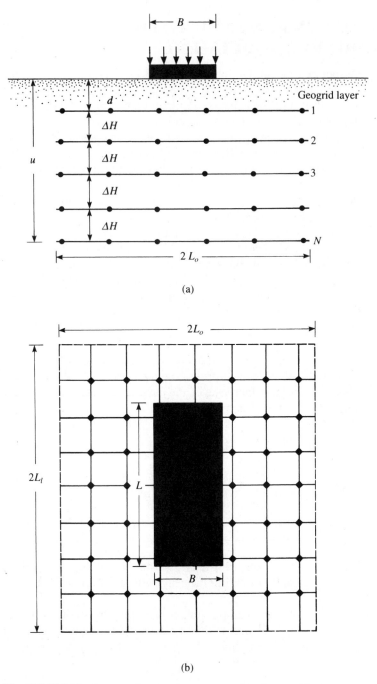

(a)

(b)

▼ **FIGURE 3.34** Rectangular foundation on sand with geogrid reinforcement

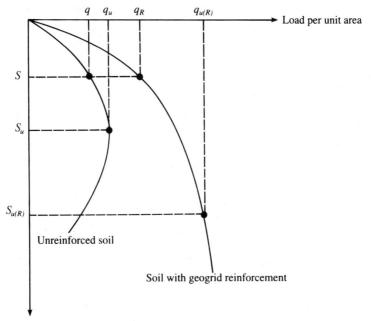

FIGURE 3.35 General form of load–settlement curves for unreinforced soil and soil with geogrid reinforcement supporting a foundation

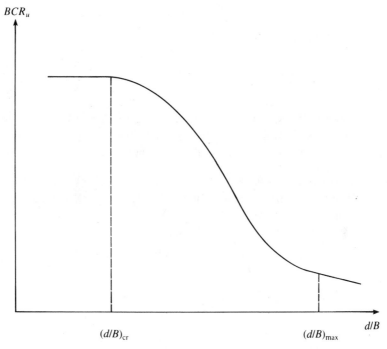

FIGURE 3.36 Nature of variation of BCR_u with d/B for given values of L_o/B, L_1/B, $\Delta H/B$, and N

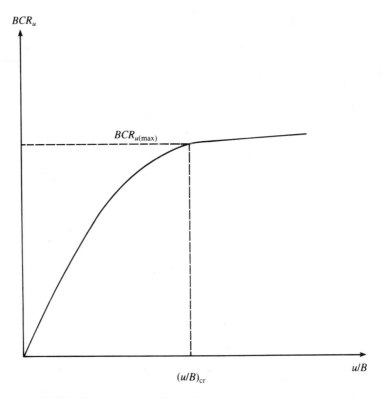

▼   **FIGURE 3.37**   Variation of BCR_u with u/B

geogrid layers, N, is increased (thus increasing u/B), the value of BCR_u will increase to a maximum at $u/B = (u/B)_{cr}$ and remain virtually constant thereafter (Figure 3.37). Similarly, there is a critical value of $L_o/B = (L_o/B)_{cr}$ and $L_i/B = (L_i/B)_{cr}$ at which the magnitudes of BCR_u will nearly reach a maximum. Figure 3.38 shows the variation of BCR_u with u/B for various magnitudes of the B/L ratio of the foundation. Based on their experimental results, Omar et al. (1993a) provided the following empirical relationships:

$$\left(\frac{u}{B}\right)_{cr} = 2 - 1.4\left(\frac{B}{L}\right) \qquad \left(\text{for } 0 \le \frac{B}{L} \le 0.5\right) \tag{3.90}$$

$$\left(\frac{u}{B}\right)_{cr} = 1.43 - 0.26\left(\frac{B}{L}\right) \qquad \left(\text{for } 0.5 \le \frac{B}{L} \le 1.0\right) \tag{3.91}$$

$$\left(\frac{d}{B}\right)_{max} \approx 0.9 - 1.0 \tag{3.92}$$

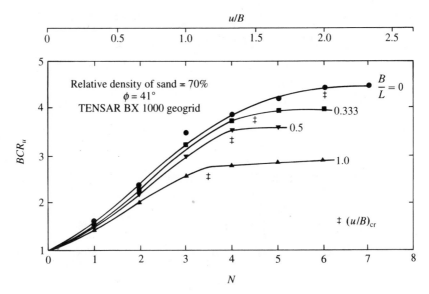

FIGURE 3.38 Variation of BCR_u with u/B for $d/B = \Delta H/B = 0.333$ (after Omar et al., 1993a)

$$\left(\frac{L_o}{B}\right)_{cr} = 4 - 1.75 \left(\frac{B}{L}\right)^{0.51}$$

(3.93)

$$\left(\frac{L_l}{B}\right)_{cr} = 1.75 \left(\frac{B}{L}\right) + \frac{L}{2B}$$

(3.94)

Omar et al. (1993b) also showed that, for similar soil and geogrid reinforcement systems, $d/B = 0.25$ to about 0.4:

$$BCR_u \approx 1.7 \text{ to } 1.8(BCR_s) \qquad \left(\text{for } \frac{B}{L} = 0\right)$$

(3.95)

and

$$BCR_u \approx 1.4 \text{ to } 1.45(BCR_s) \qquad \left(\text{for } \frac{B}{L} = 1\right)$$

(3.96)

Based on the preliminary model test results, geogrids apparently can be used as soil reinforcement to increase the ultimate and allowable bearing capacities of shallow foundations. Design methodologies are expected to be developed soon for field applications. Additional information on this topic may also be found in the works of Yetimoglu et al. (1994) and Adams and Collin (1997).

3.16   STRIP FOUNDATIONS ON SATURATED CLAY ($\phi = 0$) WITH GEOGRID REINFORCEMENT

Shin et al. (1993) reported laboratory model test results for the ultimate bearing capacity of a surface strip foundation on saturated clay ($\phi = 0$) with geogrid reinforcement. Unlike the results of the tests conducted in sand (Section 3.15), in this case

$$BCR_u = BCR_s = BCR \tag{3.97}$$

Also, the ultimate bearing capacities with and without reinforcement, $q_{u(R)}$ and q_u, occurred at similar settlement levels ($S_u/B \approx S_{u(R)}/B$ — see Figure 3.35). Figures 3.39, 3.40, and 3.41 show the variations of $BCR_u = BCR_s = BCR$ with d/B, $2L_o/B$, and N (or u/B). Based on these model test results, Shin et al. (1993) determined the following parameters:

$$\left(\frac{d}{B}\right)_{cr} \approx 0.4 \qquad \left(\frac{2L_o}{B}\right)_{cr} \approx 4.0 \text{ to } 4.5$$

$$\left(\frac{d}{B}\right)_{max} \approx 0.9 - 1.0 \qquad \left(\frac{u}{B}\right)_{cr} \approx 1.8$$

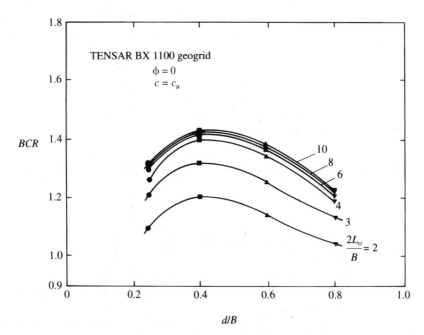

▼ **FIGURE 3.39**   Variation of BCR with d/B for $c_u = 3.14$ kN/m², $\Delta H/B = 0.333$, and $N = 4$ (after Shin et al., 1993)

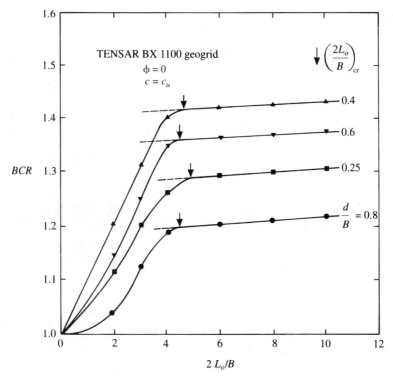

▼ **FIGURE 3.40**   Variation of *BCR* with $2L_o/B$ for $c_u = 3.14$ kN/m², $\Delta H/B = 0.333$, and $N = 4$ (after Shin et al., 1993)

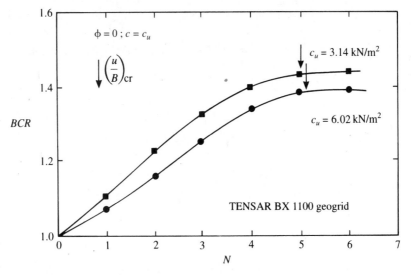

▼ **FIGURE 3.41**   Variation of *BCR* with N (i.e., u/B) for $2L_o/B = 4$, $d/B = 0.4$, and $\Delta H/B = 0.333$ (after Shin et al., 1993)

3.17 GENERAL REMARKS

The theories on the ultimate bearing capacity presented in this chapter are based on idealized conditions of soil profiles. This is, in most field conditions, not true. Soil profiles are not always homogeneous and isotropic. Hence, experience and judgment are always necessary in adopting proper soil parameters to use in the calculation of ultimate bearing capacity.

PROBLEMS

3.1 For the following cases, determine the allowable gross vertical load-bearing capacity of the foundation. Use Terzaghi's equation and assume general shear failure in soil. Use $FS = 4$.

Part	B	D_f	ϕ	c	γ	Foundation type
a.	3 ft	3 ft	28°	400 lb/ft²	110 lb/ft³	Continuous
					17.8 kN/m³	
b.	1.5 m	1.2 m	35°	0		Continuous
c.	3 m	2 m	30°	0	16.5 kN/m³	Square

3.2 A square column foundation has to carry a gross allowable load of 1805 kN ($FS = 3$). Given: $D_f = 1.5$ m, $\gamma = 15.9$ kN/m³, $\phi = 34°$, and $c = 0$. Use Terzaghi's equation to determine the size of the foundation (B). Assume general shear failure.

3.3 Use the general bearing capacity equation [Eq. (3.25)] to solve the following:
a. Problem 3.1(a)
b. Problem 3.1(b)
c. Problem 3.1(c)

3.4 The applied load on a shallow square foundation makes an angle of 15° with the vertical. Given: $B = 5.5$ ft, $D_f = 4$ ft, $\gamma = 107$ lb/ft³, $\phi = 25°$, $c = 350$ lb/ft². Use $FS = 4$ and determine the gross allowable load. Use Eq. (3.25).

3.5 A column foundation (Figure P3.5) is 3 m × 2 m in plan. Given: $D_f = 1.5$ m, $\phi = 25°$, $c = 50$ kN/m². Using Eq. (3.25) and $FS = 4$, determine the net allowable load [see Eq. (3.20)] the foundation could carry.

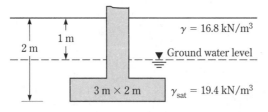

▼ **FIGURE P3.5**

3.6 For a square foundation that is $B \times B$ in plan, $D_f = 3$ ft; vertical gross allowable load, $Q_{all} = 150,000$ lb; $\gamma = 115$ lb/ft³; $\phi = 40°$; $c = 0$; and $FS = 3$. Determine the size of the foundation. Use Eq. (3.25).

3.7 A foundation measuring 8 ft × 8 ft has to be constructed in a granular soil deposit. Given: $D_f = 5$ ft and $\gamma = 110$ lb/ft³. Following are the results of a standard penetration test in that soil:

Depth (ft)	Field standard penetration number, N_F
5	11
10	14
15	16
20	21
25	24

a. Use Eq. (2.10) to estimate an average friction angle, ϕ, for the soil.

b. Using Eq. (3.25), estimate the gross ultimate load the foundation can carry.

3.8 For the design of a shallow foundation, given the following:

Soil: $\phi = 20°$
    $c = 72$ kN/m²
    Unit weight, $\gamma = 17$ kN/m³
    Modulus of elasticity, $E = 1020$ kN/m²
    Poisson's ratio, $\mu = 0.35$

Foundation: $L = 1.5$ m
          $B = 1$ m
          $D_f = 1$ m

calculate the ultimate bearing capacity. Use Eq. (3.32).

3.9 An eccentrically loaded foundation is shown in Figure P3.9. Use an *FS* of 4 and determine the maximum allowable load that the foundation can carry.

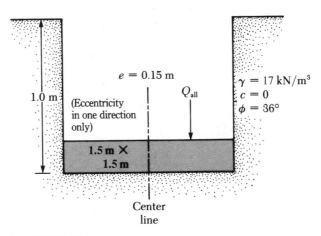

▼ **FIGURE P3.9**

3.10 An eccentrically loaded foundation is shown in Figure P3.10. Determine the ultimate load, Q_u, that the foundation can carry.

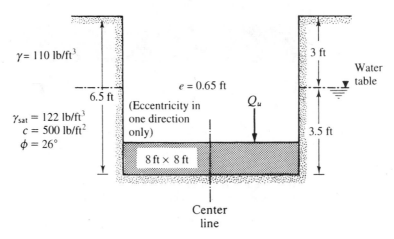

$\gamma = 110$ lb/ft^3

6.5 ft

$\gamma_{sat} = 122$ lb/ft^3
$c = 500$ lb/ft^2
$\phi = 26°$

$e = 0.65$ ft
(Eccentricity in one direction only)

Q_u

8 ft × 8 ft

3 ft

Water table

3.5 ft

Center line

▼ **FIGURE P3.10**

3.11 A square footing is shown in Figure P3.11. Use an *FS* of 6 and determine the size of the footing.

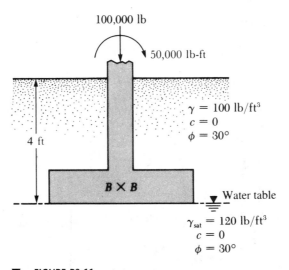

100,000 lb

50,000 lb-ft

$\gamma = 100$ lb/ft^3
$c = 0$
$\phi = 30°$

4 ft

B × B

Water table

$\gamma_{sat} = 120$ lb/ft^3
$c = 0$
$\phi = 30°$

▼ **FIGURE P3.11**

3.12 Refer to Figure 3.13. The shallow foundation measuring 4 ft × 6 ft is subjected to a centric load and a moment. If $e_B = 0.4$ ft and $e_L = 1.2$ ft and the depth of the foundation is 3 ft, determine the allowable load the foundation can carry. Use a factor of safety of 4. For the soil, given: unit weight, $\gamma = 115$ lb/ft^3; friction angle, $\phi = 35°$; cohesion, $c = 0$.

3.13 Redo Problem 3.12 with $e_L = 0.06$ ft and $e_B = 1.5$ ft.

3.14 A strip foundation in a two-layered clay is shown in Figure P3.14. Find the gross allowable bearing capacity. Factor of safety = 3.

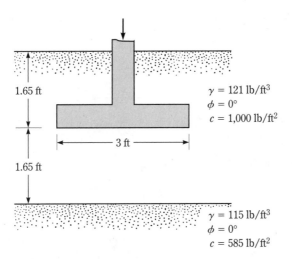

$\gamma = 121$ lb/ft^3
$\phi = 0°$
$c = 1{,}000$ lb/ft^2

3 ft

1.65 ft

1.65 ft

$\gamma = 115$ lb/ft^3
$\phi = 0°$
$c = 585$ lb/ft^2

▼ **FIGURE P3.14**

3.15 Find the gross ultimate load that the footing shown in Figure P3.15 can carry.

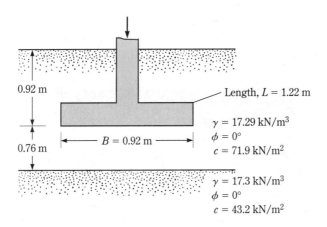

0.92 m

0.76 m

Length, $L = 1.22$ m

$B = 0.92$ m

$\gamma = 17.29$ kN/m^3
$\phi = 0°$
$c = 71.9$ kN/m^2

$\gamma = 17.3$ kN/m^3
$\phi = 0°$
$c = 43.2$ kN/m^2

▼ **FIGURE P3.15**

3.16 Figure P3.16 shows a continuous foundation.
  a. If $H = 1.5$ m, determine the ultimate bearing capacity, q_u.
  b. At what minimum value of H/B will the clay layer not have any effect on the ultimate bearing capacity of the foundation?

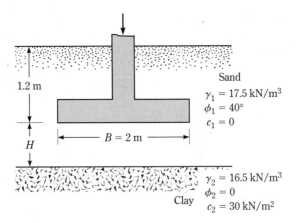

1.2 m

Sand
$\gamma_1 = 17.5 \text{ kN/m}^3$
$\phi_1 = 40°$
$c_1 = 0$

$B = 2$ m

H

$\gamma_2 = 16.5 \text{ kN/m}^3$
$\phi_2 = 0$
Clay $c_2 = 30 \text{ kN/m}^2$

▼ **FIGURE P3.16**

3.17 A continuous foundation having a width of 1 m is located on a slope made of clay soil. Referring to Figure 3.23, $D_f = 1$ m, $H = 4$ m, $b = 2$ m, $\gamma = 16.8 \text{ kN/m}^3$, $c = 68 \text{ kN/m}^3$, $\phi = 0$, and $\beta = 60°$.

  a. Determine the allowable bearing capacity of the foundation. Use $FS = 3$.

  b. Plot a graph of the ultimate bearing capacity, q_u, if b is changed from 0 to 6 m.

3.18 Refer to Figure 3.23. A continuous foundation is to be constructed near a slope made of granular soil. Given: $B = 4$ ft, $b = 6$ ft, $H = 15$ ft, $D_f = 4$ ft, $\beta = 30°$, $\phi = 40°$, and $\gamma = 110 \text{ lb/ft}^3$. Estimate the allowable bearing capacity of the foundation. Use $FS = 4$.

3.19 Following are the average values of cone penetration resistance in a granular soil deposit:

Depth (m)	Cone penetration resistance, q_c (MN/m²)
2	1.73
4	3.6
6	4.9
8	6.8
10	8.7
15	13

For the soil deposit, assume γ to be 16.5 kN/m³ and estimate the seismic ultimate bearing capacity (q_{uE}) for a continuous foundation with the following: $B = 1.5$ m, $D_f = 1.0$ m, $k_h = 0.2$, $k_v = 0$. Use Eqs. (2.26) and (3.85).

3.20 Refer to Problem 3.19. If the design earthquake parameters are $V = 0.35$ m/sec and $A = 0.3$, determine the seismic settlement of the foundation. Assume $FS = 4$ for obtaining static allowable bearing capacity.

REFERENCES

Adams, M. T., and Collin, J. G. (1997). "Large Model Spread Footing Load Tests on Geosynthetic Reinforced Soil Foundations," *Journal of Geotechnical and Geoenvironmental Engineering,* American Society of Civil Engineers, Vol. 123, No. 1, pp. 66–72.

Bjerrum, L. (1972). "Embankments on Soft Ground," *Proceedings of the Specialty Conference,* American Society of Civil Engineers, Vol. 2, pp. 1–54.

Brand, E. W., Muktabhant, C., and Taechanthummarak, A. (1972). "Load Test on Small Foundations in Soft Clay," *Proceedings,* Specialty Conference on Performance of Earth and Earth-Supported Structures, American Society of Civil Engineers, Vol. 1, Part 2, pp. 903–928.

Caquot, A., and Kerisel, J. (1953). "Sur le terme de surface dans le calcul des fondations en milieu pulverulent," *Proceedings,* Third International Conference on Soil Mechanics and Foundation Engineering, Zürich, Vol. I, pp. 336–337.

De Beer, E. E. (1970). "Experimental Determination of the Shape Factors and Bearing Capacity Factors of Sand," *Geotechnique,* Vol. 20, No. 4, pp. 387–411.

Guido, V. A., Biesiadecki, G. L., and Sullivan, M. J. (1985). "Bearing Capacity of a Geotextile Reinforced Foundation," *Proceedings,* Eleventh International Conference on Soil Mechanics and Foundation Engineering, San Francisco, Vol. 3, pp. 1777–1780.

Guido, V. A., Chang, D. K., and Sweeny, M. A. (1986). "Comparison of Geogrid and Geotextile Reinforced Slabs," *Canadian Geotechnical Journal,* Vol. 23, pp. 435–440.

Guido, V. A., Knueppel, J. D., and Sweeny, M. A. (1987). "Plate Load Tests on Geogrid-Reinforced Earth Slabs," *Proceedings,* Geosynthetics '87, pp. 216–225.

Hanna, A. M., and Meyerhof, G. G. (1981). "Experimental Evaluation of Bearing Capacity of Footings Subjected to Inclined Loads," *Canadian Geotechnical Journal,* Vol. 18, No. 4, pp. 599–603.

Hansen, J. B. (1970). "A Revised and Extended Formula for Bearing Capacity," Danish Geotechnical Institute, *Bulletin 28,* Copenhagen.

Higter, W. H., and Anders, J. C. (1985). "Dimensioning Footings Subjected to Eccentric Loads," *Journal of Geotechnical Engineering,* American Society of Civil Engineers, Vol. 111, No. GT5, pp. 659–665.

Khing, K. H., Das, B. M., Puri, V. K., Cook, E. E., and Yen, S. C. (1993). "The Bearing Capacity of a Strip Foundation on Geogrid-Reinforced Sand," *Geotextiles and Geomembranes,* Vol. 12, No. 4, pp. 351–361.

Kumbhojkar, A. S. (1993). "Numerical Evaluation of Terzaghi's N_γ," *Journal of Geotechnical Engineering,* American Society of Civil Engineers, Vol. 119, No. 3, pp. 598–607.

Liao, S., and Whitman, R. V. (1986). "Overburden Correction Factor for SPT in Sand," *Journal of Geotechnical Engineering,* ASCE, Vol. 112, No. 3, pp. 373–377.

Lundgren, H., and Mortensen, K. (1953). "Determination by the Theory of Plasticity on the Bearing Capacity of Continuous Footings on Sand," *Proceedings,* Third International Conference on Soil Mechanics and Foundation Engineering, Zürich, Vol. 1, pp. 409–412.

Meyerhof, G. G. (1953). "The Bearing Capacity of Foundations Under Eccentric and Inclined Loads," *Proceedings,* Third International Conference on Soil Mechanics and Foundation Engineering, Zürich, Vol. 1, pp. 440–445.

Meyerhof, G. G. (1957). "The Ultimate Bearing Capacity of Foundations on Slopes," *Proceedings,* Fourth International Conference on Soil Mechanics and Foundation Engineering, London, Vol. 1, pp. 384–387.

Meyerhof, G. G. (1963). "Some Recent Research on the Bearing Capacity of Foundations," *Canadian Geotechnical Journal,* Vol. 1, No. 1, pp. 16–26.

Meyerhof, G. G. (1974). "Ultimate Bearing Capacity of Footings on Sand Layer Overlying Clay," *Canadian Geotechnical Journal,* Vol. 11, No. 2, pp. 224–229.

Meyerhof, G. G., and Hanna, A. M. (1978). "Ultimate Bearing Capacity of Foundations on Layered Soil Under Inclined Load," *Canadian Geotechnical Journal,* Vol. 15, No. 4, pp. 565–572.

Omar, M. T., Das, B. M., Yen, S. C., Puri, V. K., and Cook, E. E. (1993a). "Ultimate Bearing Capacity of Rectangular Foundations on Geogrid-Reinforced Sand," *Geotechnical Testing Journal,* American Society for Testing and Materials, Vol. 16, No. 2, pp. 246–252.

Omar, M. T., Das, B. M., Yen, S. C., Puri, V. K., and Cook, E. E. (1993b). "Shallow Foundations on Geogrid-Reinforced Sand," *Transportation Research Record No. 1414,* National Academy of Sciences, National Research Council, pp. 59–64.

Prandtl, L. (1921). "Über die Eindringungsfestigkeit (Härte) plastischer Baustoffe und die Festigkeit von Schneiden," *Zeitschrift für angewandte Mathematik und Mechanik,* Vol. 1, No. 1, pp. 15–20.

Reissner, H. (1924). "Zum Erddruckproblem," *Proceedings,* First International Congress of Applied Mechanics, Delft, pp. 295–311.

Richards, R., Jr., Elms, D. G., and Budhu, M. (1993). "Seismic Bearing Capacity and Settlement of Foundations," *Journal of Geotechnical Engineering,* American Society of Civil Engineers, Vol. 119, No. 4, pp. 662–674.

Sakti, J., and Das, B. M. (1987). "Model Tests for Strip Foundation on Clay Reinforced with Geotextile Layers," *Transportation Research Record No. 1153,* National Academy of Sciences, Washington, D.C., pp. 40–45.

Shin, E. C., Das, B. M., Puri, V. K., Yen, S. C., and Cook, E. E. (1993). "Bearing Capacity of Strip Foundation on Geogrid-Reinforced Clay," *Geotechnical Testing Journal,* American Society for Testing and Materials, Vol. 17, No. 4, pp. 534–541.

Skempton, A. W. (1951). "The Bearing Capacity of Clays," *Proceedings,* Building Research Congress, London, pp. 180–189.

Terzaghi, K. (1943). *Theoretical Soil Mechanics,* Wiley, New York.

Terzaghi, K., and Peck, R. B. (1967). *Soil Mechanics in Engineering Practice,* 2nd ed., Wiley, New York.

Vesic, A. S. (1963). "Bearing Capacity of Deep Foundations in Sand," *Highway Research Record No. 39,* National Academy of Sciences, pp. 112–153.

Vesic, A. S. (1973). "Analysis of Ultimate Loads of Shallow Foundations," *Journal of the Soil Mechanics and Foundations Division,* American Society of Civil Engineers, Vol. 99, No. SM1, pp. 45–73.

Yetimoglu, T., Wu, J. T. H., and Saglamer, A. (1994). "Bearing Capacity of Rectangular Footings on Geogrid-Reinforced Soil," *Journal of Geotechnical Engineering,* American Society of Civil Engineers, Vol. 120, No. 12, pp. 2083–2099.

CHAPTER FOUR

SHALLOW FOUNDATIONS: ALLOWABLE BEARING CAPACITY AND SETTLEMENT

4.1 INTRODUCTION

It was mentioned in Chapter 3 that, in many cases, the allowable settlement of a shallow foundation may control the allowable bearing capacity. The allowable settlement may be controlled by local building codes. Thus the allowable bearing capacity will be the smaller of the following two conditions:

$$q_{all} = \begin{cases} \dfrac{q_u}{FS} \\ \text{or} \\ q_{\text{allowable settlement}} \end{cases}$$

The settlement of a foundation can be divided into two major categories: (a) elastic, or immediate, settlement, and (b) consolidation settlement. Immediate, or elastic, settlement of a foundation takes place during or immediately after the construction of the structure. Consolidation settlement occurs over time. Pore water is extruded from the void spaces of saturated clayey soils submerged in water. The total settlement of a foundation is the sum of the elastic settlement and the consolidation settlement.

Consolidation settlement comprises two phases: *primary* and *secondary*. The fundamentals of primary consolidation settlement have been explained in detail in Chapter 1. Secondary consolidation settlement occurs after completion of primary consolidation caused by slippage and reorientation of soil particles under sustained load. Primary consolidation settlement is more significant than secondary settlement in inorganic clays and silty soils. However, in organic soils, secondary consolidation settlement is more significant.

For calculation of foundation settlement (elastic and consolidation), it is required that we estimate the vertical stress increase in the soil mass due to the net load applied on the foundation. Hence, this chapter is divided into four parts.

They are:

1. Procedure for calculation of vertical stress increase
2. Settlement calculation (elastic and consolidation)
3. Allowable bearing capacity based on elastic settlement
4. Foundation with soil reinforcement.

VERTICAL STRESS INCREASE IN A SOIL MASS CAUSED BY FOUNDATION LOAD

4.2 STRESS DUE TO A CONCENTRATED LOAD

In 1885, Boussinesq developed the mathematical relationships for determining the normal and shear stresses at any point inside *homogeneous, elastic,* and *isotropic* mediums due to a *concentrated point load* located at the surface, as shown in Figure 4.1. According to his analysis, the *vertical stress increase* (Δp) at point A (Figure 4.1) caused by the point load of magnitude P is

$$\Delta p = \frac{3P}{2\pi z^2 \left[1 + \left(\dfrac{r}{z} \right)^2 \right]^{5/2}} \tag{4.1}$$

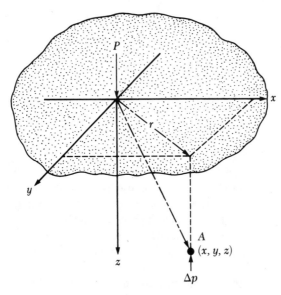

▼ **FIGURE 4.1**   Vertical stress at a point, A, caused by a point load on the surface

where $\qquad r = \sqrt{x^2 + y^2}$

x, y, z = coordinates of the point A

Note that Eq. (4.1) is not a function of the Poisson's ratio of the soil.

4.3 STRESS DUE TO A CIRCULARLY LOADED AREA

The Boussinesq equation [Eq. (4.1)] can also be used to determine the vertical stress below the center of a flexible circularly loaded area, as shown in Figure 4.2. Let the radius of the loaded area be $B/2$, and q_o be the uniformly distributed load per unit area. To determine the stress increase at a point A, located at depth z below the center of the circular area, consider an elemental area on the circle, as shown in Figure 4.2. The load on this elemental area may be considered as a point load and expressed as $q_o r \, d\theta \, dr$. The stress increase at point A caused by this load can be determined from Eq. (4.1):

$$dp = \frac{3(q_o r \, d\theta \, dr)}{2\pi z^2 \left[1 + \left(\dfrac{r}{z}\right)^2\right]^{5/2}} \tag{4.2}$$

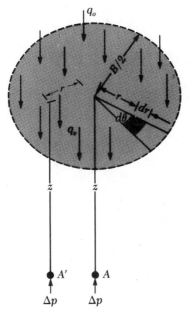

▼ **FIGURE 4.2** Increase of pressure under a uniformly loaded flexible circular area

Thus the total increase of stress caused by the entire loaded area may be obtained by integration of Eq. (4.2), or

$$\Delta p = \int dp = \int_{\theta=0}^{\theta=2\pi} \int_{r=0}^{r=B/2} \frac{3(q_0 r \, d\theta \, dr)}{2\pi z^2 \left[1 + \left(\dfrac{r}{z} \right)^2 \right]^{5/2}}$$

$$= q_0 \left\{ 1 - \frac{1}{\left[1 + \left(\dfrac{B}{2z} \right)^2 \right]^{3/2}} \right\} \tag{4.3}$$

Similar integrations could be performed to obtain the vertical stress increase at A' located at a distance r from the center of the loaded area at a depth z (Ahlvin and Ulery, 1962). Table 4.1 gives the variation of $\Delta p/q_0$ with $r/(B/2)$ and $z/(B/2)$ [for $0 \le r/(B/2) \le 1$]. Note that the variation of $\Delta p/q_0$ with depth at $r/(B/2) = 0$ can be obtained from Eq. (4.3).

4.4   STRESS BELOW A RECTANGULAR AREA

The integration technique of Boussinesq's equation also allows evaluation of the vertical stress at any point A below the corner of a flexible rectangular loaded area (Figure 4.3). To do that, consider an elementary area $dA = dx \, dy$ on the flexible

▼ **TABLE 4.1** Variation of $\Delta p/q_0$ for Uniformly Loaded Flexible Circular Area

$z/(B/2)$	$r/(B/2)$					
	0	0.2	0.4	0.6	0.8	1.0
0	1.000	1.000	1.000	1.000	1.000	1.000
0.1	0.999	0.999	0.998	0.996	0.976	0.484
0.2	0.992	0.991	0.987	0.970	0.890	0.468
0.3	0.976	0.973	0.963	0.922	0.793	0.451
0.4	0.949	0.943	0.920	0.860	0.712	0.435
0.5	0.911	0.902	0.869	0.796	0.646	0.417
0.6	0.864	0.852	0.814	0.732	0.591	0.400
0.7	0.811	0.798	0.756	0.674	0.545	0.367
0.8	0.756	0.743	0.699	0.619	0.504	0.366
0.9	0.701	0.688	0.644	0.570	0.467	0.348
1.0	0.646	0.633	0.591	0.525	0.434	0.332
1.2	0.546	0.535	0.501	0.447	0.377	0.300
1.5	0.424	0.416	0.392	0.355	0.308	0.256
2.0	0.286	0.286	0.268	0.248	0.224	0.196
2.5	0.200	0.197	0.191	0.180	0.167	0.151
3.0	0.146	0.145	0.141	0.135	0.127	0.118
4.0	0.087	0.086	0.085	0.082	0.080	0.075

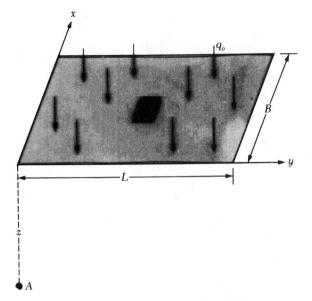

▼ **FIGURE 4.3**   Determination of stress below the corner
                     of a flexible rectangular loaded area

loaded area. If the load per unit area is q_o, the total load on the elemental area is

$$dP = q_o \, dx \, dy \tag{4.4}$$

This elemental load, dP, may be treated as a point load. The increase of vertical
stress at point A caused by dP may be evaluated by using Eq. (4.1). Note, however,
the need to substitute $dP = q_o \, dx \, dy$ for P, and $x^2 + y^2$ for r^2, in Eq. (4.1). Thus

$$\text{The stress increase at } A \text{ caused by } dP = \frac{3q_o \, (dx \, dy) \, z^3}{2\pi \, (x^2 + y^2 + z^2)^{5/2}}$$

The total stress increase caused by the entire loaded area at point A may now be
obtained by integrating the preceding equation:

$$\Delta p = \int_{y=0}^{L} \int_{x=0}^{B} \frac{3q_o \, (dx \, dy) \, z^3}{2\pi \, (x^2 + y^2 + z^2)^{5/2}} = q_o I \tag{4.5}$$

where Δp = stress increase at A

$$I = \text{influence factor} = \frac{1}{4\pi} \left(\frac{2mn\sqrt{m^2 + n^2 + 1}}{m^2 + n^2 + m^2 n^2 + 1} \cdot \frac{m^2 + n^2 + 2}{m^2 + n^2 + 1} \right.$$
$$\left. + \tan^{-1} \frac{2mn\sqrt{m^2 + n^2 + 1}}{m^2 + n^2 + 1 - m^2 n^2} \right) \tag{4.6}$$

When m and n are small, the argument of $\tan^{-1}$ becomes negative. In that case,

$$I = \text{influence factor} = \frac{1}{4\pi}\left[\frac{2mn\sqrt{m^2 + n^2 + 1}}{m^2 + n^2 + m^2n^2 + 1} \cdot \frac{m^2 + n^2 + 2}{m^2 + n^2 + 1}\right.$$
$$\left. + \tan^{-1}\left(\pi - \frac{2mn\sqrt{m^2 + n^2 + 1}}{m^2 + n^2 + 1 - m^2n^2}\right)\right] \tag{4.6a}$$

$$m = \frac{B}{z} \tag{4.7}$$

$$n = \frac{L}{z} \tag{4.8}$$

The variations of the influence values with m and n are given in Table 4.2. For convenience, they are also plotted in Figure 4.4.

▼ **TABLE 4.2** Variation of Influence Value, I [Eq. (4.6)][a]

						n						
m	0.1	0.2	0.3	0.4	0.5	0.6	0.7	0.8	0.9	1.0	1.2	1.4
0.1	0.00470	0.00917	0.01323	0.01678	0.01978	0.02223	0.02420	0.02576	0.02698	0.02794	0.02926	0.03007
0.2	0.00917	0.01790	0.02585	0.03280	0.03866	0.04348	0.04735	0.05042	0.05283	0.05471	0.05733	0.05894
0.3	0.01323	0.02585	0.03735	0.04742	0.05593	0.06294	0.06858	0.07308	0.07661	0.07938	0.08323	0.08561
0.4	0.01678	0.03280	0.04742	0.06024	0.07111	0.08009	0.08734	0.09314	0.09770	0.10129	0.10631	0.10941
0.5	0.01978	0.03866	0.05593	0.07111	0.08403	0.09473	0.10340	0.11035	0.11584	0.12018	0.12626	0.13003
0.6	0.02223	0.04348	0.06294	0.08009	0.09473	0.10688	0.11679	0.12474	0.13105	0.13605	0.14309	0.14749
0.7	0.02420	0.04735	0.06858	0.08734	0.10340	0.11679	0.12772	0.13653	0.14356	0.14914	0.15703	0.16199
0.8	0.02576	0.05042	0.07308	0.09314	0.11035	0.12474	0.13653	0.14607	0.15371	0.15978	0.16843	0.17389
0.9	0.02698	0.05283	0.07661	0.09770	0.11584	0.13105	0.14356	0.15371	0.16185	0.16835	0.17766	0.18357
1.0	0.02794	0.05471	0.07938	0.10129	0.12018	0.13605	0.14914	0.15978	0.16835	0.17522	0.18508	0.19139
1.2	0.02926	0.05733	0.08323	0.10631	0.12626	0.14309	0.15703	0.16843	0.17766	0.18508	0.19584	0.20278
1.4	0.03007	0.05894	0.08561	0.10941	0.13003	0.14749	0.16199	0.17389	0.18357	0.19139	0.20278	0.21020
1.6	0.03058	0.05994	0.08709	0.11135	0.13241	0.15028	0.16515	0.17739	0.18737	0.19546	0.20731	0.21510
1.8	0.03090	0.06058	0.08804	0.11260	0.13395	0.15207	0.16720	0.17967	0.18986	0.19814	0.21032	0.21836
2.0	0.03111	0.06100	0.08867	0.11342	0.13496	0.15326	0.16856	0.18119	0.19152	0.19994	0.21235	0.22058
2.5	0.03138	0.06155	0.08948	0.11450	0.13628	0.15483	0.17036	0.18321	0.19375	0.20236	0.21512	0.22364
3.0	0.03150	0.06178	0.08982	0.11495	0.13684	0.15550	0.17113	0.18407	0.19470	0.20341	0.21633	0.22499
4.0	0.03158	0.06194	0.09007	0.11527	0.13724	0.15598	0.17168	0.18469	0.19540	0.20417	0.21722	0.22600
5.0	0.03160	0.06199	0.09014	0.11537	0.13737	0.15612	0.17185	0.18488	0.19561	0.20440	0.21749	0.22632
6.0	0.03161	0.06201	0.09017	0.11541	0.13741	0.15617	0.17191	0.18496	0.19569	0.20449	0.21760	0.22644
8.0	0.03162	0.06202	0.09018	0.11543	0.13744	0.15621	0.17195	0.18500	0.19574	0.20455	0.21767	0.22652
10.0	0.03612	0.06202	0.09019	0.11544	0.13745	0.15622	0.17196	0.18502	0.19576	0.20457	0.21769	0.22654
∞	0.03162	0.06202	0.09019	0.11544	0.13745	0.15623	0.17197	0.18502	0.19577	0.20458	0.21770	0.22656

[a] After Newmark (1935)

The stress increase at any point below a rectangular loaded area can also be found by using Eq. (4.5) in conjunction with Figure 4.5. To determine the stress at depth z below point O, divide the loaded area into four rectangles. Point O is the corner common to each rectangle. Then use Eq. (4.5) to calculate the increase of stress at depth z below point O caused by each rectangular area. The total stress increase caused by the entire loaded area may now be expressed as

$$\Delta p = q_o(I_1 + I_2 + I_3 + I_4) \tag{4.9}$$

where I_1, I_2, I_3, and I_4 = the influence values of rectangles 1, 2, 3, and 4, respectively

In most cases, the vertical stress below the center of a rectangular area is of importance. This can be given by the following relationship:

$$\Delta p = q_o I_c \tag{4.10}$$

▼ **TABLE 4.2** (Continued)

m	1.6	1.8	2.0	2.5	3.0	4.0	5.0	6.0	8.0	10.0	∞
0.1	0.03058	0.03090	0.03111	0.03138	0.03150	0.03158	0.03160	0.03161	0.03162	0.03162	0.03162
0.2	0.05994	0.06058	0.06100	0.06155	0.06178	0.06194	0.06199	0.06201	0.06202	0.06202	0.06202
0.3	0.08709	0.08804	0.08867	0.08948	0.08982	0.09007	0.09014	0.09017	0.09018	0.09019	0.09019
0.4	0.11135	0.11260	0.11342	0.11450	0.11495	0.11527	0.11537	0.11541	0.11543	0.11544	0.11544
0.5	0.13241	0.13395	0.13496	0.13628	0.13684	0.13724	0.13737	0.13741	0.13744	0.13745	0.13745
0.6	0.15028	0.15207	0.15236	0.15483	0.15550	0.15598	0.15612	0.15617	0.15621	0.15622	0.15623
0.7	0.16515	0.16720	0.16856	0.17036	0.17113	0.17168	0.17185	0.17191	0.17195	0.17196	0.17197
0.8	0.17739	0.17967	0.18119	0.18321	0.18407	0.18469	0.18488	0.18496	0.18500	0.18502	0.18502
0.9	0.18737	0.18986	0.19152	0.19375	0.19470	0.19540	0.19561	0.19569	0.19574	0.19576	0.19577
1.0	0.19546	0.19814	0.19994	0.20236	0.20341	0.20417	0.20440	0.20449	0.20455	0.20457	0.20458
1.2	0.20731	0.21032	0.21235	0.21512	0.21633	0.21722	0.21749	0.21760	0.21767	0.21769	0.21770
1.4	0.21510	0.21836	0.22058	0.22364	0.22499	0.22600	0.22632	0.22644	0.22652	0.22654	0.22656
1.6	0.22025	0.22372	0.22610	0.22940	0.23088	0.23200	0.23236	0.23249	0.23258	0.23261	0.23263
1.8	0.22372	0.22736	0.22986	0.23334	0.23495	0.23617	0.23656	0.23671	0.23681	0.23684	0.23686
2.0	0.22610	0.22986	0.23247	0.23614	0.23782	0.23912	0.23954	0.23970	0.23981	0.23985	0.23987
2.5	0.22940	0.23334	0.23614	0.24010	0.24196	0.24344	0.24392	0.24412	0.24425	0.24429	0.24432
3.0	0.23088	0.23495	0.23782	0.24196	0.24394	0.24554	0.24608	0.24630	0.24646	0.24650	0.24654
4.0	0.23200	0.23617	0.23912	0.24344	0.24554	0.24729	0.24791	0.24817	0.24836	0.24842	0.24846
5.0	0.23236	0.23656	0.23954	0.24392	0.24608	0.24791	0.24857	0.24885	0.24907	0.24914	0.24919
6.0	0.23249	0.23671	0.23970	0.24412	0.24630	0.24817	0.24885	0.24916	0.24939	0.24946	0.24952
8.0	0.23258	0.23681	0.23981	0.24425	0.24646	0.24836	0.24907	0.24939	0.24964	0.24973	0.24980
10.0	0.23261	0.23684	0.23985	0.24429	0.24650	0.24842	0.24914	0.24946	0.24973	0.24981	0.24989
∞	0.23263	0.23686	0.23987	0.24432	0.24654	0.24846	0.24919	0.24952	0.24980	0.24989	0.25000

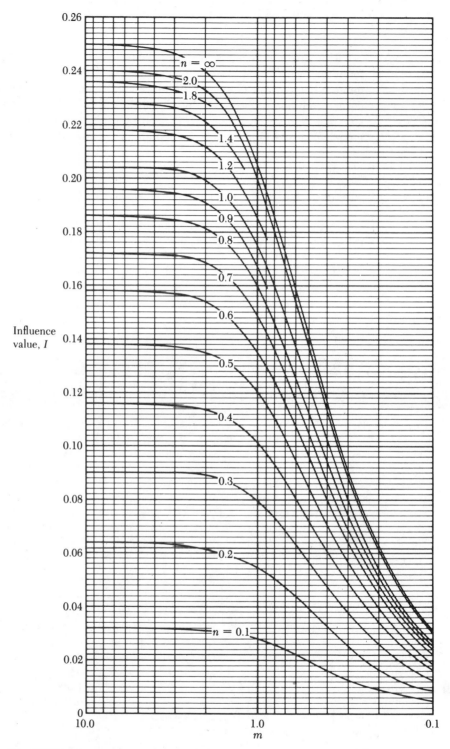

▼ **FIGURE 4.4** Variation of I with m and n — Eqs. (4.5), (4.6), and (4.6a)

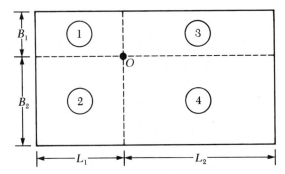

▼ **FIGURE 4.5** Stress below any point of a loaded flexible rectangular area

where

$$I_c = \frac{2}{\pi}\left[\frac{m_1 n_1}{\sqrt{1 + m_1^2 + n_1^2}}\frac{1 + m_1^2 + 2n_1^2}{(1 + n_1^2)(m_1^2 + n_1^2)}\right] + \sin^{-1}\frac{m_1}{\sqrt{m_1^2 + n_1^2}\sqrt{1 + n_1^2}} \qquad (4.11)$$

$$m_1 = \frac{L}{B} \qquad (4.12)$$

$$n_1 = \frac{z}{\left(\dfrac{B}{2}\right)} \qquad (4.13)$$

The variation of I_c with m_1 and n_1 is given in Table 4.3.

▼ **TABLE 4.3** Variation of I_c with m_1 and n_1

	m_1									
n_1	1	2	3	4	5	6	7	8	9	10
0.20	0.994	0.997	0.997	0.997	0.997	0.997	0.997	0.997	0.997	0.997
0.40	0.960	0.976	0.977	0.977	0.977	0.977	0.977	0.977	0.977	0.977
0.60	0.892	0.932	0.936	0.936	0.937	0.937	0.937	0.937	0.937	0.937
0.80	0.800	0.870	0.878	0.880	0.881	0.881	0.881	0.881	0.881	0.881
1.00	0.701	0.800	0.814	0.817	0.818	0.818	0.818	0.818	0.818	0.818
1.20	0.606	0.727	0.748	0.753	0.754	0.755	0.755	0.755	0.755	0.755
1.40	0.522	0.658	0.685	0.692	0.694	0.695	0.695	0.696	0.696	0.696
1.60	0.449	0.593	0.627	0.636	0.639	0.640	0.641	0.641	0.641	0.642
1.80	0.388	0.534	0.573	0.585	0.590	0.591	0.592	0.592	0.593	0.593
2.00	0.336	0.481	0.525	0.540	0.545	0.547	0.548	0.549	0.549	0.549
3.00	0.179	0.293	0.348	0.373	0.384	0.389	0.392	0.393	0.394	0.395
4.00	0.108	0.190	0.241	0.269	0.285	0.293	0.298	0.301	0.302	0.303
5.00	0.072	0.131	0.174	0.202	0.219	0.229	0.236	0.240	0.242	0.244
6.00	0.051	0.095	0.130	0.155	0.172	0.184	0.192	0.197	0.200	0.202
7.00	0.038	0.072	0.100	0.122	0.139	0.150	0.158	0.164	0.168	0.171
8.00	0.029	0.056	0.079	0.098	0.113	0.125	0.133	0.139	0.144	0.147
9.00	0.023	0.045	0.064	0.081	0.094	0.105	0.113	0.119	0.124	0.128
10.00	0.019	0.037	0.053	0.067	0.079	0.089	0.097	0.103	0.108	0.112

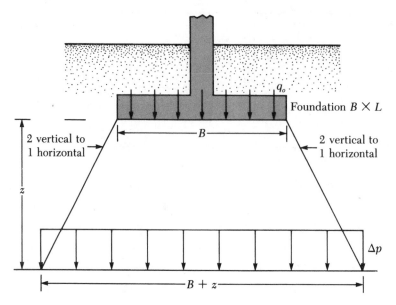

▼ **FIGURE 4.6**   2:1 method of finding stress increase under a foundation

Foundation engineers often use an approximate method to determine the in-crease of stress with depth caused by the construction of a foundation. It is referred to as the *2:1 method* (Figure 4.6). According to this method, the increase of stress at depth z is

$$\Delta p = \frac{q_o \times B \times L}{(B + z)(L + z)} \qquad (4.14)$$

Note that Eq. (4.14) is based on the assumption that the stress from the foundation spreads out along lines with a *2 vertical to 1 horizontal slope.*

▼ **EXAMPLE 4.1**

A flexible rectangular area measures 5 ft × 10 ft in plan. It supports a load of 2000 lb/ft². Determine the vertical stress increase due to the load at a depth of 12.5 ft below the center of the rectangular area.

Solution   Refer to Figure 4.5. For this case

$$B_1 = B_2 = \frac{5}{2} = 2.5 \text{ ft}$$

$$L_1 = L_2 = \frac{10}{2} = 5 \text{ ft}$$

From Eqs. (4.7) and (4.8)

$$m = \frac{B_1}{z} = \frac{B_2}{z} = \frac{2.5}{12.5} = 0.2$$

$$n = \frac{L_1}{z} = \frac{L_2}{z} = \frac{5}{12.5} = 0.4$$

From Table 4.2, for $m = 0.2$ and $n = 0.4$, the value of $I = 0.0328$. Thus

$$\Delta p = q_o(4I) = (2000)(4)(0.0328) = \mathbf{262.4\ lb/ft^2}$$

Alternate Solution   From Eq. (4.10)

$$\Delta p = q_o I_c$$

$$m_1 = \frac{L}{B} = \frac{10}{5} = 2$$

$$n_1 = \frac{z}{\left(\dfrac{B}{2}\right)} = \frac{12.5}{\left(\dfrac{5}{2}\right)} = 5$$

From Table 4.3, for $m_1 = 2$ and $n_1 = 5$, the value of $I_c = 0.131$. Thus

$$\Delta p = (2000)(0.131) = \mathbf{262\ lb/ft^2} \qquad \blacktriangle$$

4.5   AVERAGE VERTICAL STRESS INCREASE DUE TO A RECTANGULARLY LOADED AREA

In Section 4.4, the vertical stress increase below the corner of a uniformly loaded rectangular area was given as (Figure 4.7)

$$\Delta p = q_o I$$

In many cases it is required to determine the average stress increase, Δp_{av}, below the corner of a uniformly loaded rectangular area with limits of $z = 0$ to $z = H$, as shown in Figure 4.7. This can be evaluated as

$$\Delta p_{av} = \frac{1}{H}\int_0^H (q_o I)\, dz = q_o I_a \qquad (4.15)$$

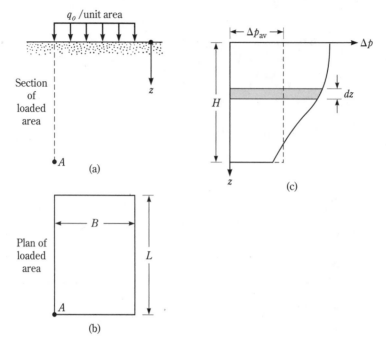

▼ **FIGURE 4.7**   Average vertical stress increase due to a rectangular loaded flexible area

where

$$I_a = f(m, n) \tag{4.16}$$

$$m = \frac{B}{H} \tag{4.17}$$

$$n = \frac{L}{H} \tag{4.18}$$

The variation of I_a is shown in Figure 4.8 as proposed by Griffiths (1984).

In the estimation of the consolidation settlement under a foundation, it may be required to determine the average vertical stress increase in only a given layer; that is, between $z = H_1$ to $z = H_2$ as shown in Figure 4.9. This can be done as (Griffiths, 1984)

$$\Delta p_{av(H_2/H_1)} = q_o \left[\frac{H_2 I_{a(H_2)} - H_1 I_{a(H_1)}}{H_2 - H_1} \right] \tag{4.19}$$

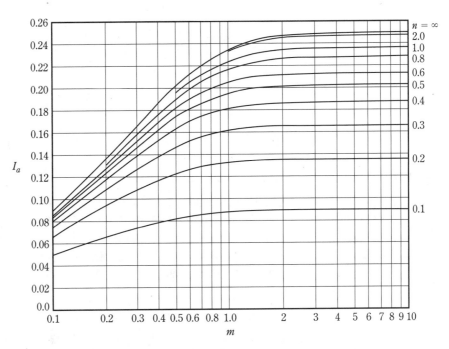

▼   **FIGURE 4.8**   Griffiths' influence factor I_a

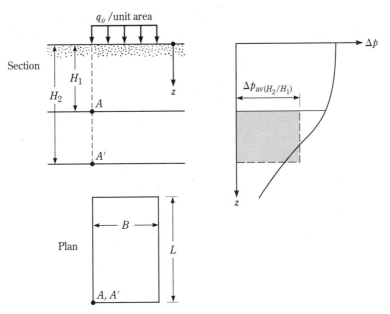

▼   **FIGURE 4.9**   Average pressure increase between $z = H_1$ to $z = H_2$ below the corner of a uniformly loaded rectangular area

where $\Delta p_{av(H_2/H_1)}$ = average stress increase immediately below the corner of a uniformly loaded rectangular area between depths $z = H_1$ to $z = H_2$

$$I_{a(H_2)} = I_a \text{ for } z = 0 \text{ to } z = H_2 = f\left(m = \frac{B}{H_2}, n = \frac{L}{H_2}\right)$$

$$I_{a(H_1)} = I_a \text{ for } z = 0 \text{ to } z = H_1 = f\left(m = \frac{B}{H_1}, n = \frac{L}{H_1}\right)$$

▼ **EXAMPLE 4.2** _____

Refer to Figure 4.10. Determine the *average* stress increase below the center of the loaded area between $z = 3$ m to $z = 5$ m (that is, between points A and A').

Solution   Refer to Figure 4.10. The loaded area can be divided into four rectangular areas, each measuring 1.5 m × 1.5 m ($L \times B$). Using Eq. (4.19), the average stress increase (between the required depths) below the corner of each rectangular area can be given as

$$\Delta p_{av(H_2/H_1)} = q_o\left[\frac{H_2 I_{a(H_2)} - H_1 I_{a(H_1)}}{H_2 - H_1}\right] = 100\left[\frac{(5)I_{a(H_2)} - (3)I_{a(H_1)}}{5 - 3}\right]$$

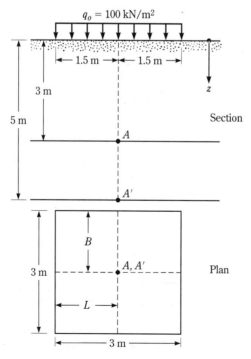

▼ **FIGURE 4.10**

For $I_{a(H_2)}$:

$$m = \frac{B}{H_2} = \frac{1.5}{5} = 0.3$$

$$n = \frac{L}{H_2} = \frac{1.5}{5} = 0.3$$

Referring to Figure 4.8, for $m = 0.3$ and $n = 0.3$, $I_{a(H_2)} = 0.136$. For $I_{a(H_1)}$:

$$m = \frac{B}{H_1} = \frac{1.5}{3} = 0.5$$

$$n = \frac{L}{H_1} = \frac{1.5}{3} = 0.5$$

Referring to Figure 4.8, $I_{a(H_1)} = 0.175$, so

$$\Delta p_{av(H_2/H_1)} = 100 \left[\frac{(5)(0.136) - (3)(0.175)}{5 - 3} \right] = \mathbf{7.75 \, kN/m^2}$$

The stress increase between $z = 3$ m to $z = 5$ m below the center of the loaded area is equal to

$$4\Delta p_{av(H_2/H_1)} = (4)(7.75) = \mathbf{31 \, kN/m^2}$$

▲

4.6 STRESS INCREASE UNDER AN EMBANKMENT

Figure 4.11 shows the cross section of an embankment of height H. For this two-dimensional loading condition the vertical stress increase may be expressed as

$$\Delta p = \frac{q_o}{\pi} \left[\left(\frac{B_1 + B_2}{B_2} \right)(\alpha_1 + \alpha_2) - \frac{B_1}{B_2}(\alpha_2) \right] \tag{4.20}$$

where $\quad q_o = \gamma H$
$\qquad \gamma = $ unit weight of the embankment soil
$\qquad H = $ height of the embankment

$$\alpha_1 \text{ (radians)} = \tan^{-1}\left(\frac{B_1 + B_2}{z}\right) - \tan^{-1}\left(\frac{B_1}{z}\right) \tag{4.21}$$

$$\alpha_2 = \tan^{-1}\left(\frac{B_1}{z}\right) \tag{4.22}$$

For a detailed derivation of the equation, see Das (1997). A simplified form of Eq. (4.20) is

$$\Delta p = q_o I' \tag{4.23}$$

where $\quad I' = $ a function of B_1/z and B_2/z

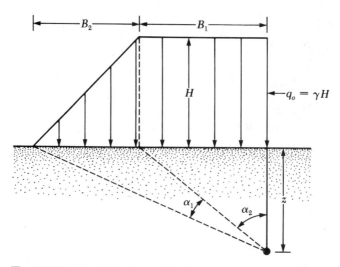

▼ **FIGURE 4.11**   Embankment loading

The variation of I' with B_1/z and B_2/z is shown in Figure 4.12. Application of this diagram is shown in Example 4.3.

▼ **EXAMPLE 4.3** _____

An embankment is shown in Figure 4.13a. Determine the stress increase under the embankment at points A_1 and A_2.

Solution

$$\gamma H = (17.5)(7) = 122.5 \text{ kN/m}^2$$

Stress Increase at A_1

The left side of Figure 4.13b indicates that $B_1 = 2.5$ m and $B_2 = 14$ m, so

$$\frac{B_1}{z} = \frac{2.5}{5} = 0.5$$

$$\frac{B_2}{z} = \frac{14}{5} = 2.8$$

According to Figure 4.12, in this case, $I' = 0.445$. Because the two sides in Figure 4.13b are symmetrical, the value of I' for the right side will also be 0.445, so

$$\Delta p = \Delta p_1 + \Delta p_2 = q_o[I'_{\text{(left side)}} + I'_{\text{(right side)}}]$$
$$= 122.5[0.445 + 0.445] = \mathbf{109.03 \text{ kN/m}^2}$$

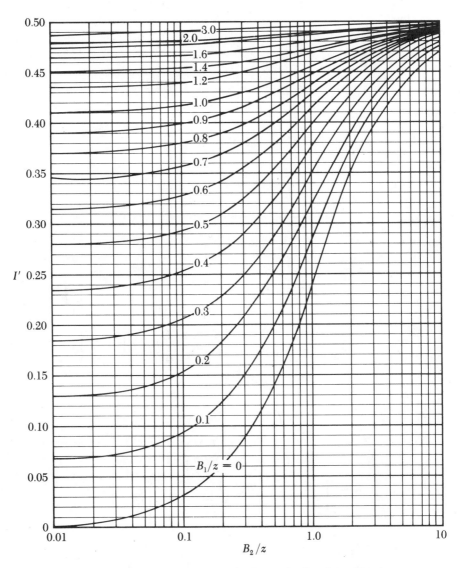

▼ **FIGURE 4.12** Influence value of I' for embankment loading (after Osterberg, 1957)

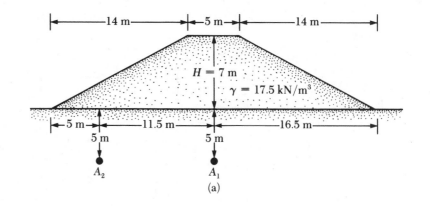

(a)

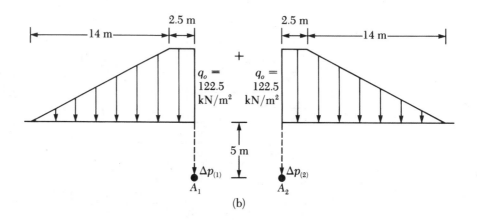

(b)

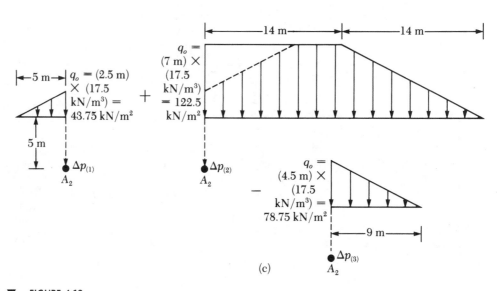

(c)

▼ FIGURE 4.13

Stress Increase at A_2

Refer to Figure 4.13c. For the left side, $B_2 = 5$ m and $B_1 = 0$, so

$$\frac{B_2}{z} = \frac{5}{5} = 1$$

$$\frac{B_1}{z} = \frac{0}{5} = 0$$

According to Figure 4.12, for these values of B_2/z and B_1/z, $I' = 0.25$, so

$$\Delta p_1 = 43.75(0.25) = 10.94 \text{ kN/m}^2$$

For the middle section,

$$\frac{B_2}{z} = \frac{14}{5} = 2.8$$

$$\frac{B_1}{z} = \frac{14}{5} = 2.8$$

Thus, $I' = 0.495$, so

$$\Delta p_2 = 0.495(122.5) = 60.64 \text{ kN/m}^2$$

For the right side,

$$\frac{B_2}{z} = \frac{9}{5} = 1.8$$

$$\frac{B_1}{z} = \frac{0}{5} = 0$$

and $I' = 0.335$, so

$$\Delta p_3 = (78.75)(0.335) = 26.38 \text{ kN/m}^2$$

Total stress increase at point A_2 is

$$\Delta p = \Delta p_1 + \Delta p_2 - \Delta p_3 = 10.94 + 60.64 - 26.38 = \textbf{45.2 kN/m}^2 \qquad \blacktriangle$$

4.7 STRESS INCREASE DUE TO ANY TYPE OF LOADING

The increase of vertical stress under any type of flexible loaded area can be easily determined by the use of Newmark's (1942) *influence chart*. The chart, in principle, is based on Eq. (4.3) for the estimation of vertical stress increase under the center of a circularly loaded area. According to Eq. (4.3),

$$\Delta p = q_o \left\{ 1 - \frac{1}{\left[1 + \left(\dfrac{B}{2z} \right)^2 \right]^{3/2}} \right\}$$

where $B/2$ = radius of the loaded area = R

The preceding equation can be rewritten as

$$\frac{R}{z} = \left[\left(1 - \frac{\Delta p}{q_o} \right)^{-2/3} - 1 \right]^{1/2} \tag{4.24}$$

We now substitute various values of $\Delta p / q_o$ into Eq. (4.24) to obtain corresponding values of R/z. Table 4.4 shows the calculated values of R/z for $\Delta p / q_o$ = 0, 0.1, 0.2, . . . , 1.

Using the nondimensional values of R/z shown in Table 4.4, we can draw the concentric circles having radii equal to R/z, as shown in Figure 4.14. Note that the distance AB in Figure 4.14 is unity. The first circle is a point having a radius of zero. Similarly, the second circle has a radius of 0.2698 $(\overline{AB})$. The last circle has a radius of infinity. These circles have been divided by equally spaced radial lines, producing what is referred to as *Newmark's chart*. The influence value, IV, of this chart is

$$IV = \frac{1}{\text{number of elements on the chart}} \tag{4.25}$$

For the chart shown in Figure 4.14, $IV = 1/200 = 0.005$.

Following is a step-by-step procedure for using Newmark's chart to determine vertical stress under a loaded area of any shape:

1. Identify the depth z below the loaded area at which the stress is to be determined.

▼ **TABLE 4.4** Values of R/z for Various Values of $\Delta p/sq_o$
[Eq. (4.24)]

$\Delta p / q_o$	R/z
0	0
0.1	0.2698
0.2	0.4005
0.3	0.5181
0.4	0.6370
0.5	0.7664
0.6	0.9174
0.7	1.1097
0.8	1.3871
0.9	1.9084
1.0	∞

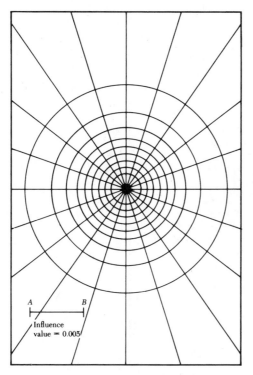

▼ **FIGURE 4.14** Influence chart for vertical pressure calculation (after Newmark, 1942)

2. Adopt a scale $z = \overline{AB}$ (that is, unit length according to Newmark's chart).
3. Draw the plan of the loaded area based on the scale adopted in Step 2.
4. Place the plan drawn in Step 3 on the Newmark's chart so that the point under which the stress is to be determined is directly above the center of the chart.
5. Count the number of elements of the chart that fall inside the plan. Let it equal N.
6. Calculate the stress increase as

$$\Delta p = (IV)(N)(q_o) \qquad (4.26)$$

where q_o = load per unit area on the loaded area

▼ **EXAMPLE 4.4**

A flexible rectangular area, 2.5 m × 5 m, is located on the ground surface and loaded with $q_o = 145$ kN/m². Determine the stress increase caused by this loading at a depth of 6.25 m below the center of the rectangular area. Use Newmark's chart.

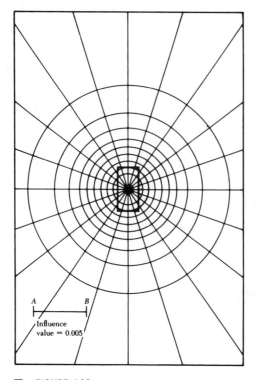

▼   **FIGURE 4.15**

Solution   Here, $z = 6.25$ m, so length $\overline{AB}$ in Figure 4.14 is 6.25 m. With this scale, the plan of the loaded rectangular area can be drawn. Figure 4.15 shows this plan placed over the Newmark's chart with the center of the loaded area above the center of the chart. The reason for this placement is that the stress increase is required at a point immediately below the center of the rectangular area. The number of elements from the influence chart that are inside the plan is about 26, so

$$\Delta p = (IV)(N)(q_o) = (0.005)(26)(145) = \mathbf{18.85 \ kN/m^2} \qquad \blacktriangle$$

SETTLEMENT CALCULATION

4.8   ELASTIC SETTLEMENT BASED ON THE THEORY OF ELASTICITY

The elastic settlement of a shallow foundation can be estimated by using the theory of elasticity. Referring to Figure 4.16 and using Hooke's law,

$$S_e = \int_0^H \varepsilon_z \, dz = \frac{1}{E_s} \int_0^H (\Delta p_z - \mu_s \, \Delta p_x - \mu_s \, \Delta p_y) \, dz \qquad (4.27)$$

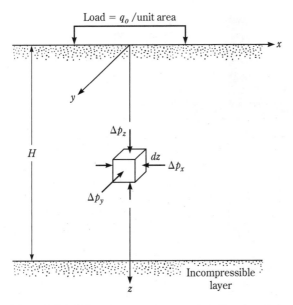

FIGURE 4.16   Elastic settlement of shallow foundation

where

$\qquad S_e$ = elastic settlement
$\qquad E_s$ = modulus of elasticity of soil
$\qquad H$ = thickness of the soil layer
$\qquad \mu_s$ = Poisson's ratio of the soil
$\Delta p_x, \Delta p_y, \Delta p_z$ = stress increase due to the net applied foundation load in the x, y, and z directions, respectively

Theoretically, if the depth of foundation $D_f = 0$, $H = \infty$, and the foundation is perfectly flexible, according to Harr (1966) the settlement may be expressed as (Figure 4.17)

$$S_e = \frac{Bq_0}{E_s}(1 - \mu_s^2)\frac{\alpha}{2} \qquad \text{(corner of the flexible foundation)} \tag{4.28}$$

$$S_e = \frac{Bq_0}{E_s}(1 - \mu_s^2)\alpha \qquad \text{(center of the flexible foundation)} \tag{4.29}$$

where $\quad \alpha = \frac{1}{\pi}\left[\ln\left(\frac{\sqrt{1 + m_1^2} + m_1}{\sqrt{1 + m_1^2} - m_1}\right) + m\ln\left(\frac{\sqrt{1 + m_1^2} + 1}{\sqrt{1 + m_1^2} - 1}\right)\right] \tag{4.30}$

$\qquad m_1 = L/B \tag{4.31}$
$\qquad B$ = width of foundation
$\qquad L$ = length of foundation

The values of α for various length-to-width (L/B) ratios are shown in Figure 4.18. The average immediate settlement for a flexible foundation also may be expressed as

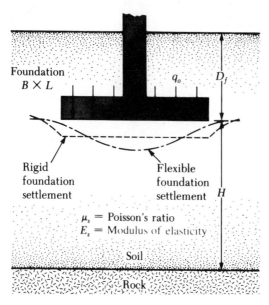

▼ **FIGURE 4.17**   Elastic settlement of flexible and rigid foundations

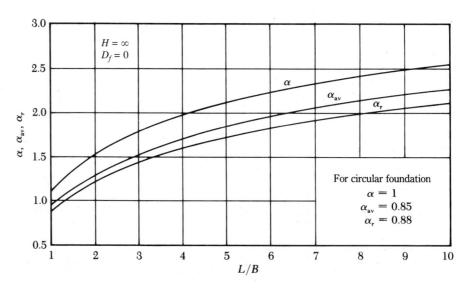

▼ **FIGURE 4.18**   Values of α, α_{av}, and α_r — Eqs. (4.28), (4.29), (4.32), and (4.32a)

$$S_e = \frac{Bq_o}{E_s}(1 - \mu_s^2)\alpha_{av} \qquad \text{(average for flexible foundation)}$$

$$\qquad (4.32)$$

Figure 4.18 also shows the values of α_{av} for various L/B ratios of foundation.

However, if the foundation shown in Figure 4.17 is rigid, the immediate settlement will be different and may be expressed as

$$S_e = \frac{Bq_o}{E_s}(1 - \mu_s^2)\alpha_r \qquad \text{(rigid foundation)}$$

$$\qquad (4.32a)$$

The values of α_r for various L/B ratios of foundation are shown in Figure 4.18.

If $D_f = 0$ and $H < \infty$ due to the presence of a rigid (incompressible) layer as shown in Figure 4.17,

$$S_e = \frac{Bq_o}{E_s}(1 - \mu_s^2)\frac{[(1 - \mu_s^2)F_1 + (1 - \mu_s - 2\mu_s^2)F_2]}{2}$$

$$\text{(corner of flexible foundation)}$$

$$\qquad (4.33a)$$

and

$$S_e = \frac{Bq_o}{E_s}(1 - \mu_s^2)[(1 - \mu_s^2)F_1 + (1 - \mu_s - 2\mu_s^2)F_2]$$

$$\text{(corner of flexible foundation)}$$

$$\qquad (4.33b)$$

The variations of F_1 and F_2 with H/B are given in Figures 4.19 and 4.20, respectively (Steinbrenner, 1934).

It is also important to realize that the preceding relationships for S_e assume that the depth of the foundation is equal to zero. For $D_f > 0$, the magnitude of S_e will decrease.

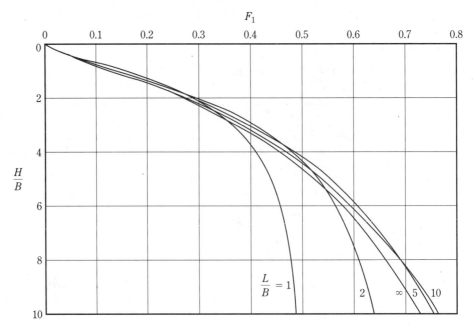

▼ **FIGURE 4.19** Variation of F_1 with H/B (based on Steinbrenner, 1934)

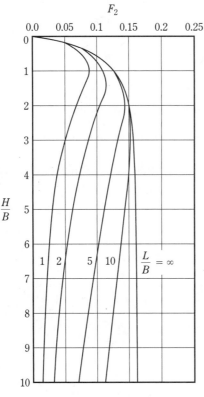

▼ **FIGURE 4.20** Variation of F_2 with H/B
(based on Steinbrenner, 1934)

▼   **EXAMPLE 4.5**

A foundation is 1 m × 2 m in plan and carries a net load per unit area, $q_o = 150$ kN/m². Given, for the soil, $E_s = 10,000$ kN/m²; $\mu_s = 0.3$. Assuming the foundation to be flexible, estimate the elastic settlement at the center of the foundation for the following conditions:

   a. $D_f = 0; H = \infty$

   b. $D_f = 0; H = 5$ m

Solution

Part a

From Eq. (4.29)

$$S_e = \frac{Bq_o}{E_s}(1 - \mu_s^2)\alpha$$

For $L/B = 2/1 = 2$, from Figure 4.18, $\alpha \approx 1.53$, so

$$S_e = \frac{(1)(150)}{10,000}(1 - 0.3^2)(1.53) = 0.0209 \text{ m} = \textbf{20.9 mm}$$

Part b

From Eq. (4.33b)

$$S_e = \frac{Bq_o}{E_s}(1 - \mu_s^2)[(1 - \mu_s^2)F_1 + (1 - \mu_s - 2\mu_s^2)F_2]$$

For $L/B = 2$ and $H/B = 5$, from Figures 4.19 and 4.20, $F_1 \approx 0.525$ and $F_2 \approx 0.06$

$$S_e = \frac{(1)(150)}{10,000}(1 - 0.3^2)[(1 - 0.3^2)(0.525) + (1 - 0.3 - 2 \times 0.3^2)(0.06)]$$
$$= 0.007 \text{ m} = \textbf{7.0 mm}$$

▲

4.9   ELASTIC SETTLEMENT OF FOUNDATIONS ON SATURATED CLAY

Janbu et al. (1956) proposed an equation for evaluating the average settlement of flexible foundations on saturated clay soils (Poisson's ratio, $\mu_s = 0.5$). For the notation used in Figure 4.21, this equation is

$$S_e = A_1 A_2 \frac{q_o B}{E_s} \tag{4.34}$$

where A_1 is a function of H/B and L/B and A_2 is a function of D_f/B

Christian and Carrier (1978) modified the values of A_1 and A_2 to some extent, as presented in Figure 4.21.

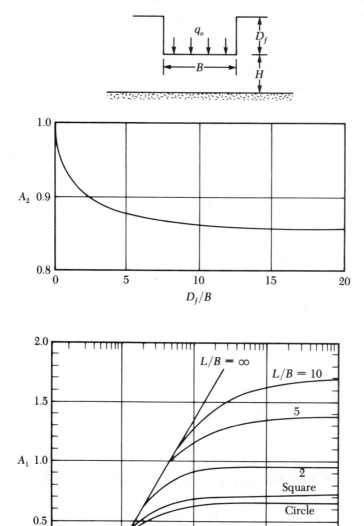

▼ **FIGURE 4.21**   Values of A_1 and A_2 for elastic settlement calculation – Eq. (4.34) (after Christian and Carrier, 1978)

4.10   SETTLEMENT OF SANDY SOIL: USE OF STRAIN INFLUENCE FACTOR

Settlement of granular soils can also be evaluated by use of a semi-empirical *strain influence factor* (Figure 4.22) proposed by Schmertmann and Hartman (1978). According to this method, the settlement is

$$S_e = C_1 C_2 (\bar{q} - q) \sum_0^{z_2} \frac{I_z}{E_s} \Delta z \qquad (4.35)$$

where I_z = strain influence factor
   C_1 = a correction factor for the depth of foundation
      embedment = $1 - 0.5[q/(\bar{q} - q)]$
   C_2 = a correction factor to account for creep in soil
      = $1 + 0.2 \log (\text{time in years}/0.1)$
   $\bar{q}$ = stress at the level of the foundation
   $q = \gamma D_f$

The variation of the strain influence factor with depth below the foundation is shown in Figure 4.22a. Note that, for square or circular foundations,

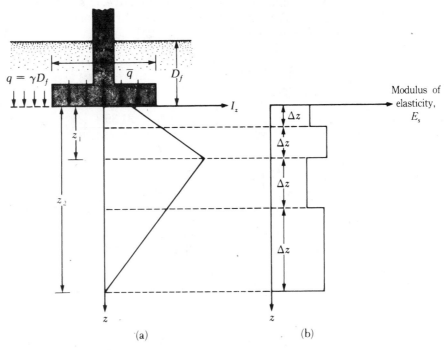

▼ **FIGURE 4.22**   Elastic settlement calculation by using strain influence factor

$I_z = 0.1$ at $z = 0$

$I_z = 0.5$ at $z = z_1 = 0.5B$

$I_z = 0$ at $z = z_2 = 2B$

Similarly, for foundations with $L/B \geq 10$,

$I_z = 0.2$ at $z = 0$

$I_z = 0.5$ at $z = z_1 = B$

$I_z = 0$ at $z = z_2 = 4B$

where B = width of the foundation and L = length of the foundation

For values of L/B between 1 and 10, necessary interpolations can be made.

To use Eq. (4.35) first requires evaluation of the approximate variation of the modulus of elasticity with depth (Figure 4.22). This evaluation can be made by using the standard penetration numbers or cone penetration resistances (Chapter 2). The soil layer can be divided into several layers to a depth of $z = z_2$, and the elastic settlement of each layer can be estimated. The sum of the settlement of all layers equals S_e. Schmertmann (1970) provided a case history of a rectangular foundation (Belgian bridge pier) having $L = 23$ m and $B = 2.6$ m and being supported by a granular soil deposit. For this foundation we may assume that $L/B \approx 10$ for plotting the strain influence factor diagram. Figure 4.23 shows the details of the foundation along with the approximate variation of the cone penetration resistance, q_c, with depth. For this foundation [Eq. (4.35)], note that

$\bar{q} = 178.54 \text{ kN/m}^2$

$q = 31.39 \text{ kN/m}^2$

$C_1 = 1 - 0.5 \dfrac{q}{\bar{q} - q} = 1 - (0.5) \left(\dfrac{31.39}{178.54 - 31.39} \right) = 0.893$

$C_2 = 1 + 0.2 \log \left(\dfrac{t\,\text{yr}}{0.1} \right)$

For $t = 5$ yr

$C_2 = 1 + 0.2 \log \left(\dfrac{5}{0.1} \right) = 1.34$

The following table shows the calculation of $\sum_0^{z_2} (I_z/E_s) \, \Delta z$ in conjunction with Figure 4.23.

Layer	Δz (m)	q_c (kN/m²)	E_s^a (kN/m²)	z to the center of the layer (m)	I_z at the center of the layer	$(I_z/E_s)\,\Delta z$ (m²/kN)
1	1	2,450	8,575	0.5	0.258	3.00×10^{-5}
2	1.6	3,430	12,005	1.8	0.408	5.43×10^{-5}
3	0.4	3,430	12,005	2.8	0.487	1.62×10^{-5}
4	0.5	6,870	24,045	3.25	0.458	0.95×10^{-5}
5	1.0	2,950	10,325	4.0	0.410	3.97×10^{-5}
6	0.5	8,340	29,190	4.75	0.362	0.62×10^{-5}
7	1.5	14,000	49,000	5.75	0.298	0.91×10^{-5}
8	1	6,000	21,000	7.0	0.247	1.17×10^{-5}
9	1	10,000	35,000	8.0	0.154	0.44×10^{-5}
10	1.9	4,000	14,000	9.45	0.062	0.84×10^{-5}
	Σ 10.4 m = 4B					Σ 18.95 × 10⁻⁵

$^a E_s \approx 3.5 q_c$ [Eq. (4.40)]

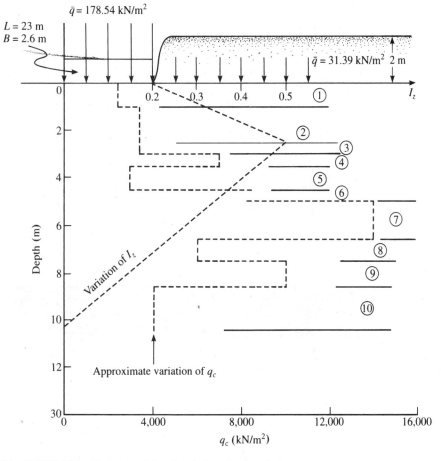

▼ **FIGURE 4.23** Variation of I_z and q_c below the foundation

Hence the immediate settlement is calculated as

$$S_e = C_1 C_2 (\bar{q} - q) \sum \frac{I_z}{E_s} \Delta z$$
$$= (0.893)(1.34)(178.54 - 31.39)(18.95 \times 10^{-5})$$
$$= 0.03336 \approx 33 \text{ mm}$$

After five years, the actual *maximum* settlement observed for the foundation was about 39 mm.

4.11 RANGE OF MATERIAL PARAMETERS FOR COMPUTING ELASTIC SETTLEMENT

Sections 4.8–4.10 presented the equations for calculating immediate settlement of foundations. These equations contain the elastic parameters, such as E_s and μ_s. If the laboratory test results for these parameters are not available, certain realistic assumptions have to be made. Table 4.5 shows the approximate range of the elastic parameters for various soils.

Several investigators have correlated the values of the modulus of elasticity, E_s, with the field standard penetration number, N_F and the cone penetration resistance, q_c. Mitchell and Gardner (1975) compiled a list of these correlations. Schmertmann (1970) indicated that the modulus of elasticity of sand may be given by

$$E_s(\text{kN/m}^2) = 766 N_F \tag{4.36}$$

where N_F = field standard penetration number

In English units

$$E (\text{U.S. ton/ft}^2) = 8 N_F \tag{4.37}$$

▼ **TABLE 4.5** Elastic Parameters of Various Soils

Type of soil	Modulus of elasticity, E_s		Poisson's ratio, μ_s
	lb/in²	MN/m²	
Loose sand	1,500–3,500	10.35–24.15	0.20–0.40
Medium dense sand	2,500–4,000	17.25–27.60	0.25–0.40
Dense sand	5,000–8,000	34.50–55.20	0.30–0.45
Silty sand	1,500–2,500	10.35–17.25	0.20–0.40
Sand and gravel	10,000–25,000	69.00–172.50	0.15–0.35
Soft clay	600–3,000	4.1–20.7	
Medium clay	3,000–6,000	20.7–41.4	0.20–0.50
Stiff clay	6,000–14,000	41.4–96.6	

Similarly

$$E_s = 2q_c \tag{4.38}$$

where q_c = static cone penetration resistance

Schmertmann and Hartman (1978) further suggested that the following correlations may be used with the strain influence factors described in Section 4.10:

$$E_s = 2.5q_c \qquad \text{(for square and circular foundations)} \tag{4.39}$$

and

$$E_s = 3.5q_c \qquad \text{(for strip foundations)} \tag{4.40}$$

Note: Any consistent set of units may be used in Eqs. (4.38)–(4.40).

The modulus of elasticity of normally consolidated clays may be estimated as

$$E_s = 250c \text{ to } 500c \tag{4.41}$$

and for overconsolidated clays as

$$E_s = 750c \text{ to } 1000c \tag{4.42}$$

where c = undrained cohesion of clay soil

4.12 CONSOLIDATION SETTLEMENT

As mentioned before, consolidation settlement occurs over time, and it occurs in saturated clayey soils when they are subjected to increased load caused by foundation construction (Figure 4.24). Based on the one-dimensional consolidation settlement equations given in Chapter 1, we write

$$S_c = \int \varepsilon_z \, dz$$

where ε_z = vertical strain

$$= \frac{\Delta e}{1 + e_o}$$

Δe = change of void ratio

$= f(p_o, p_c, \text{ and } \Delta p)$

So

$$S_c = \frac{C_c H_c}{1 + e_o} \log \frac{p_o + \Delta p_{av}}{p_o} \qquad \begin{array}{l}\text{(for normally consolidated}\\ \text{clays)}\end{array} \tag{1.64}$$

$$S_c = \frac{C_s H_c}{1 + e_o} \log \frac{p_o + \Delta p_{av}}{p_o} \qquad \begin{array}{l}\text{(for overconsolidated clays}\\ \text{with } p_o + \Delta p_{av} < p_c)\end{array} \tag{1.66}$$

$$S_c = \frac{C_s H_c}{1 + e_o} \log \frac{p_c}{p_o} + \frac{C_c H_c}{1 + e_o} \log \frac{p_o + \Delta p_{av}}{p_c} \qquad \begin{array}{l}\text{(for overconsolidated clays}\\ \text{with } p_o < p_c < p_o + \Delta p_{av})\end{array} \tag{1.68}$$

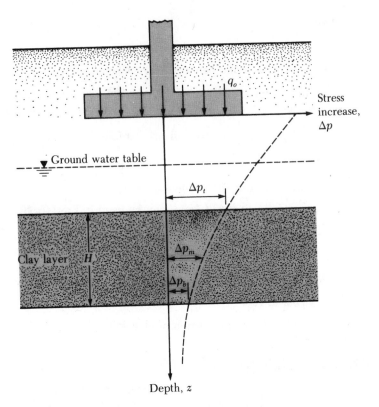

Depth, z

▼ **FIGURE 4.24**   Consolidation settlement calculation

where p_o = average effective pressure on the clay layer before the
construction of the foundation

Δp_{av} = average increase of pressure on the clay layer caused by the
foundation construction

p_c = preconsolidation pressure

e_o = initial void ratio of the clay layer

C_c = compression index

C_s = swelling index

H_c = thickness of the clay layer

The procedures for determining the compression and swelling indexes were discussed in Chapter 1.

Note that the increase of pressure, Δp, on the clay layer is not constant with
depth. The magnitude of Δp will decrease with the increase of depth measured from
the bottom of the foundation. However, the average increase of pressure may be
approximated by

$$\Delta p_{av} = \tfrac{1}{6}(\Delta p_t + 4\Delta p_m + \Delta p_b) \qquad\qquad (4.43)$$

where Δp_t, Δp_m, and Δp_b are the pressure increases at the *top, middle,* and *bottom* of the clay layer that are caused by the foundation construction.

The method of determining the pressure increase caused by various types of foundation load is discussed in Sections 4.2–4.7. Δp_{av} can also be directly obtained from the method presented in Section 4.5.

▼ **EXAMPLE 4.6**

A foundation 1 m × 2 m in plan is shown in Figure 4.25. Estimate the consolidation settlement of the foundation.

Solution   The clay is normally consolidated. Thus

$$S_c = \frac{C_c H}{1 + e_o} \log \frac{p_o + \Delta p_{av}}{p_o}$$

$$p_o = (2.5)(16.5) + (0.5)(17.5 - 9.81) + (1.25)(16 - 9.81)$$
$$= 41.25 + 3.85 + 7.74 = 52.84 \text{ kN/m}^2$$

From Eq. (4.43),

$$\Delta p_{av} = \tfrac{1}{6}(\Delta p_t + 4\Delta p_m + \Delta p_b)$$

Now the following table can be prepared (*note:* $L = 2$ m; $B = 1$ m):

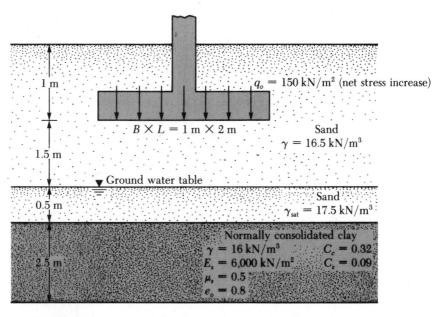

1 m

$q_o = 150 \text{ kN/m}^2$ (net stress increase)

$B \times L = 1 \text{ m} \times 2 \text{ m}$

Sand
$\gamma = 16.5 \text{ kN/m}^3$

1.5 m

▼ Ground water table

0.5 m

Sand
$\gamma_{sat} = 17.5 \text{ kN/m}^3$

Normally consolidated clay
$\gamma = 16 \text{ kN/m}^3$ $C_c = 0.32$
$E_s = 6,000 \text{ kN/m}^2$ $C_s = 0.09$
$\mu_s = 0.5$
$e_o = 0.8$

2.5 m

▼ **FIGURE 4.25**

$m_1 = L/B$	$z(m)$	$z/(B/2) = n_1$	I_c^a	$\Delta p = q_o I_c^b$
2	2	4	0.190	$28.5 = \Delta p_t$
2	$2 + 2.5/2 = 3.25$	6.5	≈ 0.085	$12.75 = \Delta p_m$
2	$2 + 2.5 = 4.5$	9	0.045	$6.75 = \Delta p_b$

[a] Table 4.3
[b] Eq. (4.10)

$$\Delta p_{av} = \tfrac{1}{6}(28.5 + 4 \times 12.75 + 6.75) = 14.38 \text{ kN/m}^2$$

so

$$S_c = \frac{(0.32)(2.5)}{1 + 0.8} \log\left(\frac{52.84 + 14.38}{52.84}\right) = 0.0465 = \textbf{46.5 mm} \qquad \blacktriangle$$

4.13 SKEMPTON–BJERRUM MODIFICATION FOR CONSOLIDATION SETTLEMENT

The consolidation settlement calculation presented in the preceding section is based on Eqs. (1.64), (1.66), and (1.68). These equations, as shown in Chapter 1, are based on one-dimensional laboratory consolidation tests. The underlying assumption for these equations is that the increase of pore water pressure, Δu, immediately after the load application equals the increase of stress, Δp, at any depth. For this case

$$S_{c(oed)} = \int \frac{\Delta e}{1 + e_o} \, dz = \int m_v \, \Delta p_{(1)} \, dz$$

where $S_{c(oed)}$ = consolidation settlement calculated by using Eqs. (1.64), (1.66), and (1.68)

$\Delta p_{(1)}$ = vertical stress increase (note the change of notation from Δp)

m_v = volume coefficient of compressibility (see Chapter 1)

In the field, however, when load is applied over a limited area on the ground surface, this assumption will not be correct. Consider the case of a circular foundation on a clay layer as shown in Figure 4.26. The vertical and the horizontal stress increases at a point in the clay layer immediately below the center of the foundation are $\Delta p_{(1)}$ and $\Delta p_{(3)}$, respectively. For a saturated clay, the pore water pressure increase at that depth (Chapter 1) is

$$\Delta u = \Delta p_{(3)} + A[\Delta p_{(1)} - \Delta p_{(3)}] \qquad (4.44)$$

where A = pore water pressure parameter

For this case

$$S_c = \int m_v \, \Delta u \, dz = \int (m_v)\{\Delta p_{(3)} + A[\Delta p_{(1)} - \Delta p_{(3)}]\} \, dz \qquad (4.45)$$

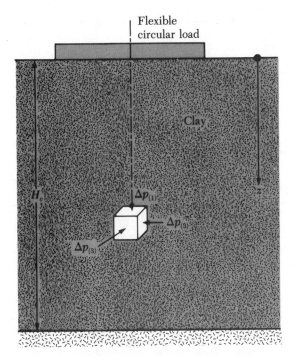

▼   **FIGURE 4.26**   Circular foundation on a clay layer

Thus we can write

$$K_{\mathrm{cir}} = \frac{S_c}{S_{c(oed)}} = \frac{\int_0^{H_c} m_v \, \Delta u \, dz}{\int_0^{H_c} m_v \, \Delta p_{(1)} \, dz} = A + (1 - A) \left[\frac{\int_0^{H_c} \Delta p_{(3)} \, dz}{\int_0^{H_c} \Delta p_{(1)} \, dz} \right] \tag{4.46}$$

where K_{cir} = settlement ratio for circular foundations

The settlement ratio for a continuous foundation (K_{str}) can be determined in a manner similar to that for a circular foundation. The variation of K_{cir} and K_{str} with A and H_c/B is given in Figure 4.27. (*Note: B* = diameter of a circular foundation, and B = width of a continuous foundation.)

Following is the procedure for determining consolidation settlement according to Skempton and Bjerrum (1957).

1. Determine the consolidation settlement, $S_{c(oed)}$, using the procedure outlined in Section 4.12. (Note the change of notation from S_c.)
2. Determine the pore water pressure parameter, A.
3. Determine H_c/B.
4. Obtain the settlement ratio — in this case, from Figure 4.27.
5. Calculate the actual consolidation settlement:

$$S_c = S_{c(oed)} \times \text{settlement ratio} \tag{4.47}$$

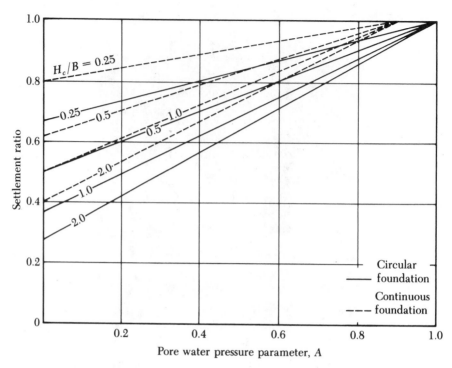

▼ **FIGURE 4.27** Settlement ratios for circular (K_{cir}) and continuous (K_{str}) foundations

This technique is generally referred to as the *Skempton–Bjerrum modification* for consolidation settlement calculation.

4.14 CONSOLIDATION SETTLEMENT— GENERAL COMMMENTS AND A CASE HISTORY

In predicting the consolidation settlement and the time rate of settlement for actual field conditions, an engineer has to make several simplifying assumptions. They include the compression index, coefficient of consolidation, preconsolidation pressure, drainage conditions, and thickness of the clay layer. Soil layering is not always uniform with ideal properties; hence field performance may deviate from the prediction, requiring adjustments during construction. The following case history on consolidation, as reported by Schnabel (1972), illustrates this reality.

Figure 4.28 shows the subsoil conditions for the construction of a school building in Waldorf, Maryland. Upper Pleistocene sand and gravel soils are underlain by deposits of very loose fine silty sand, soft silty clay, and clayey silt. The softer surface layers are underlain by various layers of stiff to firm silty clay, clayey silt, and sandy silt to a depth of 50 ft. Before construction of the building began, a compacted fill having a thickness of 8–10 ft was placed on the ground surface. This fill initiated the consolidation settlement in the soft silty clay and clayey silt.

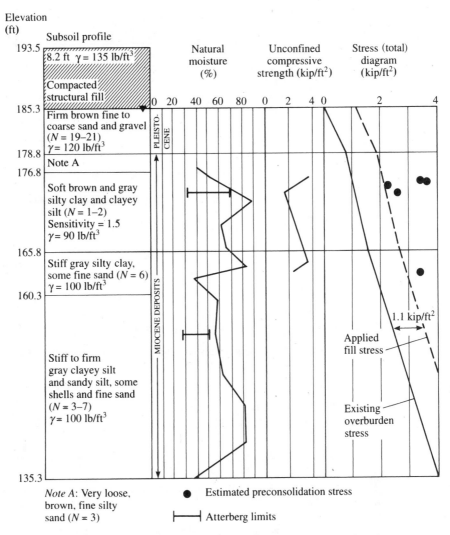

Elevation (ft)

Subsoil profile

Natural moisture (%)

Unconfined compressive strength (kip/ft²)

Stress (total) diagram (kip/ft²)

▼ FIGURE 4.28 Subsoil conditions for the construction of school building (*note:* SPT *N* values are uncorrected, i.e. N_F; after Schnabel, 1972)

To predict the time rate of settlement, based on the laboratory test results, the engineers made the following approximations:

a. The preconsolidation pressure, p_c, was 1600 to 2800 lb/ft² *in excess* of the existing overburden pressure.

b. The swell index, C_s, was 0.01 to 0.03.

c. For the more compressive layers $C_v \approx 0.36$ ft²/day, and for the stiffer soil layers $C_v \approx 3.1$ ft²/day.

The total consolidation settlement was estimated to be about 3 in. Under double drainage conditions, 90% settlement was expected to occur in 114 days.

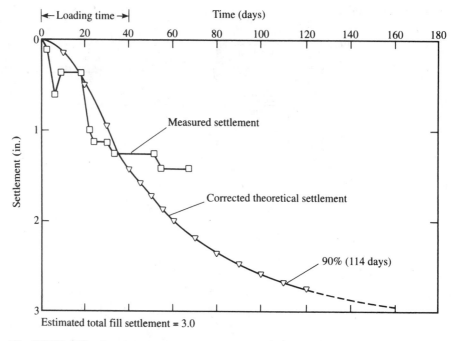

▼   **FIGURE 4.29**   Comparison of measured and predicted consolidation settlement with time (after Schnabel, 1972)

Figure 4.29 shows a comparison of measured and predicted settlement with time, which indicates that

a. $\dfrac{S_{c(observed)}}{S_{c(estimated)}} \approx 0.47$

b.   90% of the settlement occurred in about 70 days; hence $t_{90(observed)}/t_{90(estimated)} \approx 0.58$.

The relatively rapid settlement in the field is believed to be due to the presence of a fine sand layer within the Miocene deposits.

ALLOWABLE BEARING CAPACITY

4.15   ALLOWABLE BEARING PRESSURE IN SAND BASED ON SETTLEMENT CONSIDERATION

Meyerhof (1956) proposed a correlation for the *net allowable bearing pressure* for foundations with the corrected standard penetration resistance, N_{cor}. The net pressure has been defined as

$$q_{net(all)} = q_{all} - \gamma D_f$$

According to Meyerhof's theory, for 1 in. (25.4 mm) of estimated maximum settlement

$$q_{net(all)} (kN/m^2) = 11.98 N_{cor} \quad (\text{for } B \le 1.22 \text{ m}) \tag{4.48}$$

$$q_{net(all)} (kN/m^2) = 7.99 N_{cor} \left(\frac{3.28B + 1}{3.28B} \right)^2 \quad (\text{for } B > 1.22 \text{ m}) \tag{4.49}$$

where N_{cor} = corrected standard penetration number

Note that in Eqs. (4.48) and (4.49) B is in meters.
In English units

$$q_{net(all)} (kip/ft^2) = \frac{N_{cor}}{4} \quad (\text{for } B \le 4 \text{ ft}) \tag{4.50}$$

and

$$q_{net(all)} (kip/ft^2) = \frac{N_{cor}}{6} \left(\frac{B + 1}{B} \right)^2 \quad (\text{for } B > 4 \text{ ft}) \tag{4.51}$$

Since Meyerhof proposed his original correlation, researchers have observed that its results are rather conservative. Later, Meyerhof (1965) suggested that the net allowable bearing pressure should be increased by about 50%. Bowles (1977) proposed that the modified form of the bearing pressure equations be expressed as

$$q_{net(all)} (kN/m^2) = 19.16 N_{cor} F_d \left(\frac{S_e}{25.4} \right) \quad (\text{for } B \le 1.22 \text{ m}) \tag{4.52}$$

$$q_{net(all)} (kN/m^2) = 11.98 N_{cor} \left(\frac{3.28B + 1}{3.28B} \right)^2 F_d \left(\frac{S_e}{25.4} \right) \quad (\text{for } B > 1.22 \text{ m}) \tag{4.53}$$

where F_d = depth factor = $1 + 0.33(D_f/B) \le 1.33$ (4.54)
        S_e = tolerable settlement, in mm

Again, the unit of B is meters.
In English units

$$q_{net(all)} (kip/ft^2) = \frac{N_{cor}}{2.5} F_d S_e \quad (\text{for } B \le 4 \text{ ft}) \tag{4.55}$$

$$q_{net(all)} (kip/ft^2) = \frac{N_{cor}}{4} \left(\frac{B + 1}{B} \right)^2 F_d S_e \quad (\text{for } B > 4 \text{ ft}) \tag{4.56}$$

where F_d is given by Eq. (4.54)
        S_e = tolerable settlement, in in.

Based on Eqs. (4.55) and (4.56), the variations of $q_{net(all)}/(F_d S_e)$ with B and N_{cor} are given in Figure 4.30.

The empirical relations just presented may raise some questions. For example, which value of the standard penetration number should be used, and what is the effect of the water table on the net allowable bearing capacity? The design value of N_{cor} should be determined by taking into account the N_{cor} values for a depth of $2B$ to $3B$, measured from the bottom of the foundation. Many engineers are also of the opinion that the N_{cor} value should be reduced somewhat if the water table is close to the foundation. However, the author believes that this reduction is not required because the penetration resistance reflects the location of the water table.

Meyerhof (1956) also prepared empirical relations for the net allowable bearing capacity of foundations based on the cone penetration resistance, q_c:

$$q_{net(all)} = \frac{q_c}{15} \qquad \text{(for } B \leq 1.22 \text{ m and settlement of 25.4 mm)} \qquad (4.57)$$

and

$$q_{net(all)} = \frac{q_c}{25}\left(\frac{3.28B + 1}{3.28B}\right)^2 \qquad \text{(for } B > 1.22 \text{ m and settlement of 25.4 mm)} \qquad (4.58)$$

Note that in Eqs. (4.57) and (4.58) the unit of B is meters, and the units of $q_{net(all)}$ and q_c are kN/m².

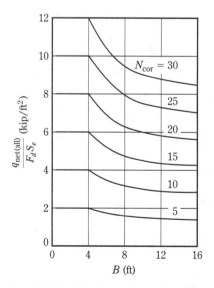

▼ **FIGURE 4.30** Plot of $q_{net(all)}/F_d S_e$ vs. B—
Eqs. (4.55) and (4.56)

In English units

$$q_{\text{net(all)}} \ (\text{lb/ft}^2) = \frac{q_c \ (\text{lb/ft}^2)}{15} \qquad (\text{for } B \le 4 \text{ ft and settlement of 1 in.}) \qquad (4.59)$$

and

$$q_{\text{net(all)}} \ (\text{lb/ft}^2) = \frac{q_c \ (\text{lb/ft}^2)}{25} \left(\frac{B+1}{B}\right)^2$$

$$(\text{for } B > 4 \text{ ft and settlement of 1 in.}) \quad (4.60)$$

Note that in Eqs. (4.59) and (4.60), the unit of B is feet.

The basic philosophy behind the development of these correlations is that, if the maximum settlement is no more than 1 in. (25.4 mm) for any foundation, the differential settlement would be no more than 0.75 in. (19 mm). These are probably the allowable limits for most building foundation designs.

4.16 FIELD LOAD TEST

The ultimate load-bearing capacity of a foundation, as well as the allowable bearing capacity based on tolerable settlement considerations, can be effectively determined from the field load test. It is generally referred to as the *plate load test* (ASTM, 1982; Test Designation D-1194-72). The plates that are used for tests in the field are usually made of steel and are 25 mm (1 in.) thick and 150 mm to 762 mm (6 in. to 30 in.) in diameter. Occasionally, square plates that are 305 mm × 305 mm (12 in. × 12 in.) are also used.

To conduct a plate load test, a hole is excavated with a minimum diameter $4B$ (B = diameter of the test plate) to a depth of D_f (D_f = depth of the proposed foundation). The plate is placed at the center of the hole. Load is applied to the plate in steps — about one-fourth to one-fifth of the estimated ultimate load — by means of a jack. A schematic diagram of the test arrangement is shown in Figure 4.31a. During each step load application, the settlement of the plate is observed on dial gauges. At least one hour is allowed to elapse between each load application step. The test should be conducted until failure, or at least until the plate has gone through 25 mm (1 in.) of settlement. Figure 4.32 shows the nature of the load–settlement curve obtained from such tests, from which the ultimate load per unit area can be determined.

For tests in clay,

$$q_{u(F)} = q_{u(P)} \tag{4.61}$$

where $q_{u(F)}$ = ultimate bearing capacity of the proposed foundation
$q_{u(P)}$ = ultimate bearing capacity of the test plate

Equation (4.61) implies that the ultimate bearing capacity in clay is virtually independent of the size of the plate.

For tests in sandy soils,

$$q_{u(F)} = q_{u(P)} \frac{B_F}{B_P} \tag{4.62}$$

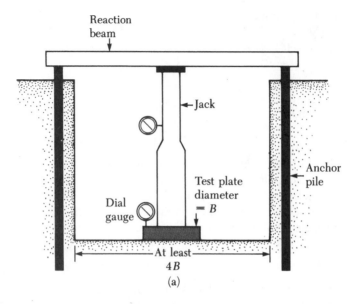

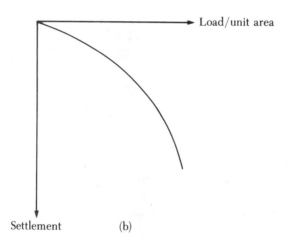

▼ **FIGURE 4.31** Plate load test: (a) test arrangement; (b) nature of load–settlement curve

where B_F = width of the foundation
B_P = width of the test plate

The allowable bearing capacity of a foundation, based on settlement considerations and for a given intensity of load, q_o, is

$$S_F = S_P \frac{B_F}{B_P} \qquad \text{(for clayey soil)} \qquad (4.63)$$

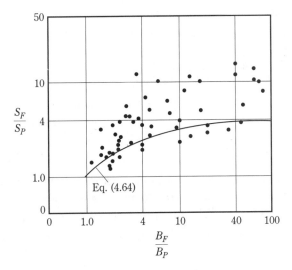

▼ **FIGURE 4.32** Comparison of field test results with Eq. (4.64) (after D'Appolonia et al., 1970)

and

$$S_F = S_P \left(\frac{2B_F}{B_F + B_P} \right)^2 \quad \text{(for sandy soil)} \tag{4.64}$$

The preceding relationship is based on the work of Terzaghi and Peck (1967). Figure 4.32 shows a comparison of several large-scale field test results with Eq. (4.64). Based on this comparison, it can be said that Eq. (4.64) is fairly approximate.

Housel (1929) proposed a different technique for determining the load-bearing capacity of shallow foundations based on settlement considerations:

1. Requirement is to find the dimensions of a foundation that will carry a load of Q_o with an allowable settlement of $S_{e(all)}$.
2. Conduct two plate load tests with plates of diameters B_1 and B_2.
3. From the load–settlement curves obtained in Step 2, determine the total loads on the plates (Q_1 and Q_2) that correspond to the settlement of $S_{e(all)}$. For plate no. 1, the total load can be expressed as

$$Q_1 = A_1 m + P_1 n \tag{4.65}$$

Similarly, for plate no. 2

$$Q_2 = A_2 m + P_2 n \tag{4.66}$$

where A_1, A_2 = areas of the plates no. 1 and no. 2, respectively
P_1, P_2 = perimeters of the plates no. 1 and no. 2, respectively
m, n = two constants that correspond to the bearing pressure and perimeter shear, respectively

The values of m and n can be determined by solving Eqs. (4.65) and (4.66).

4. For the foundation to be designed,

$$Q_o = Am + Pn \qquad (4.67)$$

where A = area of the foundation
        P = perimeter of the foundation

Because Q_o, m, and n are known, Eq. (4.67) can be solved to determine foundation width.

▼ EXAMPLE 4.7 _____

The results of a plate load test in a sandy soil are shown in Figure 4.33. The size of the plate is 0.305 m × 0.305 m. Determine the size of a square column foundation that should carry a load of 2500 kN with a maximum settlement of 25 mm.

Solution   The problem has to be solved by trial and error. Use the following table and Eq. (4.64):

Q_o (kN) (1)	Assume width B_F (m) (2)	$q_o = \dfrac{Q_o}{B_F^2}$ (kN/m²) (3)	S_P corresponding to q_o in col. 3 (mm) (4)	S_F from Eq. (4.64) (mm) (5)
2500	4.0	156.25	4.0	13.81
2500	3.0	277.80	8.0	26.37
2500	3.2	244.10	6.8	22.67

So, a column footing with dimensions of **3.2 m × 3.2 m** will be appropriate.

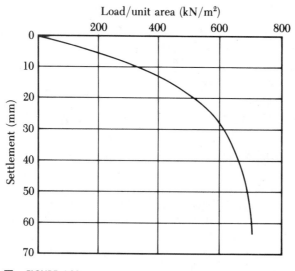

Load/unit area (kN/m²)

▼   FIGURE 4.33

▼ **EXAMPLE 4.8**

The results of two plate load tests are given in the following table:

Plate diameter, B (m)	Total load, Q (kN)	Settlement (mm)
0.305	32.2	20
0.610	71.8	20

A square column foundation has to be constructed to carry a total load of 715 kN. The tolerable settlement is 20 mm. Determine the size of the foundation.

Solution Use Eqs. (4.65) and (4.66):

$$32.2 = \frac{\pi}{4}(0.305)^2 m + \pi(0.305)n \qquad \text{(a)}$$

$$71.8 = \frac{\pi}{4}(0.610)^2 m + \pi(0.610)n \qquad \text{(b)}$$

From (a) and (b),

$$m = 50.68 \text{ kN/m}^2$$

$$n = 29.75 \text{ kN/m}$$

For the foundation to be designed [Eq. (4.67)],

$$Q_o = Am + Pn$$

or

$$Q_o = B_F^2 m + 4B_F n$$

For $Q_o = 715$ kN,

$$715 = B_F^2(50.68) + 4B_F(29.75)$$

or

$$50.68B_F^2 + 119B_F - 715 = 0$$

$$B_f \approx \mathbf{2.8\ m} \qquad \blacktriangle$$

▼ **EXAMPLE 4.9**

A shallow square foundation for a column is to be constructed. It must carry a net vertical load of 1000 kN. The foundation soil is sand. The standard penetration numbers obtained from field exploration are given in Figure 4.34. Assume that the depth of the foundation will be 1.5 m and the tolerable settlement is 25.4 mm. Determine the size of the foundation.

Solution   The field standard penetration numbers need to be corrected by using the Liao and Whitman relationship (Table 2.4). This is done in the following table:

Depth (m)	Field value of N_F	σ'_v (kN/m²)	Corrected N_{cor}^a
2	3	31.4	7
4	7	62.8	9
6	12	94.2	12
8	12	125.6	11
10	16	157.0	13
12	13	188.4	9
14	12	206.4	8
16	14	224.36	9
18	18	242.34	11

ª Rounded off

From the table, it appears that a corrected average N_{cor} value of about 10 would be appropriate. Using Eq. (4.53)

$$q_{net(all)} = 11.98 N_{cor}\left(\frac{3.28B + 1}{3.28B}\right)^2 F_d\left(\frac{S_e}{25.4}\right)$$

Allowable $S_e = 25.4$ mm and $N_{cor} = 10$, so

$$q_{net(all)} = 119.8\left(\frac{3.28B + 1}{3.28B}\right)^2 F_d$$

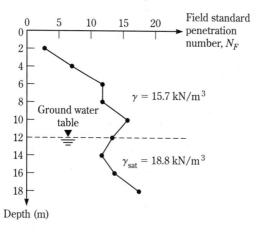

▼ FIGURE 4.34

The following table can now be prepared for trial calculations:

B (m)	$F_d^{\,a}$	$q_{net(all)}$ (kN/m^2)	$Q_o = q_{net(all)} \times B^2$ (kN)
2	1.248	197.24	788.96
2.25	1.22	187.19	947.65
2.3	1.215	185.46	981.1
2.4	1.206	182.29	1050.0
2.5	1.198	179.45	1121.56
[a] $D_f = 1.5$ m			

Because Q_o required is 1000 kN, B will be approximately equal to **2.4 m**. ▲

4.17 PRESUMPTIVE BEARING CAPACITY

Several building codes (for example, Uniform Building Code, Chicago Building Code, New York City Building Code) specify the allowable bearing capacity of foundations on various types of soil. For minor construction, they often provide fairly acceptable guidelines. However, these bearing capacity values are based primarily on the *visual* classification of near-surface soils. They generally do not take into consideration factors such as the stress history of the soil, water table location, depth of the foundation, and tolerable settlement. So, for large construction projects, the codes' presumptive values should be used only as guides.

4.18 TOLERABLE SETTLEMENT OF BUILDINGS

As has been emphasized in this chapter, settlement analysis is an important part of the design and construction of foundations. Large settlements of various components of a structure may lead to considerable damage and/or may interfere with the proper functioning of the structure. Limited studies have been made to evaluate the conditions for tolerable settlement of various types of structures (for example, Bjerrum, 1963; Burland and Worth, 1974; Grant et al. 1974; Polshin and Tokar, 1957; and Wahls, 1981). Wahls (1981) has provided an excellent review of these studies.

Figure 4.35 gives the parameters for defintion of tolerable settlement. Figure 4.35a is for a structure that has undergone settlement without tilt; Figure 4.35b is for a structure that has undergone settlement with tilt.

The parameters are

ρ_i = total vertical displacement at point i

δ_{ij} = differential settlement between points i and j

Δ = relative deflection

ω = tilt

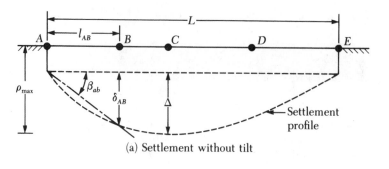

(a) Settlement without tilt

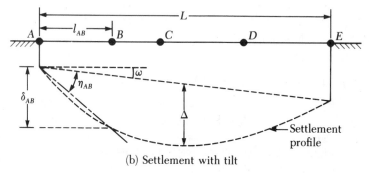

(b) Settlement with tilt

▼ **FIGURE 4.35**   Parameters for definition of tolerable settlement
(redrawn after Wahls, 1981)

$$\eta_{ij} = \frac{\delta_{ij}}{l_{ij}} - \omega = \text{angular distortion}$$

$$\frac{\Delta}{L} = \text{deflection ratio}$$

$$L = \text{lateral dimension of the structure}$$

Bjerrum (1963) provided the conditions of *limiting* angular distortion, η, for various
structures (see Table 4.6).

Polshin and Tokar (1957) presented the settlement criteria of the 1955 U.S.S.R.
Building Code. These criteria were based on experience gained from observations
of foundation settlement over 25 years. Tables 4.7 and 4.8 present the criteria.

FOUNDATION WITH SOIL REINFORCEMENT

4.19   SHALLOW FOUNDATION ON SOIL
WITH REINFORCEMENT

It was discussed in Chapter 3 that the ultimate bearing capacity of shallow foundations
can be improved by including tensile reinforcement such as metallic strips, geotextiles, and geogrids in the soil supporting the foundation. The procedure for designing

▼ **TABLE 4.6** Limiting Angular Distortion as Recommended by Bjerrum[a]

Category of potential damage	η
Danger to machinery sensitive to settlement	1/750
Danger to frames with diagonals	1/600
Safe limit for no cracking of buildings[b]	1/500
First cracking of panel walls	1/300
Difficulties with overhead cranes	1/300
Tilting of high rigid buildings becomes visible	1/250
Considerable cracking of panel and brick walls	1/150
Danger of structural damage to general buildings	1/150
Safe limit for flexible brick walls, $L/H > 4$[b]	1/150

[a] After Wahls (1981)
[b] Safe limits include a factor of safety. H = height of building

▼ **TABLE 4.7** Allowable Settlement Criteria: 1955 U.S.S.R. Building Code[a]

Type of structure	Sand and hard clay	Plastic clay
(a) η		
Civil- and industrial-building column foundations:		
For steel and reinforced concrete structures	0.002	0.002
For end rows of columns with brick cladding	0.007	0.001
For structures where auxiliary strain does not arise during nonuniform settlement of foundations	0.005	0.005
Tilt of smokestacks, towers, silos, and so on	0.004	0.004
Craneways	0.003	0.003
(b) Δ/L		
Plain brick walls: For multistory dwellings and civil buildings		
at $L/H \leq 3$	0.0003	0.0004
at $L/H \geq 5$	0.0005	0.0007
For one-story mills	0.0010	0.0010

[a] After Wahls (1981). H = height of building

shallow foundations for limiting settlement condition (that is, allowable bearing capacity) with layers of geogrid as reinforcement is still in the research and development stages. However, the problem of allowable bearing capacity of shallow foundations resting on granular soil reinforced with metallic strips was studied in detail by Binquet and Lee (1975a, b), who proposed the rational design method presented in the following sections.

▼ **TABLE 4.8** Allowable Average Settlement for Different Building Types[a]

Type of building	Allowable average settlement, in. (mm)
Building with plain brick walls	
$L/H \geq 2.5$	3 (80)
$L/H \leq 1.5$	4 (100)
Building with brick walls, reinforced with reinforced concrete or reinforced brick	6 (150)
Framed building	4 (100)
Solid reinforced concrete foundations of smokestacks, silos, towers, and so on	12 (300)

[a] After Wahls (1981). H = height of building

4.20 STRIP FOUNDATION ON GRANULAR SOIL REINFORCED BY METALLIC STRIPS

Mode of Failure

The nature of bearing capacity failure of a shallow strip foundation resting on a compact and homogeneous soil mass was presented in Figure 3.1a. In contrast, if layers of reinforcing strips, or *ties*, are placed in the soil under a shallow strip foundation, the nature of failure in the soil mass will be like that shown in Figure 4.36a, b, and c.

The type of failure in the soil mass shown in Figure 4.36a generally occurs when the first layer of reinforcement is placed at a depth, d, greater than about $\frac{2}{3}B$ (B = width of the foundation). If the reinforcements in the first layer are strong and they are sufficiently concentrated, they may act as a rigid base at a limited depth. The bearing capacity of foundations in such cases can be evaluated by the theory presented by Mandel and Salencon (1972). Experimental laboratory results for the bearing capacity of shallow foundations resting on a sand layer with a rigid rough base at a limited depth have also been provided by Meyerhof (1974), Pfeifle and Das (1979), and Das (1981).

The type of failure shown in Figure 4.36b could occur if d/B is less than about $\frac{2}{3}$ and the number of layers of reinforcement, N, is less than about 2–3. In this type of failure, reinforcement tie pullout occurs.

The most beneficial effect of reinforced earth is obtained when d/B is less than about $\frac{2}{3}$ and the number of reinforcement layers is greater than 4 but no more than 6–7. In this case, the soil mass fails when the upper ties break (see Figure 4.36c).

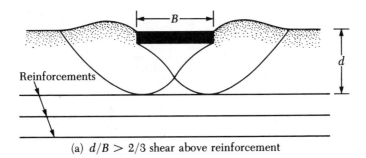

(a) $d/B > 2/3$ shear above reinforcement

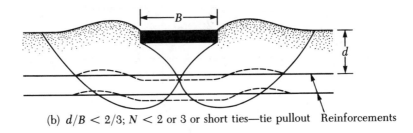

(b) $d/B < 2/3$; $N < 2$ or 3 or short ties—tie pullout   Reinforcements

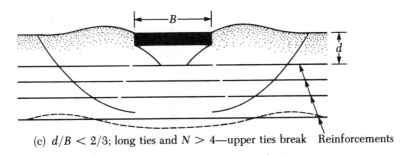

(c) $d/B < 2/3$; long ties and $N > 4$—upper ties break   Reinforcements

▼ **FIGURE 4.36**   Three modes of bearing capacity failure in reinforced earth (redrawn after Binquet and Lee, 1975b)

Location of Failure Surface

Figure 4.37 shows an idealized condition for development of the failure surface in soil for the condition shown in Figure 4.36c. It consists of a central zone — zone I — immediately below the foundation that settles along with the foundation with the application of load. On each side of zone I, the soil is pushed outward and upward — this is zone II. The points A', A'', A''', . . . , and B', B'', B''', . . . , which define the limiting lines between zones I and II, can be obtained by considering the shear stress distribution, τ_{xz}, in the soil caused by the foundation load. The term τ_{xz} refers to the shear stress developed at a depth z below the foundation at a distance x measured from the center line of the foundation. If integration of Boussinesq's

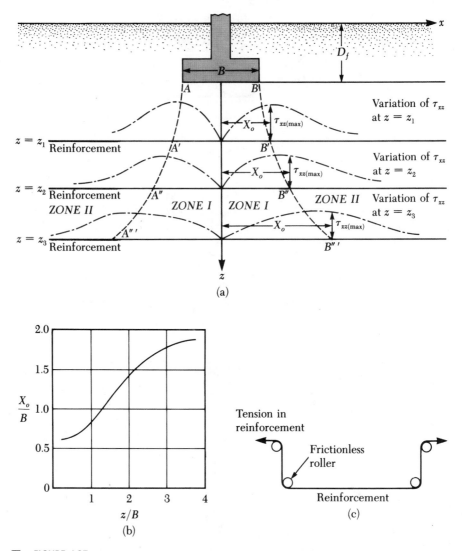

▼   **FIGURE 4.37**   Failure mechanism under a foundation supported by reinforced earth [part (b) after Binquet and Lee, 1975b]

equation is performed, τ_{xz} is given by the relation

$$\tau_{xz} = \frac{4bq_R x z^2}{\pi[(x^2 + z^2 - b^2)^2 + 4b^2 z^2]} \tag{4.68}$$

where b = half-width of the foundation = $B/2$
        B = width of foundation
        q_R = load per unit area on the foundation

The variation of τ_{xz} at any depth, z, is shown by the broken lines in Figure 4.37a. Points A' and B' refer to the points at which the value of τ_{xz} is maximum at $z = z_1$.

Similarly, A'' and B'' refer to the points at which τ_{xz} is maximum at $z = z_2$. The distances $x = X_o$ at which the maximum value of τ_{xz} occurs take a nondimensional form and are shown in Figure 4.37b.

Force Induced in Reinforcement Ties

Assumptions needed to obtain the tie force at any given depth are as follows:

1. Under the application of bearing pressure by the foundation, the reinforcing ties at points $A', A'', A''', \ldots$, and $B', B'', B''', \ldots$, take the shape shown in Figure 4.37c. That is, the ties take two right-angle turns on each side of zone I around two frictionless rollers.

2. For N reinforcing layers, the ratio of the load per unit area on the foundation supported by reinforced earth, q_R, to the load per unit area on the foundation supported by unreinforced earth, q_o, is constant irrespective of the settlement level, S (see Figure 4.38). Binquet and Lee (1975a) proved this relation in laboratory experiments.

Figure 4.39a shows a continuous foundation supported by unreinforced soil and subjected to a load of q_o per unit area. Similarly, Figure 4.39b shows a continuous foundation supported by a reinforced soil layer (one layer of reinforcement, or $N = 1$) and subjected to a load of q_R per unit area. (Due to symmetry only one-half of the foundation is shown in Figure 4.39). In both cases — that is, in Figures 4.39a and 4.39b — let the settlement equal S_e. For one-half of each foundation under consideration, the following are the forces per unit length on a soil element of thickness ΔH located at a depth z.

Unreinforced Case F_1 and F_2 are the vertical forces and S_1 is the shear force. Hence, for equilibrium,

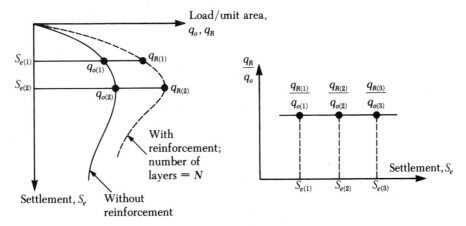

▼ **FIGURE 4.38**   Relationship between load per unit area and settlement for foundations resting on reinforced and unreinforced soil

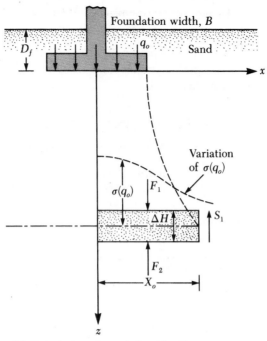

(a) Foundation of unreinforced soil

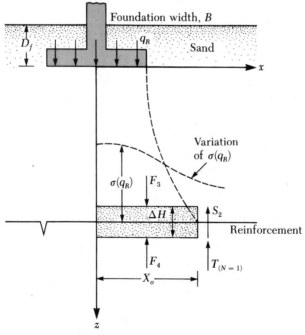

(b) Foundation of reinforced soil
(one layer reinforcement)

▼ **FIGURE 4.39** Derivation of Eq. (4.87)

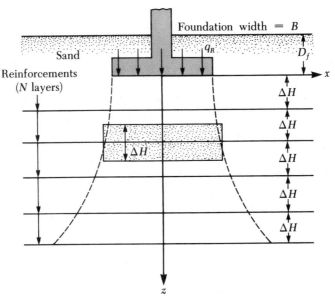

(c) Foundation on reinforced soil (N layers of reinforcement)

▼ **FIGURE 4.39** (Continued)

$$F_1 - F_2 - S_1 = 0 \tag{4.69}$$

Reinforced Case Here, F_3 and F_4 are the vertical forces, S_2 is the shear force, and $T_{(N=1)}$ is the tensile force developed in the reinforcement. The force $T_{(N=1)}$ is vertical because of the assumption made for the deformation of reinforcement as shown in Figure 4.37c. So

$$F_3 - F_4 - S_2 - T_{(N=1)} = 0 \tag{4.70}$$

If the foundation settlement, S_e, is the same in both cases,

$$F_2 = F_4 \tag{4.71}$$

Subtracting Eq. (4.69) from Eq. (4.70) and using the relationship given in Eq. (4.71), we obtain

$$T_{(N=1)} = F_3 - F_1 - S_2 + S_1 \tag{4.72}$$

Note that the force F_1 is caused by the vertical stress, σ, on the soil element under consideration as a result of the load q_o on the foundation. Similarly, F_3 is caused by the vertical stress imposed on the soil element as a result of the load q_R. Hence

$$F_1 = \int_0^{X_o} \sigma(q_o) \, dx \tag{4.73}$$

$$F_3 = \int_0^{X_o} \sigma(q_R) \, dx \tag{4.74}$$

$$S_1 = \tau_{xz}(q_o) \, \Delta H \tag{4.75}$$

$$S_2 = \tau_{xz}(q_R) \, \Delta H \tag{4.76}$$

where $\sigma(q_o)$ and $\sigma(q_R)$ are the vertical stresses at a depth z caused by the loads q_o and q_R on the foundation

   $\tau_{xz}(q_o)$ and $\tau_{xz}(q_R)$ are the shear stresses at a depth z and at a distance X_o from the center line caused by the loads q_o and q_R

Integrating Boussinesq's solution yields

$$\sigma(q_o) = \frac{q_o}{\pi} \left[\tan^{-1} \frac{z}{x-b} - \tan^{-1} \frac{z}{x+b} - \frac{2bz\,(x^2 - z^2 - b^2)}{(x^2 + z^2 - b^2)^2 + 4b^2 z^2} \right] \tag{4.77}$$

$$\sigma(q_R) = \frac{q_R}{\pi} \left[\tan^{-1} \frac{z}{x-b} - \tan^{-1} \frac{z}{x+b} - \frac{2bz\,(x^2 - z^2 - b^2)}{(x^2 + z^2 - b^2)^2 + 4b^2 z^2} \right] \tag{4.78}$$

$$\tau_{xz}(q_o) = \frac{4bq_o X_o z^2}{\pi [(X_o^2 + z^2 - b^2)^2 + 4b^2 z^2]} \tag{4.79}$$

$$\tau_{xz}(q_R) = \frac{4bq_R X_o z^2}{\pi [(X_o^2 + z^2 - b^2)^2 + 4b^2 z^2]} \tag{4.80}$$

where $b = B/2$

The procedure for derivation of Eqs. (4.77) to (4.80) is not presented here; for this information see a soil mechanics textbook (for example, Das, 1997). Proper substitution of Eqs. (4.77) to (4.80) into Eqs. (4.73) to (4.76) and simplification yields

$$F_1 = A_1 q_o B \tag{4.81}$$

$$F_3 = A_1 q_R B \tag{4.82}$$

$$S_1 = A_2 q_o \, \Delta H \tag{4.83}$$

$$S_2 = A_2 q_R \, \Delta H \tag{4.84}$$

where A_1 and $A_2 = f(z/B)$

The variations of A_1 and A_2 with nondimensional depth z are given in Figure 4.40. Substituting Eqs. (4.81)–(4.84) into Eq. (4.72) gives

$$\begin{aligned} T_{(N=1)} &= A_1 q_R B - A_1 q_o B - A_2 q_R \, \Delta H + A_2 q_o \, \Delta H \\ &= A_1 B(q_R - q_o) - A_2 \, \Delta H(q_R - q_o) \\ &= q_o \left(\frac{q_R}{q_o} - 1 \right)(A_1 B - A_2 \, \Delta H) \end{aligned} \tag{4.85}$$

Note that the derivation of Eq. (4.85) was based on the assumption that there is only one layer of reinforcement under the foundation shown in Figure 4.39b. However, if there are N layers of reinforcement under the foundation with center-to-center

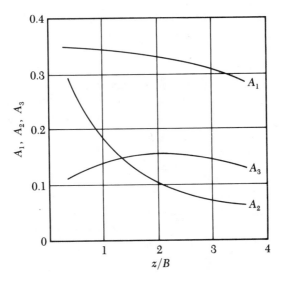

▼ **FIGURE 4.40** Variation of A_1, A_2, and A_3 with z/B (after Binquet and Lee, 1975b)

spacing of ΔH, as shown in Figure 4.39c, the assumption can be made that

$$T_{(N)} = \frac{T_{(N=1)}}{N} \tag{4.86}$$

Combining Eqs. (4.85) and (4.86) gives

$$T_{(N)} = \frac{1}{N}\left[q_o\left(\frac{q_R}{q_o} - 1\right)(A_1 B - A_2\,\Delta H) \right] \tag{4.87}$$

The unit of $T_{(N)}$ in Eq. (4.87) is lb/ft (or kN/m) per unit length of foundation.

4.21 FACTOR OF SAFETY OF TIES AGAINST BREAKING AND PULLOUT

Once the tie forces that develop in each layer as the result of the foundation load are determined from Eq. (4.87), an engineer must determine whether the ties at any depth z will fail either by *breaking* or by *pullout*. The factor of safety against tie breaking at any depth z below the foundation can be calculated as

$$FS_{(B)} = \frac{wtnf_y}{T_{(N)}} \tag{4.88}$$

where $FS_{(B)}$ = factor of safety against tie breaking
w = width of a single tie
t = thickness of each tie
n = number of ties per unit length of the foundation
f_y = yield or breaking strength of the tie material

The term wn may be defined as the *linear density ratio*, LDR, so

$$FS_{(B)} = \left[\frac{tf_y}{T_{(N)}}\right](LDR) \tag{4.89}$$

The resistance against the tie being pulled out derives from the frictional resistance between the soil and the ties at any depth. From the fundamental principles of statics, we know that the frictional force per unit length of the foundation resisting tie pullout at any depth z (Figure 4.41) is

$$F_B = 2\tan\phi_\mu[\text{normal force}] \tag{4.90}$$

$$= 2\tan\phi_\mu\left[\underbrace{(LDR)\int_{X_o}^{L_o}\sigma(q_R)\,dx}_{\substack{\text{Due to foundation}\\\text{load} = F_5}} + \underbrace{(LDR)(\gamma)(L_o - X_o)(z + D_f)}_{\substack{\text{Due to effective}\\\text{overburden}\\\text{pressure} = F_6}}\right]$$

↑
Two sides
of tie
(i.e., top
and bottom)

where γ = unit weight of soil
D_f = depth of foundation
ϕ_μ = tie–soil friction angle

The relation for $\sigma(q_R)$ was defined in Eq. (4.78). The value of $x = L_o$ is generally

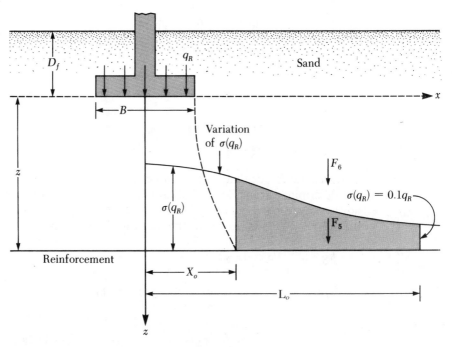

▼   **FIGURE 4.41**   Derivation of Eq. (4.91)

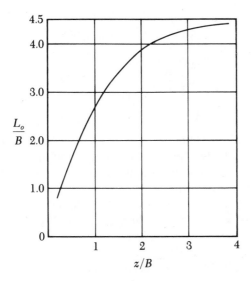

▼   **FIGURE 4.42**   Variation of L_o/B with z/B
(after Binquet and Lee, 1975b)

assumed to be the distance at which $\sigma(q_R)$ equals $0.1q_R$. The value of L_o as a function of depth z is given in Figure 4.42. Equation (4.90) may be simplified as

$$F_B = 2\tan\phi_\mu\,(LDR)\left[A_3 Bq_o\left(\frac{q_R}{q_o}\right) + \gamma(L_o - X_o)(z + D_f)\right] \tag{4.91}$$

where A_3 is a nondimensional quantity that may be expressed as a function of depth (z/B) (see Figure 4.40)

The factor of safety against tie pullout, $FS_{(P)}$, is

$$FS_{(P)} = \frac{F_B}{T_{(N)}} \tag{4.92}$$

4.22   DESIGN PROCEDURE FOR STRIP FOUNDATION ON REINFORCED EARTH

Following is a step-by-step procedure for the design of a strip foundation supported by granular soil reinforced by metallic strips:

1. Obtain the total load to be supported per unit length of the foundation. Also obtain the quantities
    a. Soil–friction angle, ϕ
    b. Soil–tie friction angle, ϕ_μ
    c. Factor for safety against bearing capacity failure
    d. Factor of safety against tie breaking, $FS_{(B)}$

    e. Factor of safety against tie pullout, $FS_{(P)}$
    f. Breaking strength of reinforcement ties, f_y
    g. Unit weight of soil, γ
    h. Modulus of elasticity of soil, E_s
    i. Poisson's ratio of soil, μ_s
    j. Allowable settlement of foundations, S_e
    k. Depth of foundation, D_f

2. Assume a width of foundation, B, and also d and N. The value of d should be less than $\frac{2}{3}B$. Also, the distance from the bottom of the foundation to the lowest layer of the reinforcement should be about $2B$ or less. Calculate ΔH.

3. Assume a value of LDR.

4. For width B (Step 2) determine the ultimate bearing capacity, q_u, for unreinforced soil [Eq. (3.3); *note:* $c = 0$]. Determine $q_{\text{all}(1)}$:

$$q_{\text{all}(1)} = \frac{q_u}{FS \text{ against bearing capacity failure}} \tag{4.93}$$

5. Calculate the allowable load, $q_{\text{all}(2)}$, based on the tolerable settlement, S_e, assuming that the soil is not reinforced [Eq. (4.32a)]:

$$S_e = \frac{Bq_{\text{all}(2)}}{E_s}(1 - \mu_s^2)\alpha_r$$

For $L/B = \infty$, the value of α_r may be taken as 2, or

$$q_{\text{all}(2)} = \frac{E_s S_e}{B(1 - \mu_s^2)\alpha_r} \tag{4.94}$$

(The allowable load for a given settlement, S_e, could have also been determined from equations that relate to standard penetration resistances.)

6. Determine the lower of the two values of q_{all} obtained from Steps 4 and 5. The lower value of q_{all} equals q_o.

7. Calculate the magnitude of q_R for the foundation supported by reinforced earth:

$$q_R = \frac{\text{load on foundation per unit length}}{B} \tag{4.95}$$

8. Calculate the tie force, $T_{(N)}$, in each layer of reinforcement by using Eq. (4.87) (*note:* unit of $T_{(N)}$ as kN/m of foundation).

9. Calculate the frictional resistance of ties for each layer per unit length of foundation, F_B, by using Eq. (4.91). For each layer, determine whether $F_B/T_{(N)} \geq FS_{(P)}$. If $F_B/T_{(N)} < FS_{(P)}$, the length of the reinforcing strips for a layer may be increased. That will increase the value of F_B and thus $FS_{(P)}$,

and so Eq. (4.91) must be rewritten as

$$F_B = 2 \tan \phi_\mu (LDR) \left[A_3 Bq_o \left(\frac{q_R}{q_o} \right) + \gamma (L - X_o)(z + D_f) \right] \qquad (4.96)$$

where L = the required length to obtain the desired value of F_B

10. Use Eq. (4.89) to obtain the tie thickness for each layer. Some allowance should be made for the corrosion effect of the reinforcements during the life of the structure.
11. If the design is unsatisfactory, repeat Steps 2–10.

The following example demonstrates the application of these steps.

▼ **EXAMPLE 4.10** _____

Design a strip foundation that will carry a load of 1.8 MN/m. Use the following parameters:

Soil: $\gamma = 17.3$ kN/m³; $\phi = 35°$; $E_s = 3 \times 10^4$ kN/m²; $\mu_s = 0.35$
Reinforcement ties: $f_y = 2.5 \times 10^5$ kN/m²; $\phi_\mu = 28°$; $FS_{(B)} = 3$; $FS_{(P)} = 2.5$
Foundation: $D_f = 1$ m; factor of safety against bearing capacity failure = 3, tolerable settlement = $S_e = 25$ mm; desired life of structure = 50 years

Solution   Let

$$B = 1 \text{ m}$$

d = depth from the bottom of the foundation to the first reinforcing layer = 0.5 m

$$\Delta H = 0.5 \text{ m}$$

$$N = 5$$

$$LDR = 65\%$$

If the reinforcing strips used are 75 mm wide, then

$$wn = LDR$$

or

$$n = \frac{LDR}{w} = \frac{0.65}{0.075 \text{ m}} = \textbf{8.67/m}$$

Hence each layer will contain 8.67 strips per meter length of the foundation.

Determination of q_o
For an unreinforced foundation

$$q_u = \gamma D_f N_q + \frac{1}{2} \gamma B N_\gamma$$

From Table 3.4 for $\phi = 35°$, $N_q = 33.30$ and $N_\gamma = 48.03$. Thus

$$q_u = (17.3)(1)(33.3) + \frac{1}{2}(17.3)(1)(48.03)$$

$$= 576.09 + 415.46 = 991.55 \approx 992 \text{ kN/m}^2$$

$$q_{all(1)} = \frac{q_u}{FS} = \frac{992}{3} = 330.7 \text{ kN/m}^2$$

From Eq. (4.94)

$$q_{all(2)} = \frac{(E_s)(S_e)}{B(1 - \mu_s^2)\alpha_r} = \frac{(30{,}000 \text{ kN/m}^2)(0.025 \text{ m})}{(1 \text{ m})(1 - 0.35^2)(2)} = 427.35 \text{ kN/m}^2$$

As $q_{all(1)} < q_{all(2)}$, $q_o = q_{all(1)} = \textbf{330.7 kN/m}^2$

Determination of q_R
From Eq. (4.95),

$$q_R = \frac{1.8 \text{ MN/m}}{B} = \frac{1.8 \times 10^3}{1} = \textbf{1.8} \times \textbf{10}^3 \textbf{ kN/m}^2$$

Calculation of Tie Force
From Eq. (4.87),

$$T_{(N)} = \left(\frac{q_o}{N}\right)\left(\frac{q_R}{q_o} - 1\right)(A_1 B - A_2 \Delta H)$$

The tie forces for each layer are given in the following table:

Layer no.	$\left(\dfrac{q_o}{N}\right)\left(\dfrac{q_R}{q_o} - 1\right)$	z (m)	$\dfrac{z}{B}$	$A_1 B$	$A_2 \Delta H$	$A_1 B - A_2 \Delta H$	$T_{(N)}$ (kN/m)
1	293.7	0.5	0.5	0.35	0.125	0.225	66.08
2	293.7	1.0	1.0	0.34	0.09	0.25	73.43
3	293.7	1.5	1.5	0.34	0.065	0.275	80.77
4	293.7	2.0	2.0	0.33	0.05	0.28	82.24
5	293.7	2.5	2.5	0.32	0.04	0.28	82.24

Note: A_1 is from Figure 4.40; $B = 1$ m; $\Delta H = 0.5$ m; A_2 is from Figure 4.40; $q_R/q_o = 1.8 \times 10^3/330.7 \approx 5.44$

Calculation of Tie Resistance Due to Friction, F_B
Use Eq. (4.91):

$$F_B = 2 \tan \phi_\mu (LDR) \left[A_3 B q_o \left(\frac{q_R}{q_o} \right) + \gamma (L_o - X_o)(z + D_f) \right]$$

The following table shows the magnitude of F_B for each layer:

	Layer number				
Quantity	1	2	3	4	5
$2 \tan \phi_\mu (LDR)$	0.691	0.691	0.691	0.691	0.691
A_3	0.125	0.14	0.15	0.15	0.15
$A_3 B q_o (q_R/q_o)$	225.0	252.0	270.0	270.0	270.0
z (m)	0.5	1.0	1.5	2.0	2.5
z/B	0.5	1.0	1.5	2.0	2.5
L_o (m)	1.55	2.6	3.4	3.85	4.2
X_o (m)	0.55	0.8	1.1	1.4	1.65
$L_o - X_o$ (m)	1.0	1.8	2.3	2.45	2.55
$z + D_f$ (m)	1.5	2.0	2.5	3.0	3.5
$\gamma (L_o - X_o)(z + D_f)$	25.95	62.28	99.48	127.16	154.4
F_B (kN/m)	173.4	217.2	255.1	274.4	293.3
$FS_{(P)} = F_B/T_{(N)}$	2.62	2.96	3.16	3.34	3.57

Note: A_3 is from Figure 4.40; X_o is from Figure 4.37; L_o is from Figure 4.42; $T_{(N)}$ is from the preceding table.

The minimum factor of safety is greater than the required value of $FS_{(P)}$, which is 2.5.

Calculation of Tie Thickness to Resist Tie Breaking
From Eq. (4.89),

$$FS_{(B)} = \frac{t f_y}{T_{(N)}} (LDR)$$

$$t = \frac{FS_{(B)} T_{(N)}}{(LDR)(f_y)}$$

Here, $f_y = 2.5 \times 10^5$ kN/m², $LDR = 0.65$, and $FS_{(B)} = 3$, so

$$t = \left[\frac{3}{(2.5 \times 10^5)(0.65)} \right] T_{(N)} = (1.846 \times 10^{-5}) T_{(N)}$$

So, for layer 1

$$t = (1.846 \times 10^{-5})(66.08) = 0.00122 \text{ m} = \textbf{1.22 mm}$$

For layer 2

$$t = (1.846 \times 10^{-5})(73.43) = 0.00136 \text{ m} = \textbf{1.36 mm}$$

Similarly, for layer 3

$$t = 0.00149 = \textbf{1.49 mm}$$

For layer 4

$$t = \textbf{1.52 mm}$$

For layer 5

$$t = \textbf{1.52 mm}$$

Thus in each layer ties with a thickness of 1.6 mm will be sufficient. However, if galvanized steel is used, the rate of corrosion is about 0.025 mm/yr, so t should be $1.6 + (0.025)(50) = \textbf{2.85 mm}$.

Calculation of Minimum Length of Ties

The minimum length of ties in each layer should equal $2L_o$. Following is the length of ties in each layer:

Layer no.	Minimum length of the tie, $2L_o$ (m)
1	3.1
2	5.2
3	6.8
4	7.7
5	8.4

Figure 4.43 is a diagram of the foundation with the ties. The design could be changed by varying B, d, N, and ΔH to determine the most economical combination.

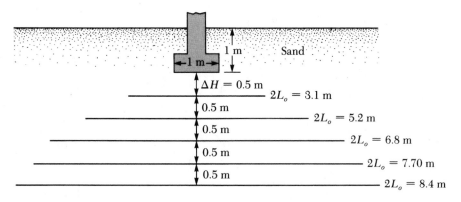

▼ **FIGURE 4.43**   ▲

▼ **EXAMPLE 4.11**

Refer to Example 4.10. For the loading given, determine the width of the foundation that is needed for unreinforced earth. Note that the factor of safety against bearing capacity failure is 3 and that the tolerable settlement is 25 mm.

Solution

Bearing Capacity Consideration
For a continuous foundation,

$$q_u = \gamma D_f N_q + \frac{1}{2} \gamma B N_\gamma$$

For $\phi = 35°$, $N_q = 33.3$ and $N_\gamma = 48.03$, so

$$q_{all} = \frac{q_u}{FS} = \frac{1}{FS}\left[\gamma D_f N_q + \frac{1}{2}\gamma B N_\gamma\right]$$

or

$$q_{all} = \frac{1}{3}\left[(17.3)(1)(33.3) + \frac{1}{2}(17.3)(B)(48.03)\right] \tag{a}$$

$$= 192.03 + 138.5B$$

However,

$$q_{all} = \frac{1.8 \times 10^3 \text{ kN}}{(B)(1)} \tag{b}$$

Equating the right-hand sides of Eqs. (a) and (b) yields

$$\frac{1800}{(B)(1)} = 192.03 + 138.5B$$

Solving the preceding equation gives $B \approx 3$ m, so, with $B = 3$ m, $q_{all} = $ **600 kN/m²**.

Settlement Consideration
For a friction angle of $\phi = 35°$, the corrected average standard penetration number is about 10–15 (Eq. 2.11). From Eq. (4.53) for the higher value, $N_{cor} = 15$,

$$q_{all} = 11.98 N_{cor} \left(\frac{3.28B + 1}{3.28}\right)^2 \left(1 + \frac{0.33D_f}{B}\right)$$

for a settlement of about 25 mm. Now, we can make a few trials:

Assumed B (m) (1)	$q_{all} = 11.98N_{cor}\left(\dfrac{3.28B+1}{3.28B}\right)^2\left(1+\dfrac{0.33D_f}{B}\right)$ (kN/m²) (2)	$Q = (B)(q_{all})$ = Col. 1 × Col. 2 (kN/m)
6	209	1254
9	199	1791[a]

Note: D_f = 1 m
[a] Required 1800 kN/m

For N_{cor} = 15, the width of the foundation should be 9 m or more. Based on the consideration of bearing capacity failure and tolerable settlement, the latter criteria will control, so B is about **9 m**.

Note: At first, the results of this calculation may show the use of reinforced earth for foundation construction to be desirable. However, several factors must be considered before a final decision is made. For example, reinforced earth needs overexcavation and backfilling. Hence, under many circumstances, proper material selection and compaction may make the construction of foundations on *unreinforced* soils more economical.   ▲

PROBLEMS

4.1 A flexible circular area is subjected to a uniformly distributed load of 3000 lb/ft². The diameter of the loaded area is 9.5 ft. Determine the stress increase in a soil mass at a point located 7.5 ft below the center of the loaded area.

4.2 Refer to Figure 4.5, which shows a flexible rectangular area. Given: B_1 = 1.2 m, B_2 = 3 m, L_1 = 3 m, and L_2 = 6 m. If the area is subjected to a uniform load of 110 kN/m², determine the stress increase at a depth of 8 m located immediately below point O.

4.3 Repeat Problem 4.2 with the following:

B_1 = 5 ft B_2 = 10 ft

L_1 = 7 ft L_2 = 12 ft

Uniform load on the flexible area = 2500 lb/ft²

Determine the stress increase below point O at a depth of 20 ft.

4.4 Using Eq. (4.10), determine the stress increase (Δp) from z = 0 to z = 5 m below the center of the area described in Problem 4.2.

4.5 Using Eq. (4.10), determine the stress increase (Δp) from z = 0 to z = 20 ft below the center of the area described in Problem 4.3.

4.6 Refer to Figure P4.6. Using the procedure outlined in Section 4.5, determine the average stress increase in the clay layer below the center of the foundation due to the net foundation load of 900 kN.

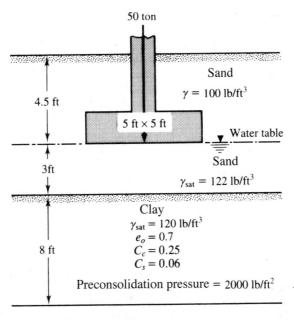

50 ton

Sand
$\gamma = 100 \ lb/ft^3$

4.5 ft

5 ft × 5 ft

▼ Water table

3ft

Sand
$\gamma_{sat} = 122 \ lb/ft^3$

Clay
$\gamma_{sat} = 120 \ lb/ft^3$
$e_o = 0.7$
$C_c = 0.25$
$C_s = 0.06$

8 ft

Preconsolidation pressure = 2000 lb/ft²

▼ **FIGURE P4.6**

4.7 Solve Problem 4.6 using the 2:1 method [Eq. (4.14) and Eq. (4.43)].

4.8 Figure P4.8 shows an embankment load on a silty clay soil layer. Determine the stress increase at points A, B, and C, which are located at a depth of 15 ft below the ground surface.

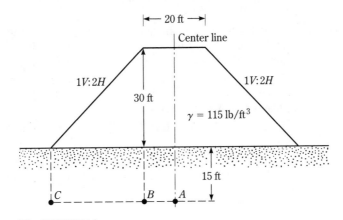

|← 20 ft →|

Center line

1V:2H

30 ft

1V:2H

$\gamma = 115 \ lb/ft^3$

15 ft

C   B A

▼ **FIGURE P4.8**

4.9 Solve Problem 4.2 using Newmark's chart.

4.10 Solve Problem 4.3 using Newmark's chart.

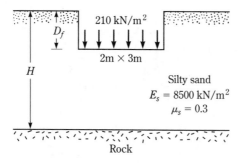

$210 \, kN/m^2$

D_f

$2m \times 3m$

H

Silty sand
$E_s = 8500 \, kN/m^2$
$\mu_s = 0.3$

Rock

▼ **FIGURE P4.11**

4.11 A flexible load area (Figure P4.11) is 2 m × 3 m in plan and carries a uniformly distributed load of 210 kN/m². Estimate the elastic settlement below the center of the loaded area. Assume $D_f = 0$ and $H = \infty$.

4.12 Redo Problem 4.11 assuming $D_f = 0$ and $H = 4$ m.

4.13 Refer to Figure 4.17. A foundation that is 10 ft × 6.5 ft in plan is resting on a sand deposit. The net load per unit area at the level of the foundation, q_o, is 3200 lb/ft². For the sand, $\mu_s = 0.3$, $E_s = 3200$ lb/in.², $D_f = 2.95$ ft, and $H = 32$ ft. Assume that the foundation is rigid and determine the elastic settlement that the foundation would undergo. Use Eq. (4.32a). Ignore the effect of depth of embedment.

4.14 Repeat Problem 4.13 for foundation criteria of size = 1.8 m × 1.8 m, $q_o = 190$ kN/m², $D_f = 1$ m, and $H = 15$ m; and soil conditions of $\mu_s = 0.35$, $E_s = 16{,}500$ kN/m², and $\gamma = 16.5$ kN/m³.

4.15 Refer to Figure 4.21. A foundation measuring 1.5 m × 3 m is supported by a saturated clay. Given: $D_f = 1.2$ m, $H = 3$ m, E_s (clay) = 600 kN/m², and $q_o = 150$ kN/m². Determine the elastic settlement of the foundation.

4.16 Solve Problem 4.13 with Eq. (4.35). For the correction factor, C_2, use a time of 5 yr for creep and, for the unit weight of soil, γ, use 110 lb/ft³. Assume an I_z plot the same as that for a square foundation.

4.17 Solve Problem 4.14 with Eq. (4.35). For the correction factor, C_2, use a time of 4 yr for creep.

4.18 A continuous foundation on a deposit of sand layer is shown in Figure P4.18 along with the variation of the modulus of elasticity of the soil (E_s). Assuming $\gamma = 115$ lb/ft³ and $C_2 = 10$ yr, calculate the elastic settlement of the foundation using the strain influence factor.

4.19 Estimate the consolidation settlement of the clay layer shown in Figure P4.6 using the results of Problem 4.6.

4.20 Estimate the consolidation settlement of the clay layer shown in Figure P4.6 using the results of Problem 4.7.

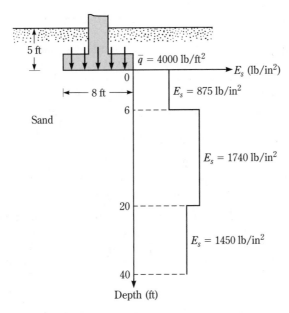

▼   **FIGURE P4.18**

4.21 Following are the results of standard penetration tests in a granular soil deposit:

Depth (ft)	Field standard penetration number, N_F
5	11
10	10
15	12
20	9
25	14

a.   Use Skempton's relationship given in Table 2.4 to obtain corrected standard penetration numbers. Use $\gamma = 115 \ lb/ft^3$.
b.   What will be the net allowable bearing capacity of a foundation 5 ft × 5 ft in plan? Given: $D_f = 3$ ft and allowable settlement = 1 in. Use the relationships presented in Section 4.15.

4.22 Two plate load tests with square plates were conducted in the field. At 1-in. settlement, the results were

Width of plate (in.)	Load (lb)
12	8,070
24	25,800

What size of square footing is required to carry a net load of 150,000 lb at a settlement of 1 in.?

4.23 Figure 4.39c shows a continuous foundation on reinforced soil. Here, $B = 0.9$ m; $D_f = 1$ m; number of layers of reinforcement, $N = 5$; $\Delta H = 0.4$ m. Make the necessary calculations and plot the lines on both sides of the foundation that define the point of maximum shear stress, $\tau_{xz(max)}$, on the reinforcements.

4.24 The tie forces under a continuous foundation are given by Eq. (4.87). For the foundation described in Problem 4.23, $q_o = 200$ kN/m² and $q_R/q_o = 4.5$. Determine the tie forces, $T_{(N)}$, in kN/m for each layer of reinforcement.

4.25 Repeat Problem 4.24 with $q_o = 300$ kN/m² and $q_R/q_o = 6$.

4.26 A continuous foundation (see Figure 4.39c) is to be built on reinforced earth to carry a load of 82.3 kip/ft. Use the following parameters:

Foundation: $B = 4$ ft, $D_f = 2.6$ ft, factor of safety against bearing capacity failure = 3, and tolerable settlement = 0.8 in.
Soil: $\gamma = 116$ lb/ft³, $\phi = 37°$, $E_s = 5200$ lb/in², and $\mu_s = 0.30$.
Reinforcement: $\Delta H = 1.3$ ft, $N = 5$, $LDR = 70\%$, and width of reinforcement strips = 0.23 ft

Calculate:
a. The number of reinforcement strips per foot length of the foundation
b. The allowable load per unit area of the foundation, q_o, without reinforcement
c. The ratio q_R/q_o
d. The tie forces for each layer of reinforcement under the foundation (kip/ft)

4.27 Refer to Problem 4.26. For the reinforcements, $f_y = 38,000$ lb/in², $\phi_\mu = 25°$, factor of safety against tie breaking = 2.5, and factor of safety against tie pullout = 2.5, calculate:
a. The minimum thickness of ties needed to resist tie breaking
b. The minimum length of ties necessary for each layer of reinforcement

REFERENCES

Ahlvin, R. G., and Ulery, H. H. (1962). "Tabulated Values of Determining the Composite Pattern of Stresses, Strains, and Deflections Beneath a Uniform Load on a Homogeneous Half Space," *Highway Research Board Bulletin 342,* pp. 1–13.

American Society for Testing and Materials (1982). *Annual Book of ASTM Standards,* Part 19, Philadelphia.

Binquet, J., and Lee, K. L. (1975a). "Bearing Capacity Tests on Reinforced Earth Mass," *Journal of the Geotechnical Engineering Division,* American Society of Civil Engineers, Vol. 101, No. GT12, PP. 1241–1255.

Binquet, J., and Lee, K. L. (1975b). "Bearing Capacity Analysis of Reinforced Earth Slabs," *Journal of the Geotechnical Engineering Division,* American Society of Civil Engineers, Vol. 101, No. GT12, pp. 1257–1276.

Bjerrum, L. (1963). "Allowable Settlement of Structures," *Proceedings,* European Conference on Soil Mechanics and Foundation Engineering, Wiesbaden, Germany, Vol. III, pp. 135–137.

Boussinesq, J. (1983). *Application des Potentials à L'Etude de L'Equilibre et du Mouvement des Solides Elastiques,* Gauthier–Villars, Paris.

Bowles, J. E. (1977). *Foundation Analysis and Design,* 2nd ed., McGraw-Hill, New York.

Burland, J. B., and Worth, C. P. (1974). "Allowable and Differential Settlement of Structures

Including Damage and Soil-Structure Interaction," *Proceedings,* Conference on Settlement of Structures, Cambridge University, England, pp. 611–654.

Christian, J. T., and Carrier, W. D. (1978). "Janbu, Bjerrum, and Kjaernsli's Chart Reinterpreted," *Canadian Geotechnical Journal,* Vol. 15, pp. 124–128.

D'Appolonia, D. J., D'Appolonia, E., and Brissettee, R. F. (1970). "Settlement of Spread Footings on Sand: Closure," *Journal of the Soil Mechanics and Foundation Engineering Division,* ASCE, Vol. 96, No. 2, pp. 754–762.

Das, B. M. (1981). "Bearing Capacity of Eccentrically Loaded Surface Footings on Sand," *Soils and Foundations,* Vol. 21, No. 1, pp. 115–119.

Das, B. (1997). *Advanced Soil Mechanics,* 2nd ed., Taylor and Francis, Washington, D.C.

Grant, R. J., Christian, J. T., and Vanmarcke, E. H. (1974). "Differential Settlement of Buildings," *Journal of the Geotechnical Engineering Division,* American Society of Civil Engineers, Vol. 100, No. GT9, pp. 973–991.

Griffiths, D. V. (1984). "A Chart for Estimating the Average Vertical Stress Increase in an Elastic Foundation Below a Uniformly Loaded Rectangular Area," *Canadian Geotechnical Journal,* Vol. 21, No. 4, 710–713.

Harr, M. E. (1966). *Fundamentals of Theoretical Soil Mechanics,* McGraw-Hill, New York.

Housel, W. S. (1929). "A Practical Method for the Selection of Foundations Based on Fundamental Research in Soil Mechanics," *Research Bulletin No. 13,* University of Michigan, Ann Arbor.

Janbu, N., Bjerrum, L., and Kjaernsli, B. (1956). "Veiledning ved losning av fundamentering—soppgaver," *Publication No. 16,* Norwegian Geotechnical Institute, pp. 30–32.

Liao, S. S. C., and Whitman, R. V. (1986). "Overburden Correction Factors for SPT in Sand," *Journal of Geotechnical Engineering,* American Society of Civil Engineers, Vol. 112, No. 3, pp. 373–377.

Mandel, J., and Salencon, J. (1972). "Force portante d'un sol sur une assise rigide (étude theorizué)," *Geotechnique,* Vol. 22, No. 1, pp. 79–93.

Meyerhof, G. G. (1956). "Penetration Tests and Bearing Capacity of Cohesionless Soils," *Journal of the Soil Mechanics and Foundations Division,* American Society of Civil Engineers, Vol. 82, No. SM1, pp. 1–19.

Meyerhof, G. G. (1965). "Shallow Foundations," *Journal of the Soil Mechanics and Foundations Division,* ASCE, Vol. 91, No. SM2, pp. 21–31.

Meyerhof, G. G. (1974). "Ultimate Bearing Capacity of Footings on Sand Layer Overlying Clay," *Canadian Geotechnical Journal,* Vol. 11, No. 2, pp. 223–229.

Mitchell, J. K., and Gardner, W. S. (1975). "*In Situ* Measurement of Volume Change Characteristics," *Proceedings,* Specialty Conference, American Society of Civil Engineers, Vol. 2, pp. 279–345.

Newmark, N. M. (1935). "Simplified Computation of Vertical Pressure in Elastic Foundation," *Circular 24,* University of Illinois Engineering Experiment Station, Urbana.

Newmark, N. M. (1942). "Influence Charts for Computation of Stresses in Elastic Foundations," *Bulletin No. 338,* University of Illinois Engineering Experiment Station, Urbana.

Osterberg, J. O. (1957). "Influence Values for Vertical Stresses in Semi-Infinite Mass Due to Embankment Loading," *Proceedings,* Fourth International Conference on Soil Mechanics and Foundation Engineering, London, Vol. 1, pp. 393–396.

Pfeifle, T. W., and Das, B. M. (1979). "Bearing Capacity of Surface Footings on Sand Layer Resting on a Rigid Rough Base," *Soils and Foundations,* Vol. 19, No. 1, pp. 1–11.

Polshin, D. E., and Tokar, R. A. (1957). "Maximum Allowable Nonuniform Settlement of Structures," *Proceedings,* Fourth International Conference on Soil Mechanics and Foundation Engineering, London, Vol. 1, pp. 402–405.

Schmertmann, J. H. (1970). "Static Cone to Compute Settlement Over Sand," *Journal of the*

Soil Mechanics and Foundations Division, American Society of Civil Engineers, Vol. 96, No. SM3, pp. 1011–1043.

Schmertmann, J. H. (1978). *Guidelines for Cone Penetrations: Performance and Design,* FHWA-TS-78-209, U.S. Department of Transportation, Washington, D.C.

Schmertmann, J. H., and Hartman, J. P. (1978). "Improved Strain Influence Factor Diagrams," *Journal of the Geotechnical Engineering Division,* American Society of Civil Engineers, Vol. 104, No. GT8, pp. 1131–1135.

Schnabel, J. J. (1972). "Foundation Construction on Compacted Structural Fill in the Washington, D.C., Area," *Proceedings,* Specialty Conference on Performance of Earth and Earth-Supported Structures, American Society of Civil Engineers, Vol. 1, Part 2, pp. 1019–1036.

Skempton, A. W., and Bjerrum, L. (1957). "A Contribution to Settlement Analysis of Foundations in Clay," *Geotechnique,* London, Vol. 7, p. 178.

Steinbrenner, W. (1934). "Tafeln zur Setzungsberechnung," *Die Strasse,* Vol. 1, pp. 121–124.

Terzaghi, K., and Peck, R. B. (1967). *Soil Mechanics in Engineering Practice,* 2nd ed., Wiley, New York.

Wahls, H. E. (1981). "Tolerable Settlement of Buildings," *Journal of the Geotechnical Engineering Division,* American Society of Civil Engineers, Vol. 107, No. GT11, pp. 1489–1504.

CHAPTER FIVE

MAT FOUNDATIONS

5.1 INTRODUCTION

Mat foundations are primarily shallow foundations. They are one of four major types of *combined footing* (see Figure 5.1a). A brief overview of combined footings and the methods used to calculate their dimensions follows:

1. *Rectangular Combined Footing:* In several instances, the load to be carried by a column and the soil bearing capacity are such that the standard spread footing design will require extension of the column foundation beyond the property line. In such a case, two or more columns can be supported on a single rectangular foundation, as shown in Figure 5.1b. If the net allowable soil pressure is known, the size of the foundation ($B \times L$) can be determined in the following manner.

    a. Determine the area of the foundation, A:

    $$A = \frac{Q_1 + Q_2}{q_{\text{all(net)}}} \tag{5.1}$$

    where Q_1, Q_2 = column loads
    $q_{\text{all(net)}}$ = net allowable soil bearing capacity

    b. Determine the location of the resultant of the column loads. From Figure 5.1b,

    $$X = \frac{Q_2 L_3}{Q_1 + Q_2} \tag{5.2}$$

    c. For uniform distribution of soil pressure under the foundation, the resultant of the column loads should pass through the centroid of the foundation. Thus

    $$L = 2(L_2 + X) \tag{5.3}$$

    where L = length of the foundation

    d. Once the length L is determined, the value of L_1 can be obtained:

    $$L_1 = L - L_2 - L_3 \tag{5.4}$$

    Note that the magnitude of L_2 will be known and depends on the location of the property line.

    e. The width of the foundation then is

    $$B = \frac{A}{L} \tag{5.5}$$

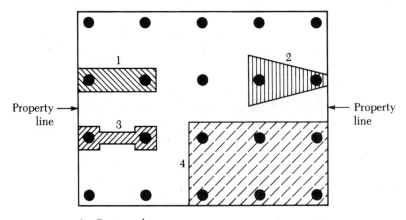

1 Rectangular
  combined footing
2 Trapezoidal
  combined footing
3 Cantilever footing
4 Mat foundation

(a)

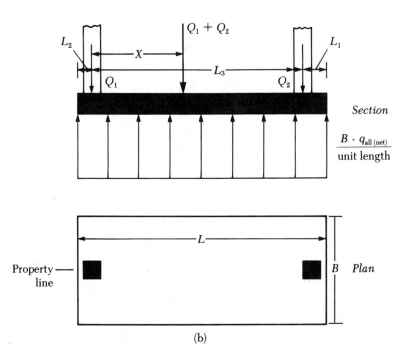

(b)

▼ **FIGURE 5.1** (a) Combined footing; (b) rectangular combined footing

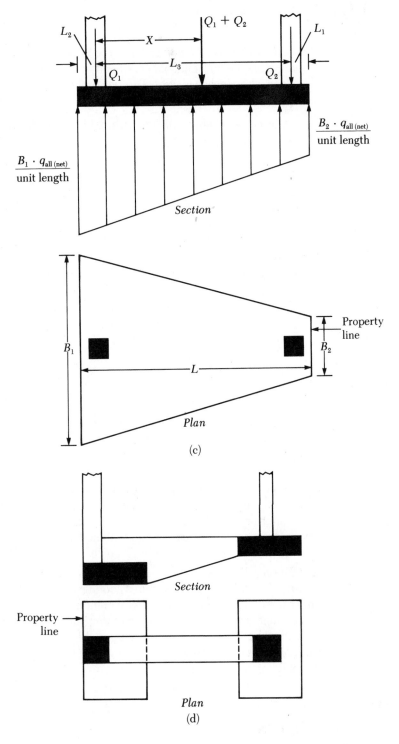

▼ **FIGURE 5.1** (Continued) (c) Trapezoidal combined footing; (d) cantilever footing

2. *Trapezoidal Combined Footing:* This type of combined footing (Figure 5.1c) is sometimes used as an isolated spread foundation of a column carrying a large load where space is tight. The size of the foundation that will uniformly distribute pressure on the soil can be obtained in the following manner.

   a. If the net allowable soil pressure is known, determine the area of the foundation:

   $$A = \frac{Q_1 + Q_2}{q_{all(net)}}$$

   From Figure 5.1c,

   $$A = \frac{B_1 + B_2}{2} L \tag{5.6}$$

   b. Determine the location of the resultant for the column loads:

   $$X = \frac{Q_2 L_3}{Q_1 + Q_2}$$

   c. From the property of a trapezoid,

   $$X + L_2 = \left(\frac{B_1 + 2B_2}{B_1 + B_2}\right)\frac{L}{3} \tag{5.7}$$

   With known values of A, L, X, and L_2, solve Eqs. (5.6) and (5.7) to obtain B_1 and B_2. Note that for a trapezoid

   $$\frac{L}{3} < X + L_2 < \frac{L}{2}$$

3. *Cantilever Footing:* This type of combined footing construction uses a *strap beam* to connect an eccentrically loaded column foundation to the foundation of an interior column (Figure 5.1d). Cantilever footings may be used in place of trapezoidal or rectangular combined footings when the allowable soil bearing capacity is high and the distances between the columns are large.

4. *Mat Foundation:* This type of foundation, which is sometimes referred to as a *raft foundation,* is a combined footing that may cover the entire area under a structure supporting several columns and walls (Figure 5.1a). Mat foundations are sometimes preferred for soils that have low load-bearing capacities but that will have to support high column and/or wall loads. Under some conditions, spread footings would have to cover more than half the building area, and mat foundations might be more economical.

5.2 COMMON TYPES OF MAT FOUNDATIONS

Several types of mat foundations are used currently. Some of the common types are shown schematically in Figure 5.2 and include:

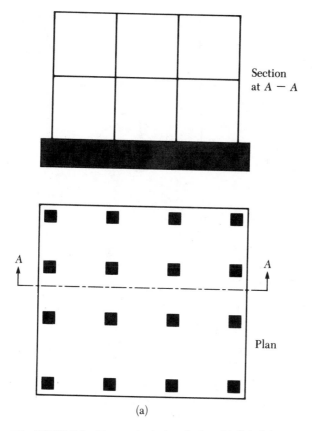

Section
at $A - A$

A　　　　　　　　　　　　　　A

Plan

(a)

▼ **FIGURE 5.2** Types of mat foundation: (a) flat plate

1. Flat plate (Figure 5.2a). The mat is of uniform thickness.
2. Flat plate thickened under columns (Figure 5.2b, p. 298).
3. Beams and slab (Figure 5.2c, p. 299). The beams run both ways, and the columns are located at the intersection of the beams.
4. Slab with basement walls as a part of the mat (Figure 5.2d, p. 300). The walls act as stiffeners for the mat.

Mats may be supported by piles. The piles help in reducing the settlement of a structure built over highly compressible soil. Where the water table is high, mats are often placed over piles to control buoyancy.

5.3 BEARING CAPACITY OF MAT FOUNDATIONS

The *gross ultimate bearing capacity* of a mat foundation can be determined by the same equation used for shallow foundations (see Section 3.7), or

$$q_u = cN_cF_{cs}F_{cd}F_{ci} + qN_qF_{qs}F_{qd}F_{qi} + \tfrac{1}{2}\gamma BN_\gamma F_{\gamma s}F_{\gamma d}F_{\gamma i} \tag{3.25}$$

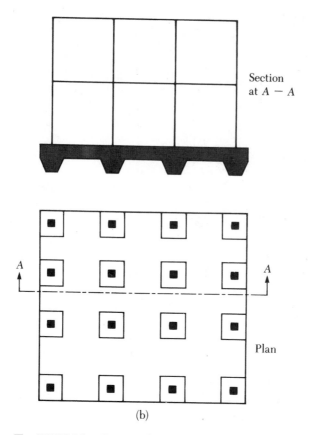

Section
at $A - A$

Plan

(b)

▼ **FIGURE 5.2**   (Continued) (b) Flat plate thickened under column

(Chapter 3 gives the proper values of the bearing capacity factors, and the shape, depth, and load inclination factors.) The term B in Eq. (3.25) is the smallest dimension of the mat. The *net ultimate capacity* is

$$q_{net(u)} = q_u - q \tag{3.19}$$

A suitable factor of safety should be used to calculate the net *allowable* bearing capacity. For rafts on clay, the factor of safety should not be less than 3 under dead load and maximum live load. However, under the most extreme conditions, the factor of safety should be at least 1.75 to 2. For rafts constructed over sand, a factor of safety of 3 should normally be used. Under most working conditions, the factor of safety against bearing capacity failure of rafts on sand is very large.

For saturated clays with $\phi = 0$ and vertical loading condition, Eq. (3.25) gives

$$q_u = c_u N_c F_{cs} F_{cd} + q \tag{5.8}$$

where c_u = undrained cohesion

(*Note:* $N_c = 5.14$, $N_q = 1$, and $N_\gamma = 0$.)

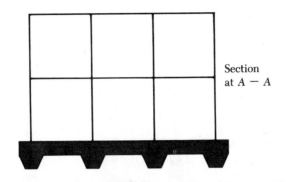

Section at A − A

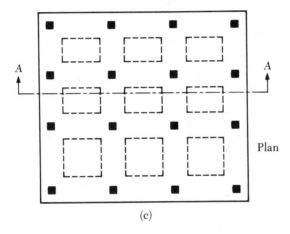

Plan

(c)

▼ **FIGURE 5.2** (Continued) (c) Beams and slab

From Table 3.5, for $\phi = 0$,

$$F_{cs} = 1 + \frac{B}{L}\left(\frac{N_q}{N_c}\right) = 1 + \left(\frac{B}{L}\right)\left(\frac{1}{5.14}\right) = 1 + \frac{0.195B}{L}$$

and

$$F_{cd} = 1 + 0.4\left(\frac{D_f}{B}\right)$$

Substitution of the preceding shape and depth factors into Eq. (5.8) yields

$$q_u = 5.14c_u\left(1 + \frac{0.195B}{L}\right)\left(1 + 0.4\frac{D_f}{B}\right) + q \qquad (5.9)$$

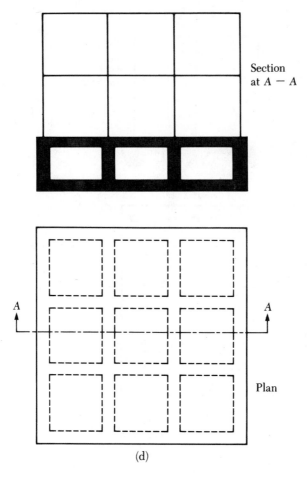

Section at $A - A$

Plan

(d)

▼ **FIGURE 5.2**   (Continued) (d) Slab with basement wall

Hence the net ultimate bearing capacity is

$$q_{\text{net}(u)} = q_u - q = 5.14c_u\left(1 + \frac{0.195B}{L}\right)\left(1 + 0.4\frac{D_f}{B}\right) \tag{5.10}$$

For $FS = 3$, the net allowable soil bearing capacity becomes

$$q_{\text{all(net)}} = \frac{q_{u(\text{net})}}{FS} = 1.713c_u\left(1 + \frac{0.195B}{L}\right)\left(1 + 0.4\frac{D_f}{B}\right) \tag{5.11}$$

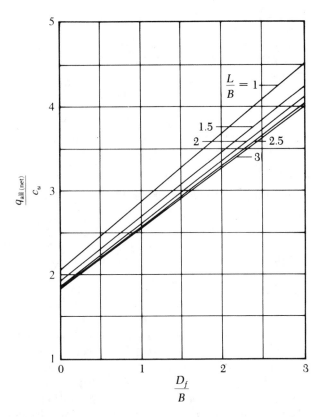

▼ **FIGURE 5.3**   Plot of $q_{all(net)}/c_u$ against D_f/B [Eq. (5.11)]
(*note:* factor of safety = 3)

Figure 5.3 shows a plot of $q_{all(net)}/c_u$ for various values of L/B and D_f/B, based on Eq. (5.11).

The net allowable bearing capacity for mats constructed over granular soil deposits can be adequately determined from the standard penetration resistance numbers. From Eq. (4.53), for shallow foundations,

$$q_{all(net)} \, (kN/m^2) = 11.98 N_{cor} \left(\frac{3.28B + 1}{3.28B} \right)^2 F_d \left(\frac{S_e}{25.4} \right)$$

where N_{cor} = corrected standard penetration resistance
B = width (m)
$F_d = 1 + 0.33(D_f/B) \le 1.33$
S_e = settlement, in mm

When the width, B, is large, the preceding equation can be approximated (assuming $3.28B + 1 \approx 3.28B$) as

$$q_{all(net)}(kN/m^2) \approx 11.98 N_{cor} F_d \left(\frac{S_e}{25.4}\right)$$
$$= 11.98 N_{cor} \left[1 + 0.33\left(\frac{D_f}{B}\right)\right]\left[\frac{S_e(mm)}{25.4}\right]$$
$$\leq 15.93 N_{cor}\left[\frac{S_e(mm)}{25.4}\right]$$

(5.12)

In English units, Eq. (5.12) may be expressed as

$$q_{all(net)}(kip/ft^2) = 0.25 N_{cor}\left[1 + 0.33\left(\frac{D_f}{B}\right)\right][S_e(in.)]$$
$$\leq 0.33 N_{cor}[S_e(in.)]$$

(5.13)

Note that Eq. (5.13) could have been derived from Eqs. (4.54) and (4.56).

Note that the original Eqs. (4.53) and (4.56) were for a settlement of 1 in. (25.4 mm) with a differential settlement of about 0.75 in. (19 mm). However, the widths of the raft foundations are larger than the isolated spread footings. As Table 4.3 shows, the depth of significant stress increase in the soil below a foundation depends on the foundation width. Hence, for a raft foundation, the depth of the zone of influence is likely to be much larger than that of a spread footing. Thus the loose soil pockets under a raft may be more evenly distributed, resulting in a smaller differential settlement. Hence the customary assumption is that, for a maximum raft settlement of 2 in. (50.8 mm), the differential settlement would be 0.75 in. (19 mm). Using this logic and conservatively assuming that F_d equals 1, we can approximate Eqs. (5.12) and (5.13) as

$$q_{all(net)}(kN/m^2) \approx 23.96 N_{cor}$$

(5.14)

and

$$q_{all(net)}(kip/ft^2) = 0.5 N_{cor}$$

(5.15)

The net pressure applied on a foundation (Figure 5.4) may be expressed as

$$q = \frac{Q}{A} - \gamma D_f$$

(5.16)

where Q = dead weight of the structure and the live load
         A = area of the raft

In all cases, q should be less than or equal to $q_{all(net)}$.

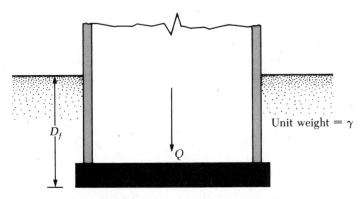

▼ **FIGURE 5.4** Definition of net pressure on soil caused by a mat foundation

▼ **EXAMPLE 5.1**

Determine the net ultimate bearing capacity of a mat foundation measuring 45 ft × 30 ft on a saturated clay with c_u = 1950 lb/ft², ϕ = 0, and D_f = 6.5 ft.

Solution From Eq. (5.10)

$$q_{net(u)} = 5.14c_u \left[1 + \left(\frac{0.195B}{L} \right) \right] \left[1 + 0.4 \frac{D_f}{B} \right]$$

$$= (5.14)(1950) \left[1 + \left(\frac{0.195 \times 30}{45} \right) \right] \left[1 + \left(\frac{0.4 \times 6.5}{30} \right) \right]$$

$$= \mathbf{12{,}307 \ lb/ft^2}$$ ▲

▼ **EXAMPLE 5.2**

What will be the net allowable bearing capacity of a mat foundation with dimensions of 45 ft × 30 ft constructed over a sand deposit? Here, D_f = 6 ft, allowable settlement = 1 in., and corrected average penetration number N_{cor} = 10.

Solution From Eq. (5.13)

$$q_{all(net)} = 0.25N_{cor} \left(1 + \frac{0.33D_f}{B} \right) S_e \le 0.33N_{cor}S_e$$

$$q_{all(net)} = 0.25(10) \left[1 + \frac{0.33(6)}{30} \right] (1) \approx \mathbf{2.67 \ kip/ft^2}$$ ▲

5.4 DIFFERENTIAL SETTLEMENT OF MATS

The American Concrete Institute Committee 336 (1988) suggested the following method for calculating the differential settlement of mat foundations. According to this method, the rigidity factor (K_r) is calculated as

$$K_r = \frac{E'I_b}{E_s B^3} \tag{5.17}$$

where E' = modulus of elasticity of the material used in the structure
E_s = modulus of elasticity of the soil
B = width of foundation
I_b = moment of inertia of the structure per unit length at right angles to B

The term $E'I_b$ can be expressed as

$$E'I_b = E'\left(I_F + \sum I_{b'} + \sum \frac{ah^3}{12}\right) \tag{5.18}$$

where $E'I_b$ = flexural rigidity of the superstructure and foundation per unit
length at right angles to B
$\sum E'I_b'$ = flexural rigidity of the framed members at right angles to B
$\sum (E'ah^3/12)$ = flexural rigidity of the shear walls
a = shear wall thickness
h = shear wall height
$E'I_F$ = flexibility of the foundation

Based on the value of K_r, the ratio (δ) of the differential settlement to the total settlement can be estimated in the following manner:

1. If $K_r > 0.5$, it can be treated as a rigid mat, and $\delta = 0$.
2. If $K_r = 0.5$, then $\delta \approx 0.1$.
3. If $K_r = 0$, then $\delta = 0.35$ for square mats ($B/L = 1$) and $\delta = 0.5$ for long foundations ($B/L = 0$).

5.5 FIELD SETTLEMENT OBSERVATIONS FOR MAT FOUNDATIONS

Several field settlement observations for mat foundations are currently available in the literature. In this section we compare the observed settlements for some mat foundations constructed over granular soil deposits with those obtained from Eqs. (5.12) and (5.13).

Meyerhof (1965) compiled the observed maximum settlements for mat foundations constructed on sand and gravel, as listed in Table 5.1. In Eq. (5.13), if the depth factor, $1 + 0.33(D_f/B)$, is assumed to be approximately 1,

$$S_e = \frac{q_{all(net)}}{0.25N_{cor}} \tag{5.19}$$

Table 5.2 shows a comparison of the observed maximum settlements in Table 5.1 and the settlements obtained from Eq. (5.19). For the cases considered, the ratio of $S_{e_{calculated}}/S_{e_{observed}}$ varies from 0.84 to 3.6. Thus calculation of the net allowable bearing capacity with Eq. (5.12) or (5.13) will yield a safe and conservative value.

Stuart and Graham (1975) reported the case history of the 13-story Ashby Institute building of Queens University, Belfast, Ireland, construction of which began

▼ **TABLE 5.1** Observed Maximum Settlement of Mat Foundations on Sand and Gravel[a]

Case no.	Structure	Reference	B (ft)	N_{cor} (avg)	$q_{all(net)}$ (kip/ft^2)	Observed maximum settlement, S_e (in.)
1	T. Edison Sao Paulo, Brazil	Rios and Silva (1948)	60	15	4.8	0.6
2	Banco do Brasil Sao Paulo, Brazil	Rios and Silva (1948); Vargas (1961)	75	18	5.0	1.1
3	Iparanga Sao Paulo, Brazil	Vargas (1948)	30	9	6.4	1.4
4	C.B.I., Esplanada Sao Paulo, Brazil	Vargas (1961)	48	22	8.0	1.1
5	Riscala Sao Paulo, Brazil	Vargas (1948)	13	20	4.8	0.5
6	Thyssen Dusseldorf, Germany	Schultze (1962)	74	25	5.0	0.95
7	Ministry Dusseldorf, Germany	Schultze (1962)	52	20	4.6	0.8
8	Chimney Cologne, Germany	Schultze (1962)	67	10	3.6	0.4

[a] After Meyerhof (1965)

in August 1960. It was supported by a mat foundation 180 ft (length) × 65 ft (width). Figure 5.5a shows a schematic diagram of the building cross section. The nature of the subsoil along with the field standard penetration resistance values at the south end of the building are shown in Figure 5.5b. The base of the mat was constructed about 20 ft below the ground surface.

The variation of the corrected standard penetration number with depth is shown in Table 5.3. Note that the average N_{cor} value between the bottom of the mat and a

▼ **TABLE 5.2** Comparison of Settlements Observed and Calculated

Case 1[a]	Maximum observed settlement, S_e(in.)	Calculated settlement, S_e [Eq. (5.19)]	$\dfrac{S_{e_{calculated}}}{S_{e_{observed}}}$
1	0.6	1.28	2.1
2	1.1	1.11	1.0
3	1.4	2.84	2.03
4	1.1	1.45	1.32
5	0.5	0.96	1.92
6	0.95	0.8	0.84
7	0.8	0.92	1.15
8	0.4	1.44	3.6

[a] Refer to Table 5.1

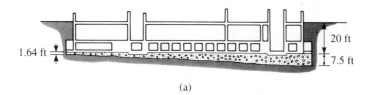

(a)

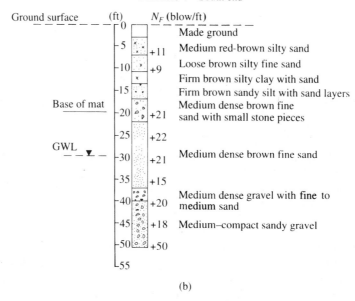

(b)

▼ **FIGURE 5.5**   Ashby Institute Building of Queens University, as reported by Stuart and Graham (1975): (a) building cross section; (b) subsoil conditions at south end

▼ **TABLE 5.3**   Determination of Corrected Standard Penetration Resistance

Depth below ground surface (ft)	Field standard penetration number, N_F	$\sigma_v'^{a}$ (ton/ft²)	$C_N = \sqrt{\dfrac{1}{\sigma_v'}}^{(b)}$	N_{cor} [Eq. (2.7)]
20	21	1.2	0.91	19
25	22	1.5	0.82	18
30	21	1.8	0.75	16
35	15	2.1	0.69	10
40	20	2.4	0.65	13
45	18	2.7	0.61	11
50	50	3.0	0.58	29
$^a \sigma_v' = $ (depth) (γ); $\gamma = 120$ lb/ft³ (assumed)				
^{b} Table 2.4				

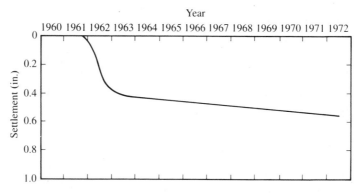

▼ **FIGURE 5.6** Mean settlement at the south end of the mat foundation, as reported by Stuart and Graham (1975)

depth of 30 ft ($\approx B/2$) is about 17. The engineers estimated the average net *dead
and live load* [Eq. (5.16)] at the level of the mat foundation to be about 3360 lb/ft^2.
From Eq. (5.13)

$$S_e = \frac{q_{\text{all(net)}}}{0.25N_{\text{cor}}\left[1 + 0.33\left(\dfrac{D_f}{B}\right)\right]} \tag{5.20}$$

Substituting appropriate values into Eq. (5.20) yields the settlement at the south
end of the building:

$$S_e = \frac{(3360/1000)}{(0.25)(17)[1 + 0.33(20/65)]} = 0.72 \text{ in.}$$

The construction of the building was completed in February 1964. Figure 5.6
shows the variation of the mean settlement of the mat at the south end. In 1972
(eight years after completion of the building) the mean settlement was about 0.55
in. Thus the estimated settlement of 0.72 in. is about 30% higher than that actually observed.

5.6 COMPENSATED FOUNDATIONS

The settlement of a mat foundation can be reduced by decreasing the net pressure
increase on soil, which can be done by increasing the depth of embedment, D_f. This
increase is particularly important for mats on soft clays, where large consolidation
settlements are expected. From Eq. (5.16), the net average applied pressure on soil is

$$q = \frac{Q}{A} - \gamma D_f$$

For no increase of the net soil pressure on soil below a raft foundation, q should
be zero. Thus

$$D_f = \frac{Q}{A\gamma} \tag{5.21}$$

This relation for D_f is usually referred to as the depth of a *fully compensated foundation*.

The factor of safety against bearing capacity failure for partially compensated foundations (that is, $D_f < Q/A\gamma$) may be given as

$$FS = \frac{q_{\text{net}(u)}}{q} = \frac{q_{\text{net}(u)}}{\dfrac{Q}{A} - \gamma D_f} \tag{5.22}$$

For saturated clays, the factor of safety against bearing capacity failure can thus be obtained by substituting Eq. (5.10) into Eq. (5.22):

$$FS = \frac{5.14c_u\left(1 + \dfrac{0.195B}{L}\right)\left(1 + 0.4\dfrac{D_f}{B}\right)}{\dfrac{Q}{A} - \gamma D_f} \tag{5.23}$$

▼ **EXAMPLE 5.3**

Refer to Figure 5.4. The mat has dimensions of 60 ft × 100 ft. The total dead and live load on the mat is 25×10^3 kip. The mat is placed over a saturated clay having a unit weight of 120 lb/ft³ and $c_u = 2800$ lb/ft². Given $D_f = 5$ ft, determine the factor of safety against bearing capacity failure.

Solution   From Eq. (5.23), the factor of safety

$$FS = \frac{5.14c_u\left(1 + \dfrac{0.195B}{L}\right)\left(1 + 0.4\dfrac{D_f}{B}\right)}{\dfrac{Q}{A} - \gamma D_f}$$

Given: $c_u = 2800$ lb/ft², $D_f = 5$ ft, $B = 60$ ft, $L = 100$ ft, and $\gamma = 120$ lb/ft³. Hence

$$FS = \frac{(5.14)(2800)\left[1 + \dfrac{(0.195)(60)}{100}\right]\left[1 + 0.4\left(\dfrac{5}{60}\right)\right]}{\left(\dfrac{25 \times 10^6\ \text{lb}}{60 \times 100}\right) - (120)(5)} = \mathbf{4.66}$$

▲

▼ **EXAMPLE 5.4**_____

Consider a mat foundation 90 ft × 120 ft in plan, as shown in Figure 5.7. The total dead load and live load on the raft is 45 × 10³ kip. Estimate the consolidation settlement at the center of the foundation.

Solution From Eq. (1.64)

$$S_c = \frac{C_c H_c}{1 + e_o} \log \left(\frac{p_o + \Delta p_{av}}{p_o} \right)$$

$$p_o = (11)(100) + (40)(121.5 - 62.4) + \frac{18}{2}(118 - 62.4) \approx 3964 \text{ lb/ft}^2$$

$$H_c = 18 \times 12 \text{ in.}$$

$$C_c = 0.28$$

$$e_o = 0.9$$

For $Q = 45 \times 10^6$ lb, the net load per unit area

$$q = \frac{Q}{A} - \gamma D_f = \frac{45 \times 10^6}{90 \times 120} - (100)(6) \approx 3567 \text{ lb/ft}^2$$

In order to calculate Δp_{av}, we refer to Section 4.5. The loaded area can be divided into four areas, each measuring 45 ft × 60 ft. Now using Eq. (4.19), we can calculate the average stress increase in the clay layer below the corner of each rectangular area, or

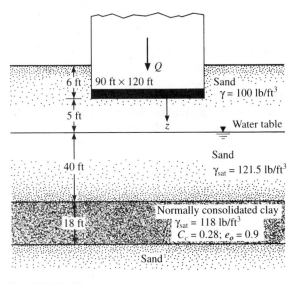

▼ **FIGURE 5.7**

$$\Delta p_{av(H_2/H_1)} = q \left[\frac{H_2 I_{a(H_2)} - H_1 I_{a(H_1)}}{H_2 - H_1} \right]$$

$$= 3567 \left[\frac{(5 + 40 + 18) I_{a(H_2)} - (5 + 40) I_{a(H_1)}}{18} \right]$$

For $I_{a(H_2)}$,

$$m = \frac{B}{H_2} = \frac{45}{5 + 40 + 18} = 0.71$$

$$n = \frac{L}{H_2} = \frac{60}{63} = 0.95$$

From Fig. 4.8, for $m = 0.71$ and $n = 0.95$, the value of $I_{a(H_2)}$ is 0.21. Again, for $I_{a(H_1)}$,

$$m = \frac{B}{H_1} = \frac{45}{45} = 1$$

$$n = \frac{L}{H_1} = \frac{60}{45} = 1.33$$

From Figure 4.8, $I_{a(H_1)} = 0.225$, so

$$\Delta p_{av(H_2/H_1)} = 3567 \left[\frac{(63)(0.21) - (45)(0.225)}{18} \right] = 615.3 \text{ lb/ft}^2$$

So, the stress increase below the center of the 90 ft $\times$ 120 ft area is $(4)(615.3) = 2461.2$ lb/ft^2. Thus

$$S_c = \frac{(0.28)(18 \times 12)}{1 + 0.9} \log \left(\frac{3964 + 2461.2}{3964} \right) = \textbf{6.68 in.}$$

▲

5.7   STRUCTURAL DESIGN OF MAT FOUNDATIONS

The structural design of mat foundations can be carried out by two conventional methods: the conventional rigid method and the approximate flexible method. Finite difference and finite element methods can also be used, but this section covers only the basic concepts of the first two design methods.

Conventional Rigid Method

The *conventional rigid method* of mat foundation design can be explained step by step with reference to Figure 5.8.

1.   Figure 5.8a shows mat dimensions of $L \times B$ and column loads of Q_1, Q_2, Q_3, ... Calculate the total column load as

$$Q = Q_1 + Q_2 + Q_3 + \cdots \tag{5.24}$$

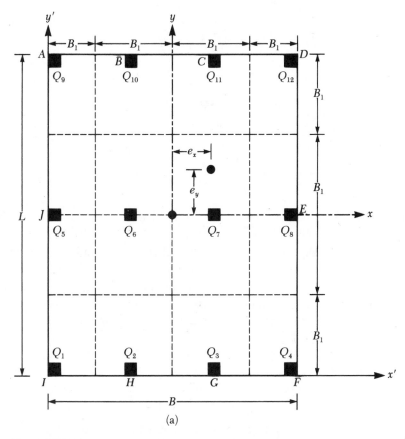

▼ **FIGURE 5.8** Conventional rigid mat foundation design

2. Determine the pressure on the soil, q, below the mat at points A, B, C, D, ..., by using the equation

$$q = \frac{Q}{A} \pm \frac{M_y x}{I_y} \pm \frac{M_x y}{I_x}$$ (5.25)

where $A = BL$

$I_x = (1/12)BL^3$ = moment of inertia about the x axis
$I_y = (1/12)LB^3$ = moment of inertia about the y axis
M_x = moment of the column loads about the x axis = $Q e_y$
M_y = moment of the column loads about the y axis = $Q e_x$

The load eccentricities, e_x and e_y, in the x and y directions can be determined by using (x', y') coordinates:

$$x' = \frac{Q_1 x_1' + Q_2 x_2' + Q_3 x_3' + \cdots}{Q}$$ (5.26)

and

$$e_x = x' - \frac{B}{2} \tag{5.27}$$

Similarly

$$y' = \frac{Q_1 y_1' + Q_2 y_2' + Q_3 y_3' + \cdots}{Q} \tag{5.28}$$

and

$$e_y = y' - \frac{L}{2} \tag{5.29}$$

3. Compare the values of the soil pressures determined in Step 2 with the net allowable soil pressure to determine whether $q \leq q_{\text{all(net)}}$.

4. Divide the mat into several strips in x and y directions (see Figure 5.8a). Let the width of any strip be B_1.

5. Draw the shear, V, and the moment, M, diagrams for each individual strip (in the x and y directions). For example, the average soil pressure of the bottom strip in the x direction of Figure 5.8a is

$$q_{\text{av}} \approx \frac{q_I + q_F}{2} \tag{5.30}$$

where q_I and q_F = soil pressures at points I and F as determined from Step 2.

The total soil reaction is equal to $q_{\text{av}} B_1 B$. Now obtain the total column load on the strip as $Q_1 + Q_2 + Q_3 + Q_4$. The sum of the column loads on the strip will not equal $q_{\text{av}} B_1 B$ because the shear between the adjacent strips has not been taken into account. For this reason, the soil reaction and the column loads need to be adjusted, or

$$\text{Average load} = \frac{q_{\text{av}} B_1 B + (Q_1 + Q_2 + Q_3 + Q_4)}{2} \tag{5.31}$$

Now, the modified average soil reaction becomes

$$q_{\text{av(modified)}} = q_{\text{av}} \left(\frac{\text{average load}}{q_{\text{av}} B_1 B} \right) \tag{5.32}$$

and the column load modification factor is

$$F = \frac{\text{average load}}{Q_1 + Q_2 + Q_3 + Q_4} \tag{5.33}$$

So, the modified column loads are FQ_1, FQ_2, FQ_3, and FQ_4. This modified loading on the strip under consideration is shown in Figure 5.8b. The shear and the moment diagram for this strip can now be drawn. This procedure is repeated for all strips in the x and y directions.

6. Determine the effective depth of the mat d by checking for diagonal tension shear near various columns. According to ACI Code 318-95 (Section 11.12.2.1c, American Concrete Institute, 1995), for the critical section,

$$U = b_o d[\phi(0.34)\sqrt{f_c'}] \tag{5.34}$$

where U = factored column load (MN), or (column load) $\times$ (load factor)
ϕ = reduction factor = 0.85
f_c' = compressive strength of concrete at 28 days (MN/m²)

The units of b_o and d in Eq. (5.34) are in meters. In English units, Eq. (5.34) may be expressed as

$$U = b_o d(4\phi\sqrt{f_c'}) \tag{5.35}$$

where U is in lb, b_o and d are in in., and f_c' is in lb/in²

The expression for b_o in terms of d, which depends on the location of the column with respect to the plan of the mat, can be obtained from Figure 5.8c.

7. From the moment diagrams of all strips *in one direction* (x or y), obtain the *maximum* positive and negative moments per unit width (that is, $M' = M/B_1$).

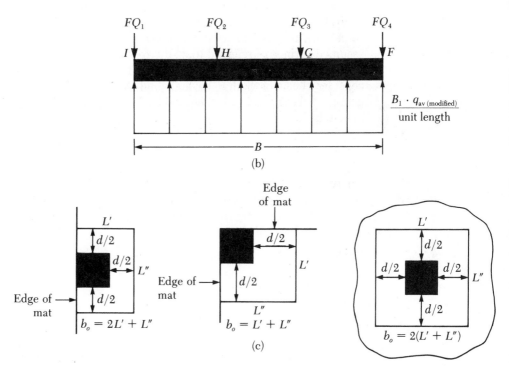

(b)

(c)

▼ **FIGURE 5.8** (Continued)

8. Determine the areas of steel per unit width for positive and negative reinforcement in the x and y directions.

$$M_u = (M')\,(\text{load factor}) = \phi A_s f_y \left(d - \frac{a}{2} \right) \tag{5.36}$$

and

$$a = \frac{A_s f_y}{0.85 f'_c b} \tag{5.37}$$

where A_s = area of steel per unit width
   f_y = yield stress of reinforcement in tension
   M_u = factored moment
   $\phi = 0.9$ = reduction factor

Examples 5.5 and 5.6 illustrate the use of the conventional rigid method of mat foundation design.

Approximate Flexible Method

In the conventional rigid method of design, the mat is assumed to be infinitely rigid. Also, the soil pressure is distributed in a straight line, and the centroid of the soil pressure is coincidental with the line of action of the resultant column loads (see Figure 5.9). In the *approximate flexible method* of design, the soil is assumed to be equivalent to infinite number of elastic springs, as shown in Figure 5.9b. It is sometimes referred to as the *Winkler foundation*. The elastic constant of these assumed springs is referred to as the *coefficient of subgrade reaction, k*.

To understand the fundamental concepts behind flexible foundation design, consider a beam of width B_1 having infinite length, as shown in Figure 5.9c. The beam is subjected to a single concentrated load Q. From the fundamentals of mechanics of materials,

$$M = E_F I_F \frac{d^2 z}{dx^2} \tag{5.38}$$

where M = moment at any section
   E_F = modulus of elasticity of foundation material
   I_F = moment of inertia of the cross section of the beam = $\left(\frac{1}{12}\right) B_1 h^3$
       (see Figure 5.9c).

However

$$\frac{dM}{dx} = \text{shear force} = V$$

and

$$\frac{dV}{dx} = q = \text{soil reaction}$$

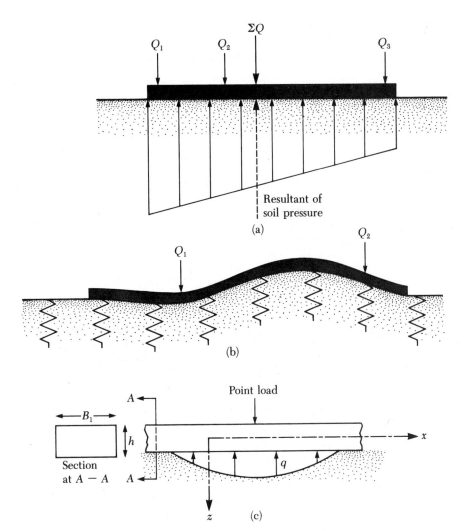

▼   **FIGURE 5.9**   (a) Principles of design by conventional rigid method; (b) principles of approximate flexible method; (c) derivation of Eq. (5.42)
for beams on elastic foundation

Hence

$$\frac{d^2M}{dx^2} = q \tag{5.39}$$

Combining Eqs. (5.38) and (5.39) yields

$$E_F I_F \frac{d^4z}{dx^4} = q \tag{5.40}$$

However, the soil reaction is

$$q = -zk'$$

where z = deflection

$k' = kB_1$

k = coefficient of subgrade reaction (kN/m³ or lb/in³)

So

$$E_F I_F = \frac{d^4 z}{dx^4} = -zkB_1 \tag{5.41}$$

Solution of Eq. (5.41) yields

$$z = e^{-\alpha x}(A' \cos \beta x + A'' \sin \beta x) \tag{5.42}$$

where A' and A'' are constants and

$$\beta = \sqrt[4]{\frac{B_1 k}{4 E_F I_F}} \tag{5.43}$$

The unit of the term β as defined by the preceding equation is (length)$^{-1}$. This parameter is very important in determining whether a mat foundation should be designed by conventional rigid method or approximate flexible method. According to the American Concrete Institute Committee 336 (1988), mats should be designed by the conventional rigid method if the spacing of columns in a strip is less than $1.75/\beta$. If the spacing of columns is larger than $1.75/\beta$, the approximate flexible method may be used.

To perform the analysis for the structural design of a flexible mat, you must know the principles of evaluating the *coefficient of subgrade reaction, k*. Before proceeding with the discussion of the approximate flexible design method, let us discuss this coefficient in more detail.

If a foundation of width B (Figure 5.10) is subjected to a load per unit area of q, it will undergo a settlement, Δ. The coefficient of subgrade modulus, k, can be defined as

$$k = \frac{q}{\Delta} \tag{5.44}$$

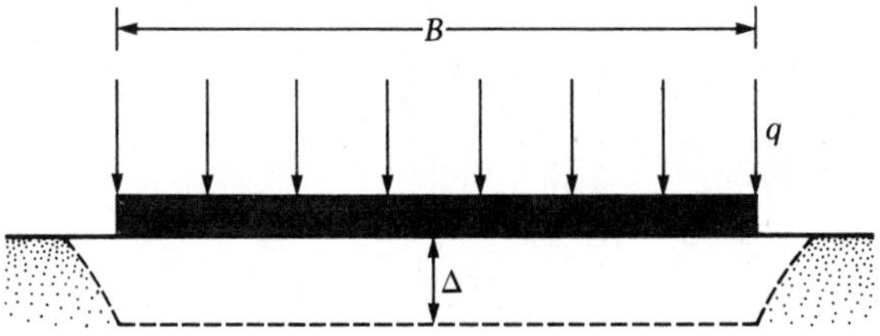

▼ **FIGURE 5.10**   Definition of coefficient of subgrade
                    reaction, k

The unit of k is kN/m^3 (or lb/in^3). The value of the coefficient of subgrade reaction is not a constant for a given soil. It depends on several factors, such as the length, L, and width, B, of the foundation and also the depth of embedment of the foundation. Terzaghi (1955) made a comprehensive study of the parameters affecting the coefficient of subgrade reaction. It indicated that the value of the coefficient of subgrade reaction decreases with the width of the foundation. In the field, load tests can be carried out by means of square plates measuring 1 ft × 1 ft (0.3 m × 0.3 m), and values of k can be calculated. The value of k can be related to large foundations measuring $B \times B$ in the following ways.

Foundations on Sandy Soils

$$k = k_{0.3} \left(\frac{B + 0.3}{2B} \right)^2 \tag{5.45}$$

where $k_{0.3}$ and k = coefficients of subgrade reaction of foundations measuring
0.3 m × 0.3 m and B (m) × B (m), respectively (unit is kN/m^3)
In English units, Eq. (5.45) may be expressed as

$$k = k_1 \left(\frac{B + 1}{2B} \right)^2 \tag{5.46}$$

where k_1 and k = coefficients of subgrade reaction of foundations measuring
1 ft × 1 ft and B (ft) × B (ft), respectively (unit is lb/in^3)

Foundations on Clays

$$k(kN/m^3) = k_{0.3}(kN/m^3) \left[\frac{0.3 \, (m)}{B \, (m)} \right] \tag{5.47}$$

The definition of k in Eq. (5.47) is the same as in Eq. (5.45).
    In English units,

$$k(lb/in^3) = k_1 \, (lb/in^3) \left[\frac{1 \, (ft)}{B \, (ft)} \right] \tag{5.48}$$

The definitions of k and k_1 are the same as in Eq. (5.46).

For rectangular foundations having dimensions of $B \times L$ (for similar soil and q),

$$k = \frac{k_{(B \times B)} \left(1 + 0.5 \dfrac{B}{L} \right)}{1.5} \tag{5.49}$$

where k = coefficient of subgrade modulus of the rectangular foundation $(L \times B)$

   $k_{(B \times B)}$ = coefficient of subgrade modulus of a square foundation having dimension of $B \times B$

Equation (5.49) indicates that the value of k of a very long foundation with a width B is approximately $0.67 k_{(B \times B)}$.

The modulus of elasticity of granular soils increases with depth. Because the settlement of a foundation depends on the modulus of elasticity, the value of k increases as the depth of the foundation increases.

Following are some typical ranges of value for the coefficient of subgrade reaction k_1 for sandy and clayey soils.

Sand (dry or moist)

Loose:	29–92 lb/in^3 (8–25 MN/m^3)
Medium:	91–460 lb/in^3 (25–125 MN/m^3)
Dense:	460–1380 lb/in^3 (125–375 MN/m^3)

Sand (saturated)

Loose:	38–55 lb/in^3 (10–15 MN/m^3)
Medium:	128–147 lb/in^3 (35–40 MN/m^3)
Dense:	478–552 lb/in^3 (130–150 MN/m^3)

Clay

Stiff:	44–92 lb/in^3 (12–25 MN/m^3)
Very stiff:	92–184 lb/in^3 (25–50 MN/m^3)
Hard:	>184 lb/in^3 (>50 MN/m^3)

Scott (1981) proposed that for sandy soils the value of $k_{0.3}$ can be obtained from standard penetration resistance at any given depth, or

$$k_{0.3} \, (\text{MN/m}^3) = 18 N_{cor} \tag{5.50}$$

where N_{cor} = *corrected* standard penetration resistance.

In English units,

$$k_1 \, (\text{U.S. ton/ft}^3) = 6 N_{cor} \tag{5.51}$$

For long beams, Vesic (1961) proposed an equation for estimating subgrade reaction:

$$k' = Bk = 0.65 \sqrt[12]{\frac{E_s B^4}{E_F I_F}} \frac{E_s}{1 - \mu_s^2}$$

or

$$k = 0.65 \sqrt[12]{\frac{E_s B^4}{E_F I_F}} \frac{E_s}{B(1 - \mu_s^2)} \qquad (5.52)$$

where E_s = modulus of elasticity of soil
  B = foundation width
  E_F = modulus of elasticity of foundation material
  I_F = moment of inertia of the cross section of the foundation
  μ_s = Poisson's ratio of soil

For most practical purposes, Eq. (5.52) can be approximated as

$$k = \frac{E_s}{B(1 - \mu_s^2)} \qquad (5.53)$$

The coefficient of subgrade reaction is also very useful parameter in the design of rigid highway and airfield pavements. The pavement with a concrete wearing surface is generally referred to as a *rigid pavement,* and the pavement with an asphaltic wearing surface is called a *flexible pavement.* For a surface load acting on a rigid pavement, the maximum tensile stress occurs at the base of the slab. For estimating the magnitude of the maximum horizontal tensile stress developed at the base of the rigid pavement, elastic solutions involving slabs on Winkler foundations are extremely useful. Some of the early work in this area was done by Westergaard (1926, 1939, 1947).

Now that we have discussed the coefficient of subgrade reaction, we will proceed with the discussion of the approximate flexible method of designing mat foundations. This method, as proposed by the American Concrete Institute Committee 336 (1988), is described step by step. The design procedure is based primarily on the theory of plates. Its use allows the effects (that is, moment, shear, and deflection) of a concentrated column load in the area surrounding it to be evaluated. If the zones of influence of two or more columns overlap, superposition can be used to obtain the net moment, shear, and deflection at any point.

1. Assume a thickness, h, for the mat, according to Step 6 as outlined for the conventional rigid method. (*Note:* h is the *total* thickness of the mat.)

2. Determine the flexural ridigity R of the mat:

$$R = \frac{E_F h^3}{12(1 - \mu_F^2)} \qquad (5.54)$$

where E_F = modulus of elasticity of foundation material
  μ_F = Poisson's ratio of foundation material

3. Determine the radius of effective stiffness:

$$L' = \sqrt[4]{\frac{R}{k}} \tag{5.55}$$

where k = coefficient of subgrade reaction

The zone of influence of any column load will be on the order of 3 to 4 L'.

4. Determine the moment (in polar coordinates at a point) caused by a column load (Figure 5.11a):

$$M_r = \text{radial moment} = -\frac{Q}{4}\left[A_1 - \frac{(1 - \mu_F)A_2}{\dfrac{r}{L'}}\right] \tag{5.56}$$

$$M_t = \text{tangential moment} = -\frac{Q}{4}\left[\mu_F A_1 + \frac{(1 - \mu_F)A_2}{\dfrac{r}{L'}}\right] \tag{5.57}$$

where r = radial distance from the column load
             Q = column load
       A_1, A_2 = functions of r/L'

The variations of A_1 and A_2 with r/L' are shown in Figure 5.11b (for details, see Hetenyi, 1946).

In the cartesian coordinate system (Figure 5.11a),

$$M_x = M_t \sin^2 \alpha + M_r \cos^2 \alpha \tag{5.58}$$

$$M_y = M_t \cos^2 \alpha + M_r \sin^2 \alpha \tag{5.59}$$

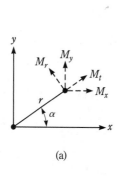

(a)

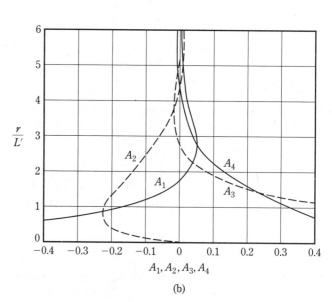

(b)

▼ **FIGURE 5.11**   Approximate flexible method of mat design

5. For the unit width of the mat, determine the shear force, V, caused by a column load:

$$V = \frac{Q}{4L'} A_3 \tag{5.60}$$

The variation of A_3 with r/L' is shown in Figure 5.11b.

6. If the edge of the mat is located in the zone of influence of a column, determine the moment and shear along the wedge (assume that the mat is continuous). Moment and shear opposite in sign to those determined are applied at the edges to satisfy the known conditions.

7. Deflection (δ) at any point is given by

$$\delta = \frac{QL'^2}{4R} A_4 \tag{5.61}$$

The variation of A_4 is given in Figure 5.11.

▼ **EXAMPLE 5.5**

The plan of a mat foundation with column loads is shown in Figure 5.12. Use Eq. (5.25) to calculate the soil pressures at points A, B, C, D, E, F, G, H, I, J, K, L, M, and N. The size of the mat is 76 ft $\times$ 96 ft, all columns are 24 in. $\times$ 24 in. in section, and $q_{all(net)} = 1.5$ kip/ft^2. Verify that the soil pressures are less than the net allowable bearing capacity.

Solution   From Figure 5.12,

$$
\begin{aligned}
\text{Column dead load } (DL) &= 100 + 180 + 190 + 110 + 180 + 360 + 400 + 200 \\
&\quad + 190 + 400 + 440 + 200 + 120 + 180 + 180 + 120 \\
&= 3550 \text{ kip}
\end{aligned}
$$

$$
\begin{aligned}
\text{Column live load } (LL) &= 60 + 120 + 120 + 70 + 120 + 200 + 250 + 120 + 130 \\
&\quad + 240 + 300 + 120 + 70 + 120 + 120 + 70 \\
&= 2230 \text{ kip}
\end{aligned}
$$

So

Service load $= 3550 + 2230 = 5780$ kip

According to ACI 318-95 (Section 9.2), factored load, $U = (1.4)$ (Dead load) $+ (1.7)$ (Live load). So

Factored load $= (1.4)(3550) + (1.7)(2230) = 8761$ kip

The moments of inertia of the foundation are

$$I_x = \tfrac{1}{12}(76)(96)^3 = 5603 \times 10^3 \text{ ft}^4$$

$$I_y = \tfrac{1}{12}(96)(76)^3 = 3512 \times 10^3 \text{ ft}^4$$

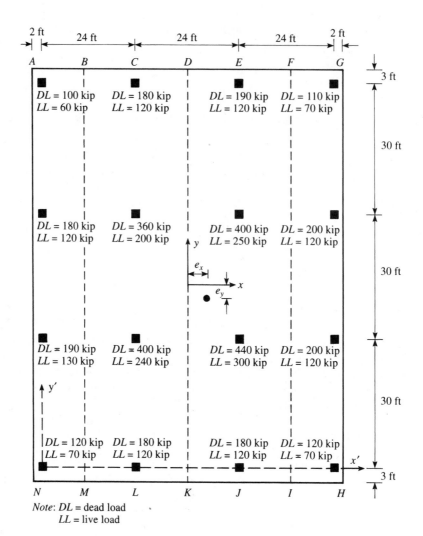

▼   **FIGURE 5.12**   Plan of a mat foundation

and

$$\sum M_{y'} = 0$$

So

$$5780x' = (24)(300 + 560 + 640 + 300) + (48)(310 + 650 + 740 + 300) \\ + (72)(180 + 320 + 320 + 190)$$

$$x' = 36.664 \text{ ft}$$

and

$$e_x = 36.664 - 36.0 = 0.664 \text{ ft}$$

Similarly,

$$\Sigma M_{x'} = 0$$

So

$$5780y' = (30)(320 + 640 + 740 + 320) + (60)(300 + 560 + 650 + 320)$$
$$+ (90)(160 + 300 + 310 + 180)$$

$$y' = 44.273 \text{ ft}$$

and

$$e_y = 44.273 - \tfrac{90}{2} = -0.727 \text{ ft}$$

The moments caused by eccentricity are

$$M_x = Qe_y = (8761)(0.727) = 6369 \text{ kip-ft}$$

$$M_y = Qe_x = (8761)(0.664) = 5817 \text{ kip-ft}$$

From Eq. (5.25)

$$q = \frac{Q}{A} \pm \frac{M_y x}{I_y} \pm \frac{M_x y}{I_x}$$
$$= \frac{8761}{(76)(96)} \pm \frac{(5817)(x)}{3512 \times 10^3} \pm \frac{(6369)(y)}{5603 \times 10^3}$$

or

$$q = 1.20 \pm 0.0017x \pm 0.0011y \ (\text{kip/ft}^2)$$

Now the following table can be prepared.

Point	$\dfrac{Q}{A}$ (kip/ft²)	x (ft)	$\pm 0.0017x$ (ft)	y (ft)	$\pm 0.0011y$ (ft)	q (kip/ft²)
A	1.2	−38	−0.065	48	−0.053	**1.082**
B	1.2	−24	−0.041	48	−0.053	**1.106**
C	1.2	−12	−0.020	48	−0.053	**1.127**
D	1.2	0	0.0	48	−0.053	**1.147**
E	1.2	12	0.020	48	−0.053	**1.167**
F	1.2	24	0.041	48	−0.053	**1.188**
G	1.2	38	0.065	48	−0.053	**1.212**
H	1.2	38	0.065	−48	0.053	**1.318**
I	1.2	24	0.041	−48	0.053	**1.294**
J	1.2	12	0.020	−48	0.053	**1.273**
K	1.2	0	0.0	−48	0.053	**1.253**
L	1.2	−12	−0.020	−48	0.053	**1.233**
M	1.2	−24	−0.041	−48	0.053	**1.212**
N	1.2	−38	−0.065	−48	0.053	**1.188**

The soil pressures at all points are less than the given value of $q_{\text{all(net)}} = 1.5 \text{ kip/ft}^2$.

▲

▼   **EXAMPLE 5.6**

Use the results of Example 5.5 and the conventional rigid method.

a. Determine the thickness of the slab.
b. Divide the mat into four strips (that is, *ABMN, BCDKLM, DEFIJK,* and *FGHI*) and determine the average soil reactions at the ends of each strip.
c. Determine the reinforcement requirements in the y direction for $f'_c = 3000$ lb/in² and $f_y = 60,000$ lb/in².

Solution

Part a: Determination of Mat Thickness

For the critical perimeter column as shown in Figure 5.13 (ACI 318-95; Section 9.2.1),

$$U = 1.4(DL) + 1.7(LL) = (1.4)(190) + (1.7)(130) = 487 \text{ kip}$$

$$b_o = 2(36 + d/2) + (24 + d) = 96 + 2d \text{ (in.)}$$

From ACI 318-95

$$\phi V_c \geq V_u$$

where V_c = nominal shear strength of concrete
   V_u = factored shear strength

$$\phi V_c = \phi(4)\sqrt{f'_c}\,b_o d = (0.85)(4)(\sqrt{3000})(96 + 2d)d$$

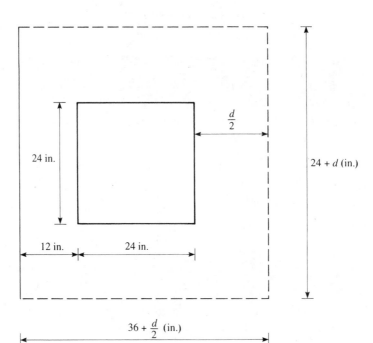

▼   **FIGURE 5.13**   Critical perimeter column

So

$$\frac{(0.85)\,(4)\,(\sqrt{3000})\,(96 + 2d)d}{1000} \geq 487$$

$$(96 + 2d)d \geq 2615.1$$

$$d \approx 19.4 \text{ in.}$$

For the critical internal column shown in Figure 5.14,

$$b_o = 4\,(24 + d) = 96 + 4d \text{ (in.)}$$

$$U = (1.4)\,(440) + (1.7)\,(300) = 1126 \text{ kip}$$

and

$$\frac{(0.85)\,(4)\,(\sqrt{3000})\,(96 + 4d)d}{1000} \geq 1126$$

$$(96 + 4d)d \geq 6046.4$$

$$d \approx 28.7 \text{ in.}$$

Use $d = 29$ in.

With a minimum cover of 3 in. over the steel reinforcement and 1-in. diameter steel bars, the total slab thickness is

$$h = 29 + 3 + 1 = \textbf{33 in.}$$

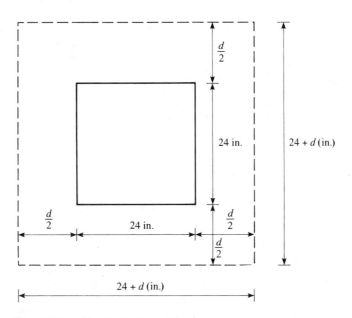

▼   **FIGURE 5.14**   Critical internal column

Part b: Average Soil Reaction

Refer to Figure 5.12. For strip *ABMN* (width = 14 ft),

$$q_1 = \frac{q_{(at\,A)} + q_{(at\,B)}}{2} = \frac{1.082 + 1.106}{2} = \textbf{1.094 kip/ft}^2$$

$$q_2 = \frac{q_{(at\,M)} + q_{(at\,N)}}{2} = \frac{1.212 + 1.188}{2} = \textbf{1.20 kip/ft}^2$$

For strip *BCDKLM* (width = 24 ft),

$$q_1 = \frac{1.106 + 1.127 + 1.147}{3} = \textbf{1.127 kip/ft}^2$$

$$q_2 = \frac{1.253 + 1.233 + 1.212}{3} = = \textbf{1.233 kip/ft}^2$$

For strip *DEFIJK* (width = 24 ft),

$$q_1 = \frac{1.147 + 1.167 + 1.188}{3} = \textbf{1.167 kip/ft}^2$$

$$q_2 = \frac{1.294 + 1.273 + 1.253}{3} = \textbf{1.273 kip/ft}^2$$

For strip *FGHI* (width = 14 ft),

$$q_1 = \frac{1.188 + 1.212}{2} = \textbf{1.20 kip/ft}^2$$

$$q_2 = \frac{1.318 + 1.294}{2} = \textbf{1.306 kip/ft}^2$$

Check for $\Sigma F_V = 0$:

Soil reaction for strip *ABMN* = $\frac{1}{2}$ (1.094 + 1.20) (14) (96)   = 1541.6 kip

Soil reaction for strip *BCDKLM* = $\frac{1}{2}$ (1.127 + 1.233) (24) (96) = 2718.7 kip

Soil reaction for strip *DEFIJK* = $\frac{1}{2}$ (1.167 + 1.273) (24) (96) = 2810.9 kip

Soil reaction for strip *FGHJ* = $\frac{1}{2}$ (1.20 + 1.306) (14) (96)   = <u>1684.0 kip</u>

Σ 8755.2 kip $\approx \Sigma$ Column load = 8761 kip—OK

Part c: Reinforcement Requirements

Refer to Figure 5.15 on page 328 for the design of strip $BCDKLM$. Figure 5.15 shows the load diagram, in which

$$Q_1 = (1.4)(180) + (1.7)(120) = 456 \text{ kip}$$

$$Q_2 = (1.4)(360) + (1.7)(200) = 844 \text{ kip}$$

$$Q_3 = (1.4)(400) + (1.7)(240) = 968 \text{ kip}$$

$$Q_4 = (1.4)(180) + (1.7)(120) = 456 \text{ kip}$$

The shear and moment diagrams are shown in Figures 5.15b and c, respectively. From Figure 5.15c, the maximum positive moment at the bottom of the foundation = 2281.1/24 = 95.05 kip-ft/ft.

For the design concepts of a rectangular section in bending refer to Figure 5.16.

$$\sum \text{Compressive force, } C = 0.85 f_c' a b$$

$$\sum \text{Tensile force, } T = A_s f_y$$

$$C = T$$

Note that for this case $b = 1$ ft = 12 in.

$$(0.85)(3)(12)a = A_s(60)$$

$$A_s = 0.51a$$

From Eq. (5.36),

$$M_u = \phi A_s f_y \left(d - \frac{a}{2} \right)$$

$$(95.05)(12) = (0.9)(0.51a)(60)\left(29 - \frac{a}{2} \right)$$

$$a = 1.47 \text{ in.}$$

Thus

$$A_s = (0.51)(1.47) = 0.75 \text{ in}^2$$

▶ Minimum reinforcement, s_{min} (ACI 318-95, Section 10.5) = $200/f_y$ = 200/60,000 = 0.00333
▶ Minimum A_s = (0.00333)(12)(29) = 1.16 in²/ft. Hence use minimum reinforcement with A_s = 1.16 in²/ft.
▶ **Use No. 9 bars at 10 in. center-to-center (A_s = 1.2 in²/ft) at the bottom of the foundation.**

From Figure 5.15c, the maximum negative moment = 2447.8 kip-ft/24 = 102 kip-ft/ft. By observation, $A_s \leq A_{s(min)}$.

▶ **Use No. 9 bars at 10 in. center-to-center at the top of the foundation.**

▲

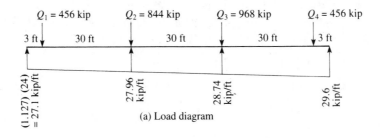

(a) Load diagram

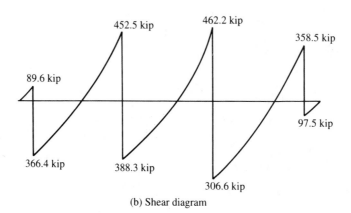

(b) Shear diagram

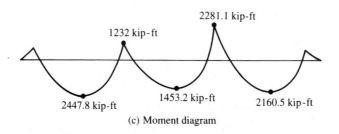

(c) Moment diagram

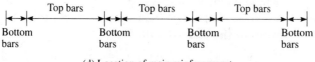

(d) Location of main reinforcement

▼   FIGURE 5.15

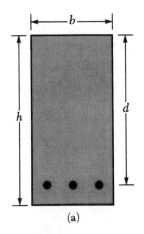
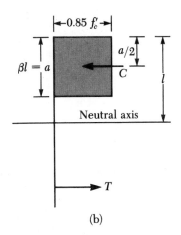

▼ **FIGURE 5.16**   Rectangular section in bending; (a) section, (b) assumed stress distribution across the section

▼ **EXAMPLE 5.7**

From the plate load test (plate dimension 1 ft × 1 ft) in the field, the coefficient of subgrade reaction of a sandy soil was determined to be 80 lb/in³. (a) What will be the value of the coefficient of subgrade reaction on the same soil for a foundation with dimensions of 30 ft × 30 ft? (b) If the full-sized foundation has dimensions of 45 ft × 30 ft, what will be the value of the coefficient of subgrade reaction?

Solution

Part a
From Eq. (5.46),

$$k = k_1 \left(\frac{B+1}{2B} \right)^2$$

where $k_1 = 80 \, \text{lb/in}^2$
            $B = 30 \, \text{ft}$

So

$$k = 80 \left[\frac{30+1}{(2)(30)} \right]^2 = \mathbf{21.36 \, in^3}$$

Part b

From Eq. (5.49),

$$k = \frac{k_{(B \times B)}\left(1 + 0.5\dfrac{B}{L}\right)}{1.5}$$

$$k_{(30\,\text{ft} \times 30\,\text{ft})} = 21.36\,\text{lb/in}^3$$

So

$$k = \frac{(21.36)\,(1 + 0.5\frac{30}{45})}{1.5} = \mathbf{19\,lb/in^3}$$

▲

PROBLEMS

5.1 Determine the net ultimate bearing capacity of mat foundations with the following characteristics:

    a. $c_u = 120$ kN/m², $\phi = 0$, $B = 8$ m, $L = 18$ m, $D_f = 3$ m

    b. $c_u = 2500$ lb/ft², $\phi = 0$, $B = 20$ ft, $L = 30$ ft, $D_f = 6.2$ ft

5.2 Following are the results of a standard penetration test in the field (sandy soil):

Depth (m)	Field value of N_F
1.5	9
3.0	12
4.5	11
6.0	7
7.5	13
9.0	11
10.5	13

Estimate the net allowable bearing capacity of a mat foundation 6.5 m × 5 m in plan. Here, $D_f = 1.5$ m, and allowable settlement = 50.8 mm. Assume that the unit weight of soil $\gamma = 16.5$ kN/m³.

5.3 Repeat Problem 5.2 for an allowable settlement of 30 mm.

5.4 A mat foundation on a saturated clay soil has dimensions of 20 m × 20 m. Given: dead and live load = 48 MN, $c_u = 30$ kN/m², $\gamma_{clay} = 18.5$ kN/m³.

    a. Find the depth, D_f, of the mat for a fully compensated foundation.

    b. What will be the depth of the mat (D_f) for a factor of safety of 2 against bearing capacity failure?

5.5 Repeat Problem 5.4 part b for $c_u = 20$ kN/m².

5.6 A mat foundation is shown in Figure P5.6. The design considerations are $L = 12$ m, $B = 10$ m, $D_f = 2.2$ m, $Q = 30$ MN, $x_1 = 2$ m, $x_2 = 2$ m, $x_3 = 5.2$ m, and preconsolidation pressure $p_c = 105$ kN/m². Calculate the consolidation settlement under the center of the mat.

5.7 For the mat foundation in Problem 5.6, estimate the consolidation settlement under the corner of the mat.

5.8 Refer to Figure P5.8. For the mat, Q_1, $Q_3 = 40$ tons, Q_4, Q_5, $Q_6 = 60$ tons, Q_2, $Q_9 =$

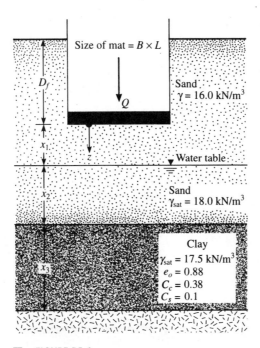

▼ **FIGURE P5.6**

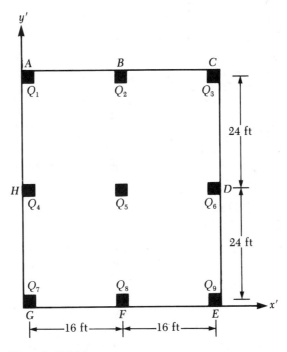

▼ **FIGURE P5.8**

45 tons, and Q_7, Q_8 = 50 tons. All columns are 20 in. × 20 in. in cross section. Use the procedure outlined in Section 5.7 to determine the pressure on the soil at A, B, C, D, E, F, G, and H.

5.9 The plan of a mat foundation with column loads is shown in Figure P5.9. Calculate the soil pressure at points A, B, C, D, E, and F. *Note:* All columns are 0.5 m × 0.5 m in plan.

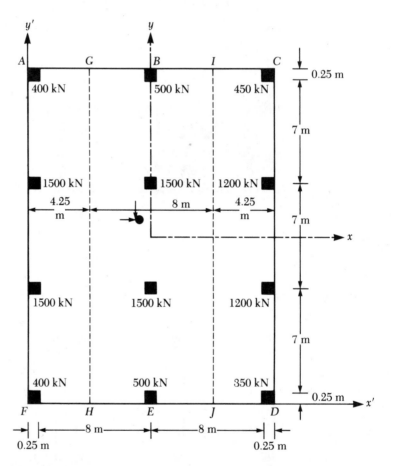

▼ **FIGURE P5.9**

5.10 Divide the mat shown in Figure P5.9 into three strips, such as $AGHF$ (B_1 = 4.25 m), $GIJH$ (B_1 = 8 m), and $ICDJ$ (B_1 = 4.25 m). Use the results of Problem 5.9 and determine the reinforcement requirements in the y direction. Here, f'_c = 20.7 MN/m², f_y = 413.7 MN/m², and the load factor is 1.7.

5.11 From the plate load test (plate dimension 1 ft × 1 ft) in the field, the coefficient of subgrade reaction of a sandy soil is determined to be 55 lb/in³. What will be the value of the coefficient of subgrade reaction on the same soil for a foundation with dimensions of 25 ft × 25 ft?

5.12 Refer to Problem 5.11. If the full-sized foundation had dimensions of 70 ft $\times$ 30 ft, what will be the value of the coefficient of the subgrade reaction?

5.13 The subgrade reaction of a sandy soil obtained from a plate load test (plate dimensions 1 m $\times$ 0.7 m) is 18 kN/m². What will be the value of k on the same soil for a foundation measuring 5m $\times$ 3.5 m?

REFERENCES

American Concrete Institute (1995). *ACI Standard Building Code Requirements for Reinforced Concrete,* ACI 318-95, Farmington Hills, Michigan.

American Concrete Institute Committee 336 (1988). "Suggested Design Procedures for Combined Footings and Mats," *Journal of the American Concrete Institute,* Vol. 63, No. 10, pp. 1041–1077.

Hetenyi, M. (1946). *Beams of Elastic Foundations,* University of Michigan Press, Ann Arbor.

Meyerhof, G. G. (1965). "Shallow Foundations," *Journal of the Soil Mechanics and Foundations Division,* American Society of Civil Engineers, Vol. 91, No. SM2, pp. 21–31.

Rios, L., and Silva, F. P. (1948). "Foundations in Downtown Sao Paulo (Brazil)," *Proceedings,* Second International Conference on Soil Mechanics and Foundation Engineering, Rotterdam, Vol. 4, p. 69.

Schultze, E. (1962). "Probleme bei der Auswertung von Setzungsmessungen," *Proceedings,* Baugrundtagung, Essen, Germany, p. 343.

Scott, R. F. (1981). *Foundation Analysis,* Prentice-Hall, Englewood Cliffs, N.J.

Stuart, J. G., and Graham, J. (1975). "Settlement Performance of a Raft Foundation on Sand," in *Settlement of Structures,* Halsted Press, New York, pp. 62–67.

Terzaghi, K. (1955). "Evaluation of the Coefficient of Subgrade Reactions," *Geotechnique,* Institute of Engineers, London, Vol. 5, No. 4, pp. 197–226.

Vargas, M. (1948). "Building Settlement Observations in Sao Paulo," *Proceedings,* Second International Conference on Soil Mechanics and Foundation Engineering, Rotterdam, Vol. 4, p. 13.

Vargas, M. (1961). "Foundations of Tall Buildings on Sand in Sao Paulo (Brazil)," *Proceedings,* Fifth International Conference on Soil Mechanics and Foundation Engineering, Paris, Vol. 1, p. 841.

Vesic, A. S. (1961). "Bending of Beams Resting on Isotropic Solid," *Journal of the Engineering Mechanics Division,* American Society of Civil Engineers, Vol. 87, No. EM2, pp. 35–53.

Westgaard, H. M. (1926). "Stresses in Concrete Pavements Computed by Theoretical Analysis," *Public Roads,* Vol. 7, No. 12, pp. 23–35.

Westergaard, H. M. (1939). "Stresses in Concrete Runways of Airports," *Proceedings,* Highway Research Board, Vol. 19, pp. 197–205.

Westergaard, H. M. (1947). "New Formulas for Stresses in Concrete Pavements of Airfields," *Proceedings,* American Society of Civil Engineers, Vol. 73, pp. 687–701.

CHAPTER SIX

LATERAL EARTH PRESSURE

6.1 INTRODUCTION

Vertical or near vertical slopes of soil are supported by retaining walls, cantilever sheet-pile walls, sheet-pile bulkheads, braced cuts, and other similar structures. The proper design of those structures requires estimation of lateral earth pressure, which is a function of several factors, such as (a) type and amount of wall movement, (b) shear strength parameters of the soil, (c) unit weight of the soil, and (d) drainage conditions in the backfill. Figure 6.1 shows a retaining wall of height H. For similar types of backfill:

a. The wall may be restrained from moving (Figure 6.1a). The lateral earth pressure on the wall at any depth is called the *at-rest earth pressure.*

b. The wall may tilt away from the soil retained (Figure 6.1b). With sufficient wall tilt, a triangular soil wedge behind the wall will fail. The lateral pressure for this condition is referred to as *active earth pressure.*

c. The wall may be pushed into the soil retained (Figure 6.1c). With sufficient wall movement, a soil wedge will fail. The lateral pressure for this condition is referred to as *passive earth pressure.*

Figure 6.2 shows the nature of variation of the lateral pressure (σ_h) at a certain depth of the wall with the magnitude of wall movement.

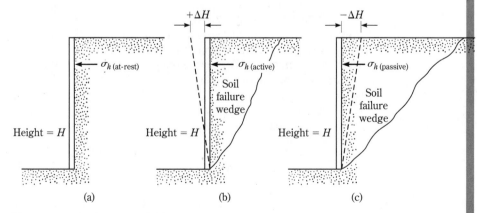

▼ FIGURE 6.1 Nature of lateral earth pressure on a retaining wall

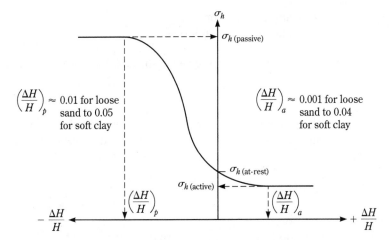

▼ **FIGURE 6.2**   Nature of variation of lateral earth pressure at a certain depth

In the following sections we will discuss various relationships to determine the at-rest, active, and passive pressures on a retaining wall. It is assumed that the readers have been exposed to lateral earth pressure in the past, so this chapter will serve as a review.

6.2   LATERAL EARTH PRESSURE AT REST

Consider a vertical wall of height H, as shown in Figure 6.3, retaining a soil having a unit weight of γ. A uniformly distributed load, q/unit area, is also applied at the ground surface. The shear strength, s, of the soil is

$$s = c + \sigma' \tan \phi$$

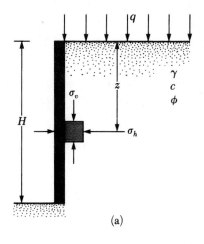

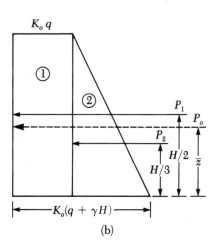

▼ **FIGURE 6.3**   At-rest earth pressure

where c = cohesion
ϕ = angle of friction
σ' = effective normal stress

At any depth z below the ground surface, the vertical subsurface stress is

$$\sigma_v = q + \gamma z \tag{6.1}$$

If the *wall is at rest and is not allowed to move at all* either away from the soil mass or into the soil mass (e.g., zero horizontal strain), the lateral pressure at a depth z is

$$\sigma_h = K_o \sigma_v' + u \tag{6.2}$$

where u = pore water pressure
K_o = coefficient of at-rest earth pressure

For normally consolidated soil, the relation for K_o (Jaky, 1944) is

$$\boxed{K_o \approx 1 - \sin \phi} \tag{6.3}$$

Equation (6.3) is an empirical approximation.

For normally consolidated clays, the coefficient of earth pressure at rest can be approximated (Brooker and Ireland, 1965) as

$$\boxed{K_o \approx 0.95 - \sin \phi} \tag{6.4}$$

where ϕ = drained peak friction angle

Based on Brooker and Ireland's (1965) experimental results, the value of K_o for normally consolidated clays may be approximately correlated with the plasticity index (PI):

$$K_o = 0.4 + 0.007(PI) \qquad \text{(for } PI \text{ between 0 and 40)} \tag{6.5}$$

and

$$K_o = 0.64 + 0.001(PI) \qquad \text{(for } PI \text{ between 40 and 80)} \tag{6.6}$$

For overconsolidated clays,

$$K_{o(\text{overconsolidated})} \approx K_{o(\text{normally consolidated})} \sqrt{OCR} \qquad (6.7)$$

where OCR = overconsolidation ratio

Mayne and Kulhawy (1982) analyzed the results of 171 different laboratory-tested soils. Based on this study, they proposed a general empirical relationship to estimate the magnitude of K_o for sand and clay:

$$K_o = (1 - \sin \phi) \left[\frac{OCR}{OCR_{\max}^{(1 - \sin \phi)}} + \frac{3}{4} \left(1 - \frac{OCR}{OCR_{\max}} \right) \right] \qquad (6.8)$$

where OCR = present overconsolidation ratio
            $OCR_{\max}$ = maximum overconsolidation ratio

In Figure 6.4, $OCR_{\max}$ is the value of OCR at point B.

With a properly selected value of the at-rest earth pressure coefficient, Eq. (6.2) can be used to determine the variation of lateral earth pressure with depth z. Figure 6.3b shows the variation of σ_h with depth for the wall shown in Figure 6.3a. Note that if the surcharge $q = 0$ and the pore water pressure $u = 0$, the pressure diagram will be a triangle. The total force, P_o, *per unit length* of the wall given in Figure 6.3a can now be obtained from the area of the pressure diagram given in Figure 6.3b as

$$P_o = P_1 + P_2 = qK_oH + \tfrac{1}{2}\gamma H^2 K_o \qquad (6.9)$$

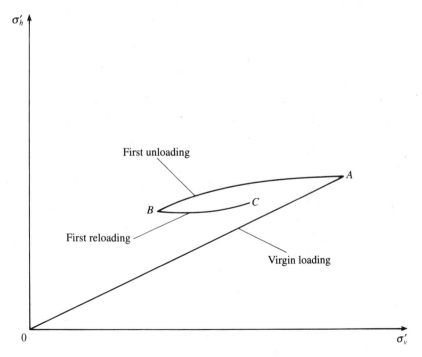

▼ **FIGURE 6.4**   Stress history for soil under K_o condition

where P_1 = area of rectangle 1
P_2 = area of triangle 2

The location of the line of action of the resultant force, P_o, can be obtained by taking the moment about the bottom of the wall. Thus

$$\bar{z} = \frac{P_1\left(\dfrac{H}{2}\right) + P_2\left(\dfrac{H}{3}\right)}{P_o} \tag{6.10}$$

If the water table is located at depth $z < H$, the at-rest pressure diagram shown in Figure 6.3b will have to be somewhat modified, as shown in Figure 6.5. If the effective unit weight of soil below the water table equals γ' (that is, $\gamma_{sat} - \gamma_w$),

At $z = 0$, $\sigma'_h = K_o\sigma'_v = K_o q$

At $z = H_1$, $\sigma'_h = K_o\sigma'_v = K_o(q + \gamma H_1)$

At $z = H_2$, $\sigma'_h = K_o\sigma'_v = K_o(q + \gamma H_1 + \gamma'H_2)$

Note that in the preceding equations, σ'_v and σ'_h are effective vertical and horizontal pressures. Determining the total pressure distribution on the wall requires adding the hydrostatic pressure. The hydrostatic pressure, u, is zero from $z = 0$ to $z = H_1$; at $z = H_2$, $u = H_2\gamma_w$. The variation of σ'_h and u with depth is shown in Figure 6.5b. Hence the total force per unit length of the wall can be determined from the area of the pressure diagram. Thus

$$P_o = A_1 + A_2 + A_3 + A_4 + A_5$$

where A = area of the pressure diagram

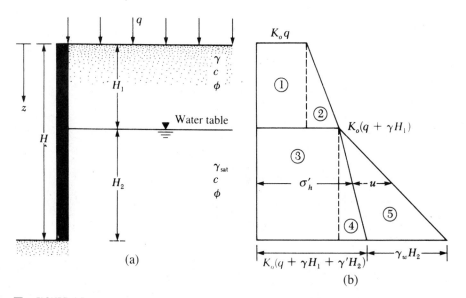

(a)

(b)

▼   FIGURE 6.5

So

$$P_o = K_o q H_1 + \tfrac{1}{2} K_o \gamma H_1^2 + K_o (q + \gamma H_1) H_2 + \tfrac{1}{2} K_o \gamma' H_2^2 + \tfrac{1}{2} \gamma_w H_2^2 \qquad (6.11)$$

Sherif et al. (1984) showed by several laboratory model tests that Eq. (6.3) gives good results for estimating the lateral earth pressure at rest for loose sands. However, for compacted dense sand, it grossly underestimates the value of K_o. For that reason, they proposed a modified relationship for K_o:

$$K_o = (1 - \sin \phi) + \left(\frac{\gamma_d}{\gamma_{d(min)}} - 1 \right) 5.5 \qquad (6.12)$$

where γ_d = *in situ* unit weight of sand
$\gamma_{d(min)}$ = minimum possible dry unit weight of sand (see Chapter 1)

▼ **EXAMPLE 6.1**

For the retaining wall shown in Figure 6.6(a), determine the lateral earth force at rest per unit length of the wall. Also determine the location of the resultant force.

Solution

$$K_o = 1 - \sin \phi = 1 - \sin 30° = 0.5$$

At $z = 0$, $\sigma_v' = 0$; $\sigma_h' = 0$

At $z = 2.5$ m, $\sigma_v' = (16.5)(2.5) = 41.25$ kN/m²;

$$\sigma_h' = K_o \sigma_v' = (0.5)(41.25) = 20.63 \text{ kN/m}^2$$

At $z = 5$ m, $\sigma_v' = (16.5)(2.5) + (19.3 - 9.81)2.5 = 64.98$ kN/m²;

$$\sigma_h' = K_o \sigma_v' = (0.5)(64.98) = 32.49 \text{ kN/m}^2$$

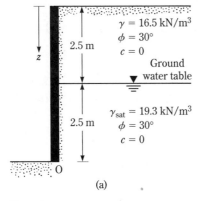

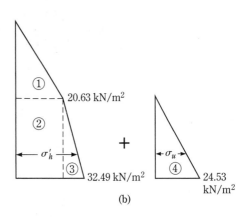

(a)   (b)

▼ **FIGURE 6.6**

The hydrostatic pressure distribution is as follows:

From $z = 0$ to $z = 2.5$ m, $u = 0$. At $z = 5$ m, $u = \gamma_w(2.5) = (9.81)(2.5) = 24.53$ kN/m². The pressure distribution for the wall is shown in Figure 6.6b.

The total force per unit length of the wall can be determined from the area of the pressure diagram, or

$$
\begin{aligned}
P_o &= \text{Area } 1 + \text{Area } 2 + \text{Area } 3 + \text{Area } 4 \\
&= \tfrac{1}{2}(2.5)(20.63) + (2.5)(20.63) + \tfrac{1}{2}(2.5)(32.49 - 20.63) \\
&\quad + \tfrac{1}{2}(2.5)(24.53) = \mathbf{122.85 \ kN/m}
\end{aligned}
$$

The location of the center of pressure measured from the bottom of the wall (point O) =

$$
\begin{aligned}
\bar{z} &= \frac{(\text{Area } 1)\left(2.5 + \dfrac{2.5}{3}\right) + (\text{Area } 2)\left(\dfrac{2.5}{2}\right) + (\text{Area } 3 + \text{Area } 4)\left(\dfrac{2.5}{3}\right)}{P_o} \\
&= \frac{(25.788)(3.33) + (51.575)(1.25) + (14.825 + 30.663)(0.833)}{122.85} \\
&= \frac{85.87 + 64.47 + 37.89}{122.85} = \mathbf{1.53 \ m}
\end{aligned}
$$

▲

ACTIVE PRESSURE

6.3   RANKINE ACTIVE EARTH PRESSURE

The lateral earth pressure condition described in Section 6.2 involves walls that do not yield at all. However, if a wall tends to move away from the soil a distance Δx, as shown in Figure 6.7a, the soil pressure on the wall at any depth will decrease. For a wall that is *frictionless*, the horizontal stress, σ_h, at depth z will equal $K_o \sigma_v (= K_o \gamma z)$ when Δx is zero. However, with $\Delta x > 0$, σ_h will be less than $K_o \sigma_v$.

The Mohr's circles corresponding to wall displacements of $\Delta x = 0$ and $\Delta x > 0$ are shown as circles a and b, respectively, in Figure 6.7b. If the displacement of the wall, Δx, continues to increase, the corresponding Mohr's circle eventually will just touch the Mohr–Coulomb failure envelope defined by the equation

$$s = c + \sigma \tan \phi$$

This circle is marked c in Figure 6.7b. It represents the failure condition in the soil mass; the horizontal stress then equals σ_a. This horizontal stress, σ_a, is referred to as the *Rankin active pressure*. The *slip lines* (failure planes) in the soil mass will then make angles of $\pm(45 + \phi/2)$ with the horizontal, as shown in Figure 6.7a.

Refer back to Eq. (1.84), the equation relating the principal stresses for a Mohr's circle that touches the Mohr–Coulomb failure envelope:

$$\sigma_1 = \sigma_3 \tan^2\left(45 + \frac{\phi}{2}\right) + 2c \tan\left(45 + \frac{\phi}{2}\right)$$

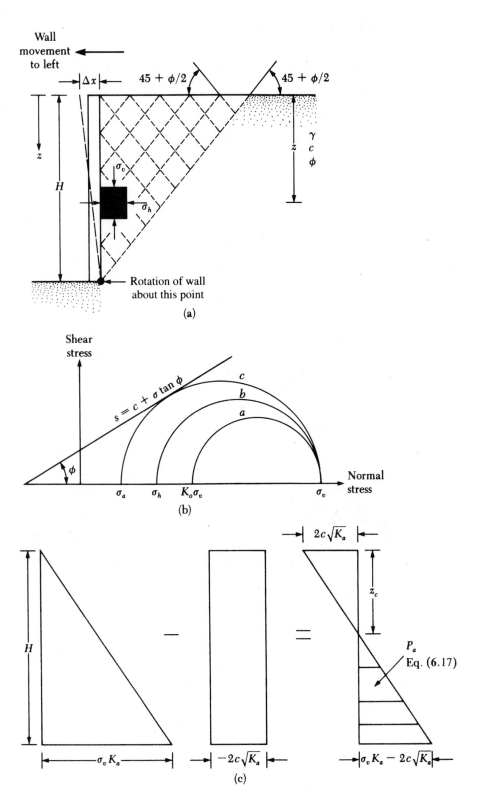

▼ FIGURE 6.7   Rankine active pressure

341

For the Mohr's circle c in Figure 6.7b,

Major principal stress, $\sigma_1 = \sigma_v$

and

Minor principal stress, $\sigma_3 = \sigma_a$

Thus

$$\sigma_v = \sigma_a \tan^2\left(45 + \frac{\phi}{2}\right) + 2c\tan\left(45 + \frac{\phi}{2}\right)$$

$$\sigma_a = \frac{\sigma_v}{\tan^2\left(45 + \dfrac{\phi}{2}\right)} - \frac{2c}{\tan\left(45 + \dfrac{\phi}{2}\right)}$$

or

$$\begin{aligned} \sigma_a &= \sigma_v \tan^2\left(45 - \frac{\phi}{2}\right) - 2c\tan\left(45 - \frac{\phi}{2}\right) \\ &= \sigma_v K_a - 2c\sqrt{K_a} \end{aligned} \qquad (6.13)$$

where $K_a = \tan^2(45 - \phi/2) =$ Rankine active pressure coefficient (Table 6.1)

The variation of the active pressure with depth for the wall shown in Figure 6.7a is given in Figure 6.7c. Note that $\sigma_v = 0$ at $z = 0$ and $\sigma_v = \gamma H$ at $z = H$. The pressure distribution shows that at $z = 0$ the active pressure equals $-2c\sqrt{K_a}$, indicating tensile stress. This tensile stress decreases with depth and becomes zero at a depth $z = z_c$, or

$$\gamma z_c K_a - 2c\sqrt{K_a} = 0$$

and

$$z_c = \frac{2c}{\gamma\sqrt{K_a}} \qquad (6.14)$$

The depth z_c is usually referred to as the *depth of tensile crack,* because the tensile stress in the soil will eventually cause a crack along the soil–wall interface. Thus the total Rankine active force per unit length of the wall before the tensile crack occurs is

$$\begin{aligned} P_a &= \int_0^H \sigma_a\,dz = \int_0^H \gamma z K_a\,dz - \int_0^H 2c\sqrt{K_a}\,dz \\ &= \tfrac{1}{2}\gamma H^2 K_a - 2cH\sqrt{K_a} \end{aligned} \qquad (6.15)$$

▼ **TABLE 6.1** Variation of Rankine K_a

Soil friction angle, ϕ (deg)	$K_a = \tan^2 (45 - \phi/2)$
20	0.490
21	0.472
22	0.455
23	0.438
24	0.422
25	0.406
26	0.395
27	0.376
28	0.361
29	0.347
30	0.333
31	0.320
32	0.307
33	0.295
34	0.283
35	0.271
36	0.260
37	0.249
38	0.238
39	0.228
40	0.217
41	0.208
42	0.198
43	0.189
44	0.180
45	0.172

After the occurrence of the tensile crack, the force on the wall will be caused only by the pressure distribution between depths $z = z_c$ and $z = H$, as shown by the hatched area in Figure 6.7c. It may be expressed as

$$P_a = \tfrac{1}{2}(H - z_c)(\gamma H K_a - 2c\sqrt{K_a}) \tag{6.16}$$

or

$$P_a = \frac{1}{2}\left(H - \frac{2c}{\gamma\sqrt{K_a}}\right)(\gamma H K_a - 2c\sqrt{K_a}) \tag{6.17}$$

For calculation purposes in some retaining wall design problems, a cohesive soil backfill is replaced by an assumed granular soil with a triangular Rankine active pressure diagram with $\sigma_a = 0$ at $z = 0$ and $\sigma_a = \sigma_v K_a - 2c\sqrt{K_a}$ at $z = H$ (see Figure 6.8). In such a case, the assumed active force per unit length of the wall is

$$P_a = \tfrac{1}{2}H(\gamma H K_a - 2c\sqrt{K_a}) = \tfrac{1}{2}\gamma H^2 K_a - cH\sqrt{K_a} \tag{6.18}$$

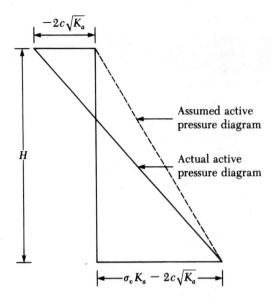

▼ **FIGURE 6.8**   Assumed active pressure diagram
for clay backfill behind a retaining
wall

However, the active earth pressure condition will be reached only if the wall is
allowed to "yield" sufficiently. The amount of outward displacement of the wall
necessary is about $0.001H$ to $0.004H$ for granular soil backfills and about $0.01H$ to
$0.04H$ for cohesive soil backfills.

▼ **EXAMPLE 6.2**

A 6-m-high retaining wall is to support a soil with unit weight $\gamma = 17.4 \ \text{kN/m}^3$, soil
friction angle $\phi = 26°$, and cohesion $c = 14.36 \ \text{kN/m}^2$. Determine the Rankine active
force per unit length of the wall both before and after the tensile crack occurs, and
determine the line of action of the resultant in both cases.

Solution   For $\phi = 26°$,

$$K_a = \tan^2\left(45 - \frac{\phi}{2}\right) = \tan^2(45 - 13) = 0.39$$

$$\sqrt{K_a} = 0.625$$

$$\sigma_a = \gamma H K_a - 2c\sqrt{K_a}$$

Refer to Figure 6.7c:

At $z = 0$, $\sigma_a = -2c\sqrt{K_a} = -2(14.36)(0.625) = -17.95 \ \text{kN/m}^2$

At $z = 6 \ \text{m}$, $\sigma_a = (17.4)(6)(0.39) - 2(14.36)(0.625)$
$= 40.72 - 17.95 = 22.77 \ \text{kN/m}^2$

Active Force Before the Occurrence of Tensile Crack: Eq. (6.15)

$$P_a = \tfrac{1}{2}\gamma H^2 K_a - 2cH\sqrt{K_a}$$
$$= \tfrac{1}{2}(6)(40.72) - (6)(17.95) = 122.16 - 107.7 = 14.46 \text{ kN/m}$$

The line of action of the resultant can be determined by taking the moment of the area of the pressure diagrams about the bottom of the wall, or

$$P_a \bar{z} = (122.16)\left(\tfrac{6}{3}\right) - (107.7)\left(\tfrac{6}{2}\right)$$

or

$$\bar{z} = \frac{244.32 - 323.1}{14.46} = -\mathbf{5.45 \; m}$$

Active Force After the Occurrence of Tensile Crack: Eq. (6.14)

$$z_c = \frac{2c}{\gamma\sqrt{K_a}} = \frac{2(14.36)}{(17.4)(0.625)} = 2.64 \text{ m}$$

Using Eq. (6.16) gives

$$P_a = \tfrac{1}{2}(H - z_c)(\gamma H K_a - 2c\sqrt{K_a}) = \tfrac{1}{2}(6 - 2.64)(22.77) = 38.25 \text{ kN/m}$$

Figure 6.7c shows that the force $P_a = 38.25$ kN/m is the area of the hatched triangle. Hence the line of action of the resultant will be located at a height of $\bar{z} = (H - z_c)/3$ above the bottom of the wall, or

$$\bar{z} = \frac{6 - 2.64}{3} = \mathbf{1.12 \; m}$$

▲

For most retaining wall construction, a granular backfill is used and $c = 0$. Thus Example 6.2 is an academic problem; however, it illustrates the basic principles of the Rankine active earth pressure calculation.

▼ **EXAMPLE 6.3**

For the retaining wall shown in Figure 6.9a, assume that the wall can yield sufficiently to develop active state. Determine the Rankine active force per unit length of the wall and the location of the resultant line of action.

Solution   If the cohesion, c, is equal to zero

$$\sigma'_a = \sigma'_v K_a$$

For the top soil layer, $\phi_1 = 30°$, so

$$K_{a(1)} = \tan^2\left(45 - \frac{\phi_1}{2}\right) = \tan^2(45 - 15) = \frac{1}{3}$$

Similarly, for the bottom soil layer, $\phi_2 = 36°$, and

$$K_{a(2)} = \tan^2\left(45 - \frac{36}{2}\right) = 0.26$$

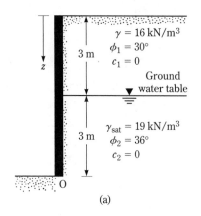

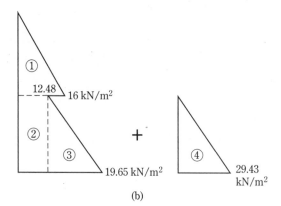

(a)   (b)

▼ **FIGURE 6.9**

Because of the presence of the water table, the effective lateral pressure and the hydrostatic pressure have to be calculated separately.

At $z = 0$, $\sigma'_v = 0$, $\sigma'_a = 0$

At $z = 3$ m, $\sigma'_v = \gamma z = (16)(3) = 48 \text{ kN/m}^2$

At this depth, for the top soil layer

$$\sigma'_a = K_{a(1)}\sigma'_v = (\tfrac{1}{3})(48) = 16 \text{ kN/m}^2$$

Similarly, for the bottom soil layer

$$\sigma'_a = K_{a(2)}\sigma'_v = (0.26)(48) = 12.48 \text{ kN/m}^2$$

$$\text{At } z = 6 \text{ m}, \sigma'_v = (\gamma)(3) + (\gamma_{\text{sat}} - \gamma_w)(3) = (16)(3) + (19 - 9.81)(3)$$
$$= 48 + 27.57 = 75.57 \text{ kN/m}^2$$

$$\sigma'_a = K_{a(2)}\sigma'_v = (0.26)(75.57) = 19.65 \text{ kN/m}^2$$

The hydrostatic pressure, u, is zero from $z = 0$ to $z = 3$ m. At $z = 6$ m, $u = 3\gamma_w = 3(9.81) = 29.43 \text{ kN/m}^2$. The pressure distribution diagram is plotted in Figure 6.9b. The force per unit length

$$P_a = \text{Area 1} + \text{Area 2} + \text{Area 3} + \text{Area 4}$$
$$= \tfrac{1}{2}(3)(16) + (3)(12.48) + \tfrac{1}{2}(3)(19.65 - 12.48) + \tfrac{1}{2}(3)(29.43)$$
$$= 24 + 37.44 + 10.76 + 44.15 = 116.35 \text{ kN/m}$$

The distance of the line of action of the resultant from the bottom of the wall ($\bar{z}$) can be determined by taking the moments about the bottom of the wall (point

O in Figure 6.9a), or

$$\bar{z} = \frac{(24)\left(3+\dfrac{3}{3}\right) + (37.44)\left(\dfrac{3}{2}\right) + (10.76)\left(\dfrac{3}{3}\right) + (44.15)\left(\dfrac{3}{3}\right)}{116.35}$$

$$= \frac{96 + 56.16 + 10.76 + 44.15}{116.35} = \mathbf{1.78\,m}$$

▲

▼ **EXAMPLE 6.4** _____

Refer to Example 6.3. Other quantities remaining the same, assume that, in the top layer, $c_1 = 24\ \mathrm{kN/m^2}$ (not zero as in Example 6.3). Determine P_a after the occurrence of the tensile crack.

Solution   From Eq. (6.14)

$$z_c = \frac{2c_1}{\gamma\sqrt{K_{a(1)}}} = \frac{(2)(24)}{(16)\sqrt{\left(\frac{1}{3}\right)}} = 5.2\ \mathrm{m}$$

Since the depth of the top layer is only 3 m, the depth of the tensile crack will be only 3 m. So the pressure diagram up to $z = 3$ m will be zero. For $z > 3$ m, the pressure diagram will be the same as shown in Figure 6.9, or

$$P_a = \underbrace{\mathrm{Area\ 2 + Area\ 3 + Area\ 4}}_{\mathrm{Figure\ 6.9}}$$

$$= 37.44 + 10.76 + 44.15 = \mathbf{92.35\ kN/m}$$

▲

6.4   RANKINE ACTIVE EARTH PRESSURE FOR INCLINED BACKFILL

If the backfill of a frictionless retaining wall is a *granular soil* ($c = 0$) and rises at an angle α with respect to the horizontal (Figure 6.10), the *active earth pressure coefficient, K_a,* may be expressed in the form

$$K_a = \cos\alpha\,\frac{\cos\alpha - \sqrt{\cos^2\alpha - \cos^2\phi}}{\cos\alpha + \sqrt{\cos^2\alpha - \cos^2\phi}} \tag{6.19}$$

where ϕ = angle of friction of soil

At any depth, z, the *Rankine active pressure* may be expressed as

$$\sigma_a = \gamma z K_a \tag{6.20}$$

Also, the total force per unit length of the wall is

$$P_a = \tfrac{1}{2}\gamma H^2 K_a \tag{6.21}$$

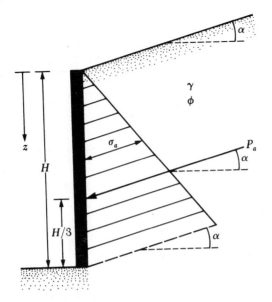

▼   **FIGURE 6.10**   Notations for active pressure —
Eqs. (6.19), (6.20), (6.21)

Note that, in this case, the direction of the resultant force, P_a, is *inclined at an angle α with the horizontal* and intersects the wall at a distance of $H/3$ from the base of the wall. Table 6.2 presents the values of K_a (active earth pressure) for various values of α and ϕ.

The preceding analysis can be extended for an inclined backfill with a $c-\phi$ soil. The details of the mathematical derivation are given by Mazindrani and Ganjali (1997). As in Eq. (6.20), for this case

$$\sigma_a = \gamma z K_a = \gamma z K_a' \cos \alpha \tag{6.22}$$

▼   **TABLE 6.2**   Active Earth Pressure Coefficient, K_a [Eq. (6.19)]

↓ α (deg)	ϕ (deg)→						
	28	30	32	34	36	38	40
0	0.361	0.333	0.307	0.283	0.260	0.238	0.217
5	0.366	0.337	0.311	0.286	0.262	0.240	0.219
10	0.380	0.350	0.321	0.294	0.270	0.246	0.225
15	0.409	0.373	0.341	0.311	0.283	0.258	0.235
20	0.461	0.414	0.374	0.338	0.306	0.277	0.250
25	0.573	0.494	0.434	0.385	0.343	0.307	0.275

▼ **TABLE 6.3** Values of K'_a

ϕ (deg)	α (deg)	$\dfrac{c}{\gamma z}$			
		0.025	0.05	0.1	0.5
15	0	0.550	0.512	0.435	−0.179
	5	0.566	0.525	0.445	−0.184
	10	0.621	0.571	0.477	−0.186
	15	0.776	0.683	0.546	−0.196
20	0	0.455	0.420	0.350	−0.210
	5	0.465	0.429	0.357	−0.212
	10	0.497	0.456	0.377	−0.218
	15	0.567	0.514	0.417	−0.229
25	0	0.374	0.342	0.278	−0.231
	5	0.381	0.348	0.283	−0.233
	10	0.402	0.366	0.296	−0.239
	15	0.443	0.401	0.321	−0.250
30	0	0.305	0.276	0.218	−0.244
	5	0.309	0.280	0.221	−0.246
	10	0.323	0.292	0.230	−0.252
	15	0.350	0.315	0.246	−0.263

where

$$K'_a = \frac{1}{\cos^2 \phi} \left\{ \frac{2 \cos^2 \alpha + 2 \left(\dfrac{c}{\gamma z} \right) \cos \phi \sin \phi}{- \sqrt{\left[4 \cos^2 \alpha (\cos^2 \alpha - \cos^2 \phi) + 4 \left(\dfrac{c}{\gamma z} \right)^2 \cos^2 \phi + 8 \left(\dfrac{c}{\gamma z} \right) \cos^2 \alpha \sin \phi \cos \phi \right]}} \right\} - 1 \quad (6.23)$$

Some values of K'_a are given in Table 6.3. For a problem of this type, the depth of tensile crack, z_c, is given as

$$z_c = \frac{2c}{\gamma} \sqrt{\frac{1 + \sin \phi}{1 - \sin \phi}} \quad (6.24)$$

▼ **EXAMPLE 6.5**

Refer to the retaining wall shown in Figure 6.10. Given: $H = 7.5$ m, $\gamma = 18$ kN/m³, $\phi = 20°$, $c = 13.5$ kN/m², and $\alpha = 10°$. Calculate the Rankine active force, P_a, per unit length of the wall and the location of the resultant after the occurrence of the tensile crack.

Solution   From Eq. (6.24),

$$z_c = \frac{2c}{\gamma} \sqrt{\frac{1 + \sin \phi}{1 - \sin \phi}} = \frac{(2)(13.5)}{18} \sqrt{\frac{1 + \sin 20}{1 - \sin 20}} = 2.14 \text{ m}$$

At $z = 7.5$ m

$$\frac{c}{\gamma z} = \frac{13.5}{(18)(7.5)} = 0.1$$

From Table 6.3, for 20°, $c/\gamma z = 0.1$ and $\alpha = 10°$, the value of K_a' is 0.377, so at $z = 7.5$ m

$$\sigma_a = \gamma z K_a' \cos \alpha = (18)(7.5)(0.377)(\cos 10) = 50.1 \text{ kN/m}^2$$

After the occurrence of the tensile crack, the pressure distribution on the wall will be as shown in Figure 6.11, so

$$P_a = \left(\frac{1}{2}\right)(50.1)(7.5 - 2.14) = \textbf{134.3 kN/m}$$

$$\bar{z} = \frac{7.5 - 2.14}{3} = \textbf{1.79 m}$$

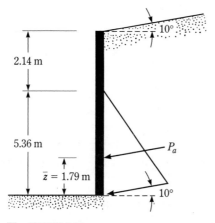

▼   FIGURE 6.11   ▲

6.5   COULOMB'S ACTIVE EARTH PRESSURE

The Rankine active earth pressure calculations discussed in the preceding sections were based on the assumption that the wall is frictionless. In 1776, Coulomb proposed a theory to calculate the lateral earth pressure on a retaining wall with granular soil backfill. This theory takes wall friction into consideration.

To apply Coulomb's active earth pressure theory, let us consider a retaining wall with its back face inclined at an angle β with the horizontal, as shown in Figure 6.12a. The backfill is a granular soil that slopes at an angle α with the horizontal. Also, let δ be the angle of friction between the soil and the wall (that is, angle of wall friction).

Under active pressure the wall will move away from the soil mass (to the left in Figure 6.12a). Coulomb assumed that, in such a case, the failure surface in the soil mass would be a plane (e.g., BC_1, BC_2, ...). So, to find the active force in our

$$P_a = \frac{1}{2}K_a\gamma H^2$$

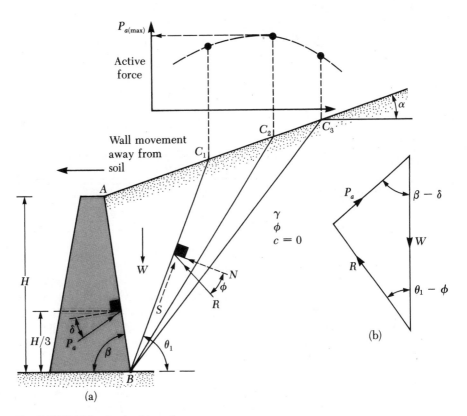

▼ **FIGURE 6.12**   Coulomb's active pressure

example, consider a possible soil failure wedge ABC_1. The forces acting on this wedge, ABC_1 (per unit length at right angles to the cross section shown), are as follows:

1. Weight of the wedge, W.
2. The resultant, R, of the normal and resisting shear forces along the surface, BC_1. The force R will be inclined at an angle ϕ to the normal drawn to the surface BC_1.
3. The active force per unit length of the wall, P_a. The force P_a will be inclined at an angle δ to the normal drawn to the back face of the wall.

For equilibrium purposes, a force triangle can be drawn, as shown in Figure 6.12b. Note that θ_1 is the angle that BC_1 makes with the horizontal. Because the magnitude of W as well as the directions of all three forces are known, the value of P_a can now be determined. Similarly, the active forces of other trial wedges, such as ABC_2, ABC_3, . . . can be determined. The maximum value of P_a thus determined is Coulomb's active force (see top part of Figure 6.12), which may be expressed as

$$P_a = \tfrac{1}{2}K_a\gamma H_s^2 \qquad\qquad (6.25)$$

where

$$
\begin{aligned}
K_a &= \text{Coulomb's active earth pressure coefficient}\\[4pt]
&= \frac{\sin^2(\beta + \phi)}{\sin^2\beta\,\sin(\beta - \delta)\left[1 + \sqrt{\dfrac{\sin(\phi + \delta)\sin(\phi - \alpha)}{\sin(\beta - \delta)\sin(\alpha + \beta)}}\right]^2}
\end{aligned}
\qquad (6.26)
$$

and H = height of the wall

The values of the active earth pressure coefficient, K_a, for a vertical retaining wall ($\beta = 90°$) with horizontal backfill ($\alpha = 0°$) are given in Table 6.4. Note that the line of action of the resultant (P_a) will act at a distance of $H/3$ above the base of the wall and will be inclined at an angle δ to the normal drawn to the back of the wall.

In the actual design of retaining walls, the value of the wall friction angle, δ, is assumed to be between $\phi/2$ and $\tfrac{2}{3}\phi$. The active earth pressure coefficients for various values of ϕ, α, and β with $\delta = \tfrac{1}{2}\phi$ and $\tfrac{2}{3}\phi$ are given in Tables 6.5 and 6.6. These coefficients are very useful design considerations.

If a uniform surcharge of intensity q is located above the backfill, as shown in

▼ **TABLE 6.4** Values of K_a [Eq. (6.26)] for $\beta = 90°$, $\alpha = 0°$

ϕ (deg)	δ (deg)					
	0	5	10	15	20	25
28	0.3610	0.3448	0.3330	0.3251	0.3203	0.3186
30	0.3333	0.3189	0.3085	0.3014	0.2973	0.2956
32	0.3073	0.2945	0.2853	0.2791	0.2755	0.2745
34	0.2827	0.2714	0.2633	0.2579	0.2549	0.2542
36	0.2596	0.2497	0.2426	0.2379	0.2354	0.2350
38	0.2379	0.2292	0.2230	0.2190	0.2169	0.2167
40	0.2174	0.2098	0.2045	0.2011	0.1994	0.1995
42	0.1982	0.1916	0.1870	0.1841	0.1828	0.1831

Figure 6.13, the active force, P_a, can be calculated as

$$P_a = \tfrac{1}{2}K_a\gamma_{eq}H^2$$
$$\uparrow$$
$$\text{Eq. (6.26)}$$

(6.27)

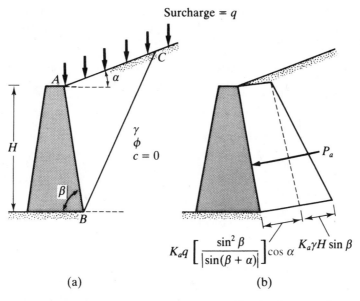

(a)   (b)

▼ **FIGURE 6.13** Coulomb's active pressure with a surcharge on the backfill

▼ TABLE 6.5 Values of K_a [Eq. (6.26)]. Note: $\delta = \frac{2}{3}\phi$

α (deg)	ϕ (deg)	β (deg) 90	85	80	75	70	65
0	28	0.3213	0.3588	0.4007	0.4481	0.5026	0.5662
	29	0.3091	0.3467	0.3886	0.4362	0.4908	0.5547
	30	0.2973	0.3349	0.3769	0.4245	0.4794	0.5435
	31	0.2860	0.3235	0.3655	0.4133	0.4682	0.5326
	32	0.2750	0.3125	0.3545	0.4023	0.4574	0.5220
	33	0.2645	0.3019	0.3439	0.3917	0.4469	0.5117
	34	0.2543	0.2916	0.3335	0.3813	0.4367	0.5017
	35	0.2444	0.2816	0.3235	0.3713	0.4267	0.4919
	36	0.2349	0.2719	0.3137	0.3615	0.4170	0.4824
	37	0.2257	0.2626	0.3042	0.3520	0.4075	0.4732
	38	0.2168	0.2535	0.2950	0.3427	0.3983	0.4641
	39	0.2082	0.2447	0.2861	0.3337	0.3894	0.4553
	40	0.1998	0.2361	0.2774	0.3249	0.3806	0.4468
	41	0.1918	0.2278	0.2689	0.3164	0.3721	0.4384
	42	0.1840	0.2197	0.2606	0.3080	0.3637	0.4302
5	28	0.3431	0.3845	0.4311	0.4843	0.5461	0.6190
	29	0.3295	0.3709	0.4175	0.4707	0.5325	0.6056
	30	0.3165	0.3578	0.4043	0.4575	0.5194	0.5926
	31	0.3039	0.3451	0.3916	0.4447	0.5067	0.5800
	32	0.2919	0.3329	0.3792	0.4324	0.4943	0.5677
	33	0.2803	0.3211	0.3673	0.4204	0.4823	0.5558
	34	0.2691	0.3097	0.3558	0.4088	0.4707	0.5443
	35	0.2583	0.2987	0.3446	0.3975	0.4594	0.5330
	36	0.2479	0.2881	0.3338	0.3866	0.4484	0.5221
	37	0.2379	0.2778	0.3233	0.3759	0.4377	0.5115
	38	0.2282	0.2679	0.3131	0.3656	0.4273	0.5012
	39	0.2188	0.2582	0.3033	0.3556	0.4172	0.4911
	40	0.2098	0.2489	0.2937	0.3458	0.4074	0.4813
	41	0.2011	0.2398	0.2844	0.3363	0.3978	0.4718
	42	0.1927	0.2311	0.2753	0.3271	0.3884	0.4625
10	28	0.3702	0.4164	0.4686	0.5287	0.5992	0.6834
	29	0.3548	0.4007	0.4528	0.5128	0.5831	0.6672
	30	0.3400	0.3857	0.4376	0.4974	0.5676	0.6516
	31	0.3259	0.3713	0.4230	0.4826	0.5526	0.6365
	32	0.3123	0.3575	0.4089	0.4683	0.5382	0.6219
	33	0.2993	0.3442	0.3953	0.4545	0.5242	0.6078
	34	0.2868	0.3314	0.3822	0.4412	0.5107	0.5942
	35	0.2748	0.3190	0.3696	0.4283	0.4976	0.5810
	36	0.2633	0.3072	0.3574	0.4158	0.4849	0.5682
	37	0.2522	0.2957	0.3456	0.4037	0.4726	0.5558
	38	0.2415	0.2846	0.3342	0.3920	0.4607	0.5437
	39	0.2313	0.2740	0.3231	0.3807	0.4491	0.5321
	40	0.2214	0.2636	0.3125	0.3697	0.4379	0.5207
	41	0.2119	0.2537	0.3021	0.3590	0.4270	0.5097
	42	0.2027	0.2441	0.2921	0.3487	0.4164	0.4990

▼ **TABLE 6.5** Continued

α (deg)	ϕ (deg)	β (deg)					
		90	85	80	75	70	65
15	28	0.4065	0.4585	0.5179	0.5868	0.6685	0.7670
	29	0.3881	0.4397	0.4987	0.5672	0.6483	0.7463
	30	0.3707	0.4219	0.4804	0.5484	0.6291	0.7265
	31	0.3541	0.4049	0.4629	0.5305	0.6106	0.7076
	32	0.3384	0.3887	0.4462	0.5133	0.5930	0.6895
	33	0.3234	0.3732	0.4303	0.4969	0.5761	0.6721
	34	0.3091	0.3583	0.4150	0.4811	0.5598	0.6554
	35	0.2954	0.3442	0.4003	0.4659	0.5442	0.6393
	36	0.2823	0.3306	0.3862	0.4513	0.5291	0.6238
	37	0.2698	0.3175	0.3726	0.4373	0.5146	0.6089
	38	0.2578	0.3050	0.3595	0.4237	0.5006	0.5945
	39	0.2463	0.2929	0.3470	0.4106	0.4871	0.5805
	40	0.2353	0.2813	0.3348	0.3980	0.4740	0.5671
	41	0.2247	0.2702	0.3231	0.3858	0.4613	0.5541
	42	0.2146	0.2594	0.3118	0.3740	0.4491	0.5415
20	28	0.4602	0.5205	0.5900	0.6714	0.7689	0.8880
	29	0.4364	0.4958	0.5642	0.6445	0.7406	0.8581
	30	0.4142	0.4728	0.5403	0.6195	0.7144	0.8303
	31	0.3935	0.4513	0.5179	0.5961	0.6898	0.8043
	32	0.3742	0.4311	0.4968	0.5741	0.6666	0.7799
	33	0.3559	0.4121	0.4769	0.5532	0.6448	0.7569
	34	0.3388	0.3941	0.4581	0.5335	0.6241	0.7351
	35	0.3225	0.3771	0.4402	0.5148	0.6044	0.7144
	36	0.3071	0.3609	0.4233	0.4969	0.5856	0.6947
	37	0.2925	0.3455	0.4071	0.4799	0.5677	0.6759
	38	0.2787	0.3308	0.3916	0.4636	0.5506	0.6579
	39	0.2654	0.3168	0.3768	0.4480	0.5342	0.6407
	40	0.2529	0.3034	0.3626	0.4331	0.5185	0.6242
	41	0.2408	0.2906	0.3490	0.4187	0.5033	0.6083
	42	0.2294	0.2784	0.3360	0.4049	0.4888	0.5930

where

$$\gamma_{eq} = \gamma + \left[\frac{\sin \beta}{\sin(\beta + \alpha)}\right]\left(\frac{2q}{H}\right)\cos \alpha \qquad (6.28)$$

The derivations of Eqs. (6.27) and (6.28) are contained in other soil mechanics texts (e.g., Das, 1987).

▼ **TABLE 6.6** Values of K_a [Eq. (6.26)]. Note: $\delta = \phi/2$

α (deg)	ϕ (deg)	β (deg)					
		90	85	80	75	70	65
0	28	0.3264	0.3629	0.4034	0.4490	0.5011	0.5616
	29	0.3137	0.3502	0.3907	0.4363	0.4886	0.5492
	30	0.3014	0.3379	0.3784	0.4241	0.4764	0.5371
	31	0.2896	0.3260	0.3665	0.4121	0.4645	0.5253
	32	0.2782	0.3145	0.3549	0.4005	0.4529	0.5137
	33	0.2671	0.3033	0.3436	0.3892	0.4415	0.5025
	34	0.2564	0.2925	0.3327	0.3782	0.4305	0.4915
	35	0.2461	0.2820	0.3221	0.3675	0.4197	0.4807
	36	0.2362	0.2718	0.3118	0.3571	0.4092	0.4702
	37	0.2265	0.2620	0.3017	0.3469	0.3990	0.4599
	38	0.2172	0.2524	0.2920	0.3370	0.3890	0.4498
	39	0.2081	0.2431	0.2825	0.3273	0.3792	0.4400
	40	0.1994	0.2341	0.2732	0.3179	0.3696	0.4304
	41	0.1909	0.2253	0.2642	0.3087	0.3602	0.4209
	42	0.1828	0.2168	0.2554	0.2997	0.3511	0.4117
5	28	0.3477	0.3879	0.4327	0.4837	0.5425	0.6115
	29	0.3337	0.3737	0.4185	0.4694	0.5282	0.5972
	30	0.3202	0.3601	0.4048	0.4556	0.5144	0.5833
	31	0.3072	0.3470	0.3915	0.4422	0.5009	0.5698
	32	0.2946	0.3342	0.3787	0.4292	0.4878	0.5566
	33	0.2825	0.3219	0.3662	0.4166	0.4750	0.5437
	34	0.2709	0.3101	0.3541	0.4043	0.4626	0.5312
	35	0.2596	0.2986	0.3424	0.3924	0.4505	0.5190
	36	0.2488	0.2874	0.3310	0.3808	0.4387	0.5070
	37	0.2383	0.2767	0.3199	0.3695	0.4272	0.4954
	38	0.2282	0.2662	0.3092	0.3585	0.4160	0.4840
	39	0.2185	0.2561	0.2988	0.3478	0.4050	0.4729
	40	0.2090	0.2463	0.2887	0.3374	0.3944	0.4620
	41	0.1999	0.2368	0.2788	0.3273	0.3840	0.4514
	42	0.1911	0.2276	0.2693	0.3174	0.3738	0.4410
10	28	0.3743	0.4187	0.4688	0.5261	0.5928	0.6719
	29	0.3584	0.4026	0.4525	0.5096	0.5761	0.6549
	30	0.3432	0.3872	0.4368	0.4936	0.5599	0.6385
	31	0.3286	0.3723	0.4217	0.4782	0.5442	0.6225
	32	0.3145	0.3580	0.4071	0.4633	0.5290	0.6071
	33	0.3011	0.3442	0.3930	0.4489	0.5143	0.5920
	34	0.2881	0.3309	0.3793	0.4350	0.5000	0.5775
	35	0.2757	0.3181	0.3662	0.4215	0.4862	0.5633
	36	0.2637	0.3058	0.3534	0.4084	0.4727	0.5495
	37	0.2522	0.2938	0.3411	0.3957	0.4597	0.5361
	38	0.2412	0.2823	0.3292	0.3833	0.4470	0.5230
	39	0.2305	0.2712	0.3176	0.3714	0.4346	0.5103
	40	0.2202	0.2604	0.3064	0.3597	0.4226	0.4979
	41	0.2103	0.2500	0.2956	0.3484	0.4109	0.4858
	42	0.2007	0.2400	0.2850	0.3375	0.3995	0.4740

▼ **TABLE 6.6** Continued

α (deg)	ϕ (deg)	β (deg)					
		90	85	80	75	70	65
15	28	0.4095	0.4594	0.5159	0.5812	0.6579	0.7498
	29	0.3908	0.4402	0.4964	0.5611	0.6373	0.7284
	30	0.3730	0.4220	0.4777	0.5419	0.6175	0.7080
	31	0.3560	0.4046	0.4598	0.5235	0.5985	0.6884
	32	0.3398	0.3880	0.4427	0.5059	0.5803	0.6695
	33	0.3244	0.3721	0.4262	0.4889	0.5627	0.6513
	34	0.3097	0.3568	0.4105	0.4726	0.5458	0.6338
	35	0.2956	0.3422	0.3953	0.4569	0.5295	0.6168
	36	0.2821	0.3282	0.3807	0.4417	0.5138	0.6004
	37	0.2692	0.3147	0.3667	0.4271	0.4985	0.5846
	38	0.2569	0.3017	0.3531	0.4130	0.4838	0.5692
	39	0.2450	0.2893	0.3401	0.3993	0.4695	0.5543
	40	0.2336	0.2773	0.3275	0.3861	0.4557	0.5399
	41	0.2227	0.2657	0.3153	0.3733	0.4423	0.5258
	42	0.2122	0.2546	0.3035	0.3609	0.4293	0.5122
20	28	0.4614	0.5188	0.5844	0.6608	0.7514	0.8613
	29	0.4374	0.4940	0.5586	0.6339	0.7232	0.8313
	30	0.4150	0.4708	0.5345	0.6087	0.6968	0.8034
	31	0.3941	0.4491	0.5119	0.5851	0.6720	0.7772
	32	0.3744	0.4286	0.4906	0.5628	0.6486	0.7524
	33	0.3559	0.4093	0.4704	0.5417	0.6264	0.7289
	34	0.3384	0.3910	0.4513	0.5216	0.6052	0.7066
	35	0.3218	0.3736	0.4331	0.5025	0.5851	0.6853
	36	0.3061	0.3571	0.4157	0.4842	0.5658	0.6649
	37	0.2911	0.3413	0.3991	0.4668	0.5474	0.6453
	38	0.2769	0.3263	0.3833	0.4500	0.5297	0.6266
	39	0.2633	0.3120	0.3681	0.4340	0.5127	0.6085
	40	0.2504	0.2982	0.3535	0.4185	0.4963	0.5912
	41	0.2381	0.2851	0.3395	0.4037	0.4805	0.5744
	42	0.2263	0.2725	0.3261	0.3894	0.4653	0.5582

▼ **EXAMPLE 6.6**

Consider the retaining wall shown in Figure 6.12a. Given: $H = 4.6$ m; unit weight of soil = 16.5 kN/m³; angle of friction of soil = 30°; wall friction-angle, $\delta = \frac{2}{3}\phi$; soil cohesion, $c = 0$; $\alpha = 0$, and $\beta = 90°$. Calculate the Coulomb's active force per unit length of the wall.

Solution From Eq. (6.25)

$$P_a = \frac{1}{2}\gamma H^2 K_a$$

From Table 6.5, for $\alpha = 0°$, $\beta = 90°$, $\phi = 30°$, and $\delta = \frac{2}{3}\phi = 20°$, $K_a = 0.297$. Hence

$$P_a = \frac{1}{2}(16.5)(4.6)^2(0.297) = \mathbf{51.85\ kN/m} \quad \blacktriangle$$

6.6 ACTIVE EARTH PRESSURE FOR EARTHQUAKE CONDITIONS

Coulomb's active earth pressure theory (see Section 6.5) can be extended to take into account the forces caused by an earthquake. Figure 6.14 shows a condition of active pressure with a granular backfill ($c = 0$). Note that the forces acting on the soil failure wedge in Figure 6.14 are essentially the same as those shown in Figure 6.12a, with the addition of k_hW and k_vW in the horizontal and vertical directions, respectively; k_h and k_v may be defined as

$$k_h = \frac{\text{horizontal earthquake acceleration component}}{\text{acceleration due to gravity}, g} \quad (6.29)$$

$$k_v = \frac{\text{vertical earthquake acceleration component}}{\text{acceleration due to gravity}, g} \quad (6.30)$$

As in Section 6.5, the relation for the active force per unit length of the wall (P_{ae}) can be determined as

$$P_{ae} = \frac{1}{2}\gamma H^2(1 - k_v)K_{ae} \quad (6.31)$$

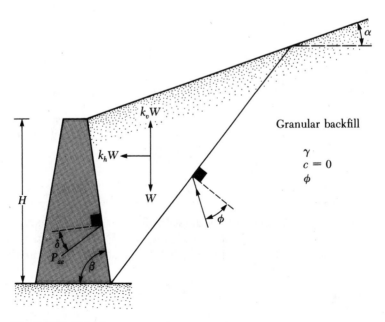

▼ **FIGURE 6.14**   Derivation of Eq. (6.31)

where

$$K_{ae} = \text{active earth pressure coefficient}$$
$$= \frac{\sin^2(\phi + \beta - \theta')}{\cos\theta' \sin^2\beta \sin(\beta - \theta' - \delta)\left[1 + \sqrt{\dfrac{\sin(\phi + \delta)\sin(\phi - \theta' - \alpha)}{\sin(\beta - \delta - \theta')\sin(\alpha + \beta)}}\right]^2}$$

(6.32)

$$\theta' = \tan^{-1}\left[\frac{k_h}{1 - k_v}\right]$$

(6.33)

Note that for no earthquake condition

$$k_h = 0, \qquad k_v = 0, \qquad \text{and} \qquad \theta' = 0$$

Hence $K_{ae} = K_a$ [as given by Eq. (6.26)].

The variation of $K_{ae}\cos\delta$ with k_h for the case of $k_v = 0$, $\beta = 90°$, $\alpha = 0°$, and $\delta = \phi/2$ is shown in Figure 6.15. Some values of K_{ae} for $\beta = 90°$ and $k_v = 0$ are given in Table 6.7.

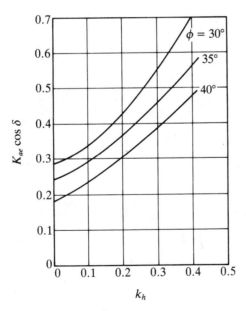

▼ **FIGURE 6.15**   Variation of $K_{ae}\cos\delta$ with k_h (*note:*
$k_v = 0$, $\beta = 90°$, $\alpha = 0°$, and $\delta = \phi/2$). (Note: $K_{ae}\cos\delta$ is the component of earth pressure coefficient at right angles to the back face of the wall.)

▼ TABLE 6.7 Values of K_{ae} [Eq. (6.32)] for $\beta = 90°$ and $k_v = 0$

k_h	δ (deg)	α (deg)	ϕ (deg) 28	30	35	40	45
0.1	0	0	0.427	0.397	0.328	0.268	0.217
0.2			0.508	0.473	0.396	0.382	0.270
0.3			0.611	0.569	0.478	0.400	0.334
0.4			0.753	0.697	0.581	0.488	0.409
0.5			1.005	0.890	0.716	0.596	0.500
0.1	0	5	0.457	0.423	0.347	0.282	0.227
0.2			0.554	0.514	0.424	0.349	0.285
0.3			0.690	0.635	0.522	0.431	0.356
0.4			0.942	0.825	0.653	0.535	0.442
0.5			—	—	0.855	0.673	0.551
0.1	0	10	0.497	0.457	0.371	0.299	0.238
0.2			0.623	0.570	0.461	0.375	0.303
0.3			0.856	0.748	0.585	0.472	0.383
0.4			—	—	0.780	0.604	0.486
0.5			—	—	—	0.809	0.624
0.1	$\phi/2$	0	0.396	0.368	0.306	0.253	0.207
0.2			0.485	0.452	0.380	0.319	0.267
0.3			0.604	0.563	0.474	0.402	0.340
0.4			0.778	0.718	0.599	0.508	0.433
0.5			1.115	0.972	0.774	0.648	0.552
0.1	$\phi/2$	5	0.428	0.396	0.326	0.268	0.218
0.2			0.537	0.497	0.412	0.342	0.283
0.3			0.699	0.640	0.526	0.438	0.367
0.4			1.025	0.881	0.690	0.568	0.475
0.5			—	—	0.962	0.752	0.620
0.1	$\phi/2$	10	0.472	0.433	0.352	0.285	0.230
0.2			0.616	0.562	0.454	0.371	0.303
0.3			0.908	0.780	0.602	0.487	0.400
0.4			—	—	0.857	0.656	0.531
0.5			—	—	—	0.944	0.722
0.1	$\frac{2}{3}\phi$	0	0.393	0.366	0.306	0.256	0.212
0.2			0.486	0.454	0.384	0.326	0.276
0.3			0.612	0.572	0.486	0.416	0.357
0.4			0.801	0.740	0.622	0.533	0.462
0.5			1.177	1.023	0.819	0.693	0.600
0.1	$\frac{2}{3}\phi$	5	0.427	0.395	0.327	0.271	0.224
0.2			0.541	0.501	0.418	0.350	0.294
0.3			0.714	0.655	0.541	0.455	0.386
0.4			1.073	0.921	0.722	0.600	0.509
0.5			—	—	1.034	0.812	0.679
0.1	$\frac{2}{3}\phi$	10	0.472	0.434	0.354	0.290	0.237
0.2			0.625	0.570	0.463	0.381	0.317
0.3			0.942	0.807	0.624	0.509	0.423
0.4			—	—	0.909	0.699	0.573
0.5			—	—	—	1.037	0.800

Equation (6.31) is usually referred to as the *Mononobe–Okabe* solution. Unlike the case shown in Figure 6.12a, the resultant earth pressure in this situation, as calculated by Eq. (6.31) *does not act* at a distance of $H/3$ from the bottom of the wall. The following procedure may be used to obtain the location of the resultant force P_{ae}:

1. Calculate P_{ae} by using Eq. (6.31)
2. Calculate P_a by using Eq. (6.25)
3. Calculate

$$\Delta P_{ae} = P_{ae} - P_a \tag{6.34}$$

4. Assume that P_a acts at a distance of $H/3$ from the bottom of the wall (Figure 6.16).
5. Assume that ΔP_{ae} acts at a distance of $0.6H$ from the bottom of the wall (Figure 6.16).
6. Calculate the location of the resultant as

$$\bar{z} = \frac{(0.6H)(\Delta P_{ae}) + \left(\dfrac{H}{3}\right)(P_a)}{P_{ae}} \tag{6.35}$$

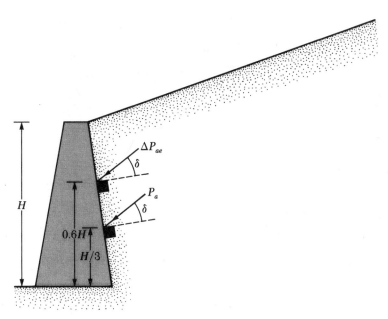

▼ **FIGURE 6.16**   Determining the line of action of P_{ae}

▼ **EXAMPLE 6.7**

Refer to Figure 6.17. For $k_v = 0$ and $k_h = 0.3$, determine:

a. P_{ae}
b. The location of the resultant, $\bar{z}$, from the bottom of the wall

Solution

Part a

From Eq. (6.31),

$$P_{ae} = \tfrac{1}{2}\gamma H^2(1 - k_v)K_{ae}$$

Here, $\gamma = 105$ lb/ft³, $H = 10$ ft, and $k_v = 0$. As $\delta = \phi/2$, we can use Figure 6.15 to determine K_{ae}. For $k_h = 0.3$, $K_{ae} \approx 0.472$, so

$$P_{ae} = \tfrac{1}{2}(105)(10)^2(1 - 0)(0.472) = \textbf{2478 lb/ft}$$

Part b

From Eq. (6.25),

$$P_a = \tfrac{1}{2}\gamma H^2 K_a$$

From Eq. (6.26) with $\delta = 17.5°$, $\beta = 90°$, and $\alpha = 0°$, $K_a \approx 0.246$ (Table 6.6), so

$$P_a = \tfrac{1}{2}(105)(10)^2(0.246) = 1292 \text{ lb/ft}$$

$$\Delta P_{ae} = P_{ae} - P_a = 2478 - 1292 = 1186 \text{ lb/ft}$$

From Eq. (6.35),

$$\bar{z} = \frac{(0.6H)(\Delta P_{ae}) + (H/3)(P_a)}{P_{ae}}$$

$$= \frac{[(0.6)(10)](1292) + (10/3)(1186)}{2478} = \textbf{4.72 ft}$$

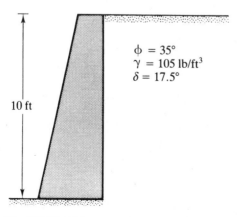

$\phi = 35°$
$\gamma = 105$ lb/ft³
$\delta = 17.5°$

10 ft

▼ **FIGURE 6.17**

6.7 LATERAL EARTH PRESSURE DUE TO SURCHARGE

In several instances, the theory of elasticity is used to determine the lateral earth pressure on retaining structures caused by various types of surcharge loading, such as *line loading* (Figure 6.18a) and *strip loading* (Figure 6.18b).

According to the theory of elasticity, the stress at any depth, z, on a retaining structure caused by a line load of intensity q/unit length (Figure 6.18a) may be given as

$$\sigma = \frac{2q}{\pi H} \frac{a^2 b}{(a^2 + b^2)^2} \tag{6.36}$$

where σ = horizontal stress at depth $z = bH$

(See Figure 6.18a for explanations of the terms a and b.)

However, because soil is not a perfectly elastic medium, some deviations from Eq. (6.36) may be expected. The modified forms of this equation generally accepted for use with soils are as follows:

$$\sigma = \frac{4q}{\pi H} \frac{a^2 b}{(a^2 + b^2)^2} \qquad \text{for } a > 0.4 \tag{6.37}$$

and

$$\sigma = \frac{q}{H} \frac{0.203b}{(0.16 + b^2)^2} \qquad \text{for } a \le 0.4 \tag{6.38}$$

Figure 6.18b shows a strip load with an intensity of q/unit area located at a distance b' from a wall of height H. Based on the theory of elasticity, the horizontal stress, σ, at any depth z on a retaining structure is

$$\sigma = \frac{q}{\pi} (\beta - \sin \beta \cos 2\alpha) \tag{6.39}$$

(The angles α and β are defined in Figure 6.18b.)

However, in the case of soils, the right-hand side of Eq. (6.39) is doubled to account for the yielding soil continuum, or

$$\sigma = \frac{2q}{\pi} (\beta - \sin \beta \cos 2\alpha) \tag{6.40}$$

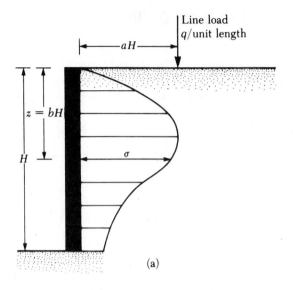

(a)

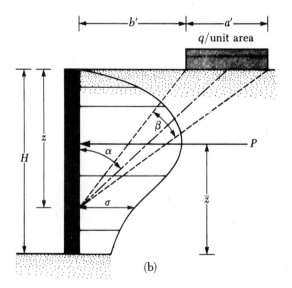

(b)

▼   **FIGURE 6.18**   Lateral earth pressure caused by (a) line load and (b) strip load

The total force per unit length (*P*) due to the *strip loading only* (Jarquio, 1981) may be expressed as

$$P = \frac{q}{90} [H(\theta_2 - \theta_1)] \qquad (6.41)$$

where

$$\theta_1 = \tan^{-1}\left(\frac{b'}{H}\right) \quad \text{(deg)} \qquad (6.42)$$

$$\theta_2 = \tan^{-1}\left(\frac{a' + b'}{H}\right) \quad \text{(deg)} \qquad (6.43)$$

▼ **EXAMPLE 6.8**

Refer to Figure 6.18b. Here, $a' = 2$ m, $b' = 1$ m, $q = 40$ kN/m², and $H = 6$ m. Determine the total pressure on the wall caused by the strip loading only.

Solution   From Eqs. (6.42) and (6.43),

$$\theta_1 = \tan^{-1}\left(\frac{1}{6}\right) = 9.46°$$

$$\theta_2 = \tan^{-1}\left(\frac{2 + 1}{6}\right) = 26.57°$$

From Eq. (6.41),

$$P = \frac{q}{90} [H(\theta_2 - \theta_1)] = \frac{40}{90} [6(26.57 - 9.46)] = \textbf{45.63 kN/m} \qquad ▲$$

6.8 ACTIVE PRESSURE FOR WALL ROTATION ABOUT TOP—BRACED CUT

In the preceding sections, we have seen that a retaining wall rotates about its bottom (Figure 6.19a). With sufficient yielding of the wall, the lateral earth pressure is approximately equal to that obtained by Rankine's theory or Coulomb's theory. In contrast to retaining walls, braced cuts show a different type of wall yielding (see Figure 6.19b). In this case, deformation of the wall gradually increases with the depth of excavation. The variation of the amount of deformation depends on several factors, such as the type of soil, the depth of excavation, and the workmanship. However, with very little wall yielding at the top of the cut, the lateral earth pressure will be close to the at-rest pressure. At the bottom of the wall, with a much larger degree of yielding, the lateral earth pressure will be substantially lower than the Rankine active earth pressure. As a result, the distribution of lateral earth pressure

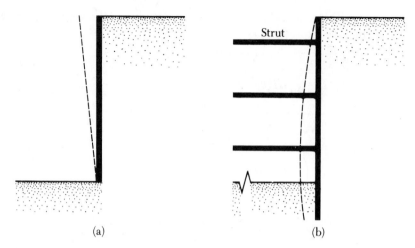

▼   **FIGURE 6.19**   Nature of yielding of walls: (a) retaining wall; (b) braced cut

will vary substantially in comparison to the linear distribution assumed in the case of retaining walls.

The total lateral force per unit length of the wall, P_a, imposed on a wall may be evaluated theoretically by using Terzaghi's (1943) general wedge theory (Figure 6.20). The failure surface is assumed to be the arc of a logarithmic spiral, defined as

$$r = r_0 e^{\theta \tan \phi} \tag{6.44}$$

where ϕ = angle of friction of soil

In Figure 6.20, H is the height of the cut. The unit weight, angle of friction, and cohesion of the soil are equal to γ, ϕ, and c, respectively. Following are the forces per unit length of the cut acting on the trial failure wedge:

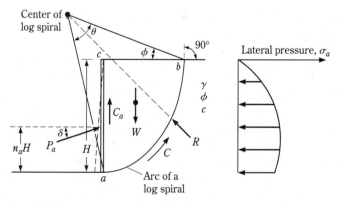

▼   **FIGURE 6.20**   Braced cut analysis by general wedge theory — wall rotation about top

1. Weight of the wedge, W
2. Resultant of the normal and shear forces along ab, R
3. Cohesive force along ab, C
4. Adhesive force along ac, C_a
5. P_a, which is the active force acting a distance $n_a H$ from the bottom of the wall and is inclined at an angle δ to the horizontal

The adhesive force is

$$C_a = c_a H \tag{6.45}$$

where c_a = unit adhesion

A detailed outline for the evaluation of P_a is beyond the scope of this text; those interested should check a soil mechanics text for more information (for example, Das, 1998). Kim and Preber (1969) provided tabulated values of $P_a/\frac{1}{2}\gamma H^2$ determined by using the principles of general wedge theory, and these values are given in Table 6.8. In developing the theoretical values in Table 6.8, it was assumed that

$$\frac{c_a}{c} = \frac{\tan \delta}{\tan \phi} \tag{6.46}$$

6.9 ACTIVE EARTH PRESSURE FOR TRANSLATION OF RETAINING WALL—GRANULAR BACKFILL

Under certain circumstances, retaining walls may undergo lateral translation, as shown in Figure 6.21. A solution to the distribution of active pressure for this case was provided by Dubrova (1963) and was also described by Harr (1966). The solution of Dubrova assumes the validity of Coulomb's solution [Eqs. (6.25) and (6.26)]. In order to understand this procedure, let us consider a vertical wall with a horizontal granular backfill (Figure 6.22). For rotation about the top of the wall, the resultant R of the normal and shear forces along the rupture line AC is inclined at an angle

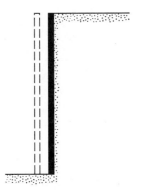

▼ **FIGURE 6.21**   Lateral translation of retaining wall

ϕ, in degrees (1)	δ, in degrees (2)	$n_a = 0.3$ $c/\gamma H$ 0 (3)	0.1 (4)	0.2 (5)	$n_a = 0.4$ $c/\gamma H$ 0 (6)	0.1 (7)	0.2 (8)	$n_a = 0.5$ $c/\gamma H$ 0 (9)	0.1 (10)	0.2 (11)	$n_a = 0.6$ $c/\gamma H$ 0 (12)	0.1 (13)	0.2 (14)
0	0	0.952	0.558	0.164	—	0.652	0.192	—	0.782	0.230	—	0.978	0.288
5	0	0.787	0.431	0.076	0.899	0.495	0.092	1.050	0.580	0.110	1.261	0.697	0.134
	5	0.756	0.345	−0.066	0.863	0.399	−0.064	1.006	0.474	−0.058	1.209	0.573	−0.063
10	0	0.653	0.334	0.015	0.734	0.378	0.021	0.840	0.434	0.027	0.983	0.507	0.032
	5	0.623	0.274	−0.074	0.700	0.312	−0.077	0.799	0.358	−0.082	0.933	0.420	−0.093
	10	0.610	0.242	−0.125	0.685	0.277	−0.131	0.783	0.324	−0.135	0.916	0.380	−0.156
15	0	0.542	0.254	−0.033	0.602	0.285	−0.033	0.679	0.322	−0.034	0.778	0.370	−0.039
	5	0.518	0.214	−0.089	0.575	0.240	−0.094	0.646	0.270	−0.106	0.739	0.310	−0.118
	10	0.505	0.187	−0.131	0.559	0.210	−0.140	0.629	0.238	−0.153	0.719	0.273	−0.174
	15	0.499	0.169	−0.161	0.554	0.191	−0.171	0.623	0.218	−0.187	0.714	0.251	−0.212
20	0	0.499	0.191	−0.067	0.495	0.210	−0.074	0.551	0.236	−0.080	0.622	0.266	−0.090
	5	0.430	0.160	−0.110	0.473	0.179	−0.116	0.526	0.200	−0.126	0.593	0.225	−0.142
	10	0.419	0.140	−0.139	0.460	0.156	−0.149	0.511	0.173	−0.165	0.575	0.196	−0.184
	15	0.413	0.122	−0.169	0.454	0.137	−0.179	0.504	0.154	−0.195	0.568	0.174	−0.219
	20	0.413	0.113	−0.188	0.454	0.124	−0.206	0.504	0.140	−0.223	0.569	0.160	−0.250
25	0	0.371	0.138	−0.095	0.405	0.150	−0.104	0.447	0.167	−0.112	0.499	0.187	−0.125
	5	0.356	0.116	−0.125	0.389	0.128	−0.132	0.428	0.141	−0.146	0.477	0.158	−0.162
	10	0.347	0.099	−0.149	0.378	0.110	−0.158	0.416	0.122	−0.173	0.464	0.136	−0.192
	15	0.342	0.085	−0.172	0.373	0.095	−0.182	0.410	0.106	−0.198	0.457	0.118	−0.221
	20	0.341	0.074	−0.193	0.372	0.083	−0.205	0.409	0.093	−0.222	0.456	0.104	−0.248
	25	0.344	0.065	−0.215	0.375	0.074	−0.228	0.413	0.083	−0.247	0.461	0.093	−0.275
30	0	0.304	0.093	−0.117	0.330	0.103	−0.124	0.361	0.113	−0.136	0.400	0.125	−0.150
	5	0.293	0.078	−0.137	0.318	0.086	−0.145	0.347	0.094	−0.159	0.384	0.105	−0.175
	10	0.286	0.066	−0.154	0.310	0.073	−0.164	0.339	0.080	−0.179	0.374	0.088	−0.198
	15	0.282	0.056	−0.171	0.306	0.060	−0.185	0.334	0.067	−0.199	0.368	0.074	−0.220
	20	0.281	0.047	−0.188	0.305	0.051	−0.204	0.332	0.056	−0.220	0.367	0.062	−0.242
	25	0.284	0.036	−0.211	0.307	0.042	−0.223	0.335	0.047	−0.241	0.370	0.051	−0.267
	30	0.289	0.029	−0.230	0.313	0.033	−0.246	0.341	0.038	−0.265	0.377	0.042	−0.294
35	0	0.247	0.059	−0.129	0.267	0.064	−0.139	0.290	0.069	−0.151	0.318	0.076	−0.165
	5	0.239	0.047	−0.145	0.258	0.052	−0.154	0.280	0.057	−0.167	0.307	0.062	−0.183
	10	0.234	0.038	−0.157	0.252	0.041	−0.170	0.273	0.046	−0.182	0.300	0.050	−0.200
	15	0.231	0.030	−0.170	0.249	0.033	−0.183	0.270	0.035	−0.199	0.296	0.039	−0.218
	20	0.231	0.022	−0.187	0.248	0.025	−0.198	0.269	0.027	−0.215	0.295	0.030	−0.235
	25	0.232	0.015	−0.202	0.250	0.016	−0.218	0.271	0.019	−0.234	0.297	0.020	−0.256
	30	0.236	0.006	−0.224	0.254	0.008	−0.238	0.276	0.011	−0.255	0.302	0.011	−0.281
	35	0.243	0	−0.243	0.262	0.001	−0.260	0.284	0.002	−0.279	0.312	0.002	−0.307
40	0	0.198	0.030	−0.138	0.213	0.032	−0.148	0.230	0.036	−0.159	0.252	0.038	−0.175
	5	0.192	0.021	−0.150	0.206	0.024	−0.158	0.223	0.026	−0.171	0.244	0.029	−0.186
	10	0.189	0.015	−0.158	0.202	0.016	−0.170	0.219	0.018	−0.182	0.238	0.020	−0.199
	15	0.187	0.008	−0.171	0.200	0.010	−0.180	0.216	0.011	−0.195	0.236	0.012	−0.212
	20	0.187	0.003	−0.181	0.200	0.003	−0.195	0.216	0.004	−0.208	0.235	0.004	−0.227
	25	0.188	−0.005		0.202	−0.003		0.218	−0.003		0.237	−0.003	
	30	0.192	−0.010		0.205	−0.010		0.222	−0.011		0.241	−0.012	
	35	0.197	−0.018		0.211	−0.018		0.228	−0.018		0.248	−0.020	
	40	0.205	−0.025		0.220	−0.025		0.237	−0.027		0.259	−0.030	

▼ **TABLE 6.8** Continued

ϕ, in degrees (1)	δ, in degrees (2)	$n_a = 0.3$ $c/\gamma H$			$n_a = 0.4$ $c/\gamma H$			$n_a = 0.5$ $c/\gamma H$			$n_a = 0.6$ $c/\gamma H$		
		0 (3)	0.1 (4)	0.2 (5)	0 (6)	0.1 (7)	0.2 (8)	0 (9)	0.1 (10)	0.2 (11)	0 (12)	0.1 (13)	0.2 (14)
45	0	0.156	0.007	−0.142	0.167	0.008	−0.150	0.180	0.009	−0.162	0.196	0.010	−0.177
	5	0.152	0.002	−0.148	0.163	0.002	−0.158	0.175	0.002	−0.170	0.190	0.003	−0.185
	10	0.150	−0.003		0.160	−0.004		0.172	−0.003		0.187	−0.004	
	15	0.148	−0.009		0.159	−0.008		0.171	−0.009		0.185	−0.010	
	20	0.149	−0.013		0.159	−0.014		0.171	−0.014		0.185	−0.016	
	25	0.150	−0.018		0.160	−0.020		0.173	−0.020		0.187	−0.022	
	30	0.153	−0.025		0.164	−0.026		0.176	−0.026		0.190	−0.029	
	35	0.158	−0.030		0.168	−0.031		0.181	−0.034		0.196	−0.037	
	40	0.164	−0.038		0.175	−0.040		0.188	−0.042		0.204	−0.045	
	45	0.173	−0.046		0.184	−0.048		0.198	−0.052		0.215	−0.057	

* After Kim and Preber (1969)

ϕ to the normal drawn to AC. According to Dubrova there exists infinite number of quasi-rupture lines such as $A'C'$, $A''C''$, ... for which the resultant force R is inclined at an angle ψ, where

$$\psi = \frac{\phi z}{H} \tag{6.47}$$

Now, refer to Eqs. (6.25) and (6.26) for Coulomb's active pressure. For $\beta = 90°$ and $\alpha = 0$, the relationship for Coulomb's active force can also be rewritten as

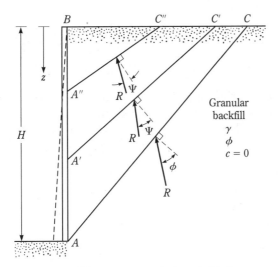

▼ **FIGURE 6.22** Quasi-rupture lines behind a retaining wall

$$P_a = \frac{\gamma}{2 \cos \delta} \left[\frac{H}{\dfrac{1}{\cos \phi} + (\tan^2 \phi + \tan \phi \tan \delta)^{0.5}} \right]^2 \tag{6.48}$$

The force against the wall at any z is then given as

$$P_a = \frac{\gamma}{2 \cos \delta} \left[\frac{z}{\dfrac{1}{\cos \psi} + (\tan^2 \psi + \tan \psi \tan \delta)^{0.5}} \right]^2 \tag{6.49}$$

The active pressure at any depth z for *wall rotation about the top* is

$$\sigma_a(z) = \frac{dP_a}{dz} \approx \frac{\gamma}{\cos \delta} \left[\frac{z \cos^2 \psi}{(1 + m \sin \psi)^2} - \frac{z^2 \phi \cos^2 \psi}{H(1 + m \sin \psi)} (\sin \psi + m) \right] \tag{6.50}$$

where $m = \left(1 + \dfrac{\tan \delta}{\tan \psi} \right)^{0.5}$ \qquad (6.51)

For frictionless walls, $\delta = 0$ and Eq. (6.50) simplifies to

$$\sigma_a(z) = \gamma \tan^2 \left(45 - \frac{\psi}{2} \right) \left(z - \frac{\phi z^2}{H \cos \psi} \right) \tag{6.52}$$

For wall rotation about the bottom, a similar expression can be found in the form

$$\sigma_a(z) = \frac{\gamma z}{\cos \delta} \left(\frac{\cos \phi}{1 + m \sin \phi} \right)^2 \tag{6.53}$$

For translation of the wall, the active pressure can then be taken as

$$\sigma_a(z)_{\text{translation}} = \tfrac{1}{2} [\sigma_a(z)_{\text{rotation about top}} + \sigma_a(z)_{\text{rotation about bottom}}] \tag{6.54}$$

An experimental verification of this procedure was provided by Matsuzawa and Hazazika (1996). The results were obtained from large-scale model tests and are shown in Figure 6.23. The theory and experimental results show good agreement.

▼ EXAMPLE 6.9

Consider a frictionless wall 16 ft high. For the granular backfill, $\gamma = 110$ lb/ft³ and $\phi = 36°$. Calculate and plot the variation of $\sigma_a(z)$ for a translation mode of the wall movement.

Solution   For a frictionless wall, $\delta = 0$. Hence, m is equal to one [Eq. (6.51)]. So, for rotation about the top, from Eq. (6.52),

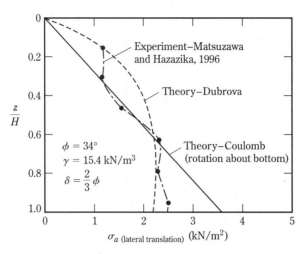

▼ **FIGURE 6.23**   Experimental verification of Dubrova's
theory for lateral wall translation by
large-scale model tests

$$\sigma_a(z) = \sigma_a'(z) = \gamma \tan^2\left(45 - \frac{\phi z}{2H}\right)\left[z - \frac{\phi z^2}{H \cos\left(\dfrac{\phi z}{H}\right)}\right]$$

For rotation about the bottom, from Eq. (6.53),

$$\sigma_a(z) = \sigma_a''(z) = \gamma z \left(\frac{\cos \phi}{1 + \sin \phi}\right)^2$$

$$\sigma_a(z)_{\text{translation}} = \frac{\sigma_a'(z) + \sigma_a''(z)}{2}$$

The following table can now be prepared with $\gamma = 110$ lb/ft³, $\phi = 36°$, and $H = 16$ ft.

z (ft)	$\sigma_a'(z)$ (lb/ft²)	$\sigma_a''(z)$ (lb/ft²)	$\sigma_a(z)_{\text{translation}}$ (lb/ft²)
0	0	0	0
4	269.9	112.6	191.25
8	311.2	225.3	268.25
12	233.6	337.9	285.75
16	102.2	450.6	276.4

The plot of $\sigma_a(z)$ versus z is shown in Figure 6.24.

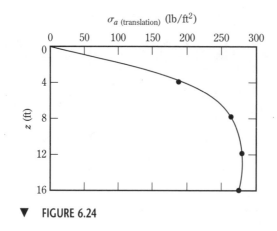

σ_a (translation) (lb/ft^2)

▼   FIGURE 6.24

▼   FIGURE 6.24   ▲

PASSIVE PRESSURE

6.10 RANKINE PASSIVE EARTH PRESSURE

Figure 6.25a shows a vertical frictionless retaining wall with a horizontal backfill. At depth z, the vertical pressure on a soil element is $\sigma_v = \gamma z$. Initially, if the wall does not yield at all, the lateral stress at that depth will be $\sigma_h = K_o \sigma_v$. This state of stress is illustrated by the Mohr's circle a in Figure 6.25b. Now, if the wall is pushed into the soil mass by an amount Δx, as shown in Figure 6.25a, the vertical stress at depth z will stay the same; however, the horizontal stress will increase. Thus σ_h will be greater than $K_o \sigma_v$. The state of stress can now be represented by the Mohr's circle b in Figure 6.25b. If the wall moves farther inward (that is, Δx is increased still more), the stresses at depth z will ultimately reach the state represented by Mohr's circle c (Figure 6.25b). Note that this Mohr's circle touches the Mohr–Coulomb failure envelope, which implies that the soil behind the wall will fail by being pushed upward. The horizontal stress, σ_h, at this point is referred to as the *Rankine passive pressure*, or $\sigma_h = \sigma_p$.

For Mohr's circle c in Figure 6.25b, the major principal stress is σ_p, and the minor principal stress is σ_v. Substituting them into Eq. (1.84) yields

$$\sigma_p = \sigma_v \tan^2\left(45 + \frac{\phi}{2}\right) + 2c \tan\left(45 + \frac{\phi}{2}\right) \tag{6.55}$$

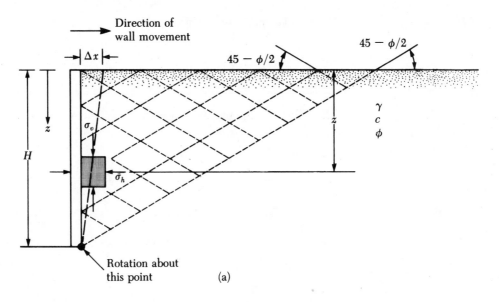

Direction of
wall movement

Δx

$45 - \phi/2$

$45 - \phi/2$

z

σ_v

γ
c
ϕ

z

H

σ_h

Rotation about
this point

(a)

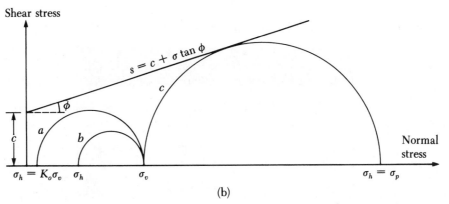

Shear stress

$s = c + \sigma \tan \phi$

c

ϕ

a

b

c

Normal
stress

$\sigma_h = K_o \sigma_v$ σ_h σ_v

$\sigma_h = \sigma_p$

(b)

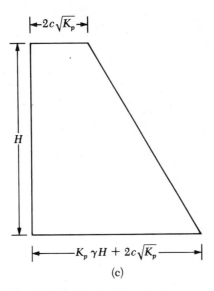

$2c\sqrt{K_p}$

H

$K_p \gamma H + 2c\sqrt{K_p}$

(c)

▼ **FIGURE 6.25**   Rankine passive pressure

Now, let

$$K_p = \text{Rankine passive earth pressure coefficient}$$
$$= \tan^2\left(45 + \frac{\phi}{2}\right) \tag{6.56}$$

(see Table 6.9). Hence, from Eq. (6.55),

$$\sigma_p = \sigma_v K_p + 2c\sqrt{K_p} \tag{6.57}$$

▼ **TABLE 6.9** Variation of Rankine K_p

Soil friction angle, ϕ (deg)	$K_p = \tan^2(45 + \phi/2)$
20	2.040
21	2.117
22	2.198
23	2.283
24	2.371
25	2.464
26	2.561
27	2.663
28	2.770
29	2.882
30	3.000
31	3.124
32	3.255
33	3.392
34	3.537
35	3.690
36	3.852
37	4.023
38	4.204
39	4.395
40	4.599
41	4.815
42	5.045
43	5.289
44	5.550
45	5.828

Equation (6.57) produces Figure 6.25c, the passive pressure diagram for the wall shown in Figure 6.25a. Note that at $z = 0$,

$$\sigma_v = 0 \quad \text{and} \quad \sigma_p = 2c\sqrt{K_p}$$

and at $z = H$,

$$\sigma_v = \gamma H \quad \text{and} \quad \sigma_p = \gamma H K_p + 2c\sqrt{K_p}$$

The passive force per unit length of the wall can be determined from the area of the pressure diagram, or

$$P_p = \tfrac{1}{2}\gamma H^2 K_p + 2cH\sqrt{K_p} \tag{6.58}$$

The approximate magnitudes of the wall movements, Δx, required to develop failure under passive conditions are

Soil type	Wall movement for passive condition, Δx
Dense sand	$0.005H$
Loose sand	$0.01H$
Stiff clay	$0.01H$
Soft clay	$0.05H$

▼ **EXAMPLE 6.10**

A 3-m-high wall is shown in Figure 6.26a. Determine the Rankine passive force per unit length of the wall.

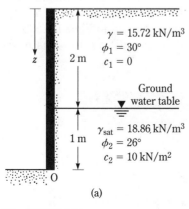
(a)

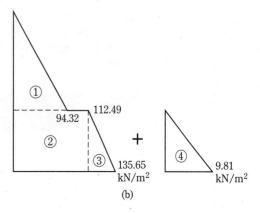
(b)

▼ **FIGURE 6.26**

Solution   For the top layer

$$K_{p(1)} = \tan^2\left(45 + \frac{\phi_1}{2}\right) = \tan^2(45 + 15) = 3$$

From the bottom soil layer

$$K_{p(2)} = \tan^2\left(45 + \frac{\phi_2}{2}\right) = \tan^2(45 + 13) = 2.56$$

$$\sigma_p = \sigma_v' K_p + 2c\sqrt{K_p}$$

where σ_v' = effective vertical stress
 at $z = 0$, $\sigma_v' = 0$, $c_1 = 0$, $\sigma_p = 0$
 at $z = 2$ m, $\sigma_v' = (15.72)(2) = 31.44$ kN/m², $c_1 = 0$

So, for the top soil layer

$$\sigma_p = 31.44 K_{p(1)} + 2(0)\sqrt{K_{p(1)}} = 31.44(3) = 94.32 \text{ kN/m}^2$$

At this depth, that is, $z = 2$ m, for the bottom soil layer

$$\sigma_p = \sigma_v' K_{p(2)} + 2c\sqrt{K_{p(2)}} = 31.44(2.56) + 2(10)\sqrt{2.56}$$

$$= 80.49 + 32 = 112.49 \text{ kN/m}^2$$

Again, at $z = 3$ m, $\sigma_v' = (15.72)(2) + (\gamma_{\text{sat}} - \gamma_w)(1)$

$$= 31.44 + (18.86 - 9.81)(1) = 40.49 \text{ kN/m}^2$$

Hence

$$\sigma_p = \sigma_v' K_{p(2)} + 2c\sqrt{K_{p(2)}} = 40.49(2.56) + (2)(10)(1.6)$$

$$= 135.65 \text{ kN/m}^2$$

Note that, because a water table is present, the hydrostatic stress, u, also has to be taken into consideration. For $z = 0$ to 2 m, $u = 0$; $z = 3$ m, $u = (1)(\gamma_w) = 9.81$ kN/m².

The passive pressure diagram is plotted in Figure 6.26b. The passive force per unit length of the wall can be determined from the area of the pressure diagram as follows:

Area no.	Area	
1	$(\tfrac{1}{2})(2)(94.32)$	= 94.32
2	$(112.49)(1)$	= 112.49
3	$(\tfrac{1}{2})(1)(135.65 - 112.49)$	= 11.58
4	$(\tfrac{1}{2})(9.81)(1)$	= 4.905
		$P_p \approx$ **223.3 kN/m**

6.11   RANKINE PASSIVE EARTH PRESSURE—INCLINED BACKFILL

For a frictionless vertical retaining wall (Figure 6.10) with a *granular backfill* ($c = 0$), the Rankine passive pressure at any depth can be determined in a manner similar to that done in the case of active pressure in Section 6.4, or

$$\sigma_p = \gamma z K_p \tag{6.59}$$

and the passive force

$$P_p = \tfrac{1}{2}\gamma H^2 K_p \tag{6.60}$$

where

$$K_p = \cos\alpha \frac{\cos\alpha + \sqrt{\cos^2\alpha - \cos^2\phi}}{\cos\alpha - \sqrt{\cos^2\alpha - \cos^2\phi}} \tag{6.61}$$

As in the case of the active force, the resultant force, P_p, is inclined at an angle α with the horizontal and intersects the wall at a distance of $H/3$ from the bottom of the wall. The values of K_p (passive earth pressure coefficient) for various values of α and ϕ are given in Table 6.10.

If the backfill of the frictionless vertical retaining wall is a $c - \phi$ soil (Figure 6.10), then

$$\sigma_a = \gamma z K_p = \gamma z K_p' \cos\alpha \tag{6.62}$$

where

$$K_p' = \frac{1}{\cos^2\phi}\left\{ \frac{2\cos^2\alpha + 2\left(\frac{c}{\gamma z}\right)\cos\phi\sin\phi}{+\sqrt{4\cos^2\alpha(\cos^2\alpha - \cos^2\phi) + 4\left(\frac{c}{\gamma z}\right)^2\cos^2\phi + 8\left(\frac{c}{\gamma z}\right)\cos^2\alpha\sin\phi\cos\phi}} \right\} - 1 \tag{6.63}$$

▼ **TABLE 6.10**   Passive Earth Pressure Coefficient, K_p [Eq. (6.61)]

↓ α (deg)	ϕ (deg) →						
	28	30	32	34	36	38	40
0	2.770	3.000	3.255	3.537	3.852	4.204	4.599
5	2.715	2.943	3.196	3.476	3.788	4.136	4.527
10	2.551	2.775	3.022	3.295	3.598	3.937	4.316
15	2.284	2.502	2.740	3.003	3.293	3.615	3.977
20	1.918	2.132	2.362	2.612	2.886	3.189	3.526
25	1.434	1.664	1.894	2.135	2.394	2.676	2.987

▼   **TABLE 6.11** Values of K_p'

ϕ (deg)	α (deg)	$c/\gamma z$			
		0.025	0.050	0.100	0.500
15	0	1.764	1.829	1.959	3.002
	5	1.716	1.783	1.917	2.971
	10	1.564	1.641	1.788	2.880
	15	1.251	1.370	1.561	2.732
20	0	2.111	2.182	2.325	3.468
	5	2.067	2.140	2.285	3.435
	10	1.932	2.010	2.162	3.339
	15	1.696	1.786	1.956	3.183
25	0	2.542	2.621	2.778	4.034
	5	2.499	2.578	2.737	3.999
	10	2.368	2.450	2.614	3.895
	15	2.147	2.236	2.409	3.726
30	0	3.087	3.173	3.346	4.732
	5	3.042	3.129	3.303	4.674
	10	2.907	2.996	3.174	4.579
	15	2.684	2.777	2.961	4.394

The variation of K_p' with ϕ, α, $c/\gamma z$ is given in Table 6.11 (Mazindrani and Ganjali, 1997).

6.12 COULOMB'S PASSIVE EARTH PRESSURE

Coulomb (1776) also presented an analysis for determining the passive earth pressure (that is, when the wall moves *into* the soil mass) for walls possessing friction (δ = angle of wall friction) and retaining a granular backfill material similar to that discussed in Section 6.5.

To understand the determination of Coulomb's passive force, P_p, consider the wall shown in Figure 6.27a. As in the case of active pressure, Coulomb assumed that the potential failure surface in soil is a plane. For a trial failure wedge of soil, such as ABC_1, the forces per unit length of the wall acting on the wedge are

1. The weight of the wedge, W
2. The resultant, R, of the normal and shear forces on the plane BC_1
3. The passive force, P_p

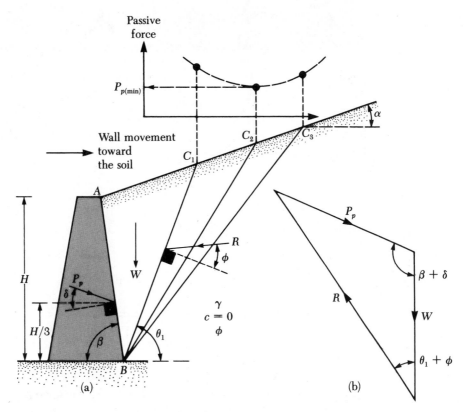

▼ FIGURE 6.27 Coulomb's passive pressure

Figure 6.27 shows the force triangle at equilibrium for the trial wedge ABC_1. From this force triangle, the value of P_p can be determined because the direction of all three forces and the magnitude of one force are known.

Similar force triangles for several trial wedges, such as $ABC_1, ABC_2, ABC_3, \ldots$ can be constructed, and the corresponding values of P_p can be determined. The top part of Figure 6.27a shows the nature of variation of the P_p values for different wedges. The *minimum value of P_p* in this diagram is *Coulomb's passive force.* Mathematically, this can be expressed as

$$P_p = \tfrac{1}{2}\gamma H^2 K_p$$

(6.64)

▼ **TABLE 6.12** Values of K_p [Eq. (6.65)] for $\beta = 90°$ and $\alpha = 0°$

	δ (deg)				
ϕ (deg)	0	5	10	15	20
15	1.698	1.900	2.130	2.405	2.735
20	2.040	2.313	2.636	3.030	3.525
25	2.464	2.830	3.286	3.855	4.597
30	3.000	3.506	4.143	4.977	6.105
35	3.690	4.390	5.310	6.854	8.324
40	4.600	5.590	6.946	8.870	11.772

where

$$K_p = \text{Coulomb's passive pressure coefficient}$$

$$= \frac{\sin^2(\beta - \phi)}{\sin^2\beta \sin(\beta + \delta)\left[1 - \sqrt{\dfrac{\sin(\phi + \delta)\sin(\phi + \alpha)}{\sin(\beta + \delta)\sin(\beta + \alpha)}}\right]^2} \qquad (6.65)$$

The values of the passive pressure coefficient, K_p, for various values of ϕ and δ are given in Table 6.12 ($\beta = 90°$, $\alpha = 0°$).

Note that the resultant passive force, P_p, will act at a distance of $H/3$ from the bottom of the wall and will be inclined at an angle δ to the normal drawn to the back face of the wall.

6.13   COMMENTS ON THE FAILURE SURFACE ASSUMPTION FOR COULOMB'S PRESSURE CALCULATIONS

Coulomb's pressure calculation methods for active and passive pressure have been discussed in Sections 6.5 and 6.12. The fundamental assumption for these analyses is the acceptance of *plane failure surfaces*. However, for walls with friction, this assumption does not hold in practice. The nature of *actual* failure surfaces in the soil mass for active and passive pressure is shown in Figure 6.28a and b, respectively (for a vertical wall with a horizontal backfill). Note that the failure surfaces BC are curved and that the failure surfaces CD are planes.

Although the actual failure surface in soil for the case of active pressure is somewhat different from that assumed in the calculation of the Coulomb pressure, the results are not greatly different. However, in the cases of passive pressure, as

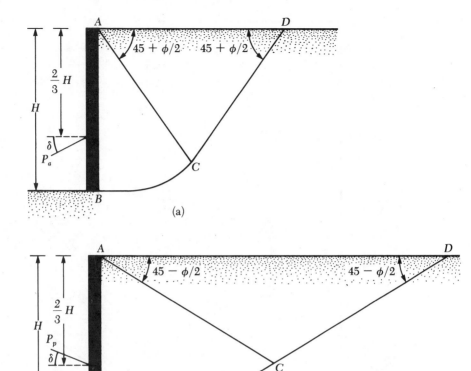

▼ **FIGURE 6.28** Nature of failure surface in soil with wall friction for (a) active pressure case and (b) passive pressure case

the value of δ increases, Coulomb's method of calculation gives increasingly errone-ous values of P_p. This factor of error could lead to an unsafe condition because the values of P_p would become higher than the soil resistance.

Caquot and Kerisel (1948) developed a chart (Figure 6.29) for estimating the value of the passive pressure coefficient (K_p) with curved failure surface in granular soil ($c = 0$), such as that shown in Figure 6.28b. In their solution, the portion BC of the failure surface was assumed to be an arc of a logarithmic spiral. While using Figure 6.29, the following points should be kept in mind:

1. The curves are for $\phi = \delta$.
2. If δ/ϕ is less than one, then

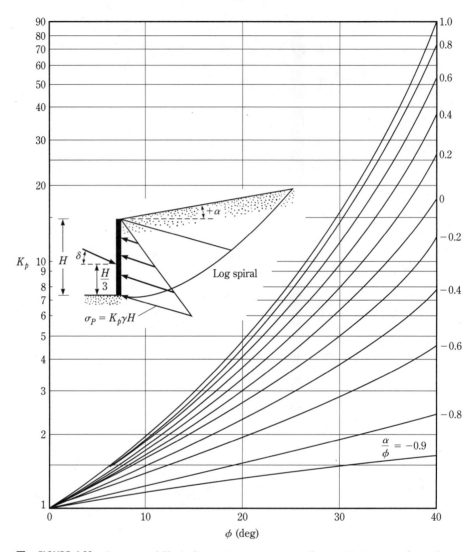

▼ **FIGURE 6.29** Caquot and Kerisel's passive pressure coefficient, K_p, for granular soil

$$K_{p(\delta)} = R\,K_{p(\delta=\phi)}$$

where R = reduction factor

The reduction factor R is given in Table 6.13.

3. The passive pressure is

$$P_p = \tfrac{1}{2}\gamma H^2\,K_{p(\delta)}$$

▼ **TABLE 6.13** Reduction factor, R, for use in conjunction with Figure 6.29

ϕ (deg)	δ/ϕ							
	0.7	0.6	0.5	0.4	0.3	0.2	0.1	0.0
10	0.978	0.962	0.946	0.929	0.912	0.898	0.880	0.864
15	0.961	0.934	0.907	0.881	0.854	0.830	0.803	0.775
20	0.939	0.901	0.862	0.824	0.787	0.752	0.716	0.678
25	0.912	0.860	0.808	0.759	0.711	0.666	0.620	0.574
30	0.878	0.811	0.746	0.686	0.627	0.574	0.520	0.467
35	0.836	0.752	0.674	0.603	0.536	0.475	0.417	0.362
40	0.783	0.682	0.592	0.512	0.439	0.375	0.316	0.262
45	0.718	0.600	0.500	0.414	0.339	0.276	0.221	0.174

PROBLEMS

6.1 Refer to Figure 6.3a. Given: $H = 12$ ft, $q = 0$, $\gamma = 108$ lb/ft³, $c = 0$, and $\phi = 30°$. Determine the at-rest lateral earth force per foot length of the wall. Also find the location of the resultant. Use Eq. (6.3).

6.2 Repeat Problem 6.1 with the following: $H = 3.5$ m, $q = 20$ kN/m², $\gamma = 18.2$ kN/m³, $c = 0$, and $\phi = 35°$.

6.3 Use Eq. (6.3), Figure P6.3, and the following values to determine the at-rest lateral earth force per unit length of the wall. Also find the location of the resultant. $H = 10$ ft, $H_1 = 4$ ft, $H_2 = 6$ ft, $\gamma = 105$ lb/ft³, $\gamma_{sat} = 122$ lb/ft³, $\phi = 30°$, $c = 0$, $q = 300$ lb/ft².

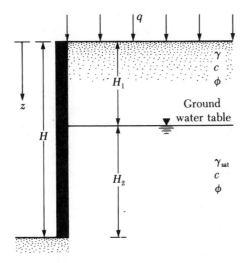

▼ **FIGURE P6.3**

6.4 Repeat Problem 6.3 with the following: $H = 5$ m, $H_1 = 2$ m, $H_2 = 3$ m, $\gamma = 15.5$ kN/m^3, $\gamma_{sat} = 18.5$ kN/m^3, $\phi = 34°$, $c = 0$, $q = 20$ kN/m^2.

6.5 Refer to Figure P6.3. Given; $H_1 = 4.5$ m, $H_2 = 0$, $q = 0$, and $\gamma = 17$ kN/m^3. The backfill is an overconsolidated clay with a plasticity index of 23. If the overconsolidation ratio is 2.2, determine the at-rest lateral earth pressure per meter length of the wall. Also find the location of the resultant. Use Eqs. (6.5) and (6.7).

6.6 Refer to Figure 6.7a. Given: the height of the retaining wall, H is 18 ft; the backfill is a saturated clay with $\phi = 0°$, $c = 500$ lb/ft^2, $\gamma_{sat} = 120$ lb/ft^3.
   a. Determine the Rankine active pressure distribution diagram behind the wall.
   b. Determine the depth of the tensile crack, z_c.
   c. Estimate the Rankine active force per foot of the wall before and after the occurrence of the tensile crack.

6.7 A vertical retaining wall (Figure 6.7a) is 6.3 m high with a horizontal backfill. For the backfill, assume that $\gamma = 17.9$ kN/m^3, $\phi = 26°$, and $c = 15$ kN/m^2. Determine the Rankine active force per unit length of the wall after the occurrence of the tensile crack.

6.8 Refer to the retaining wall described in Problem 6.3. Determine the Rankine active force per unit length of wall and the location of the line of action of the resultant measured from the bottom of the wall.

6.9 Refer to Problem 6.4. For the retaining wall, determine the Rankine active force per unit length of the wall and the location of the line of action of the resultant.

6.10 Refer to Figure 6.10. For the retaining wall, $H = 7.5$ m, $\phi = 32°$, $\alpha = 5°$, $\gamma = 18.2$ kN/m^3, and $c = 0$.
   a. Determine the intensity of the Rankine active force at $z = 2, 4, 6$, and 7.5 m.
   b. Determine the Rankine active force per meter of the wall and also the location and direction of the resultant.

6.11 Refer to Figure 6.10. Given: $H = 22$ ft, $\gamma = 115$ lb/ft^3, $\phi = 25°$, $c = 250$ lb/ft^2, and $\alpha = 10°$. Calculate the Rankine active force per unit length of the wall after the occurrence of the tensile crack.

6.12 Refer to Figure 6.12a. Given: $H = 12$ ft, $\gamma = 105$ lb/ft^3, $\phi = 30°$, $c = 0$, and $\beta = 85°$. Determine the Coulomb's active force per foot length of the wall and the location and direction of the resultant for the following cases:
   a. $\alpha = 10°$ and $\delta = 20°$
   b. $\alpha = 20°$ and $\delta = 15°$

6.13 Refer to Figure 6.14. Here, $H = 5$ m, $\gamma = 18.2$ kN/m^3, $\phi = 30°$, $\delta = 20°$, $c = 0$, $\alpha = 10°$, and $\beta = 85°$. Determine the Coulomb's active force of earthquake conditions (P_{ae}) per meter length of the wall and the location and direction of the resultant. Given $k_h = 0.2$ and $k_v = 0$.

6.14 Refer to Figure 6.18b. Given: $H = 10$ ft, $a' = 3$ ft, $b' = 4.5$ ft, and $q = 525$ lb/ft^2. Determine the lateral force per unit length of the wall caused by surcharge loading only.

6.15 A retaining wall is shown in Figure P6.15. If the wall rotates about its top, determine the magnitude of the active force per unit length of the wall for $n_a = 0.3, 0.4$, and 0.5. Assume unit adhesion, $c_a = c(\tan \delta / \tan \phi)$.

6.16 A vertical frictionless retaining wall is 6 m high with a horizontal granular backfill. Given: $\gamma = 16$ kN/m^3, $\phi = 30°$. For the translation mode of the wall, calculate the active pressure at depths $z = 1.5$ m, 3 m, 4.5 m, and 6 m.

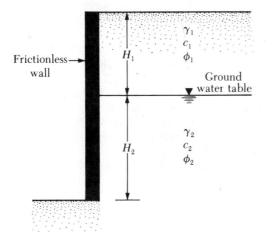

$\gamma = 17.5$ kN/m^3
$\phi = 25°$
$\delta = 15°$
$c = 14$ kN/m^2

8 m

δ

P_a

▼ **FIGURE P6.15**

6.17   Refer to Problem 6.6.
  a.   Draw the Rankine passive pressure distribution diagram behind the wall.
  b.   Estimate the Rankine passive force per foot length of the wall and also the location
    of the resultant.
6.18–6.19   Use Figure P6.18 and the following data to determine the Rankine passive force
    per unit length of the wall:

Frictionless → wall

H_1

γ_1
c_1
ϕ_1

Ground
▼ water table

H_2

γ_2
c_2
ϕ_2

▼ **FIGURE P6.18**

Prob.	H_1	H_2	γ_1	γ_2	ϕ_1	ϕ_2	c_1	c_2
6.18	8 ft	16 ft	110 lb/ft^3	140 lb/ft^3	38°	25°	0	209 lb/ft^2
6.19	8.2 ft	14.8 ft	107 lb/ft^3	125 lb/ft^3	28°	20°	350 lb/ft^2	100 lb/ft^2

6.20   Refer to the retaining wall with a granular backfill shown in Figure 6.29. Given: $H =$
    20 ft, $\alpha = +10°$, $\phi = 36°$, $\gamma = 110$ lb/ft^3, $\delta/\phi = 0.4$. Calculate the passive force per
    unit length of the wall assuming curved failure surface in the soil.

REFERENCES

Brooker, E. W., and Ireland, H. O. (1965). "Earth Pressure at Rest Related to Stress History," *Canadian Geotechnical Journal,* Vol. 2, No. 1, pp. 1–15.

Caquot, A., and Kerisel, J. (1948). *Tables for Calculation of Passive Pressure, Active Pressure, and Bearing Capacity of Foundations,* Gauthier-Villars, Paris, France.

Coulomb, C. A. (1776). *Essai sur une Application des Regles de Maximis et Minimum à quelques Problemes de Statique Relatifs à l'Architecture,* Mem. Acad. Roy. des Sciences, Paris, Vol. 3, p. 38.

Das, B. M. (1987). *Theoretical Foundation Engineering,* Elsevier, Amsterdam.

Das, B. M. (1998). *Principles of Geotechnical Engineering,* Fourth Edition, PWS Publishing Company, Boston.

Dubrova, G. A. (1963). "Interaction of Soil and Structures," Izd. *Rechnoy Transport,* Moscow.

Harr, M. E. (1966). *Fundamentals of Theoretical Soil Mechanics,* McGraw-Hill, New York.

Jaky, J. (1944). "The Coefficient of Earth Pressure at Rest," *Journal for the Society of Hungarian Architects and Engineers,* October, pp. 355–358.

Jarquio, R. (1981). "Total Lateral Surcharge Pressure Due to Strip Load," *Journal of the Geotechnical Engineering Division,* American Society of Civil Engineers, Vol. 107, No. GT10, pp. 1424–1428.

Kim, J. S., and Preber, T. (1969). "Earth Pressure Against Braced Excavations," *Journal of the Soil Mechanics and Foundations Division,* ASCE, Vol. 96, No. 6, pp. 1581–1584.

Matsuzawa, H., and Hazarika, H. (1996). "Analysis of Active Earth Pressure Against Rigid Retaining Wall Subjected to Different Modes of Movement," *Soils and Foundations,* Tokyo, Japan, Vol. 36, No. 3, pp. 51–66.

Mayne, P. W., and Kulhawy, F. H. (1982). "K_o–OCR Relationships in Soil," *Journal of the Geotechnical Engineering Division,* ASCE, Vol. 108, No. GT6, pp. 851–872.

Mazindrani, Z. H., and Ganjali, M. H. (1997). "Lateral Earth Pressure Problem of Cohesive Backfill with Inclined Surface," *Journal of Geotechnical and Geoenvironmental Engineering,* ASCE, Vol. 123, No. 2, pp. 110–112.

Sherif, M. A., Fang, Y. S., and Sherif, R. I. (1984). "k_a and k_o Behind Rotating and Non-Yielding Walls," *Journal of Geotechnical Engineering,* American Society of Civil Engineers, Vol. 110, No. GT1, pp. 41–56.

Terzaghi, K. (1943). *Theoretical Soil Mechanics,* Wiley, New York.

CHAPTER SEVEN

RETAINING WALLS

7.1 INTRODUCTION

In Chapter 6 you were introduced to various types of lateral earth pressure. Those theories will be used in this chapter to design various types of retaining walls. In general, retaining walls can be divided into two major categories: (a) conventional retaining walls, and (b) mechanically stabilized earth walls.

Conventional retaining walls can generally be classified as

1. Gravity retaining walls
2. Semigravity retaining walls
3. Cantilever retaining walls
4. Counterfort retaining walls

Gravity retaining walls (Figure 7.1a) are constructed with plain concrete or stone masonry. They depend on their own weight and any soil resting on the masonry for stability. This type of construction is not economical for high walls.

In many cases, a small amount of steel may be used for the construction of gravity walls, thereby minimizing the size of wall sections. Such walls are generally referred to as *semigravity walls* (Figure 7.1b).

Cantilever retaining walls (Figure 7.1c) are made of reinforced concrete that consists of a thin stem and a base slab. This type of wall is economical to a height of about 25 ft (8 m).

Counterfort retaining walls (Figure 7.1d) are similar to cantilever walls. At regular intervals, however, they have thin vertical concrete slabs known as *counterforts* that tie the wall and the base slab together. The purpose of the counterforts is to reduce the shear and the bending moments.

To design retaining walls properly, an engineer must know the basic soil parameters — that is, the *unit weight, angle of friction,* and *cohesion* — for the soil retained behind the wall and the soil below the base slab. Knowing the properties of the soil behind the wall enables the engineer to determine the lateral pressure distribution that has to be designed for.

There are two phases in the design of a conventional retaining wall. First, with the lateral earth pressure known, the structure as a whole is checked for *stability*. That includes checking for possible *overturning, sliding,* and *bearing capacity* failures.

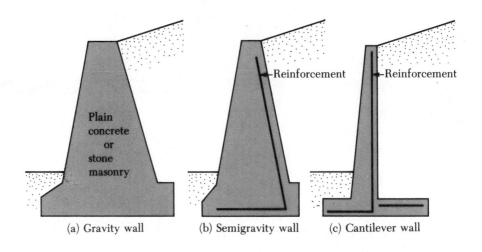

(a) Gravity wall　　(b) Semigravity wall　　(c) Cantilever wall

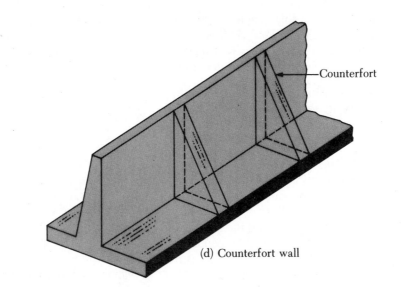

(d) Counterfort wall

▼　**FIGURE 7.1**　Types of retaining wall

Second, each component of the structure is checked for *adequate strength,* and the *steel reinforcement* of each component is determined.

This chapter presents the procedures for determining retaining-wall stability. Checks for adequate strength of each component of the structures can be found in any textbook on reinforced concrete.

Mechanically stabilized retaining walls have their backfills stabilized by inclusion of reinforcing elements such as metal strips, bars, welded wire mats, geotextiles, and geogrids. These walls are relatively flexible and can sustain large horizontal and vertical displacement without much damage.

In this chapter the gravity and cantilever retaining walls will be described first, followed by mechanically stabilized walls with metal strips, geotextiles, and geogrid-reinforced backfills.

GRAVITY AND CANTILEVER WALLS

7.2 PROPORTIONING RETAINING WALLS

When designing retaining walls, an engineer must assume some of the dimensions, called *proportioning,* which allows the engineer to check trial sections for stability. If the stability checks yield undesirable results, the sections can be changed and rechecked. Figure 7.2 shows the general proportions of various retaining-wall components that can be used for initial checks.

Note that the top of the stem of any retaining wall should not be less than about 12 in. ($\approx$0.3 m) for proper placement of concrete. The depth, D, to the bottom of the base slab should be a minimum of 2 ft ($\approx$0.6 m). However, the bottom of the base slab should be positioned below the seasonal frost line.

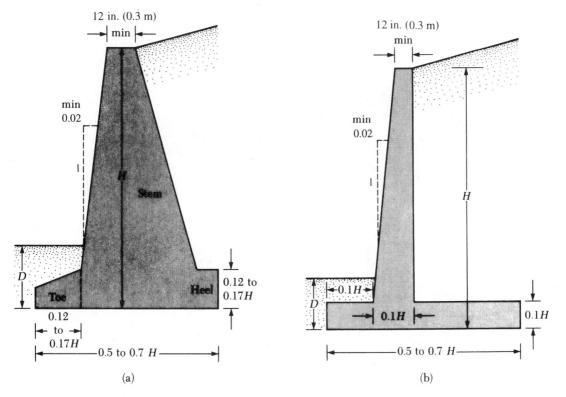

(a)   (b)

▼ **FIGURE 7.2** Approximate dimensions for various components of retaining wall for initial stability checks: (a) gravity wall; (b) cantilever wall [*note:* minimum dimension of D is 2 ft ($\approx$ 0.6 m)]

For counterfort retaining walls, the general proportion of the stem and the base slab is the same as for cantilever walls. However, the counterfort slabs may be about 12 in. ($\approx$0.3 m) thick and spaced at center-to-center distances of 0.3H to 0.7H.

7.3   APPLICATION OF LATERAL EARTH PRESSURE THEORIES TO DESIGN

The fundamental theories for calculating lateral earth pressure have been presented in Chapter 6. To use these theories in design, an engineer must make several simple assumptions. In the case of cantilever walls, use of the Rankine earth pressure theory for stability checks involves drawing a vertical line AB through point A, as shown in Figure 7.3a, (which is located at the edge of the heel of the base slab. The Rankine active condition is assumed to exist along the vertical plane AB. Rankine active earth pressure equations may then be used to calculate the lateral pressure on the face AB. In the analysis of stability for the wall, the force $P_{a(Rankine)}$, the weight of soil above the heel, W_s, and the weight of the concrete, W_c, all should be taken into consideration. The assumption for the development of Rankine active pressure along the soil face AB is theoretically correct if the shear zone bounded by the line AC is not obstructed by the stem of the wall. The angle, η, that the line AC makes

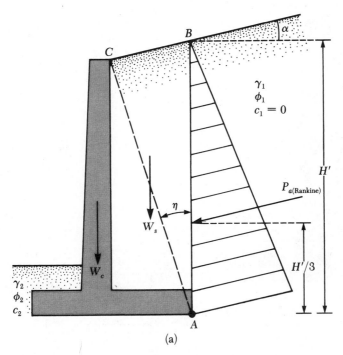

(a)

▼ **FIGURE 7.3**   Assumption for the determination of lateral earth pressure: (a) cantilever wall; (b) and (c) gravity wall

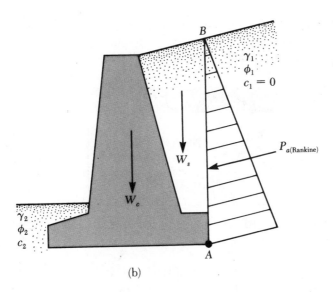

(b)

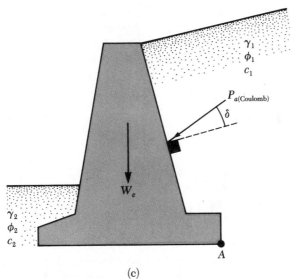

(c)

▼ **FIGURE 7.3** (Continued)

with the vertical is

$$\eta = 45 + \frac{\alpha}{2} - \frac{\phi}{2} - \sin^{-1}\left(\frac{\sin \alpha}{\sin \phi}\right) \tag{7.1}$$

A similar type of analysis may be used for gravity walls, as shown in Figure 7.3b. However, Coulomb's theory also may be used, as shown in Figure 7.3c. If *Coulomb's active pressure theory* is used, the only forces to be considered are $P_{a(\text{Coulomb})}$ and the weight of the wall, W_c.

If Coulomb's earth pressure theory is used, it will be necessary to know the range of the wall friction angle δ with various types of backfill material. Following are some ranges of wall friction angle for masonry or mass concrete walls:

Backfill material	Range of δ (deg)
Gravel	27–30
Coarse sand	20–28
Fine sand	15–25
Stiff clay	15–20
Silty clay	12–16

In the case of ordinary retaining walls, water table problems and hence hydrostatic pressure are not encountered. Facilities for drainage from the soils retained are always provided.

In several instances, for small retaining walls, *semiempirical charts* are used to evaluate lateral earth pressure. Figures 7.4 and 7.5 show two semiempirical charts given by Terzaghi and Peck (1967). Figure 7.4 is for backfills with plane surfaces, and Figure 7.5 is for backfills that slope upward from the crest of the wall for a limited distance and then become horizontal. Note that $\frac{1}{2}K_vH'^2$ is the vertical component of the active force on plane AB; similarly, $\frac{1}{2}K_hH'^2$ is the horizontal force. The numerals on the curves indicate the types of soil described in Table 7.1.

7.4   STABILITY CHECKS

To check the stability of a retaining wall, the following steps are necessary:

1. Check for *overturning* about its toe
2. Check for *sliding failure* along its base
3. Check for *bearing capacity failure* of the base
4. Check for *settlement*
5. Check for *overall stability*

This section describes the procedure for checking for overturning and sliding and bearing capacity failure. The principles of investigation for settlement were covered in Chapter 4 and will not be repeated here. Some problems regarding the overall stability of retaining walls are discussed in Section 7.5.

Check for Overturning

Figure 7.6 (p. 396) shows the forces acting on a cantilever and a gravity retaining wall, based on the assumption that the Rankine active pressure is acting along a vertical plane AB drawn through the heel. P_p is the Rankine passive pressure; recall that its magnitude is

$$P_p = \tfrac{1}{2}K_p\gamma_2D^2 + 2c_2\sqrt{K_p}D \tag{6.58}$$

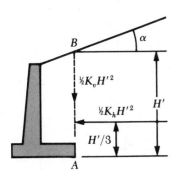

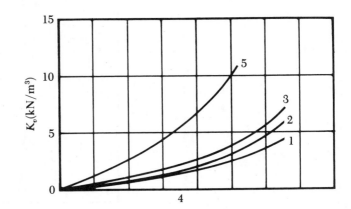

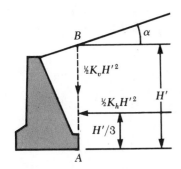

Note: Numerals on curves indicate soil types as described in Table 7.1 For materials, type-5 computations of pressure may be based on value of H' 1.3 m less than actual value.

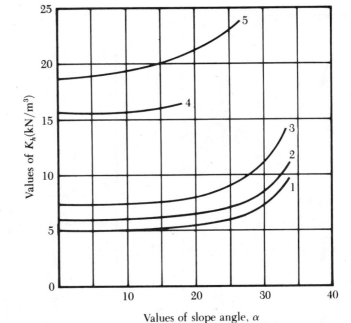

Values of slope angle, α

▼ **FIGURE 7.4** Chart for estimating pressure of backfill against retaining walls supporting backfills with plane surface (after *Soil Mechanics in Engineering Practice*, Second Edition, by K. Terzaghi and R. B. Peck. Copyright 1967 by John Wiley and Sons. Reprinted with permission) (*note:* 1 kN/m³ = 6.361 lb/ft³)

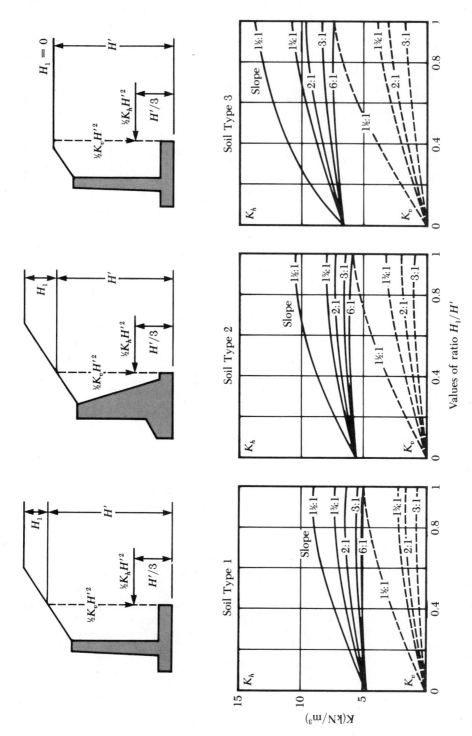

▶ **FIGURE 7.4** (Continued)

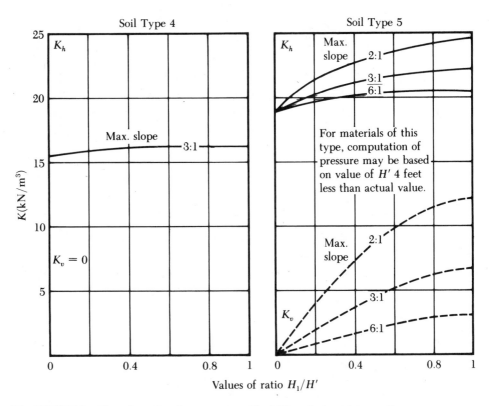

▼ **FIGURE 7.5** Chart for estimating pressure of backfill against retaining walls supporting backfills with surface that slopes upward from crest of wall for limited distance and then becomes horizontal (after *Soil Mechanics in Engineering Practice,* Second Edition, by K. Terzaghi and R. B. Peck. Copyright 1967 by John Wiley and Sons. Reprinted with permission) (*note:* 1 kN/m³ = 6.361 lb/ft³)

▼ **TABLE 7.1** Types of Backfill for Retaining Walls[a]

1. Coarse-grained soil without admixture of fine soil particles, very permeable (clean sand or gravel).
2. Coarse-grained soil of low permeability due to admixture of particles of silt size.
3. Residual soil with stones, fine silty sand, and granular materials with conspicuous clay content.
4. Very soft or soft clay, organic silts, or silty clays.
5. Medium or stiff clay, deposited in chunks and protected in such a way that a negligible amount of water enters the spaces between the chunks during floods or heavy rains. If this condition of protection cannot be satisfied, the clay should not be used as backfill material. With increasing stiffness of the clay, danger to the wall due to infiltration of water increases rapidly.

[a] From *Soil Mechanics in Engineering Practice,* Second Edition, by K. Terzaghi and R. B. Peck. Copyright 1967 by John Wiley and Sons. Reprinted with permission.

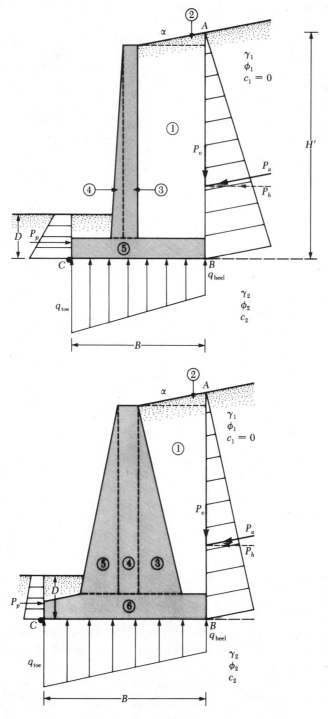

▼ FIGURE 7.6 Check for overturning; assume that Rankine pressure is valid

where $\quad \gamma_2 =$ unit weight of soil in front of the heel and under the base slab
$\qquad K_p =$ Rankine passive earth pressure coefficient $= \tan^2(45 + \phi_2/2)$
$\quad c_2, \phi_2 =$ cohesion and soil friction angle, respectively

The factor of safety against overturning about the toe — that is, about point C in Figure 7.6 — may be expressed as

$$FS_{(overturning)} = \frac{\Sigma M_R}{\Sigma M_O} \tag{7.2}$$

where $\quad \Sigma M_O =$ sum of the moments of forces tending to overturn about point C
$\qquad \Sigma M_R =$ sum of the moments of forces tending to resist overturning about point C

The overturning moment is

$$\Sigma M_O = P_h \left(\frac{H'}{3} \right) \tag{7.3}$$

where $\quad P_h = P_a \cos \alpha$

For calculation of the resisting moment, ΣM_R (neglecting P_p), a table (such as Table 7.2) can be prepared. The weight of the soil above the heel and the weight of the concrete (or masonry) are both forces that contribute to the resisting moment. Note that the force P_v also contributes to the resisting moment. P_v is the vertical component of the active force P_a, or

$$P_v = P_a \sin \alpha$$

The moment of the force P_v about C is

$$M_v = P_v B = P_a \sin \alpha B \tag{7.4}$$

where $\quad B =$ width of the base slab

▼ **TABLE 7.2** Procedure for Calculation of ΣM_R

Section (1)	Area (2)	Weight/unit length of wall (3)	Moment arm measured from C (4)	Moment about C (5)
1	A_1	$W_1 = \gamma_1 \times A_1$	X_1	M_1
2	A_2	$W_2 = \gamma_2 \times A_2$	X_2	M_2
3	A_3	$W_3 = \gamma_c \times A_3$	X_3	M_3
4	A_4	$W_4 = \gamma_c \times A_4$	X_4	M_4
5	A_5	$W_5 = \gamma_c \times A_5$	X_5	M_5
6	A_6	$W_6 = \gamma_c \times A_6$	X_6	M_6
		P_v	B	M_v
		ΣV		ΣM_R

Note: $\gamma_1 =$ unit weight of backfill
$\qquad \gamma_c =$ unit weight of concrete

Once ΣM_R is known, the factor of safety can be calculated as

$$FS_{(overturning)} = \frac{M_1 + M_2 + M_3 + M_4 + M_5 + M_6 + M_v}{P_a \cos \alpha (H'/3)} \qquad (7.5)$$

The usual minimum desirable value of the factor of safety with respect to overturning is 2 to 3.

Some designers prefer to determine the factor of safety against overturning with

$$FS_{(overturning)} = \frac{M_1 + M_2 + M_3 + M_4 + M_5 + M_6}{P_a \cos \alpha (H'/3) - M_v} \qquad (7.6)$$

Check for Sliding Along the Base

The factor of safety against sliding may be expressed by the equation

$$FS_{(sliding)} = \frac{\Sigma F_{R'}}{\Sigma F_d} \qquad (7.7)$$

where $\Sigma F_{R'}$ = sum of the horizontal resisting forces
        ΣF_d = sum of the horizontal driving forces

Figure 7.7 indicates that the shear strength of the soil immediately below the base slab may be represented as

$$s = \sigma \tan \delta + c_a$$

where δ = angle of friction between the soil and the base slab
        c_a = adhesion between the soil and the base slab

Thus the maximum resisting force that can be derived from the soil per unit length of the wall along the bottom of the base slab is

$$R' = s(\text{area of cross section}) = s(B \times 1) = B\sigma \tan \delta + Bc_a$$

However,

$$B\sigma = \text{sum of the vertical force} = \Sigma V \text{ (see Table 7.2)}$$

so

$$R' = (\Sigma V)\tan \delta + Bc_a$$

Figure 7.7 shows that the passive force P_p is also a horizontal resisting force. The expression for P_p is given in Eq. (6.58). Hence

$$\Sigma F_{R'} = (\Sigma V)\tan \delta + Bc_a + P_p \qquad (7.8)$$

The only horizontal force that will tend to cause the wall to slide (*driving force*) is the horizontal component of the active force P_a, so

$$\Sigma F_d = P_a \cos \alpha \qquad (7.9)$$

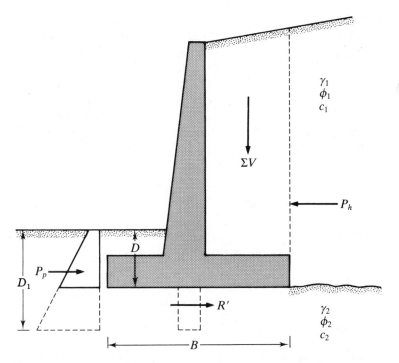

▼ **FIGURE 7.7**   Check for sliding along the base

Combining Eqs. (7.7), (7.8), and (7.9) yields

$$FS_{\text{(sliding)}} = \frac{(\Sigma V)\tan \delta + Bc_a + P_p}{P_a \cos \alpha} \tag{7.10}$$

A minimum factor of safety of 1.5 against sliding is generally required.

In many cases, the passive force P_p is ignored for calculation of the factor of safety with respect to sliding. In general, we can write $\delta = k_1\phi_2$ and $c_a = k_2c_2$. In most cases, k_1 and k_2 are in the range of $\frac{1}{2}$ to $\frac{2}{3}$. Thus

$$FS_{\text{(sliding)}} = \frac{(\Sigma V)\tan(k_1\phi_2) + Bk_2c_2 + P_p}{P_a \cos \alpha} \tag{7.11}$$

In some instances, certain walls may not yield a desired factor of safety of 1.5. To increase their resistance to sliding, a base key may be used. Base keys are illustrated by broken lines in Figure 7.7. It indicates that the passive force at the toe *without the key* is

$$P_p = \tfrac{1}{2}\gamma_2 D^2 K_p + 2c_2 D\sqrt{K_p}$$

However, if a key is included, the passive force per unit length of the wall becomes

$$P_p = \tfrac{1}{2}\gamma_2 D_1^2 K_p + 2c_2 D_1\sqrt{K_p}$$

where $K_p = \tan^2(45 + \phi_2/2)$

Because $D_1 > D$, a key obviously will help increase the passive resistance at the toe and hence the factor of safety against sliding. Usually the base key is constructed below the stem and some main steel is run into the key.

Another possible way to increase the value of $FS_{(sliding)}$ is to consider reducing the value of P_a [see Eq. (7.11)]. One possible way to do so is to use the method developed by Elman and Terry (1988). The discussion here is limited to the case in which the retaining wall has a horizontal granular backfill (Figure 7.8). In Figure 7.8, the active force, P_a, is horizontal ($\alpha = 0$) so that

$$P_a \cos \alpha = P_h = P_a$$

and

$$P_a \sin \alpha = P_v = 0$$

However

$$P_a = P_{a(1)} + P_{a(2)} \tag{7.12}$$

The magnitude of $P_{a(2)}$ can be reduced if the heel of the retaining wall is sloped as shown in Figure 7.8b. For this case,

$$P_a = P_{a(1)} + AP_{a(2)} \tag{7.13}$$

The magnitude of A, as shown in Figure 7.9, is valid for $\alpha' = 45°$. However, note that in Figure 7.8a

$$P_{a(1)} = \tfrac{1}{2}\gamma_1 K_a (H' - D')^2$$

and

$$P_a = \tfrac{1}{2}\gamma_1 K_a H'^2$$

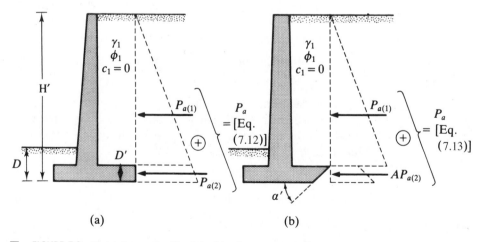

▼ **FIGURE 7.8**   Retaining wall with sloped heel

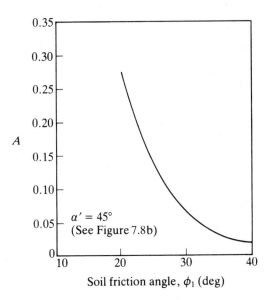

▼ FIGURE 7.9 Variation of A with friction angle of backfill [Eq. (7.14)] (after Elman and Terry, 1988)

Hence

$$P_{a(2)} = \tfrac{1}{2}\gamma_1 K_a [H'^2 - (H' - D')^2]$$

So, for the active pressure diagram shown in Figure 7.8b,

$$P_a = \frac{1}{2}\gamma_1 K_a (H' - D')^2 + \frac{A}{2}\gamma_1 K_a [H'^2 - (H' - D')^2] \qquad (7.14)$$

Sloping the heel of a retaining wall can thus be extremely helpful in some cases.

Check for Bearing Capacity Failure

The vertical pressure as transmitted to the soil by the base slab of the retaining wall should be checked against the ultimate bearing capacity of the soil. The nature of variation of the vertical pressure transmitted by the base slab into the soil is shown in Figure 7.10. Note that q_{toe} and q_{heel} are the *maximum* and the *minimum* pressures occurring at the ends of the toe and heel sections, respectively. The magnitudes of q_{toe} and q_{heel} can be determined in the following manner.

The sum of the vertical forces acting on the base slab is ΣV (see column 3, Table 7.2), and the horizontal force is $P_a \cos \alpha$. Let R be the resultant force, or

$$\vec{R} = \overrightarrow{\Sigma V} + \overrightarrow{(P_a \cos \alpha)} \qquad (7.15)$$

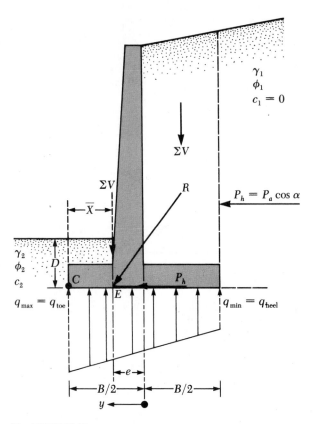

▼ **FIGURE 7.10**   Check for bearing capacity failure

The net moment of these forces about point C (Figure 7.10) is

$$M_{net} = \Sigma\, M_R - \Sigma\, M_O \tag{7.16}$$

Note that the values of $\Sigma\, M_R$ and $\Sigma\, M_O$ have been previously determined [see column 5, Table 7.2 and Eq. (7.3)]. Let the line of action of the resultant, R, intersect the base slab at E, as shown in Figure 7.10. The distance CE then is

$$\overline{CE} = \overline{X} = \frac{M_{net}}{\Sigma\, V} \tag{7.17}$$

Hence the eccentricity of the resultant, R, may be expressed as

$$e = \frac{B}{2} - \overline{CE} \tag{7.18}$$

The pressure distribution under the base slab may be determined by using the simple principles of mechanics of materials:

$$q = \frac{\Sigma\, V}{A} \pm \frac{M_{net}\, y}{I} \tag{7.19}$$

where M_{net} = moment = $(\Sigma V)e$
$\quad\quad\quad I$ = moment of inertia per unit length of the base section
$\quad\quad\quad\quad = \frac{1}{12}(1)(B^2)$

For maximum and minimum pressures, the value of y in Eq. (7.19) equals $B/2$. Substituting the preceding values into Eq. (7.19) gives

$$q_{max} = q_{toe} = \frac{\Sigma V}{(B)(1)} + \frac{e(\Sigma V)\frac{B}{2}}{\left(\frac{1}{12}\right)(B^3)} = \frac{\Sigma V}{B}\left(1 + \frac{6e}{B}\right) \quad\quad (7.20)$$

Similarly,

$$q_{min} = q_{heel} = \frac{\Sigma V}{B}\left(1 - \frac{6e}{B}\right) \quad\quad (7.21)$$

Note that ΣV includes the soil weight, as shown in Table 7.2, and that, when the value of the eccentricity, e, becomes greater than $B/6$, q_{min} becomes negative [Eq. (7.21)]. Thus, there will be some tensile stress at the end of the heel section. This stress is not desirable because the tensile strength of soil is very small. If the analysis of a design shows that $e > B/6$, the design should be reproportioned and calculations redone.

The relationships for the ultimate bearing capacity of a shallow foundation were discussed in Chapter 3. Recall that

$$q_u = c_2 N_c F_{cd} F_{ci} + q N_q F_{qd} F_{qi} + \tfrac{1}{2}\gamma_2 B' N_\gamma F_{\gamma d} F_{\gamma i} \quad\quad (7.22)$$

where $\quad q = \gamma_2 D$

$\quad\quad\quad B' = B - 2e$

$\quad\quad\quad F_{cd} = 1 + 0.4\dfrac{D}{B'}$

$\quad\quad\quad F_{qd} = 1 + 2\tan\phi_2(1 - \sin\phi_2)^2\dfrac{D}{B'}$

$\quad\quad\quad F_{\gamma d} = 1$

$\quad\quad\quad F_{ci} = F_{qi} = \left(1 - \dfrac{\psi^\circ}{90^\circ}\right)^2$

$\quad\quad\quad F_{\gamma i} = \left(1 - \dfrac{\psi^\circ}{\phi_2^\circ}\right)^2$

$\quad\quad\quad \psi^\circ = \tan^{-1}\left(\dfrac{P_a\cos\alpha}{\Sigma V}\right)$

Note that the shape factors F_{cs}, F_{qs}, and $F_{\gamma s}$ given in Chapter 3 are all equal to 1 because they can be treated as a continuous foundation. For this reason, the shape factors are not shown in Eq. (7.22).

Once the ultimate bearing capacity of the soil has been calculated by using Eq. (7.22), the factor of safety against bearing capacity failure can be determined:

$$FS_{\text{(bearing capacity)}} = \frac{q_u}{q_{max}} \tag{7.23}$$

Generally, a factor of safety of 3 is required. In Chapter 3 we noted that the ultimate bearing capacity of shallow foundations occurs at a settlement of about 10% of the foundation width. In the case of retaining walls, the width B is large. Hence the ultimate load q_u will occur at a fairly large foundation settlement. A factor of safety of 3 against bearing capacity failure may not ensure, in all cases, that settlement of the structure will be within the tolerable limit. Thus this situation needs further investigation.

▼ **EXAMPLE 7.1** _____

The cross section of a cantilever retaining wall is shown in Figure 7.11 Calculate the factors of safety with respect to overturning and sliding and bearing capacity.

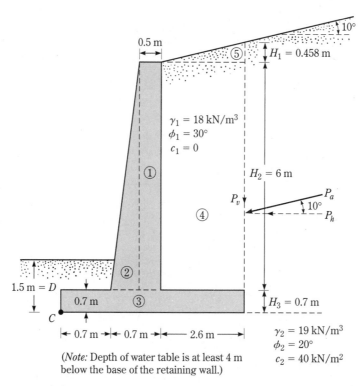

(*Note:* Depth of water table is at least 4 m below the base of the retaining wall.)

▼ FIGURE 7.11

Solution Referring to Figure 7.11,

$$H' = H_1 + H_2 + H_3 = 2.6 \tan 10° + 6 + 0.7$$
$$= 0.458 + 6 + 0.7 = 7.158 \text{ m}$$

The Rankine active force per unit length of wall $= P_p = \frac{1}{2}\gamma_1 H'^2 K_a$. For $\phi_1 = 30°$, $\alpha = 10°$, K_a is equal to 0.350 (Table 6.2). Thus,

$$P_a = \frac{1}{2}(18)(7.158)^2(0.35) = 161.4 \text{ kN/m}$$

$$P_v = P_a \sin 10° = 161.4(\sin 10°) = 28.03 \text{ kN/m}$$

$$P_h = P_a \cos 10° = 161.4(\cos 10°) = 158.95 \text{ kN/m}$$

Factor of Safety Against Overturning

The following table can now be prepared for determination of the resisting moment:

Section no.[a]	Area (m²)	Weight/unit length (kN/m)	Moment arm from point C (m)	Moment (kN-m)
1	6 × 0.5 = 3	70.74	1.15	81.35
2	$\frac{1}{2}$(0.2)6 = 0.6	14.15	0.833	11.79
3	4 × 0.7 = 2.8	66.02	2.0	132.04
4	6 × 2.6 = 15.6	280.80	2.7	758.16
5	$\frac{1}{2}$(2.6)(0.458) = 0.595	10.71	3.13	33.52
		$P_v = 28.03$	4.0	112.12
		$\Sigma V = 470.45$		$\Sigma 1128.98$ $= \Sigma M_R$

[a] For section numbers, refer to Figure 7.11.
$\gamma_{\text{concrete}} = 23.58 \text{ kN/m}^3$

The overturning moment, M_O

$$M_O = P_h\left(\frac{H'}{3}\right) = 158.95\left(\frac{7.158}{3}\right) = 379.25 \text{ kN-m}$$

$$FS_{(\text{overturning})} = \frac{\Sigma M_R}{M_O} = \frac{1128.98}{379.25} = \textbf{2.98} > \textbf{2} - \textbf{OK}$$

Factor of Safety Against Sliding

From Eq. (7.11)

$$FS_{(\text{sliding})} = \frac{(\Sigma V)\tan(k_1\phi_2) + Bk_2c_2 + P_p}{P_a \cos \alpha}$$

Let $k_1 = k_2 = \frac{2}{3}$

Also

$$P_p = \tfrac{1}{2}K_p\gamma_2 D^2 + 2c_2\sqrt{K_p}D$$

$$K_p = \tan^2\left(45 + \frac{\phi_2}{2}\right) = \tan^2(45 + 10) = 2.04$$

$$D = 1.5\,\text{m}$$

So

$$P_p = \tfrac{1}{2}(2.04)(19)(1.5)^2 + 2(40)(\sqrt{2.04})(1.5)$$
$$= 43.61 + 171.39 = 215\,\text{kN/m}$$

Hence

$$FS_{(\text{sliding})} = \frac{(470.45)\tan\left(\dfrac{2 \times 20}{3}\right) + (4)\left(\dfrac{2}{3}\right)(40) + 215}{158.95}$$

$$= \frac{111.5 + 106.67 + 215}{158.95} = 2.73 > 1.5\text{—OK}$$

Note: For some designs, the depth D for passive pressure calculation may be taken to be *equal to the thickness of the base slab.*

Factor of Safety Against Bearing Capacity Failure

Combining Eqs. (7.16), (7.17), and (7.18),

$$e = \frac{B}{2} - \frac{\Sigma M_R - \Sigma M_O}{\Sigma V} = \frac{4}{2} - \frac{1128.98 - 379.25}{470.45}$$

$$= 0.406\,\text{m} < \frac{B}{6} = \frac{4}{6} = 0.666\,\text{m}$$

Again, from Eqs. (7.20) and (7.21)

$$q_{\substack{\text{toe} \\ \text{heel}}} = \frac{\Sigma V}{B}\left(1 \pm \frac{6e}{B}\right) = \frac{470.45}{4}\left(1 \pm \frac{6 \times 0.406}{4}\right) = 189.2\,\text{kN/m}^2 \ (\text{toe})$$

$$= 45.99\,\text{kN/m}^2 \ (\text{heel})$$

The ultimate bearing capacity of the soil can be determined from Eq. (7.22):

$$q_u = c_2 N_c F_{cd}F_{ci} + qN_q F_{qd}F_{qi} + \tfrac{1}{2}\gamma_2 B' N_\gamma F_{\gamma d}F_{\gamma i}$$

For $\phi_2 = 20°$ (Table 3.4), $N_c = 14.83$, $N_q = 6.4$, and $N_\gamma = 5.39$. Also

$$q = \gamma_2 D = (19)(1.5) = 28.5\,\text{kN/m}^2$$

$$B' = B - 2e = 4 - 2(0.406) = 3.188\,\text{m}$$

$$F_{cd} = 1 + 0.4\left(\frac{D}{B'}\right) = 1 + 0.4\left(\frac{1.5}{3.188}\right) = 1.188$$

$$F_{qd} = 1 + 2 \tan \phi_2 (1 - \sin \phi_2)^2 \left(\frac{D}{B'}\right) = 1 + 0.315 \left(\frac{1.5}{3.188}\right) = 1.148$$

$$F_{\gamma d} = 1$$

$$F_{ci} = F_{qi} = \left(1 - \frac{\psi^\circ}{90^\circ}\right)^2$$

$$\psi = \tan^{-1}\left(\frac{P_a \cos \alpha}{\Sigma V}\right) = \tan^{-1}\left(\frac{158.95}{470.45}\right) = 18.67^\circ$$

So

$$F_{ci} = F_{qi} = \left(1 - \frac{18.67}{90}\right)^2 = 0.628$$

$$F_{\gamma i} = \left(1 - \frac{\psi}{\phi}\right)^2 = \left(1 - \frac{18.67}{20}\right)^2 \approx 0$$

Hence

$$\begin{aligned}
q_u &= (40)(14.83)(1.188)(0.628) + (28.5)(6.4)(1.148)(0.628) \\
&\quad + \tfrac{1}{2}(19)(5.93)(3.188)(1)(0) \\
&= 442.57 + 131.50 + 0 = 574.07 \text{ kN/m}^2
\end{aligned}$$

$$FS_{\text{(bearing capacity)}} = \frac{q_u}{q_{\text{toe}}} = \frac{574.07}{189.2} = 3.03 > 3 - \text{OK}$$

▲

▼ **EXAMPLE 7.2**

A concrete gravity retaining wall is shown in Figure 7.12. Determine

a. The factor of safety against overturning
b. The factor of safety against sliding
c. The pressure on the soil at the toe and heel

(*Note:* Unit weight of concrete $= \gamma_c = 150$ lb/ft³.)

Solution

$$H' = 15 + 2.5 = 17.5 \text{ ft}$$

$$K_a = \tan^2\left(45 - \frac{\phi_1}{2}\right) = \tan^2\left(45 - \frac{30}{2}\right) = \frac{1}{3}$$

$$P_a = \tfrac{1}{2}\gamma(H')^2 K_a = \tfrac{1}{2}(121)(17.5)^2 \left(\tfrac{1}{3}\right) = 6176 \text{ lb/ft}$$
$$= 6.176 \text{ kip/ft}$$

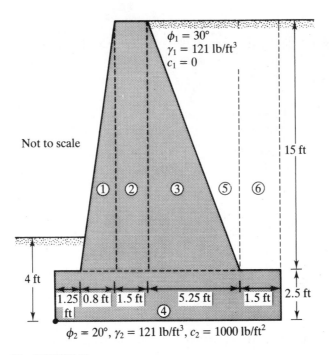

Not to scale

$\phi_1 = 30°$
$\gamma_1 = 121 \text{ lb/ft}^3$
$c_1 = 0$

15 ft

① ② ③ ⑤ ⑥

4 ft

1.25 ft | 0.8 ft | 1.5 ft | 5.25 ft | 1.5 ft | 2.5 ft

④

$\phi_2 = 20°$, $\gamma_2 = 121 \text{ lb/ft}^3$, $c_2 = 1000 \text{ lb/ft}^2$

▼ **FIGURE 7.12**

Since $\alpha = 0$

$$P_h = P_a = 6.176 \text{ kip/ft}$$

$$P_v = 0$$

Part a: Factor of Safety Against Overturning
The following table can now be prepared to obtain $\Sigma \, M_R$:

Area (from Figure 7.12)	Weight (kip)	Moment arm from C (ft)	Moment about C (kip/ft)
1	$\frac{1}{2}(0.8)(15)(\gamma_c) = 0.9$	$1.25 + \frac{2}{3}(0.8) = 1.783$	1.605
2	$(1.5)(15)(\gamma_c) = 3.375$	$1.25 + 0.8 + 0.75 = 2.8$	9.45
3	$\frac{1}{2}(5.25)(15)(\gamma_c) = 5.906$	$1.25 + 0.8 + 1.5 + \dfrac{5.25}{3} = 5.3$	31.30
4	$(10.3)(2.5)(\gamma_c) = 3.863$	$\dfrac{10.3}{2} = 5.15$	19.89
5	$\frac{1}{2}(5.25)(15)(0.121) = 4.764$	$1.25 + 0.8 + 1.5 + \frac{2}{3}(5.25) = 7.05$	33.59
6	$(1.5)(15)(0.121) = 2.723$	$1.25 + 0.8 + 1.5 + 5.25 + 0.75 = 9.55$	26.0
	$\overline{21.531}$		$\overline{121.84} = M_R$

The overturning moment

$$M_O = \frac{H'}{3} P_a = \left(\frac{17.5}{3}\right)(6.176) = 36.03 \text{ kip/ft}$$

$$FS_{(\text{overturning})} = \frac{121.84}{36.03} = \mathbf{3.38}$$

Part b: Factor of Safety Against Sliding

From Eq. (7.11), with $k_1 = k_2 = \frac{2}{3}$ and assuming that $P_p = 0$,

$$FS_{(\text{sliding})} = \frac{\Sigma V \tan\left(\frac{2}{3}\right)\phi_2 + B\left(\frac{2}{3}\right)c_2}{P_a}$$

$$= \frac{21.531 \tan\left(\frac{2 \times 20}{3}\right) + 10.3 \left(\frac{2}{3}\right)(1.0)}{6.176}$$

$$= \frac{5.1 + 6.87}{6.176} = \mathbf{1.94}$$

Part c: Pressure on the Soil at the Toe and Heel

From Eqs. (7.16), (7.17), and (7.18),

$$e = \frac{B}{2} - \frac{\Sigma M_R - \Sigma M_O}{\Sigma V} = \frac{10.3}{2} - \frac{121.84 - 36.03}{21.531} = 5.15 - 3.99 = 1.16 \text{ ft}$$

$$q_{\text{toe}} = \frac{\Sigma V}{B}\left[1 + \frac{6e}{B}\right] = \frac{21.531}{10.3}\left[1 + \frac{(6)(1.16)}{10.3}\right] = \mathbf{3.5 \text{ kip/ft}^2}$$

$$q_{\text{heel}} = \frac{\Sigma V}{B}\left[1 - \frac{6e}{B}\right] = \frac{21.531}{10.3}\left[1 - \frac{(6)(1.16)}{10.3}\right] = \mathbf{0.678 \text{ kip/ft}^2}$$

▲

▼ **EXAMPLE 7.3**

Repeat Example 7.2 and use Coulomb's active pressure for calculation and $\delta = 2\phi/3$.

Solution Refer to Figure 7.13 for the pressure calculation:

$$\delta = \tfrac{2}{3}\phi = (\tfrac{2}{3})(30) = 20°$$

From Table 6.5, $K_a = 0.4794(\alpha = 0°, \beta = 70°)$, so

$$P_a = \tfrac{1}{2}(0.121)(17.5)^2(0.4794) = 8.882 \text{ kip/ft}$$

$$P_h = P_a \cos 40 = (8.882)(\cos 40) = 6.8 \text{ kip/ft}$$

$$P_v = P_a \sin 40 = 5.71 \text{ kip/ft}$$

Part a: Factor of Safety Against Overturning

Refer to Figures 7.14 and 7.12.

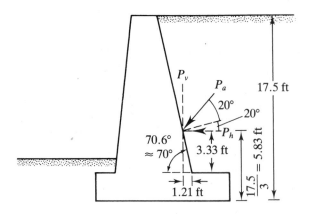

Not to scale

▼   **FIGURE 7.13**

Area (from Figures 7.12 and 7.14)	Weight (kip)	Moment arm from C (ft)	Moment about C (kip/ft)
1	0.9[a]	1.783[a]	1.605
2	3.375[a]	2.8[a]	9.46
3	5.906[a]	5.3[a]	31.30
4	3.863[a]	5.15[a]	19.89
	$P_v = 5.71$	1.25 + 0.8 + 1.5 +5.25 − 1.21 = 7.59	43.34
	$\overline{19.75}$		$\overline{105.6}$

[a] Same as in Example 7.2

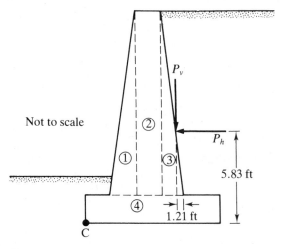

Not to scale

▼   **FIGURE 7.14**

The overturning moment is

$$M_O = P_h \frac{H'}{3} = (6.8)\left(\frac{17.5}{3}\right) = 39.67 \text{ kip/ft}$$

Hence

$$FS_{\text{(overturning)}} = \frac{105.6}{39.67} = \mathbf{2.66}$$

Part b: Factor of Safety Against Sliding

$$FS_{\text{(sliding)}} = \frac{\Sigma V \tan\left(\frac{2}{3}\right)\phi_2 + B\left(\frac{2}{3}\right)c_2}{P_h}$$

$$= \frac{19.75 \tan\left(\frac{2}{3}\right)(20) + 10.3\left(\frac{2}{3}\right)(1.0)}{6.8} = \mathbf{1.7}$$

Part c: Pressure on the Soil at the Toe and Heel

$$e = \frac{B}{2} - \frac{\Sigma M_R - \Sigma M_O}{\Sigma V} = \frac{10.3}{2} - \frac{(105.6 - 39.67)}{19.67} = 1.8 \text{ ft}$$

$$q_{\text{toe}} = \frac{19.75}{10.3}\left[1 + \frac{(6)(1.8)}{10.3}\right] = \mathbf{3.93 \text{ kip/ft}^2}$$

$$q_{\text{heel}} = \frac{19.75}{10.3}\left[1 - \frac{(6)(1.8)}{10.3}\right] = \mathbf{-0.093 \text{ kip/ft}^2 \approx 0}$$

▲

7.5   OTHER TYPES OF POSSIBLE RETAINING-WALL FAILURE

In addition to the three types of possible failure for retaining walls discussed in Section 7.4, two other types of failure could occur: shallow shear failure and deep shear failure.

Shallow shear failure in soil below the base of a retaining wall takes place along a cylindrical surface *abc* passing through the heel, as shown in Figure 7.15a. The center of the arc of the circle *abc* is located at *O*, which is found by trial and error (corresponds to the minimum factor of safety). This type of failure can occur as the result of excessive induced shear stress along the cylindrical surface in soil. In general, the factor of safety against horizontal sliding is lower than the factor of safety obtained by shallow shear failure. So, if $FS_{\text{(sliding)}}$ is greater than about 1.5, shallow shear failure under the base may not occur.

Deep shear failure can occur along a cylindrical surface *abc*, as shown in Figure 7.15b, as the result of the existence of a weak layer of soil underneath the wall at a depth of about 1.5 times the width of the retaining wall. In such cases, the critical cylindrical failure surface *abc* has to be determined by trial and error with various

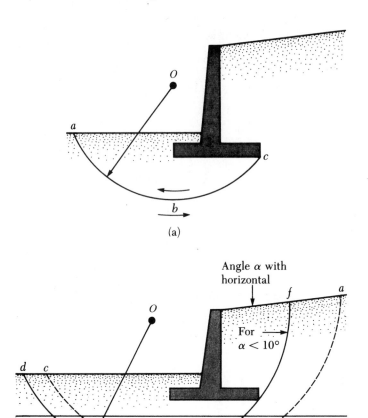

▼ FIGURE 7.15   (a) Shallow shear failure; (b) deep shear failure

centers, such as O (Figure 7.15b). The failure surface along which the minimum factor of safety is obtained is the *critical surface of sliding*. For the backfill slope with α less than about 10°, the critical failure circle apparently passes through the edge of the heel slab (such as *def* in Figure 7.15b). In this situation, the minimum factor of safety also has to be determined by trial and error by changing the center of the trial circle.

The following is an approximate procedure for determining the factor of safety against deep-seated shear failure for a gently sloping backfill ($\alpha < 10°$) developed by Teng (1962). Refer to Figure 7.16.

1.   Draw the retaining wall and the underlying soil layer to a convenient scale.

2.   For a trial center O, draw an arc of a circle *abcd*. For all practical purposes,

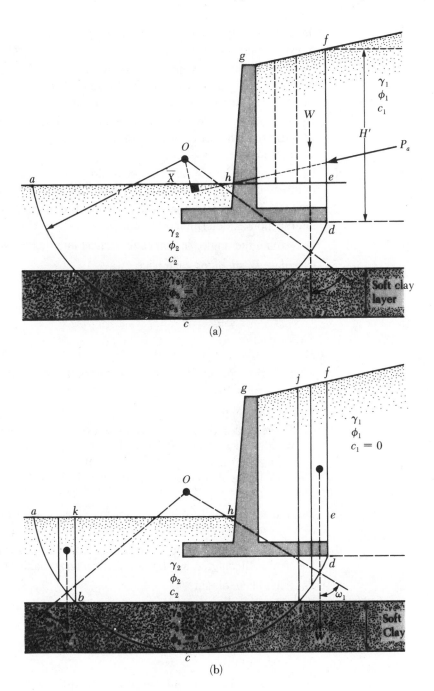

▼ **FIGURE 7.16** Deep shear failure analysis

the weight of the soil in the area *abcde* is symmetrical about a vertical line drawn through point O. Let the radius of the trial circle be r.

3. To determine the driving force on the failure surface causing instability (Figure 7.16a), divide the area in the zone *efgh* into several slices. These slices can be treated as rectangles or triangles, as the case may be.

4. Determine the area of each of these slices and then determine the weight W of the soil (and/or concrete) contained inside each slice (per unit length of the wall).

5. Draw a vertical line through the centroid of each slice, and locate the point of intersection of each vertical line with the trial failure circle.

6. Join point O (that is, the center of the trial circles) with the points of intersection as determined in Step 5.

7. Determine the angle, ω, that each vertical line makes with the radial line.

8. Calculate $W \sin \omega$ for each slice.

9. Determine the active force P_a on the face *df*, $\frac{1}{2}\gamma_1 H'^2 K_a$.

10. Calculate the total driving force:

$$\Sigma \, (W \sin \omega) + \frac{P_a \overline{X}}{r} \tag{7.24}$$

where $\overline{X}$ = perpendicular distance between the line of action of P_a and the center O

11. To determine the resisting force on the failure surface (Figure 7.16b), divide the area in the zones *abk* and *ideff* into several slices, and determine the weight of each slice, W_1 (per unit length of the wall). Note that points *b* and *i* are on top of the soft clay layer; the weight of each slice shown in Figure 7.16b is W_1 in contrast to the weight of each slice W, as shown in Figure 7.16a.

12. Draw a vertical line through the centroid of each slice and locate the point of intersection of each line with the trial failure circle.

13. Join point O with the points of intersection as determined in Step 12. Determine the angles, ω_1, that the vertical lines make with the radial lines.

14. For each slice, obtain

$W_1 \tan \phi_2 \cos \omega_1$

15. Calculate

$c_2 l_1 + c_3 l_2 + c_2 l_3$

where l_1, l_2, and l_3 are the lengths of the arcs *ab, bi,* and *id*

16. The maximum resisting force that can be derived along the failure surface is

$$\Sigma \, (W_1 \tan \phi_2 \cos \omega_1) + c_2 l_1 + c_3 l_2 + c_2 l_3 \tag{7.25}$$

17. Determine the factor of safety against deep shear failure for this trial failure surface:

$$FS_{\text{(deep shear failure)}} = \frac{\Sigma\,(W_1 \tan \phi_2 \cos \omega_1) + c_2 l_1 + c_3 l_2 + c_2 l_3}{\Sigma\,(W \sin \omega) + \dfrac{P_a \overline{X}}{r}} \tag{7.26}$$

Several other trial failure surfaces may be drawn, and the factor of safety can be determined in a similar manner. The lowest value of the factor of safety obtained from all trial surfaces is the desired factor of safety.

7.6 COMMENTS RELATING TO STABILITY

When a weak soil layer is located at a shallow depth — that is, within a depth of about 1.5 times the width of the retaining wall — the bearing capacity of the weak layer should be carefully investigated. The possibility of excessive settlement also should be considered. In some cases, the use of lightweight backfill material behind the retaining wall may solve the problem.

In many instances, piles are used to transmit the foundation load to a firmer layer. However, often the thrust of the sliding wedge of soil, in the case of deep shear failure, bends the piles and eventually causes them to fail. Careful attention should be given to this possibility when considering the option of pile foundations for retaining walls. (Pile foundations may be required for bridge abutments to avoid the problem of scouring.)

As illustrated in Examples 7.1, 7.2, and 7.3, the *active earth pressure coefficient* is used to determine the lateral force of the backfill. The active state of the backfill can be established only if the wall yields sufficiently, which does not happen in all cases. The degree of wall yielding will depend on its height and the section modulus. Furthermore, the lateral force of the backfill will depend on several factors, as identified by Casagrande (1973):

a. Effect of temperature
b. Groundwater fluctuation
c. Readjustment of the soil particles due to creep and prolonged rainfall
d. Tidal changes
e. Heavy wave action
f. Traffic vibration
g. Earthquakes

Insufficient wall yielding when combined with other unforeseen factors may generate a larger lateral force on the retaining structure compared to that obtained from the active earth pressure theory. Casagrande (1973) investigated the distribution of lateral earth pressure behind a bridge abutment (in Germany) with a slag backfill, as shown in Figure 7.17. Laboratory tests on the slag backfill gave angles of friction between 37° and 45°, depending on the degree of compaction. For purposes of comparison, the variation of the Rankine active earth pressure with $\phi = 37°$ and $\phi = 45°$ is also shown in Figure 7.17. Comparing the actual and theoretical pressure distribution diagrams indicates:

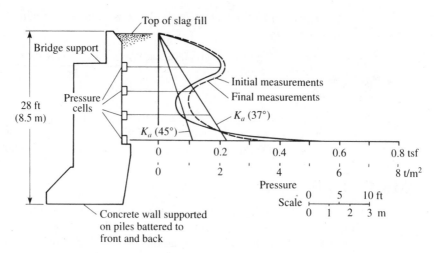

▼ **FIGURE 7.17**   Bridge abutment on piles backfilled with granulated slag (after Casagrande, 1973)

a. The actual lateral earth pressure distribution may not be triangular.
b. The lateral earth pressure distribution may change with time.
c. The actual active force is greater than the minimum theoretical active force.

The primary reason that many retaining walls designed with theoretical active earth pressure perform satisfactorily is the use of a large factor of safety. Recently, Goh (1993) analyzed the behavior of a retaining wall using the finite element method and proposed the simplified earth pressure distribution shown in Figure 7.18.

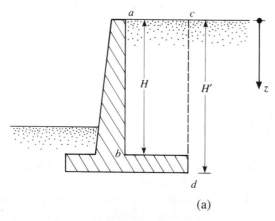

(a)

▼ **FIGURE 7.18**   Simplified lateral earth pressure (σ_h) profile: (a) retaining wall; (b) pressure distribution behind wall stem; (c) pressure distribution behind virtual wall (after Goh, 1993)

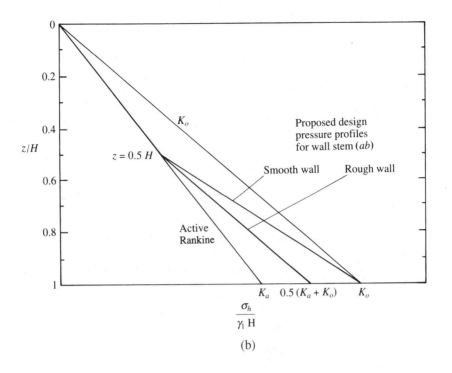

(b)

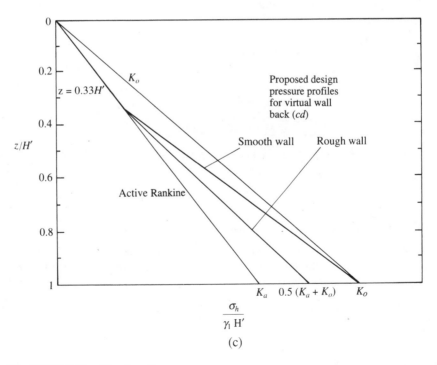

(c)

▼ FIGURE 7.18 (Continued)

7.7 DRAINAGE FROM THE BACKFILL OF THE RETAINING WALL

As the result of rainfall or other wet conditions, the backfill material for a retaining wall may become saturated. Saturation will increase the pressure on the wall and may create an unstable condition. For this reason, adequate drainage must be provided by means of *weepholes* and/or *perforated drainage pipes* (see Figure 7.19).

The *weepholes,* if provided, should have a minimum diameter of about 4 in. (0.1 m) and be adequately spaced. Note that there is always a possibility that the backfill material may be washed into weepholes or drainage pipes and ultimately clog them. Thus a filter material needs to be placed behind the weepholes or around the drainage pipes, as the case may be; geotextiles now serve that purpose. Whenever granular soil is used as a filter, the principles outlined in Section 1.10 should be followed. Example 7.4 gives the procedure for designing a filter.

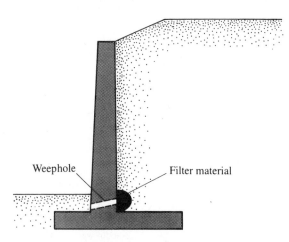

▼ **FIGURE 7.19**   Drainage provisions for the backfill of a retaining wall

▼ **EXAMPLE 7.4**

Figure 7.20 shows the grain-size distribution of a backfill material. Using the conditions outlined in Section 1.10, determine the range of the grain-size distribution for the filter material.

Solution   From the grain-size distribution curve given in Figure 7.20, the following values can be determined:

$$D_{15(B)} = 0.04 \text{ mm}$$

$$D_{85(B)} = 0.25 \text{ mm}$$

$$D_{50(B)} = 0.13 \text{ mm}$$

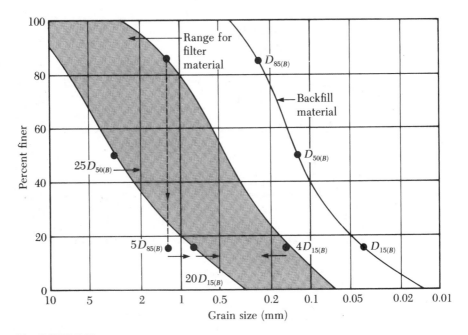

▼   FIGURE 7.20

Conditions of Filter

1. $D_{15(F)}$ should be less than $5D_{85(F)}$ — that is, $5 \times 0.25 = 1.25$ mm.
2. $D_{15(F)}$ should be greater than $4D_{15(B)}$ — that is, $4 \times 0.04 = 0.16$ mm.
3. $D_{50(F)}$ should be less than $25D_{50(B)}$ — that is, $25 \times 0.13 = 3.25$ mm.
4. $D_{15(F)}$ should be less than $20D_{15(B)}$ — that is, $20 \times 0.04 = 0.8$ mm.

These limiting points are plotted in Figure 7.20. Through these points two curves can be drawn that are similar in nature to the grain-size distribution curve of the backfill material. These curves define the range for the filter material to be used.

▲

7.8   PROVISION OF JOINTS IN RETAINING-WALL CONSTRUCTION

A retaining wall may be constructed with one or more of the following joints:

1.   *Construction joints* (Figure 7.21a) are vertical and horizontal joints that are placed between two successive pours of concrete. To increase the shear at the joints, keys may be used. If keys are not used, the surface of the first pour is cleaned and roughened before the next pour of concrete.
2.   *Contraction joints* (Figure 7.21b) are vertical joints (grooves) placed in the face of a wall (from the top of the base slab to the top of the wall) that allow the concrete to shrink without noticeable harm. The grooves may be about 0.25 to 0.3 in. ($\approx$6 to 8 mm) wide and 0.5 to 0.6 in. ($\approx$12 to 16 mm) deep.

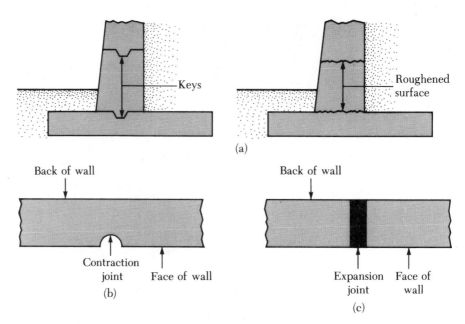

Keys

Roughened surface

Back of wall

Back of wall

Contraction joint   Face of wall

(b)

Expansion joint   Face of wall

(c)

▼ **FIGURE 7.21** (a) Construction joints; (b) contraction joint; (c) expansion joint

3. Expansion joints (Figure 7.21c) allow for the expansion of concrete caused by temperature changes; vertical expansion joints from the base to the top of the wall may also be used. These joints may be filled with flexible joint fillers. In most cases, horizontal reinforcing steel bars running across the stem are continuous through all joints. The steel is greased to allow the concrete to expand.

7.9 GRAVITY RETAINING-WALL DESIGN FOR EARTHQUAKE CONDITIONS

Even in mild earthquakes, most retaining walls undergo limited lateral displacement. Richards and Elms (1979) proposed a procedure for designing gravity retaining walls for earthquake conditions that allows limited lateral displacement. This procedure takes into consideration the wall inertia effect. Figure 7.22 shows a retaining wall with various forces acting on it, which are as follows (per unit length of the wall):

a. W_w = weight of the wall
b. P_{ae} = active force with earthquake condition taken into consideration (Section 6.6)

The backfill of the wall and the soil on which the wall is resting are assumed cohesionless. Considering the equilibrium of the wall, it can be shown that

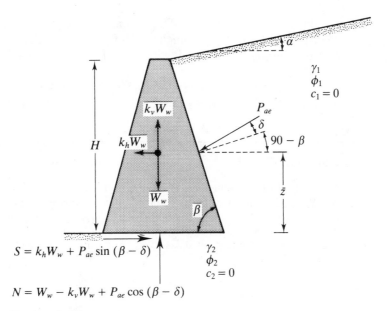

▼ **FIGURE 7.22**   Stability of a retaining wall under earthquake forces

$$W_w = [\tfrac{1}{2}\gamma_1 H^2 (1 - k_v) K_{ae}] C_{IE} \tag{7.27}$$

where γ_1 = unit weight of the backfill;

$$C_{IE} = \frac{\sin(\beta - \delta) - \cos(\beta - \delta)\tan\phi_2}{(1 - k_v)(\tan\phi_2 - \tan\theta')} \tag{7.28}$$

and $\theta' = \tan^{-1}\left(\dfrac{k_h}{1 - k_v}\right)$

For a detailed derivation of Eq. (7.28), see Das (1983).

Based on Eqs. (7.27) and (7.28), the following procedure may be used to determine the weight of the retaining wall, W_w, for tolerable displacement that may take place during an earthquake.

1. Determine the tolerable displacement of the wall, Δ.
2. Obtain a design value of k_h from

$$k_h = A_a \left(\frac{0.2 A_v^2}{A_a \Delta} \right)^{0.25} \tag{7.29}$$

In Eq. (7.29), A_a and A_v are effective acceleration coefficients and Δ is displacement in inches. The magnitudes of A_a and A_v are given by the Applied Technology Council (1978) for various regions of the United States.

3. Assume that $k_v = 0$, and, with the value of k_h obtained, calculate K_{ae} from Eq. (6.32).
4. Use the value of K_{ae} determined in Step 3 to obtain the weight of the wall (W_w).
5. Apply a factor of safety to the value of W_w obtained in Step 4.

▼ **EXAMPLE 7.5**

Refer to Figure 7.23. For $k_v = 0$ and $k_h = 0.3$, determine:

a. Weight of the wall for static condition
b. Weight of the wall for zero displacement during an earthquake
c. Weight of the wall for lateral displacement of 1.5 in. during an earthquake

For Part c, assume that $A_a = 0.2$ and $A_v = 0.2$. For Parts a, b, and c, use a factor of safety of 1.5.

Solution

Part a

For static conditions, $\theta' = 0$ and Eq. (7.28) becomes

$$C_{IE} = \frac{\sin(\beta - \delta) - \cos(\beta - \delta)\tan \phi_2}{\tan \phi_2}$$

For $\beta = 90°$, $\delta = 24°$ and $\phi = 36°$,

$$C_{IE} = \frac{\sin(90 - 24) - \cos(90 - 24)\tan 36}{\tan 36} = 0.85$$

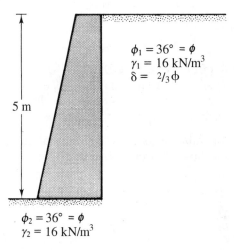

$\phi_1 = 36° = \phi$
$\gamma_1 = 16$ kN/m^3
$\delta = {}^{2}/_{3}\phi$

5 m

$\phi_2 = 36° = \phi$
$\gamma_2 = 16$ kN/m^3

▼ **FIGURE 7.23**

For static conditions, $K_{ae} = K_a$, so

$$W_w = \tfrac{1}{2}\gamma H^2 K_a C_{IE}$$

For $K_a \approx 0.2349$ [Table 6.5],

$$W_w = \tfrac{1}{2}(16)(5)^2(0.2349)(0.85) = 39.9 \text{ kN/m}$$

With a factor of safety of 1.5,

$$W_w = (39.9)(1.5) = \textbf{59.9 kN/m}$$

Part b

For zero displacement, $k_v = 0$,

$$C_{IE} = \frac{\sin(\beta - \delta) - \cos(\beta - \delta)\tan\phi_2}{\tan\phi_2 - \tan\theta'}$$

$$\tan\theta' = \frac{k_h}{1 - k_v} = \frac{0.3}{1 - 0} = 0.3$$

$$C_{IE} = \frac{\sin(90 - 24) - \cos(90 - 24)\tan 36}{\tan 36 - 0.3} = 1.45$$

For $k_h = 0.3$, $\phi_1 = 36°$ and $\delta = 2\phi/3$, the value of $K_{ae} \approx 0.48$ (Table 6.7).

$$W_w = \tfrac{1}{2}\gamma_1 H^2(1 - k_v)K_{ae}C_{IE} = \tfrac{1}{2}(16)(5)^2(1 - 0)(0.48)(1.45) = 139.2 \text{ kN/m}$$

With a factor of safety of 1.5, $W_w = \textbf{208.8 kN/m}$

Part c

For a lateral displacement of 1.5 in.,

$$k_h = A_a\left(\frac{0.2A_v^2}{A_a\Delta}\right)^{0.25} = (0.2)\left[\frac{(0.2)(0.2)^2}{(0.2)(1.5)}\right]^{0.25} = 0.081$$

$$\tan\theta' = \frac{k_h}{1 - k_v} = \frac{0.081}{1 - 0} = 0.081$$

$$C_{IE} = \frac{\sin(90 - 24) - \cos(90 - 24)\tan 36}{\tan 36 - 0.081} = 0.957$$

$$W_w = \tfrac{1}{2}\gamma_1 H^2 K_{ae} C_{IE}$$
$$\uparrow$$
$$\approx 0.29 \text{ [Table 6.7]}$$

$$W_w = \tfrac{1}{2}(16)(5)^2(0.29)(0.957) = 55.5 \text{ kN/m}$$

With a factor of safety of 1.5, $W_w = \textbf{83.3 kN/m}$ ▲

MECHANICALLY STABILIZED RETAINING WALLS

7.10 GENERAL DESIGN CONSIDERATIONS

The general design procedure of any mechanically stabilized retaining wall can be divided into two parts:

1. Satisfying *internal stability* requirements
2. Checking the *external stability* of the wall

The internal stability checks involve determining tension and pullout resistance in the reinforcing elements and the integrity of facing elements. The external stability checks include checks for overturning, sliding, and bearing capacity failure (Figure 7.24). The following sections will discuss the retaining-wall design procedures with metallic strips, geotextiles, and geogrids.

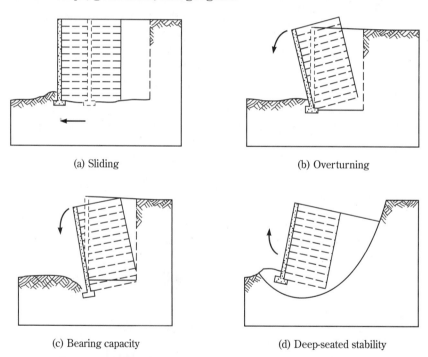

(a) Sliding

(b) Overturning

(c) Bearing capacity

(d) Deep-seated stability

▼   **FIGURE 7.24**   External stability checks (after Transportation Research Board, 1995)

7.11 RETAINING WALLS WITH METALLIC STRIP REINFORCEMENT

Reinforced earth walls are flexible walls. Their main components are

1. *Backfill,* which is granular soil

2. *Reinforcing strips,* which are thin, wide strips placed at regular intervals
3. *A cover* on the front face, which is referred to as the *skin*

Figure 7.25 is a diagram of a reinforced earth wall. Note that, at any depth, the reinforcing stripes or ties are placed with a horizontal spacing of S_H center-to-center; the vertical spacing of the strips or ties is S_V center-to-center. The skin can be constructed with sections of relatively flexible thin material. Lee et al. (1973) showed that, with a conservative design, a 0.2-in.-thick ($\approx$5 mm) galvanized steel skin would be enough to hold a wall about 45–50 ft (14–15 m) high. In most cases, precast concrete slabs can also be used as skin. The slabs are grooved to fit into each other so that soil cannot flow out between the joints. When metal skins are used, they are bolted together, and reinforcing strips are placed between the skins.

Figures 7.26 and 7.27 show a reinforced earth retaining wall under construction; its skin (facing) is a precast concrete slab. Figure 7.28 (p. 428) shows a metallic reinforcement tie attached to the concrete slab.

The simplest and most common method for the design of ties is the *Rankine method.* The following is a detailed discussion of this procedure.

Calculation of Active Horizontal and Vertical Pressure

Figure 7.29a (p. 429) shows a retaining wall with a granular backfill having a unit weight of γ_1 and a friction angle of ϕ_1. Below the base of the retaining wall, the *in*

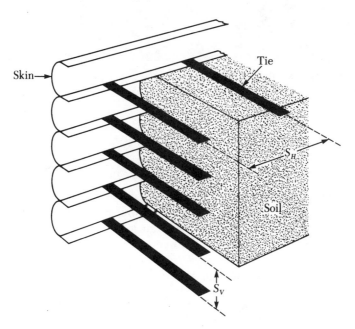

▼ **FIGURE 7.25**   Reinforced earth retaining wall

▼ **FIGURE 7.26**   Reinforced earth retaining wall (with metallic strip) under construction

situ soil has been excavated and recompacted, with granular soil used as backfill. Below the backfill, the *in situ* soil has a unit weight of γ_2, friction angle of ϕ_2, and cohesion of c_2. A surcharge having an intensity of q per unit area lies atop the retaining wall. The wall has reinforcement ties at depths $z = 0$, S_V, $2S_V$, ..., NS_V. The height of the wall is $NS_V = H$.

According to the Rankine active pressure theory (Section 6.3).

$$\sigma_a = \sigma_v K_a - 2c\sqrt{K_a}$$

where σ_a = Rankine active pressure at any depth z

▼ **FIGURE 7.27**   Another view of the retaining wall shown in Figure 7.26

For dry granular soils with no surcharge at the top, $c = 0$, $\sigma_v = \gamma_1 z$, and $K_a = \tan^2(45 - \phi_1/2)$. Thus

$$\sigma_{a(1)} = \gamma_1 z K_a \tag{7.30}$$

When a surcharge is added at the top, as shown in Figure 7.29,

$$\sigma_v = \underset{\substack{\uparrow \\ = \gamma_1 z \\ \text{Due to} \\ \text{soil only}}}{\sigma_{v(1)}} + \underset{\substack{\uparrow \\ \text{Due to the} \\ \text{surcharge}}}{\sigma_{v(2)}} \tag{7.31}$$

The magnitude of $\sigma_{v(2)}$ can be calculated by using the 2:1 method of stress distribution described in Eq. (4.14) and Figure 4.6. It is shown in Figure 7.30a (p. 430). According to Laba and Kennedy (1986),

▼ **FIGURE 7.28**   Metallic strip attachment to the precast concrete slab used as the skin

$$\sigma_{v(2)} = \frac{qa'}{a' + z} \qquad (\text{for } z \le 2b')$$

(7.32)

and

$$\sigma_{v(2)} = \frac{qa'}{a' + \dfrac{z}{2} + b'} \qquad (\text{for } z > 2b')$$

(7.33)

Also, when a surcharge is added at the top, the lateral pressure at any depth is

$$
\begin{array}{ccc}
\sigma_a = & \sigma_{a(1)} & + \sigma_{a(2)} \\
& \uparrow & \uparrow \\
& = K_a \gamma_1 z & \text{Due to the} \\
& \text{Due to} & \text{surcharge} \\
& \text{soil only} &
\end{array}
$$

(7.34)

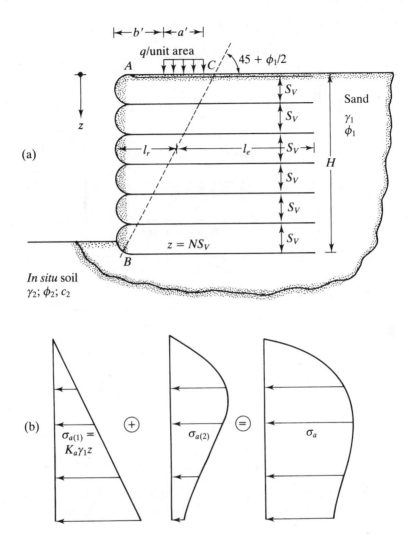

▼ **FIGURE 7.29** Analysis of a reinforced earth retaining wall

According to Laba and Kennedy (1986), $\sigma_{a(2)}$ may be expressed (Figure 7.30b) as

$$\sigma_{a(2)} = M\left[\frac{2q}{\pi}\;(\beta - \sin\beta\cos2\alpha)\right]$$

$$\uparrow$$
$$\text{(in radians)}$$

(7.35)

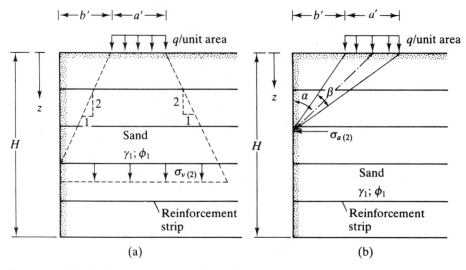

▼ **FIGURE 7.30**   (a) Notation for the relationship of $\sigma_{v(2)}$ — Eqs. (7.32) and (7.33);
(b) notation for the relationship of $\sigma_{a(2)}$ — Eqs. (7.35) and (7.36)

where

$$M = 1.4 - \frac{0.4b'}{0.14H} \geq 1 \qquad\qquad (7.36)$$

The net active (lateral) pressure distribution on the retaining wall calculated by using Eqs. (7.34), (7.35), and (7.36) is shown in Figure 7.29b.

Tie Force

Refer again to Figure 7.29. The tie force *per unit length of the wall* developed at any depth z is

T = active earth pressure at depth z
  × area of the wall to be supported by the tie   (7.37)

  $= (\sigma_a)(S_V S_H)$

Factor of Safety Against Tie Failure

The reinforcement ties at each level and thus the walls could fail by either (a) tie breaking or (b) tie pullout.

The factor of safety against *tie breaking* may be determined as

$$
FS_{(B)} = \frac{\text{yield or breaking strength of each tie}}{\text{maximum tie force in any tie}}
$$

$$
= \frac{wtf_y}{\sigma_a S_V S_H}
$$

(7.38)

where w = width of each tie
t = thickness of each tie
f_y = yield or breaking strength of the tie material

A factor of safety of about 2.5–3 is generally recommended for ties at all levels.

Reinforcing ties at any depth, z, will fail by pullout if the frictional resistance developed along their surfaces is less than the force to which the ties are being subjected. The *effective length* of the ties along which the frictional resistance is developed may be conservatively taken as the length that extends *beyond the limits of the Rankine active failure zone,* which is the zone *ABC* in Figure 7.29. Line *BC* in Figure 7.29 makes an angle of $45 + \phi_1/2$ with the horizontal. Now, the maximum friction force F_R that can be realized for a tie at depth z is

$$
F_R = 2l_e w \sigma_v \tan \phi_\mu
$$

(7.39)

where l_e = effective length
σ_v = effective vertical pressure at a depth z
ϕ_μ = soil–tie friction angle

Thus the factor of safety against *tie pullout* at any depth z is

$$
FS_{(P)} = \frac{F_R}{T}
$$

(7.40)

where $FS_{(P)}$ = factor of safety against tie pullout

Substituting Eqs. (7.37) and (7.39) into Eq. (7.40) yields

$$
FS_{(P)} = \frac{2l_e w \sigma_v \tan \phi_\mu}{\sigma_a S_V S_H}
$$

(7.41)

Total Length of Tie

The total length of ties at any depth is

$$
L = l_r + l_e
$$

(7.42)

where l_r = length within the Rankine failure zone
l_e = effective length

For a given $FS_{(P)}$ from Eq. (7.41),

$$l_e = \frac{FS_{(P)} \, \sigma_a S_V S_H}{2w\sigma_v \tan \phi_\mu} \tag{7.43}$$

Again, at any depth z,

$$l_r = \frac{(H - z)}{\tan\left(45 + \dfrac{\phi_1}{2}\right)} \tag{7.44}$$

So, combining Eqs. (7.42), (7.43), and (7.44) gives

$$L = \frac{(H - z)}{\tan\left(45 + \dfrac{\phi_1}{2}\right)} + \frac{FS_{(P)} \sigma_a S_V S_H}{2w\sigma_v \tan \phi_\mu} \tag{7.45}$$

7.12 STEP-BY-STEP-DESIGN PROCEDURE (METALLIC STRIP REINFORCEMENT)

Following is a step-by-step procedure for the design of reinforced earth retaining walls.

General:

1. Determine the height of the wall, H, and the properties of the granular backfill material, such as unit weight (γ_1) and angle of friction (ϕ_1).
2. Obtain the soil–tie friction angle, ϕ_μ, and the required values of $FS_{(B)}$ and $FS_{(P)}$.

Internal Stability:

3. Assume values for horizontal and vertical tie spacing. Also assume the width of reinforcing strip, w, to be used.
4. Calculate σ_a from Eqs. (7.34), (7.35), and (7.36).
5. Calculate the tie forces at various levels from Eq. (7.37).
6. For the known values of $FS_{(B)}$, calculate the thickness of ties, t, to resist the tie breakout:

$$T = \sigma_a S_V S_H = \frac{wtf_y}{FS_{(B)}}$$

or

$$t = \frac{(\sigma_a S_V S_H)\,[FS_{(B)}]}{w f_y}$$

(7.46)

The convention is to keep the magnitude of t the same at all levels, so σ_a in Eq. (7.46) should equal $\sigma_{a(\max)}$.

7. For the known values of ϕ_μ and $FS_{(P)}$, determine the length, L, of the ties at various levels from Eq. (7.45).

8. The magnitudes of S_V, S_H, t, w, and L may be changed to obtain the most economical design.

External Stability:

9. A check for *overturning* can be done as follows with reference to Figure 7.31. Taking the moment about B yields the overturning moment for the unit length of the wall:

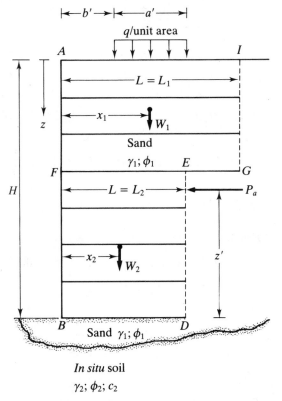

▼ **FIGURE 7.31** Stability check for the retaining wall

$$M_O = P_a z' \tag{7.47}$$

where P_a = active force = $\int_0^H \sigma_a\, dz$

The resisting moment per unit length of the wall is

$$M_R = W_1 x_1 + W_2 x_2 + \cdots + q a' \left(b' + \frac{a'}{2} \right) \tag{7.48}$$

where W_1 = (area $AFEGI$) (1) (γ_1)
$\qquad\;\; W_2$ = (area $FBDE$) (1) (γ_1)
$\qquad\qquad \vdots$

So

$$\boxed{\begin{aligned} FS_{(\text{overturning})} &= \frac{M_R}{M_O} \\[2mm] &= \frac{W_1 x_1 + W_2 x_2 + \cdots + q a' \left(b' + \dfrac{a'}{2} \right)}{\left(\displaystyle\int_0^H \sigma_a\, dz \right) z'} \end{aligned}} \tag{7.49}$$

10.   The check for *sliding* can be done by using Eq. (7.11), or

$$\boxed{FS_{(\text{sliding})} = \frac{(W_1 + W_2 + \cdots + q a')\, [\tan(k\phi_1)]}{P_a}} \tag{7.50}$$

where $k \approx \frac{2}{3}$.

11.   Check for ultimate bearing capacity failure. The ultimate bearing capacity can be given as

$$q_u = c_2 N_c + \tfrac{1}{2} \gamma_2 L_2' N_\gamma \tag{7.51a}$$

The bearing capacity factors N_c and N_γ correspond to the soil friction angle, ϕ_2 (Table 3.4). In Eq. (7.51a), L_2' is the effective length, or

$$L_2' = L_2 - 2e \tag{7.51b}$$

where e = eccentricity

$$e = \frac{L_2}{2} - \frac{M_R - M_O}{\Sigma V} \tag{7.51c}$$

where $\Sigma V = W_1 + W_2 \ldots + qa'$

The vertical stress at $z = H$, from Eq. (7.31), is

$$\sigma_{v(H)} = \gamma_1 H + \sigma_{v(2)} \tag{7.52}$$

So the factor of safety against bearing capacity failure is

$$FS_{(\text{bearing capacity})} = \frac{q_{\text{ult}}}{\sigma_{v(H)}} \tag{7.53}$$

Generally, minimum values of $FS_{(\text{overturning})} = 3$, $FS_{(\text{sliding})} = 3$, and $FS_{(\text{bearing capacity failure})} = 3$ to 5 are recommended.

▼ EXAMPLE 7.6 _____

A 30-ft-high retaining wall with galvanized steel-strip reinforcement in a granular backfill has to be constructed. Referring to Figure 7.29, given:

Granular backfill:	ϕ_1	$= 36°$
	γ_1	$= 105\ \text{lb/ft}^3$
Foundation soil:	ϕ_2	$= 28°$
	γ_2	$= 110\ \text{lb/ft}^3$
	c_2	$= 1000\ \text{lb/ft}^2$

Galvanized steel reinforcement:

Width of strip,	w	$= 3$ in.
	S_V	$= 2$ ft center-to-center
	S_H	$= 3$ ft center-to-center
	f_y	$= 35,000\ \text{lb/in}^2$
	ϕ_μ	$= 20°$
Required	$FS_{(B)}$	$= 3$
Required	$FS_{(P)}$	$= 3$

Check for the external and internal stability. Assume the corrosion rate of the galvanized steel to be 0.001 in./year and the life span of the structure to be 50 years.

Solution

Internal Stability Check

a. **Tie thickness:** Maximum tie force, $T_{max} = \sigma_{a(max)} S_V S_H$

$$\sigma_{a(max)} = \gamma H K_a = \gamma H \tan^2\left(45 - \frac{\phi_1}{2}\right)$$

so

$$T_{max} = \gamma H \tan^2\left(45 - \frac{\phi_1}{2}\right) S_V S_H$$

From Eq. (7.46), for *tie break*,

$$t = \frac{(\sigma_a S_V S_H)\,[FS_{(B)}]}{w f_y} = \frac{\left[\gamma H \tan^2\left(45 - \dfrac{\phi_1}{2}\right) S_V S_H\right]}{w f_y}$$

or

$$t = \frac{\left[(105)\,(30)\tan^2\left(45 - \dfrac{36}{2}\right)(2)\,(3)\right](3)}{\left(\dfrac{3}{12}\,\text{ft}\right)(35{,}000 \times 144\ \text{lb/ft}^2)} = 0.0117\ \text{ft} = 0.14\ \text{in.}$$

If the rate of corrosion is 0.001 in./yr and the life span of the structure is 50 yr, then the actual thickness, t, of the ties will be

$$t = 0.14 + (0.001)\,(50) = 0.19\ \text{in.}$$

So a **tie thickness of 0.2 in.** would be enough.

b. **Tie length:** Refer to Eq. (7.45). For this case, $\sigma_a = \gamma_1 z K_a$ and $\sigma_v = \gamma_1 z$, so

$$L = \frac{(H - z)}{\tan\left(45 + \dfrac{\phi_1}{2}\right)} + \frac{FS_{(P)}\gamma_1 z K_a S_V S_H}{2w\gamma_1 z \tan \phi_\mu}$$

Now the following table can be prepared. (Note: $FS_{(P)} = 3$, $H = 30$ ft, $w = 3$ in., and $\phi_\mu = 20°$.)

z (ft)	Tie length, L (ft) [Eq. (7.45)]
5	38.45
10	35.89
15	33.34
20	30.79
25	28.25
30	25.7

So use a **tie length of L = 40 ft.**

External Stability Check

a. **Check for overturning:** Refer to Figure 7.32. For this case, using Eq. (7.49),

$$FS_{(overturning)} = \frac{W_1 x_1}{\left[\int_0^H \sigma_a \, dz \right] z'}$$

$$W_1 = \gamma_1 HL = (105)(30)(40) = 126,000 \text{ lb}$$

$$x_1 = 20 \text{ ft}$$

$$P_a = \int_0^H \sigma_a \, dz = \tfrac{1}{2} \gamma_1 K_a H^2 = (\tfrac{1}{2})(105)(0.26)(30)^2 = 12,285 \text{ lb/ft}$$

$$z' = \frac{30}{3} = 10 \text{ ft}$$

$$FS_{(overturning)} = \frac{(126,000)(20)}{(12,285)(10)} = \textbf{20.5} > \textbf{3—OK}$$

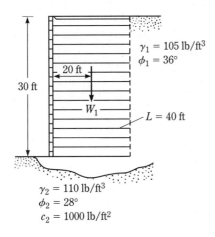

$$\gamma_1 = 105 \text{ lb/ft}^3$$
$$\phi_1 = 36°$$

20 ft

30 ft

W_1

$L = 40 \text{ ft}$

$$\gamma_2 = 110 \text{ lb/ft}^3$$
$$\phi_2 = 28°$$
$$c_2 = 1000 \text{ lb/ft}^2$$

▼ FIGURE 7.32

b. **Check for sliding:** From Eq. (7.50)

$$FS_{(sliding)} = \frac{W_1 \tan(k\phi_1)}{P_a} = \frac{126{,}000 \tan\left[\left(\frac{2}{3}\right)(36)\right]}{12{,}285} = \mathbf{4.57 > 3—OK}$$

c. **Check for bearing capacity:** For $\phi_2 = 28°$, $N_c = 25.8$, $N_\gamma = 16.78$ (Table 3.4). From Eq. (7.51a),

$$q_{ult} = c_2 N_c + \tfrac{1}{2}\gamma_2 L' N_\gamma$$

$$e = \frac{L}{2} - \frac{M_R - M_O}{\Sigma V} = \frac{40}{2} - \left[\frac{(126{,}000 \times 20) - (12{,}285 \times 10)}{126{,}000}\right] = 0.975 \text{ ft}$$

$$L' = 40 - (2 \times 0.975) = 38.05 \text{ ft}$$

$$q_{ult} = (1000)(25.8) + (\tfrac{1}{2})(110)(38.05)(16.72) = 60{,}791 \text{ lb/ft}^2$$

From Eq. (7.52),

$$\sigma_{V(H)} = \gamma_1 H = (105)(30) = 3150 \text{ lb/ft}^2$$

$$FS_{(bearing\ capacity)} = \frac{q_{ult}}{q_{v(H)}} = \frac{60{,}791}{3150} = \mathbf{19.3 > 5—OK}$$

▲

7.13 RETAINING WALLS WITH GEOTEXTILE REINFORCEMENT

Figure 7.33 shows a retaining wall in which layers of geotextile have been used as reinforcement. As in Figure 7.30, the backfill is a granular soil. In this type of retaining wall, the facing of the wall is formed by lapping the sheets as shown with a lap length of l_l. When construction of the wall is finished, the exposed face of the wall must be covered; otherwise, the geotextile will deteriorate from exposure to ultraviolet light. *Bitumen emulsion* or *Gunite* is sprayed on the wall face. A wire mesh anchored to the geotextile facing may be necessary to keep the coating on the face of the wall.

The design of this type of retaining wall is similar to that presented in Section 7.11. Following is a step-by-step procedure for design based on the recommendations of Bell et al. (1975) and Koerner (1990).

Internal Stability:

1. Determine the active pressure distribution on the wall from

$$\sigma_a = K_a \sigma_v = K_a \gamma_1 z \tag{7.54}$$

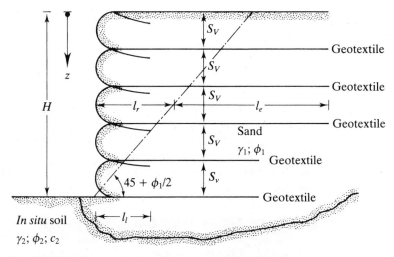

▼ FIGURE 7.33 Retaining wall with geotextile reinforcement

where K_a = Rankine earth pressure coefficient = $\tan^2(45 - \phi_1/2)$
γ_1 = unit weight of the granular backfill
ϕ_1 = friction angle of the granular backfill

2. Select a geotextile fabric that has an allowable strength of σ_G (lb/ft or kN/m).

3. Determine the vertical spacing of the layers at any depth z from

$$S_V = \frac{\sigma_G}{\sigma_a FS_{(B)}} = \frac{\sigma_G}{(\gamma_1 z K_a)[FS_{(B)}]} \tag{7.55}$$

Note that Eq. (7.55) is similar to Eq. (7.38). The magnitude of $FS_{(B)}$ is generally 1.3–1.5.

4. Determine the length of each layer of geotextile from

$$L = l_r + l_e \tag{7.56}$$

where

$$l_r = \frac{H - z}{\tan\left(45 + \dfrac{\phi_1}{2}\right)} \tag{7.57}$$

and

$$l_e = \frac{S_V \sigma_a [FS_{(P)}]}{2\sigma_v \tan \phi_F} \tag{7.58}$$

$$\sigma_a = \gamma_1 z K_a$$

$$\sigma_v = \gamma_1 z$$

$$FS_{(P)} = 1.3 \text{ to } 1.5$$

$$\phi_F = \text{friction angle at geotextile–soil interface}$$

$$\approx \tfrac{2}{3}\,\phi_1$$

Note that Eqs. (7.56), (7.57), and (7.58) are similar to Eqs. (7.42), (7.44), and (7.43), respectively.

Based on the published results, the assumption of $\phi_F/\phi_1 \approx \tfrac{2}{3}$ is reasonable and appears to be conservative. Martin et al. (1984) presented the following laboratory test results for ϕ_F/ϕ_1 between various types of geotextiles and sand.

Type	ϕ_F/ϕ_1
Woven — monofilament/concrete sand	0.87
Woven — silt film/concrete sand	0.8
Woven — silt film/rounded sand	0.86
Woven — silt film/silty sand	0.92
Nonwoven — melt-bonded/concrete sand	0.87
Nonwoven — needle-punched/concrete sand	1.0
Nonwoven — needle-punched/rounded sand	0.93
Nonwoven — needle-punched/silty sand	0.91

5. Determine the lap length, l_l, from

$$l_l = \frac{S_V \sigma_a FS_{(P)}}{4\sigma_v \tan \phi_F} \tag{7.59}$$

The minimum lap length should be 3 ft (1 m).

External Stability:

6. Check the factors of safety against overturning, sliding, and bearing capacity failure as described in Section 7.12 (Steps 9, 10, and 11).

▼ **EXAMPLE 7.7**

A geotextile-reinforced retaining wall 16 ft high is shown in Figure 7.34. For the granular backfill, $\gamma_1 = 110 \text{ lb/ft}^3$ and $\phi_1 = 36°$. For the geotextile, $\sigma_G = 80 \text{ lb/in.}$ For the design of the wall, determine S_V, L, and l_l.

Solution

$$K_a = \tan^2\left(45 - \frac{\phi_1}{2}\right) = 0.26$$

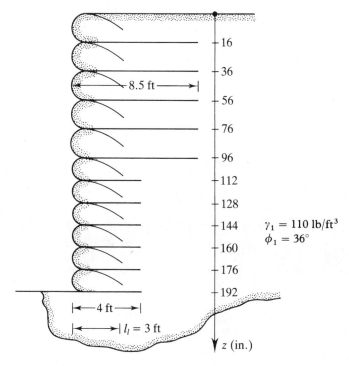

$\gamma_1 = 110 \text{ lb/ft}^3$
$\phi_1 = 36°$

▼ **FIGURE 7.34**

Determination of S_V

To find S_V, we make a few trials. From Eq. (7.55),

$$S_V = \frac{\sigma_G}{(\gamma_1 z K_a)[FS_{(B)}]}$$

With $FS_{(B)} = 1.5$ at $z = 8$ ft,

$$S_V = \frac{(80 \times 12 \text{ lb/ft})}{(110)(8)(0.26)(1.5)} = 2.8 \text{ ft} \approx 33.6 \text{ in.}$$

At $z = 12$ ft,

$$S_V = \frac{(80 \times 12 \text{ lb/ft})}{(110)(12)(0.26)(1.5)} = 1.87 \text{ ft} \approx 22 \text{ in.}$$

At $z = 16$ ft,

$$S_V = \frac{(80 \times 12 \text{ lb/ft})}{(110)(16)(0.26)(1.5)} = 1.4 \text{ ft} \approx 16.8 \text{ in.}$$

So, **use $S_V = 20$ in. for $z = 0$ to $z = 8$ ft and $S_V = 16$ in. for $z > 8$ ft.** This is shown in Figure 7.34.

Determination of L

From Eqs. (7.56), (7.57), and (7.58),

$$L = \frac{(H - z)}{\tan\left(45 + \dfrac{\phi_1}{2}\right)} + \frac{S_V K_a [FS_{(P)}]}{2 \tan \phi_F}$$

For $FS_{(P)} = 1.5$, $\tan \phi_F = \tan[(\tfrac{2}{3})(36)] = 0.445$ and

$$L = (0.51)(H - z) + 0.438 S_V$$

Now the following table can be prepared.

z (in.)	(ft)	S_V (ft)	$(0.51)(H - z)$ (ft)	$0.438 S_V$ (ft)	L (ft)
16	1.33	1.67	7.48	0.731	8.21
56	4.67	1.67	5.78	0.731	6.51
76	6.34	1.67	4.93	0.731	5.66
96	8.0	1.67	4.08	0.731	4.81
112	9.34	1.33	3.40	0.582	3.982
144	12.0	1.33	2.04	0.582	2.662
176	14.67	1.33	0.68	0.582	1.262

Based on the preceding calculations, **use $L = 8.5$ ft for $z \leq 8$ ft and $L = 4$ ft for $z > 8$ ft.**

Determination of l_l

From Eq. (7.59),

$$l_l = \frac{S_V \sigma_a [FS_{(P)}]}{4\sigma_v \tan \phi_F}$$

With $\sigma_a = \gamma_1 z K_a$, $FS_{(P)} = 1.5$; with $\sigma_v = \gamma_1 z$, $\phi_F = \tfrac{2}{3}\phi_1$. So

$$l_l = \frac{S_V K_a [FS_{(P)}]}{4 \tan \phi_F} = \frac{S_V (0.26)(1.5)}{4 \tan[(\tfrac{2}{3})(36)]} = 0.219 S_V$$

At $z = 16$ in.,

$$l_l = 0.219 S_V = (0.219)\left(\frac{20}{12}\right) = 0.365 \text{ ft} \leq 3 \text{ ft}$$

So, use $l_l = 3$ ft.   ▲

7.14 RETAINING WALLS WITH GEOGRID REINFORCEMENT

Geogrids can also be used as reinforcement in granular backfill for the construction of retaining walls. Figure 7.35 shows typical schematic diagrams of retaining walls with geogrid reinforcement.

Relatively few field measurements are available for lateral earth pressure on retaining walls constructed with geogrid reinforcement. Figure 7.36 shows a comparison of measured and design lateral pressures (Berg et al., 1986) for two retaining walls constructed with precast panel facing (see Figure 7.35c). It indicates that the

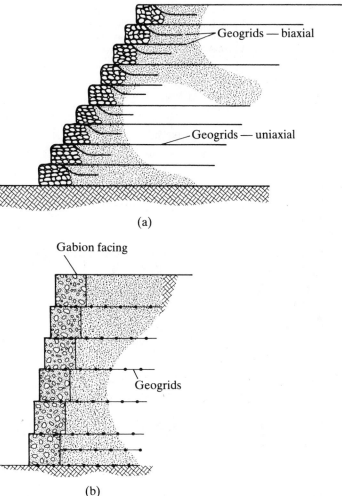

(a)

(b)

▼ **FIGURE 7.35** Typical schematic diagrams of retaining walls with geogrid reinforcement: (a) geogrid wraparound wall; (b) wall with gabion facing; (c) concrete panel–faced wall (after The Tensar Corporation, 1986)

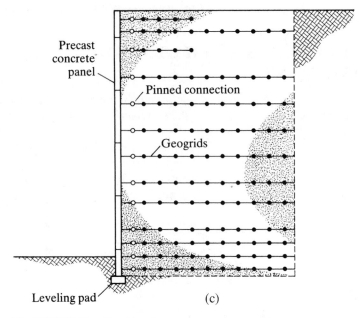

Precast
concrete
panel

Pinned connection

Geogrids

Leveling pad

(c)

▼ **FIGURE 7.35**   (Continued)

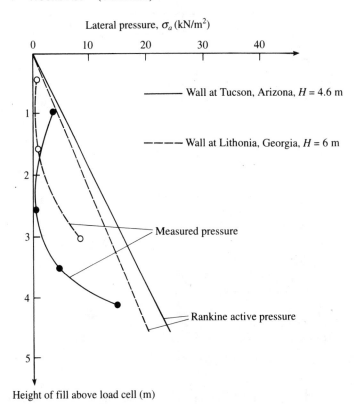

Lateral pressure, σ_a (kN/m^2)

Wall at Tucson, Arizona, $H = 4.6$ m

Wall at Lithonia, Georgia, $H = 6$ m

Measured pressure

Rankine active pressure

Height of fill above load cell (m)

▼ **FIGURE 7.36**   Comparison of theoretical and measured lateral pressures in geogrid reinforced retaining
walls (based on Berg et al., 1986)

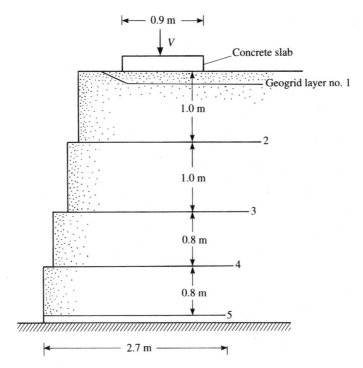

▼ **FIGURE 7.37** Schematic diagram of the retaining wall tested by
Thamm et al. (1990)

measured earth pressures were substantially smaller than those calculated for the
Rankine active case.

The results of another interesting full-scale test on a retaining wall with geogrid
reinforcement, granular backfill, and a height of 3.6 m was reported by Thamm
et al. (1990). The main reinforcement for the wall was TENSAR SR2 geogrid. Figure
7.37 shows a schematic diagram of the retaining wall. Failure in the wall was caused
by applying load to a concrete slab measuring 2.4 m × 0.9 m. The wall failed when
the vertical load, V, on the concrete slab reached 1065 kN. Figure 7.38 shows the
variation of the wall face displacement and the distribution of lateral pressure as
the loading progressed, from which the following conclusions can be drawn:

1. The shape of the lateral earth pressure distribution on the wall face is similar
    to that shown in Figure 7.29b.
2. At failure load, the magnitude of $\Delta L/H$ (ΔL = facing displacement) at the
    top of the wall was about 1.7%, which is considerably higher than may be
    encountered for a rigid retaining wall.

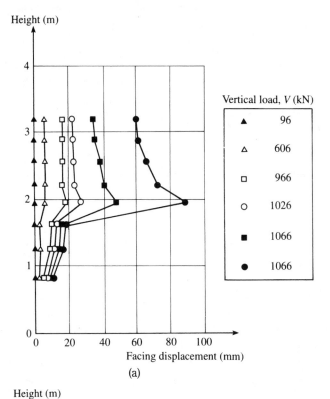

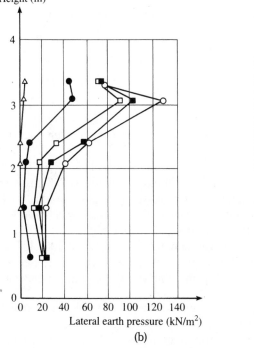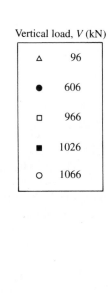

▼ **FIGURE 7.38** Observations from tests on the retaining wall shown in Figure 7.37: (a) facing displacement with loading; (b) lateral earth pressure with loading (based on Thamm et al., 1990)

7.15 GENERAL COMMENTS

Great progress is being made in the development of rational design procedures for mechanically stabilized earth (MSE) retaining walls. Readers are directed to *Transportation Research Circular No. 444* (1995) and *Federal Highway Administration Publication No. FHWA-SA-96-071* (1996) for further information. However, a few recent developments are summarized below.

1. Rankine's active pressure was used in the design of MSE retaining walls in this chapter. The appropriate value of the earth pressure coefficient to be used in the design depends, however, on the degree of restraint that the reinforcing elements impose on the soil. If the wall can yield substantially, the Rankine active earth pressure may be appropriate, which is not the case for all types of MSE walls. Figure 7.39 shows the recommended design values for lateral earth pressure coefficient, K. Note that

$$\sigma_h = K\sigma_v = K\gamma_1 z$$

where σ_h = lateral earth pressure
σ_v = effective vertical stress
γ_1 = unit weight of granular backfill

In Figure 7.39, $K_a = \tan^2(45 - \phi_1/2)$, where ϕ_1 is the angle of friction of the backfill.

2. In Sections 7.11 and 7.13, the effective length (l_e) against tie pullout was calculated behind the Rankine failure surface (for example, see Figue 7.29a).

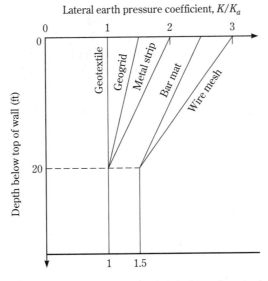

▼ **FIGURE 7.39** Recommended design values for lateral earth pressure coefficient, K (after Transportation Research Board, 1995)

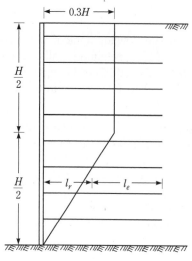

(a) Inextensible reinforcement

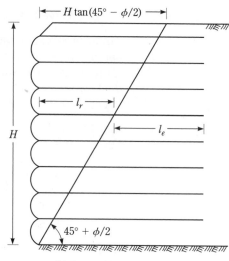

(b) Extensible reinforcement

▼ **FIGURE 7.40**   Location of potential failure surface (after Transportation Research Board, 1995)

Recent field measurements and theoretical analysis show that the potential failure surface may depend on the type of reinforcement. Figure 7.40 shows the potential failure plane locations for walls with inextensible and extensible reinforcement in the granular backfill.

New developments in the design of MSE walls will be incorporated into future editions of the text.

PROBLEMS   For Problems 7.1 through 7.7, use $k_1 = k_2 = \frac{2}{3}$ and $P_p = 0$ while using Eq. (7.11).

7.1   For the cantilever retaining wall shown in Figure P7.1., the wall dimensions are $H = 8$ m, $x_1 = 0.4$ m, $x_2 = 0.6$ m, $x_3 = 1.5$ m, $x_4 = 3.5$ m, $x_5 = 0.96$ m, $D = 1.75$ m, and $\alpha = 10°$. The soil properties are $\gamma_1 = 16.8$ kN/m³, $\phi_1 = 32°$, $\gamma_2 = 17.6$ kN/m³, $\phi_2 = 28°$, and $c_2 = 30$ kN/m². Calculate the factors of safety with respect to overturning, sliding, and bearing capacity. Use unit weight of concrete, $\gamma_c = 23.58$ kN/m³.

7.2   Repeat Problem 7.1 with the following:

Wall dimensions: $H = 18$ ft, $x_1 = 18$ in., $x_2 = 30$ in., $x_3 = 4$ ft, $x_4 = 6$ ft, $x_5 = 2.75$ ft, $D = 4$ ft, $\alpha = 10°$

Soil properties: $\gamma_1 = 117$ lb/ft³, $\phi_1 = 34°$, $\gamma_2 = 110$ lb/ft³, $\phi_2 = 18°$, $c_2 = 800$ lb/ft²

Use unit weight of concrete, $\gamma_c = 150$ lb/ft³.

7.3   Repeat Problem 7.1 with the following:

Wall dimensions: $H = 22$ ft, $x_1 = 12$ in., $x_2 = 27$ in., $x_3 = 4.5$ ft, $x_4 = 8$ ft, $x_5 = 2.75$ ft, $D = 4$ ft, $\alpha = 5°$

Soil properties: $\gamma_1 = 110$ lb/ft³, $\phi_1 = 36°$, $\gamma_2 = 120$ lb/ft³, $\phi_2 = 15°$, $c_2 = 1000$ lb/ft²

Use unit weight of concrete, $\gamma_c = 150$ lb/ft³.

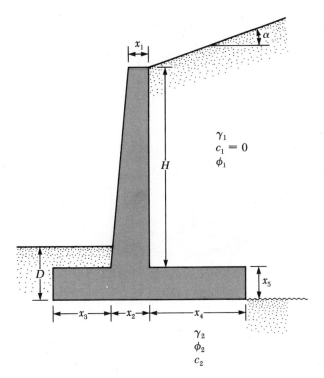

▼ FIGURE P7.1

7.4 Repeat Problem 7.1 with the following:

Wall dimensions: $H = 6.5$ m, $x_1 = 0.3$ m, $x_2 = 0.6$ m, $x_3 = 0.8$ m, $x_4 = 2$ m, $x_5 = 0.8$ m, $D = 1.5$ m, $\alpha = 0°$

Soil properties: $\gamma_1 = 18.08$ kN/m³, $\phi_1 = 36°$, $\gamma_2 = 19.65$ kN/m³, $\phi_2 = 15°$, $c_2 = 30$ kN/m²

Use unit weight of concrete, $\gamma_c = 23.58$ kN/m³.

7.5 Refer to Problem 7.4. What will be the factor of safety against sliding if the heel of the wall is sloped with $\alpha' = 45°$? (See Figure 7.8b for the definition of α'.)

7.6 A gravity retaining wall is shown in Figure P7.6. Calculate the factor of safety with respect to overturning and sliding. Given:

Wall dimensions: $H = 6$ m, $x_1 = 0.6$ m, $x_2 = 2$ m, $x_3 = 2$ m, $x_4 = 0.5$ m, $x_5 = 0.75$ m, $x_6 = 0.8$ m, $D = 1.5$ m

Soil properties: $\gamma_1 = 16.5$ kN/m³, $\phi_1 = 32°$, $\gamma_2 = 18$ kN/m³, $\phi_2 = 22°$, $c_2 = 40$ kN/m²

Use Rankine active pressure for calculation and $\gamma_{concrete} = 23.58$ kN/m³.

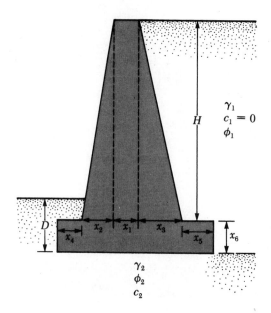

$$H \qquad \begin{array}{l} \gamma_1 \\ c_1 = 0 \\ \phi_1 \end{array}$$

$$\begin{array}{l} \gamma_2 \\ \phi_2 \\ c_2 \end{array}$$

▼   **FIGURE P7.6**

7.7   Repeat Problem 7.6 using Coulomb's active pressure for calculation and $\delta = \frac{2}{3}\phi_1$.

7.8   Refer to Figure P7.8 for the design of a gravity retaining wall for earthquake conditions.
Given: $k_v = 0$ and $k_h = 0.3$.
   a.   What should be the weight of the wall for a zero displacement condition? Use a
   factor of safety of 2.
   b.   What should be the weight of the wall for an allowable displacement of 50.8 mm?

Given: $A_v = 0.15$ and $A_a = 0.25$. Use a factor of safety of 2.

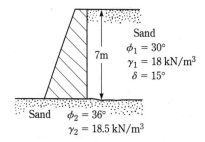

7m

Sand
$\phi_1 = 30°$
$\gamma_1 = 18 \text{ kN/m}^3$
$\delta = 15°$

Sand $\phi_2 = 36°$
$\gamma_2 = 18.5 \text{ kN/m}^3$

▼   **FIGURE P7.8**

7.9   Refer to Figure 7.29. Use the following parameters:

Wall: $H = 6$ m

Soil: $\gamma_1 = 16.5 \text{ kN/m}^3$ and $\phi_1 = 35°$

Reinforcement: $S_V = 1$ m and $S_H = 1.5$ m

Surcharge: $q = 50$ kN/m², $a' = 1.5$ m, and $b' = 2$ m

Calculate the vertical stress, σ_v [Eqs. (7.31), (7.32), and (7.33)] at $z = 1$ m, 2 m, 3 m, 4 m, 5 m, and 6 m.

7.10 For Problem 7.9, calculate the lateral pressure σ_a at $z = 1$ m, 2 m, 3 m, 4 m, 5 m, and 6 m. Use Eqs. (7.34), (7.35), and (7.36).

7.11 A reinforced earth retaining wall (Figure 7.29) is to be 30 ft high. Here,

Backfill: unit weight, $\gamma_1 = 119$ lb/ft³, soil friction angle, $\phi_1 = 34°$

Reinforcement: vertical spacing, $S_V = 3$ ft; horizontal spacing, $S_H = 4$ ft; width of reinforcement = 4.75 in., $f_y = 38,000$ lb/in.², $\phi_\mu = 25°$, factor of safety against tie pullout = 3, and factor of safety against tie breaking = 3

Determine:
a. The required thickness of ties
b. The required maximum length of ties

7.12 In Problem 7.11, assume that the ties at all depths are the length determined in part (b). For the *in situ* soil, $\phi_2 = 25°$, $\gamma_2 = 116$ lb/ft³, and $c_2 = 650$ lb/ft². Calculate the factor of safety against (a) overturning, (b) sliding, and (c) bearing capacity failure.

7.13 Redo Problem 7.11 for a retaining wall with a height of 24 ft.

7.14 In Problem 7.13, assume that the ties at all depths are the length determined in part (b). For the *in situ* soil, $\phi_2 = 28°$, $\gamma_2 = 121$ lb/ft³, and $c_2 = 500$ lb/ft². Calculate the factor of safety against (a) overturning, (b) sliding, and (c) bearing capacity failure.

7.15 Redo Problem 7.11 and change S_V to 1.5 ft.

7.16 A retaining wall with geotextile reinforcement is 6 m high. For the granular backfill, $\gamma_1 = 15.9$ kN/m³ and $\phi_1 = 30°$. For the geotextile, $\sigma_G = 16$ kN/m. For the design of the wall, determine S_V, L, and l_l. Use $FS_{(B)} = FS_{(P)} = 1.5$.

7.17 For the S_V, L, and l_l determined in Problem 7.16, check the overall stability (that is, factor of safety against overtuning, sliding, and bearing capacity failure). For the *in situ* soil, $\gamma_2 = 16.8$ kN/m³, $\phi_2 = 20°$, $c_2 = 55$ kN/m².

7.18 Check the overall stability (that is, factor of safety against overturning, sliding, and bearing capacity failure) of the retaining wall with geotextile reinforcement given in Example 7.7. Use $\gamma_2 = 108$ lb/ft³, $\phi_2 = 20°$, $c_2 = 1200$ lb/ft².

REFERENCES

Applied Technology Council (1978). "Tentative Provisions for the Development of Seismic Regulations for Buildings," *Publication ATC-3-06.*

Bell, J. R., Stilley, A. N., and Vandre, B. (1975). "Fabric Retaining Earth Walls," *Proceedings, Thirteenth Engineering Geology and Soils Engineering Symposium,* Moscow, Idaho.

Berg, R. R., Bonaparte, R., Anderson, R. P., and Chouery, V. E. (1986). "Design Construction and Performance of Two Tensar Geogrid Reinforced Walls," *Proceedings,* Third International Conference on Geotextiles, Vienna, pp. 401–406.

Casagrande, L. (1973). "Comments on Conventional Design of Retaining Structure," *Journal of the Soil Mechanics and Foundations Division,* ASCE, Vol. 99, No. SM2, pp. 181–198.

Das, B. M. (1983). *Fundamentals of Soil Dynamics,* Elsevier, New York.

Elman, M. T., and Terry, C. F. (1988). "Retaining Walls with Sloped Heel," *Journal of Geotechnical Engineering,* American Society of Civil Engineers, Vol. 114, No. GT10, pp. 1194–1199.

Federal Highway Administration (1996). *Mechanically Stabilized Earth Walls and Reinforced Soil Slopes Design and Construction Guidelines, Publication No. FHWA-SA-96-071,* Washington, D.C.

Goh, A. T. C. (1993). "Behavior of Cantilever Retaining Walls," *Journal of Geotechnical Engineering,* ASCE, Vol. 119, No. 11, pp. 1751–1770.

Koerner, R. B. (1990). *Design with Geosynthetics,* 2nd ed., Prentice-Hall, Englewood Cliffs, N.J.

Laba, J. T., and Kennedy, J. B. (1986). "Reinforced Earth Retaining Wall Analysis and Design," *Canadian Geotechnical Journal,* Vol. 23, No. 3, pp. 317–326.

Lee, K. L., Adams, B. D., and Vagneron, J. J. (1973). "Reinforced Earth Retaining Walls," *Journal of the Soil Mechanics and Foundations Division,* American Society of Civil Engineers, Vol. 99, No. SM10, pp. 745–763.

Martin, J. P., Koerner, R. M., and Whitty, J. E. (1984). "Experimental Friction Evaluation of Slippage Between Geomembranes, Geotextiles, and Soils," *Proceedings,* International Conference on Geomembranes, Denver, pp. 191–196.

Richards, R., and Elms, D. G. (1979). "Seismic Behavior of Gravity Retaining Walls," *Journal of the Geotechnical Engineering Division,* American Society of Civil Engineers, Vol. 105, No. GT4, pp. 449–464.

Teng, W. C. (1962). *Foundation Design,* Prentice-Hall, Englewood Cliffs, N. J.

Tensar Corporation (1986). *The Tensar Technical Note. No. TTN:RW1,* August.

Terzaghi, K., and Peck, R. B. (1967). *Soil Mechanics in Engineering Practice,* Wiley, New York.

Thamm, B. R., Krieger, B., and Krieger, J. (1990). "Full-Scale Test on a Geotextile-Reinforced Retaining Structure," *Proceedings,* Fourth International Conference on Geotextiles, Geomembranes, and Related Products, The Hague, Vol. 1, pp. 3–8.

Transportation Research Board (1995). *Transportation Research Circular No. 444,* National Research Council, Washington, D.C.

CHAPTER EIGHT

SHEET PILE STRUCTURES

8.1 INTRODUCTION

Connected or semiconnected sheet piles are often used to build continuous walls for waterfront structures that may range from small waterfront pleasure boat launching facilities to large dock facilities (Figure 8.1a). In contrast to the construction of other types of retaining wall, the building of sheet pile walls does not usually require dewatering of the site. Sheet piles are also used for some temporary structures, such as braced cuts (Figure 8.1b). The principles of sheet pile wall design and the design used in braced cuts are discussed in this chapter.

Several types of sheet pile are commonly used in construction: (a) wooden sheet piles, (b) precast concrete sheet piles, and (c) steel sheet piles. Aluminum sheet piles are also marketed.

Wooden sheet piles are used only for temporary light structures that are above the water table. The most common types are ordinary wooden planks and *Wakefield piles*. The wooden planks are about 2 in. × 12 in. (50 mm × 300 mm) in cross section and are driven edge to edge (Figure 8.2a). Wakefield piles are made by nailing three planks together with the middle plank offset by 2–3 in. (50–75 mm) (Figure 8.2b). Wooden planks can also be milled to form *tongue-and-groove piles,* as shown in Figure 8.2c. Figure 8.2d shows another type of wooden sheet pile that has precut grooves. Metal *splines* are driven into the grooves of the adjacent sheetings to hold them together after they are driven into the ground.

Precast concrete sheet piles are heavy and are designed with reinforcements to withstand the permanent stresses to which the structure will be subjected after construction and also to handle the stresses produced during construction. In cross section, these piles are about 20–32 in. (500–800 mm) wide and 6–10 in. (150–250 mm) thick. Figure 8.2e shows schematic diagrams of the elevation and the cross section of a reinforced concrete sheet pile.

Steel sheet piles in the United States are about 0.4–0.5 in. (10–13 mm) thick. European sections may be thinner and wider. Sheet pile sections may be *Z, deep arch, low arch,* or *straight web* sections. The interlocks of the sheet pile sections are shaped like a *thumb-and-finger* or a *ball-and-socket* for watertight connections. Figure 8.3a shows schematic diagrams of the thumb-and-finger type of interlocking for straight web sections. The ball-and-socket type of interlocking for Z section piles is

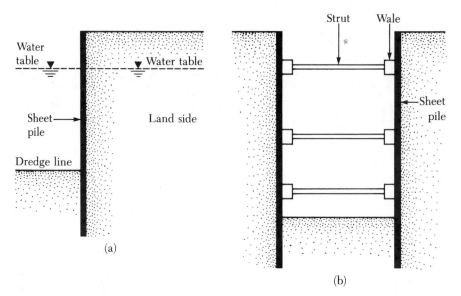

▼ **FIGURE 8.1** Examples of uses of sheet piles: (a) waterfront sheet pile wall;
(b) braced cut

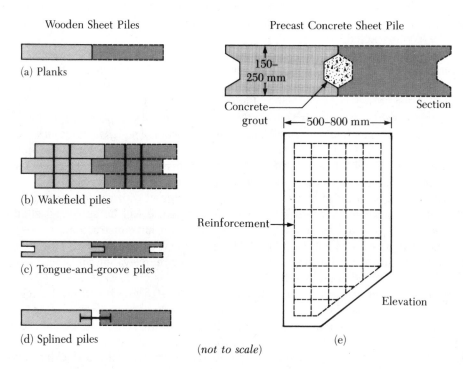

▼ **FIGURE 8.2** Various types of wooden and concrete sheet pile

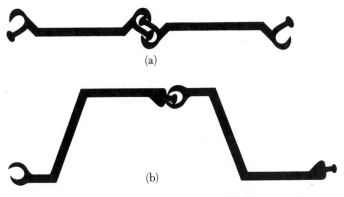

(a)

(b)

▼ **FIGURE 8.3** Nature of sheet pile connections: (a) thumb-and-finger type; (b) ball-and-socket type

shown in Figure 8.3b. Table C-1 (Appendix C) shows the properties of the sheet pile sections produced by the Bethlehem Steel Corporation. The allowable design flexural stress for the steel sheet piles is as follows:

Type of steel	Allowable stress (lb/in^2)
ASTM A-328	25,000 lb/in^2 (170 MN/m^2)
ASTM A-572	30,000 lb/in^2 (210 MN/m^2)
ASTM A-690	30,000 lb/in^2 (210 MN/m^2)

Steel sheet piles are convenient to use because of their resistance to high driving stress developed when being driven into hard soils. They are also lightweight and reusable.

SHEET PILE WALLS

8.2 CONSTRUCTION METHODS

Sheet pile walls may be divided into two basic categories: (a) cantilever and (b) anchored.

In the construction of sheet pile walls, sheet piles may be driven into the ground and then the backfill is placed on the land side, or the sheet pile may first be driven into the ground and the soil in front of the sheet pile dredged. In any case, the soil used for backfill behind the sheet pile wall is usually granular. The soil below the dredge line may be sandy or clayey soil. The surface of soil on the water side is referred to as the *mud line* or *dredge line*.

Thus construction methods generally can be divided into two categories (Tsinker, 1983):

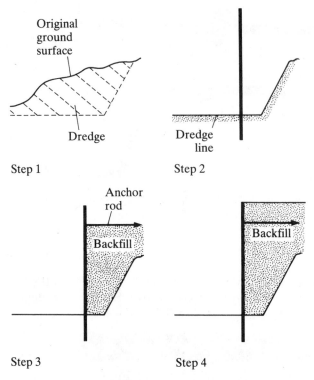

▼ FIGURE 8.4 Sequence of construction for a backfilled structure

1. Backfilled structure
2. Dredged structure

The sequence of construction for a *backfilled structure* is as follows (Figure 8.4):

Step 1. Dredge the *in situ* soil in front and back of the proposed structure.
Step 2. Drive the sheet piles.
Step 3. Backfill up to the level of the anchor and place the anchor system.
Step 4. Backfill up to the top of the wall.

For a cantilever type of wall, only Steps 1, 2, and 4 apply. The sequence of construction for a *dredged structure* is as follows (Figure 8.5):

Step 1. Drive the sheet piles.
Step 2. Backfill up to the anchor level and place the anchor system.
Step 3. Backfill up to the top of the wall.
Step 4. Dredge the front side of the wall.

For cantilever sheet pile walls, Step 2 is not required.

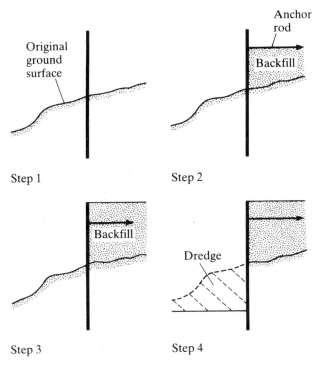

Step 1

Step 2

Step 3

Step 4

▼ **FIGURE 8.5** Sequence of construction for a dredged structure

8.3 CANTILEVER SHEET PILE WALLS—GENERAL

Cantilever sheet pile walls are usually recommended for walls of moderate height—about 20 ft ($\approx$6 m) or less, measured above the dredge line. In such walls, the sheet piles act as a wide cantilever beam above the dredge line. The basic principles for estimating net lateral pressure distribution on a cantilever sheet pile wall can be explained with the aid of Figure 8.6. It shows the nature of lateral yielding of a cantilever wall penetrating a sand layer below the dredge line. The wall rotates about point O. Because the hydrostatic pressures at any depth from both sides of the wall will cancel each other, we consider only the effective lateral soil pressures. In zone A, the lateral pressure is only the active pressure from the land side. In zone B, because of the nature of yielding of the wall, there will be active pressure from the land side and passive pressure from the water side. The condition is reversed in zone C — that is, below the point of rotation, O. The net actual pressure distribution on the wall is like that shown in Figure 8.6b. However, for design purposes, Figure 8.6c shows a simplified version.

Sections 8.4–8.7 present the mathematical formulation of the analysis of cantilever sheet pile walls. Note that, in some waterfront structures, the water level may

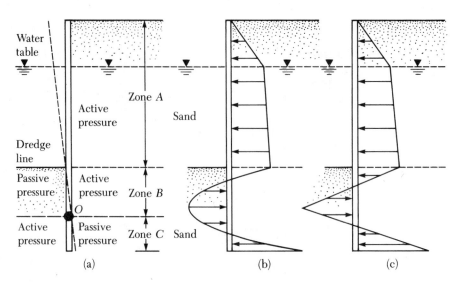

▼ **FIGURE 8.6**   Cantilever sheet pile penetrating sand

fluctuate as the result of tidal effects. Care should be taken in determining the water level that will affect the net pressure diagram.

8.4 CANTILEVER SHEET PILING PENETRATING SANDY SOILS

To develop the relationships for the proper depth of embedment of sheet piles driven into a granular soil, we refer to Figure 8.7a. The soil retained by the sheet piling above the dredge line also is sand. The water table is at depth L_1 below the top of the wall. Let the angle of friction of the sand be ϕ. The intensity of the active pressure at a depth $z = L_1$ is

$$p_1 = \gamma L_1 K_a \qquad (8.1)$$

where K_a = Rankine active pressure coefficient = $\tan^2 (45 - \phi/2)$
        γ = unit weight of soil above the water table

Similarly, the active pressure at depth $z = L_1 + L_2$ (that is, at the level of the dredge line) is

$$p_2 = (\gamma L_1 + \gamma' L_2) K_a \qquad (8.2)$$

where γ' = effective unit weight of soil = $\gamma_{sat} - \gamma_w$

Note that, at the level of the dredge line, the hydrostatic pressures from both sides of the wall are the same magnitude and cancel each other.

To determine the net lateral pressure below the dredge line up to the point of rotation O, as shown in Figure 8.6a, an engineer has to consider the passive pressure acting from the left side (water side) toward the right side (land side) and also the

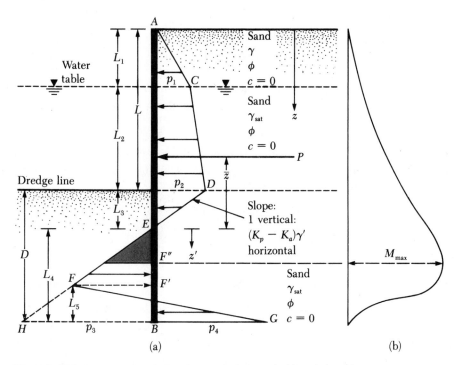

▼ **FIGURE 8.7** Cantilever sheet pile penetrating sand: (a) variation of net pressure diagram; (b) variation of moment

active pressure acting from the right side toward the left side of the wall. For such cases, ignoring the hydrostatic pressure from both sides of the wall, the active pressure at depth z is

$$p_a = [\gamma L_1 + \gamma' L_2 + \gamma'(z - L_1 - L_2)]K_a \tag{8.3}$$

Also, the passive pressure at depth z is

$$p_p = \gamma'(z - L_1 - L_2)K_p \tag{8.4}$$

where K_p = Rankine passive pressure coefficient = $\tan^2(45 + \phi/2)$

Hence, combining Eqs. (8.3) and (8.4) yields the net lateral pressure:

$$\begin{aligned} p = p_a - p_p &= (\gamma L_1 + \gamma' L_2)K_a - \gamma'(z - L_1 - L_2)(K_p - K_a) \\ &= p_2 - \gamma'(z - L)(K_p - K_a) \end{aligned} \tag{8.5}$$

where $L = L_1 + L_2$

The net pressure, p, equals zero at depth L_3 below the dredge line, so

$$p_2 - \gamma'(z - L)(K_p - K_a) = 0$$

or

$$(z - L) = L_3 = \frac{p_2}{\gamma'(K_p - K_a)} \tag{8.6}$$

Equation (8.6) indicates that the slope of the net pressure distribution line *DEF* is 1 vertical to $(K_p - K_a)\gamma'$ horizontal, so, in the pressure diagram

$$\overline{HB} = p_3 = L_4(K_p - K_a)\gamma' \tag{8.7}$$

At the bottom of the sheet pile, passive pressure, p_p, acts from the right toward the left side and active pressure acts from the left toward the right side of the sheet pile, so, at $z = L + D$,

$$p_p = (\gamma L_1 + \gamma' L_2 + \gamma' D)K_p \tag{8.8}$$

At the same depth

$$p_a = \gamma' D K_a \tag{8.9}$$

Hence the net lateral pressure at the bottom of the sheet pile is

$$\begin{aligned}
p_p - p_a = p_4 &= (\gamma L_1 + \gamma' L_2)K_p + \gamma' D(K_p - K_a) \\
&= (\gamma L_1 + \gamma' L_2)K_p + \gamma' L_3(K_p - K_a) + \gamma' L_4(K_p - K_a) \\
&= p_5 + \gamma' L_4(K_p - K_a)
\end{aligned} \tag{8.10}$$

where
$$p_5 = (\gamma L_1 + \gamma' L_2)K_p + \gamma' L_3(K_p - K_a) \tag{8.11}$$
$$D = L_3 + L_4 \tag{8.12}$$

For the stability of the wall, the principles of statics can now be applied:

Σ horizontal forces per unit length of wall $= 0$

and

Σ moment of the forces per unit length of wall about point $B = 0$

For summation of the horizontal forces,

Area of the pressure diagram *ACDE* $-$ area of *EFHB* $+$ area of *FHBG* $= 0$

or

$$P - \tfrac{1}{2}p_3 L_4 + \tfrac{1}{2}L_5(p_3 + p_4) = 0 \tag{8.13}$$

where $P = $ area of the pressure diagram *ACDE*

Summing the moment of all the forces about point B yields

$$P(L_4 + \bar{z}) - \left(\frac{1}{2}L_4 p_3\right)\left(\frac{L_4}{3}\right) + \frac{1}{2}L_5(p_3 + p_4)\left(\frac{L_5}{3}\right) = 0 \tag{8.14}$$

From Eq. (8.13),

$$L_5 = \frac{p_3 L_4 - 2P}{p_3 + p_4} \tag{8.15}$$

Combining Eqs. (8.7), (8.10), (8.14), and (8.15) and simplifying them further, we obtain the following fourth-degree equation in terms of L_4:

$$L_4^4 + A_1 L_4^3 - A_2 L_4^2 - A_3 L_4 - A_4 = 0 \qquad (8.16)$$

where

$$A_1 = \frac{p_5}{\gamma'(K_p - K_a)} \qquad (8.17)$$

$$A_2 = \frac{8P}{\gamma'(K_p - K_a)} \qquad (8.18)$$

$$A_3 = \frac{6P[2\bar{z}\gamma'(K_p - K_a) + p_5]}{\gamma'^2(K_p - K_a)^2} \qquad (8.19)$$

$$A_4 = \frac{P(6\bar{z}p_5 + 4P)}{\gamma'^2(K_p - K_a)^2} \qquad (8.20)$$

Step-by-Step Procedure for Obtaining the Pressure Diagram

Based on the preceding theory, the step-by-step procedure for obtaining the pressure diagram for a cantilever sheet pile wall penetrating a granular soil is as follows:

1. Calculate K_a and K_p.
2. Calculate p_1 [Eq. (8.1)] and p_2 [Eq. (8.2)]. *Note:* L_1 and L_2 will be given.
3. Calculate L_3 [Eq. (8.6)].
4. Calculate P.
5. Calculate $\bar{z}$ (that is, the center of pressure for the area *ACDE*) by taking the moment about E.
6. Calculate p_5 [Eq. (8.11)].
7. Calculate A_1, A_2, A_3, and A_4 [Eqs. (8.17) to (8.20)].
8. Solve Eq. (8.16) by trial and error to determine L_4.
9. Calculate p_4 [Eq. (8.10)].
10. Calculate p_3 [Eq. (8.7)].
11. Obtain L_5 from Eq. (8.15).
12. Draw the pressure distribution diagram like the one shown in Figure 8.7a.
13. Obtain the theoretical depth [Eq. (8.12)] of penetration as $L_3 + L_4$. The actual depth of penetration is increased by about 20%–30%.

Note: Some designers prefer to use a factor of safety on the passive earth pressure coefficient at the beginning. In that case, in Step 1

$$K_{p(\text{design})} = \frac{K_p}{FS}$$

where FS = factor of safety (usually between 1.5 to 2)

For this type of analysis, follow Steps 1–12 with the value of $K_a = \tan^2(45 - \phi/2)$ and $K_{p(\text{design})}$ (instead of K_p). The actual depth of penetration can now be determined by adding L_3, obtained from Step 3, and L_4, obtained from Step 8.

Calculation of Maximum Bending Moment

The nature of variation of the moment diagram for a cantilever sheet pile wall is shown in Figure 8.7b. The maximum moment will occur between points E and F'. To obtain the maximum moment ($M_{\max}$) per unit length of the wall requires determining the point of zero shear. For a new axis z' (with origin at point E) for zero shear,

$$P = \tfrac{1}{2}(z')^2(K_p - K_a)\gamma'$$

or

$$z' = \sqrt{\frac{2P}{(K_p - K_a)\gamma'}} \tag{8.21}$$

Once the point of zero shear force is determined (point F'' in Figure 8.7a), the magnitude of the maximum moment can be obtained as

$$M_{\max} = P(\bar{z} + z') - [\tfrac{1}{2}\gamma'z'^2(K_p - K_a)](\tfrac{1}{3})z' \tag{8.22}$$

The necessary profile of the sheet piling is then sized according to the allowable flexural stress of the sheet pile material, or

$$S = \frac{M_{\max}}{\sigma_{\text{all}}} \tag{8.23}$$

where S = section modulus of the sheet pile required per unit length of the structure
   σ_{all} = allowable flexural stress of the sheet pile

▼ **EXAMPLE 8.1** _____

Figure 8.8 shows a cantilever sheet pile wall penetrating a granular soil. Here, $L_1 = 2$ m, $L_2 = 3$ m, $\gamma = 15.9$ kN/m³, $\gamma_{\text{sat}} = 19.33$ kN/m³, and $\phi = 32°$.

   a.   What is the theoretical depth of embedment, D?
   b.   For a 30% increase in D, what should be the total length of the sheet piles?
   c.   What should be the minimum section modulus of the sheet piles? Use $\sigma_{\text{all}} = 172$ MN/m².

Solution

Part a

The following table can now be prepared for a step-by-step calculation. Refer to Figure 8.7a for the pressure distribution diagram.

Quantity required	Eq. no.	Equation and calculation
K_a	—	$\tan^2\left(45-\dfrac{\phi}{2}\right)=\tan^2\left(45-\dfrac{32}{2}\right)=0.307$
K_p	—	$\tan^2\left(45+\dfrac{\phi}{2}\right)=\tan^2\left(45+\dfrac{32}{2}\right)=3.25$
p_1	8.1	$\gamma L_1 K_a = (15.9)(2)(0.307) = 9.763 \text{ kN/m}^2$
p_2	8.2	$(\gamma L_1 + \gamma' L_2)K_a = [(15.9)(2) + (19.33 - 9.81)(3)](0.307) = 18.53 \text{ kN/m}^2$
L_3	8.6	$\dfrac{p_2}{\gamma'(K_p - K_a)} = \dfrac{18.53}{(19.33-9.81)(3.25-0.307)} = 0.66 \text{ m}$
P	—	$\frac{1}{2}p_1L_1 + p_1L_2 + \frac{1}{2}(p_2-p_1)L_2 + \frac{1}{2}p_2L_3$
		$\quad = \left(\frac{1}{2}\right)(9.763)(2) + (9.763)(3) + \left(\frac{1}{2}\right)(18.53-9.763)(3) + \left(\frac{1}{2}\right)(18.53)(0.66)$
		$\quad = 9.763 + 29.289 + 13.151 + 6.115 = 58.32 \text{ kN/m}$
$\bar{z}$	—	$\dfrac{\Sigma M_E}{P} = \dfrac{1}{58.32}\left[\begin{array}{l}9.763(0.66+3+\frac{2}{3})+29.289(0.66+\frac{3}{2})\\ +13.151(0.66+\frac{3}{3})+6.115(0.66\times\frac{2}{3})\end{array}\right] = 2.23 \text{ m}$
p_5	8.11	$(\gamma L_1 + \gamma' L_2)K_p + \gamma' L_3(K_p - K_a) = [(15.9)(2) + (19.33-9.81)(3)](3.25)$
		$\quad + (19.33-9.81)(0.66)(3.25-0.307) = 214.66 \text{ kN/m}^2$
A_1	8.17	$\dfrac{p_5}{\gamma'(K_p-K_a)} = \dfrac{214.66}{(19.33-9.81)(3.25-0.307)} = 7.66$
A_2	8.18	$\dfrac{8P}{\gamma'(K_p-K_a)} = \dfrac{(8)(58.32)}{(19.33-9.81)(3.25-0.307)} = 16.65$
A_3	8.19	$\dfrac{6P[2\bar{z}\gamma'(K_p-K_a)+p_5]}{\gamma'^2(K_p-K_a)^2}$
		$\quad = \dfrac{(6)(58.32)[(2)(2.23)(19.33-9.81)(3.25-0.307)+214.66]}{(19.33-9.81)^2(3.25-0.307)^2} = 151.93$
A_4	8.20	$\dfrac{P(6\bar{z}p_5+4P)}{\gamma'^2(K_p-K_a)^2} = \dfrac{58.32[(6)(2.23)(214.66)+(4)(58.32)]}{(19.33-9.81)^2(3.25-0.307)^2} = 230.72$
L_4	8.16	$L_4^4 + A_1L_4^3 - A_2L_4^2 - A_3L_4 - A_4 = 0$
		$L_4^4 + 7.66L_4^3 - 16.55L_4^2 - 151.93L_4 - 230.72 = 0; \quad L_4 \approx 4.8 \text{ m}$

$$D_{\text{theory}} = L_3 + L_4 = 0.66 + 4.8 = \textbf{5.46 m}$$

Part b

Total length of sheet piles,

$$L_1 + L_2 + 1.3(L_3 + L_4) = 2 + 3 + 1.3(5.46) = \textbf{12.1 m}$$

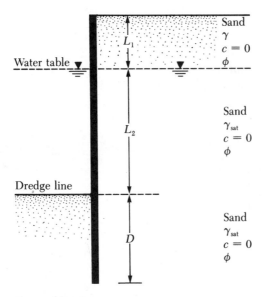

Water table ▼

Sand
γ
$c = 0$
ϕ

L_1

Sand
γ_{sat}
$c = 0$
ϕ

L_2

Dredge line

Sand
γ_{sat}
$c = 0$
ϕ

D

▼   FIGURE 8.8

Part c

Quantity required	Eq. no.	Equation and calculation
z'	8.21	$\sqrt{\dfrac{2P}{(K_p - K_a)\gamma'}} = \sqrt{\dfrac{(2)(58.32)}{(3.25 - 0.307)(19.33 - 9.81)}} = 2.04 \text{ m}$
M_{max}	8.22	$P(\bar{z} + z') - \left[\dfrac{1}{2}\gamma' z'^2 (K_p - K_a)\right]\dfrac{z'}{3} = (58.32)(2.23 + 2.04)$ $- \left[\left(\dfrac{1}{2}\right)(19.33 - 9.81)(2.04)^2(3.25 - 0.307)\right]\dfrac{2.04}{3} = 209.39 \text{ kN} \cdot \text{m/m}$
S	8.29	$\dfrac{M_{max}}{\sigma_{all}} = \dfrac{209.39 \text{ kN} \cdot \text{m}}{172 \times 10^3 \text{ kN/m}^2} = \mathbf{1.217 \times 10^{-3} \text{ m}^3/\text{m of wall}}$

▲

8.5   SPECIAL CASES FOR CANTILEVER WALLS (PENETRATING A SANDY SOIL)

Following are two special cases of the mathematical formulation shown in Section 8.4.

Case 1: Sheet Pile Wall in the Absence of Water Table

In the absence of the water table, the net pressure diagram on the cantilever sheet pile wall will be as shown in Figure 8.9, which is a modified version of Figure 8.7. In this case,

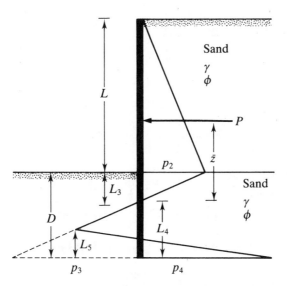

FIGURE 8.9   Sheet piling penetrating a sandy soil
in the absence of the water table

$$p_2 = \gamma L K_a \tag{8.24}$$

$$p_3 = L_4(K_p - K_a)\gamma \tag{8.25}$$

$$p_4 = p_5 + \gamma L_4(K_p - K_a) \tag{8.26}$$

$$p_5 = \gamma L K_p + \gamma L_3(K_p - K_a) \tag{8.27}$$

$$L_3 = \frac{p_2}{\gamma(K_p - K_a)} = \frac{LK_a}{(K_p - K_a)} \tag{8.28}$$

$$P = \tfrac{1}{2}p_2 L + \tfrac{1}{2}p_2 L_3 \tag{8.29}$$

$$\bar{z} = L_3 + \frac{L}{3} = \frac{LK_a}{K_p - K_a} + \frac{L}{3} = \frac{L(2K_a + K_p)}{3(K_p - K_a)} \tag{8.30}$$

and Eq. (8.16) transforms to

$$L_4^4 + A_1' L_4^3 - A_2' L_4^2 - A_3' L_4 - A_4' = 0 \tag{8.31}$$

where

$$A_1' = \frac{p_5}{\gamma(K_p - K_a)}$$ (8.32)

$$A_2' = \frac{8P}{\gamma(K_p - K_a)}$$ (8.33)

$$A_3' = \frac{6P[2\bar{z}\gamma(K_p - K_a) + p_5]}{\gamma^2(K_p - K_a)^2}$$ (8.34)

$$A_4' = \frac{P(6\bar{z}p_5 + 4P)}{\gamma^2(K_p - K_a)^2}$$ (8.35)

Case 2: Free Cantilever Sheet Piling

Figure 8.10 shows a free cantilever sheet pile wall penetrating a sandy soil and subjected to a line load of P per unit length of the wall. For this case,

$$D^4 - \left[\frac{8P}{\gamma(K_p - K_a)}\right]D^2 - \left[\frac{12PL}{\gamma(K_p - K_a)}\right]D - \left[\frac{2P}{\gamma(K_p - K_a)}\right]^2 = 0$$ (8.36)

and

$$L_5 = \frac{\gamma(K_p - K_a)D^2 - 2P}{2D(K_p - K_a)\gamma}$$ (8.37)

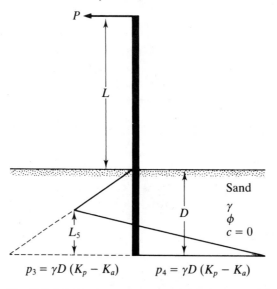

$$p_3 = \gamma D(K_p - K_a) \qquad p_4 = \gamma D(K_p - K_a)$$

▼ **FIGURE 8.10**   Free cantilever sheet piling penetrating a sand layer

$$M_{\text{max}} = P(L + z') - \frac{\gamma z'^3 (K_p - K_a)}{6} \tag{8.38}$$

$$z' = \sqrt{\frac{2P}{\gamma'(K_p - K_a)}} \tag{8.39}$$

▼ **EXAMPLE 8.2**

Redo Parts a and b of Example 8.1 assuming the absence of the water table. Use $\gamma = 15.9$ kN/m³ and $\phi = 32°$. Note: $L = 5$ m.

Solution

Part a

Quantity required	Eq. no.	Equation and calculation
K_a	—	$\tan^2\left(45 - \dfrac{\phi}{2}\right) = \tan^2\left(45 - \dfrac{32}{2}\right) = 0.307$
K_p	—	$\tan^2\left(45 + \dfrac{\phi}{2}\right) = \tan^2\left(45 + \dfrac{32}{2}\right) = 3.25$
p_2	8.24	$\gamma L K_a = (15.9)(5)(0.307) = 24.41$ kN/m²
L_3	8.28	$\dfrac{LK_a}{K_p - K_a} = \dfrac{(5)(0.307)}{3.25 - 0.307} = 0.521$ m
p_5	8.27	$\gamma L K_p + \gamma L_3(K_p - K_a) = (15.9)(5)(3.25) + (15.9)(0.521)(3.25 - 0.307)$ $= 282.76$ kN/m²
P	8.29	$\frac{1}{2}p_2 L + \frac{1}{2}p_2 L_3 = \frac{1}{2}p_2(L + L_3) = (\frac{1}{2})(24.41)(5 + 0.521) = 67.38$ kN/m
$\bar{z}$	8.30	$\dfrac{L(2K_a - K_p)}{3(K_p - K_a)} = \dfrac{5[(2)(0.307) + 3.25]}{3(3.25 - 0.307)} = 2.188$ m
A_1'	8.32	$\dfrac{p_5}{\gamma(K_p - K_a)} = \dfrac{282.76}{(15.9)(3.25 - 0.307)} = 6.04$
A_2'	8.33	$\dfrac{8P}{\gamma(K_p - K_a)} = \dfrac{(8)(67.38)}{(15.9)(3.25 - 0.307)} = 11.52$
A_3'	8.34	$\dfrac{6P[2\bar{z}\gamma(K_p - K_a) + p_5]}{\gamma^2(K_p - K_a)^2}$ $= \dfrac{(6)(67.38)[(2)(2.188)(15.9)(3.25 - 0.307) + 282.76]}{(15.9)^2(3.25 - 0.307)^2} = 90.01$
A_4'	8.35	$\dfrac{P(6\bar{z}p_5 + 4P)}{\gamma^2(K_p - K_a)^2} = \dfrac{(67.38)[(6)(2.188)(282.76) + (4)(67.38)]}{(15.9)^2(3.25 - 0.307)^2} = 122.52$
L_4	8.31	$L_4^4 + A_1'L_4^3 - A_2'L_4^2 - A_3'L_4 - A_4' = 0$ $L_4^4 + 6.04L_4^3 - 11.52L_4^2 - 90.01L_4 - 122.52 = 0; \quad L_4 \approx 4.1$ m

$$D_{\text{theory}} = L_3 + L_4 = 0.521 + 4.1 = \mathbf{4.7\ m}$$

Part b

Total length, $L + 1.3(D_{\text{theory}}) = 5 + 1.3(4.7) = \mathbf{11.11\ m}$ ▲

▼ **EXAMPLE 8.3**

Refer to Figure 8.10. For $L = 15$ ft, $\gamma = 110$ lb/ft³, $\phi = 30°$, and $P = 2000$ lb/ft, determine:

    a.   The theoretical depth of penetration, D
    b.   The maximum moment, M_{max} (lb-ft/ft)

Solution

$$K_p = \tan^2\left(45 + \frac{\phi}{2}\right) = \tan^2\left(45 + \frac{30}{2}\right) = 3$$

$$K_a = \tan^2\left(45 - \frac{\phi}{2}\right) = \tan^2\left(45 - \frac{30}{2}\right) = \frac{1}{3}$$

$$K_p - K_a = 3 - 0.333 = 2.667$$

Part a

From Eq. (8.36),

$$D^4 - \left[\frac{8P}{\gamma(K_p - K_a)}\right]D^2 - \left[\frac{12PL}{\gamma(K_p - K_a)}\right]D - \left[\frac{2P}{\gamma(K_p - K_a)}\right]^2 = 0$$

and

$$\frac{8P}{\gamma(K_p - K_a)} = \frac{(8)(2000)}{(110)(2.667)} = 54.54$$

$$\frac{12PL}{\gamma(K_p - K_a)} = \frac{(12)(2000)(15)}{(110)(2.667)} = 1227.1$$

$$\frac{2P}{\gamma(K_p - K_a)} = \frac{(2)(2000)}{(110)(2.667)} = 13.63$$

so

$$D^4 - 54.54D^2 - 1227.1D - (13.63)^2 = 0$$

From the preceding equation, $D \approx \mathbf{12.5\ ft}$

Part b

From Eq. (8.39),

$$z' = \sqrt{\frac{2P}{\gamma(K_p - K_a)}} = \sqrt{\frac{(2)(2000)}{(110)(2.667)}} = 3.69\ \text{ft}$$

From Eq. (8.38),

$$M_{max} = P(L + z') - \frac{\gamma z'^3 (K_p - K_a)}{6}$$

$$= (2000)(15 + 3.69) - \frac{(110)(3.69)^3(2.667)}{6}$$

$$= 37,380 - 2456.65 \approx \mathbf{34,923\ lb\text{-}ft/ft}$$

▲

8.6 CANTILEVER SHEET PILING PENETRATING CLAY

At times, cantilever sheet piles must be driven into a clay layer possessing an undrained cohesion, c ($\phi = 0$ concept). The net pressure diagram will be somewhat different from that shown in Figure 8.7a. Figure 8.11 shows a cantilever sheet pile wall driven into clay with a backfill of granular soil above the level of the dredge line. The water table is at depth L_1 below the top of the wall. As before, Eqs. (8.1) and (8.2) give the intensity of the net pressures p_1 and p_2, and the diagram for pressure distribution above the level of the dredge line can be drawn. The diagram for net pressure distribution below the dredge line can now be determined as follows.

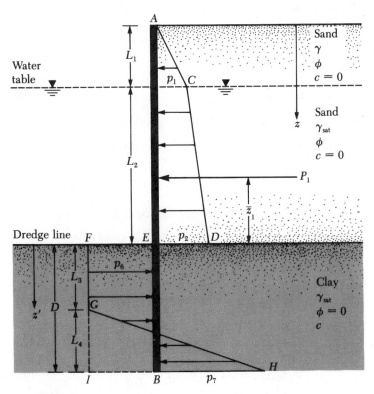

▼ **FIGURE 8.11** Cantilever sheet pile penetrating clay

At any depth greater than $L_1 + L_2$, for $\phi = 0$ condition, the Rankine active earth pressure coefficient $K_a = 1$. Similarly, for $\phi = 0$ condition, the Rankine passive earth pressure (K_p) is equal to 1. Thus, above the point of rotation (point O in Figure 8.6a), the active pressure, p_a, from right to left is

$$p_a = [\gamma L_1 + \gamma' L_2 + \gamma_{sat}(z - L_1 - L_2)] - 2c \qquad (8.40)$$

Similarly, the passive pressure, p_p, from left to right may be expressed as

$$p_p = \gamma_{sat}(z - L_1 - L_2) + 2c \qquad (8.41)$$

Thus the net pressure is

$$\begin{aligned}
p_6 = p_p - p_a &= [\gamma_{sat}(z - L_1 - L_2) + 2c] \\
&\quad - [\gamma L_1 + \gamma' L_2 + \gamma_{sat}(z - L_1 - L_2)] + 2c \\
&= 4c - (\gamma L_1 + \gamma' L_2)
\end{aligned} \qquad (8.42)$$

At the bottom of the sheet pile, the passive pressure from right to left is

$$p_p = (\gamma L_1 + \gamma' L_2 + \gamma_{sat} D) + 2c \qquad (8.43)$$

Similarly, the active pressure from left to right is

$$p_a = \gamma_{sat} D - 2c \qquad (8.44)$$

Hence the net pressure is

$$p_7 = p_p - p_a = 4c + (\gamma L_1 + \gamma' L_2) \qquad (8.45)$$

For equilibrium analysis, $\Sigma F_H = 0$ — that is, area of pressure diagram $ACDE$ − area of $EFIB$ + area of $GIH = 0$, or

$$P_1 - [4c - (\gamma L_1 + \gamma' L_2)]D + \tfrac{1}{2} L_4 [4c - (\gamma L_1 + \gamma' L_2) + 4c + (\gamma L_1 + \gamma' L_2)] = 0$$

where P_1 = area of the pressure diagram $ACDE$

Simplifying the preceding equation produces

$$L_4 = \frac{D[4c - (\gamma L_1 + \gamma' L_2)] - P_1}{4c} \qquad (8.46)$$

Now, taking the moment about point B, $\Sigma M_B = 0$, yields

$$P_1(D + \bar{z}_1) - [4c - (\gamma L_1 + \gamma' L_2)]\frac{D^2}{2} + \frac{1}{2}L_4(8c)\left(\frac{L_4}{3}\right) = 0 \qquad (8.47)$$

where $\bar{z}_1$ = distance of the center of pressure of the pressure diagram $ACDE$ measured from the level of the dredge line

Combining Eqs. (8.46) and (8.47) yields

$$D^2[4c - (\gamma L_1 + \gamma' L_2)] - 2DP_1 - \frac{P_1(P_1 + 12c\bar{z}_1)}{(\gamma L_1 + \gamma' L_2) + 2c} = 0 \qquad (8.48)$$

Equation (8.48) may be solved to obtain D, the theoretical depth of penetration of the clay layer by the sheet pile.

Step-by-Step Procedure to Obtain the Pressure Diagram

1. Calculate $K_a = \tan^2 (45 - \phi/2)$ for the granular soil (backfill).
2. Obtain p_1 and p_2 [Eqs. (8.1) and (8.2)].
3. Calculate P_1 and $\bar{z}_1$.
4. Use Eq. (8.48) to obtain the theoretical value of D.
5. Using Eq. (8.46), calculate L_4.
6. Calculate p_6 and p_7 [Eqs. (8.42) and (8.45)].
7. Draw the pressure distribution diagram as shown in Figure 8.11.
8. The actual depth of penetration is

$$D_{\text{actual}} = 1.4 \text{ to } 1.6 (D_{\text{theoretical}})$$

Maximum Bending Moment

According to Figure 8.11, the maximum moment (zero shear) will occur between $L_1 + L_2 < z < L_1 + L_2 + L_3$. Using a new coordinate system z' ($z' = 0$ at dredge line) for zero shear gives

$$P_1 - p_6 z' = 0$$

or

$$z' = \frac{P_1}{p_6} \tag{8.49}$$

The magnitude of the maximum moment may now be obtained:

$$M_{\text{max}} = P_1(z' + \bar{z}_1) - \frac{p_6 z'^2}{2} \tag{8.50}$$

Knowing the maximum bending moment, we determine the section modulus of the sheet pile section from Eq. (8.23).

▼ **EXAMPLE 8.4**

Refer to Figure 8.12. For the sheet pile wall, determine the

a. Theoretical and actual depth of penetration. Use $D_{\text{actual}} = 1.5 D_{\text{theory}}$.
b. Minimum size of sheet pile section necessary. Use $\sigma_{\text{all}} = 172 \text{ MN/m}^2$.

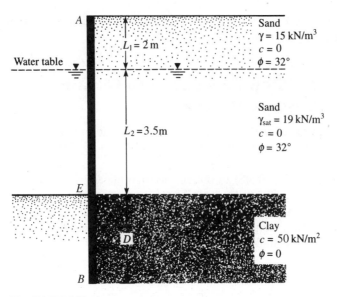

Solution

Part a

Refer to the pressure diagram shown in Figure 8.11.

Quantity required	Eq. no.	Equation and calculation
K_a	—	$\tan^2\left(45 - \dfrac{\phi}{2}\right) = \tan^2\left(45 - \dfrac{32}{2}\right) = 0.307$
p_1	8.1	$\gamma L_1 K_a = (15)(2)(0.307) = 9.21\ \text{kN/m}^2$
p_2	8.2	$(\gamma L_1 + \gamma' L_2)K_a = [(15)(2) + (19 - 9.81)(3.5)](0.307) = 19.08\ \text{kN/m}^2$
P_1	—	$\tfrac{1}{2}p_1 L_1 + p_1 L_2 + \tfrac{1}{2}(p_2 - p_1)L_2$
		$= (\tfrac{1}{2})(9.21)(2) + (9.21)(3.5) + (\tfrac{1}{2})(19.08 - 9.21)(3.5)$
		$= 9.21 + 32.24 + 17.27 = 58.72\ \text{kN/m}$
$\bar{z}_1$	—	$\dfrac{\Sigma M_E}{P_1} = \dfrac{1}{58.72}\left[9.21\left(3.5 + \dfrac{2}{3}\right) + 32.24\left(\dfrac{3.5}{2}\right) + (17.27)\left(\dfrac{3.5}{3}\right)\right] = 1.957\ \text{m}$
D_{theory}	8.48	$D^2[4c - (\gamma L_1 + \gamma' L_2)] - 2DP_1 - \dfrac{P_1(P_1 + 12c\bar{z}_1)}{(\gamma L_1 + \gamma' L_2) + 2c}$
		$D^2\{(4)(50) - [(15)(2) + (19 - 9.81)(3.5)]\} - (2)(D)(58.72)$
		$\quad - \dfrac{58.72[58.72 + (12)(50)(1.957)]}{[(15)(2) + (19 - 9.81)(3.5)] + (2)(50)} = 0$
		$137.84D^2 - 117.44D - 446.44 = 0;\quad D \approx \mathbf{2.3\ m}$
D_{actual}	—	$1.5 D_{\text{theory}} = (1.5)(2.3) = \mathbf{3.45\ m}$

Part b

Quantity required	Eq. no.	Equation and calculation
z'	8.49	$\dfrac{P_1}{p_6} = \dfrac{P_1}{4c - (\gamma L_1 + \gamma' L_2)} = \dfrac{58.72}{(4)(50) - [(15)(2) + (19 - 9.81)(3.5)]} = 0.426 \text{ m}$
M_{max}	8.50	$P_1(z' + \bar{z}_1) - \dfrac{p_6 z'^2}{2} = P_1(z' + \bar{z}_1) - \dfrac{[4c - (\gamma L_1 + \gamma' L_2)]z'^2}{2}$
		$= (58.72)(0.426 + 1.957)$
		$\quad - \dfrac{\{(4)(50) - [(15)(2) + (19 - 9.81)(3.5)]\}(0.426)^2}{2}$
		$= 127.42 \text{ kN} \cdot \text{m/m}$
S	—	$\dfrac{M_{max}}{\sigma_{all}} = \dfrac{127.42}{172 \times 10^3} = \mathbf{0.741 \times 10^{-3}} \ \mathbf{m^3/m \ of \ wall}$

▲

8.7   SPECIAL CASES FOR CANTILEVER WALLS (PENETRATING CLAY)

As in Section 8.5, relationships for special cases for cantilever walls penetrating clay may also be derived.

Case 1: Sheet Pile Wall in the Absence of Water Table

Referring to Figure 8.13, we can write

$$p_2 = \gamma L K_a \tag{8.51}$$

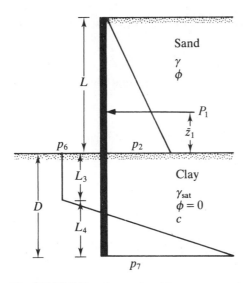

▼   **FIGURE 8.13**   Sheet pile wall penetrating clay

$$p_6 = 4c - \gamma L \tag{8.52}$$

$$p_7 = 4c + \gamma L \tag{8.53}$$

$$P_1 = \tfrac{1}{2} L p_2 = \tfrac{1}{2} \gamma L^2 K_a \tag{8.54}$$

$$L_4 = \frac{D(4c - \gamma L) - \tfrac{1}{2}\gamma L^2 K_a}{4c} \tag{8.55}$$

The theoretical depth of penetration, D, can be calculated [similar to Eq. (8.48)] as

$$D^2(4c - \gamma L) - 2DP_1 - \frac{P_1(P_1 + 12c\bar{z}_1)}{\gamma L + 2c} = 0 \tag{8.56}$$

where $\quad \bar{z}_1 = \dfrac{L}{3}$ $\tag{8.57}$

The magnitude of the maximum moment in the wall is

$$M_{max} = P_1(z' + \bar{z}_1) - \frac{p_6 z'^2}{2} \tag{8.58}$$

where $\quad z' = \dfrac{P_1}{p_6} = \dfrac{\tfrac{1}{2}\gamma L^2 K_a}{4c - \gamma L}$ $\tag{8.59}$

Case 2: Free Cantilever Sheet Pile Wall Penetrating Clay

Figure 8.14 shows a free cantilever sheet pile wall penetrating a clay layer. The wall is being subjected to a line load of P per unit length. For this case,

$$p_6 = p_7 = 4c \tag{8.60}$$

The depth of penetration, D, may be obtained from

$$4D^2c - 2PD - \frac{P(P + 12cL)}{2c} = 0 \tag{8.61}$$

Also note that, for pressure diagram construction,

$$L_4 = \frac{4cD - P}{4c} \tag{8.62}$$

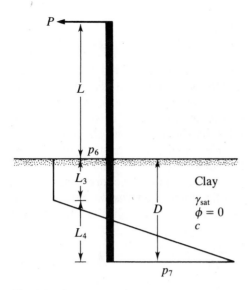

▼ **FIGURE 8.14**   Free cantilever sheet piling penetrating clay

The maximum moment in the wall is

$$M_{\max} = P(L + z') - \frac{4cz'^2}{2} \qquad (8.63)$$

where $z' = \dfrac{P}{4c}$ (8.64)

▼ **EXAMPLE 8.5**

Refer to the free cantilever sheet pile wall shown in Figure 8.14, for which $P = 32$ kN/m, $L = 3.5$ m, and $c = 12$ kN/m². Calculate the theoretical depth of penetration.

Solution   From Eq. (8.61),

$$4D^2c - 2PD - \frac{P(P + 12cL)}{2c} = 0$$

$$(4)(D^2)(12) - (2)(32)(D) - \frac{32[32 + (12)(12)(3.5)]}{(2)(12)} = 0$$

$$48D^2 - 64D - 714.7 = 0$$

Hence $D \approx$ **4.6 m.**   ▲

8.8   ANCHORED SHEET PILE WALL—GENERAL

When the height of the backfill material behind a cantilever sheet pile wall exceeds about 20 ft ($\approx$6 m), tying the sheet pile wall near the top to anchor plates, anchor walls, or anchor piles becomes more economical. This type of construction is referred to as *anchored sheet pile wall* or an *anchored bulkhead*. Anchors minimize the depth of required penetration by the sheet piles and also reduce the cross-sectional area and weight of the sheet piles needed for construction. However, the tie rods and anchors must be carefully designed.

The two basic methods of designing anchored sheet pile walls are (a) the *free earth support* method and (b) the *fixed earth support* method. Figure 8.15 shows the assumed nature of deflection of the sheet piles for the two methods.

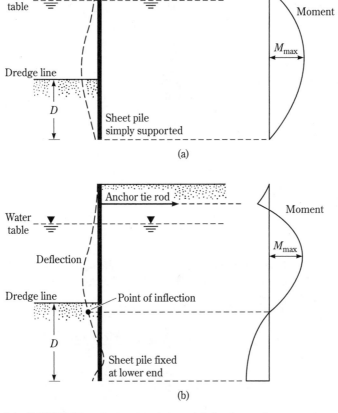

▼ **FIGURE 8.15**   Nature of variation of deflection and moment for anchored sheet piles: (a) free earth support method; (b) fixed earth support method

The free earth support method involves minimum penetration depth. Below the dredge line, no pivot point exists for the static system. The nature of variation of the bending moment with depth for both methods is also shown in Figure 8.15. Note that

$$D_{\text{free earth}} < D_{\text{fixed earth}}$$

8.9   FREE EARTH SUPPORT METHOD FOR PENETRATION OF SANDY SOIL

Figure 8.16 shows an anchor sheet pile wall with a granular soil backfill; the wall has been driven into a granular soil. The tie rod connecting the sheet pile and the anchor is located at depth l_1 below the top of the sheet pile wall.

The diagram of net pressure distribution above the dredge line is similar to that shown in Figure 8.7. At depth $z = L_1$, $p_1 = \gamma L_1 K_a$; and, at $z = L_1 + L_2$, $p_2 = (\gamma L_1 + \gamma' L_2) K_a$. Below the dredge line, the net pressure will be zero at $z = L_1 + L_2 + L_3$. The relation for L_3 is given by Eq. (8.6), or

$$L_3 = \frac{p_2}{\gamma' (K_p - K_a)}$$

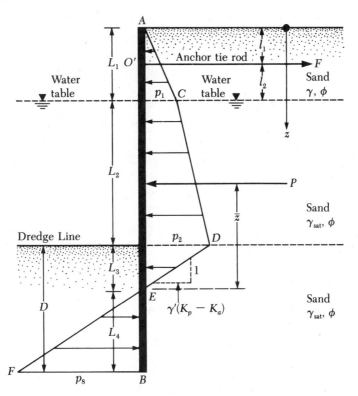

▼ **FIGURE 8.16**   Anchored sheet pile wall penetrating sand

At $z = (L_1 + L_2 + L_3 + L_4)$, the net pressure is given by

$$p_8 = \gamma'(K_p - K_a)L_4 \tag{8.65}$$

Note that the slope of the line DEF is 1 vertical to $\gamma'(K_p - K_a)$ horizontal.

For equilibrium of the sheet pile, Σ horizontal forces $= 0$, and Σ moment about $O' = 0$. (*Note:* Point O' is located at the level of the tie rod.)

Summing the forces in the horizontal direction (per unit length of the wall) gives

Area of the pressure diagram $ACDE$ − area of EBF − $F = 0$

where $F =$ tension in the tie rod/unit length of the wall, or

$$P - \tfrac{1}{2}p_8L_4 - F = 0$$

or

$$\boxed{F = P - \tfrac{1}{2}[\gamma'(K_p - K_a)]L_4^2} \tag{8.66}$$

where $P =$ area of the pressure diagram $ACDE$

Now, taking the moment about point O' gives

$$-P[(L_1 + L_2 + L_3) - (\bar{z} + l_1)] + \tfrac{1}{2}[\gamma'(K_p - K_a)]L_4^2(l_2 + L_2 + L_3 + \tfrac{2}{3}L_4) = 0$$

or

$$\boxed{L_4^3 + 1.5L_4^2(l_2 + L_2 + L_3) - \frac{3P[(L_1 + L_2 + L_3) - (\bar{z} + l_1)]}{\gamma'(K_p - K_a)} = 0} \tag{8.67}$$

Equation (8.67) may be solved by trial and error to determine the theoretical depth, L_4:

$$D_{\text{theoretical}} = L_3 + L_4$$

The theoretical depth is increased by about 30%–40% for actual construction, or

$$D_{\text{actual}} = 1.3 \text{ to } 1.4D_{\text{theoretical}} \tag{8.68}$$

The step-by-step procedure in Section 8.4 indicated that a factor of safety can be applied to K_p at the beginning [that is, $K_{p(\text{design})} = K_p/FS$]. If done, there is no

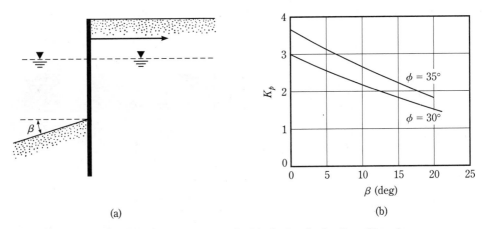

▼ **FIGURE 8.17**   (a) Anchored sheet pile wall with sloping dredge line; (b) variation K_p with β and ϕ

need to increase the theoretical depth by 30%–40%. This approach is often more conservative.

The maximum theoretical moment to which the sheet pile will be subjected
occurs at a depth between $z = L_1$ and $z = L_1 + L_2$. The depth, z, for zero shear and
hence maximum moment, may be evaluated from

$$\tfrac{1}{2}p_1L_1 - F + p_1(z - L_1) + \tfrac{1}{2}K_a\gamma'(z - L_1)^2 = 0 \qquad (8.69)$$

Once the value of z is determined, the magnitude of the maximum moment is easily
obtained. The procedure for determining the holding capacity of anchors is treated
in Sections 8.16 and 8.17.

In some cases, the dredge line may be sloping at an angle β with respect to
the horizontal, as shown in Figure 8.17a. In that case, the passive pressure coefficient
will not be equal to $\tan^2 (45 + \phi/2)$. The variations of K_p (Coulomb—for wall friction
angle of zero) with β for $\phi = 30°$ and 35° are shown in Figure 8.17b. With these
values of K_p, the procedure described above may be used to determine the depth
of penetration, D.

▼ **EXAMPLE 8.6** _____

Refer to Figure 8.16. Here, $L_1 = 3.05$ m, $L_2 = 6.1$ m, $l_1 = 1.53$ m, $l_2 = 1.52$ m, $c =$
0, $\phi = 30°$, $\gamma = 16$ kN/m³, and $\gamma_{sat} = 19.5$ kN/m³.

    a. Determine the theoretical and actual depths of penetration. Note: $D_{actual} =$
        $1.3D_{theory}$.
    b. Find the anchor force per unit length of the wall.

Solution

Part a

Quantity required	Eq. no.	Equation and calculation
K_a	—	$\tan^2\left(45 - \dfrac{\phi}{2}\right) = \tan^2\left(45 - \dfrac{30}{2}\right) = \dfrac{1}{3}$
K_p	—	$\tan^2\left(45 + \dfrac{\phi}{2}\right) = \tan^2\left(45 + \dfrac{30}{2}\right) = 3$
$K_a - K_p$	—	$3 - 0.333 = 2.667$
γ'	—	$\gamma_{sat} - \gamma_w = 19.5 - 9.81 = 9.69 \ \text{kN/m}^3$
p_1	8.1	$\gamma L_1 K_a = (16)(3.05)\left(\tfrac{1}{3}\right) = 16.27 \ \text{kN/m}^2$
p_2	8.2	$(\gamma L_1 + \gamma' L_2)K_a = [(16)(3.05) + (9.69)(6.1)]\tfrac{1}{3} = 35.97 \ \text{kN/m}^2$
L_3	8.6	$\dfrac{p_2}{\gamma'(K_p - K_a)} = \dfrac{35.97}{(9.69)(2.667)} = 1.39 \ \text{m}$
P	—	$\tfrac{1}{2}p_1 L_1 + p_1 L_2 + \tfrac{1}{2}(p_2 - p_1)L_2 + \tfrac{1}{2}p_2 L_3 = \left(\tfrac{1}{2}\right)(16.27)(3.05)$ $+ (16.27)(6.1) + \left(\tfrac{1}{2}\right)(35.97 - 16.27)(6.1) + \left(\tfrac{1}{2}\right)(35.97)(1.39)$ $= 24.81 + 99.25 + 60.01 + 25.0 = 209.07 \ \text{kN/m}$
$\bar{z}$	—	$\dfrac{\sum M_E}{P} = \left[\begin{array}{l}(24.81)\left(1.39 + 6.1 + \dfrac{3.05}{3}\right) + (99.25)\left(1.39 + \dfrac{6.1}{2}\right) \\ + (60.01)\left(1.39 + \dfrac{6.1}{3}\right) + (25.0)\left(\dfrac{2 \times 1.39}{3}\right)\end{array}\right]\dfrac{1}{209.07} = 4.21 \ \text{m}$
L_4	8.67	$L_4^3 + 1.5 L_4^2(l_2 + L_2 + L_3) - \dfrac{3P[(L_1 + L_2 + L_3) - (\bar{z} + l_1)]}{\gamma'(K_p - K_a)} = 0$ $L_4^3 + 1.5 L_4^2(1.52 + 6.1 + 1.39)$ $- \dfrac{(3)(209.07)[(3.05 + 6.1 + 1.39) - (4.21 + 1.53)]}{(9.69)(2.667)} = 0$ $L_4 = 2.7 \ \text{m}$
D_{theory}	—	$L_3 + L_4 = 1.39 + 2.7 = 4.09 \approx \mathbf{4.1 \ m}$
D_{actual}	—	$1.3 D_{\text{theory}} = (1.3)(4.1) = \mathbf{5.33 \ m}$

Part b

$$F = P - \tfrac{1}{2}\gamma'(K_p - K_a)L_4^2$$
$$= 209.07 - \left(\tfrac{1}{2}\right)(9.69)(2.667)(2.7)^2 = 114.87 \ \text{kN/m} \approx \mathbf{115 \ kN/m} \quad \blacktriangle$$

8.10 DESIGN CHARTS FOR FREE EARTH SUPPORT METHOD (PENETRATION INTO SANDY SOIL)

Using the free earth support method, Hagerty and Nofal (1992) provided simplified design charts for quick estimation of the depth of penetration, D, anchor force, F, and maximum moment, $M_{\max}$, for anchored sheet pile walls penetrating into sandy soil, as shown in Figure 8.16. They made the following assumptions for their analysis.

a. The soil friction angle, ϕ, above and below the dredge line is the same.
b. The angle of friction between the sheet pile wall and the soil is $\phi/2$.
c. The passive earth pressure below the dredge line has a logarithmic spiral failure surface.
d. For active earth pressure calculation, Coulomb's theory is valid.

The magnitudes of D, F, and M_{max} may be calculated from the following relationships:

$$\frac{D}{L_1 + L_2} = (GD)(CDL_1) \tag{8.70}$$

$$\frac{F}{\gamma_a (L_1 + L_2)^2} = (GF)(CFL_1) \tag{8.71}$$

$$\frac{M_{max}}{\gamma_a (L_1 + L_2)^3} = (GM)(CML_1) \tag{8.72}$$

where γ_a = average unit weight of soil

$$= \frac{\gamma L_1^2 + (\gamma_{sat} - \gamma_w)L_2^2 + 2\gamma L_1 L_2}{(L_1 + L_2)^2} \tag{8.73}$$

GD = generalized nondimensional embedment

$$= \frac{D}{L_1 + L_2} \quad \text{(for } L_1 = 0 \text{ and } L_2 = L_1 + L_2\text{)}$$

GF = generalized nondimensional anchor force

$$= \frac{F}{\gamma_a (L_1 + L_2)^2} \quad \text{(for } L_1 = 0 \text{ and } L_2 = L_1 + L_2\text{)}$$

GM = generalized nondimensional moment

$$= \frac{M_{max}}{\gamma_a (L_1 + L_2)^3} \quad \text{(for } L_1 = 0 \text{ and } L_2 = L_1 + L_2\text{)}$$

CDL_1, CFL_1, CML_1 = correction factors for $L_1 \neq 0$

The variations of GD, GF, GM, CDL_1, CFL_1, and CML_1 are shown in Figures 8.18, 8.19, 8.20, 8.21, 8.22, and 8.23, respectively.

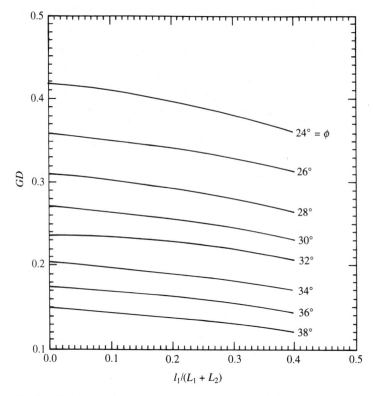

▼ **FIGURE 8.18** Variation of GD with $l_1/(L_1 + L_2)$ and ϕ (after Hagerty and Nofal, 1992)

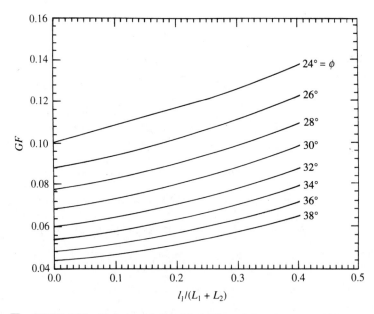

▼ **FIGURE 8.19** Variation of GF with $l_1/(L_1 + L_2)$ and ϕ (after Hagerty and Nofal, 1992)

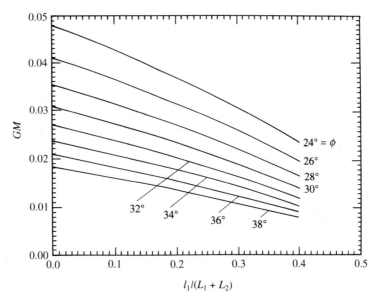

▼ **FIGURE 8.20** Variation of GM with $l_1/(L_1 + L_2)$ and ϕ (after Hagerty and Nofal, 1992)

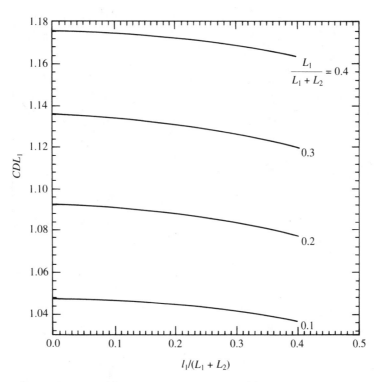

▼ **FIGURE 8.21** Variation of CDL_1 with $L_1/(L_1 + L_2)$ and $l_1/(L_1 + L_2)$ (after Hagerty and Nofal, 1992)

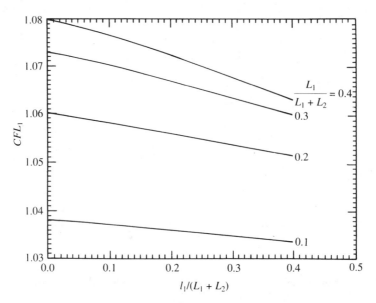

▼ **FIGURE 8.22** Variation of CFL_1 with $L_1/(L_1 + L_2)$ and $l_1/(L_1 + L_2)$ (after Hagerty and Nofal, 1992)

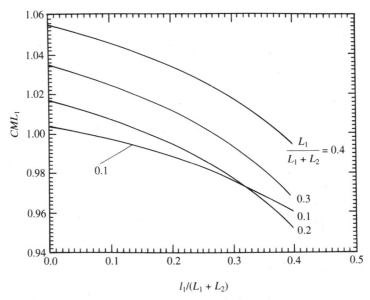

▼ **FIGURE 8.23** Variation of CML_1 with $L_1/(L_1 + L_2)$ and $l_1/(L_1 + L_2)$ (after Hagerty and Nofal, 1992)

▼ **EXAMPLE 8.7** _____

Refer to Figure 8.16. Given: $L_1 = 2$ m, $L_2 = 3$ m, $l_1 = l_2 = 1$ m, $c = 0$, $\phi = 32°$, $\gamma = 15.9$ kN/m³, $\gamma_{sat} = 19.33$ kN/m³. Determine:

a. Theoretical and actual depth of penetration. Note: $D_{actual} = 1.4D_{theory}$.
b. Anchor force per unit length of wall.
c. Maximum moment, M_{max}.

Use the charts presented in Section 8.10.

Solution

Part a

From Eq. (8.70),

$$\frac{D}{L_1 + L_2} = (GD)(CDL_1)$$

$$\frac{l_1}{L_1 + L_2} = \frac{1}{2 + 3} = 0.2$$

From Figure 8.18 for $l_1/(L_1 + L_2) = 0.2$ and $\phi = 32°$, $GD = 0.22$. From Figure 8.21, for

$$\frac{L_1}{L_1 + L_2} = \frac{2}{2 + 3} = 0.4 \quad \text{and} \quad \frac{l_1}{L_1 + L_2} = 0.2$$

$CDL_1 \approx 1.172$. So

$$D_{theory} = (L_1 + L_2)(GD)(CDL_1) = (5)(0.22)(1.172) \approx 1.3$$

$$D_{actual} \approx (1.4)(1.3) = 1.82 \approx \textbf{2 m}$$

Part b

From Figure 8.19 for $l_1/(L_1 + L_2) = 0.2$ and $\phi = 32°$, $GF \approx 0.074$. Also, from Figure 8.22, for

$$\frac{L_1}{L_1 + L_2} = \frac{2}{2 + 3} = 0.4, \quad \frac{l_1}{L_1 + L_2} = 0.2, \quad \text{and} \quad \phi = 32°$$

$CFL_1 = 1.073$. From Eq. (8.73),

$$\gamma_a = \frac{\gamma L_1^2 + \gamma' L_2^2 + 2\gamma L_1 L_2}{(L_1 + L_2)^2}$$

$$= \frac{(15.9)(2)^2 + (19.33 - 9.81)(3)^2 + (2)(15.9)(2)(3)}{(2 + 3)^2} = 13.6 \, \text{kN/m}^3$$

Using Eq. (8.71) yields

$$F = \gamma_a(L_1 + L_2)^2(GF)(CFL_1) = (13.6)(5)^2(0.074)(1.073) \approx \textbf{27 kN/m}$$

Part c

From Figure 8.20, for $l_1/(L_1 + L_2) = 0.2$ and $\phi = 32°$, $GM = 0.021$. Also, from Figure 8.23, for

$$\frac{L_1}{L_1 + L_2} = \frac{2}{2 + 3} = 0.4, \quad \frac{l_1}{L_1 + L_2} = 0.2, \quad \text{and} \quad \phi = 32°$$

$CML_1 = 1.036$. Hence from Eq. (8.72),

$$M_{max} = \gamma_a (L_1 + L_2)^3 (GM)(CML_1) = (13.6)(5)^3(0.021)(1.036) = \mathbf{36.99 \ kN \cdot m/m}$$

Note: If this problem had been solved by the procedure described in Section 8.9, the following answers would have been obtained:

$$D_{actual} = 2.9 \ m$$

$$F = 30.86 \ kN/m$$

$$M_{max} = 43.72 \ kN/m$$

The difference between the results is primarily due to the assumed wall friction angle and the method used to calculate passive earth pressure.   ▲

8.11 MOMENT REDUCTION FOR ANCHORED SHEET PILE WALLS

Sheet piles are flexible and hence sheet pile walls yield (that is, displace laterally), which redistributes the lateral earth pressure. This change tends to reduce the maximum bending moment, M_{max}, as calculated by the procedure outlined in Sections 8.9 and 8.10. For that reason, Rowe (1952, 1957) suggested a procedure to reduce the maximum design moment on the sheet pile walls *obtained from the free earth support method*. This section discusses the procedure of moment reduction for sheet piles *penetrating into sand*.

In Figure 8.24, which is valid for the case of a sheet pile penetrating sand, the following notations are used.

1. H' = total height of pile driven (that is, $L_1 + L_2 + D_{actual}$)

$$2. \quad \text{Relative flexibility of pile} = \rho = 10.91 \times 10^{-7} \left(\frac{H'^4}{EI} \right) \tag{8.74}$$

where H' is in meters

$\qquad E$ = modulus of elasticity of the pile material (MN/m^2)

$\qquad I$ = moment of inertia of the pile section per meter of the wall $(m^4/m$ of wall)

3. M_d = design moment
4. M_{max} = maximum theoretical moment

In English units, Eq. (8.74) takes the form

$$\rho = \frac{H'^4}{EI} \tag{8.75}$$

where H' is in ft, E is in lb/in^2, and I is in in^4/ft of the wall

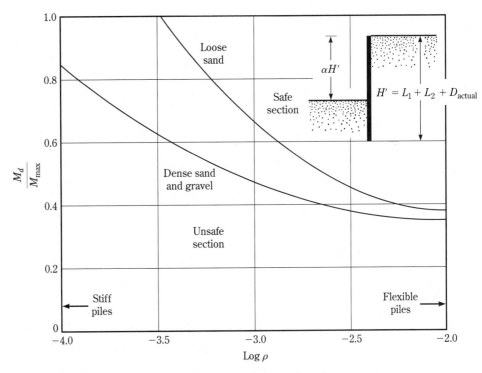

FIGURE 8.24   Plot of log ρ against M_d/M_{max} for sheet pile walls penetrating sand (after Rowe, 1952)

The procedure for the use of the moment reduction diagram (Figure 8.24) is as follows:

Step 1.   Choose a sheet pile section (such as those given in Table B-1 in Appendix B).

Step 2.   Find the section modulus, S, of the selected section (Step 1) per unit length of the wall.

Step 3.   Determine the moment of inertia of the section (Step 1) per unit length of the wall.

Step 4.   Obtain H' and calculate ρ [Eq. (8.74) or Eq. (8.75)].

Step 5.   Find log ρ.

Step 6.   Find the moment capacity of the pile section chosen in Step 1 as $M_d = \sigma_{all}S$.

Step 7.   Determine M_d/M_{max}. Note that M_{max} is the maximum theoretical moment determined before.

Step 8.   Plot log ρ (Step 5) and M_d/M_{max} in Figure 8.24.

Step 9.   Repeat Steps 1–8 for several sections. The points that fall above the curve (loose sand or dense sand, as the case may be) are *safe sections*.

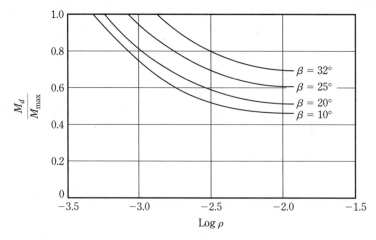

▼ **FIGURE 8.25**   Plot of ρ against M_d/M_{max} for sheet pile walls penetrating into sand with a sloping dredge line (after
Schroeder and Roumillac, 1983)

Those points that fall below the curve are *unsafe sections*. The cheapest
section may now be chosen from those points that fall above the proper
curve. Note that the section chosen will have an $M_d < M_{max}$.

For anchor sheet pile walls penetrating into sand with a sloping dredge line
(Figure 8.17), a moment reduction procedure similar to that outlined above may be
adopted. For this procedure, Figure 8.25 (which was developed by Schroeder and
Roumillac, 1983) should be used.

▼ **EXAMPLE 8.8** _____

Refer to Example 8.6.

    a. Determine the maximum moment, M_{max}.
    b. Use Rowe's moment reduction technique and find a suitable sheet pile
        section. Use $\sigma_{all} = 172{,}500$ kN/m².

Given: $E = 207 \times 10^3$ MN/m².

Solution

Part a

From Eq. (8.69), for zero shear,

$$\tfrac{1}{2}p_1L_1 - F + p_1(z - L_1) + \tfrac{1}{2}K_a\gamma'(z - L_1)^2 = 0$$

Let $z - L_1 = x$, so

$$\tfrac{1}{2}p_1l_1 - F + p_1x + \tfrac{1}{2}K_a\gamma'x^2 = 0$$

$$(\tfrac{1}{2})(16.27)(3.05) - 115 + (16.27)(x) + (\tfrac{1}{2})(\tfrac{1}{3})(9.69)x^2 = 0$$

$$x^2 + 10.07x - 55.84 = 0$$

$x = 4$ m; $z = x + L_1 = 4 + 3.05 = 7.05$ m. Taking the moment about the point of zero shear,

$$M_{max} = -\frac{1}{2} p_1 L_1 \left(x + \frac{3.05}{3} \right) + F(x + 1.52) - p_1 \frac{x^2}{2} - \frac{1}{2} K_a \gamma' x^2 \left(\frac{x}{3} \right)$$

or

$$M_{max} = -\left(\frac{1}{2} \right)(16.27)(3.05)\left(4 + \frac{3.05}{3} \right) + (115)(4 + 1.52) - (16.27)\left(\frac{4^2}{2} \right)$$

$$-\left(\frac{1}{2} \right)\left(\frac{1}{3} \right)(9.69)(4)^2 \left(\frac{4}{3} \right) = \textbf{344.9 kN} \cdot \textbf{m/m}$$

Part b

$$H' = L_1 + L_2 + D_{actual} = 3.05 + 6.1 + 5.33 = 14.48 \text{ m}$$

Section	I (m⁴/m)	H' (m)	$\rho = 10.91$ $\times 10^{-7} \left(\dfrac{H'^4}{EI} \right)$	$\log \rho$	S (m³/m)	$M_d = S\sigma_{all}$ (kN·m/m)	$\dfrac{M_d}{M_{max}}$
PZ-22	115.2×10^{-6}	14.48	20.11×10^{-4}	-2.7	97×10^{-5}	167.33	0.485
PZ-27	251.5×10^{-6}	14.48	9.21×10^{-4}	-3.04	162.3×10^{-5}	284.84	0.826

Figure 8.26 shows the plot of M_d/M_{max} versus ρ. It can be seen that **PZ-27** will be sufficient.

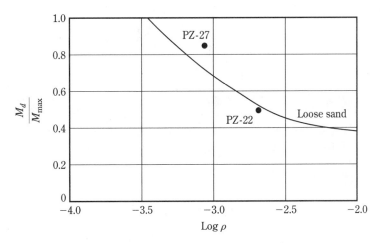

▼ **FIGURE 8.26**

8.12   FREE EARTH SUPPORT METHOD FOR PENETRATION OF CLAY

Figure 8.27 shows an anchored sheet pile wall penetrating a clay soil and having a granular soil backfill. The diagram of pressure distribution above the dredge line is similar to that shown in Figure 8.11. From Eq. (8.42), the net pressure distribution below the dredge line (from $z = L_1 + L_2$ to $z = L_1 + L_2 + D$) is

$$p_6 = 4c - (\gamma L_1 + \gamma' L_2)$$

For static equilibrium, the sum of the forces in the horizontal direction is

$$P_1 - p_6 D = F \tag{8.76}$$

where P_1 = area of the pressure diagram ACD
        F = anchor force per unit length of the sheet pile wall

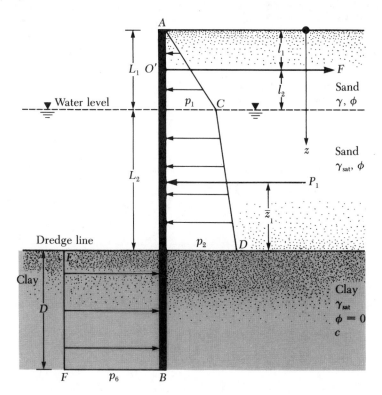

▼   **FIGURE 8.27**   Anchored sheet pile wall penetrating clay

Again, taking the moment about O' produces

$$P_1(L_1 + L_2 - l_1 - \bar{z}_1) - p_6 D\left(l_2 + L_2 + \frac{D}{2}\right) = 0$$

Simplification yields

$$p_6 D^2 + 2p_6 D(L_1 + L_2 - l_1) - 2P_1(L_1 + L_2 - l_1 - \bar{z}_1) = 0 \tag{8.77}$$

Equation (8.77) gives the theoretical depth of penetration, D.

As in Section 8.9, the maximum moment in this case occurs at depth $L_1 < z < L_1 + L_2$. The depth of zero shear (and thus the maximum moment) may be determined from Eq. (8.69).

A moment reduction technique similar to that in Section 8.11 for anchored sheet piles penetrating into clay has also been developed by Rowe (1952, 1957). This technique is presented in Figure 8.28. In this figure, the notations are as follows:

1.   The stability number is

$$S_n = 1.25 \frac{c}{(\gamma L_1 + \gamma' L_2)} \tag{8.78}$$

where c = undrained cohesion ($\phi = 0$)

For the definition of γ, γ', L_1, and L_2, see Figure 8.27.
2.   The nondimensional wall height is

$$\alpha = \frac{L_1 + L_2}{L_1 + L_2 + D_{actual}} \tag{8.79}$$

3.   Flexibility number, ρ [see Eqs. (8.74) or Eq. (8.75)]
4. M_d = design moment
     M_{max} = maximum theoretical moment

The procedure for moment reduction using Figure 8.28 is as follows:

Step 1.   Obtain $H' = L_1 + L_2 + D_{actual}$
Step 2.   Determine $\alpha = (L_1 + L_2)/H'$.
Step 3.   Determine S_n [Eq. (8.78)].
Step 4.   For the magnitudes of α and S_n obtained (Steps 2 and 3), determine M_d/M_{max} for various values of log ρ from Figure 8.28 and plot M_d/M_{max} against log ρ.
Step 5.   Follow Steps 1–9 as outlined for the case of moment reduction of sheet pile walls penetrating granular soil (Section 8.11).

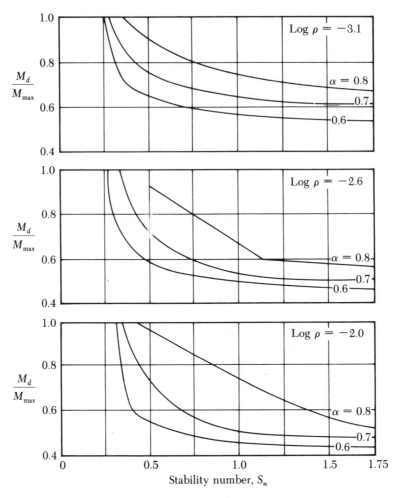

▼ **FIGURE 8.28**   Plot of M_d/M_{max} against stability number for sheet pile wall penetrating clay (after Rowe, 1957)

▼ **EXAMPLE 8.9** _____

Refer to Figure 8.27, which shows that $L_1 = 10.8$ ft, $L_2 = 21.6$ ft, and $l_1 = 5.4$ ft. Also, $\gamma = 108$ lb/ft³, $\gamma_{sat} = 127.2$ lb/ft³, $\phi = 35°$, and $c = 850$ lb/ft².

  a.   Determine the theoretical depth of embedment.
  b.   Calculate the anchor force per unit length of the sheet pile wall.

Solution

Part a

Quantity to be determined	Eq. no.	Equation and calculation
K_a	—	$\tan^2\left(45 - \dfrac{\phi}{2}\right) = \tan^2\left(45 - \dfrac{35}{2}\right) = 0.271$
γ'	—	$\gamma_{sat} - \gamma_w = 127.2 - 62.4 = 64.8 \text{ lb/ft}^3$
p_1	8.1	$\gamma L_1 K_a = (0.108)(10.8)(0.271) = 0.316 \text{ kip/ft}^2$
p_2	8.2	$(\gamma L_1 + \gamma' L_2)K_a = [(0.108)(10.8) + (0.0648)(21.6)](0.271) = 0.695 \text{ kip/ft}^2$
P_1	—	$\frac{1}{2}p_1 L_1 + p_1 L_2 + \frac{1}{2}(p_2 - p_1)L_2$ $= (\frac{1}{2})(0.316)(10.8) + (0.316)(21.6) + (\frac{1}{2})(0.695 - 0.316)(21.6)$ $= 1.706 + 6.826 + 4.093 = 12.625 \text{ kip/ft}$
$\bar{z}_1$	—	$\dfrac{\Sigma M_{\text{about dredge line}}}{P_1} = \dfrac{(1.706)\left(21.6 + \dfrac{10.8}{3}\right) + (6.826)(10.8) + (4.093)\left(\dfrac{21.6}{3}\right)}{12.625}$ $\qquad = 11.58 \text{ ft}$
p_6	8.42	$4c - (\gamma L_1 + \gamma' L_2) = (4)(0.850) - [(0.108)(10.8) + (0.0648)(21.6)]$ $\qquad = 0.834 \text{ kip/ft}^2$
D	8.77	$p_6 D^2 + 2p_6 D(L_1 + L_2 - l_1) - 2P_1(L_1 + L_2 - l_1 - \bar{z}_1) = 0$ $0.834 D^2 + (2)(0.834)(D)(27) - (2)(12.625)(15.42) = 0$ $D^2 + 54D - 466.85 = 0; \quad D = \textbf{7.6 ft}$

Part b

From Eq. (8.76)

$$F = P_1 - p_6 D = 12.625 - (0.834)(7.6) = \textbf{6.29 kip/ft} \qquad \blacktriangle$$

8.13 COMPUTATIONAL PRESSURE DIAGRAM METHOD (FOR PENETRATION INTO SANDY SOIL)

The computational pressure diagram method (CPD method) for sheet pile penetrating a sandy soil is a simplified method of design and an alternative to the free earth method described in Sections 8.9, 8.10, and 8.11 (Nataraj and Hoadley, 1984). In this method, the net pressure diagram shown in Figure 8.16 is replaced by rectangular pressure diagrams, as shown in Figure 8.29. Note that $\bar{p}_a$ is the width of the net active pressure diagram above the dredge line and $\bar{p}_p$ is the width of the net passive pressure diagram below the dredge line. The magnitudes of $\bar{p}_a$ and $\bar{p}_p$ may be

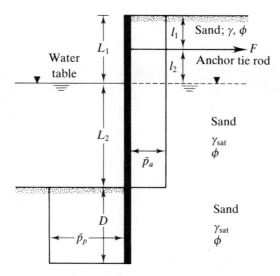

▼ **FIGURE 8.29**   Computational pressure diagram method (*note:* $L_1 + L_2 = L$)

expressed as

$$\overline{p}_a = CK_a\gamma_{av}L \tag{8.80}$$

$$\overline{p}_p = RCK_a\gamma_{av}L = R\overline{p}_a \tag{8.81}$$

where γ_{av} = average effective unit weight of sand

$$\approx \frac{\gamma L_1 + \gamma' L_2}{L_1 + L_2} \tag{8.82}$$

C = coefficient

$$R = \text{coefficient} = \frac{L(L - 2l_1)}{D(2L + D - 2l_1)} \tag{8.83}$$

The range of values for C and R is given in Table 8.1.

The depth of penetration, D, anchor force per unit length of the wall, F, and maximum moment in the wall, M_{max}, are obtained from the following relationships.

▼ **TABLE 8.1** Range of Values for C and R [Eqs. (8.80) and (8.81)]

Soil type	C[a]	R
Loose sand	0.8–0.85	0.3–0.5
Medium sand	0.7–0.75	0.55–0.65
Dense sand	0.55–0.65	0.60–0.75

[a] Valid for the case in which there is no surcharge above the granular backfill (that is, on the right side of the wall as shown in Figure 8.29)

Depth of Penetration

$$D^2 + 2DL\left[1 - \left(\frac{l_1}{L}\right)\right] - \left(\frac{L^2}{R}\right)\left[1 - 2\left(\frac{l_1}{L}\right)\right] = 0 \qquad (8.84)$$

Anchor Force

$$F = \bar{p}_a(L - RD) \qquad (8.85)$$

Maximum Moment

$$M_{\text{max}} = 0.5\bar{p}_a L^2\left[\left(1 - \frac{RD}{L}\right)^2 - \left(\frac{2l_1}{L}\right)\left(1 - \frac{RD}{L}\right)\right] \qquad (8.86)$$

Note the following qualifications.

1. The magnitude of D obtained from Eq. (8.84) is about 1.25 to 1.5 times the value of D_{theory} obtained by the conventional free earth support method (Section 8.9), so
$$D \approx D_{\text{actual}}$$
$$\uparrow \quad \uparrow$$
Eq. (8.84) Eq. (8.68)

2. The magnitude of F obtained by using Eq. (8.85) is about 1.2 to 1.6 times the value obtained by using Eq. (8.66). Thus an additional factor of safety for actual design of anchors need not be used.

3. The magnitude of M_{max} obtained from Eq. (8.86) is about 0.6 to 0.75 times the value of M_{max} obtained by the conventional free earth support method. Hence this value of M_{max} can be used as the actual design value, and Rowe's moment reduction need not be applied.

▼ **EXAMPLE 8.10**

For the anchored sheet pile wall shown in Figure 8.30, determine (a) D, (b) F, and (c) M_{max}. Use the CPD method; assume that $C = 0.68$ and $R = 0.6$.

Solution

Part a

$$\gamma' = \gamma_{\text{sat}} - \gamma_w = 122.4 - 62.4 = 60 \text{ lb/ft}^3$$

From Eq. (8.82),

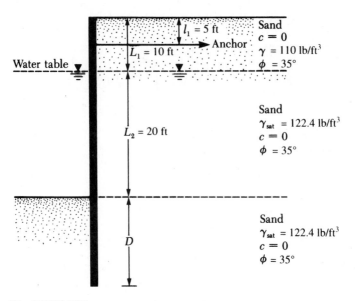

▼ FIGURE 8.30

$$\gamma_{av} = \frac{\gamma L_1 + \gamma' L_2}{L_1 + L_2} = \frac{(110)(10) + (60)(20)}{10 + 20} = 76.67 \text{ lb/ft}^3$$

$$K_a = \tan^2\left(45 - \frac{\phi}{2}\right) = \tan^2\left(45 - \frac{35}{2}\right) = 0.271$$

$$\bar{p}_a = CK_a\gamma_{av}L = (0.68)(0.271)(76.67)(30) = 423.9 \text{ lb/ft}^2$$

$$\bar{p}_p = R\bar{p}_a = (0.6)(423.9) = 254.3 \text{ lb/ft}^2$$

From Eq. (8.84):

$$D^2 + 2DL\left[1 - \left(\frac{l_1}{L}\right)\right] - \frac{L^2}{R}\left[1 - 2\left(\frac{l_1}{L}\right)\right] = 0$$

or

$$D^2 + 2(D)(30)\left[1 - \left(\frac{5}{30}\right)\right] - \frac{(30)^2}{0.6}\left[1 - 2\left(\frac{5}{30}\right)\right] = D^2 + 50D - 1000 = 0$$

Hence $D \approx$ **15.3 ft.**

Check for the assumption of R:

$$R = \frac{L(L - 2l_1)}{D(2L + D - 2l_1)} = \frac{30[30 - (2)(5)]}{15.3[(2)(30) + 15.3 - (2)(5)]} \approx \textbf{0.6—OK}$$

Part b

From Eq. (8.85),

$$F = \bar{p}_a(L - RD) = 423.9[30 - (0.6)(15.3)] = \mathbf{8826\ lb/ft}$$

Part c

From Eq. (8.86),

$$M_{\max} = 0.5\bar{p}_aL^2\left[\left(1 - \frac{RD}{L}\right)^2 - \left(\frac{2l_1}{L}\right)\left(1 - \frac{RD}{L}\right)\right]$$

$$1 - \frac{RD}{L} = 1 - \frac{(0.6)(15.3)}{30} = 0.694$$

So

$$M_{\max} = (0.5)(423.9)(30)^2\left[(0.694)^2 - \frac{(2)(5)(0.694)}{30}\right] = \mathbf{47{,}810\ lb\text{-}ft/ft} \qquad \blacktriangle$$

8.14 FIXED EARTH SUPPORT METHOD FOR PENETRATION INTO SANDY SOIL

When using the fixed earth support method we assume that the toe of the pile is restrained from rotating as shown in Figure 8.31a. In the fixed earth support solution,

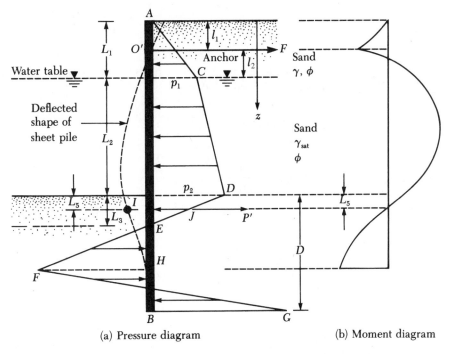

(a) Pressure diagram   (b) Moment diagram

▼ **FIGURE 8.31**   Fixed earth support method for penetration of sandy soil a simplified method called the *equivalent beam solution* is generally used to calculate L_3 and, thus, D. The development of the equivalent beam method is generally attributed to Blum (1931).

In order to understand this method, compare the sheet pile to a loaded cantilever beam *RSTU,* as shown in Figure 8.32. Note that the support at T for the beam is equivalent to the *anchor load reaction* (F) on the sheet pile (Figure 8.31). It can be seen that the point S of the beam *RSTU* is the inflection point of the elastic line of the beam, which is equivalent to point I in Figure 8.31. If the beam is cut at S and a free support (reaction P_s) is provided at that point, the bending moment diagram for portion *STU* of the beam will remain unchanged. This beam *STU* will be equivalent to the section *STU* of the beam *RSTU.* The force P' shown in Figure 8.31a at I will be equivalent to the reaction P_s on the beam (Figure 8.32).

Following is an approximate procedure for the design of an anchored sheet pile wall (Cornfield, 1975). Refer to Figure 8.31.

1. Determine L_5, which is a function of the soil friction angle ϕ below the dredge line, from the following:

ϕ (deg)	$\dfrac{L_5}{L_1 + L_2}$
30	0.08
35	0.03
40	0

2. Calculate the span of the equivalent beam as $l_2 + L_2 + L_5 = L'$.
3. Calculate the total load of the span, W. This is the area of the pressure diagram between O' and I.

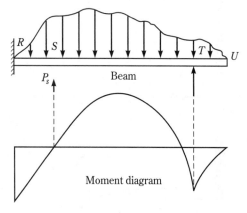

▼ **FIGURE 8.32**   Equivalent cantilever beam concept

4. Calculate the maximum moment, M_{max}, as $WL'/8$.
5. Calculate P' by taking the moment about O', or

$$P' = \frac{1}{L'} \text{ (moment of area } ACDJI \text{ about } O')$$

6. Calculate D as

$$D = L_5 + 1.2\sqrt{\frac{6P'}{(K_p - K_a)\gamma'}} \tag{8.87}$$

7. Calculate the anchor force per unit length, F, by taking the moment about I, or

$$F = \frac{1}{L'} \text{ (moment of area } ACDJI \text{ about } I)$$

▼ **EXAMPLE 8.11** _____

Consider the anchored sheet pile structure described in Example 8.6. Using the equivalent beam method described in Section 8.14, determine

a. Maximum moment
b. Theoretical depth of penetration
c. Anchor force per unit length of the structure

Solution

Part a

Determination of L_5:   For $\phi = 30°$,

$$\frac{L_5}{L_1 + L_2} = 0.08$$

$$\frac{L_5}{3.05 + 6.1} = 0.08$$

$$L_5 = 0.73$$

Net Pressure Diagram: From Example 8.6, $K_a = \frac{1}{3}$, $K_p = 3$, $\gamma = 16 \text{ kN/m}^3$, $\gamma' = 9.69 \text{ kN/m}^3$, $p_1 = 16.27 \text{ kN/m}^2$, $p_2 = 35.97 \text{ kN/m}^2$. The net active pressure at a

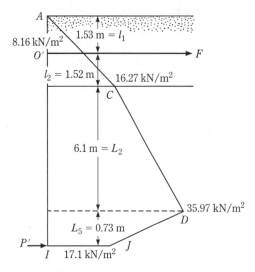

▼   FIGURE 8.33

depth L_5 below the dredge line can be calculated as

$$p_2 - \gamma' (K_p - K_a)L_5 = 35.97 - (9.69)(3 - 0.333)(0.73) = 17.1 \text{ kN/m}^2$$

The net pressure diagram from $z = 0$ to $z = L_1 + L_2 + L_5$ is shown in Figure 8.33.

Maximum Moment:

$$W = \left(\frac{1}{2}\right)(8.16 + 16.27)(1.52) + \left(\frac{1}{2}\right)(6.1)(16.27 + 35.97)$$

$$+ \left(\frac{1}{2}\right)(0.73)(35.97 + 17.1)$$

$$= 197.2 \text{ kN/m}$$

$$L' = l_2 + L_2 + L_5 = 1.52 + 6.1 + 0.73 = 8.35 \text{ m}$$

$$M_{\text{max}} = \frac{WL'}{8} = \frac{(197.2)(8.35)}{8} = \textbf{205.8 kN} \cdot \textbf{m/m}$$

Part b

$$P' = \frac{1}{L'} (\text{moment of area } ACDJI \text{ about } O')$$

$$P' = \frac{1}{8.35} \begin{bmatrix} \left(\frac{1}{2}\right)(16.27)(3.05)\left(\frac{2}{3} \times 3.05 - 1.53\right) + (16.27)(6.1)\left(1.52 + \frac{6.1}{2}\right) \\ + \left(\frac{1}{2}\right)(6.1)(35.97 - 16.27)\left(1.52 + \frac{2}{3} \times 6.1\right) + \left(\frac{1}{2}\right)(35.97 + 17.1) \\ \times (0.73)\left(1.52 + 6.1 + \frac{0.73}{2}\right) \end{bmatrix}$$

↑
Approximate

$= 114.48 \text{ kN/m}$

From Eq. (8.87)

$$D = L_5 + 1.2\sqrt{\frac{6P'}{(K_p - K_a)\gamma'}} = 0.73 + 1.2\sqrt{\frac{(6)(114.48)}{(3 - 0.333)(9.69)}} = \mathbf{6.92 \text{ m}}$$

Part c

Taking the moment about I (Figure 8.33)

$$F = \frac{1}{8.35} \begin{bmatrix} \left(\frac{1}{2}(16.27)(3.05)\left(0.73 + 6.1 + \frac{3.05}{3}\right) + (16.27)(6.1)\left(0.73 + \frac{6.1}{2}\right)\right) \\ + \left(\frac{1}{2}\right)(6.1)(35.97 - 16.27)\left(0.73 + \frac{6.1}{3}\right) + \left(\frac{1}{2}\right)(35.97 + 17.1)(0.73)\left(\frac{0.73}{2}\right) \end{bmatrix}$$

↑
Approximate

$= \mathbf{88.95 \text{ kN/m}}$ ▲

8.15 FIELD OBSERVATIONS FOR ANCHORED SHEET PILE WALLS

In the preceding sections, large factors of safety were used for the depth of penetration, D. In most cases, designers use smaller magnitudes of soil friction angle, ϕ, thereby ensuring a built-in factor of safety for the active earth pressure. This procedure is followed primarily because of the uncertainties involved in predicting the actual earth pressure a sheet pile wall will be subjected to in the field. In addition, Casagrande (1973) observed that if the soil behind the sheet pile wall has grain sizes that are predominantly smaller than those of coarse sand, the active earth pressure after construction sometimes increases to an at-rest earth pressure condition. Such an increase causes a large increase in the anchor force, F. The following two case histories are given by Casagrande (1973).

Bulkhead of Pier C—Long Beach Harbor, California (1949)

A typical cross section of the Pier C bulkhead of the Long Beach harbor is shown in Figure 8.34. Except for a rockfill dike constructed with 3 in. (76.2 mm) maximum-size quarry wastes, the backfill of the sheet pile wall consisted of fine sand.

Figure 8.35a–f shows the lateral earth pressure variation between May 17, 1949, and August 6, 1949 at Station 27 + 30. The fine sand backfill reached the design grade on May 24, 1949. The following general observations are based on Figure 8.35.

1. On May 17 (Figure 8.35a), the backfill was several feet below the design grade. However, the earth pressure was greater on this date than on May 24. This is probably because of the fact that, due to lack of lateral yielding of the wall, the earth pressure was closer to at-rest state than the active state.

2. Due to wall yielding on May 24, the earth pressure reached an active state (Figure 8.35b).

3. Between May 24 and June 3, the anchor resisted further yielding and the lateral earth pressure increased to the at-rest state (Figure 8.35c).

4. Figures 8.35d, e, and f show how the flexibility of sheet piles resulted in a gradual decrease in the lateral earth pressure distribution on the sheet piles.

These observations show that the magnitude of the active earth pressure may vary with time and depends greatly on the flexibility of the sheet piles. Also, the actual

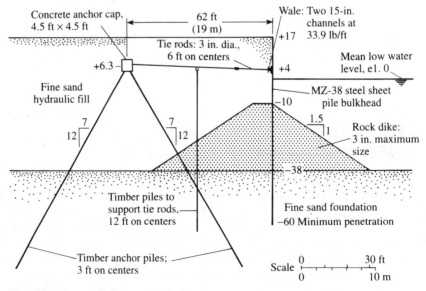

Note: Elevations are in feet

▼ **FIGURE 8.34**   Pier C bulkhead, Long Beach harbor (after Casagrande, 1973)

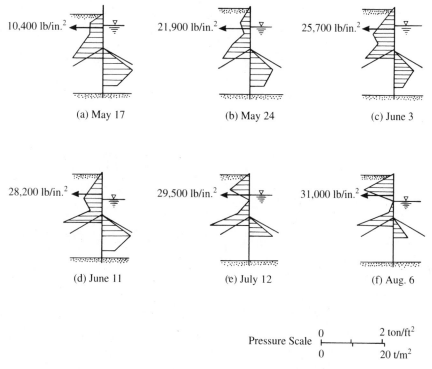

10,400 lb/in.²

(a) May 17

21,900 lb/in.²

(b) May 24

25,700 lb/in.²

(c) June 3

28,200 lb/in.²

(d) June 11

29,500 lb/in.²

(e) July 12

31,000 lb/in.²

(f) Aug. 6

Pressure Scale

0   2 ton/ft²

0   20 t/m²

▼ **FIGURE 8.35**   Measured stresses at Station 27 + 30 during construction of Pier C bulkhead, Long Beach (after Casagrande, 1973)

variations in the lateral earth pressure diagram may not be identical to those used for design.

Bulkhead—Toledo, Ohio (1961)

A typical cross section of a toledo bulkhead completed in 1961 is shown in Figure 8.36. The foundation soil was primarily fine to medium sand, but the dredge line did cut into highly overconsolidated clay. Figure 8.36 also shows the actual measured values of stress and bending moment along the sheet pile wall. Casagrande (1973) used the Rankine active earth pressure distribution to calculate the maximum bending moment according to the free earth support method with and without Rowe's moment reduction.

Design method	Maximum predicted bending moment, M_{max}
Free earth support method	108 kip-ft/ft
Free earth support method with Rowe's moment reduction	58 kip-ft/ft

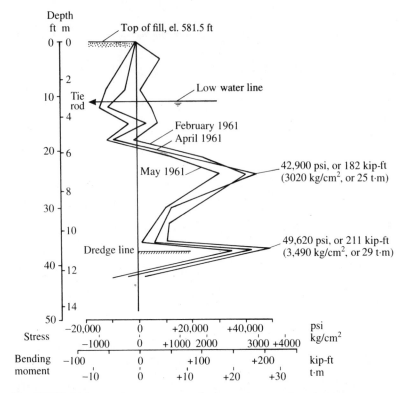

▼ **FIGURE 8.36** Bending moment and sresses from strain gage measurements at test location 3, Toledo bulkhead (after Casagrande, 1973)

Comparisons of these magnitudes of M_{max} with those actually observed show that the field values are substantially larger. The reason probably is that the backfill was primarily fine sand the measured active earth pressure distribution was larger than that predicted theoretically.

8.16 ANCHORS—GENERAL

Sections 8.8–8.14 presented the analysis of anchored sheet pile walls. Those sections also discussed how to obtain the force, F, per unit length of the sheet pile wall that has to be taken by the anchors. This section covers in more detail the various types of anchor generally used and the procedures for evaluating their ultimate holding capacities.

The general types of anchor used in sheet pile walls are

1. Anchor plates and beams (deadman)
2. Tie backs
3. Vertical anchor piles
4. Anchor beams supported by batter (compression and tension) piles

Anchor plates and beams are generally made of cast concrete blocks (Figure 8.37a). The anchors are attached to the sheet pile by *tie rods*. A *wale* is placed at the front or back face of a sheet pile for the purpose of conveniently attaching the tie rod to the wall. To protect the tie rod from corrosion, it is generally coated with paint or asphaltic materials.

In the construction of *tie backs,* bars or cables are placed in predrilled holes (Figure 8.37b) with concrete grout (cables are commonly high-strength, prestressed steel tendons). Figures 8.37c and 8.37d show a vertical anchor pile and an anchor beam with batter piles.

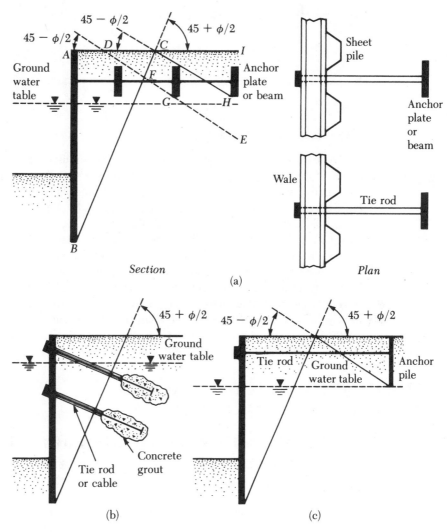

▼ **FIGURE 8.37** Various types of anchoring for sheet pile walls: (a) anchor plate or beam; (b) tie back; (c) vertical anchor pile; (d) anchor beam with batter piles

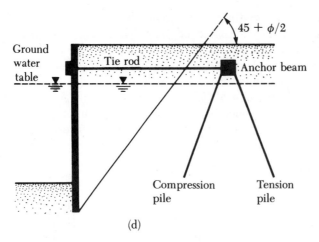

Placement of Anchors

The resistance offered by anchor plates and beams is derived primarily from the passive force of the soil located in front of them. Figure 8.37a, in which *AB* is the sheet pile wall, shows the best location for an anchor plate (for maximum efficiency). If the anchor is placed inside wedge *ABC*, which is the Rankine active zone, it would not provide any resistance to failure. Alternatively, the anchor could be placed in zone *CFEH*. Note that line *DFG* is the slip line for the Rankine passive pressure. If part of the passive wedge is located inside the active wedge *ABC*, full passive resistance of the anchor cannot be realized upon failure of the sheet pile wall. However, if the anchor is placed in zone *ICH*, the Rankine passive zone in front of the anchor slab or plate is located completely outside the Rankine active zone *ABC*. In this case, full passive resistance from the anchor can be realized.

Figures 8.37b, 8.37c, and 8.37d also show the proper locations for placement of tie backs, vertical anchor piles, and anchor beams supported by batter piles.

8.17 HOLDING CAPACITY OF ANCHOR PLATES AND BEAMS IN SAND

A. Teng's Method: Calculation of the Ultimate Resistance Offered by Anchor Plates and Beams in Sand

Teng (1962) proposed a method of determining the ultimate resistance of anchor plates or walls in granular soils located at or near the ground surface ($H/h \leq 1.5$ to 2 in Figure 8.38):

$$P_u = B(P_p - P_a) \qquad \text{(for continuous plates or beams — that is, } B/h \approx \infty) \qquad (8.88)$$

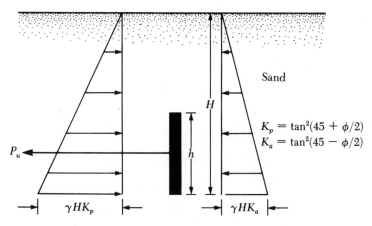

where P_u = ultimate resistance of anchor
B = length of anchor at right angle to the cross section shown
P_p and P_a = Rankine passive and active force per unit length of anchor

Note that P_p acts in front of the anchor, as shown in Figure 8.38. Also,

$$P_p = \frac{1}{2}\gamma H^2 \tan^2\left(45 + \frac{\phi}{2}\right)$$
(8.89)

and

$$P_a = \frac{1}{2}\gamma H^2 \tan^2\left(45 - \frac{\phi}{2}\right)$$
(8.90)

Equation (8.88) is valid for the plane-strain condition. For all practical cases, $B/h > 5$ may be considered to be plane-strain condition.

For $B/h <$ about 5, considering the three-dimensional failure surface (that is, accounting for the frictional resistance developed at the two ends of an anchor), Teng (1962) gave the following relation for the ultimate anchor resistance:

$$P_u = B(P_p - P_a) + \frac{1}{3}K_0\gamma(\sqrt{K_p} + \sqrt{K_a})H^3 \tan\phi \quad \left(\text{for } \frac{H}{h} \leq 1.5 \text{ to } 2\right)$$
(8.91)

where K_0 = earth pressure coefficient at rest ≈ 0.4.

B. Ovesen and Stromann's Method

Ovesen and Stromann (1972) proposed a semi-empirical method for determining the ultimate resistance of anchors in sand. The calculations are made in three steps, and they are described below.

Step 1. **Basic Case Consideration.** Determine the depth of embedment, H. Assume that the anchor slab has height H and is continuous (that is, B = length of anchor slab perpendicular to the cross section = ∞), as shown in Figure 8.39. In Figure 8.39, the following notations are used:

P_p = passive force per unit length of anchor

P_a = active force per unit length of anchor

ϕ = soil friction angle

δ = friction angle between anchor slab and soil

P_u' = ultimate resistance per unit length of anchor

W = weight per unit length of anchor slab

The magnitude of P_u' is

$$P_u' = \tfrac{1}{2}\gamma H^2 K_p \cos\delta - P_a\cos\phi = \tfrac{1}{2}\gamma H^2 K_p\cos\delta - \tfrac{1}{2}\gamma H^2 K_a\cos\phi$$
$$= \tfrac{1}{2}\gamma H^2(K_p\cos\delta - K_a\cos\phi)$$

(8.92)

where K_a = active pressure coefficient with $\delta = \phi$
        (see Figure 8.40a)
        K_p = passive pressure coefficient

To obtain $K_p\cos\delta$, first calculate $K_p\sin\delta$:

$$K_p\sin\delta = \frac{W + P_a\sin\phi}{\tfrac{1}{2}\gamma H^2} = \frac{W + \tfrac{1}{2}\gamma H^2 K_a\sin\phi}{\tfrac{1}{2}\gamma H^2}$$

(8.93)

Use the magnitude of $K_p\sin\delta$ obtained from Eq. (8.93) to estimate the magnitude of $K_p\cos\delta$ from the plots given in Figure 8.40b.

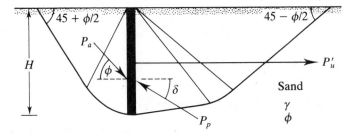

▼ **FIGURE 8.39** Basic case: continuous vertical anchor in granular soil

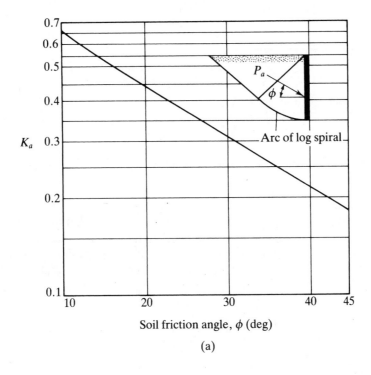

(a)

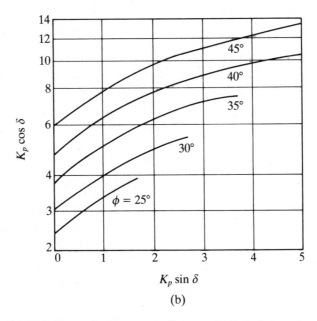

(b)

▼ **FIGURE 8.40** (a) Variation of K_a (for $\delta = \phi$); (b) variation of $K_p \cos \delta$ with $K_p \sin \delta$ (based on Ovesen and Stromann, 1972)

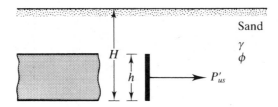

▼ **FIGURE 8.41**   Strip case: vertical anchor

Step 2. **Strip Case.** Determine the actual height of the anchor, h, to be constructed. If a continuous anchor (that is, $B = \infty$) of height h is placed in the soil so that its depth of embedment is H, as shown in Figure 8.41, the ultimate resistance per unit length is

$$P'_{us} = \left[\frac{C_{ov} + 1}{C_{ov} + \left(\dfrac{H}{h} \right)} \right] \underset{\underset{\text{Eq. 8.92}}{\uparrow}}{P'_u} \qquad (8.94)$$

where P'_{us} = ultimate resistance for the *strip case*
         C_{ov} = 19 for dense sand and 14 for loose sand

Step 3. **Actual Case.** In practice, the anchor plates are placed in a row with center-to-center spacing, S', as shown in Figure 8.42a. The ultimate resistance of each anchor, P_u, is

$$P_u = P'_{us} B_e \qquad (8.95)$$

where B_e = equivalent length

The equivalent length is a function of S', B, H, and h. Figure 8.42b shows a plot of $(B_e - B)/(H + h)$ against $(S' - B)/(H + h)$ for the cases of loose and dense sand. With known values of S', B, H, and h, the value of B_e can be calculated and used in Eq. (8.95) to obtain P_u.

C. Empirical Correlation Based on Model Tests

Ghaly (1997) used the results of 104 laboratory tests, 15 centrifugal model tests, and 9 field tests to propose an empirical correlation for the ultimate resistance of

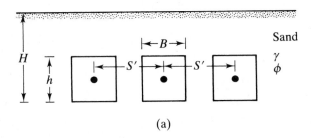

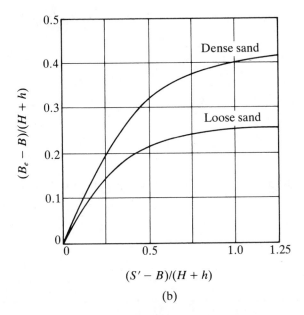

▼ **FIGURE 8.42**   (a) Actual case for row of anchors; (b) variation of $(B_e - B)/(H + h)$ with $(S' - B)/(H + h)$ (based on Ovesen and Stromann, 1972)

single anchors (Figure 8.43). The correlation can be written as

$$P_u = \frac{5.4}{\tan \phi} \left(\frac{H^2}{A}\right)^{0.28} \gamma AH \tag{8.96}$$

where A = area of the anchor = Bh

Ghaly (1997) also used the model test results of Das and Seeley (1975) to develop a load-displacement relationship for single anchors (Figure 8.44). The relationship can be given as

$$\frac{P}{P_u} = 2.2 \left(\frac{u}{H}\right)^{0.3} \tag{8.97}$$

where u = horizontal displacement of the anchor at a load level P

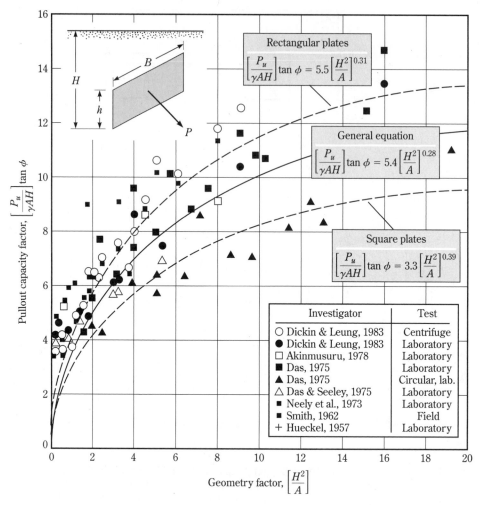

▼ **FIGURE 8.43**   Variation of $(P_u/\gamma AH)\tan\phi$ also $(P_u/\gamma AH)\tan\phi$ with H^2/A for single shallow anchors in sand (after Ghaly, 1997)

The relationship given by Eqs. (8.96) and (8.97) are for single anchors (that is, $S'/B = \infty$). For all practical purposes, when $S'/B \approx 2$, they behave as single anchors.

Factor of Safety for Anchor Plates and Beams

The allowable resistance per anchor plate, P_{all}, may be given as

$$P_{all} = \frac{P_u}{FS}$$

where FS = factor of safety

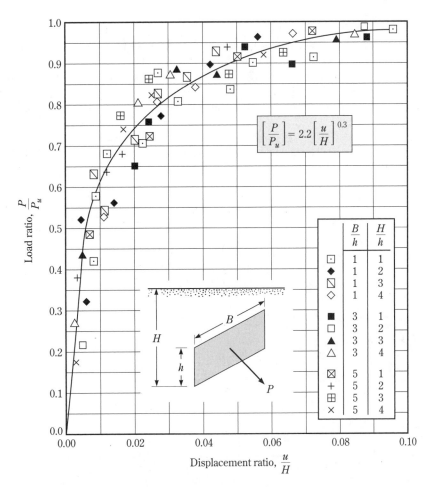

The figure contains the equation:

$$\left[\frac{P}{P_u}\right] = 2.2\left[\frac{u}{H}\right]^{0.3}$$

with axes Load ratio, $\dfrac{P}{P_u}$ versus Displacement ratio, $\dfrac{u}{H}$, and a legend table:

	$\dfrac{B}{h}$	$\dfrac{H}{h}$
⊡	1	1
◆	1	2
◰	1	3
◇	1	4
■	3	1
□	3	2
▲	3	3
△	3	4
⊠	5	1
+	5	2
⊞	5	3
×	5	4

▼ **FIGURE 8.44** Relationship of load ratio versus displacement ratio [from data reported by Das and Seeley (1975)] (after Ghaly, 1997)

Generally, a factor of safety of 2 is suggested when the method of Ovesen and Stromann is used. A factor of safety of 3 is suggested for P_u calculated by Eq. (8.96).

Spacing of Anchor Plates

The center-to-center spacing of anchors, S', may be obtained from

$$S' \doteq \frac{P_{all}}{F}$$

where F = force per unit length of the sheet pile

A row of vertical anchors embedded in sand is shown in Figure 8.45. The anchor plates are made of 6-in.-thick concrete. The design parameters are $B = h = 15$ in., $S' = 48$ in., $H = 37.5$ in., $\gamma = 105$ lb/ft³, $\phi = 35°$, and the unit weight of concrete $= 150$ lb/ft³.

Determine the allowable resistance of each anchor by using

  a.   The method of Ovesen and Stromann. Use $FS = 2$.
  b.   The empirical correlation method [Eq. (8.96)]. Use $FS = 3$.

Solution

Part a

From Figure 8.40a for $\phi = 35°$, the magnitude of K_a is about 0.26. Also,

$$W = Ht\gamma_{\text{concrete}} = \left(\frac{37.5}{12}\right)\left(\frac{6}{12}\right)(150) = 234.4 \text{ lb/ft}$$

From Eq. (8.93),

$$K_p \sin \delta = \frac{W + \dfrac{1}{2}\gamma H^2 K_a \sin \phi}{\dfrac{1}{2}\gamma H^2}$$

$$= \frac{234.4 + \dfrac{1}{2}(105)\left(\dfrac{37.5}{12}\right)^2 (0.26)(\sin 35)}{\dfrac{1}{2}(105)\left(\dfrac{37.5}{12}\right)^2} = 0.606$$

From Figure 8.40b with $\phi = 35°$ and $K_p \sin \delta = 0.606$, the magnitude of $K_p \cos \delta$ is about 4.5. Now, from Eq. (8.92),

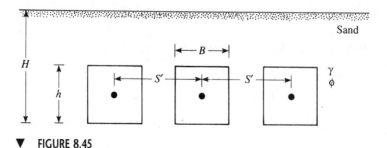

▼   **FIGURE 8.45**

$$P'_u = \frac{1}{2}\gamma H^2 (K_p \cos \delta - K_a \cos \phi)$$

$$= \frac{1}{2}(105)\left(\frac{37.5}{12}\right)^2 [4.5 - (0.26)(\cos 35)] = 2198 \, \text{lb/ft}$$

To calculate P'_{us}, we assume the sand to be loose. So, C_{ov} in Eq. (8.94) is 14. Hence

$$P'_{us} = \left[\frac{C_{ov} + 1}{C_{ov} + \left(\frac{H}{h}\right)}\right] P'_u = \left[\frac{14 + 1}{14 + \left(\frac{37.5}{15}\right)}\right](2198) = 1998 \, \text{lb/ft}$$

$$\frac{S' - B}{H + h} = \frac{48 - 15}{37.5 + 15} = 0.629$$

For $(S' - B)/(H + h) = 0.629$ and loose sand, Figure 8.42b yields

$$\frac{B_e - B}{H - h} = 0.227$$

So

$$B_e = (0.227)(H + h) + B = \frac{(0.227)(37.5 + 15) + 15}{12} = 2.24 \, \text{ft}$$

Hence from Eq. (8.95),

$$P_u = P'_{us}B_e = (1998)(2.24) = \mathbf{4476 \, lb}$$

$$P_{\text{all}} = \frac{P_u}{FS} = \frac{4476}{2} = \mathbf{2238 \, lb}$$

Part b

From Eq. (8.96),

$$P_u = \frac{5.4}{\tan \phi}\left(\frac{H^2}{A}\right)^{0.28} \gamma A H$$

$$= \frac{5.4}{\tan 35}\left(\frac{37.5^2}{15 \times 15}\right)^{0.28}(105)\left(\frac{15 \times 15}{144}\right)\left(\frac{37.5}{12}\right) = 6605 \, \text{lb}$$

$$P_{\text{all}} = \frac{P_u}{FS} = \frac{6606}{3} \approx \mathbf{2202 \, lb}$$

▲

8.18   ULTIMATE RESISTANCE OF ANCHOR PLATES AND BEAMS IN CLAY ($\phi = 0$ CONDITION)

Relatively few studies have been conducted on the ultimate resistance of anchor plates and beams in clayey soils ($\phi = 0$). Mackenzie (1955) and Tschebotarioff

(1973) identified the nature of the variation of the ultimate resistance of strip anchors and beams as a function of H, h, and c (undrained cohesion based on $\phi = 0$) in a nondimensional form based on laboratory model test results. Das et al., (1985) suggested the following procedure to obtain the ultimate resistance of an anchor embedded in clay.

When an anchor plate having dimensions of $h \times B$ is embedded at depth H, the failure surface in soil at ultimate load may extend to the ground surface, as shown in Figure 8.46a. This condition will arise when the ratio H/h is relatively small. However, for larger values of H/h, local shear failure takes place at ultimate load (Figure 8.46b). The critical value of H/h at which general shear failure changes to local shear failure in soil is

$$\left(\frac{H}{h}\right)_{cr\text{-}S} = 4.7 + 2.9 \times 10^{-3}c \leq 7$$

$$\text{(for } square \text{ anchors, that is, } B/h = 1)$$

(8.98)

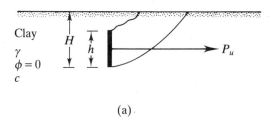

(a)

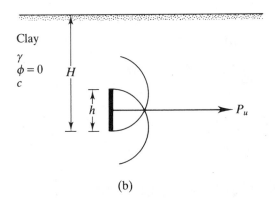

(b)

▼ **FIGURE 8.46**   Nature of failure surface in soil
around vertical anchor plate:
(a) H/h relatively small; (b) $H/h > (H/h)_{cr}$
(Note: width of anchor = B)

and

$$\left(\frac{H}{h}\right)_{\text{cr-R}} = \left(\frac{H}{h}\right)_{\text{cr-S}}\left[0.9 + 0.1\left(\frac{B}{h}\right)\right] \leq 1.3\left(\frac{H}{h}\right)_{\text{cr-S}}$$

(for *rectangular* anchors, that is, $B/h \geq 1$)

(8.99)

In Eqs. (8.98) and (8.99), the unit of the undrained cohesion is lb/ft².

The ultimate resistance of an anchor plate may be expressed in a nondimensional form as

$$F_c = \frac{P_u}{Bhc}$$

(8.100)

where F_c = breakout factor
P_u = ultimate resistance

Figure 8.47 shows the nature of variation of F_c against H/h for an anchor plate embedded in clay. Note that, for $H/h \geq (H/h)_{\text{cr}}$, the magnitude of F_c equals that of $F_{c(\text{max})}$, which is a constant. For square anchors $(B = h)$, $F_{c(\text{max})} = 9$. So, with $H/h \geq (H/h)_{\text{cr-S}}$,

$$P_u = 9h^2c \qquad \text{(for square anchors)}$$

(8.101)

For rectangular anchors with $H/h \geq (H/h)_{\text{cr-R}}$, the ultimate resistance may be given as

$$P_u = 9Bhc\left[0.825 + 0.175\left(\frac{h}{B}\right)\right]$$

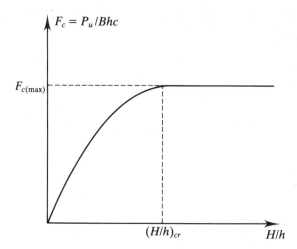

▼ **FIGURE 8.47**   Nature of variation of F_c with H/h for vertical anchor in clay

or

$$P_u = Bch \left[7.425 + 1.575 \left(\frac{h}{B} \right) \right] \tag{8.102}$$

So, for square and rectangular anchors with $H/h \leq (H/h)_{cr}$, the ultimate resistance may be calculated from the empirical relationship:

$$\frac{\left[\dfrac{H/h}{(H/h)_{cr}} \right]}{\left[\dfrac{P_u/cBh}{7.425 + 1.575(h/B)} \right]} = 0.41 + 0.59 \left[\frac{H/h}{(H/h)_{cr}} \right] \tag{8.103}$$

▼ **EXAMPLE 8.13** _____

For a vertical anchor plate in clay, the design parameters are $B = 5$ ft, $h = 2$ ft, $H = 6$ ft, and $c = 500$ lb/ft². Determine the ultimate resistance of the anchor.

Solution   From Eq. (8.98),

$$\left(\frac{H}{h} \right)_{cr\text{-}S} = 4.7 + 2.9 \times 10^{-3}c$$

$$= 4.7 + (2.9 \times 10^{-3})(500) = 6.15$$

Again, from Eq. (8.99),

$$\left(\frac{H}{h} \right)_{cr\text{-}R} = \left(\frac{H}{h} \right)_{cr\text{-}S} \left[0.9 + 0.1 \left(\frac{B}{h} \right) \right]$$

$$= (6.15) \left[0.9 + (0.1) \left(\frac{5}{2} \right) \right] = 7.07$$

As $H/h = 6/2 = 3$ is less than $(H/h)_{cr}$, we use Eq. (8.103) to obtain P_u, from which

$$\frac{\left(\dfrac{3}{7.07} \right)}{\left[\dfrac{(P_u/cBh)}{7.425 + 1.575(2/5)} \right]} = 0.41 + (0.59) \left(\frac{3}{7.07} \right)$$

or

$$\frac{0.424}{\left[\dfrac{(P_u/cBh)}{8.055} \right]} = 0.66$$

So

$$\frac{P_u}{cBh} = \left(\frac{0.424}{0.66} \right) (8.055) = 5.17$$

$$P_u = (5.17)cBh = (5.17)(500)(5)(2) = \mathbf{25{,}850\ lb}$$

▲

8.19 ULTIMATE RESISTANCE OF TIE BACKS

According to Figure 8.48, the ultimate resistance offered by a tie back in sand is

$$P_u = \pi dl \overline{\sigma}'_v K \tan \phi \qquad (8.104)$$

where P_u = ultimate resistance
ϕ = angle of friction of soil
$\overline{\sigma}'_v$ = average effective vertical stress ($= \gamma z$ in dry sand)
K = earth pressure coefficient

The magnitude of K can be taken to be equal to the earth pressure coefficient at rest (K_O) if the concrete grout is placed under pressure (Littlejohn, 1970). The lower limit of K can be taken to be equal to the Rankine active earth pressure coefficient.

In clays, the ultimate resistance of tie backs may be approximated as

$$P_u = \pi dl c_a \qquad (8.105)$$

where c_a = adhesion

The value of c_a may be approximated as $\frac{2}{3}c_u$ (where c_u = undrained cohesion). A factor of safety of 1.5–2 may be used over the ultimate resistance to obtain the allowable resistance offered by each tie back.

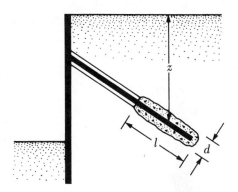

▼ **FIGURE 8.48**   Parameters for defining the ultimate resistance of tie backs

BRACED CUTS

8.20 BRACED CUTS—GENERAL

Sometimes construction work requires ground excavations with vertical or near-vertical faces — for example, basements of buildings in developed areas or underground transportation facilities at shallow depths below the ground surface (cut-

and-cover type of construction). The vertical faces of the cuts need to be protected by temporary bracing systems to avoid failure that may be accompanied by considerable settlement or by bearing capacity failure of nearby foundations.

Figure 8.49 shows two types of braced cut commonly used in construction work. One type uses the *soldier beam* (Figure 8.49a), which is driven into the ground before excavation and is a vertical steel or timber beam. *Laggings,* which are horizontal timber planks, are placed between soldier beams as the excavation proceeds. When the excavation reaches the desired depth, *wales* and *struts* (horizontal steel beams) are installed. The struts are horizontal compression members. Figure 8.49b shows another type of braced excavation. In this case, interlocking *sheet piles* are driven into the soil before excavation. Wales and struts are inserted immediately after excavation reaches the appropriate depth.

To design braced excavations (that is, to select wales, struts, sheet piles, and soldier beams), an engineer must estimate the lateral earth pressure to which the braced cuts will be subjected.

The theoretical aspects of the lateral earth pressure on a braced cut were discussed in Section 6.8. The total active force per unit length of the wall (P_a) was calculated using the general wedge theory. However, that analysis does not provide the relationships for estimation of the variation of lateral pressure with depth, which is a function of several factors such as the type of soil, experience of the construction crew, type of construction equipment used, and so forth. For that reason, empirical pressure envelopes developed from field observations are used for the design of braced cuts. This procedure is discussed in the following section.

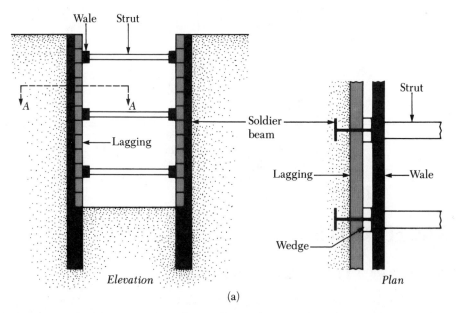

▼ **FIGURE 8.49**   Types of braced cut: (a) use of soldier beams; (b) use of sheet piles

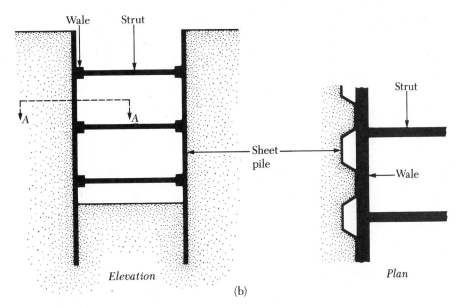

Wale Strut

Sheet pile

Wale

Strut

Elevation

Plan

(b)

▼ **FIGURE 8.49** (*Continued*)

8.21 PRESSURE ENVELOPE FOR BRACED-CUT DESIGN

After observation of several braced cuts, Peck (1969) suggested using *design pressure envelopes* for braced cuts in sand and clay. Figures 8.50, 8.51, and 8.52 show Peck's pressure envelopes, to which the following guidelines apply.

Cuts in Sand

Figure 8.50 shows the pressure envelope for cuts in sand. This pressure, p_a, may be expressed as

$$p_a = 0.65\gamma H K_a \tag{8.106}$$

where γ = unit weight
H = height of the cut
K_a = Rankine active pressure coefficient = $\tan^2 (45 - \phi/2)$

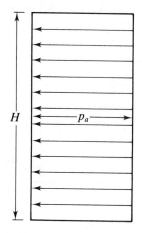

▼ **FIGURE 8.50** Peck's (1969) apparent pressure envelope for cuts in sand

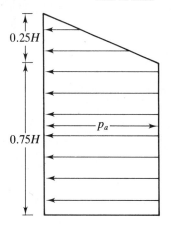

▼ **FIGURE 8.51** Peck's (1969) apparent pressure envelope for cuts in soft to medium clay

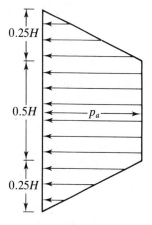

▼ **FIGURE 8.52** Peck's (1969) apparent pressure envelope for cuts in stiff clay

Cuts in Soft and Medium Clay

The pressure envelope for soft to medium clay is shown in Figure 8.51. It is applicable for the condition

$$\frac{\gamma H}{c} > 4$$

where c = undrained cohesion ($\phi = 0$)

The pressure, p_a, is the larger of

$$p_a = \gamma H \left[1 - \left(\frac{4c}{\gamma H} \right) \right]$$
or
$$p = 0.3\gamma H$$

(8.107)

where γ = unit weight of clay

Cuts in Stiff Clay

The pressure envelope shown in Figure 8.52, in which

$$p_a = 0.2\gamma H \text{ to } 0.4\gamma H \qquad \text{(with an average of } 0.3\gamma H)$$

(8.108)

is applicable to the condition $\gamma H/c \leq 4$.

Limitations of the Pressure Envelopes

When using the pressure envelopes just described, keep the following points in mind:

1. The pressure envelopes are sometimes referred to as *apparent pressure envelopes*. However, the actual pressure distribution is a function of the construction sequence and the relative flexibility of the wall.
2. They apply to excavations having depths greater than about 20 ft ($\approx$6 m).
3. They are based on the assumption that the water table is below the bottom of the cut.
4. Sand is assumed to be drained with zero pore water pressure.
5. Clay is assumed to be undrained and pore water pressure is not considered.

Cuts in Layered Soil

Sometimes, layers of both sand and clay are encountered when a braced cut is being constructed. In this case, Peck (1943) proposed that an equivalent value of cohesion ($\phi = 0$ concept) should be determined in the following manner (refer to Figure 8.53a):

$$c_{av} = \frac{1}{2H} [\gamma_s K_s H_s^2 \tan \phi_s + (H - H_s) n' q_u] \tag{8.109}$$

where H = total height of the cut
γ_s = unit weight of sand
H_s = height of the sand layer
K_s = a lateral earth pressure coefficient for the sand layer (≈ 1)
ϕ_s = angle of friction of sand
q_u = unconfined compression strength of clay
n' = a coefficient of progressive failure
(ranging from 0.5 to 1.0; average value 0.75)

The average unit weight, γ_a, of the layers may be expressed as

$$\gamma_a = \frac{1}{H} [\gamma_s H_s + (H - H_s) \gamma_c] \tag{8.110}$$

where γ_c = saturated unit weight of clay layer

Once the average values of cohesion and unit weight are determined, the pressure envelopes in clay can be used to design the cuts.

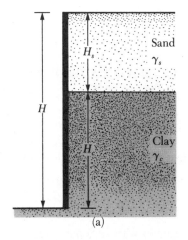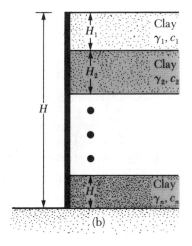

▼   **FIGURE 8.53**   Layered soils in braced cuts

Similarly, when several clay layers are encountered in the cut (Figure 8.53b), the average undrained cohesion becomes

$$c_{av} = \frac{1}{H} (c_1 H_1 + c_2 H_2 + \cdots + c_n H_n) \tag{8.111}$$

where $c_1, c_2, \ldots, c_n$ = undrained cohesion in layers $1, 2, \ldots, n$
$H_1, H_2, \ldots, H_n$ = thickness of layers $1, 2, \ldots, n$

The average unit weight, γ_a, is

$$\gamma_a = \frac{1}{H} (\gamma_1 H_1 + \gamma_2 H_2 + \gamma_3 H_3 + \cdots + \gamma_n H_n) \tag{8.112}$$

8.22 DESIGN OF VARIOUS COMPONENTS OF A BRACED CUT

Struts

In construction work, struts should have a minimum vertical spacing of about 9 ft (2.75 m) or more. The struts are actually horizontal columns subject to bending. The load-carrying capacity of columns depends on the *slenderness ratio*. The slenderness ratio can be reduced by providing vertical and horizontal supports at intermediate points. For wide cuts, splicing the struts may be necessary. For braced cuts in clayey soils, the depth of the first strut below the ground surface should be less than the depth of tensile crack, z_c. From Eq. (6.13),

$$\sigma_a = \gamma z K_a - 2c\sqrt{K_a}$$

where K_a = coefficient of Rankine active pressure

For determining the depth of tensile crack,

$$\sigma_a = 0 = \gamma z_c K_a - 2c\sqrt{K_a}$$

or

$$z_c = \frac{2c}{\sqrt{K_a}\gamma}$$

With $\phi = 0$, $K_a = \tan^2 (45 - \phi/2) = 1$, so

$$z_c = \frac{2c}{\gamma}$$

A simplified conservative procedure may be used to determine the strut loads. Although this procedure will vary, depending on the engineers involved in the project, the following is a step-by-step outline of the general procedure (refer to Figure 8.54).

1.   Draw the pressure envelope for the braced cut (see Figures 8.50, 8.51, and 8.52). Also show the proposed strut levels. Figure 8.54a shows a pressure envelope for a sandy soil; however, it could also be for a clay. The strut levels are marked A, B, C, and D. The sheet piles (or soldier beams) are assumed to be hinged at the strut levels, except for the top and bottom ones. In Figure 8.54a, the hinges are at the level of struts B and C. (Many designers also assume the sheet piles, or soldier beams, to be hinged at all strut levels, except for the top.)

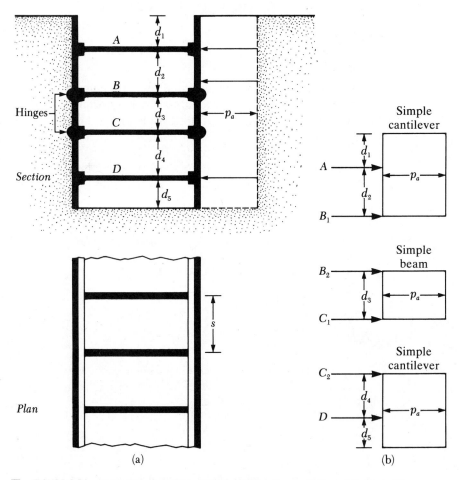

▼ **FIGURE 8.54**   Determination of strut loads; (a) section and plan of the cut; (b) method for determining strut loads

2. Determine the reactions for the two simple cantilever beams (top and bottom) and all the simple beams between. In Figure 8.54b, these reactions are A, B_1, B_2, C_1, C_2, and D.
3. The strut loads in Figure 8.54 may be calculated as follows:

$$P_A = (A)(s)$$

$$P_B = (B_1 + B_2)(s) \tag{8.113}$$

$$P_C = (C_1 + C_2)(s)$$

$$P_D = (D)(s)$$

where

$\qquad P_A, P_B, P_C, P_D$ = loads to be taken by the individual struts at levels A, B, C, and D, respectively

$\qquad A, B_1, B_2, C_1, C_2, D$ = reactions calculated in Step 2 (note unit: force/unit length of the braced cut)

$\qquad s$ = horizontal spacing of the struts (see plan in Figure 8.54a)

4. Knowing the strut loads at each level and the intermediate bracing conditions allows selection of the proper sections from the steel construction manual.

Sheet Piles

The following steps are involved in designing the sheet piles:

1. For each of the sections shown in Figure 8.54b, determine the maximum bending moment.
2. Determine the maximum value of the maximum bending moments (M_{max}) obtained in Step 1. Note that the unit of this moment will be, for example, lb-ft/ft (kN · m/m) length of the wall.
3. Obtain the required section modulus of the sheet piles:

$$S = \frac{M_{max}}{\sigma_{all}} \tag{8.114}$$

where σ_{all} = allowable flexural stress of the sheet pile material

4. Choose a sheet pile having a section modulus greater than or equal to the required section modulus from a table such as Table C.1 (Appendix C).

Wales

Wales may be treated as continuous horizontal members if they are spliced properly. Conservatively, they may also be treated as though they are pinned at the struts.

For the section shown in Figure 8.54a, the maximum moments for the wales (assuming that they are pinned at the struts) are

At level A, $M_{max} = \dfrac{(A)\,(s^2)}{8}$

At level B, $M_{max} = \dfrac{(B_1 + B_2)s^2}{8}$

At level C, $M_{max} = \dfrac{(C_1 + C_2)s^2}{8}$

At level D, $M_{max} = \dfrac{(D)\,(s^2)}{8}$

where A, B_1, B_2, C_1, C_2, and D are the reactions under the struts per unit length of the wall (Step 2 of strut design)

Determine the section modulus of the wales:

$$S = \dfrac{M_{max}}{\sigma_{all}}$$

The wales are sometimes fastened to the sheet piles at points that satisfy the lateral support requirements.

▼ **EXAMPLE 8.14** _____

Refer to the braced cut shown in Figure 8.55.

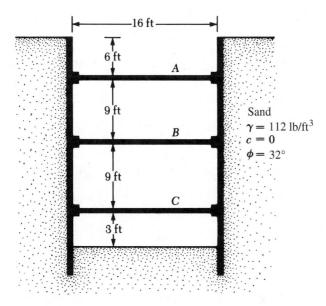

▼ **FIGURE 8.55**

a. Draw the earth pressure envelope and determine the strut loads (Note: the struts are spaced horizontally at 12 ft center-to-center).
b. Determine the sheet pile section.
c. Determine the required section modulus of the wales at level A. Given: $\sigma_{all} = 24$ kip/in².

Solution

Part a

For this case, the earth pressure envelope shown in Figure 8.50 is applicable. Hence

$$K_a = \tan^2\left(45 - \frac{\phi}{2}\right) = \tan^2\left(45 - \frac{32}{2}\right) = 0.307$$

From Equation (8.106),

$$p_a = 0.65\gamma H K_a = (0.65)(112)(27)(0.307) = 603.44 \text{ lb/ft}^2$$

Figure 8.56a shows the pressure envelope. Refer to Figure 8.56b and calculate B_1:

$$\Sigma M_{B_1} = 0$$

$$A = \frac{(603.44)(15)\left(\dfrac{15}{2}\right)}{9} = 7543 \text{ lb/ft}$$

$$B_1 = (603.44)(15) - 7543 = 1508.6 \text{ lb/ft}$$

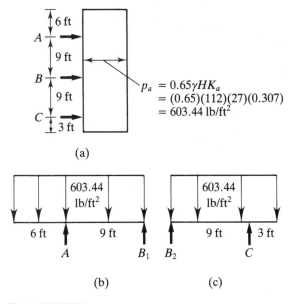

(a)

(b)   (c)

▼ FIGURE 8.56

Now, refer to Figure 8.56c and calculate B_2:

$$\Sigma M_{B_2} = 0$$

$$C = \frac{(603.44)(12)\left(\dfrac{12}{2}\right)}{9} = 4827.5 \text{ lb/ft}$$

$$B_2 = (603.44)(12) - 4827.5 = 2413.7 \text{ lb/ft}$$

The strut loads are

At A, $(7.543)(\text{spacing}) = (7.543)(12) = \textbf{90.52 kip}$

At B, $(B_1 + B_2)(\text{spacing}) = (1.509 + 2.414)(12) = \textbf{47.07 kip}$

At C, $(C)(\text{spacing}) = (4.827)(12) = \textbf{57.93 kip}$

Part b

Refer to the load diagrams shown in Figure 8.56b and 8.56c. Figure 8.57 shows the shear force diagrams based on the load diagrams. First, determine x_1 and x_2:

$$x_1 = \frac{3.923}{0.603} = 6.5 \text{ ft}$$

$$x_2 = \frac{3.017}{0.603} = 5 \text{ ft}$$

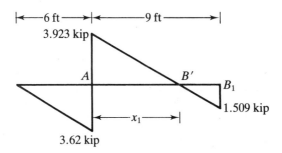

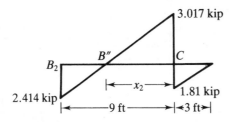

▼ FIGURE 8.57

Then the moments are

At A, $\frac{1}{2}(3.62)(6) = 10.86$ kip-ft

At C, $\frac{1}{2}(1.81)(3) = 2.715$ kip-ft

At B', $\frac{1}{2}(1.509)(2.5) = 1.89$ kip-ft

At B'', $\frac{1}{2}(2.414)(4) = 4.828$ kip-ft

M_A is maximum, so

$$S_x = \frac{M_{max}}{\sigma_{all}} = \frac{(10.86 \text{ kip-ft})(12)}{24 \text{ kip/in}^2} = 5.43 \text{ in}^3$$

From Table C.1 (Appendix C),

Use PZ-22: $S_x = 18.1$ in^3/ft of wall.

Part c:

For the wale at level A,

$$M_{max} = \frac{A(s^2)}{8}$$

$A = 7543$ lb/ft, so

$$M_{max} = \frac{(7.543)(12^2)}{8} = 135.77 \text{ kip-ft}$$

$$S_x = \frac{M_{max}}{\sigma_{all}} = \frac{(135.77)(12)}{24 \text{ kip/in}^2} = \textbf{67.9 in}^3\textbf{/ft of wall} \qquad \blacktriangle$$

▼ **EXAMPLE 8.15**

The cross section of a long braced cut is shown in Figure 8.58a.

   a. Draw the earth pressure envelope.
   b. Determine the strut loads at levels A, B, and C.
   c. Determine the section modulus of sheet pile section required. Use $\sigma_{all} = 170$ MN/m^2.

Note: The struts are placed at 3 m center-to-center in the plan.

Solution

Part a

Given: $\gamma = 18$ kN/m^3, $c = 35$ kN/m^2, and $H = 6$ m.

$$\frac{\gamma H}{c} = \frac{(18)(7)}{35} = 3.6 < 4$$

So, the pressure envelope will be like the one in Figure 8.52. This is plotted in Figure 8.58a with maximum pressure intensity, p_a, equal to $0.3\gamma H = 0.3(18)(7) = $ **37.8 kN/m^2.**

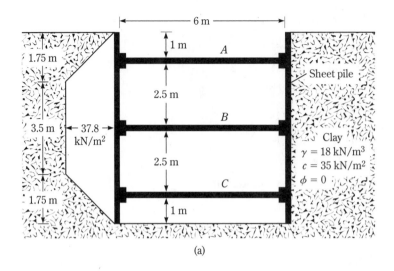

(a)

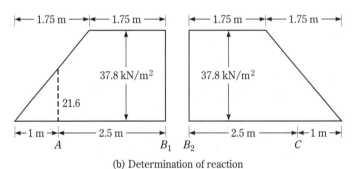

(b) Determination of reaction

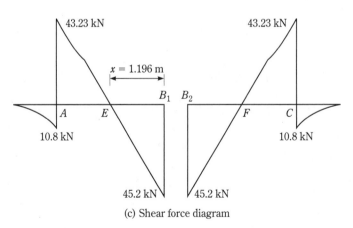

(c) Shear force diagram

▼ FIGURE 8.58

Part b

For determination of the strut loads, refer to Figure 8.58b. Taking the moment about B_1, $\Sigma M_{B_1} = 0$,

$$A(2.5) - \left(\frac{1}{2}\right)(37.8)(1.75)\left(1.75 + \frac{1.75}{3}\right) - (1.75)(37.8)\left(\frac{1.75}{2}\right) = 0$$

or

$$A = 54.02 \text{ kN/m}$$

Also, Σ vertical forces $= 0$. Thus

$$\tfrac{1}{2}(1.75)(37.8) + (37.8)(1.75) = A + B_1$$

$$33.08 + 66.15 - A = B_1$$

So

$$B_1 = 45.2 \text{ kN/m}$$

Due to symmetry

$$B_2 = 45.2 \text{ kN/m}$$

$$C = 54.02 \text{ kN/m}$$

Strut load at various levels

$$P_A = 54.02 \times \text{horizontal spacing}, s = 54.02 \times 3 = \textbf{162.06 kN}$$

$$P_B = (B_1 + B_2)3 = (45.2 + 45.2)3 = \textbf{271.2 kN}$$

$$P_C = 54.02 \times 3 = \textbf{162.06 kN}$$

Part c

Refer to the left side of Figure 8.58b. For the maximum moment, the shear force should be zero. The nature of variation of the shear force is shown in Figure 8.58c. The location of point E can be given as

$$x = \frac{\text{reaction at } B_1}{37.8} = \frac{45.2}{37.8} = 1.196 \text{ m}$$

The magnitude of moment at $A = \frac{1}{2}(1)\left(\frac{37.8}{1.75} \times 1\right)\left(\frac{1}{3}\right)$

$$= 3.6 \text{ kN-m/meter of wall}$$

The magnitude of moment at $E = (45.2 \times 1.196) - (37.8 \times 1.196)\left(\frac{1.196}{2}\right)$

$$= 54.06 - 27.03$$

$$= 27.03 \text{ kN-m/meter of wall}$$

Because the loading on the left and right sections of Figure 8.58b are the same, the magnitude of moments at F and C (Figure 8.58c) will be the same as E and A, respectively. Hence, the maximum moment = 27.03 kN-m/meter of wall.

The section modulus of the sheet piles

$$S = \frac{M_{max}}{\sigma_{all}} = \frac{27.03 \text{ kN-m}}{170 \times 10^3 \text{ kN/m}^2} = \mathbf{15.9 \times 10^{-5} \, m^3/m \, of \, the \, wall} \qquad \blacktriangle$$

8.23 BOTTOM HEAVING OF A CUT IN CLAY

Braced cuts in clay may become unstable as a result of heaving of the bottom of the excavation. Terzaghi (1943) analyzed the factor of safety of braced excavations against bottom heave. The failure surface for such a case is shown in Figure 8.59. The vertical load (per unit length of the cut) at the bottom of the cut along line bd and af is

$$Q = \gamma H B_1 - cH \qquad (8.115)$$

where $B_1 = 0.7B$
$\qquad c$ = cohesion ($\phi = 0$ concept)

This load Q may be treated as a load per unit length on a continuous foundation at the level of bd (and af) and having a width of $B_1 = 0.7B$. Based on Terzaghi's bearing capacity theory, the net ultimate load-carrying capacity per unit length of this foundation (Chapter 3) is

$$Q_u = cN_cB_1 = 5.7cB_1$$

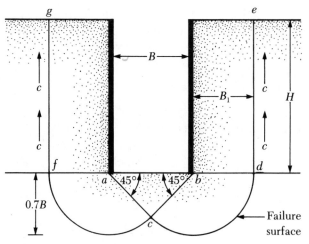

Note: cd and cf are arcs of circles with centers at b and a, respectively

▼ **FIGURE 8.59**   Factor of safety against bottom heave

Hence from Eq. (8.115), the factor of safety against bottom heave is

$$FS = \frac{Q_u}{Q} = \frac{5.7cB_1}{\gamma H B_1 - cH} = \frac{1}{H}\left(\frac{5.7c}{\gamma - \dfrac{c}{0.7B}}\right) \tag{8.116}$$

This factor of safety is based on the assumption that the clay layer is homogeneous, at least to a depth of $0.7B$ below the bottom of the cut. However, a *hard layer of rock or rocklike material at a depth of* $D < 0.7B$ will modify the failure surface to some extent. In such a case, the factor of safety becomes

$$FS = \frac{1}{H}\left(\frac{5.7c}{\gamma - c/D}\right) \tag{8.117}$$

Bjerrum and Eide (1956) also studied the problem of bottom heave for braced cuts in clay. For the factor of safety, they proposed:

$$FS = \frac{cN_c}{\gamma H} \tag{8.118}$$

The bearing capacity factor, N_c, varies with the ratios H/B and L/B (where $L = $ length of the cut). For infinitely long cuts $(B/L = 0)$, $N_c = 5.14$ at $H/B = 0$ and increases to $N_c = 7.6$ at $H/B = 4$. Beyond that — that is, for $H/B > 4$ — the value of N_c remains constant. For cuts square in plan $(B/L = 1)$, $N_c = 6.3$ at $H/B = 0$, and $N_c = 9$ for $H/B \geq 4$. In general, for any H/B,

$$N_{c(\text{rectangle})} = N_{c(\text{square})}\left(0.84 + 0.16\frac{B}{L}\right) \tag{8.119}$$

Figure 8.60 shows the variation of the value of N_c for $L/B = 1, 2, 3,$ and ∞.

When Eqs. (8.118) and (8.119) are combined, the factor of safety against heave becomes

$$FS = \frac{cN_{c(\text{square})}\left(0.84 + 0.16\dfrac{B}{L}\right)}{\gamma H} \tag{8.120}$$

Equation (8.120) and the variation of the bearing capacity factor, N_c, as shown in Figure 8.60 are based on the assumptions that the clay layer below the bottom of the cut is homogeneous and that the magnitude of the undrained cohesion in

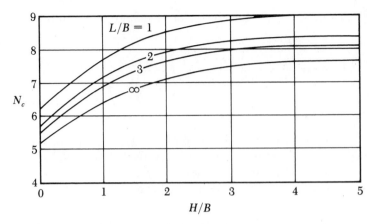

▼ **FIGURE 8.60** Variation of N_c with L/B and H/B [based on Bjerrum and Eide's equation, Eq. (8.119)]

the soil that contains the failure surface is equal to c (Figure 8.61). However, if a stronger clay layer is encountered at a shallow depth, as shown in Figure 8.62a, the failure surface below the cut will be controlled by the undrained cohesions c_1 and c_2. For this type of condition, the factor of safety is

$$FS = \frac{c_1 [N'_{c(\text{strip})} F_d] F_s}{\gamma H}$$

(8.121)

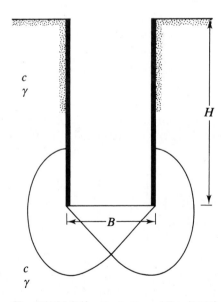

▼ **FIGURE 8.61** Derivation of Eq. (8.120)

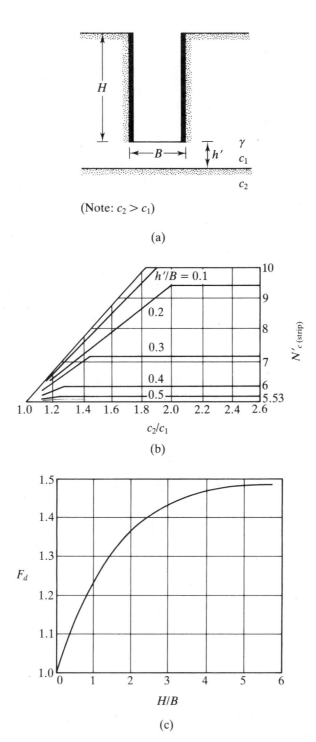

FIGURE 8.62 (a) Layered clay below the bottom of the cut; (b) variation of $N'_{c(strip)}$ with c_2/c_1 and h'/B (redrawn after Reddy and Srinivasan, 1967); and (c) variation of F_d with H/B

where $N'_{c(\text{strip})}$ = bearing capacity factor of an infinitely long cut ($B/L = 0$), which is a function of h'/B and c_2/c_1
   F_d = depth factor, which is a function of H/B
   F_s = shape factor

The variation of $N'_{c(\text{strip})}$ is shown in Figure 8.62b, and the variation of F_d as a function of H/B is given in Figure 8.62c. The shape factor, F_s, is

$$F_s = 1 + 0.2\frac{B}{L}$$

(8.122)

In most cases, a factor of safety of about 1.5 is generally recommended. If *FS* becomes less than about 1.5, the sheet pile is driven deeper (Figure 8.63). Usually the depth, d, is kept less than or equal to $B/2$. In that case, the force, P, per unit length of the buried sheet pile (aa' and bb') may be expressed as follows (U.S. Department of the Navy, 1971):

$$P = 0.7(\gamma HB - 1.4cH - \pi cB) \qquad \text{for } d > 0.47B$$

(8.123)

and

$$P = 1.5d\left(\gamma H - \frac{1.4cH}{B} - \pi c\right) \qquad \text{for } d < 0.47B$$

(8.124)

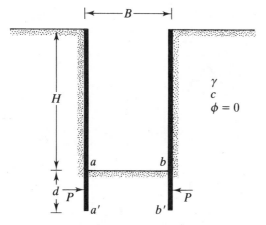

▼ **FIGURE 8.63**   Force on the buried length of sheet pile

8.24 STABILITY OF THE BOTTOM OF A CUT IN SAND

The bottom of a cut in sand is generally stable. When the water table is encountered, the bottom of the cut is stable as long as the water level inside the excavation is higher than the groundwater level. In case dewatering is needed (Figure 8.64), the factor of safety against piping should be checked. [*Piping* is another term for failure by heave, as defined in Section 1.11; see Eq. (1.51)]. Piping may occur when a high hydraulic gradient is created by water flowing into the excavation. To check the factor of safety, draw flow nets and determine the maximum exit gradient [$i_{max(exit)}$] that will occur at points A and B. Figure 8.65 shows such a flow net, for which the maximum exit gradient is

$$i_{max(exit)} = \frac{\frac{h}{N_d}}{a} = \frac{h}{N_d a} \tag{8.125}$$

where a = length of the flow element at A (or B)
N_d = number of drops (*note:* in Figure 8.65,
$N_d = 8$ — also see Section 1.9)

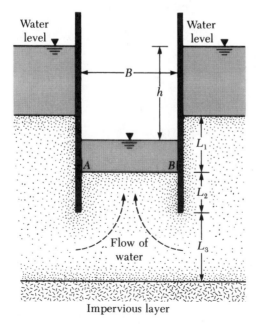

Water level

Water level

B

h

L_1

A B

L_2

Flow of water L_3

Impervious layer

▼ FIGURE 8.64

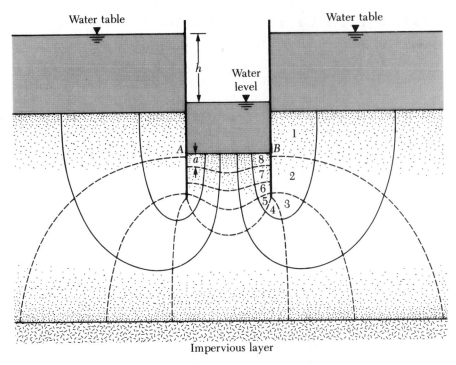

▼ **FIGURE 8.65**   Determining the factor of safety against piping by drawing flow net

The factor of safety against piping may be expressed as

$$FS = \frac{i_{cr}}{i_{max(exit)}} \tag{8.126}$$

where i_{cr} = critical hydraulic gradient

The relationship for i_{cr} was given in Chapter 1 as

$$i_{cr} = \frac{G_s - 1}{e + 1}$$

The magnitude of i_{cr} varies between 0.9 and 1.1 in most soils, with an average of about 1. A factor of safety of about 1.5 is desirable.

The maximum exit gradient for sheeted excavations in sands with $L_3 = \infty$ can also be evaluated theoretically (Harr, 1962). (Only the results of these mathematical derivations will be presented here. For further details, refer to the original work.) To calculate the maximum exit gradient, refer to Figures 8.66 and 8.67 and perform the following steps.

1. Determine the modulus, m, from Figure 8.66 by obtaining $2L_2/B$ (or $B/2L_2$) and $2L_1/B$.

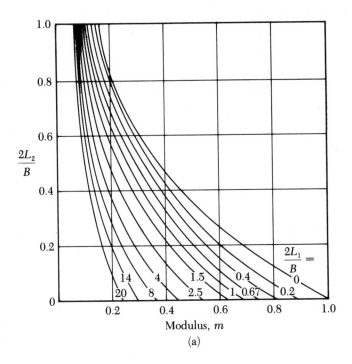

(a)

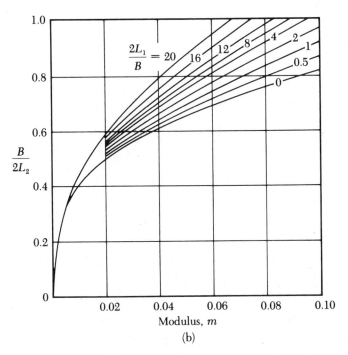

(b)

▼ **FIGURE 8.66** Variation of modulus (from *Groundwater and Seepage,* by M. E. Harr. Copyright © 1962 by McGraw-Hill. Used with permission.)

2. With the known modulus and $2L_1/B$, refer to Figure 8.67 and determine $L_2 i_{exit(max)}/h$. Because L_2 and h will be known, $i_{exit(max)}$ can be calculated.
3. The factor of safety against piping can be evaluated by using Eq. (8.126).

Marsland (1958) presented the results of model tests conducted to study the influence of seepage on the stability of sheeted excavations in sand. They were summarized by the U.S. Department of the Navy (1971) in *NAVFAC DM-7* and are given in Figure 8.68a, b, and c. Note that Figure 8.68b is for the case of determining the sheet pile penetration (L_2) needed for the required factor of safety against piping when the sand layer extends to a great depth below the excavation. However, Figure 8.68c represents the case in which an impervious layer lies at depth $L_2 + L_3$ below the bottom of the excavation.

▼ **EXAMPLE 8.16** _____

Refer to Figure 8.64, for which $h = 4.5$ m, $L_1 = 5$ m, $L_2 = 4$ m, $B = 5$ m, and $L_3 = \infty$. Determine the factor of safety against piping.

Solution

$$\frac{2L_1}{B} = \frac{2(5)}{5} = 2$$

$$\frac{B}{2L_2} = \frac{5}{2(4)} = 0.625$$

According to Figure 8.66b for $2L_1/B = 2$ and $B/2L_2 = 0.625$, $m \approx 0.033$. From Figure 8.67a for $m = 0.033$ and $2L_1/B = 2$, $L_2 i_{exit(max)}/h = 0.54$. Hence

$$i_{exit(max)} = \frac{0.54(h)}{L_2} = 0.54(4.5)/4 = 0.608$$

$$FS = \frac{i_{cr}}{i_{max(exit)}} = \frac{1}{0.608} = 1.645 \qquad\qquad ▲$$

8.25   LATERAL YIELDING OF SHEET PILES AND GROUND SETTLEMENT

In braced cuts, some lateral movement of sheet pile walls may be expected (Figure 8.69, p. 845). The amount of lateral yield depends on several factors, the most important of which is the elapsed time between excavation and placement of wales and struts. Mana and Clough (1981) analyzed the field records of several braced cuts in clay from the San Francisco, Oslo (Norway), Boston, Chicago, and Bowline Point (New York) areas. Under ordinary construction conditions, they found that the maximum lateral wall yield, $\delta_{H(max)}$, has a definite relationship with the factor of

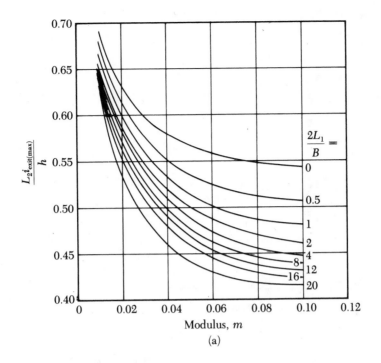

(a)

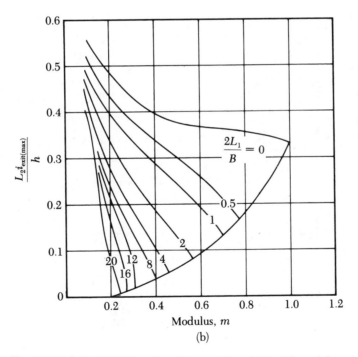

(b)

▼ **FIGURE 8.67** Variation of maximum exit gradient with modulus (from *Groundwater and Seepage,* by M. E. Harr. Copyright © 1962 by McGraw-Hill. Used with permission.)

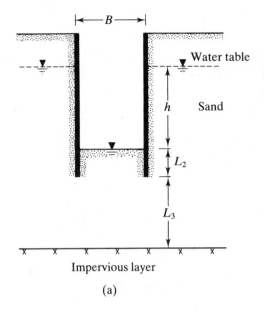

(a)

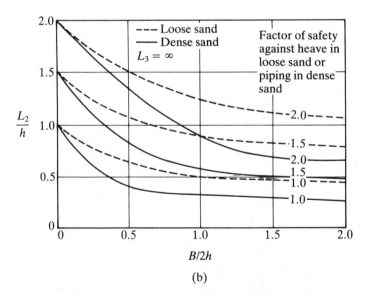

(b)

▼ **FIGURE 8.68**   Influence of seepage on the stability of sheeted excavation (after U.S. Department of the Navy, 1971)

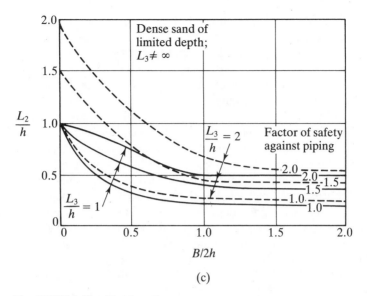

(c)

▼   **FIGURE 8.68**   (Continued)

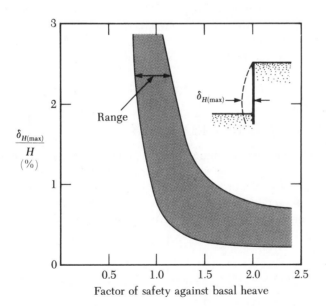

▼   **FIGURE 8.69**   Range of variation of $\delta_{H(max)}/H$ with FS against basal heave from field observation (redrawn after Mana and Clough, 1981)

safety against heave, as shown in Figure 8.69. Note that the factor of safety against heave plotted in Figure 8.69 was calculated by using Eqs. (8.116) and (8.117).

As discussed before, in several instances the sheet piles (or the soldier piles, as the case may be) are driven to a certain depth below the bottom of the excavation. The reason is to reduce the lateral yielding of the walls during the last stages of excavation. Lateral yielding of the walls will cause the ground surface surrounding the cut to settle. The degree of lateral yielding, however, depends mostly on the soil type below the bottom of the cut. If clay below the cut extends to a great depth and $\gamma H/c$ is less than about 6, extension of the sheet piles or soldier piles below the bottom of the cut will help considerably in reducing the lateral yield of the walls.

However, under similar circumstances, if $\gamma H/c$ is about 8, the extension of sheet piles into the clay below the cut does not help greatly. In such circumstances, we may expect a great degree of wall yielding that may result in the total collapse of the bracing systems. If a hard soil layer lies below a clay layer at the bottom of the cut, the piles should be embedded in the stiffer layer. This action will greatly reduce lateral yield.

The lateral yielding of walls will generally induce ground settlement, δ_V, around a braced cut, which is generally referred to as *ground loss*. Based on several field observations, Peck (1969) provided curves for predicting ground settlement in various types of soil (see Figure 8.70). The magnitude of ground loss varies extensively; however, Figure 8.70 may be used as a general guide.

Based on the field data obtained from various cuts in the areas of San Francisco, Oslo, and Chicago, Mana and Clough (1981) provided a correlation between the maximum lateral yield of sheet piles, $\delta_{H(max)}$, and the maximum ground settlement, $\delta_{V(max)}$. It is shown in Figure 8.71. Note that

$$\delta_{V(max)} \approx 0.5\delta_{H(max)} \qquad \text{to} \qquad 1.0\delta_{H(max)} \tag{8.127}$$

8.26 CASE STUDIES OF BRACED CUTS

The procedure for determining strut loads and the design of sheet piles and wales presented in the preceding sections appears to be fairly straightforward. It is, however, only possible if a proper pressure envelope is chosen for the design, which is difficult. This section describes some case studies of braced cuts and highlights the difficulties and degree of judgment needed for successful completion of various projects.

A. 12th Street Station in Oakland—San Francisco Bay Area Rapid Transit (BART) System

Armento (1972) provided the lateral pressure measurements for the braced cut constructed for the 12th Street Station of the San Francisco Bay Area Rapid Transit (BART) system. The soil conditions principally consisted of medium dense to dense brown clayey sands and stiff to very stiff sandy clays. Figures 8.72a and 8.72b show the assumed pressure envelopes and the measured average pressure in Zone 12-1

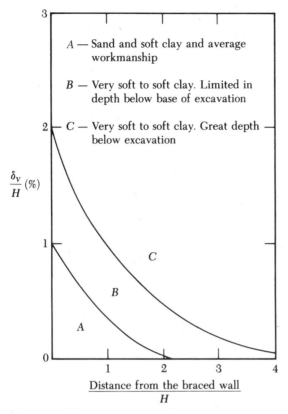

A — Sand and soft clay and average workmanship

B — Very soft to soft clay. Limited in depth below base of excavation

C — Very soft to soft clay. Great depth below excavation

▼ **FIGURE 8.70** Variation of ground settlement with distance (after Peck, 1969)

and 12-2. From the figures, it can be seen that (a) the measured pressure does not follow a pattern of linear variation with depth, and (b) at some depths, the measured pressure exceeded that assumed in the pressure envelope.

B. Subway Extension of the Massachusetts Bay Transportation Authority (MBTA)

Lambe (1970) provided data on the performance of three excavations for the subway extension of the MBTA in Boston (test sections A, B, and D), all of which were well instrumented. Figure 8.73 gives the details of test section B, where the cut was 58 ft ($\approx$ 18 m), including subsoil conditions. The subsoil consisted of gravel, sand, silt, and clay (fill) to a depth of about 26 ft (8 m), followed by a light gray, slightly organic silt to a depth of 46 ft (14 m). A layer of coarse sand and gravel with some clay was present from 46 ft to 54 ft (14 m to 16.5 m) below the ground surface.

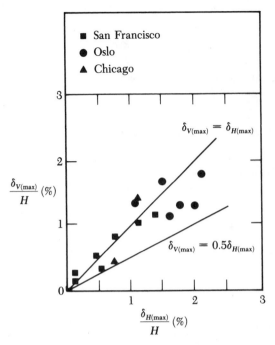

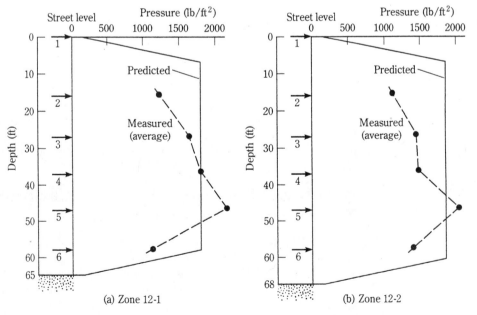

▼ FIGURE 8.71 Variation of maximum lateral yield with maximum ground settlement (after Mana and Clough, 1981)

▼ FIGURE 8.72 Lateral pressure measurements for braced cuts for construction of 12th Street Station of BART (after Armento, 1972)

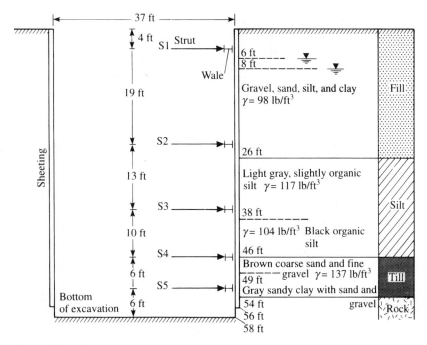

Rock was encountered below 54 ft (16.5 m). The details of the struts and wales are given below. The horizontal spacing of the struts was 12 ft (3.66 m) center-to-center.

Strut level	Depth (ft)	Section of wales	Section of struts
S1	4	14BP89	36WF150
S2	23	27WF160	14WF153
S3	36	27WF177	14WF167
S4	46	27WF160	14WF158
S5	52	27WF177	14WF167

Because the apparent pressure envelopes available (Section 8.21) are for *sand* and *clay* only, questions may arise about how to treat the fill, silt, and till. Figure 8.74 shows the apparent pressure envelopes proposed by Peck (1969), considering the soil as *sand* and also as *clay*, to overcome that problem. For the average soil parameters of the profile, the following values of p_a were used to develop the pressure envelopes shown in Figure 8.74.

Sand

$$p_a = 0.65\gamma H K_a \tag{8.106}$$

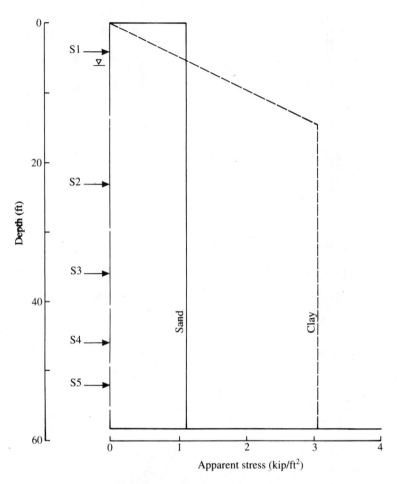

▼ **FIGURE 8.74**   Apparent pressure envelopes for test section B, MBTA (after Lambe, 1970)

For $\gamma = 114$ lb/ft³, $H = 58$ ft, and $K_a = 0.26$,

$$p_a = (0.65)(114)(58)(0.26) = 1117 \text{ lb/ft}^2 \approx 1.12 \text{ kip/ft}^2$$

Clay

$$p_a = \gamma H \left[1 - \left(\frac{4c}{\gamma H} \right) \right] \tag{8.107}$$

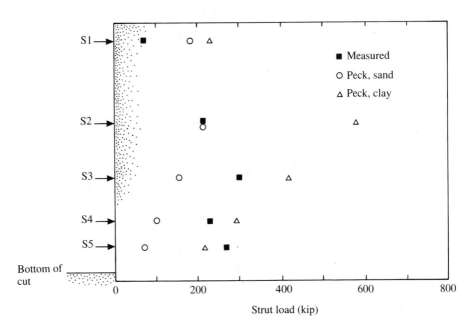

▼ **FIGURE 8.75**   Measured and predicted strut loads, test section B, MBTA (after Lambe, 1970)

For c = 890 lb/ft²,

$$p_a = (114)(58)\left[1 - \frac{(4)(890)}{(114)(58)}\right] = 3052 \text{ lb/ft}^2 \approx 3.05 \text{ kip/ft}^2$$

Figure 8.75 shows the variations of the strut load, based on the assumed pressure envelopes shown in Figure 8.74. Also shown in Figure 8.75 are the measured strut loads in the field and the design strut loads. This comparison indicates that

1. In most cases the measured strut loads differed widely from those predicted. This result is due primarily to the uncertainties involved in the assumption of the soil parameters.
2. The actual design strut loads were substantially higher than those measured.

Figure 8.76 shows test section A for the subway extension along with the soil profile. The measured and predicted strut loads for this section (similar to Figure 8.75) are shown in Figure 8.77. Considering the uncertainties involved, the general agreement between the measured and predicted values is quite good.

Figure 8.78 shows soil movements in the vicinity of sections A and B in nondimensional form (δ_V/H and δ_H/H). They appear to be in general agreement with those presented in Section 8.25.

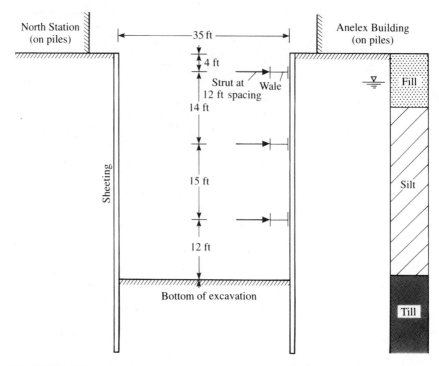

▼ **FIGURE 8.76** Test section A for subway extension, MBTA (after Lambe, 1970)

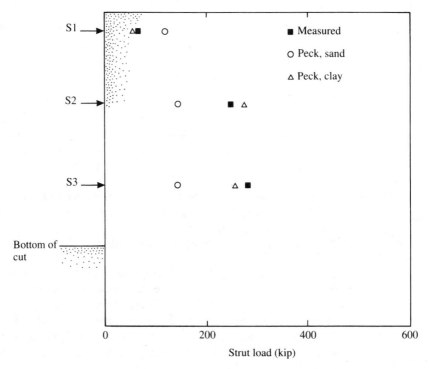

▼ **FIGURE 8.77** Measured and predicted strut loads, test section A, MBTA (after Lambe, 1970)

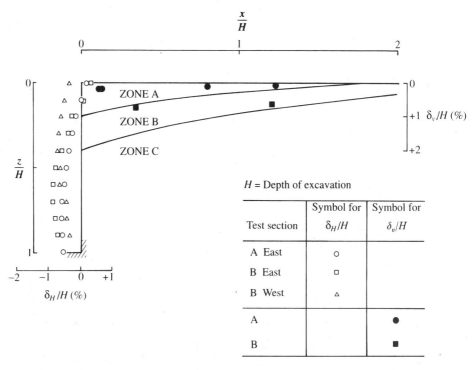

▼ **FIGURE 8.78** Variation of δ_H/H and δ_v/H in the vicinity of test sections A and B, MBTA (after Lambe, 1970)

C. Construction of National Plaza (South Half) in Chicago

The construction of the south half of the National Plaza in Chicago required a braced cut 70 ft (≈21 m) deep. Swatek et al. (1972) reported the case history for this construction. Figure 8.79 shows a schematic diagram for the braced cut and the subsoil profile. There were six levels of struts. The following table gives the actual maximum wale and strut loads.

Strut level	Elevation (ft)	Load measured (kip/ft)
A	+3	16.0
B	−6	26.5
C	−15	29.0
D	−24.5	29.0
E	−34	29.0
F	−44.5	30.7
		Σ 160.2

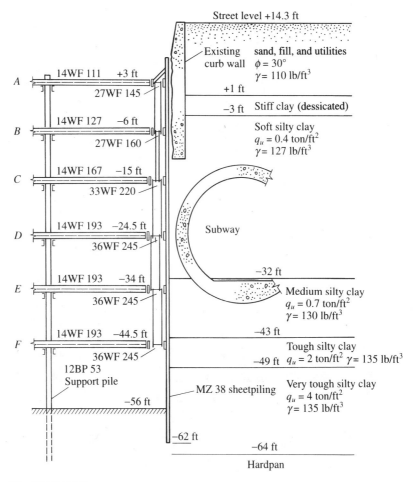

▼ FIGURE 8.79   Schematic diagram of braced cut and subsoil profile, National Plaza of Chicago (after Swatek et al., 1972)

Figure 8.80 presents a lateral earth pressure envelope based on the maximum wale loads measured. To compare the theoretical prediction to the actual observation requires making an approximate calculation. To do so, we convert the clayey soil layers from Elevation $+1$ ft to -56 ft to a single equivalent layer by using Eq. (8.111).

Elevation (ft)	Thickness, H_i (ft)	c (lb/ft²)	Equivalent c (lb/ft²)
+1 to −32 ft	33	400	$c_{av} = \dfrac{1}{57}\left[(33)(400) + (11)(700) + (6)(2000) + (7)(4000)\right]$
−32 ft to −43 ft	11	700	
− 43 ft to −49 ft	6	2000	$= 1068 \text{ lb/ft}^2$
−49 ft to −56 ft	7	4000	
	Σ 57		

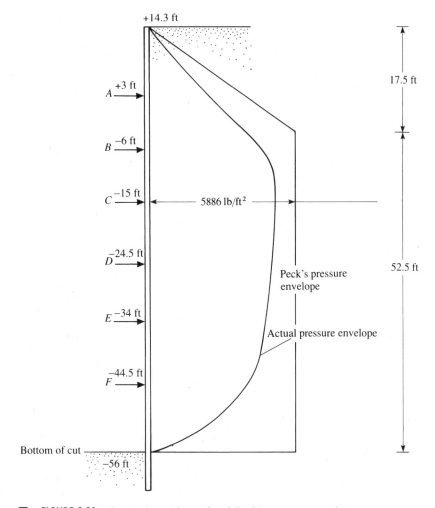

▼ FIGURE 8.80 Comparison of actual and Peck's pressure envelopes

Now, using Eq. (8.109), we can convert the sand layer located between elevations +14 ft and +1 ft and the equivalent clay layer of 57 ft to one equivalent clay layer with a thickness of 70 ft:

$$c_{av} = \frac{1}{2H} [\gamma_s K_s H_s^2 \tan \phi_s + (H - H_s) n' q_u]$$

$$= \left[\frac{1}{(2)(70)} \right] [(110)(1)(13)^2 \tan 30 + (57)(0.75)(2 \times 1068)] \approx 730 \, lb/ft^2$$

Equation (8.112) gives

$$\gamma_{av} = \frac{1}{H}[\gamma_1 H_1 + \gamma_2 H_2 + \cdots + \gamma_n H_n)$$

$$= \frac{1}{70}[(110)(13) + (127)(33) + (130)(11) + (135)(6) + (135)(7)]$$

$$= 125.8 \text{ lb/ft}^3$$

For the equivalent clay layer of 70 ft,

$$\frac{\gamma_{av} H}{c_{av}} = \frac{(125.8)(70)}{730} = 12.06 > 4$$

Hence the apparent pressure envelope will be of the type shown in Figure 8.51. From Eq. (8.107)

$$p_a = \gamma H\left[1 - \left(\frac{4c_{av}}{\gamma_{av} H}\right)\right] = (125.8)(70)\left[1 - \frac{(4)(730)}{(125.8)(70)}\right] = 5886 \text{ lb/ft}^2$$

The pressure envelope is shown in Figure 8.80. The area of this pressure diagram is 201 kip/ft. Thus Peck's pressure envelope gives a lateral earth pressure of about 1.8 times that actually observed. This result is not surprising because the pressure envelope provided by Figure 8.51 is an envelope developed considering several cuts made at different locations. Under actual field conditions, past experience with the behavior of similar soils can help reduce overdesigning substantially.

PROBLEMS

8.1   Figure P8.1 shows a cantilever sheet pile wall penetrating a granular soil. Here, $L_1 = 8$ ft, $L_2 = 15$ ft, $\gamma = 100$ lb/ft³, $\gamma_{sat} = 110$ lb/ft³, $\phi = 35°$.
   a.   What is the theoretical depth of embedment, D?
   b.   For a 30% increase in D, what should be the total length of the sheet piles?
   c.   Determine the theoretical maximum moment of the sheet pile.

8.2   Solve Problem 8.1 with the following: $L_1 = 4$ m, $L_2 = 8$ m, $\gamma = 16.1$ kN/m³, $\gamma_{sat} = 18.2$ kN/m³, $\phi = 32°$.

8.3   Redo Problem 8.1 with the following: $L_1 = 3$ m, $L_2 = 6$ m, $\gamma = 17.3$ kN/m³, $\gamma_{sat} = 19.4$ kN/m³, $\phi = 30°$.

8.4   Refer to Figure 8.9. Given: $L = 15$ ft, $\gamma = 108$ lb/ft³, and $\phi = 35°$. Calculate the theoretical depth of penetration, D, and the maximum moment.

8.5   Repeat Problem 8.4 with the following: $L = 3$ m, $\gamma = 16.7$ kN/m³, $\phi = 30°$.

8.6   Refer to Figure 8.10, which shows a free cantilever sheet pile wall penetrating a sand layer. Determine the theoretical depth of penetration and the maximum moment. Given: $\gamma = 17$ kN/m³, $\phi = 36°$, $L = 4$ m, and $P = 15$ kN/m.

8.7   Refer to Figure P8.7, for which $L_1 = 2.4$ m, $L_2 = 4.6$ m, $\gamma = 15.7$ kN/m³, $\gamma_{sat} = 17.3$ kN/m³, $c = 29$ kN/m², $\phi = 30°$.
   a.   Find the theoretical depth of penetration, D.
   b.   Increase D by 40%. What length of sheet piles is needed?
   c.   Determine the theoretical maximum moment in the sheet pile.

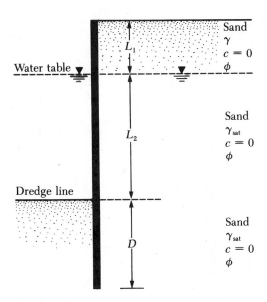

▼ **FIGURE P8.1**

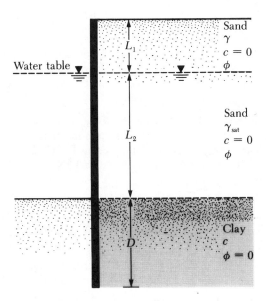

▼ **FIGURE P8.7**

8.8 Solve Problem 8.7 with the following: $L_1 = 5$ ft, $L_2 = 10$ ft, $\gamma = 108$ lb/ft³, $\gamma_{sat} = 122.4$ lb/ft³, $\phi = 36°$, and $c = 800$ lb/ft².

8.9 Refer to Figure 8.14. Here, $L = 4$ m, $P = 8$ kN/m, $c = 45$ kN/m², and $\gamma_{sat} = 19.2$ kN/m³. Find the theoretical value of D and the maximum moment.

8.10 An anchored sheet pile bulkhead is shown in Figure P8.10. Given: $L_1 = 5$ m, $L_2 = 8$ m, $l_1 = 2$ m, $\gamma = 16.5$ kN/m³, $\gamma_{sat} = 18.5$ kN/m³, and $\phi = 32°$.

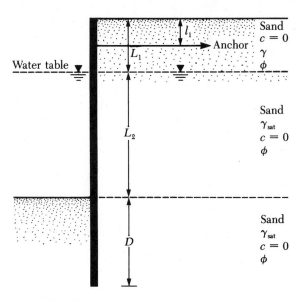

▼ **FIGURE P8.10**

a. Calculate the theoretical value of the depth of embedment, D.
b. Draw the pressure distribution diagram.
c. Determine the anchor force per unit length of the wall.

Use the free earth support method.

8.11 Refer to Problem 8.10. Assume that $D_{actual} = 1.25 D_{theory}$.
a. Determine the theoretical maximum moment.
b. Choose a sheet pile section by using Rowe's moment reduction technique. Use $E = 207 \times 10^3$ MN/m², $\sigma_{all} = 210,000$ kN/m².

8.12 Redo Problem 8.10 with the following: $L_1 = 10$ ft, $L_2 = 25$ ft, $l_1 = 4$ ft, $\gamma = 120$ lb/ft³, $\gamma_{sat} = 129.4$ lb/ft³, and $\phi = 40°$.

8.13 Refer to Problem 8.12. Assume that $D_{actual} = 1.3 D_{theoretical}$. Using Rowe's moment reduction method, choose a sheet pile section. Use $E = 29 \times 10^6$ lb/in² and $\sigma_{all} = 25$ kip/in².

8.14 Refer to Figure P8.10, for which $L_1 = 3$ m, $L_2 = 6$ m, $l_1 = 1.5$ m, $\gamma = 17.5$ kN/m³, $\gamma_{sat} = 19.5$ kN/m³, and $\phi = 35°$. Use the computational diagram method (Section 8.13) to determine D, F, and M_{max}. Assume that $C = 0.68$ and $R = 0.6$.

8.15 Refer to Figure P8.10. Given: $L_1 = 4$ m, $L_2 = 8$ m, $l_1 = l_2 = 2$ m, $\gamma = 16$ kN/m³, $\gamma_{sat} = 18.5$ kN/m³, and $\phi = 35°$.
Determine:
a. Theoretical depth of penetration
b. Anchor force per unit length
c. Maximum moment in the sheet pile. Use the charts presented in Section 8.10.

8.16 An anchored sheet pile bulkhead is shown in Figure P8.16. Given: $L_1 = 2$ m, $L_2 = 6$ m, $l_1 = 1$ m, $\gamma = 16$ kN/m³, $\gamma_{sat} = 18.86$ kN/m³, $\phi = 32°$, and $c = 27$ kN/m².

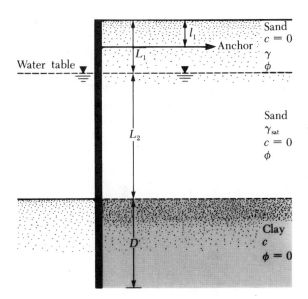

▼ FIGURE P8.16

a. Determine the theoretical depth of embedment, D.
b. Calculate the anchor force per unit length of the sheet pile wall.

Use the free earth support method.

8.17 Repeat Problem 8.16 with the following: $L_1 = 8$ ft, $L_2 = 20$ ft, $l_1 = 4$ ft, $c = 1500$ lb/ft², $\gamma = 115$ lb/ft³, $\gamma_{sat} = 128$ lb/ft³, and $\phi = 40°$.

8.18 Refer to Figure P8.10. Given: $L_1 = 9$ ft, $L_2 = 26$ ft, $l_1 = 5$ ft, $\gamma = 108.5$ lb/ft³, $\gamma_{sat} = 128.5$ lb/ft³, and $\phi = 35°$. Use the fixed earth support method to determine the following:
a. Maximum moment
b. Theoretical depth of penetration
c. Anchor force per unit length of sheet pile wall

8.19 Refer to Figure 8.42a. For the anchor slab in sand, $H = 5$ ft, $h = 3$ ft, $B = 4$ ft, $S' = 7$ ft, $\phi = 30°$, and $\gamma = 110$ lb/ft³. Calculate the ultimate holding capacity of each anchor. The anchor plates are made of concrete and have a thickness of 3 in. Use $\gamma_{concrete} = 150$ lb/ft³. Use Ovesen and Stromann's method.

8.20 A single anchor slab is shown in Figure P8.20. Here, $H = 0.9$ m, $h = 0.3$ m, $\gamma = 17$ kN/m³, and $\phi = 32°$. Calculate the ultimate holding capacity of the anchor slab if the width B is (a) 0.3 m, (b) 0.6 m, and (c) 0.9 m. (*Note:* center-to-center spacing, $S' = \infty$.) Use the empirical correlation given in Section 8.17.

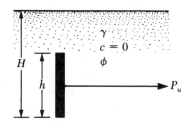

▼ FIGURE P8.20

8.21   Refer to Problem 8.20. Estimate the probable horizontal displacement of the anchors at ultimate load. Also estimate the resistance of anchors if the allowable displacement is 20 mm. Use the empirical correlation method given in Section 8.17.

8.22   A vertical anchor plate in clay has $B = 3$ ft, $H = 3$ ft, $h = 3$ ft, and $c = 700$ lb/ft^2. Estimate the ultimate resistance of the anchor.

8.23   Refer to the braced cut shown in Figure P8.23. Given: $\gamma = 17$ kN/m^3, $\phi = 35°$, and $c = 0$. The struts are located at 3 m center-to-center in the plan. Draw the earth pressure envelope and determine the strut loads at levels A, B, and C.

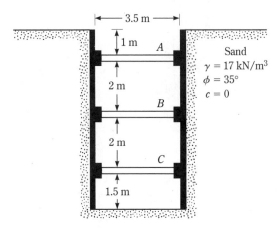

▼   **FIGURE P8.23**

8.24   For the braced cut described in Problem 8.23, determine the following:
   a.   The sheet pile section
   b.   The section modulus of the wales at level B

   Assume that $\sigma_{all} = 170$ MN/m^2.

8.25   Redo Problem 8.23 with $\gamma = 18$ kN/m^3, $\phi = 40°$, $c = 0$, and center-to-center strut spacing in the plan $= 4$ m.

8.26   Determine the sheet pile section required for the braced cut described in Problem 8.25. Given: $\sigma_{all} = 170$ MN/m^2.

8.27   Refer to Figure 8.53a. For the braced cut, given: $H = 8$ m; $H_s = 3$ m; $\gamma_s = 17.5$ kN/m^3; angle of friction of sand, $\phi_s = 34°$; $H_c = 5$ m; $\gamma_c = 18.2$ kN/m^3; and unconfined compression strength of clay layer, $q_u = 55$ kN/m^2.
   a.   Estimate the average cohesion (c_{av}) and average unit weight (γ_{av}) for the construction of the earth pressure envelope.
   b.   Plot the earth pressure envelope.

8.28   Refer to Figure 8.53b, which shows a braced cut in clay. Given: $H = 25$ ft, $H_1 = 5$ ft, $c_1 = 2125$ lb/ft^2, $\gamma_1 = 111$ lb/ft^3, $H_2 = 10$ ft, $c_2 = 1565$ lb/ft^2, $\gamma_2 = 107$ lb/ft^3, $H_3 = 10$ ft, $c_3 = 1670$ lb/ft^2, and $\gamma_3 = 109$ lb/ft^3.
   a.   Determine the average cohesion (c_{av}) and average unit weight (γ_{av}) for calculation of the earth pressure envelope.
   b.   Plot the earth pressure envelope.

8.29 Refer to Figure P8.29. Given: $\gamma = 118$ lb/ft³, $c = 800$ lb/ft², and center-to-center spacing of struts in the plan $= 12$ ft. Draw the earth pressure envelope and determine the strut loads at levels A, B, and C.

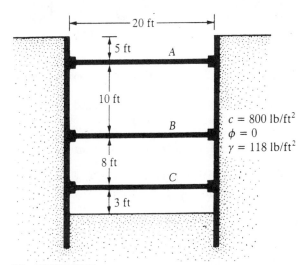

▼ **FIGURE P8.29**

8.30 Determine the sheet pile section modulus for the braced cut described in Problem 8.29. Use $\sigma_{all} = 25,000$ lb/in.².

8.31 Redo Problem 8.29 assuming that $c = 600$ lb/ft².

8.32 Determine the factor of safety against bottom heave for the braced cut described in Problem 8.29. Use Eqs. (8.116) and (8.118). For Eq. (8.118), assume the length of the cut, $L = 30$ ft.

8.33 Refer to Figure 8.62a. Given: $H = 8$ m, $B = 3$ m, $L = 8$ m, $\gamma = 17.8$ kN/m³, $c_1 = 30$ kN/m², $c_2 = 45$ kN/m², and $h' = 2$ m. Determine the factor of safety against bottom heave.

8.34 Determine the factor of safety against bottom heave for the braced cut described in Problem 8.31. Use Eq. (8.120). The length of the cut is 40 ft.

REFERENCES

Akinmusuru, J. O. (1978). "Horizontally loaded vertical plate anchor in sand," *Journal of the Geotechnical Engineering Division,* ASCE, Vol. 104, No. 2, pp. 283–286.

Armento, W. J. (1972). "Criteria for Lateral Pressures for Braced Cuts," *Proceedings,* Specialty Conference on Performance of Earth and Earth-Supported Structures, ASCE, Vol. 1, Part 2, pp. 1283–1302.

Bjerrum, L., and Eide, O. (1956). "Stability of Strutted Excavation in Clay," *Geotechnique,* Vol. 6, No. 1, pp. 32–47.

Blum, H. (1931). *Einspannungsverhältnisse bei Bohlwerken,* W. Ernst und Sohn, Berlin, Germany.

Casagrande, L. (1973). "Comments on Conventional Design of Retaining Structures," *Journal of the Soil Mechanics and Foundations Division,* ASCE, Vol. 99, No. SM2, pp. 181–198.

Cornfield, G. M. (1975). "Sheet Pile Structures," in *Foundation Engineering Handbook,* ed. H. F. Wintercorn and H. Y. Fang, Van Nostrand Reinhold, New York, pp. 418–444.

Das, B. M. (1975). "Pullout Resistance of Vertical Anchors," *Journal of the Geotechnical Engineering Division,* American Society of Civil Engineers, Vol. 101, No. GT1, pp. 87–91.

Das, B. M., and Seeley, G. R. (1975). "Load-Displacement Relationships for Vertical Anchor Plates," *Journal of the Geotechnical Engineering Division,* American Society of Civil Engineers, Vol. 101, No. GT7, pp. 711–715.

Das, B. M., Tarquin, A. J., and Moreno, R. (1985). "Model Tests for Pullout Resistance of Vertical Anchor in Clay," *Journal of Civil Engineering for Practicing and Design Engineers,* Vol. 4, No. 2, pp. 191–209.

Dickin, E. A., and Leung, C. F. (1983). "Centrifugal Model Tests on Vertical Anchor Plates," *Journal of Geotechnical Engineering,* ASCE, Vol. 109, No. 12, pp. 1503–1525.

Ghaly, A. M. (1997). "Load-Displacement Prediction for Horizontally Loaded Vertical Plates," *Journal of Geotechnical and Geoenvironmental Engineering,* ASCE, Vol. 123, No. 1, pp. 74–76.

Hagerty, D. J., and Nofal, M. M. (1992). "Design Aids: Anchored Bulkheads in Sand," *Canadian Geotechnical Journal,* Vol. 29, No. 5, pp. 789–795.

Harr, M. E. (1962). *Groundwater and Seepage,* McGraw-Hill, New York.

Hueckel, S. (1957). "Model Tests on Anchoring Capacity of Vertical and Inclined Plates," *Proceedings,* Fourth International Conference on Soil Mechanics and Foundation Engineering, Butterworth Scientific Publications, London, England, Vol. 2, pp. 203–206.

Lambe, T. W. (1970). "Braced Excavations," *Proceedings of the Specialty Conference on Lateral Stresses in the Ground and Design of Earth-Retaining Structures,* American Society of Civil Engineers, pp. 149–218.

Littlejohn, G. S. (1970). "Soil Anchors," *Proceedings,* Conference on Ground Engineering, Institute of Civil Engineers, London, pp. 33–44.

Mackenzie, T. R. (1955). *Strength of Deadman Anchors in Clay,* M. S. Thesis, Princeton University, Princeton, N.J.

Mana, A. I., and Clough, G. W. (1981). "Prediction of Movements for Braced Cuts in Clay," *Journal of the Geotechnical Engineering Division,* American Society of Civil Engineers, Vol. 107, No. GT8, pp. 759–777.

Marsland, A. (1958). "Model Experiments to Study the Influence of Seepage on the Stability of a Sheeted Excavation in Sand," *Geotechnique,* London, Vol. 3, p. 223.

Nataraj, M. S., and Hoadley, P. G. (1984). "Design of Anchored Bulkheads in Sand," *Journal of Geotechnical Engineering,* American Society of Civil Engineers, Vol. 110, No. GT4, pp. 505–515.

Neeley, W. J., Stuart, J. G., and Graham, J. (1973). "Failure Loads of Vertical Anchor Plates in Sand," *Journal of the Soil Mechanics and Foundations Division,* American Society of Civil Engineers, Vol. 99, No. SM9, pp. 669–685.

Ovesen, N. K., and Stromann, H. (1972). "Design Methods for Vertical Anchor Slabs in Sand," *Proceedings,* Specialty Conference on Performance of Earth and Earth-Supported Structures, American Society of Civil Engineers, Vol. 2.1, pp. 1481–1500.

Peck, R. B. (1943). "Earth Pressure Measurements in Open Cuts, Chicago (Ill.) Subway," *Transactions,* American Society of Civil Engineers, Vol. 108, pp. 1008–1058.

Peck, R. B. (1969). "Deep Excavation and Tunneling in Soft Ground," *Proceedings,* Seventh International Conference on Soil Mechanics and Foundation Engineering, Mexico City, State-of-the-Art Volume, pp. 225–290.

Reddy, A. S., and Srinivasan, R. J. (1967). "Bearing Capacity of Footing on Layered Clay," *Journal of the Soil Mechanics and Foundations Division,* American Society of Civil Engineers, Vol. 93, No. SM2, pp. 83–99.

Rowe, P. W. (1952). "Anchored Sheet Pile Walls," *Proceedings,* Institute of Civil Engineers, Vol. 1, Part 1, pp. 27–70.

Rowe, P. W. (1957). "Sheet Pile Walls in Clay," *Proceedings,* Institute of Civil Engineers, Vol. 7, pp. 654–692.

Schroeder, W. L., and Roumillac, P. (1983). "Anchored Bulkheads with Sloping Dredge Lines," *Journal of Geotechnical Engineering,* American Society of Civil Engineers, Vol. 109, No. 6, pp. 845–851.

Smith, J. E. (1962). "Deadman Anchorages in Sand," *Technical Report No. R-199,* U.S. Naval Civil Engineering Laboratory, Washington, D.C.

Swatek, E. P., Jr., Asrow, S. P., and Seitz, A. (1972). "Performance of Bracing for Deep Chicago Excavation," *Proceeding of the Specialty Conference on Performance of Earth and Earth Supported Structures,* American Society of Civil Engineers, Vol. 1, Part 2, pp. 1303–1322.

Teng, W. C. (1962). *Foundation Design,* Prentice-Hall, Englewood Cliffs, N.J.

Terzaghi, K. (1943). *Theoretical Soil Mechanics,* Wiley, New York.

Tschebotarioff, G. P. (1973). *Foundations, Retaining and Earth Structures,* 2nd ed., McGraw-Hill, New York.

Tsinker, G. P. (1983). "Anchored Street Pile Bulkheads: Design Practice," *Journal of Geotechnical Engineering,* American Society of Civil Engineers, Vol. 109, No. GT8, pp. 1021–1038.

U.S. Department of the Navy (1971). "Design Manual—Soil Mechanics, Foundations, and Earth Structures,"—NAVFAC DM-7, Washington, D.C.

CHAPTER NINE

PILE FOUNDATIONS

9.1 INTRODUCTION

Piles are structural members that are made of steel, concrete, and/or timber. They are used to build pile foundations, which are deep and which cost more than shallow foundations (Chapters 3 and 4). Despite the cost, the use of piles often is necessary to ensure structural safety. The following list identifies some of the conditions that require pile foundations (Vesic, 1977).

1. When the upper soil layer(s) is (are) highly compressible and too weak to support the load transmitted by the superstructure, piles are used to transmit the load to underlying bedrock or a stronger soil layer, as shown in Figure 9.1a. When bedrock is not encountered at a reasonable depth below the ground surface, piles are used to transmit the structural load to the soil gradually. The resistance to the applied structural load is derived mainly from the frictional resistance developed at the soil-pile interface (Figure 9.1b).

2. When subjected to horizontal forces (see Figure 9.1c), pile foundations resist by bending while still supporting the vertical load transmitted by the superstructure. This type of situation is generally encountered in the design and construction of earth-retaining structures and foundations of tall structures that are subjected to high wind and/or earthquake forces.

3. In many cases, expansive and collapsible soils (Chapter 11) may be present at the site of a proposed structure. These soils may extend to a great depth below the ground surface. Expansive soils swell and shrink as the moisture content increases and decreases, and the swelling pressure of such soils can be considerable. If shallow foundations are used in such circumstances, the structure may suffer considerable damage. However, pile foundations may be considered as an alternative when piles are extended beyond the active zone, which swells and shrinks (Figure 9.1d).

    Soils such as loess are collapsible in nature. When the moisture content of these soils increases, their structures may break down. A sudden decrease in the void ratio of soil induces large settlements

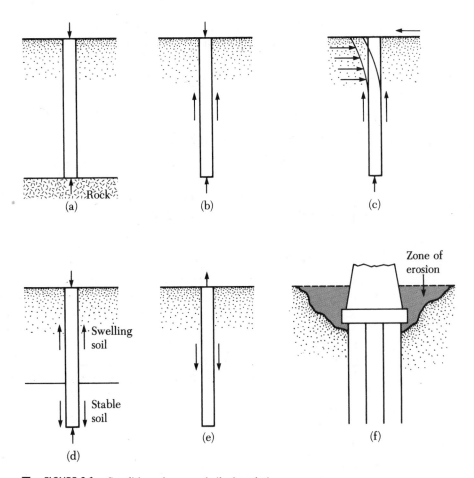

▼ **FIGURE 9.1** Conditions for use of pile foundations

of structures supported by shallow foundations. In such cases, pile foundations may be used in which piles are extended into stable soil layers beyond the zone of possible moisture change.

4. Foundations of some structures, such as transmission towers, offshore platforms, and basement mats below the water table, are subjected to uplifting forces. Piles are sometimes used for these foundations to resist the uplifting force (Figure 9.1e).

5. Bridge abutments and piers are usually constructed over pile foundations to avoid the possible loss of bearing capacity that a shallow foundation might suffer because of soil erosion at the ground surface (Figure 9.1f).

Although numerous investigations, both theoretical and experimental, have been conducted in the past to predict the behavior and the load-bearing capacity of piles in granular and cohesive soils, the mechanisms are not yet entirely understood and may never be. The design of pile foundations may be considered somewhat of an "art" as a result of the uncertainties involved in working with some subsoil conditions. This chapter discusses the present state of the art for design and analysis of pile foundations.

9.2 TYPES OF PILES AND THEIR STRUCTURAL CHARACTERISTICS

Different types of piles are used in construction work, depending on the type of load to be carried, the subsoil conditions, and the location of the water table. Piles can be divided into the following categories: (a) steel piles, (b) concrete piles, (c) wooden (timber) piles, and (d) composite piles.

Steel Piles

Steel piles generally are either *pipe piles* or *rolled steel* H-*section piles*. Pipe piles can be driven into the ground with their ends open or closed. Wide-flange and I-section steel beams can also be used as piles. However, H-section piles are usually preferred because their web and flange thicknesses are equal. In wide-flange and I-section beams, the web thicknesses are smaller than the thicknesses of the flange. Table D.1 (Appendix D) gives the dimensions of some standard H-section steel piles used in the United States. Table D.2 (Appendix D) shows selected pipe sections frequently used for piling purposes. In many cases, the pipe piles are filled with concrete after driving.

The allowable structural capacity for steel piles is

$$Q_{all} = A_s f_s \tag{9.1}$$

where A_s = cross-sectional area of the steel
f_s = allowable stress of steel

Based on geotechnical considerations (once the design load for a pile is fixed) determining whether $Q_{(design)}$ is within the allowable range as defined by Eq. (9.1) is always advisable.

When necessary, steel piles are spliced by welding or by riveting. Figure 9.2a shows a typical condition of splicing by welding for an H-pile. A typical case of splicing by welding for a pipe pile is shown in Figure 9.2b. Figure 9.2c shows a diagram of splicing an H-pile by rivets or bolts.

When hard driving conditions are expected, such as driving through dense gravel, shale, and soft rock, steel piles can be fitted with driving points or shoes. Figure 9.2d and 9.2e are diagrams of two types of shoe used for pipe piles.

Steel piles may be subject to corrosion. For example, swamps, peats, and other organic soils are corrosive. Soils that have a pH greater than 7 are not so corrosive. To offset the effect of corrosion, an additional thickness of steel (over the actual design cross-sectional area) is generally recommended. In many circumstances, factory-applied epoxy coatings on piles work satisfactorily against corrosion. These

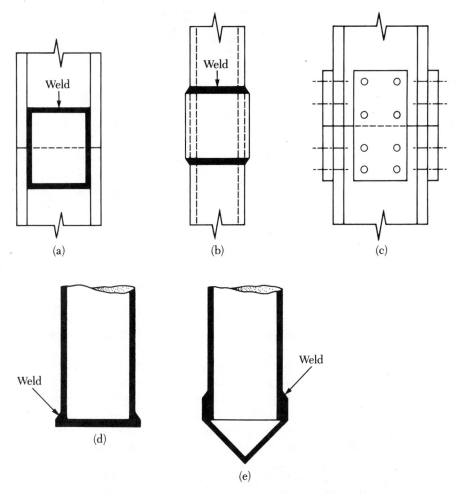

▼ **FIGURE 9.2** Steel piles: (a) splicing of H-pile by welding; (b) splicing of pipe pile by welding; (c) splicing of H-pile by rivets and bolts; (d) flat driving point of pipe pile; (e) conical driving point of pipe pile

coatings are not easily damaged by pile driving. Concrete encasement of steel piles in most corrosive zones also protects against corrosion.

Concrete Piles

Concrete piles may be divided into two basic categories: (a) precast piles and (b) cast-*in-situ* piles. *Precast piles* can be prepared by using ordinary reinforcement, and they can be square or octagonal in cross section (Figure 9.3). Reinforcement is provided to enable the pile to resist the bending moment developed during pickup and transportation, the vertical load, and the bending moment caused by lateral

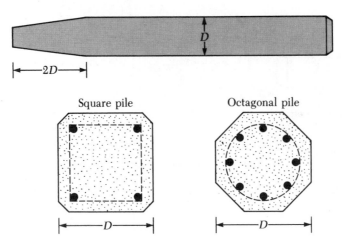

▼ **FIGURE 9.3** Precast piles with ordinary reinforcement

load. The piles are cast to desired lengths and cured before being transported to the work sites.

Precast piles can also be prestressed by the use of high-strength steel prestressing cables. The ultimate strength of these steel cables is about 260 ksi ($\approx$1800 MN/m²). During casting of the piles, the cables are pretensioned to about 130–190 ksi ($\approx$900–1300 MN/m²), and concrete is poured around them. After curing, the cables are cut, thus producing a compressive force on the pile section. Table D.3 (Appendix D) gives additional information about prestressed concrete piles with square and octagonal cross sections.

Cast-in-situ, or *cast-in-place, piles* are built by making a hole in the ground and then filling it with concrete. Various types of cast-in-place concrete pile are currently used in construction, and most of them have been patented by their manufacturers. These piles may be divided into two broad categories: (a) cased and (b) uncased. Both types may have a pedestal at the bottom.

Cased piles are made by driving a steel casing into the ground with the help of a mandrel placed inside the casing. When the pile reaches the proper depth, the mandrel is withdrawn and the casing is filled with concrete. Figure 9.4a, 9.4b, 9.4c, and 9.4d show some examples of cased piles without a pedestal. Table 9.1 gives additional information about these cased piles. Figure 9.4e shows a cased pile with a pedestal. The pedestal is an expanded concrete bulb that is formed by dropping a hammer on fresh concrete.

Figure 9.4f and 9.4g are two types of uncased pile, one with a pedestal and the other without. The uncased piles are made by first driving the casing to the desired depth and then filling it with fresh concrete. The casing is then gradually withdrawn.

The allowable loads for cast-in-place concrete piles are given by the following equations.

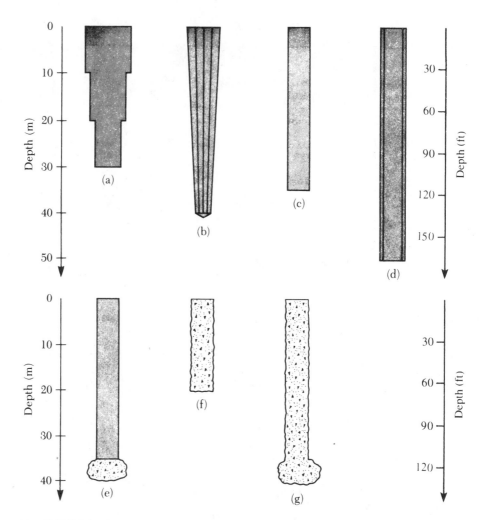

▼ **FIGURE 9.4**   Cast-in-place concrete piles (see Table 9.1 for descriptions)

Cased Pile

$$Q_{\text{all}} = A_s f_s + A_c f_c \tag{9.2a}$$

where A_s = area of cross section of steel
A_c = area of cross section of concrete
f_s = allowable stress of steel
f_c = allowable stress of concrete

Uncased Pile

$$Q_{\text{all}} = A_c f_c \tag{9.2b}$$

▼ **TABLE 9.1** Descriptions of the Cast-in-Place Piles Shown in Figure 9.4

Part in Figure 9.4	Name of pile	Type of casing	Maximum usual depth of pile (ft)	Maximum usual depth of pile (m)
a	Raymond Step-Taper	Corrugated, thin, cylindrical casing	100	30
b	Monotube or Union Metal	Thin, fluted, tapered steel casing driven without mandrel	130	40
c	Western cased	Thin sheet casing	100–130	30–40
d	Seamless pipe or Armco	Straight steel pipe casing	160	50
e	Franki cased pedestal	Thin sheet casing	100–130	30–40
f	Western uncased without pedestal	—	50–65	15–20
g	Franki uncased pedestal	—	100–130	30–40

Timber Piles

Timber piles are tree trunks that have had their branches and bark carefully trimmed off. The maximum length of most timber piles is 30–65 ft (10–20 m). To qualify for use as a pile, the timber should be straight, sound, and without any defects. The American Society of Civil Engineers' *Manual of Practice,* No. 17 (1959), divided timber piles into three classifications:

1. *Class A piles* carry heavy loads. The minimum diameter of the butt should be 14 in. (356 mm).

2. *Class B piles* are used to carry medium loads. The minimum butt diameter should be 12–13 in. (305–330 mm).

3. *Class C piles* are used in temporary construction work. They can be used permanently for structures when the entire pile is below the water table. The minimum butt diameter should be 12 in. (305 mm).

In any case, a pile tip should not have a diameter less than 6 in. (150 mm).

Timber piles cannot withstand hard driving stress; therefore, the pile capacity is generally limited to about 25–30 tons (220–270 kN). Steel shoes may be used to avoid damage at the pile tip (bottom). The tops of timber piles may also be damaged during the driving operation. The crushing of the wooden fibers caused by the impact of the hammer is referred to as *brooming*. To avoid damage to the pile top, a metal band or a cap may be used.

Splicing of timber piles should be avoided, particularly when they are expected to carry tensile load or lateral load. However, if splicing is necessary, it can be done by using *pipe sleeves* (Figure 9.5a) or *metal straps and bolts* (Figure 9.5b). The length of the pipe sleeve should be at least five times the diameter of the pile. The butting ends should be cut square so that full contact can be maintained. The spliced portions

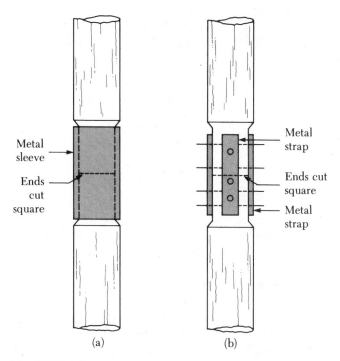

▼ FIGURE 9.5 Splicing of timber piles: (a) use of pipe sleeves;
(b) use of metal straps and bolts

should be carefully trimmed so that they fit tightly to the inside of the pipe sleeve. In the case of metal straps and bolts, the butting ends should also be cut square. Also, the sides of the spliced portion should be trimmed plane for putting the straps on.

Timber piles can stay undamaged indefinitely if they are surrounded by saturated soil. However, in a marine environment timber piles are subject to attack by various organisms and can be damaged extensively in a few months. When located above the water table, the piles are subject to attack by insects. The life of the piles may be increased by treating them with preservatives such as creosote.

The allowable load-carrying capacity of wooden piles is

$$Q_{all} = A_p f_w \tag{9.3}$$

where A_p = average area of cross section of the pile
   f_w = allowable stress for the timber

The following allowable stresses are for pressure-treated round timber piles made from Pacific Coast Douglas fir and Southern pine, when used in hydraulic structures (ASCE, 1993).

Allowable stress	Pacific Coast Douglas fir	Southern pine
Compression parallel to grain	875 lb/in² (6.04 MN/m²)	825 lb/in² (5.7 MN/m²)
Bending	1700 lb/in² (11.7 MN/m²)	1650 lb/in² (11.4 MN/m²)
Horizontal shear	95 lb/in² (0.66 MN/m²)	90 lb/in² (0.62 MN/m²)
Compression perpendicular to grain	190 lb/in² (1.31 MN/m²)	205 lb/in² (1.41 MN/m²)

Composite Piles

The upper and lower portions of *composite piles* are made of different materials. For example, composite piles may be made of steel and concrete or timber and concrete. Steel and concrete piles consist of a lower portion of steel and an upper portion of cast-in-place concrete. This type of pile is the one used when the length of the pile required for adequate bearing exceeds the capacity of simple cast-in-place concrete piles. Timber and concrete piles usually consist of a lower portion of timber pile below the permanent water table and an upper portion of concrete. In any case, forming proper joints between two dissimilar materials is difficult, and, for that reason, composite piles are not widely used.

Comparison of Pile Types

Several factors affect the selection of piles for a particular structure at a specific site. Table 9.2 gives a brief comparison of the advantages and disadvantages of the various types of pile based on the pile material.

▼ **TABLE 9.2**   Comparisons of Piles Made of Different Materials

Pile type	Usual length of piles	Maximum length of pile	Usual load	Approximate maximum load	Comments
Steel	50–200 ft (15–60 m)	Practically unlimited	67–270 kip (300–1200 kN)	Eq. (9.1)	*Advantages* a. Easy to handle with respect to cutoff and extension to the desired length b. Can stand high driving stresses c. Can penetrate hard layers such as dense gravel, soft rock d. High load-carrying capacity *Disadvantages* a. Relatively costly material b. High level of noise during pile driving c. Subject to corrosion d. H-piles may be damaged or deflected from the vertical during driving through hard layers or past major obstructions

▼ **TABLE 9.2** Continued

Pile type	Usual length of piles	Maximum length of pile	Usual load	Approximate maximum load	Comments
Precast concrete	*Precast:* 30–50 ft (10–15 m) *Prestressed:* 30–150 ft (10–35 m)	*Precast:* 100 ft (30 m) *Prestressed:* 200 ft (60 m)	67–675 kip (300–3000 kN)	*Precast:* 180–200 kip (800–900 kN) *Prestressed:* 1700–1900 kip (7500–8500 kN)	*Advantages* a. Can be subjected to hard driving b. Corrosion resistant c. Can be easily combined with concrete superstructure *Disadvantages* a. Difficult to achieve proper cutoff b. Difficult to transport
Cased cast-in-place concrete	15–50 ft (5–15 m)	100–130 ft (30–40 m)	45–115 kip (200–500 kN)	180 kip (800 kN)	*Advantages* a. Relatively cheap b. Possibility of inspection before pouring concrete c. Easy to extend *Disadvantages* a. Difficult to splice after concreting b. Thin casings may be damaged during driving
Uncased cast-in-place concrete	15–50 ft (5–15 m)	100–130 ft (30–40 m)	65–115 kip (300–500 kN)	160 kip (700 kN)	*Advantages* a. Initially economical b. Can be finished at any elevation *Disadvantages* a. Voids may be created if concrete is placed rapidly b. Difficult to splice after concreting c. In soft soils, the sides of the hole may cave in, thus squeezing the concrete
Wood	30–50 ft (10–15 m)	100 ft (30 m)	22–45 kip (100–200 kN)	60 kip (270 kN)	*Advantages* a. Economical b. Easy to handle c. Permanently submerged piles are fairly resistant to decay *Disadvantages* a. Decay above water table b. Can be damaged in hard driving c. Low load-bearing capacity d. Low resistance to tensile load when spliced

9.3 ESTIMATING PILE LENGTH

Selecting the type of pile to be used and estimating its necessary length are fairly difficult tasks that require good judgment. In addition to the classification given in Section 9.2, piles can be divided into three major categories, depending on their lengths and the mechanisms of load transfer to the soil: (a) point bearing piles, (b) friction piles, and (c) compaction piles.

Point Bearing Piles

If soil-boring records establish the presence of bedrock or rocklike material at a site within a reasonable depth, piles can be extended to the rock surface (Figure 9.6a). In this case, the ultimate capacity of the piles depends entirely on the load-bearing capacity of the underlying material; thus the piles are called *point bearing piles*. In most of these cases, the necessary length of the pile can be fairly well established.

If, instead of bedrock, a fairly compact and hard stratum of soil is encountered at a reasonable depth, piles can be extended a few meters into the hard stratum (Figure 9.6b). Piles with pedestals can be constructed on the bed of the hard stratum, and the ultimate pile load may be expressed as

$$Q_u = Q_p + Q_s \qquad (9.4)$$

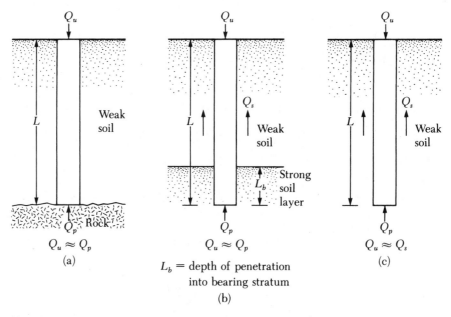

L_b = depth of penetration into bearing stratum

▼ **FIGURE 9.6** (a) and (b) Point bearing piles; (c) friction piles

where Q_p = load carried at the pile point
$\quad\quad\quad Q_s$ = load carried by skin friction developed at the side of the pile
$\quad\quad\quad\quad$ (caused by shearing resistance between the soil and the pile)

If Q_s is very small,

$$Q_u \approx Q_p \tag{9.5}$$

In this case, the required pile length may be estimated accurately if proper subsoil exploration records are available.

Friction Piles

When no layer of rock or rocklike material is present at a reasonable depth at a site, point bearing piles become very long and uneconomical. For this type of subsoil condition, piles are driven through the softer material to specified depths (Figure 9.6c). The ultimate load of these piles may be expressed by Eq. (9.4). However, if the value of Q_p is relatively small,

$$Q_u \approx Q_s \tag{9.6}$$

These piles are called *friction piles* because most of the resistance is derived from skin friction. However, the term *friction pile,* although used often in the literature, is a misnomer: in clayey soils, the resistance to applied load is also caused by *adhesion.*

The length of friction piles depends on the shear strength of the soil, the applied load, and the pile size. To determine the necessary lengths of these piles, an engineer needs a good understanding of soil-pile interaction, good judgment, and experience. Theoretical procedures for the calculation of load-bearing capacity of piles are presented later in this chapter.

Compaction Piles

Under certain circumstances, piles are driven in granular soils to achieve proper compaction of soil close to the ground surface. These piles are called *compaction piles.* The length of compaction piles depends on factors such as (a) relative density of the soil before compaction, (b) desired relative density of the soil after compaction, and (c) required depth of compaction. These piles are generally short; however, some field tests are necessary to determine a reasonable length.

9.4 INSTALLATION OF PILES

Most piles are driven into the ground by means of *hammers* or *vibratory drivers.* In special circumstances, piles can also be inserted by *jetting* or *partial augering.* The types of hammer used for pile driving include the (a) drop hammer, (b) single-acting air or steam hammer, (c) double-acting and differential air or steam hammer, and (d) diesel hammer. In the driving operation, a cap is attached to the top of the pile. A cushion may be used between the pile and the cap. This cushion has the effect of reducing the impact force and spreading it over a longer time; however, its use is optional. A hammer cushion is placed on the pile cap. The hammer drops on the cushion.

Figure 9.7 illustrates various hammers. A drop hammer (Figure 9.7a) is raised by a winch and allowed to drop from a certain height H. It is the oldest type of hammer used for pile driving. The main disadvantage of the drop hammer is the slow rate of hammer blows. The principle of the single-acting air or steam hammer is shown in Figure 9.7b. In this case, the striking part, or ram, is raised by air or steam pressure and then drops by gravity. Figure 9.7c shows the operation of the double-acting and differential air or steam hammer. For these hammers, air or steam is used both to raise the ram and to push it downward. This increases the impact velocity of the ram. The diesel hammer (Figure 9.7d) essentially consists of a ram, an anvil block, and a fuel-injection system. During the operation, the ram is first raised and fuel is injected near the anvil. Then the ram is released. When the ram drops, it compresses the air–fuel mixture, which ignites it. This action, in effect, pushes the pile downward and raises the ram. Diesel hammers work well under hard driving conditions. In soft soils, the downward movement of the pile is rather large, and the upward movement of the ram is small. This differential may not be sufficient to ignite the air–fuel system, so the ram may have to be lifted manually.

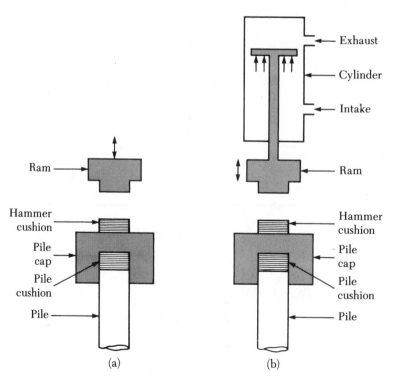

▼ **FIGURE 9.7**   Pile-driving equipment: (a) drop hammer; (b) single-acting air or steam hammer; (c) double-acting and differential air or steam hammer; (d) diesel hammer; (e) vibratory pile driver

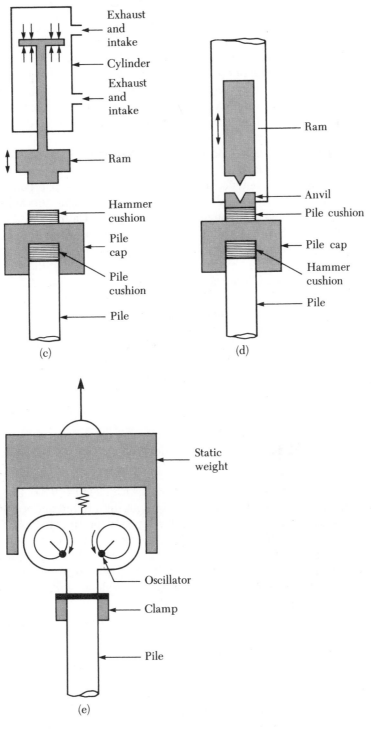

Exhaust and intake

Cylinder

Exhaust and intake

Ram

Hammer cushion

Pile cap

Pile cushion

Pile

(c)

Ram

Anvil

Pile cushion

Pile cap

Hammer cushion

Pile

(d)

Static weight

Oscillator

Clamp

Pile

(e)

▼ FIGURE 9.7 (Continued)

Tables D.4 and D.5 (Appendix D) list some of the commercially available diesel, single-acting, double-acting, and differential hammers.

The principles of operation of a vibratory pile driver are shown in Figure 9.7e. This driver essentially consists of two counter-rotating weights. The horizontal components of the centrifugal force generated as a result of rotating masses cancel each other. As a result, a sinusoidal dynamic vertical force is produced on the pile and helps drive the pile downward.

Jetting is a technique sometimes used in pile driving when the pile needs to penetrate a thin layer of hard soil (such as sand and gravel) overlying a softer soil layer. In this technique, water is discharged at the pile point by means of a pipe 2–3 in. (50–75 mm) in diameter to wash and loosen the sand and gravel.

Piles driven at an angle to the vertical, typically 14° to 20°, are referred to as *batter piles.* Batter piles are used in group piles when higher lateral load-bearing capacity is required. Piles also may be advanced by partial augering, with power augers (Chapter 2) being used to predrill holes part of the way. The piles can then be inserted into the holes and driven to the desired depth.

Based on the nature of their placement, piles may be divided into two categories: *displacement piles* and *nondisplacement piles.* Driven piles are displacement piles because they move some soil laterally; hence there is a tendency for densification of soil surrounding them. Concrete piles and closed-ended pipe piles are high-displacement piles. However, steel H-piles displace less soil laterally during driving, and so they are low-displacement piles. In contrast, bored piles are nondisplacement piles because their placement causes very little change in the state of stress in the soil.

9.5 LOAD TRANSFER MECHANISM

The load transfer mechanism from a pile to the soil is complicated. To understand it, consider a pile of length L, as shown in Figure 9.8a. The load on the pile is gradually increased from zero to $Q_{(z=0)}$ at the ground surface. Part of this load will be resisted by the side friction developed along the shaft, Q_1, and part by the soil below the tip of the pile, Q_2. Now, how are Q_1 and Q_2 related to the total load? If measurements are made to obtain the load carried by the pile shaft $Q_{(z)}$, at any depth z, the nature of variation will be like that shown in curve 1 of Figure 9.8b. The *frictional resistance per unit area, $f_{(z)}$,* at any depth z may be determined as

$$f_{(z)} = \frac{\Delta Q_{(z)}}{(p)(\Delta z)} \tag{9.7}$$

where p = perimeter of the pile cross section

Figure 9.8c shows the variation of $f_{(z)}$ with depth.

If the load Q at the ground surface is gradually increased, maximum frictional resistance along the pile shaft will be fully mobilized when the relative displacement between the soil and the pile is about 0.2–0.3 in. (5–10 mm) irrespective of pile size and length L. However, the maximum point resistance $Q_2 = Q_p$ will not be mobilized until the pile tip has moved about 10%–25% of the pile width (or diameter). The

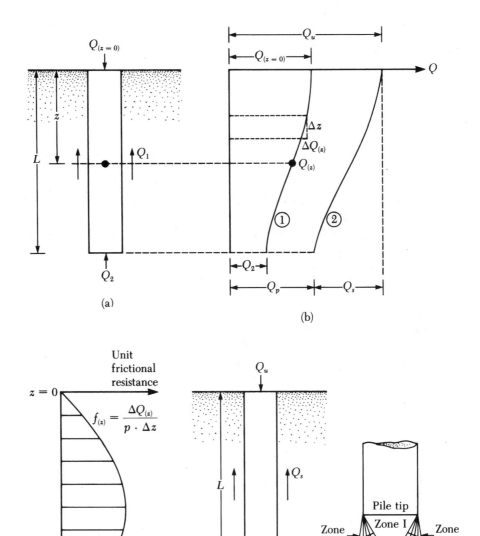

$$f_{(z)} = \frac{\Delta Q_{(z)}}{p \cdot \Delta z}$$

▼ **FIGURE 9.8** Load transfer mechanism for piles

lower limit applies to driven piles and the upper limit to bored piles. At ultimate load (Figure 9.8d and curve 2 in Figure 9.8b), $Q_{(z=0)} = Q_u$. Thus

$$Q_1 = Q_s$$

and

$$Q_2 = Q_p$$

The preceding explanation indicates that Q_s (or the unit skin friction, f, along the pile shaft) is developed at a *much smaller pile displacement compared to the point resistance, Q_p.* This condition can be seen in Vesic's (1970) pile-load test results in granular soil, shown in Figure 9.9. Note that these results are for *pipe piles in dense sand.*

At ultimate load, the failure surface in the soil at the pile tip (bearing capacity failure caused by Q_p) is like that shown in Figure 9.8e. Note that pile foundations are deep foundations and that the soil fails mostly in a *punching mode,* as illustrated previously in Figures 3.1c and 3.3. That is, a *triangular zone,* I, is developed at the pile tip, which is pushed downward without producing any other visible slip surface. In dense sands and stiff clayey soils, a *radial shear zone,* II, may partially develop. Hence the load displacement curves of piles will resemble those shown in Figure 3.1c.

Figure 9.10 shows the field load-transfer curves reported by Woo and Juang (1995) on a bored concrete pile (drilled shaft) in Taiwan. The pile was 41.7 m long.

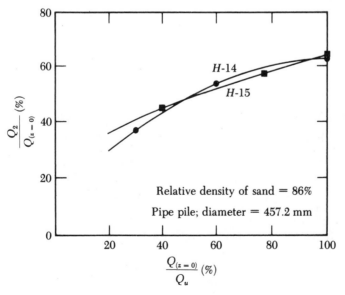

▼ **FIGURE 9.9**   Relative magnitude of point load transferred at various stages of pile loading (redrawn after Vesic, 1970)

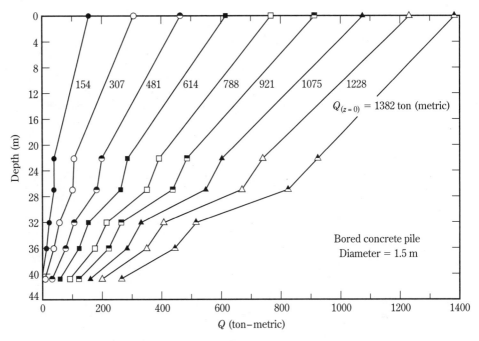

▼ **FIGURE 9.10** Load transfer curves for a pile as obtained by Woo and Juang (1975)

The subsoil conditions where the pile was bored were as follows:

Depth below ground surface (m)	Unified soil classification
0–3.7	SM
3.7–6.0	GP–GM
6.0–9.0	GM–SM
9.0–12.0	GM–SM
12.0–18.0	SM
18.0–20.0	CL–ML
20.0–33.0	ML/SM
33.0–39.0	GP–GM
39.0–41.7	GP–SM/GM

9.6 EQUATIONS FOR ESTIMATING PILE CAPACITY

The ultimate load-carrying capacity of a pile is given by a simple equation as the sum of the load carried at the pile point plus the total frictional resistance (skin friction) derived from the soil-pile interface (Figure 9.11a), or

$$Q_u = Q_p + Q_s \tag{9.8}$$

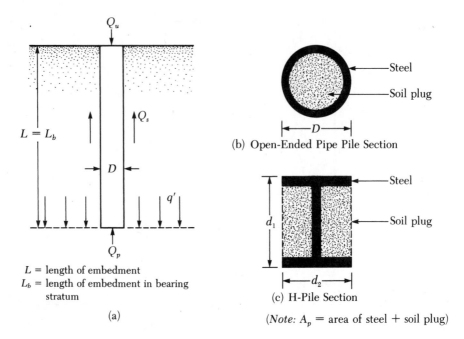

▼ FIGURE 9.11 Ultimate load-carrying capacity of pile

where Q_u = ultimate pile capacity
$\qquad$ Q_p = load-carrying capacity of the pile point
$\qquad$ Q_s = frictional resistance

Numerous published studies cover the determination of the values of Q_p and Q_s. Excellent reviews of many of these investigations have been provided by Vesic (1977), Meyerhof (1976), and Coyle and Castello (1981). These studies provide insight into the problem of determining ultimate pile capacity.

Point Bearing Capacity, Q_p

The ultimate bearing capacity of shallow foundations was discussed in Chapter 3. According to Terzaghi's equations,

$$q_u = 1.3cN_c + qN_q + 0.4\gamma BN_\gamma \qquad \text{(for shallow square foundations)}$$

and

$$q_u = 1.3cN_c + qN_q + 0.3\gamma BN_\gamma \qquad \text{(for shallow circular foundations)}$$

Similarly, the general bearing capacity equation for shallow foundations was given in Chapter 3 (for vertical loading) as

$$q_u = cN_c F_{cs} F_{cd} + qN_q F_{qs} F_{qd} + \tfrac{1}{2}\gamma BN_\gamma F_{\gamma s} F_{\gamma d}$$

Hence, in general, the ultimate load-bearing capacity may be expressed as

$$q_u = cN_c^* + qN_q^* + \gamma BN_\gamma^* \tag{9.9a}$$

where N_c^*, N_q^*, and N_γ^* are the bearing capacity factors that include the necessary shape and depth factors

Pile foundations are deep. However, the ultimate resistance per unit area developed at the pile tip, q_p, may be expressed by an equation similar in form to that shown in Eq. (9.9a), although the values of N_c^*, N_q^*, and N_γ^* will change. The notation used in this chapter for the width of a pile is D. Hence substituting D for B in Eq. (9.9a) gives

$$q_u = q_p = cN_c^* + qN_q^* + \gamma DN_\gamma^* \tag{9.9b}$$

Because the width D of a pile is relatively small, the term γDN_γ^* may be dropped from the right side of the preceding equation without introducing a serious error, or

$$q_p = cN_c^* + q'N_q^* \tag{9.10}$$

Note that the term q has been replaced by q' in Eq. (9.10) to signify effective vertical stress. Hence the point bearing of piles is

$$\boxed{Q_p = A_p q_p = A_p(cN_c^* + q'N_q^*)} \tag{9.11}$$

where A_p = area of pile tip
             c = cohesion of the soil supporting the pile tip
             q_p = unit point resistance
             q' = effective vertical stress at the level of the pile tip
        N_c^*, N_q^* = the bearing capacity factors

Frictional Resistance, Q_s

The frictional or skin resistance of a pile may be written as

$$\boxed{Q_s = \Sigma\, p\, \Delta L f} \tag{9.12}$$

where p = perimeter of the pile section
          ΔL = incremental pile length over which p and f are taken constant
          f = unit friction resistance at any depth z

There are several methods for estimating Q_p and Q_s. They are discussed in the following sections. It needs to be reemphasized that, in the field, for full mobilization of the point resistance (Q_p), the pile tip must go through a displacement of 10 to 25% of the pile width (or diameter).

9.7 MEYERHOF'S METHOD— ESTIMATION OF Q_p

Sand

The point bearing capacity, q_p, of a pile in sand generally increases with the depth of embedment in the bearing stratum and reaches a maximum value at an embedment ratio of $L_b/D = (L_b/D)_{cr}$. Note that in a homogeneous soil L_b is equal to the actual embedment length of the pile, L (see Figure 9.11a). However, in Figure 9.6b, where a pile has penetrated into a bearing stratum, $L_b < L$. Beyond the critical embedment ratio, $(L_b/D)_{cr}$, the value of q_p remains constant ($q_p = q_l$). That is, as shown in Figure 9.12 for the case of a homogeneous soil, $L = L_b$. The variation of $(L_b/D)_{cr}$ with the soil friction angle is shown in Figure 9.13. Note that the broken curve is for the determination of N_c^* and that the solid curve is for the determination of N_q^*. According to Meyerhof (1976), the bearing capacity factors increase with L_b/D and reach a maximum value at $L_b/D \approx 0.5(L_b/D)_{cr}$. Figure 9.13 indicates that $(L_b/D)_{cr}$ for $\phi = 45°$ is about 25 and that it decreases with the decrease of the friction angle, ϕ. In most cases, the magnitude of L_b/D for piles is greater than $0.5(L_b/D)_{cr}$, so the maximum values of N_c^* and N_q^* will apply for calculation of q_p for all piles. The

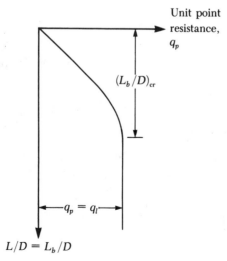

▼ **FIGURE 9.12** Nature of variation of unit point resistance in a homogeneous sand

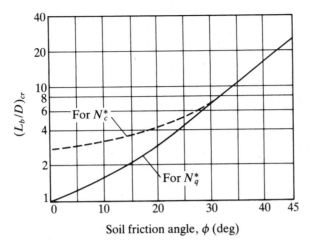

▼ **FIGURE 9.13**   Variation of $(L_b/D)_{cr}$ with soil friction angle (after Meyerhof, 1976)

variation of these maximum values of N_c^* and N_q^* with friction angle, ϕ, is shown in Figure 9.14.

For piles in sand, $c = 0$, and Eq. (9.11) simpifies to

$$Q_p = A_p q_p = A_p q' N_q^*$$
$$\uparrow$$
$$\text{Figure 9.14}$$

(9.13)

However, Q_p should not exceed the limiting value, or $A_p q_l$, so

$$Q_p = A_p q' N_q^* \leq A_p q_l$$

(9.14)

The limiting point resistance is

$$q_l (\text{kN}/\text{m}^2) = 50 N_q^* \tan \phi$$

(9.15)

where ϕ = soil friction angle in the bearing stratum

In English units, Eq. (9.15) becomes

$$q_l (\text{lb}/\text{ft}^2) = 1000 N_q^* \tan \phi$$

(9.16)

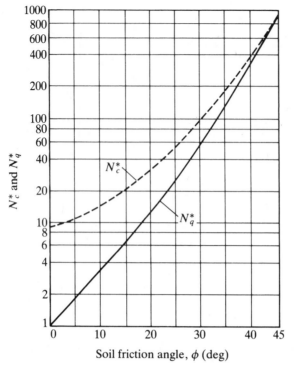

FIGURE 9.14   Variation of the maximum values of N_c^* and N_q^* with soil friction angle ϕ (after Meyerhof, 1976)

Based on field observations, Meyerhof (1976) also suggested that the ultimate point resistance, q_p, in a homogeneous granular soil ($L = L_b$) may be obtained from standard penetration numbers as

$$q_p\,(\text{kN/m}^2) = 40 N_{cor} L/D \leq 400 N_{cor} \tag{9.17}$$

where N_{cor} = average corrected standard penetration number near the pile point (about $10D$ above and $4D$ below the pile point)

In English units,

$$q_p\,(\text{lb/ft}^2) = 800 N_{cor} L/D \leq 8000 N_{cor} \tag{9.18}$$

Clay ($\phi = 0$ condition)

For piles in *saturated clays* in undrained conditions ($\phi = 0$),

$$\boxed{Q_p = N_c^* c_u A_p = 9 c_u A_p} \tag{9.19}$$

where c_u = undrained cohesion of the soil below the pile tip

9.8 VESIC'S METHOD—ESTIMATION OF Q_p

Vesic (1977) proposed a method for estimating the pile point bearing capacity based on the theory of *expansion of cavities*. According to this theory, based on effective stress parameters,

$$Q_p = A_p q_p = A_p(cN_c^* + \sigma_o' N_\sigma^*)$$ (9.20)

where σ_o' = mean normal ground stress (effective) at the level of the pile point

$$= \left(\frac{1 + 2K_o}{3}\right)q'$$ (9.21)

K_o = earth pressure coefficient at rest = $1 - \sin\phi$ (9.22)

N_c^*, N_σ^* = bearing capacity factors

Note that Eq. (9.20) is a modification of Eq. (9.11) with

$$N_\sigma^* = \frac{3N_q^*}{(1 + 2K_o)}$$ (9.23)

The relation for N_c^* given in Eq. (9.20) may be expressed as

$$N_c^* = (N_q^* - 1)\cot\phi$$ (9.24)

According to Vesic's theory,

$$N_\sigma^* = f(I_{rr})$$ (9.25)

where I_{rr} = reduced rigidity index for the soil

However,

$$I_{rr} = \frac{I_r}{1 + I_r\Delta}$$ (9.26)

where

$$I_r = \text{rigidity index} = \frac{E_s}{2(1 + \mu_s)(c + q'\tan\phi)} = \frac{G_s}{c + q'\tan\phi}$$ (9.27)

E_s = modulus of elasticity of soil
μ_s = Poisson's ratio of soil
G_s = shear modulus of soil
Δ = average volumatic strain in the plastic zone below the pile point

For conditions of no volume change (dense sand or saturated clay), $\Delta = 0$, so

$$I_r = I_{rr}$$ (9.28)

Table D.6 (Appendix D) gives the values of N_c^* and N_σ^* for various values of the soil friction angle (ϕ) and I_{rr}. For $\phi = 0$ (undrained condition),

$$N_c^* = \frac{4}{3}(\ln I_{rr} + 1) + \frac{\pi}{2} + 1 \tag{9.29}$$

The values of I_r can be estimated from laboratory consolidation and triaxial tests corresponding to the proper stress levels. However, for preliminary use the following values are recommended:

Soil type	I_r
Sand	70–150
Silts and clays (drained condition)	50–100
Clays (undrained condition)	100–200

9.9 JANBU'S METHOD—ESTIMATION OF Q_p

Janbu (1976) proposed calculating Q_p as follows:

$$Q_p = A_p(cN_c^* + q'N_q^*) \tag{9.30}$$

Note that Eq. (9.30) has the same form as Eq. (9.11). The bearing capacity factors N_c^* and N_q^* are calculated by assuming a failure surface in soil at the pile tip similar to that shown in the insert of Figure 9.15. The bearing capacity relationships then are

$$N_q^* = (\tan\phi + \sqrt{1 + \tan^2\phi})^2 (e^{2\eta'\tan\phi}) \tag{9.31}$$

(The angle η' is defined in the insert of Figure 9.15.)

$$N_c^* = \underset{\underset{\text{Eq. (9.31)}}{\uparrow}}{(N_q^* - 1)\cot\phi} \tag{9.32}$$

Figure 9.15 shows the variation of N_q^* and N_c^* with ϕ and η'. The angle η' may vary from about 70° in soft clays to about 105° in dense sandy soils.

Regardless of the theoretical procedure used to calculate Q_p, its full magnitude cannot be realized until the pile tip has penetrated at least 10%–25% of the width of the pile. This depth is critical in the case of sand.

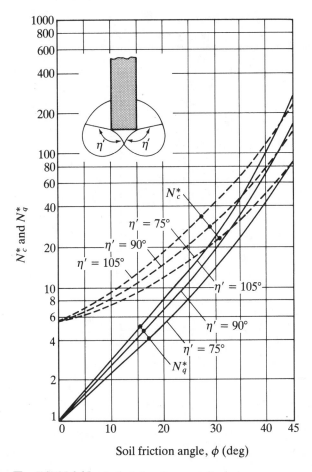

▼ **FIGURE 9.15** Janbu's bearing capacity factors

9.10 COYLE AND CASTELLO'S METHOD—ESTIMATION OF Q_p IN SAND

Coyle and Castello (1981) analyzed twenty-four large-scale field load tests of driven piles in sand. Based on the test results, they suggested that, in sand,

$$Q_p = q' N_q^* A_p \tag{9.33}$$

where q' = effective vertical stress at the pile tip
N_q^* = bearing capacity factor

Figure 9.16 shows the variation of N_q^* with L/D and the soil friction angle, ϕ.

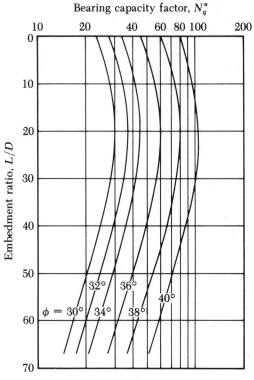

FIGURE 9.16   Variation of N_q^* with L/D (redrawn after Coyle and Castello, 1981)

9.11   FRICTIONAL RESISTANCE (Q_s) IN SAND

It was pointed out in Eq. (9.12) that the frictional resistance (Q_s) can be expressed as

$$Q_s = \Sigma \, p \, \Delta L f$$

The unit frictional resistance, f, is hard to estimate. In making an estimation of f, several important factors must be kept in mind. They are as follows:

1.   The nature of pile installation. For driven piles in sand, the vibration caused during pile driving helps densify the soil around the pile. Figure 9.17 shows the contours of the soil friction angle, ϕ, around a driven pile (Meyerhof, 1961). Note that, in this case, the original soil friction angle of the sand was 32°. The zone of sand densification is about 2.5 times the pile diameter surrounding the pile.

2.   It has been observed that the nature of variation of f in the field is approximately as shown in Figure 9.18. The unit skin friction increases with depth

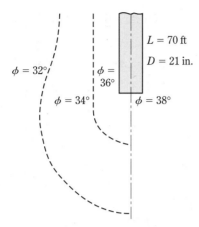

▼ **FIGURE 9.17** Compaction of sand near driven piles (after Meyerhof, 1961)

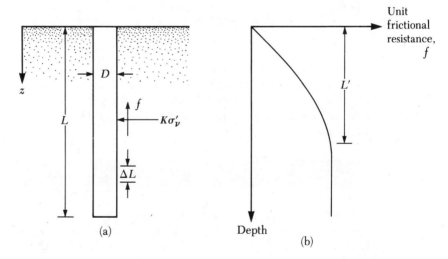

▼ **FIGURE 9.18** Unit frictional resistance for piles in sand

more or less linearly to a depth of L' and remains constant thereafter. The magnitude of the critical depth L' may be 15 to 20 pile diameters. A conservative estimate would be

$$L' \approx 15D \qquad (9.34)$$

3. At similar depths, the unit skin friction in loose sand is higher for a high-displacement pile as compared to a low-displacement pile.

4. At similar depths, bored, or jetted, piles will have a lower unit skin friction as compared to driven piles.

Considering the above factors, an approximate relationship for f can be given as follows (Figure 9.18):

For $z = 0$ to L'

$$f = K\sigma_v' \tan \delta \qquad (9.35a)$$

and for $z = L'$ to L

$$f = f_{z=L'} \qquad (9.35b)$$

where K = effective earth coefficient
σ_v' = effective vertical stress at the depth under consideration
δ = soil-pile friction angle

In reality, the magnitude of K varies with depth. It is approximately equal to the Rankine passive earth pressure coefficient, K_p, at the top of the pile and may be less than the at-rest pressure coefficient, K_o, at a greater depth. Based on the presently available results, the following average values of K are recommended for use in Eq. (9.35):

Pile type	K
Bored or jetted	$\approx K_o = 1 - \sin \phi$
Low-displacement driven	$\approx K_o = 1 - \sin \phi$ to $1.4K_o = 1.4(1 - \sin \phi)$
High-displacement driven	$\approx K_o = 1 - \sin \phi$ to $1.8K_o = 1.8(1 - \sin \phi)$

The values of δ from various investigations appear to be in the range of 0.5ϕ to 0.8ϕ. Judgment must be used in choosing the value of δ. For high-displacement driven piles, Bhusan (1982) recommended

$$K \tan \delta = 0.18 + 0.0065D_r \qquad (9.36)$$

and

$$K = 0.5 + 0.008D_r \qquad (9.37)$$

where D_r = relative density (%)

Meyerhof (1976) also indicated that the average unit frictional resistance, f_{av}, for high-displacement driven piles may be obtained from average corrected standard penetration resistance values as

$$f_{av}(\text{kN/m}^2) = 2\overline{N}_{cor} \tag{9.38}$$

where $\overline{N}_{cor}$ = average corrected value of standard penetration resistance

In English units, Eq. (9.38) becomes

$$f_{av}(\text{lb/ft}^2) = 40\overline{N}_{cor} \tag{9.39}$$

For low-displacement driven piles

$$f_{av}(\text{kN/m}^2) = \overline{N}_{cor} \tag{9.40}$$

and

$$f_{av}(\text{lb/ft}^2) = 20\overline{N}_{cor} \tag{9.41}$$

Thus

$$Q_s = pLf_{av} \tag{9.42}$$

Coyle and Castello (1981), in conjunction with the material presented in Section 9.10, proposed that

$$Q_s = f_{av}pL = (K\overline{\sigma}_v' \tan \delta)pL \tag{9.43}$$

where $\overline{\sigma}'$ = average effective overburden pressure
δ = soil–pile friction angle = 0.8ϕ

The lateral earth pressure coefficient K, which was determined from field observations, is shown in Figure 9.19. Thus, if Figure 9.19 is used,

$$Q_s = K\overline{\sigma}_v' \tan(0.8\phi)pL \tag{9.44}$$

9.12 FRICTIONAL (SKIN) RESISTANCE IN CLAY

Estimating the frictional (or skin) resistance of piles in clay is almost as difficult a task as that in sand (Section 9.11) due to the presence of several variables that cannot be easily quantified. Several methods for obtaining unit frictional resistance of piles are presently available in the literature. Three of the presently accepted procedures are described below.

1. λ *Method:* This method was proposed by Vijayvergiya and Focht (1972). It is based on the assumption that the displacement of soil caused by pile driving results in a passive lateral pressure at any depth and that the average unit skin resistance is

$$f_{av} = \lambda(\overline{\sigma}_v' + 2c_u) \tag{9.45}$$

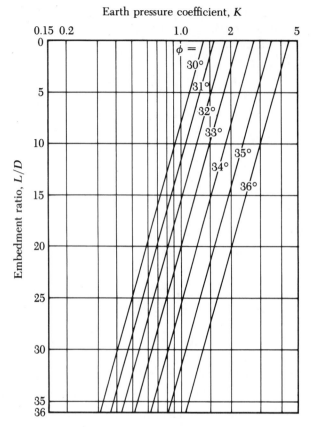

▼ **FIGURE 9.19**   Variation of K with L/D (redrawn after Coyle and Castello, 1981)

where $\overline{\sigma}'_v$ = mean effective vertical stress for the entire embedment length
c_u = mean undrained shear strength ($\phi = 0$ concept)

The value of λ changes with the depth of pile penetration (see Figure 9.20). Thus the total frictional resistance may be calculated as

$$Q_s = pLf_{av}$$

Care should be taken in obtaining the values of $\overline{\sigma}'_v$ and c_u in layered soil. Figure 9.21 helps explain the reason. According to Figure 9.21b, the mean value of c_u is $(c_{u(1)}L_1 + c_{u(2)}L_2 + \cdots)/L$. Similarly, Figure 9.21c shows the plot of the variation of effective stress with depth. The mean effective stress is

$$\overline{\sigma}'_v = \frac{A_1 + A_2 + A_3 + \cdots}{L} \tag{9.46}$$

where $A_1, A_2, A_3, \ldots$ = areas of the vertical effective stress diagrams

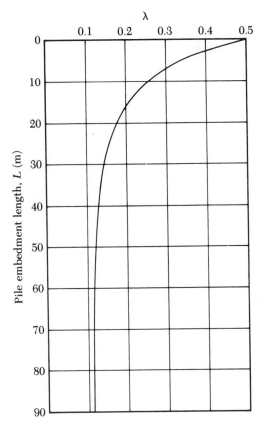

▼ FIGURE 9.20   Variation of λ with pile em-
                   bedment length (redrawn after
                   McClelland, 1974)

2. *α Method:* According to the α method, the unit skin resistance in clayey
    soils can be represented by the equation

$$f = \alpha c_u$$

(9.47)

where α = empirical adhesion factor

The approximate variation of the value of α is shown in Figure 9.22. Note
that for normally consolidated clays with $c_u \leq$ about 1 kip/ft² (50 kN/m²),
$\alpha = 1$. Thus

$$Q_s = \Sigma fp \, \Delta L = \Sigma \alpha c_u p \, \Delta L$$

(9.48)

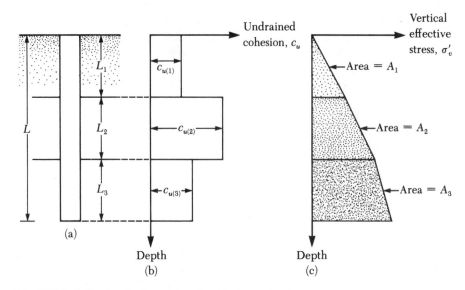

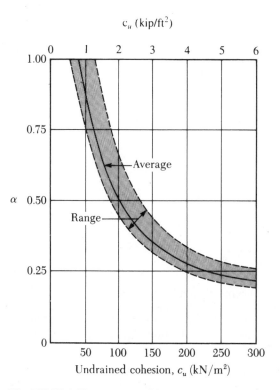

▼ **FIGURE 9.21**   Application of λ method in layered soil

▼ **FIGURE 9.22**   Variation of α with undrained
                     cohesion of clay

3.   *β Method:* When piles are driven into saturated clays, the pore water pressure in the soil around the piles increases. This excess pore water pressure in normally consolidated clays may be 4 to 6 times c_u. However, within a month or so, this pressure gradually dissipates. Hence the unit frictional resistance for the pile can be determined on the basis of the effective stress parameters of the clay in a remolded state ($c = 0$). Thus at any depth

$$\boxed{f = \beta \sigma'_v} \tag{9.49}$$

where σ'_v = vertical effective stress
$\qquad \beta = K \tan \phi_R$ $\qquad\qquad\qquad\qquad\qquad\qquad\qquad$ (9.50)
$\qquad \phi_R$ = drained friction angle of remolded clay
$\qquad K$ = earth pressure coefficient

Conservatively, the magnitude of K is the earth pressure coefficient at rest, or

$$K = 1 - \sin \phi_R \qquad \text{(for normally consolidated clays)} \tag{9.51}$$

and

$$K = (1 - \sin \phi_R)\sqrt{OCR} \qquad \text{(for overconsolidated clays)} \tag{9.52}$$

where OCR = overconsolidation ratio

Combining Eqs. (9.49), (9.50), (9.51), and (9.52), for normally consolidated clays, yields

$$f = (1 - \sin \phi_R) \tan \phi_R \sigma'_v \tag{9.53}$$

and for overconsolidated clays,

$$f = (1 - \sin \phi_R) \tan \phi_R \sqrt{OCR} \sigma'_v \tag{9.54}$$

With the value of f determined, the total frictional resistance may be evaluated as

$$Q_s = \Sigma f p \, \Delta L$$

9.13   GENERAL COMMENTS AND ALLOWABLE PILE CAPACITY

Although calculations for estimating the ultimate load-bearing capacity of a pile can be made by using the relationships presented in Sections 9.6 through 9.12, an engineer needs to keep the following points in mind:

1.   In calculating the area of cross section, A_p, and the perimeter, p, of piles with developed profiles, such as H-piles and open-ended pipe piles, the effect of soil plug should be considered. According to Figure 9.11b and 9.11c, for pipe piles

$$A_p = \left(\frac{\pi}{4}\right)D^2$$

$$p = \pi D$$

Similarly, for H-piles

$$A_p = d_1 d_2$$

$$p = 2(d_1 + d_2)$$

Also, note that for H-piles, because $d_2 > d_1$, $D = d_1$.

2. The ultimate point load relations given in Eqs. (9.11), (9.20), and (9.30) are for the gross ultimate point load; that is, they include the weight of the pile. So the net ultimate point load is approximately

$$Q_{p(\text{net})} = Q_{p(\text{gross})} - q'A_p$$

However, in practice, for soils with $\phi > 0$, the assumption is made that $Q_{p(\text{net})} = Q_{p(\text{gross})}$. In cohesive soils with $\phi = 0$, $N_q^* = 1$ (Figure 9.14). Hence from Eq. (9.11),

$$Q_{p(\text{gross})} = (c_u N_c^* + q')A_p$$

So

$$Q_{p(\text{net})} = [(c_u N_c^* + q') - q']A_p = c_u N_c^* A_p = 9c_u A_p = Q_p$$

This relation is the one given in Eq. (9.19).

After the total ultimate load-carrying capacity of a pile has been determined by summing the point bearing capacity and the frictional (or skin) resistance, a reasonable factor of safety should be used to obtain the total allowable load for each pile, or

$$Q_{\text{all}} = \frac{Q_u}{FS} \tag{9.55}$$

where Q_{all} = allowable load-carrying capacity for each pile
FS = factor of safety

The factor of safety generally used ranges from 2.5 to 4, depending on the uncertainties of ultimate load calculation.

9.14 POINT BEARING CAPACITY OF PILES RESTING ON ROCK

Sometimes piles are driven to an underlying layer of rock. In such cases, the engineer must evaluate the bearing capacity of the rock. The ultimate unit point resistance in rock (Goodman, 1980) is approximately

$$q_p = q_u(N_\phi + 1) \tag{9.56}$$

where $N_\phi = \tan^2 (45 + \phi/2)$

$\quad\quad\quad\ q_u$ = unconfined compression strength of rock

$\quad\quad\quad\ \phi$ = drained angle of friction

The unconfined compression strength of rock can be determined by laboratory tests on rock specimens collected during field investigation. However, extreme caution should be used in obtaining the proper value of q_u because laboratory specimens usually are small in diameter. As the diameter of the specimen increases, the unconfined compression strength decreases, which is referred to as the *scale effect*. For specimens larger than about 3 ft (1 m) in diameter, the value of q_u remains approximately constant. There appears to be a fourfold to fivefold reduction of the magnitude of q_u in this process. The scale effect in rock is primarily caused by randomly distributed large and small fractures and also by progressive ruptures along the slip lines. Hence, we always recommend that

$$q_{u(\text{design})} = \frac{q_{u(\text{lab})}}{5} \tag{9.57}$$

Table 9.3 lists some representative values of (laboratory) unconfined compression strengths of rock. Representative values of the rock friction angle, ϕ, are given in Table 9.4.

▼ **TABLE 9.3** Typical Unconfined Compressive Strength of Rocks

Rock type	q_u	
	lb/in²	MN/m²
Sandstone	10,000–20,000	70–140
Limestone	15,000–30,000	105–210
Shale	5,000–10,000	35–70
Granite	20,000–30,000	140–210
Marble	8,500–10,000	60–70

▼ **TABLE 9.4** Typical Values of Angle of Friction, ϕ, of Rocks

Rock type	Angle of friction, ϕ (deg)
Sandstone	27–45
Limestone	30–40
Shale	10–20
Granite	40–50
Marble	25–30

A factor of safety of at least 3 should be used to determine the allowable point bearing capacity of piles. Thus

$$Q_{p(\text{all})} = \frac{[q_{u(\text{design})}(N_\phi + 1)]A_p}{FS} \qquad (9.58)$$

▼ **EXAMPLE 9.1** _____

A concrete pile is 50 ft (L) long and 16 in. × 16 in. in cross section. The pile is fully embedded in sand for which $\gamma = 110$ lb/ft³ and $\phi = 30°$. Calculate the ultimate point load, Q_p, by using

    a. Meyerhof's method (Section 9.7).
    b. Vesic's method (Section 9.8). Use $I_r = I_{rr} = 50$.
    c. Janbu's method (Section 9.9). Use $\eta' = 90°$.

Solution

Part a. From Eq. (9.13),

$$Q_p = A_p q' N_q^* = A_p \gamma L N_q^*$$

For $\phi = 30°$, $N_q^* \approx 55$ (Figure 9.14), so

$$Q_p = \left(\frac{16 \times 16}{12 \times 12}\, \text{ft}^2\right)\left(\frac{110 \times 50}{1000}\, \text{kip/ft}^2\right)(55) = 537.8\ \text{kip}$$

Again, from Eqs. (9.14) and (9.16),

$$Q_p = A_p q_l = A_p N_q^* \tan \phi\,(\text{kip/ft}^2)$$

$$= \left(\frac{16 \times 16}{12 \times 12}\, \text{ft}^2\right)(55)\tan 30 = 56.45\ \text{kip}$$

Hence, $Q_p = $ **56.45 kip.**

Part b. From Eqs. (9.20), (9.21), and (9.22), with $c = 0$,

$$Q_p = A_p \sigma_o' N_\sigma^* = A_p \left[\frac{1 + 2(1 - \sin \phi)}{3}\right] q' N_\sigma^*$$

For $\phi = 30°$ and $I_{rr} = 50$, the value of N_σ^* is about 36 (Table D.4, Appendix D), so

$$Q_p = \left(\frac{16 \times 16}{12 \times 12}\, \text{ft}^2\right)\left[\frac{1 + 2(1 - \sin 30)}{3}\right]\left(\frac{110 \times 50}{1000}\, \text{kip/ft}^2\right)(36) = \mathbf{234.7\ kip}$$

Part c. From Eq. (9.30) with $c = 0$,

$$Q_p = A_p q' N_q^*$$

For $\phi = 30°$ and $\eta' = 90°$, the value of $N_q^* \approx 19$ (Figure 9.15).

$$Q_p = \left(\frac{16 \times 16}{12 \times 12}\, \text{ft}^2\right)\left(\frac{110 \times 50}{1000}\, \text{kip/ft}^2\right)(19) = \mathbf{185.8\ kip}$$

▲

▼ **EXAMPLE 9.2**

For the pile described in Example 9.1

a.   Determine the frictional resistance, Q_s. Use Eqs. (9.12), (9.35a), and (9.35b). Given: $K = 1.3$ and $\delta = 0.8\phi$.

b.   Using the results of Example 9.1 and Part a of this problem, estimate the allowable load-carrying capacity of the pile. Given: $FS = 4$.

Solution

Part a. From Eq. (9.34),

$$L \approx 15D = 15\left(\frac{16}{12}\,\text{ft}\right) = 20\,\text{ft}$$

From Eq. (9.35a), at $z = 0$, $\sigma_v' = 0$, so $f = 0$. Again, at $z = L' = 20$ ft,

$$\sigma_v' = \gamma L' = \frac{(110)(20)}{1000} = 2.2\,\text{kip/ft}^2$$

So

$$f = K\sigma_v'\tan\delta = (1.3)(2.2)[\tan(0.8\times30)] = 1.273\,\text{kip/ft}^2$$

Thus

$$Q_s = \left(\frac{f_{z=0} + f_{z=20\,\text{ft}}}{2}\right)pL' + f_{z=20\,\text{ft}}\,p(L - L')$$

$$= \left(\frac{0 + 1.273}{2}\right)\left(4\times\frac{16}{12}\right)(20) + (1.273)\left(4\times\frac{16}{12}\right)(50 - 20)$$

$$= 67.9 + 203.7 = \textbf{271.6 kip}$$

Part b. $Q_u = P_p + Q_s$

Average value of Q_p from Example 9.1 is

$$\frac{56.45 + 234.7 + 185.8}{3} \approx 159\,\text{kip}$$

So

$$Q_{\text{all}} = \frac{Q_u}{FS} = \frac{1}{4}\left(159 + 271.6\right) = \textbf{107.7 kip}$$

▲

▼ **EXAMPLE 9.3**

For the pile described in Example 9.1, estimate the Q_{all} using Coyle and Castello's method [Sections 9.10 and Eq. (9.44)].

Solution   From Eqs. (9.33) and (9.44),

$$Q_u = Q_p + Q_s = q'N_q^*A_p + K\sigma_v'\tan(0.8\phi)pL$$

$$\frac{L}{D} = \frac{50}{\left(\frac{16}{12}\right)} = 37.5$$

For $\phi = 30°$ and $L/D = 37.5$, $N_q^* = 25$ (Figure 9.16) and $K = 0.2$ (Figure 9.19). Thus

$$Q_u = \left(\frac{110 \times 50}{1000} \text{ kip/ft}^2\right)(25)\left(\frac{16 \times 16}{12 \times 12} \text{ ft}^2\right)$$

$$+ (0.2)\left(\frac{110 \times 50}{1000 \times 2}\right)\tan(0.8 \times 30)\left(\frac{4 \times 16}{12}\right)(50)$$

$$= 244.4 + 65.3 = 309.7 \text{ kip}$$

$$Q_{\text{all}} = \frac{Q_u}{FS} = \frac{309.7}{4} = \mathbf{77.4 \text{ kip}}$$

▲

▼ **EXAMPLE 9.4**

A driven pipe pile in clay is shown in Figure 9.23a. The pipe has an outside diameter of 406 mm and a wall thickness of 6.35 mm.

    a.   Calculate the net point bearing capacity. Use Eq. (9.19).

    b.   Calculate the skin resistance (1) by using Eqs. (9.47) and (9.48) (α method), (2) by using Eq. (9.45) (λ method), and (3) by using Eq. (9.49) (β method). For all clay layers, $\phi_R = 30°$. The top 10 m of clay is normally consolidated. The bottom clay layer has an *OCR* of 2.

    c.   Estimate the net allowable pile capacity. Use $FS = 4$.

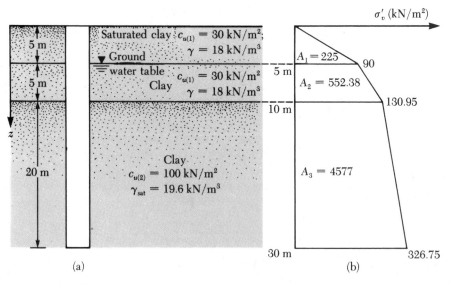

▼ **FIGURE 9.23**

Solution The area of cross section of the pile, including the soil inside the pile, is

$$A_p = \frac{\pi}{4} D^2 = \frac{\pi}{4} (0.406)^2 = 0.1295 \text{ m}^2$$

Part a. Calculation of Net Point Bearing Capacity

From Eq. (9.19)

$$Q_p = A_p q_p = A_p N_c^* c_{u(2)} = (0.1295)(9)(100) = \mathbf{116.55 \ kN}$$

Part b. Calculation of Skin Resistance

(1) Use of Eqs. (9.47) and (9.48): from Eq. (9.48),

$$Q_s = \Sigma \ \alpha c_u p \ \Delta L$$

For the top soil layer, $c_{u(1)} = 30$ kN/m². According to the average plot of Figure 9.22, $\alpha_1 = 1.0$. Similarly, for the bottom soil layer, $c_{u(2)} = 100$ kN/m²; $\alpha_2 = 0.5$. Thus

$$
\begin{aligned}
Q_s &= \alpha_1 c_{u(1)} [(\pi)(0.406)] 10 + \alpha_2 c_{u(2)} [(\pi)(0.406)] 20 \\
&= (1)(30)[(\pi)(0.406)] 10 + (0.5)(100)[(\pi)(0.406)] 20 \\
&= 382.7 + 1275.5 = \mathbf{1658.2 \ kN}
\end{aligned}
$$

(2) Use of Eq. (9.45): $f_{av} = \lambda(\overline{\sigma}'_v + 2c_u)$. The average value of c_u is

$$\frac{c_{u(1)}(10) + c_{u(2)}(20)}{30} = \frac{(30)(10) + (100)(20)}{30} = 76.7 \text{ kN/m}^2$$

To obtain the average value of $\overline{\sigma}'_v$, the diagram for vertical effective stress variation with depth is plotted in Figure 9.23b. From Eq. (9.46),

$$\overline{\sigma}'_v = \frac{A_1 + A_2 + A_3}{L} = \frac{225 + 552.38 + 4577}{30} = 178.48 \text{ kN/m}^2$$

The magnitude of λ from Figure 9.20 is 0.14. So

$$f_{av} = 0.14[178.48 + (2)(76.7)] = 46.46 \text{ kN/m}^2$$

Hence

$$Q_s = pLf_{av} = \pi (0.406)(30)(46.46) = \mathbf{1777.8 \ kN}$$

(3) Use of Eq. (9.49): The top clay layer (10 m) is normally consolidated and $\phi_R = 30°$. For $z = 0$–5 m [Eq. (9.53)],

$$
\begin{aligned}
f_{av(1)} &= (1 - \sin \phi_R) \tan \phi_R \sigma'_{v(av)} \\
&= (1 - \sin 30°)(\tan 30°)\left(\frac{0 + 90}{2}\right) = 13.0 \text{ kN/m}^2
\end{aligned}
$$

Similarly, for $z = 5\text{–}10$ m,

$$f_{av(2)} = (1 - \sin 30°)(\tan 30°)\left(\frac{90 + 130.95}{2}\right) = 31.9 \text{ kN/m}^2$$

For $z = 10\text{–}30$ m [Eq. (9.54)],

$$f_{av} = (1 - \sin \phi_R) \tan \phi_R \sqrt{OCR} \sigma'_{v(av)}$$

For $OCR = 2$,

$$f_{av(3)} = (1 - \sin 30°)(\tan 30°)\sqrt{2}\left(\frac{130.95 + 326.75}{2}\right) = 93.43 \text{ kN/m}^2$$

So

$$Q_s = p[f_{av(1)}(5) + f_{av(2)}(5) + f_{av(3)}(20)]$$
$$= (\pi)(0.406)[(13)(5) + (31.9)(5) + (93.43)(20)] = \mathbf{2669.7 \text{ kN}}$$

Part c. Calculation of Net Ultimate Capacity, Q_u

Comparing the three values shows that the α and λ methods give similar results, so we use

$$Q_s = \frac{1658.1 + 1777.8}{2} \approx 1718 \text{ kN}$$

Thus

$$Q_u = Q_p + Q_s = 116.46 + 1718 = 1834.46 \text{ kN}$$

$$Q_{all} = \frac{Q_u}{FS} = \frac{1834.46}{4} = \mathbf{458.6 \text{ kN}}$$ ▲

▼ **EXAMPLE 9.5**

An H-pile (size HP 310 × 1.226) having a length of embedment of 26 m is driven through a soft clay layer to rest on sandstone. The sandstone has a laboratory unconfined compression strength of 76 MN/m² and a friction angle of 28°. Use a factor of safety of 5 and estimate the allowable point bearing capacity.

Solution   From Eqs. (9.57) and (9.58),

$$Q_{p(all)} = \frac{\left\{\left[\dfrac{q_{u(lab)}}{5}\right]\left[\tan^2\left(45 + \dfrac{\phi}{2}\right) + 1\right]\right\} A_p}{FS}$$

From Table D.1b (Appendix D), for HP 310 × 1.226 piles, $A_p = 15.9 \times 10^{-3}$ m², so

$$Q_{p(all)} = \frac{\left\{\left[\dfrac{76 \times 10^3 \text{ kN/m}^2}{5}\right]\left[\tan^2\left(45 + \dfrac{28}{2}\right) + 1\right]\right\}(15.9 \times 10^{-3} \text{ m}^2)}{5}$$

$$= \mathbf{182 \text{ kN}}$$ ▲

9.15 PILE LOAD TESTS

In most large projects, a specific number of load tests must be conducted on piles. The primary reason is the unreliability of prediction methods. Vertical and lateral load-bearing capacity of a pile can be tested in the field. Figure 9.24a shows a schematic diagram of the pile load test arrangement for testing in *axial compression* in the field. The load is applied to the pile by a hydraulic jack. Step loads are applied

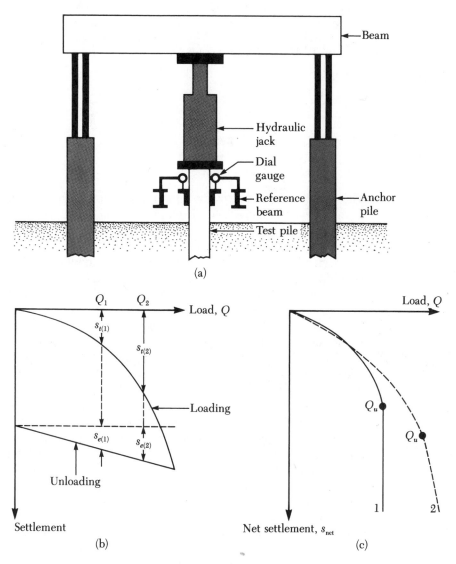

(a)

(b)

(c)

▼ **FIGURE 9.24**   (a) Schematic diagram of pile load test arrangement; (b) plot of load against total settlement; (c) plot of load against net settlement

to the pile, and sufficient time is allowed to elapse after each load so that a small amount of settlement occurs. The settlement of the pile is measured by dial gauges. The amount of load to be applied for each step will vary, depending on local building codes. Most building codes require that each step load be about one-fourth of the proposed working load. The load test should be carried out to at least a total load of two times the proposed working load. After reaching the desired pile load, the pile is gradually unloaded.

Figure 9.24b shows a load settlement diagram obtained from field loading and unloading. For any load, Q, the net pile settlement can be calculated as follows. When $Q = Q_1$,

$$\text{Net settlement, } s_{\text{net}(1)} = s_{t(1)} - s_{e(1)}$$

When $Q = Q_2$,

$$\text{Net settlement, } s_{\text{net}(2)} = s_{t(2)} - s_{e(2)}$$

$$\vdots$$

where s_{net} = net settlement
s_e = elastic settlement of the pile itself
s_t = total settlement

These values of Q can be plotted in a graph against the corresponding net settlement, s_{net}, as shown in Figure 9.24c. The ultimate load of the pile can be determined from this graph. Pile settlement may increase with load to a certain point, beyond which the load–settlement curve becomes vertical. The load corresponding to the point where the Q vs. s_{net} curve becomes vertical is the ultimate load, Q_u, for the pile; it is shown by curve 1 in Figure 9.24c. In many cases, the latter stage of the load–settlement curve is almost linear, showing a large degree of settlement for a small increment of load; it is shown by curve 2 in Figure 9.24c. The ultimate load, Q_u, for such a case is determined from the point of the Q vs. s_{net} curve where this steep linear portion starts.

The load test procedure just described requires application of step loads on the piles and measurement of settlement and is called a *load-controlled* test. Another technique used for a pile load test is the *constant-rate-of-penetration* test. In it, the load on the pile is continuously increased to maintain a constant rate of penetration, which can vary from 0.01 to 0.1 in./min (0.25 to 2.5 mm/min). This test gives a load–settlement plot similar to that obtained from the load-controlled test. Another type of pile load test is *cyclic loading,* in which an incremental load is repeatedly applied and removed.

Load tests on piles embedded in sand can be conducted immediately after the piles are driven. However, when piles are embedded in clay, care should be taken in deciding the time lapse between driving and starting the load test. When piles are driven in soft clay, a certain zone surrounding the clay becomes remolded and/or compressed, as shown in Figure 9.25. This results in a reduction of undrained shear strength, c_u (Figure 9.26). With time, the loss of undrained shear strength is partially or fully regained. This time lapse may range from 30 up to 60 days. Figure 9.27 shows the magnitude of the variation of Q_s with time for a pile driven in soft

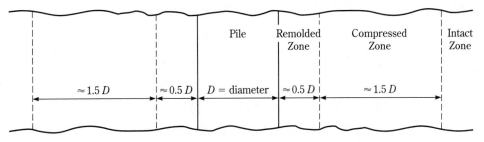

▼ **FIGURE 9.25** Remolded and/or compacted zone around a pile driven in soft clay

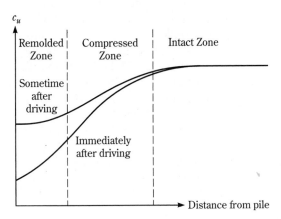

▼ **FIGURE 9.26** Nature of variation of undrained shear strength (c_u) with time around a pile driven in soft clay

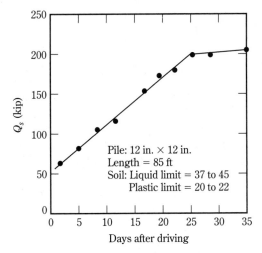

▼ **FIGURE 9.27** Variation of Q_s with time for a pile driven in soft clay (based on the load test results of Terzaghi and Peck, 1967)

clay based on the results reported by Terzaghi and Peck (1967). It can be seen from this figure that Q_s increased by about 300% with a time lapse of about 25 days.

9.16 COMPARISON OF THEORY WITH FIELD LOAD TEST RESULTS

Details of many field studies related to the estimation of the ultimate load-carrying capacity of various types of piles are available in the literature. In some cases, the results generally agree with the theoretical predictions and, in others, they vary widely. The variations between theory and field test results may be attributed to factors such as improper interpretation of subsoil properties, incorrect theoretical assumptions, erroneous acquisition of field test results, and others.

We saw from Example 9.1 that, for similar soil properties, the ultimate point load (Q_p) can vary over 400% or more depending on which theory and equation is used. Also, from the calculation of part a of Example 9.1, it is easy to see that, in most cases, for long piles embedded in sand the limiting point resistance (q_l) [Eqs. (9.15) or (9.16)] controls the unit point resistance (q_p). Meyerhof (1976) provided the results of several field load tests on long piles ($L/D \geq 10$) from which the derived values of q_p have been calculated and plotted in Figure 9.28. Also plotted

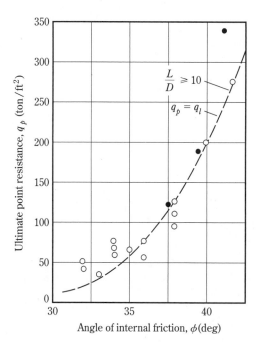

▼ **FIGURE 9.28**   Ultimate point resistance of driven piles in sand (after Meyerhof, 1976)

in this figure is the variation of q_l calculated from Eq. (9.16). It can be seen that, for a given friction angle ϕ, the magnitude of q_p can deviate substantially from the theory.

Briaud et al. (1989) reported the results of 28 axial load tests on impact-driven H-piles and pipe piles in sand performed by the U.S. Army Engineering District (St. Louis) during the construction of the New Lock and Dam No. 26 on the Mississippi River. Typical variations of field standard (uncorrected) penetration numbers with depth are shown in Figure 9.29.

The results of the load tests on four H-piles obtained from this program are

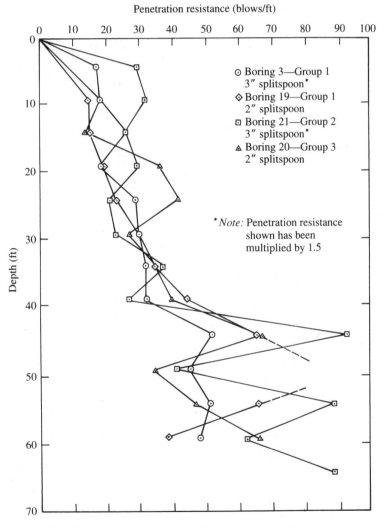

▼ **FIGURE 9.29** Results of the standard penetration test (after Briaud et al., 1989)

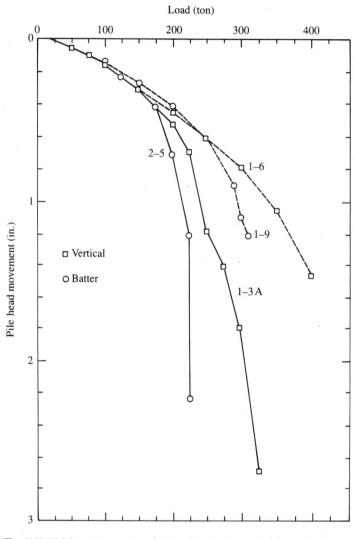

▼ **FIGURE 9.30**   Load test results for H-piles in sand (after Briaud et al., 1989)

given in Figure 9.30. Details of the H-piles and the load test results for these four piles are summarized in Table 9.5. Briaud et al. (1989) made a statistical analysis for the ratio of theoretical ultimate load to the measured ultimate load. The results of this analysis are summarized in Table 9.6 for the plugged case (Figure 9.11c). Note that a perfect prediction would have a mean = 1.0, standard deviation = 0, and a coefficient of variation = 0. Table 9.6 indicates that no method gave a perfect prediction; in general, Q_p was overestimated, and Q_s was underestimated. Again, this shows the uncertainty in predicting the load-bearing capacity of piles.

▼ **TABLE 9.5** Pile Load Test Results

Pile no.	Pile type	Batter	Q_p (ton)	Q_s (ton)	Q_u (ton)	Pile length (ft)
1–3A	HP14 × 73	Vertical	152	161	313	54
1–6	HP14 × 73	Vertical	75	353	428	53
1–9	HP14 × 73	1 : 2.5	85	252	337	58
2–5	HP14 × 73	1 : 2.5	46	179	225	59

Sharma and Joshi (1988) reported the results of field load tests on two cast-in-place concrete piles in a granular soil deposit in Alberta, Canada. The length of these piles (TP-1 and TP-2) was about 12.3 m. Figure 9.31 shows the general soil conditions, pile dimensions, and load–settlement curves. The load transfer mechanism for the two test piles is shown in Figure 9.32. The average skin friction, f_{av}, is calculated as

$$f_{av} = \frac{Q_{top} - Q_{base}}{\pi D_s L} \tag{9.59}$$

where Q_{top} and Q_{base} = loads at the top and base of the pile, respectively
$\qquad\qquad D_s$ = diameter of the pile shaft
$\qquad\qquad L$ = pile length

The variations of f_{av} with load, Q, for the two piles are plotted in Figure 9.33. Note that, for test pile TP-1, the maximum value of f_{av} appears to be about 85 kN/m² at a load of about 4000 kN. In Figure 9.31a, it corresponds to a relative displacement of about 7 mm between the soil and the pile. This result confirms that frictional resistance between the pile and the shaft is fully mobilized in about 5–10 mm of

▼ **TABLE 9.6** Summary of Briaud et al.'s Statistical Analysis for H-Piles — Plugged Case

Theoretical method	Q_p			Q_s			Q_u		
	Mean	Standard deviation	Coefficient of variation	Mean	Standard deviation	Coefficient of variation	Mean	Standard deviation	Coefficient of variation
Coyle and Castello (1981)	2.38	1.31	0.55	0.87	0.36	0.41	1.17	0.44	0.38
Briaud and Tucker (1984)	1.79	1.02	0.59	0.81	0.32	0.40	0.97	0.39	0.40
Meyerhof (1976)	4.37	2.76	0.63	0.92	0.43	0.46	1.68	0.76	0.45
API (1984)	1.62	1.00	0.62	0.59	0.25	0.43	0.79	0.34	0.43

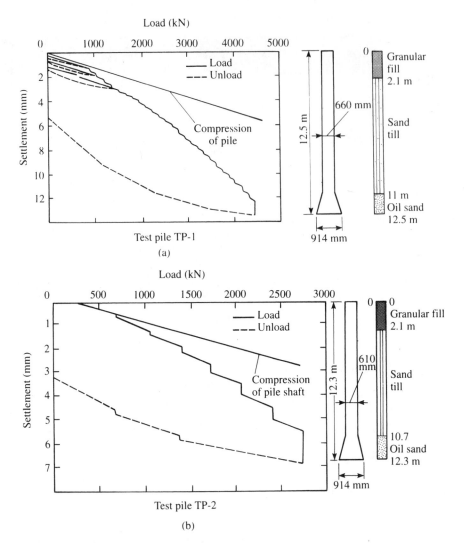

▼ **FIGURE 9.31**   General soil condition, pile dimensions, and load-settlement
curves (after Sharma and Joshi, 1988)

pile head movement (Section 9.5). Again, referring to Eqs. (9.38) and (9.40), we
can say that, in general,

$$f_{av}(kN/m^2) = m\overline{N} \tag{9.60}$$

where m = constant and varies between 1 and 2

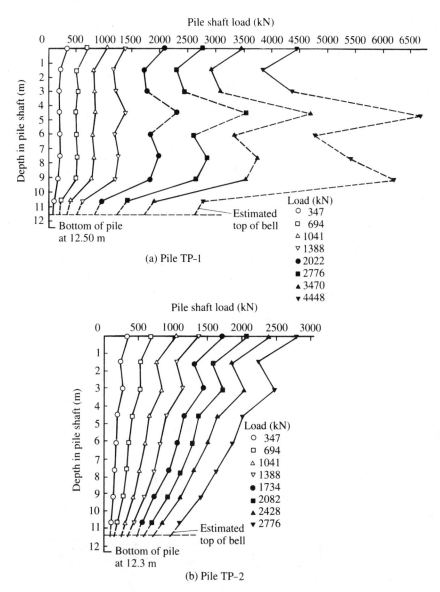

▼ **FIGURE 9.32** Load transfer mechanism for two test piles (after Sharma and Joshi, 1988)

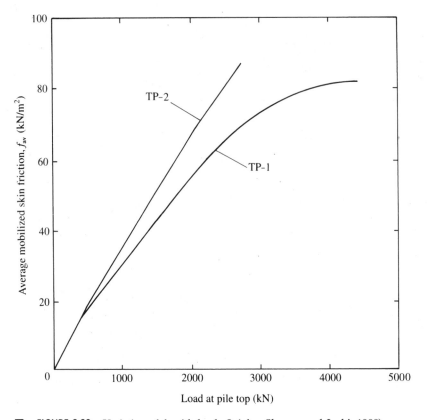

▼ **FIGURE 9.33**   Variation of f_{av} with load, Q (after Sharma and Joshi, 1988)

For test pile TP-1, the shaft length (not including the bell) is about 11 m. Hence the following calculations may be made to determine f_{av}.

Soil	Thickness (m)	$\overline{N}_{cor}$[a]	Average $\overline{N}_{cor}$
Sand and gravel	2.1	15	$\dfrac{(15)(2.1) + (39)(8.9)}{11} = 34.4$
Sand till	8.9	39	
[a] From Sharma and Joshi (1988)			

The experimental value of f_{av} is about 85 kN/m², so from Eq. (9.60),

$$m = \frac{f_{av}}{\overline{N}_{cor}} = \frac{85}{34.4} = 2.47$$

This magnitude is somewhat higher than that given by either Eq. (9.38) or (9.40).

Lessons from the case studies above and others available in the literature show that previous experience and good practical judgment are required along with the knowledge of theoretical developments to design safe pile foundations.

9.17 SETTLEMENT OF PILES

The settlement of a pile under a vertical working load, Q_w, is caused by three factors:

$$s = s_1 + s_2 + s_3 \tag{9.61}$$

where s = total pile settlement
s_1 = elastic settlement of pile
s_2 = settlement of pile caused by the load at the pile tip
s_3 = settlement of pile caused by the load transmitted along the pile shaft

If the pile material is assumed to be elastic, the deformation of the pile shaft can be evaluated using the fundamental principles of mechanics of materials:

$$\boxed{s_1 = \frac{(Q_{wp} + \xi Q_{ws})L}{A_p E_p}} \tag{9.62}$$

where Q_{wp} = load carried at the pile point under working load condition
Q_{ws} = load carried by frictional (skin) resistance under working load condition
A_p = area of pile cross section
L = length of pile
E_p = modulus of elasticity of the pile material

The magnitude of ξ will depend on the nature of unit friction (skin) resistance distribution along the pile shaft. If the distribution of f is uniform or parabolic, as shown in Figure 9.34a and 9.34b, $\xi = 0.5$. However, for triangular distribution of f (Figure 9.34c), the magnitude of ξ is about 0.67 (Vesic, 1977).

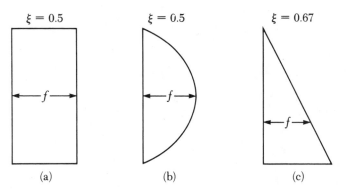

$\xi = 0.5$ $\xi = 0.5$ $\xi = 0.67$

(a) (b) (c)

▼ FIGURE 9.34 Various types of unit friction (skin) resistance distribution along the pile shaft

The settlement of a pile caused by the load carried at the pile point may be expressed in a form similar to that given for shallow foundations [Eq. (4.33)]:

$$s_2 = \frac{q_{wp}D}{E_s}(1 - \mu_s^2)I_{wp}$$

(9.63)

where D = width or diameter of pile
q_{wp} = point load per unit area at the pile point = Q_{wp}/A_p
E_s = modulus of elasticity of soil at or below the pile point
μ_s = Poisson's ratio of soil
I_{wp} = influence factor ≈ 0.85

Vesic (1977) also proposed a semi-empirical method to obtain the magnitude of the settlement, s_2:

$$s_2 = \frac{Q_{wp}C_p}{Dq_p}$$

(9.64)

where q_p = ultimate point resistance of the pile
C_p = an empirical coefficient

Representative values of C_p for various soils are given in Table 9.7.

The settlement of a pile caused by the load carried by the pile shaft is given by a relation similar to Eq. (9.63), or

$$s_3 = \left(\frac{Q_{ws}}{pL}\right)\frac{D}{E_s}(1 - \mu_s^2)I_{ws}$$

(9.65)

where p = perimeter of the pile
L = embedded length of pile
I_{ws} = influence factor

▼ **TABLE 9.7** Typical Values of C_p [Eq. (9.64)]

Soil type	Driven pile	Bored pile
Sand (dense to loose)	0.02–0.04	0.09–0.18
Clay (stiff to soft)	0.02–0.03	0.03–0.06
Silt (dense to loose)	0.03–0.05	0.09–0.12

From "Design of Pile Foundations," by A. S. Vesic, in NCHRP *Synthesis of Highway Practice 42*, Transportation Research Board, 1977. Reprinted by permission.

Note that the term Q_{ws}/pL in Eq. (9.65) is the average value of f along the pile shaft. The influence factor, I_{ws}, has a simple empirical relation (Vesic, 1977):

$$I_{ws} = 2 + 0.35 \sqrt{\frac{L}{D}} \tag{9.66}$$

Vesic (1977) also proposed a simple empirical relation similar to Eq. (9.64) for obtaining s_3:

$$s_3 = \frac{Q_{ws}C_s}{Lq_p} \tag{9.67}$$

where C_s = an empirical constant = $(0.93 + 0.16\sqrt{L/D})C_p$ (9.68)

The values of C_p for use in Eq. (9.67) may be estimated from Table 9.7.

Sharma and Joshi (1988) used Eqs. (9.61), (9.62), (9.64), and (9.67) to estimate the settlement of two concrete piles in sand, as shown previously in Figure 9.31, and compared them to observed values from the field. For these calculations, they used: $\xi = 0.5$ and 0.67, $C_p = 0.02$, and $C_s = 0.02$. Table 9.8 shows the comparison of s values. Note the fairly good agreement between estimated and observed values of settlement.

▼ **TABLE 9.8** Comparison of Observed and Estimated Values of Settlement of Two Concrete Piles (Figure 9.31)

| Pile | Load on pile (kN) | Measured s (mm) | Calculated s | |
			$\xi = 0.5$	$\xi = 0.67$
TP-1	694	1.08	1.456	1.571
	1388	2.91	3.350	3.55
	2776	6.67	7.195	7.535
	4448	13.41	11.67	13.651
TP-2	694	0.65	1.467	1.610
	1388	2.11	3.118	3.387
	2776	6.72	6.889	7.365

▼ **EXAMPLE 9.6**

The allowable working load on a prestressed concrete pile 21 m long that has been driven into sand is 502 kN. The pile is octagonal in shape with $D = 356$ mm (see Table D.3, Appendix D). Skin resistance carries 350 kN of the allowable load, and point bearing carries the rest. Use $E_p = 21 \times 10^6$ kN/m², $E_s = 25 \times 10^3$ kN/m², $\mu_s = 0.35$, and $\xi = 0.62$. Determine the settlement of the pile.

Solution   From Eq. (9.62),

$$s_1 = \frac{(Q_{wp} + \xi Q_{ws})L}{A_p E_p}$$

From Table D.3 (Appendix D) for $D = 356$ mm, the area of pile cross section $A_p = 1045$ cm². Also, perimeter $p = 1.168$ m. Given: $Q_{ws} = 350$ kN, so

$$Q_{wp} = 502 - 350 = 152 \text{ kN}$$

$$s_1 = \frac{[152 + 0.62(350)](21)}{(0.1045 \text{ m}^2)(21 \times 10^6)} = 0.00353 \text{ m} = 3.35 \text{ mm}$$

From Eq. (9.63),

$$s_2 = \frac{q_{wp}D}{E_s}(1 - \mu_s^2)I_{wp} = \left(\frac{152}{0.1045}\right)\left(\frac{0.356}{25 \times 10^3}\right)(1 - 0.35^2)(0.85)$$

$$= 0.0155 \text{ m} = 15.5 \text{ mm}$$

Again, from Eq. (9.65),

$$s_3 = \left(\frac{Q_{ws}}{pL}\right)\left(\frac{D}{E_s}\right)(1 - \mu_s^2)I_{ws}$$

$$I_{ws} = 2 + 0.35\sqrt{\frac{L}{D}} = 2 + 0.35\sqrt{\frac{21}{0.356}} = 4.69$$

$$s_3 = \left[\frac{350}{(1.168)(21)}\right]\left(\frac{0.356}{25 \times 10^3}\right)(1 - 0.35^2)(4.69)$$

$$= 0.00084 \text{ m} = 0.84 \text{ mm}$$

Hence, total settlement is

$$s = s_1 + s_2 + s_3 = 3.35 + 15.5 + 0.84 = \textbf{19.69 mm} \qquad \blacktriangle$$

9.18 PULLOUT RESISTANCE OF PILES

In Section 9.1 we noted that, under certain construction conditions, piles are subjected to uplifting forces. The ultimate resistance of piles subjected to such force did not receive much attention among researchers until recently. The gross ultimate resistance of a pile subjected to uplifting force (Figure 9.35) is

$$T_{ug} = T_{un} + W \tag{9.69}$$

where $T_{ug} = $ gross uplift capacity
    $T_{un} = $ net uplift capacity
    $W = $ effective weight of the pile

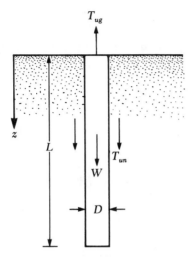

D = diameter or width
of pile

▼ **FIGURE 9.35** Uplift capacity of
piles

Piles in Clay

The net ultimate uplift capacity of piles embedded in saturated clays was studied by Das and Seeley (1982). According to that study,

$$T_{un} = Lp\alpha' c_u$$
(9.70)

where L = length of the pile
p = perimeter of pile section
α' = adhesion coefficient at soil–pile interface
c_u = undrained cohesion of clay

For cast-*in-situ* concrete piles,

$\alpha' = 0.9 - 0.00625c_u$ (for $c_u \leq 80 \, \text{kN/m}^2$)
(9.71)

and

$\alpha' = 0.4$ (for $c_u > 80 \, \text{kN/m}^2$)
(9.72)

Similarly, for pipe piles,

$\alpha' = 0.715 - 0.0191c_u$ (for $c_u \leq 27 \, \text{kN/m}^2$)
(9.73)

and

$\alpha' = 0.2$ (for $c_u > 27 \, \text{kN/m}^2$)
(9.74)

Piles in Sand

When piles are embedded in granular soils ($c = 0$), the net ultimate uplift capacity (Das and Seeley, 1975) is

$$T_{un} = \int_0^L (f_u p) \, dz \tag{9.75}$$

where f_u = unit skin friction during uplift
  p = perimeter of pile cross section

The unit skin friction during uplift, f_u, usually varies as shown in Figure 9.36a. It increases linearly to a depth of $z = L_{cr}$; beyond that it remains constant. For $z \leq L_{cr}$,

$$f_u = K_u \sigma_v' \tan \delta \tag{9.76}$$

where K_u = uplift coefficient
  σ_v' = effective vertical stress at a depth of z
  δ = soil–pile friction angle

The variation of the uplift coefficient with soil friction angle ϕ is given in Figure 9.36b. Based on the author's experience, the values of L_{cr} and δ appear to depend on the relative density of soil. Figure 9.36c shows the approximate nature of these variations with the relative density of soil. For calculating the net ultimate uplift capacity of piles, the following procedure is suggested:

1. Determine the relative density of the soil and, using Figure 9.36c, obtain the value of L_{cr}.
2. If the length of the pile, L, is less than or equal to L_{cr},

$$T_{un} = p \int_0^L f_u \, dz = p \int_0^L (\sigma_v' K_u \tan \delta) \, dz \tag{9.77}$$

In dry soils, $\sigma_v' = \gamma z$ (where γ = unit weight of soil), so

$$T_{un} = p \int_0^L (\sigma_v' K_u \tan \delta) \, dz = p \int_0^L \gamma z K_u \tan \delta \, dz$$
$$= \frac{1}{2} p \gamma L^2 K_u \tan \delta \tag{9.78}$$

Obtain the values of K_u and δ from Figure 9.36b and 9.36c.

3. For $L > L_{cr}$,

$$T_{un} = p \int_0^L f_u \, dz = p \left[\int_0^{L_{cr}} f_u \, dz + \int_{L_{cr}}^L f_u \, dz \right]$$
$$= p \left\{ \int_0^{L_{cr}} \left[\sigma_v' K_u \tan \delta \right] dz + \int_{L_{cr}}^L \left[\sigma_{v(\text{at } z = L_{cr})}' K_u \tan \delta \right] dz \right\} \tag{9.79}$$

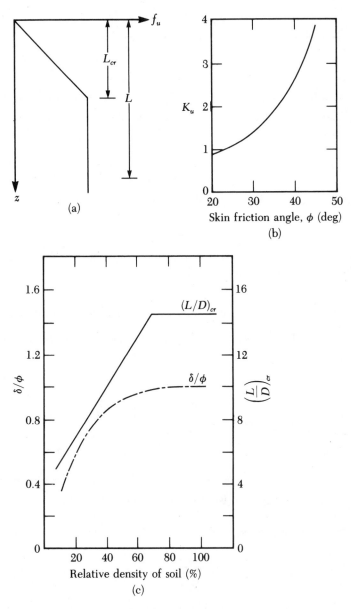

▼ **FIGURE 9.36** (a) Nature of variation of f_u; (b) uplift coefficient K_u; (c) variation of δ/ϕ and $(L/D)_{cr}$ with relative density of sand

For dry soils, Eq. (9.79) simplifies to

$$T_{un} = \tfrac{1}{2}p\gamma L_{cr}^2 K_u \tan\delta + p\gamma L_{cr}K_u \tan\delta(L - L_{cr}) \qquad (9.80)$$

Determine the values of K_u and δ from Figure 9.36b and 9.36c.

For estimating the net allowable uplift capacity, a factor of safety of 2–3 is recommended. Thus

$$T_{u(\text{all})} = \frac{T_{ug}}{FS}$$

where $T_{u(\text{all})}$ = allowable uplift capacity

▼ **EXAMPLE 9.7**

A concrete pile 50 ft long is embedded in a saturated clay with $c_u = 850$ lb/ft². The pile is 12 in. × 12 in. in cross section. Use $FS = 4$ and determine the allowable pullout capacity of the pile.

Solution   Given: $c_u = 850$ lb/ft² ≈ 40.73 kN/m². From Eq. (9.71),

$$\alpha' = 0.9 - 0.00625c_u = 0.9 - (0.00625)(40.73) = 0.645$$

From Eq. (9.70),

$$T_{un} = Lp\alpha'c_u = \frac{(50)(4 \times 1)(0.645)(850)}{1000} = 109.7 \text{ kip}$$

$$T_{un(\text{all})} = \frac{109.7}{FS} = \frac{109.7}{4} = \mathbf{27.4 \text{ kip}}$$

▲

▼ **EXAMPLE 9.8**

A precast concrete pile with a cross section of 350 mm × 350 mm is embedded in sand. The length of the pile is 15 m. Assume that $\gamma_{\text{sand}} = 15.8$ kN/m³, $\phi_{\text{sand}} = 35°$, and the relative density of sand = 70%. Estimate the allowable pullout capacity of the pile ($FS = 4$).

Solution   From Figure 9.36, for $\phi = 35°$ and relative density = 70%,

$$\left(\frac{L}{D}\right)_{cr} = 14.5; L_{cr} = (14.5)(0.35 \text{ m}) = 5.08 \text{ m}$$

$$\frac{\delta}{\phi} = 1; \delta = (1)(35) = 35°$$

$$K_u = 2$$

From Eq. (9.80),

$$T_{un} = \tfrac{1}{2}p\gamma L_{cr}^2 K_u \tan\delta + p\gamma L_{cr}K_u (L - L_{cr}) \tan\delta$$

$$= (\tfrac{1}{2})(0.35 \times 4)(15.8)(5.08)^2(2)\tan 35$$
$$+ (0.35 \times 4)(15.8)(5.08)(2)(15 - 5.08)\tan 35 = 1961 \text{ kN}$$

$$T_{un(\text{all})} = \frac{1961}{FS} = \frac{1961}{4} \approx \textbf{490 kN}$$

▲

9.19 LATERALLY LOADED PILES

A vertical pile resists lateral load by mobilizing passive pressure in the soil surrounding it (Figure 9.1c). The degree of distribution of the soil reaction depends on (a) the stiffness of the pile, (b) the stiffness of the soil, and (c) the fixity of the ends of the pile. In general, laterally loaded piles can be divided into two major categories: (1) short or rigid piles and (2) long or elastic piles. Figure 9.37a and

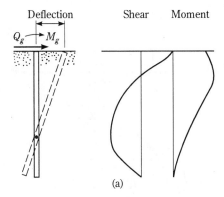

(a)

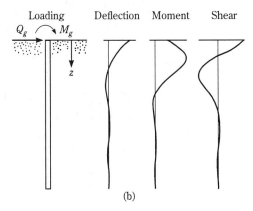

(b)

▼ **FIGURE 9.37** Nature of variation of pile deflection, moment, and shear force for (a) rigid pile, (b) elastic pile

9.37b show the nature of variation of the pile deflection and the moment and shear force distribution along the pile length when subjected to lateral loading. Following is a summary of the solutions presently available for laterally loaded piles.

Elastic Solution

A general method for determining moments and displacements of a vertical pile embedded in a *granular soil* and subjected to lateral load and moment at the ground surface was given by Matlock and Reese (1960). Consider a pile of length L subjected to a lateral force Q_g and a moment M_g at the ground surface ($z = 0$), as shown in Figure 9.38a. Figure 9.38b shows the general deflected shape of the pile and the soil resistance caused by the applied load and the moment.

According to a simpler Winkler's model, an elastic medium (soil in this case) can be replaced by a series of infinitely close independent elastic springs. Based on this assumption,

$$k = \frac{p'\,(\text{kN/m or lb/ft})}{x\,(\text{m or ft})} \qquad (9.81)$$

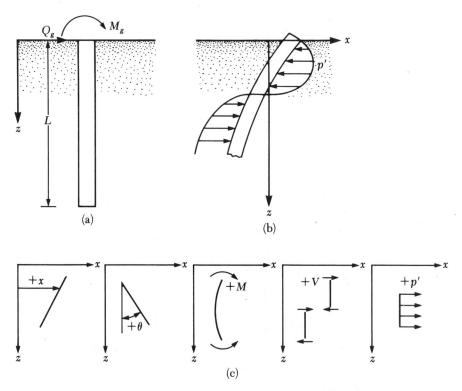

▼ **FIGURE 9.38** (a) Laterally loaded pile; (b) soil resistance on pile caused by lateral load; (c) sign conventions for displacement, slope, moment, shear, and soil reaction

where k = modulus of subgrade reaction
p' = pressure on soil
x = deflection

The subgrade modulus for *granular soils* at a depth z is defined as

$$k_z = n_h z \tag{9.82}$$

where n_h = constant of modulus of horizontal subgrade reaction

Referring to Figure 9.38b and using the theory of beams on an elastic foundation, we can write

$$E_p I_p \frac{d^4 x}{dz^4} = p' \tag{9.83}$$

where E_p = modulus of elasticity in the pile material
I_p = moment of inertia of the pile section

Based on Winkler's model

$$p' = -kx \tag{9.84}$$

The sign in Eq. (9.84) is negative because the soil reaction is in the direction opposite to the pile deflection.

Combining Eqs. (9.83) and (9.84) gives

$$E_p I_p \frac{d^4 x}{dz^4} + kx = 0 \tag{9.85}$$

The solution of Eq. (9.85) results in the following expressions:

Pile Deflection at Any Depth $[x_z(z)]$

$$x_z(z) = A_x \frac{Q_g T^3}{E_p I_p} + B_x \frac{M_g T^2}{E_p I_p} \tag{9.86}$$

Slope of Pile at Any Depth $[\theta_z(z)]$

$$\theta_z(z) = A_\theta \frac{Q_g T^2}{E_p I_p} + B_\theta \frac{M_g T}{E_p I_p} \tag{9.87}$$

Moment of Pile at Any Depth $[M_z(z)]$

$$M_z(z) = A_m Q_g T + B_m M_g \tag{9.88}$$

Shear Force on Pile at Any Depth $[V_z(z)]$

$$V_z(z) = A_v Q_g + B_v \frac{M_g}{T} \qquad (9.89)$$

Soil Reaction at Any Depth $[p_z'(z)]$

$$p_z'(z) = A_{p'} \frac{Q_g}{T} + B_{p'} \frac{M_g}{T^2} \qquad (9.90)$$

where A_x, B_x, A_θ, B_θ, A_m, B_m, A_v, B_v, $A_{p'}$, and $B_{p'}$ are coefficients

T = characteristic length of the soil–pile system

$$= \sqrt[5]{\frac{E_p I_p}{n_h}} \qquad (9.91)$$

n_h has been defined in Eq. (9.82)

When $L \geq 5T$, the pile is considered to be a *long pile*. For $L \leq 2T$, the pile is considered to be a *rigid pile*. Table 9.9 gives the values of the coefficients for long piles ($L/T \geq 5$) in Eqs. (9.86) to (9.90). Note that, in the first column of Table 9.9,

▼ **TABLE 9.9** Coefficients for Long Piles, $k_z = n_h z$

Z	A_x	A_θ	A_m	A_v	A_p'	B_x	B_θ	B_m	B_v	B_p'
0.0	2.435	−1.623	0.000	1.000	0.000	1.623	−1.750	1.000	0.000	0.000
0.1	2.273	−1.618	0.100	0.989	−0.227	1.453	−1.650	1.000	−0.007	−0.145
0.2	2.112	−1.603	0.198	0.956	−0.422	1.293	−1.550	0.999	−0.028	−0.259
0.3	1.952	−1.578	0.291	0.906	−0.586	1.143	−1.450	0.994	−0.058	−0.343
0.4	1.796	−1.545	0.379	0.840	−0.718	1.003	−1.351	0.987	−0.095	−0.401
0.5	1.644	−1.503	0.459	0.764	−0.822	0.873	−1.253	0.976	−0.137	−0.436
0.6	1.496	−1.454	0.532	0.677	−0.897	0.752	−1.156	0.960	−0.181	−0.451
0.7	1.353	−1.397	0.595	0.585	−0.947	0.642	−1.061	0.939	−0.226	−0.449
0.8	1.216	−1.335	0.649	0.489	−0.973	0.540	−0.968	0.914	−0.270	−0.432
0.9	1.086	−1.268	0.693	0.392	−0.977	0.448	−0.878	0.885	−0.312	−0.403
1.0	0.962	−1.197	0.727	0.295	−0.962	0.364	−0.792	0.852	−0.350	−0.364
1.2	0.738	−1.047	0.767	0.109	−0.885	0.223	−0.629	0.775	−0.414	−0.268
1.4	0.544	−0.893	0.772	−0.056	−0.761	0.112	−0.482	0.688	−0.456	−0.157
1.6	0.381	−0.741	0.746	−0.193	−0.609	0.029	−0.354	0.594	−0.477	−0.047
1.8	0.247	−0.596	0.696	−0.298	−0.445	−0.030	−0.245	0.498	−0.476	0.054
2.0	0.142	−0.464	0.628	−0.371	−0.283	−0.070	−0.155	0.404	−0.456	0.140
3.0	−0.075	−0.040	0.225	−0.349	0.226	−0.089	0.057	0.059	−0.213	0.268
4.0	−0.050	0.052	0.000	−0.106	0.201	−0.028	0.049	−0.042	0.017	0.112
5.0	−0.009	0.025	−0.033	0.015	0.046	0.000	−0.011	−0.026	0.029	−0.002

From *Drilled Pier Foundations*, by R. J. Woodwood, W. S. Gardner, and D. M. Greer. Copyright 1972 by McGraw-Hill. Used with the permission of McGraw-Hill Book Company.

Z is the nondimensional depth, or

$$Z = \frac{z}{T} \tag{9.92}$$

The positive sign conventions for $x_z(z)$, $\theta_z(z)$, $M_z(z)$, $V_z(z)$, and $p'_z(z)$ assumed in the derivations in Table 9.9 are shown in Figure 9.38c. Also, Figure 9.39 shows the variation of A_x, B_x, A_m, and B_m for various values of $L/T = Z_{max}$. It indicates that, when L/T is greater than about 5, the coefficients do not change, which is true of long piles only.

Calculating the characteristic length T for the pile requires assuming a proper value of n_h. Table 9.10 gives some representative values of n_h.

Elastic solutions similar to those given in Eqs. (9.86)–(9.90) for piles embedded in *cohesive soil* were developed by Davisson and Gill (1963). These relationships are given in Eqs. 9.93–9.97.

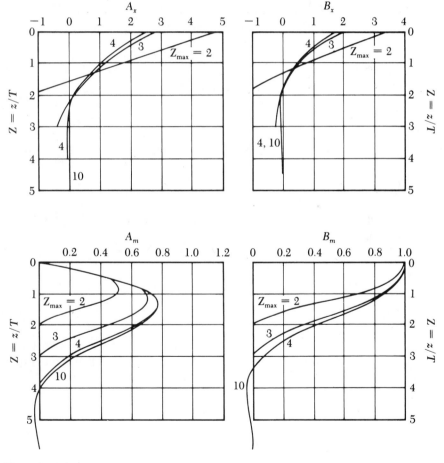

▼ **FIGURE 9.39**   Variation of A_x, B_x, A_m, and B_m with Z (after Matlock and Reese, 1960)

▼   **TABLE 9.10**   Representative Values of n_h

	n_k	
Soil	lb/in³	kN/m³
Dry or moist sand		
Loose	6.5–8.0	1800–2200
Medium	20–25	5500–7000
Dense	55–65	15,000–18,000
Submerged sand		
Loose	3.5–5.0	1000–1400
Medium	12–18	3500–4500
Dense	32–45	9000–12,000

$$x_z(z) = A'_x \frac{Q_g R^3}{E_p I_p} + B'_x \frac{M_g R^2}{E_p I_p}$$

(9.93)

and

$$M_z(z) = A'_m Q_g R + B'_m M_g$$

(9.94)

where A'_x, B'_x, A'_m, and B'_m are coefficients

$$R = \sqrt[4]{\frac{E_p I_p}{k}}$$

(9.95)

The values of the A' and B' coefficients are given in Figure 9.40. Note that

$$Z = \frac{z}{R}$$

(9.96)

and

$$Z_{max} = \frac{L}{R}$$

(9.97)

The use of Eqs. (9.93) and (9.94) requires knowing the magnitude of the characteristic length, R. It can be calculated from Eq. (9.95), provided the coefficient of the subgrade reaction is known. For sands, the coefficient of subgrade reaction was given by Eq. (9.82), which showed a linear variation with depth. However, in cohesive soils, the subgrade reaction may be assumed to be approximately constant with depth. Vesic (1961) proposed the following equation to estimate the value of k:

$$k = 0.65 \sqrt[12]{\frac{E_s D^4}{E_p I_p}} \frac{E_s}{1 - \mu_s^2}$$

(9.98)

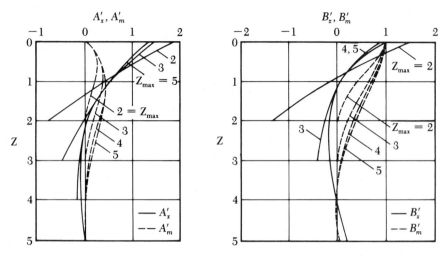

▼ FIGURE 9.40 Variation of A'_x, B'_x, A'_m, and B'_m with Z (after Davisson and Gill, 1963)

where E_s = modulus of elasticity of soil
        D = pile width (or diameter)
        μ_s = Poisson's ratio of the soil

Ultimate Load Analysis—Broms' Method

Broms (1965) developed a simplified solution for laterally loaded piles based on the assumptions of (a) shear failure in soil, which is the case for short piles, and (b) bending of the pile governed by plastic yield resistance of the pile section, which is applicable for long piles. Broms' solution for calculating the ultimate load resistance, $Q_{u(g)}$, for *short piles* is given in Figure 9.41a. A similar solution for piles embedded in cohesive soil is shown in Figure 9.41b. In using Figure 9.41a, note that

$$K_p = \text{Rankine passive earth pressure coefficient} = \tan^2\left(45 + \frac{\phi}{2}\right) \qquad (9.99)$$

Similarly, in Figure 9.41b,

$$c_u = \text{undrained cohesion} \approx \frac{0.75q_u}{FS} = \frac{0.75q_u}{2} = 0.375q_u \qquad (9.100)$$

where FS = factor of safety (=2)
        q_u = unconfined compression strength

Figure 9.42 shows Broms' analysis of long piles. In this figure, M_y is the yield moment for the pile, or

$$M_y = SF_Y \qquad (9.101)$$

where S = section modulus of the pile section

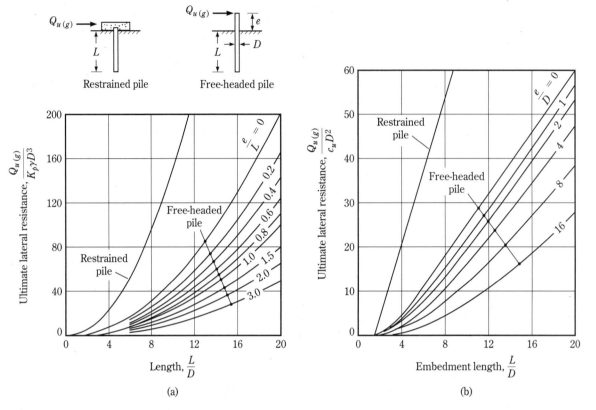

▼ FIGURE 9.41 Broms' solution for ultimate lateral resistance of short piles (a) in sand, (b) in clay

F_Y = yield stress of the pile material

In solving a given problem, both cases (that is, Figure 9.41 and Figure 9.42) should be checked.

The deflection of the pile head, x_o, under working load conditions can be estimated from Figure 9.43. In Figure 9.43a, the term η can be expressed as

$$\eta = \sqrt[5]{\frac{n_h}{E_p I_p}} \qquad (9.102)$$

The range of n_h for granular soil is given in Table 9.10. Similarly in Figure 9.43b, which is for clay, the term K is the horizontal soil modulus and can be defined as

$$K = \frac{\text{pressure (lb/in}^2 \text{ or kN/m}^2)}{\text{displacement (in. or m)}} \qquad (9.103)$$

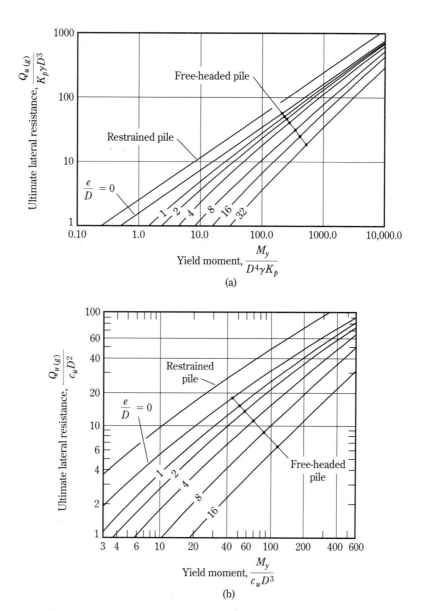

▼ **FIGURE 9.42** Broms' solution for ultimate lateral resistance of long piles
(a) in sand, (b) in clay

Also, the term β can be defined as

$$\beta = \sqrt[4]{\frac{KD}{4E_pI_p}}$$

(9.104)

Note that, in Figure 9.43, Q_g is the working load.

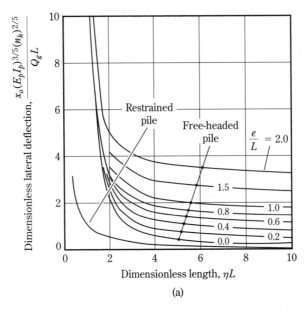

(a)

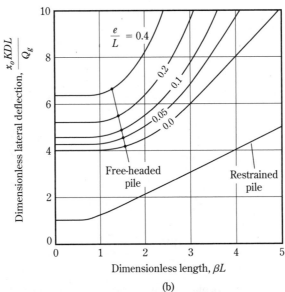

(b)

▼ **FIGURE 9.43** Broms' solution for estimating deflection of pile head (a) in sand, and (b) in clay

Ultimate Load Analysis—Meyerhof's Method

More recently, Meyerhof (1995) provided the solutions for laterally loaded rigid and flexible piles (Figure 9.44), which are summarized below. According to Meyerhof's method, a pile can be defined as flexible if

$$K_r = \text{relative stiffness of pile} = \frac{E_p I_p}{E_s L^4} < 0.01 \qquad (9.105)$$

where E_s = average horizontal soil modulus of elasticity

Piles in Sand   For short (rigid) piles in *sand*, the ultimate load resistance can be given as

$$Q_{u(g)} = 0.12\,\gamma D L^2 K_{br} \le 0.4 p_l DL \qquad (9.106)$$

where γ = unit weight of soil
   K_{br} = resultant net soil pressure coefficient (Figure 9.45)
   p_l = limit pressure obtained from pressuremeter tests (Chapter 2)

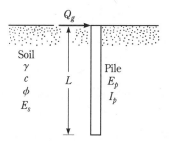

▼ **FIGURE 9.44**   Pile with lateral loading at ground level

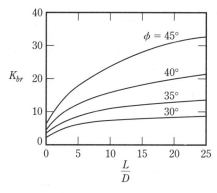

▼ **FIGURE 9.45**   Variation of resultant net soil pressure coefficient, K_{br}

The limit pressure, p_l, can be given as

$$p_l = 40N_q \tan \phi \text{ (kPa)} \qquad \text{(for Menard pressuremeter)} \tag{9.107}$$

and

$$p_l = 60N_q \tan \phi \text{ (kPa)}$$
$$\text{(for self-boring and full displacement pressuremeters)} \tag{9.108}$$

where N_q = bearing capacity factor (Table 3.4)

The maximum moment, $M_{\max}$, in the pile due to the lateral load, $Q_{u(g)}$, is

$$M_{\max} = 0.35Q_{u(g)}L \le M_y$$
$$\uparrow$$
$$\text{Eq. (9.101)} \tag{9.109}$$

For long (flexible) piles in sand, the ultimate lateral load, $Q_{u(g)}$, can be estimated from Eq. (9.106) by substituting an effective length (L_e) for L where

$$\frac{L_e}{L} = 1.65K_r^{0.12} \le 1 \tag{9.110}$$

The maximum moment in a flexible pile due to a working lateral load Q_g applied at the ground surface is

$$M_{\max} = 0.3K_r^{0.2}Q_gL \le 0.3Q_gL \tag{9.111}$$

Piles in Clay The ultimate lateral load, $Q_{u(g)}$, applied at the ground surface for short (rigid) piles embedded in clay can be given as

$$Q_{u(g)} = 0.4c_uK_{cr}DL \le 0.4p_lDL \tag{9.112}$$

where p_l = limit pressure from pressuremeter test
K_{cr} = net soil pressure coefficient (Figure 9.46)

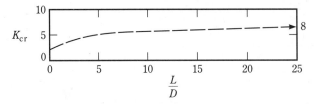

▼ **FIGURE 9.46** Variation of K_{cr}

The limit pressure in clay is

$$p_l \approx 6c_u \quad \text{(for Menard pressuremeter)} \tag{9.113}$$

$$p_l \approx 8c_u \quad \text{(for self-boring and full displacement pressuremeter)} \tag{9.114}$$

The maximum bending moment in the pile due to $Q_{u(g)}$ is

$$M_{\max} = 0.22Q_{u(g)}L \leq M_y \tag{9.115}$$
$$\uparrow$$
$$\text{Eq. (9.101)}$$

For long (flexible) piles, Eq. (9.112) can be used to estimate $Q_{u(g)}$ by substituting the effective length (L_e) in place of L.

$$\frac{L_e}{L} = 1.5K_r^{0.12} \leq 1 \tag{9.116}$$

The maximum moment in a flexible pile due to a working lateral load Q_g applied at the ground surface is

$$M_{\max} = 0.3K_r^{0.2}Q_gL \leq 0.15Q_gL \tag{9.117}$$

▼ **EXAMPLE 9.9**

Consider a steel H-pile (HP 250 × 0.834) 25 m long embedded fully in a granular soil. Assume that $n_h = 12,000 \text{ kN/m}^3$. The allowable displacement at the top of the pile is 8 mm. Determine the allowable lateral load, Q_g. Assume that $M_g = 0$. Use the elastic solution.

Solution From Table D.1b (Appendix D) for an HP 250 × 0.834 pile,

$$I_p = 123 \times 10^{-6} \text{ m}^4 \quad \text{(about the strong axis)}$$

$$E_p = 207 \times 10^6 \text{ kN/m}^2$$

From Eq. (9.91),

$$T = \sqrt[5]{\frac{E_pI_p}{n_h}} = \sqrt[5]{\frac{(207 \times 10^6)(123 \times 10^{-6})}{12,000}} = 1.16 \text{ m}$$

Here, $L/T = 25/1.16 = 21.55 > 5$, so it is a long pile. Because $M_g = 0$, Eq. (9.86) takes the form

$$x_z(z) = A_x \frac{Q_gT^3}{E_pI_p}$$

and

$$Q_g = \frac{x_z(z)E_pI_p}{A_xT^3}$$

At $z = 0$, $x_z = 8$ mm $= 0.008$ m, and $A_x = 2.435$ (Table 9.9), so

$$Q_g = \frac{(0.008)(207 \times 10^6)(123 \times 10^{-6})}{(2.435)(1.16^3)} = 53.59 \text{ kN}$$

This magnitude of Q_g is based on the *limiting displacement condition only*. However, the magnitude of Q_g based on the *moment capacity* of the pile also needs to be determined. For $M_g = 0$, Eq. (9.88) becomes

$$M_z(z) = A_mQ_gT$$

According to Table 9.9, the maximum value of A_m at any depth is 0.772. The maximum allowable moment that the pile can carry is

$$M_{z(\text{max})} = F_Y\frac{I_p}{\dfrac{d_1}{2}}$$

Let $F_Y = 248,000$ kN/m². From Table D.1b, $I_p = 123 \times 10^{-6}$ m⁴ and $d_1 = 0.254$ m, so

$$\frac{I_p}{\left(\dfrac{d_1}{2}\right)} = \frac{123 \times 10^{-6}}{\left(\dfrac{0.254}{2}\right)} = 968.5 \times 10^{-6} \text{ m}^3$$

Now

$$Q_g = \frac{M_{z(\text{max})}}{A_mT} = \frac{(968.5 \times 10^{-6})(248,000)}{(0.772)(1.16)} = 268.2 \text{ kN}$$

Because $Q_g = 268.2$ kN > 53.59 kN, the deflection criteria apply. Hence $Q_g = $ **53.59 kN.**   ▲

▼ **EXAMPLE 9.10**

Solve Example 9.9 by Broms' method. Assume that the pile is flexible and is free headed. Given: yield stress of pile material, $F_y = 248$ MN/m²; unit weight of soil, $\gamma = 18$ kN/m³; and soil friction angle, $\phi = 35°$.

Solution   Check for bending failure. From Eq. (9.101),

$$M_y = SF_y$$

From Table D.1b (Appendix D),

$$S = \frac{I_p}{\dfrac{d_1}{2}} = \frac{123 \times 10^{-6}}{\dfrac{0.254}{2}}$$

$$M_y = \left[\frac{123 \times 10^{-6}}{\dfrac{0.254}{2}}\right](248 \times 10^3) = 240.2 \text{ kN-m}$$

$$\frac{M_y}{D^4\gamma K_p} = \frac{M_y}{D^4\gamma \tan^2\left(45 + \dfrac{\phi}{2}\right)} = \frac{240.2}{(0.250)^4(18)\tan^2\left(45 + \dfrac{35}{2}\right)} = 925.8$$

From Figure 9.42a, for $M_y/D^4\gamma K_p = 925.8$, the magnitude of $Q_{u(g)}/K_p D^3\gamma$ (for free-headed pile with $e/D = 0$) is about 140, so

$$Q_{u(g)} = 140 K_p D^3 \gamma = 140 \tan^2\left(45 + \frac{35}{2}\right)(0.25)^2(18) = 581.2 \text{ kN}$$

Check for pile head deflection. From Eq. (9.102),

$$\eta = \sqrt[5]{\frac{n_h}{E_p I_p}} = \sqrt[5]{\frac{12,000}{(207 \times 10^6)(123 \times 10^{-6})}} = 0.86 \text{ m}^{-1}$$

$$\eta L = (0.86)(25) = 21.5$$

From Figure 9.43a, for $\eta L = 21.5$, $e/L = 0$ (free-headed pile),

$$\frac{x_o(E_p I_p)^{3/5}(n_h)^{2/5}}{Q_g L} \approx 0.15 \qquad \text{(by interpolation)}$$

$$Q_g = \frac{x_o(E_p I_p)^{3/5}(n_h)^{2/5}}{0.15L}$$

$$= \frac{(0.008)[(207 \times 10^6)(123 \times 10^{-6})]^{3/5}(12,000)^{2/5}}{(0.15)(25)} = 40.2 \text{ kN}$$

Hence, $Q_g = $ **40.2 kN (<581.2 kN)**. ▲

9.20 PILE-DRIVING FORMULAS

To develop the desired load-carrying capacity, a point bearing pile must penetrate the dense soil layer sufficiently or have sufficient contact with a layer of rock. This requirement cannot always be satisfied by driving a pile to a predetermined depth because soil profiles vary. For that reason, several equations have been developed to calculate the ultimate capacity of a pile during driving. These dynamic equations are widely used in the field to determine whether the pile has reached satisfactory bearing value at the predetermined depth. One of the earliest of these dynamic equations—commonly referred to as the *Engineering News Record (ENR) formula*—is derived from the work–energy theory. That is,

Energy imparted by the hammer per blow =
(pile resistance)(penetration per hammer blow)

According to the ENR formula, the pile resistance is the ultimate load Q_u, expressed as

$$Q_u = \frac{W_R h}{S + C} \qquad (9.118)$$

where W_R = weight of the ram (for example, see Table D.4, Appendix D)
   h = height of fall of the ram
   S = penetration of pile per hammer blow
   C = a constant

The pile penetration, S, is usually based on the average value obtained from the last few driving blows. In the equation's original form, the following values of C were recommended.

   For drop hammers: $C = 1$ in. (if the units of S and h are in inches)
   For steam hammers: $C = 0.1$ in. (if the units of S and h are in inches)

Also, a factor of safety, $FS = 6$, was recommended to estimate the allowable pile capacity. Note that, for single- and double-acting hammers, the term $W_R h$ can be replaced by EH_E (where E = hammer efficiency and H_E = rated energy of hammer). Thus

$$Q_u = \frac{EH_E}{S + C} \qquad (9.119)$$

The ENR formula has been revised several times over the years, and other piledriving formulas also have been suggested. Some of them are tabulated in Table 9.11.

The maximum stress developed on a pile during the driving operation can be estimated from the pile-driving formulas presented in Table 9.11. To illustrate, we use the modified ENR formula:

$$Q_u = \frac{EW_R h}{S + C} \frac{W_R + n^2 W_p}{W_R + W_p}$$

In this equation, S equals the average penetration per hammer blow, which can also be expressed as

$$S = \frac{1}{N} \qquad (9.120)$$

where S is in inches
   N = number of hammer blows per inch of penetration

▼ **TABLE 9.11** Pile-Driving Formulas

Name	Formula
Modified ENR formula	$Q_u = \dfrac{EW_R h}{S + C} \dfrac{W_R + n^2 W_p}{W_R + W_p}$ where E = hammer efficiency $C = 0.1$ in., if the units of S and h are in inches W_p = weight of the pile n = coefficient of restitution between the ram and the pile cap Typical values for E Single- and double-acting hammers 0.7–0.85 Diesel hammers 0.8–0.9 Drop hammers 0.7–0.9 Typical values for n Cast iron hammer and concrete piles (without cap) 0.4–0.5 Wood cushion on steel piles 0.3–0.4 Wooden piles 0.25–0.3
Michigan State Highway Commission formula (1965)	$Q_u = \dfrac{1.25 E H_E}{S + C} \dfrac{W_R + n^2 W_p}{W_R + W_p}$ where H_E = manufacturer's maximum rated hammer energy (lb-in.) E = hammer efficiency $C = 0.1$ in. A factor of safety of 6 is recommended.
Danish formula (Olson and Flaate, 1967)	$Q_u = \dfrac{E H_E}{S + \sqrt{\dfrac{E H_E L}{2 A_p E_p}}}$ where E = hammer efficiency H_E = rated hammer energy E_p = modulus of elasticity of the pile material L = length of pile A_p = area of the pile cross section
Pacific Coast Uniform Building Code formula (International Conference of Building Officials, 1982)	$Q_u = \dfrac{(E H_E)\left(\dfrac{W_R + n W_p}{W_R + W_p}\right)}{S + \dfrac{Q_u L}{A E_p}}$ The value of n should be 0.25 for steel piles and 0.1 for all other piles. A factor of safety of 4 is generally recommended.

(continued)

▼ **TABLE 9.11** Continued

Name	Formula
Janbu's formula (Janbu, 1953)	$Q_u = \dfrac{EH_E}{K'_u S}$ where $K'_u = C_d\left(1 + \sqrt{1 + \dfrac{\lambda'}{C_d}}\right)$ $C_d = 0.75 + 0.14\left(\dfrac{W_p}{W_R}\right)$ $\lambda' = \left(\dfrac{EH_E L}{A_p E_p S^2}\right)$
Gates formula (Gates, 1957)	$Q_u = a\sqrt{EH_E}\,(b - \log S)$ If Q_u is in kips, then S is in in., $a = 27$, $b = 1$, and $\dot{H}_E$ is in kip-ft. If Q_u is in kN, then S is in mm, $a = 104.5$, $b = 2.4$, and H_E is in kN-m. $E = 0.75$ for drop hammer; $E = 0.85$ for all other hammers Use a factor of safety of 3.
Navy-McKay formula	$Q_u = \dfrac{EH_E}{S\left(1 + 0.3\dfrac{W_P}{W_R}\right)}$ Use a factor of safety of 6.

Thus

$$Q_u = \frac{EW_R h}{(1/N) + 0.1}\,\frac{W_R + n^2 W_p}{W_R + W_p} \tag{9.121}$$

Different values of N may be assumed for a given hammer and pile and Q_u calculated. The driving stress can then be calculated for each value of N and Q_u/A_p. This procedure can be demonstrated with a set of numerical values. Assume that a prestressed concrete pile 80 ft in length has to be driven by an 11B3 (MKT) hammer. The pile sides measure 10 in. From Table D.3a (Appendix D) for this pile,

$$A_p = 100 \text{ in}^2$$

The weight of the pile is

$$A_p L \gamma_c = \left(\frac{100 \text{ in}^2}{144}\right)(80 \text{ ft})(150 \text{ lb/ft}^3) = 8.33 \text{ kip}$$

If the weight of the cap is 0.67 kip,

$$W_p = 8.33 + 0.67 = 9 \text{ kip}$$

Again, from Table D.4 (Appendix D) for an 11B3 hammer,

Rated energy = 19.2 kip-ft = $H_E = W_R h$
Weight of ram = 5 kip

Assume that the hammer efficiency is 0.85 and that $n = 0.35$. Substituting these values in Eq. (9.121) yields

$$Q_u = \left[\frac{(0.85)(19.2 \times 12)}{\dfrac{1}{N} + 0.1} \right] \left[\frac{5 + (0.35)^2(9)}{5 + 9} \right] = \frac{85.37}{\dfrac{1}{N} + 0.1} \text{ kip}$$

Now the following table can be prepared:

N	Q_u (kip)	A_p (in^2)	Q_u/A_p (kip/in^2)
0	0	100	0
2	142.3	100	1.42
4	243.9	100	2.44
6	320.1	100	3.20
8	379.4	100	3.79
10	426.9	100	4.27
12	465.7	100	4.66
20	569.1	100	5.69

Both the number of hammer blows per inch and the stress can now be plotted in a graph, as shown in Figure 9.47. If such a curve is prepared, the number of

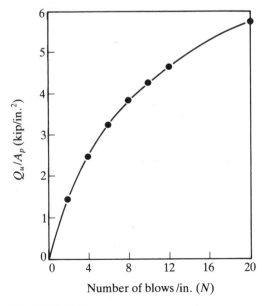

▼ FIGURE 9.47

blows per inch of pile penetration corresponding to the allowable pile-driving stress can be easily determined.

Actual driving stresses in wooden piles are limited to about $0.7f_u$. Similarly, for concrete and steel piles, driving stresses are limited to about $0.6f'_c$ and $0.85f_y$, respectively.

In most cases, wooden piles are driven with a hammer energy of less than 45 kip-ft ($\approx$60 kN·m). Driving resistances are limited mostly to 4–5 blows per inch of pile penetration. For concrete and steel piles, the usual N values are 6–8 and 12–14, respectively.

▼ **EXAMPLE 9.11** _____

A precast concrete pile 12 in. × 12 in. in cross section is driven by a hammer. Given:

> Maximum rated hammer energy = 30 kip-ft
> Hammer efficiency = 0.8
> Weight of ram = 7.5 kip
> Pile length = 80 ft
> Coefficient of restitution = 0.4
> Weight of pile cap = 550 lb
> $E_p = 3 \times 10^6$ kip/in²
> Number of blows for last 1 in. of penetration = 8

Estimate the allowable pile capacity by the

> a. Modified ENR formula (use $FS = 6$)
> b. Danish formula (use $FS = 4$)
> c. Gates formula (use $FS = 3$)

Solution

Part a

$$Q_u = \frac{EW_R h}{S + C} \frac{W_R + n^2 W_p}{W_R + W_p}$$

Weight of pile + cap = $\left(\frac{12}{12} \times \frac{12}{12} \times 80\right)(150 \text{ lb/ft}^3) + 550$
$$= 12{,}550 \text{ lb} = 12.55 \text{ kip}$$

Given: $W_R h = 30$ kip-ft

$$Q_u = \frac{(0.8)(30 \times 12 \text{ kip-in.})}{\frac{1}{8} + 0.1} \times \frac{7.5 + (0.4)^2(12.55)}{7.5 + 12.55} = 607 \text{ kip}$$

$$Q_{all} = \frac{Q_u}{FS} = \frac{607}{6} \approx \textbf{101 kip}$$

Part b

$$Q_u = \frac{EH_E}{S + \sqrt{\dfrac{EH_EL}{2A_pE_p}}}$$

Use $E_p = 3 \times 10^6$ lb/in^2.

$$\sqrt{\frac{EH_EL}{2A_pE_p}} = \sqrt{\frac{(0.8)(30 \times 12)(80 \times 12)}{2(12 \times 12)\left(\dfrac{3 \times 10^6}{1000}\,\text{kip/in}^2\right)}} = 0.566 \text{ in.}$$

$$Q_u = \frac{(0.8)(30 \times 12)}{\frac{1}{8} + 0.566} \approx 417 \text{ kip}$$

$$Q_{\text{all}} = \frac{417}{4} \approx \mathbf{104\ kip}$$

Part c

$$Q_u = a\sqrt{EH_E}\,(b - \log S) = 27\sqrt{(0.8)(30)}\,[1 - \log(\tfrac{1}{8})] \approx 252 \text{ kip}$$

$$Q_{\text{all}} = \frac{252}{3} = \mathbf{84\ kip}$$

▲

9.21   NEGATIVE SKIN FRICTION

Negative skin friction is a downward drag force exerted on the pile by the soil surrounding it. This action can occur under conditions such as the following:

1. If a fill of clay soil is placed over a granular soil layer into which a pile is driven, the fill will gradually consolidate. This consolidation process will exert a downward drag force on the pile (Figure 9.48a) during the period of consolidation.

2. If a fill of granular soil is placed over a layer of soft clay, as shown in Figure 9.48b, it will induce the process of consolidation in the clay layer and thus exert a downward drag on the pile.

3. Lowering of the water table will increase the vertical effective stress on the soil at any depth, which will induce consolidation settlement in clay. If a pile is located in the clay layer, it will be subjected to a downward drag force.

In some cases, the downward drag force may be excessive and cause foundation failure. This section outlines two tentative methods for the calculation of negative skin friction.

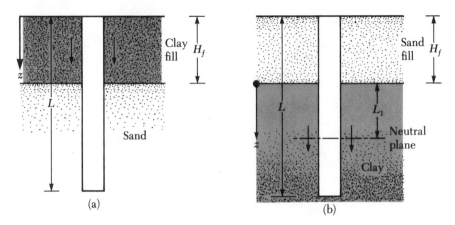

▼   **FIGURE 9.48**   Negative skin friction

Clay Fill over Granular Soil (Figure 9.48a)

Similar to the β method presented in Section 9.12, the negative (downward) skin stress on the pile is

$$f_n = K'\sigma'_v \tan \delta \tag{9.122}$$

where K' = earth pressure coefficient = $K_o = 1 - \sin \phi$

   σ'_v = vertical effective stress at any depth $z = \gamma'_f z$

   γ'_f = effective unit weight of fill

   δ = soil–pile friction angle ≈ 0.5–0.7ϕ

Hence the total downward drag force, Q_n, on a pile is

$$\boxed{Q_n = \int_0^{H_f} (pK'\gamma'_f \tan \delta)z\, dz = \frac{pK'\gamma'_f H_f^2 \tan \delta}{2}} \tag{9.123}$$

where H_f = height of the fill

If the fill is above the water table, the effective unit weight, γ'_f, should be replaced by the moist unit weight.

Granular Soil Fill over Clay (Figure 9.48b)

In this case, the evidence indicates that the negative skin stress on the pile may exist from $z = 0$ to $z = L_1$, which is referred to as the *neutral depth* (see Vesic, 1977, pp. 25–26, for discussion). The neutral depth may be given as (Bowles, 1982):

$$L_1 = \frac{(L-H_f)}{L_1}\left[\frac{L-H_f}{2}+\frac{\gamma_f' H_f}{\gamma'}\right]-\frac{2\gamma_f' H_f}{\gamma'} \tag{9.124}$$

where γ_f' and γ' = effective unit weights of the fill and the underlying clay layer, respectively

For end-bearing piles, the neutral depth may be assumed to be located at the pile tip (that is, $L_1 = L - H_f$).

Once the value of L_1 is determined, the downward drag force is obtained in the following manner. The unit negative skin friction at any depth from $z = 0$ to $z = L_1$ is

$$f_n = K'\sigma_v' \tan \delta \tag{9.125}$$

where $K' = K_o = 1 - \sin \phi$

$$\sigma_v' = \gamma_f' H_f + \gamma' z$$

$$\delta = 0.5\text{--}0.7\phi$$

$$Q_n = \int_0^{L_1} pf_n\, dz = \int_0^{L_1} pK'(\gamma_f' H_f \gamma' z) \tan \delta\, dz$$

$$= (pK'\gamma_f' H_f \tan \delta)L_1 + \frac{L_1^2 pK'\gamma' \tan \delta}{2} \tag{9.126}$$

If the soil and the fill are above the water table, the effective unit weights should be replaced by moist unit weights. In some cases, the piles can be coated with bitumen in the downdrag zone to avoid this problem. Baligh et al. (1978) summarized the results of several field tests that were conducted to evaluate the effectiveness of bitumen coating in reducing the negative skin friction. Their results are presented in Table 9.12.

A limited number of case studies on negative skin friction is available in the literature. Bjerrum et al. (1969) reported monitoring of downdrag force on a test pile at Sorenga in the harbor of Oslo, Norway (noted as pile G in the original paper). This was also discussed by Wong and Teh (1995) in terms of the pile being driven to bedrock at 40 m. Figure 9.49a shows the soil profile and the pile. Wong and Teh (1995) estimated the following:

Fill: Moist unit weight, $\gamma_f = 16\,\text{kN/m}^3$

Saturated unit weight, $\gamma_{\text{sat}(f)} = 18.5\,\text{kN/m}^3$

So

$$\gamma_f' = 18.5 - 9.81 = 8.69\,\text{kN/m}^3$$

$$H_f = 13\,\text{m}$$

Clay: $K' \tan \delta \approx 0.22$

Saturated effective unit weight, $\gamma' = 19 - 9.81 = 9.19\,\text{kN/m}^3$

▶ **TABLE 9.12** Summary of Case Studies of Bitumen-Coated Piles[a]

Case number	Downward drag				Test loadings		
	1	2	3	4	5	6	7
Soil type	Fill, sand, and clay	Fill and silty clay	Fill and clay	Sand and silty clay	Silty clay	Silty clay	Sand fill, clay, and peat
Pile type	Cast-in-place concrete	Steel pipe	Steel pipe	Steel pipe	6 RC piles	6 RC piles	Precast concrete
Pile cross section (mm)	$D = 530$	$D = 300$	$D = 500$	$D = 760$	300×300	300×300	380×450
Length in contact with settling soil (m)	25	26	40	25	7–17	9–16	24
Installation method	Predriven casing	Enlarged tip and slurry	Enlarged tip and casing	Driving	Driving	Driving	Driving
Bitumen coating							
Type (pen 25° C)	20/30	80/100	80/100	60/70	60/70	80–100 RC-0 cutback	43 special grade
Coating thickness (mm)	10	1.2	1.2	1.5	1.2	1.2	10
Measured shaft resistance							
Uncoated pile (ton)	70–80	120	300	180	31–40	31–40	160
Coated pile (ton)	5–7	10	15	3	10–33	20–42	
Coating effectiveness (%)	92	92	95	98	30–80	30–80	
Predicted downdrag							
Coated pile (ton)	0.1	2–11	5	0–23			
Coating effectiveness (%)	100	91–98	98	87–100			

* After Baligh et al. (1978)

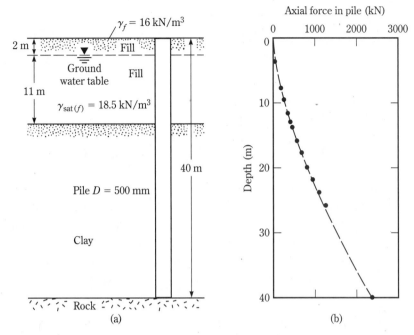

▼   **FIGURE 9.49**   Negative skin friction on a pile in the harbor of Oslo, Norway
[based on Bjerrum et al. (1969); and Wong and Teh (1995)]

Pile: $L = 40$ m

      Diameter, $D = 500$ m

      Thus, the maximum downdrag force on the pile can be estimated from Eq. (9.126). Since it is a point bearing pile, the magnitude of $L_1 = 27$ m, so

$$Q_n = (p)(K' \tan \delta)[\gamma_f \times 2 + (13 - 2)\gamma_f'](L_1) + \frac{L_1^2 p\gamma'(K' \tan \delta)}{2}$$

or

$$Q_n = (\pi \times 0.5)(0.22)[(16 \times 2) + (8.69 \times 11)](27) + \frac{(27)^2(\pi \times 0.5)(9.19)(0.22)}{2}$$

$$= 2348 \text{ kN}$$

The measured value of maximum Q_n was about 2500 kN (Figure 9.49b), which is in good agreement with the calculated value.

▼   **EXAMPLE 9.12**

      Refer to Figure 9.48a; $H_f = 3$ m. The pile is circular in cross section with a diameter of 0.5 m. For the fill that is above the water table, $\gamma_f = 17.2$ kN/m³ and $\phi = 36°$. Determine the total drag force. Use $\delta = 0.7\phi$.

Solution   From Eq. (9.123),

$$Q_n = \frac{pK'\gamma_f H_f^2 \tan \delta}{2}$$

$$p = \pi(0.5) = 1.57 \text{ m}$$

$$K' = 1 - \sin \phi = 1 - \sin 36° = 0.41$$

$$\delta = (0.7)(36) = 25.2°$$

$$Q_n = \frac{(1.57)(0.41)(17.2)(3)^2 \tan 25.2}{2} = \mathbf{23.4 \text{ kN}}$$

▲

▼ **EXAMPLE 9.13** _____

Refer to Figure 9.48b. Here, $H_f = 2$ m, pile diameter $= 0.305$ m, $\gamma_f = 16.5$ kN/m³, $\phi_{clay} = 34°$, $\gamma_{sat(clay)} = 17.2$ kN/m³, and $L = 20$ m. The water table coincides with the top of the clay layer. Determine the downward drag force. Assume $\delta = 0.6\phi_{clay}$.

Solution   The depth of the neutral plane is given in Eq. (9.124) as

$$L_1 = \frac{L - H_f}{L_1}\left(\frac{L - H_f}{2} + \frac{\gamma_f H_f}{\gamma'}\right) - \frac{2\gamma_f H_f}{\gamma'}$$

Note that γ_f' in Eq. (9.124) has been replaced by γ_f because the fill is above the water table, so

$$L_1 = \frac{(20 - 2)}{L_1}\left[\frac{(20 - 2)}{2} + \frac{(16.5)(2)}{(17.2 - 9.81)}\right] - \frac{(2)(16.5)(2)}{(17.2 - 9.81)}$$

$$L_1 = \frac{242.4}{L_1} - 8.93; L_1 = 11.75 \text{ m}$$

Now, referring to Eq. (9.126), we have

$$Q_n = (pK'\gamma_f H_f \tan \delta)L_1 + \frac{L_1^2 pK'\gamma' \tan \delta}{2}$$

$$p = \pi(0.305) = 0.958 \text{ m}$$

$$K' = 1 - \sin 34° = 0.44$$

$$Q_n = (0.958)(0.44)(16.5)(2)[\tan(0.6 \times 34)](11.75)$$

$$+ \frac{(11.75)^2(0.958)(0.44)(17.2 - 9.81)[\tan(0.6 \times 34)]}{2}$$

$$= 60.78 + 79.97 = \mathbf{140.75 \text{ kN}}$$

▲

GROUP PILES

9.22 GROUP EFFICIENCY

In most cases, piles are used in groups, as shown in Figure 9.50, to transmit the structural load to the soil. A *pile cap* is constructed over *group piles*. The pile cap

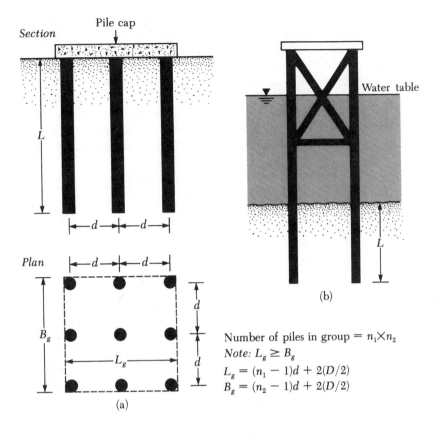

Section

Pile cap

L

d d

Plan

d d

B_g

L_g

d

d

(a)

Water table

L

(b)

Number of piles in group $= n_1 \times n_2$

Note: $L_g \geq B_g$

$L_g = (n_1 - 1)d + 2(D/2)$

$B_g = (n_2 - 1)d + 2(D/2)$

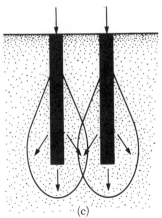

(c)

▼ **FIGURE 9.50** Pile groups

can be in contact with the ground, as in most cases (Figure 9.50a), or well above the ground, as in the case of offshore platforms (Figure 9.50b).

Determining the load-bearing capacity of group piles is extremely complicated and has not yet been fully resolved. When the piles are placed close to each other, a reasonable assumption is that the stresses transmitted by the piles to the soil will overlap (Figure 9.50c), reducing the load-bearing capacity of the piles. Ideally, the piles in a group should be spaced so that the load-bearing capacity of the group should not be less than the sum of the bearing capacity of the individual piles. In practice, the minimum center-to-center pile spacing, d, is $2.5D$, and in ordinary situations, is actually about 3–$3.5D$.

The efficiency of the load-bearing capacity of a group pile may be defined as

$$\eta = \frac{Q_{g(u)}}{\Sigma Q_u} \tag{9.127}$$

where η = group efficiency
  $Q_{g(u)}$ = ultimate load-bearing capacity of the group pile
  Q_u = ultimate load-bearing capacity of each pile without the group effect

Many structural engineers use a simplified analysis to obtain the group efficiency for *friction piles*, particularly in sand. This type of analysis can be explained with the aid of Figure 9.50a. Depending on their spacing within the group, the piles may act in one of two ways: (1) as a *block* with dimensions $L_g \times B_g \times L$, or (2) as *individual piles*. If the piles act as a block, the frictional capacity is $f_{av} p_g L \approx Q_{g(u)}$. [*Note:* p_g = perimeter of the cross section of block = $2(n_1 + n_2 - 2)d + 4D$, and f_{av} = average unit frictional resistance.] Similarly, for each pile acting individually, $Q_u \approx pLf_{av}$. (*Note:* p = perimeter of the cross section of each pile.) Thus

$$\eta = \frac{Q_{g(u)}}{\Sigma\, Q_u} = \frac{f_{av}[2(n_1 + n_2 - 2)d + 4D]L}{n_1 n_2 p L f_{av}}$$
$$= \frac{2(n_1 + n_2 - 2)d + 4D}{p n_1 n_2} \tag{9.128}$$

Hence

$$Q_{g(u)} = \left[\frac{2(n_1 + n_2 - 2)d + 4D}{p n_1 n_2}\right]\Sigma\, Q_u \tag{9.129}$$

From Eq. (9.129), if the center-to-center spacing, d, is large enough, $\eta > 1$. In that case, the piles will behave as individual piles. Thus, in practice, if $\eta < 1$,

$$Q_{g(u)} = \eta\, \Sigma\, Q_u$$

▼ **TABLE 9.13** Equations for Group Efficiency of Friction Piles

Name	Equation
Converse-Labarre equation	$\eta = 1 - \left[\dfrac{(n_1 - 1)\, n_2 + (n_2 - 1)\, n_1}{90 n_1 n_2}\right]\theta$ where θ (deg) $= \tan^{-1}(D/d)$
Los Angeles Group Action equation	$\eta = 1 - \dfrac{D}{\pi d n_1 n_2}\,[n_1(n_2 - 1)]$ $+\, n_2(n_1 - 1) + \sqrt{2}(n_1 - 1)(n_2 - 1)]$
Seiler-Keeney equation (Seiler and Keeney, 1944)	$\eta = \left\{1 - \left[\dfrac{11d}{7(d^2 - 1)}\right]\left[\dfrac{n_1 + n_2 - 2}{n_1 + n_2 - 1}\right]\right\} + \dfrac{0.3}{n_1 + n_2}$ where d is in ft

and, if $\eta \geq 1$,

$$Q_{g(u)} = \Sigma\, Q_u$$

There are several other equations like Eq. (9.129) for the group efficiency of friction piles. Some of these are given in Table 9.13.

Feld (1943) suggested a method by which the load capacity of individual piles (friction) in a group embedded in sand could be assigned. According to this method, the ultimate capacity of a pile is reduced by one-sixteenth by each adjacent diagonal or row pile. The technique can be explained by referring to Figure 9.51, which shows the plan of a group pile. For pile type A, there are eight adjacent piles; for pile type B, there are five adjacent piles; and for pile type C, there are three adjacent piles. Now the following table can be prepared:

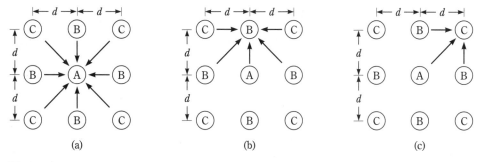

(a)   (b)   (c)

▼ **FIGURE 9.51** Feld's method for estimation of group capacity of friction piles

Pile type	No. of Piles	No. of adjacent piles/pile	Reduction factor for each pile	Ultimate capacity[a]
A	1	8	$1 - \dfrac{8}{16}$	$0.5Q_u$
B	4	5	$1 - \dfrac{5}{16}$	$2.75Q_u$
C	4	3	$1 - \dfrac{3}{16}$	$3.25Q_u$
				$\Sigma\ 6.5Q_u = Q_{g(u)}$

[a] (No of piles) (Q_u) (reduction factor)
Q_u = ultimate capacity for an isolated pile

Hence

$$\eta = \frac{Q_{g(u)}}{\Sigma\, Q_u} = \frac{6.5Q_u}{9Q_u} = 72\%$$

Figure 9.52 shows a comparison of field test results in clay with the theoretical group efficiency calculated from the Converse–Labarre equation (Table 9.13). Reported by Brand et al. (1972), these tests were conducted in soil for which the details are given in Figure 3.7. Other test details include

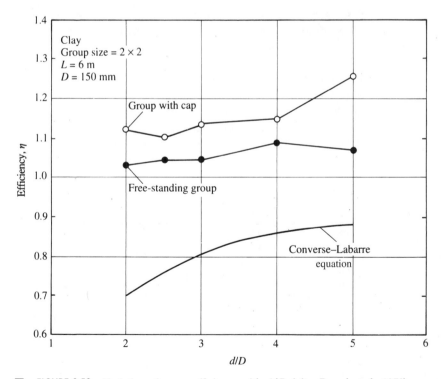

▼ **FIGURE 9.52**   Variation of group efficiency with d/D (after Brand et al., 1972)

Length of piles = 6 m
Diameter of piles = 150 mm
Pile groups tested = 2 × 2
Location of pile head = 1.5 m below the ground surface

Pile tests were conducted with and without a cap (free-standing group). Note that for $d/D \geq 2$, the magnitude of η was greater than 1.0. Also for similar values of d/D the group efficiency was greater with the pile cap than without the cap. Figure 9.53 shows the pile group settlement at various stages of the load test.

Figure 9.54 shows the variation of group efficiency (η) for a 3 × 3 group pile in sand (Kishida and Meyerhof, 1965). It can be seen that, for loose and medium sands, the magnitude of group efficiency is larger than one. This is primarily due to the densification of sand surrounding the pile.

Liu et al. (1985) reported the results of field tests on 58 pile groups and 23 single piles embedded in granular soil. Test details include

Pile length, $L = 8D$–$23D$
Pile diameter, $D = 125$ mm–330 mm
Type of pile installation = bored
Spacing of piles in group, $d = 2D$–$6D$

Figure 9.55 shows the behavior of 3 × 3 pile groups with low-set and high-set pile caps in terms of average skin friction, f_{av}. Figure 9.56 shows the variation of average skin friction based on the location of a pile in the group.

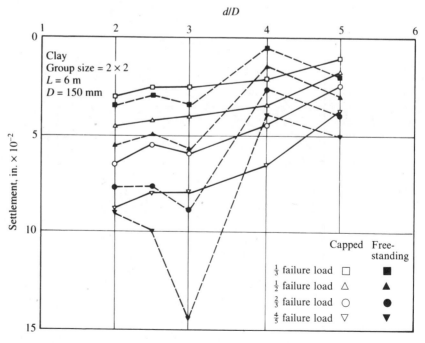

▼ **FIGURE 9.53**   Variation of group pile settlement at various stages of load (after Brand et al., 1972)

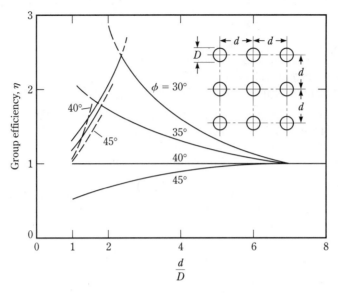

▼ **FIGURE 9.54** Variation of efficiency of pile groups in sand (based on Kishida and Meyerhof, 1965)

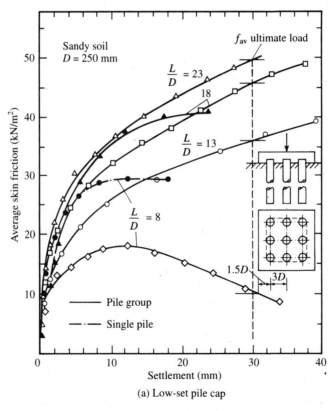

(a) Low-set pile cap

▼ **FIGURE 9.55** Behavior of low-set and high-set pile groups in terms of average skin friction (based on Liu et al., 1985)

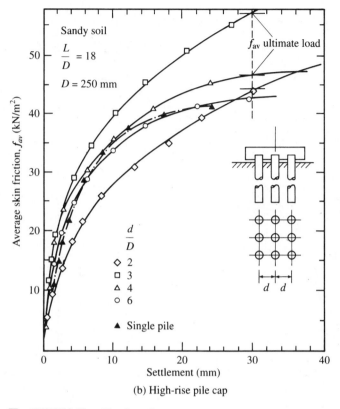

(b) High-rise pile cap

▼   **FIGURE 9.55**   (Continued)

Based on the experimental observations of the behavior of group piles in sand to date, the following general conclusions may be drawn.

1. For *driven* group piles in *sand* with $d \geq 3D$, $Q_{g(u)}$ may be taken to be $\Sigma\, Q_u$, which includes the frictional and the point bearing capacities of individual piles.

2. For *bored* group piles in *sand* at conventional spacings ($d \approx 3D$), $Q_{g(u)}$ may be taken to be $\frac{2}{3}$ to $\frac{3}{4}$ times $\Sigma\, Q_u$ (frictional and point bearing capacities of individual piles).

9.23   ULTIMATE CAPACITY OF GROUP PILES IN SATURATED CLAY

Figure 9.57 shows a group pile in saturated clay. Referring to this figure, the ultimate load-bearing capacity of group piles can be estimated in the following manner:

1. Determine $\Sigma\, Q_u = n_1 n_2 (Q_p + Q_s)$. From Eq. (9.19),

$$Q_p = A_p [9 c_{u(p)}]$$

where $c_{u(p)}$ = undrained cohesion of the clay at the pile tip

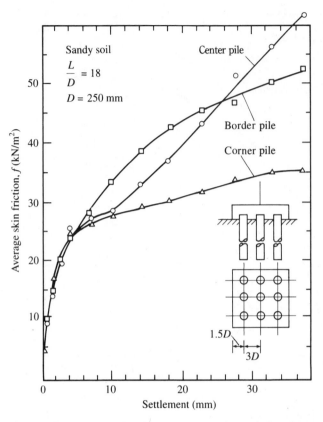

▼ **FIGURE 9.56** Average skin friction based on pile location (based on Liu et al., 1985)

Also, from Eq. (9.48),

$$Q_s = \Sigma \, \alpha p c_u \, \Delta L$$

So

$$\Sigma \, Q_u = n_1 n_2 [9 A_p c_{u(p)} + \Sigma \, \alpha p c_u \, \Delta L] \tag{9.130}$$

2. Determine the ultimate capacity by assuming that the piles in the group act as a block with dimensions of $L_g \times B_g \times L$. The skin resistance of the block is

$$\Sigma \, p_g c_u \, \Delta L = \Sigma \, 2(L_g + B_g) c_u \, \Delta L$$

Calculate the point bearing capacity:

$$A_p q_p = A_p c_{u(p)} N_c^* = (L_g B_g) c_{u(p)} N_c^*$$

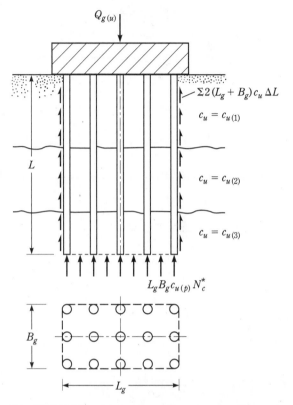

▼   **FIGURE 9.57**   Ultimate capacity of group piles in clay

Obtain the value of the bearing capacity factor, N_c^*, from Figure 9.58. Thus the ultimate load is

$$\Sigma\, Q_u = L_g B_g c_{u(p)} N_c^* + \Sigma\, 2(L_g + B_g) c_u\, \Delta L \tag{9.131}$$

3.   Compare the values obtained from Eqs. (9.130) and (9.131). The *lower* of the two values is $Q_{g(u)}$.

9.24   PILES IN ROCK

For point bearing piles resting on rock, most building codes specify that $Q_{g(u)} = \Sigma\, Q_u$, provided that the minimum center-to-center spacing of piles is $D + 300$ mm. For H-piles and piles with square cross sections, the magnitude of D is equal to the diagonal dimension of the pile cross section.

▼   **EXAMPLE 9.14** _____

The section of a 3×4 group pile in a layered saturated clay is shown in Figure 9.59. The piles are square in cross section (14 in. × 14 in.). The center-to-center

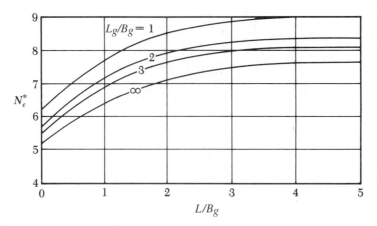

▼ **FIGURE 9.58**   Variation of N_c^* with L_g/B_g and L/B_g

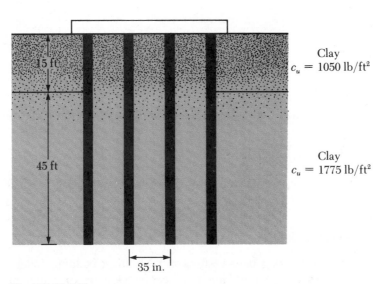

▼ **FIGURE 9.59**

spacing, d, of the piles is 35 in. Determine the allowable load-bearing capacity of the pile group. Use $FS = 4$.

Solution   From Eq. (9.130),

$$\Sigma Q_u = n_1 n_2 [9 A_p c_{u(p)} + \alpha_1 p c_{u(1)} L_1 + \alpha_2 p c_{u(2)} L_2]$$

From Figure 9.22, $c_{u(1)} = 1050$ lb/ft²; $\alpha_1 = 0.86$ and $c_{u(2)} = 1775$ lb/ft²; $\alpha_2 = 0.6$.

$$\Sigma Q_u = \frac{(3)(4)}{1000} \left[\begin{array}{c} (9)\left(\dfrac{14}{12}\right)^2 (1775) + (0.86)\left(4 \times \dfrac{14}{12}\right)(1050)(15) \\[3mm] + (0.6)\left(4 \times \dfrac{14}{12}\right)(1775)(45) \end{array} \right] = 3703 \text{ kip}$$

For piles acting as a group,

$$L_g = (3)(35) + 14 = 119 \text{ in.} = 9.92 \text{ ft}$$

$$B_g = (2)(35) + 14 = 84 \text{ in.} = 7 \text{ ft}$$

$$\frac{L_g}{B_g} = \frac{9.92}{7} = 1.42$$

$$\frac{L}{B_g} = \frac{60}{7} = 8.57$$

From Figure 9.58, $N_c^* = 8.75$. From Eq. (9.131)

$$\Sigma Q_u = L_g B_g c_{u(p)} N_c^* + \Sigma 2(L_g + B_g) c_u \Delta L$$

$$= (9.92)(7)(1775)(8.75) + (2)(9.92 + 7)[(1050)(15) + (1775)(45)]$$

$$= 4313 \text{ kip}$$

Hence, $\Sigma Q_u = 3703$ kip.

$$\Sigma Q_{all} = \frac{3703}{FS} = \frac{3703}{4} \approx \mathbf{926 \text{ kip}}$$

▲

9.25 CONSOLIDATION SETTLEMENT OF GROUP PILES

The consolidation settlement of a group pile in clay can be approximately estimated by using the 2:1 stress distribution method. The procedure of calculation involves the following steps (refer to Figure 9.60):

1. Let the depth of embedment of the piles be L. The group is subjected to a total load of Q_g. If the pile cap is below the original ground surface, Q_g

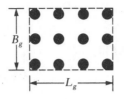

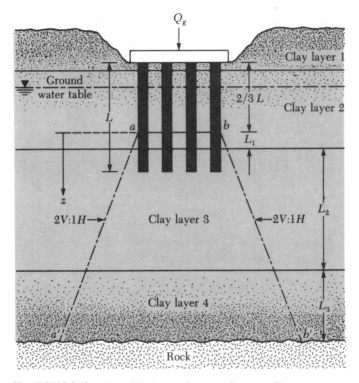

▼ **FIGURE 9.60**   Consolidation settlement of group piles

equals the total load of the superstructure on the piles minus the effective weight of soil above the pile group removed by excavation.

2. Assume that the load Q_g is transmitted to the soil beginning at a depth of $2L/3$ from the top of the pile, as shown in Figure 9.60. The load Q_g spreads out along 2 vertical : 1 horizontal lines from this depth. Lines aa' and bb' are the two 2 : 1 lines.

3. Calculate the stress increase caused at the middle of each soil layer by the load Q_g:

$$\Delta p_i = \frac{Q_g}{(B_g + z_i)(L_g + z_i)} \tag{9.132}$$

where Δp_i = stress increase at the middle of layer i
L_g, B_g = length and width of the plan of pile group, respectively
z_i = distance from $z = 0$ to the middle of the clay layer, i

For example, in Figure 9.60 for layer 2, $z_i = L_1/2$; for layer 3, $z_i = L_1 + L_2/2$; and for layer 4, $z_i = L_1 + L_2 + L_3/2$. Note, however, that there will be no stress increase in clay layer 1 because it is above the horizontal plane ($z = 0$) from which the stress distribution to the soil starts.

4. Calculate the settlement of each layer caused by the increased stress:

$$\Delta s_i = \left[\frac{\Delta e_{(i)}}{1 + e_{o(i)}}\right] H_i \qquad (9.133)$$

where Δs_i = consolidation settlement of layer i
$\Delta e_{(i)}$ = change of void ratio caused by the stress increase in layer i
e_o = initial void ratio of layer i (before construction)
H_i = thickness of layer i (*Note:* In Figure 9.60, for layer 2 $H_i = L_1$; for layer 3, $H_i = L_2$; and for layer 4, $H_i = L_3$.)

Relations for $\Delta e_{(i)}$ are given in Chapter 1.

5. Total consolidation settlement of the pile group is then

$$\Delta s_g = \Sigma \Delta s_i \qquad (9.134)$$

Note that consolidation settlement of piles may be initiated by fills placed nearby, adjacent floor loads, and lowering of water tables.

▼ **EXAMPLE 9.15**

A group pile in clay is shown in Figure 9.61. Determine the consolidation settlement of the pile groups. All clays are normally consolidated.

Solution The stress distribution pattern is shown in Figure 9.61. Hence

$$\Delta p_{(1)} = \frac{Q_g}{(L_g + z_1)(B_g + z_1)} = \frac{(500)(1000)}{\left(9 + \frac{21}{2}\right)\left(6 + \frac{21}{2}\right)} = 1554 \text{ lb/ft}^2$$

$$\Delta p_{(2)} = \frac{(500)(1000)}{(9 + 27)(6 + 27)} = 421 \text{ lb/ft}^2$$

$$\Delta p_{(3)} = \frac{(500)(1000)}{(9 + 36)(6 + 36)} = 265 \text{ lb/ft}^2$$

$$\Delta s_1 = \frac{C_{c(1)} H_1}{1 + e_{o(1)}} \log\left[\frac{p_{o(1)} + \Delta p_{(1)}}{p_{o(1)}}\right]$$

$$p_{o(1)} = (6)(105) + \left(27 + \frac{21}{2}\right)(115 - 62.4) = 2603 \text{ lb/ft}^2$$

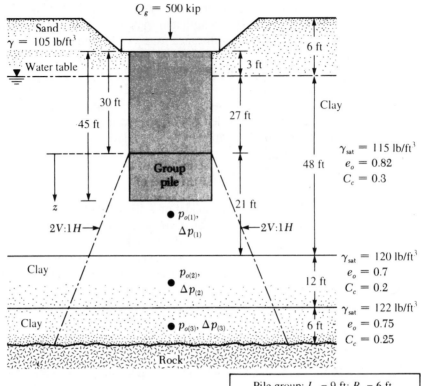

▼ FIGURE 9.61

(*not to scale*)

$$\Delta s_1 = \frac{(0.3)(21)}{1 + 0.82} \log \left(\frac{2603 + 1554}{2603} \right) = 0.703 \text{ ft} = 8.45 \text{ in.}$$

$$\Delta s_2 = \frac{C_{c(2)} H_2}{1 + e_{o(2)}} \log \left[\frac{p_{o(2)} + \Delta p_{(2)}}{p_{o(2)}} \right]$$

$$p_{o(2)} = (6)(105) + (27 + 21)(115 - 62.4) + (6)(120 - 62.4) = 3500 \text{ lb/ft}^2$$

$$\Delta s_2 = \frac{(0.2)(12)}{1 + 0.7} \log \left(\frac{3500 + 421}{3500} \right) = 0.07 \text{ ft} = 0.84 \text{ in.}$$

$$p_{o(3)} = (6)(105) + (48)(115 - 62.4) + (12)(120 - 62.4)$$
$$+ (3)(122 - 62.4) = 4025 \text{ lb/ft}^2$$

$$\Delta s_2 = \frac{(0.25)(6)}{1 + 0.75} \log \left(\frac{4025 + 265}{4025} \right) = 0.024 \text{ ft} \approx 0.29 \text{ in.}$$

Total settlement, $\Delta s_g = 8.45 + 0.84 + 0.29 = \textbf{9.58 in.}$

▲

9.26   ELASTIC SETTLEMENT OF GROUP PILES

In general, the settlement of a pile group under similar working load per pile increases with the width of the group (B_g) and the center-to-center spacing of piles (d). This fact is demonstrated in Figure 9.62 obtained from the experimental results of Meyerhof (1961) for pile groups in sand. In this figure, $s_{g(e)}$ is the settlement of the pile group and s is the settlement of isolated piles under similar working load.

Several investigations relating to the settlement of group piles with widely varying results have been reported in the literature. The simplest relation for the settlement of group piles was given by Vesic (1969) as

$$s_{g(e)} = \sqrt{\frac{B_g}{D}}\, s$$

(9.135)

where $s_{g(e)}$ = elastic settlement of group piles
   B_g = width of pile group section
   D = width or diameter of each pile in the group
   s = elastic settlement of each pile at comparable working load
      (see Section 9.17)

For pile groups in sand and gravel, Meyerhof (1976) suggested the following empirical relation for elastic settlement:

$$s_{g(e)}\,(\text{in.}) = \frac{2q\sqrt{B_g}\,I}{N_{\text{cor}}}$$

(9.136)

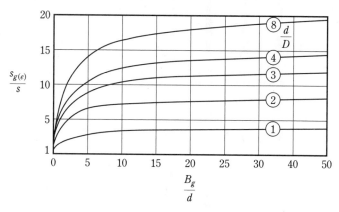

▼ **FIGURE 9.62**   Settlement of pile groups in sand (after Meyerhof, 1961)

where $\qquad q = Q_g/(L_gB_g)$ (in U.S. ton/ft^2) $\hfill$ (9.137)

L_g and B_g = length and width of the pile group section respectively (ft)

N_{cor} = average corrected standard penetration number within seat of settlement ($\approx B_g$ deep below the tip of the piles)

I = influence factor = $1 - L/8B_g \geq 0.5$ $\hfill$ (9.138)

L = length of embedment of piles

Similarly, the pile group settlement is related to the cone penetration resistance as

$$s_{g(e)} = \frac{qB_gI}{2q_c} \hspace{3cm} (9.139)$$

where $\quad q_c$ = average cone penetration resistance within the seat of settlement

In Eq. (9.139), all symbols are in consistent units.

9.27   UPLIFT CAPACITY OF GROUP PILES

The efficiency of group piles under compressive load was discussed in Section 9.21. However, under certain circumstances, group piles may be used for construction of foundations subjected to uplifting load (Figure 9.63). As in Eq. (9.127), the group

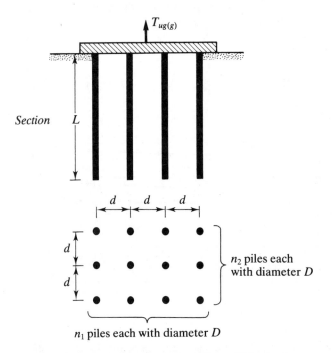

▼   FIGURE 9.63   Group piles subjected to uplifting load

efficiency under uplift may be expressed as

$$\eta_T = \frac{T_{un(g)}}{T_{un}}$$

(9.140)

where η_T = group efficiency under uplift

$T_{un(g)}$ = net ultimate uplift capacity of pile group

T_{un} = net ultimate uplift capacity of single pile (Section 9.18)

Note that

$$T_{un(g)} = T_{ug(g)} - (n_1 \times n_2)W - W_{\text{cap}}$$

(9.141)

where $T_{ug(g)}$ = gross ultimate uplift capacity of group piles

W = effective self-weight of each pile

$n_1 \times n_2$ = number of piles in the group

W_{cap} = effective weight of pile cap

At present, few field and laboratory experimental results relating to the evaluation of η_T are available in the literature. Das and Azim (1985) conducted a limited number of model tests to determine the group efficiency, η_T, of pile groups embedded in saturated clay. The results of this study are shown in Figure 9.64, from which the following general conclusions may be drawn:

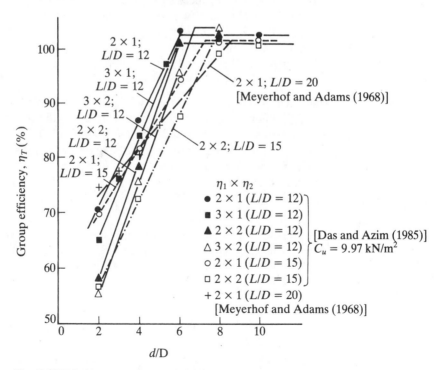

▼ **FIGURE 9.64** Efficiency of pile groups embedded in saturated clay and subjected to uplifting force

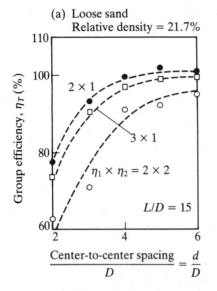 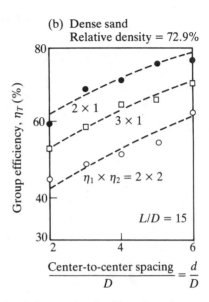

▼ **FIGURE 9.65**   Efficiency of pile groups embedded in sand and subjected to uplifting force (based on laboratory model test results of Das, 1984)

1. For a pile group, η_T increases linearly with the d/D ratio until it reaches 100%. The d/D ratio at which η_T reaches a value of 100% is about $\frac{1}{2}(L/D)$.
2. For given d/D and L/D ratios, the magnitude of η_T decreases with the increase of the number of piles in a group.
3. For a given d/D ratio and number of piles in a group, the magnitude of η_T decreases with the increase of L/D.

Group efficiency, however, may be a function of the consistency of the clay.

Figure 9.65 shows the laboratory model tests results for group efficiency of rough piles embedded in loose and dense sand (Das, 1984). The group piles in this case had an L/D ratio of 15. Note that the magnitude of η_T is a function of L/D, d/D, the number of piles in the group, and the relative density of the sand.

More studies are required to define quantitatively the parameters controlling the group efficiency, η_T.

PROBLEMS

9.1   A prestressed concrete pile is 20 m long. The cross section of the pile is 460 mm × 460 mm. The pile is fully embedded in sand. Given: for the sand: $\gamma = 18.6 \text{ kN/m}^3$ and $\phi = 30°$. Estimate the ultimate point load, Q_p, using
   a.   Meyerhof's method
   b.   Vesic's method (use $I_r = I_{rr} = 75$)
   c.   Janbu's method (use $\eta' = 90°$)

9.2   Refer to Problem 9.1. Calculate the total frictional resistance for the pile. Use Eqs. (9.12), (9.35a), and (9.35b). Given: $K = 1.5$ and $\delta = 0.6\phi$.

9.3   Refer to Problem 9.1. Calculate the allowable load-carrying capacity (that is, Q_p and Q_s) using Coyle and Castillo's method. Use $FS = 4$.

9.4   Redo Problem 9.1 with the following: length of pile = 80 ft, pile cross section = 12 in. × 12 in., $\gamma = 115 \text{ lb/ft}^3$, and $\phi = 35°$.

9.5 Refer to Problem 9.4. Calculate the frictional resistance for the pile using Eqs. (9.12), (9.35a), and (9.35b). Given: $K = 1.35$ and $\delta = 0.75\phi$.

9.6 Refer to Problem 9.4. Calculate the allowable load-carrying capacity of the pile tip (that is, Q_p) using Coyle and Castello's method. Use $FS = 4$.

9.7 A concrete pile 20 m long having a cross section of 381 mm × 381 mm is fully embedded in a saturated clay layer. For the clay, $\gamma_{sat} = 18.5$ kN/m³, $\phi = 0$, and $c_u = 70$ kN/m². Assume that the water table lies below the tip of the pile. Determine the allowable load that the pile can carry ($FS = 3$). Use the α method to estimate the skin friction.

9.8 Redo Problem 9.7 using the λ method for estimating the skin friction.

9.9 A concrete pile 60 ft long having a cross section of 15 in. × 15 in. is fully embedded in a saturated clay layer. For the clay, given: $\gamma_{sat} = 122$ lb/ft³, $\phi = 0$, and $c_u = 1450$ lb/ft². Assume that the groundwater table is located below the tip of the pile. Determine the allowable load that the pile can carry ($FS = 3$). Use the α method to estimate the skin resistance.

9.10 A concrete pile 16 in. × 16 in. in cross section is shown in Figure P9.10. Calculate the ultimate skin resistance by using the

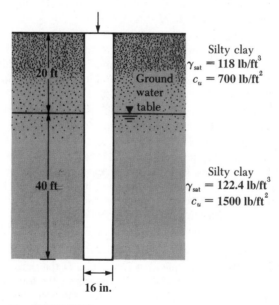

▼ **FIGURE P9.10**

   a. α method
   b. λ method
   c. β method
Use $\phi_R = 20°$ for all clays, which are normally consolidated.

9.11 A steel pile (H-section; HP 14 × 102; see Table D.1a, Appendix D) is driven to a layer of sandstone. The length of the pile is 62 ft. Following are the properties of the sandstone: unconfined compression strength = $q_{u(lab)}$ = 11,400 lb/in²; angle of friction = 36°. Using a factor of safety of 3, estimate the allowable point load that can be carried by the pile.

9.12 A concrete pile is 50 ft long and has a cross section of 16 in. × 16 in. The pile is embedded in a sand having $\gamma = 117$ lb/ft^3 and $\phi = 37°$. The allowable working load is 180 kips. If 110 kips are contributed by the frictional resistance and 70 kips are from the point load, determine the elastic settlement of the pile. Given: $E_p = 3 \times 10^6$ lb/in^2, $E_s = 5 \times 10^3$ lb/in^2, $\mu_s = 0.38$, and $\xi = 0.57$.

9.13 Solve Problem 9.12 with the following: length of pile = 12 m, pile cross section = 0.305 m × 0.305 m, allowable working load = 338 kN, contribution of frictional resistance to working load = 240 kN, $E_p = 21 \times 10^6$ kN/m^2, $E_s = 30,000$ kN/m^2, $\mu_s = 0.3$, $\xi = 0.6$.

9.14 A precast concrete pile with a cross section of 406 mm × 406 mm is embedded in sand. The length of the pile is 10.4 m. Assume that $\gamma_{sand} = 15.8$ kN/m^3, $\phi_{sand} = 30°$, and relative density of sand = 30%. Estimate the allowable pullout capacity of the pile ($FS = 4$).

9.15 Redo Problem 9.14 with the following changes: $\gamma_{sand} = 118$ lb/ft^3, $\phi_{sand} = 37°$, and relative density of sand = 80%. Give your answer in kips.

9.16 A concrete pile 60 m long is embedded in a saturated clay with $c_u = 30$ kN/m^2. The pile is 305 mm × 305 mm in cross section. Using $FS = 3$, determine the allowable pullout capacity of the pile.

9.17 Redo Problem 9.16 with the following changes: The top 5 m of the clay has a $c_u = 25$ kN/m^2 and, below that, $c_u = 55$ kN/m^2.

9.18 A 30-m-long concrete pile is 305 mm × 305 mm in cross section and is fully embedded in a sand deposit. Given: $n_h = 9200$ kN/m^2; moment at ground level, $M_g = 0$; allowable displacement of pile head = 12 mm; $E_p = 21 \times 10^6$ kN/m^2; and $F_{Y(pile)} = 21,000$ kN/m^2. Calculate the allowable lateral load, Q_g, at the ground level. Use the elastic solution method.

9.19 Solve Problem 9.18 by Broms' method. Assume that the pile is flexible and free headed. Given: soil unit weight, $\gamma = 16$ kN/m^3; soil friction angle, $\phi = 30°$; yield stress of pile material, $F_Y = 21$ MN/m^2.

9.20 A steel H-pile (Section: HP 14 × 117; see Table D.1a, Appendix D) is driven by an MKT S-20 hammer (see Table D.4, Appendix D). The length of the pile is 80 ft, the coefficient of restitution is 0.35, the weight of the pile cap is 4.8 kip, hammer efficiency is 0.84, and the number of blows for the last inch of penetration is 10. Estimate the ultimate pile capacity using Eq. (9.119). For the pile, $E_p = 30 \times 10^6$ lb/in^2.

9.21 Solve Problem 9.20 using the modified ENR formula (Table 9.11).

9.22 Solve Problem 9.20 using the Danish formula (Table 9.11).

9.23 Figure 9.48a shows a pile. Let $L = 20$ m, D (pile diameter) = 450 mm, $H_f = 4$ m, $\gamma_{fill} = 17.5$ kN/m^3, and $\phi_{fill} = 25°$. Determine the total downward drag force on the pile. Assume that the fill is located above the water table and that $\delta = 0.5\phi_{fill}$.

9.24 Redo Problem 9.23 assuming that the water table coincides with the top of the fill and that $\gamma_{sat(fill)} = 19.8$ kN/m^3. If the other quantities remain the same, what would be the downward drag force on the pile? Assume that $\delta = 0.5\phi_{fill}$.

9.25 Refer to Figure 9.48b. Let $L = 60$ ft, $\gamma_{fill} = 105$ lb/ft^3, $\gamma_{sat(clay)} = 121$ lb/ft^3, $\phi_{clay} = 28°$, $H_f = 12$ ft, and D (pile diameter) = 18 in. The water table coincides with the top of the clay layer. Determine the total downward drag force on the pile. Assume that $\delta = 0.5\phi_{clay}$.

9.26 The plan of a group pile (friction pile) in sand is shown in Figure P9.26. The piles are circular in cross section and have an outside diameter of 460 mm. The center-to-center spacings of the piles (d) are 920 mm. Determine the efficiency of the pile group by

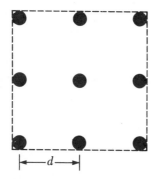

▼ **FIGURE P9.26**

a. using Eq. (9.128)

b. using the Los Angeles group action equation (Table 9.18)

9.27 Refer to Problem 9.26. If the center-to-center pile spacings are increased to 1200 mm, what will be the group efficiencies? (Solve parts a and b).

9.28 The plan of a group pile is shown in Figure P9.26. Assume that the piles are embedded in a saturated homogeneous clay having a c_u = 95.8 kN/m². Given:

Diameter of piles (D) = 406 mm

Center-to-center spacing of piles = 70 mm

Length of piles = 18.5 m

Find the allowable load-carrying capacity of the pile group. Use FS = 3.

9.29 Redo Problem 9.28 with the following:

Center-to-center spacing of piles = 30 in.

Length of piles = 45 ft

D = 12 in.

c_u = 860 lb/ft²

FS = 3

9.30 The section of a 3 × 4 group pile in a layered saturated clay is shown in Figure P9.30.

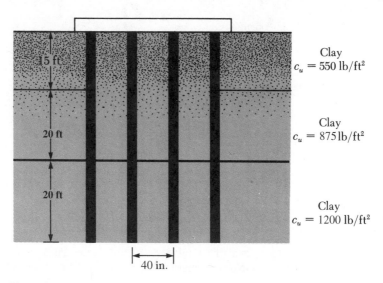

▼ **FIGURE P9.30**

The piles are square in cross section (14 in. × 14 in.). The center-to-center spacing (*d*) of the piles is 40 in. Determine the allowable load-bearing capacity of the pile group. Use *FS* = 4.

9.31 Figure P9.31 shows a group pile in clay. Determine the consolidation settlement of the group. Use the 2 : 1 method to estimate the average effective stress in the clay layers.

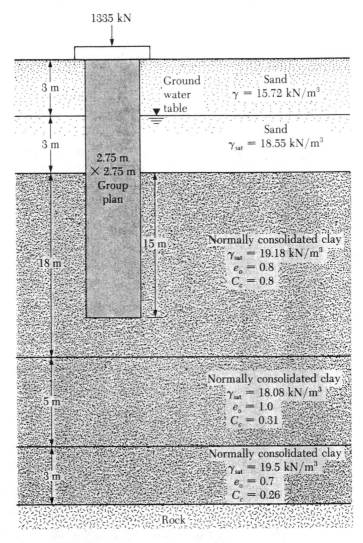

1335 kN

Ground water table

Sand
$\gamma = 15.72 \ kN/m^3$

3 m

Sand
$\gamma_{sat} = 18.55 \ kN/m^3$

3 m

2.75 m × 2.75 m Group plan

15 m

18 m

Normally consolidated clay
$\gamma_{sat} = 19.18 \ kN/m^3$
$e_o = 0.8$
$C_c = 0.8$

5 m

Normally consolidated clay
$\gamma_{sat} = 18.08 \ kN/m^3$
$e_o = 1.0$
$C_c = 0.31$

3 m

Normally consolidated clay
$\gamma_{sat} = 19.5 \ kN/m^3$
$e_o = 0.7$
$C_c = 0.26$

Rock

▼   **FIGURE P9.31**

REFERENCES

American Society of Civil Engineers (1959). "Timber Piles and Construction Timbers," *Manual of Practice,* No. 17, American Society of Civil Engineers, New York.

American Society of Civil Engineers (1993). *Design of Pile Foundations* (Technical Engineering and Design Guides as Adapted from the U.S. Army Corps of Engineers, No. 1), American Society of Civil Engineers, New York.

American Petroleum Institute (1984). *Recommended Practice of Planning, Designing, and Construction of Fixed Offshore Platforms,* Report No. API-RF-2A, Dallas, 115 pp.

Baligh, M. M., Vivatrat, V., and Pigi, H. (1978). "Downdrag on Bitumen-Coated Piles," *Journal of the Geotechnical Engineering Division,* American Society of Civil Engineers, Vol. 104, No. GT11, pp. 1355–1370.

Bhusan, K. (1982). "Discussion: New Design Correlations for Piles in Sands," *Journal of the Geotechnical Engineering Division,* American Society of Civil Engineers, Vol. 108, No. GT11, pp. 1508–1510.

Bjerrum, L., Johannessen, I. J., and Eide, O. (1969). "Reduction of Skin Friction on Steel Piles to Rock," *Proceedings,* Seventh International Conference on Soil Mechanics and Foundation Engineering, Mexico City, Vol. 2, pp. 27–34.

Bowles, J. E. (1982). *Foundation Design and Analysis,* McGraw-Hill, New York.

Brand, E. W., Muktabhant, C., and Taechathummarak, A. (1972). "Load Test on Small Foundations in Soft Clay," *Proceedings,* Performance of Earth and Earth-Supported Structures, American Society of Civil Engineers, Vol. 1, Part 2, pp. 903–928.

Briaud, J. L., Moore, B. H., and Mitchell, G. B. (1989). "Analysis of Pile Load Test at Lock and Dam 26," *Proceedings,* Foundation Engineering: Current Principles and Practices, American Society of Civil Engineers, Vol. 2, pp. 925–942.

Briaud, J. L., and Tucker, L. M. (1984). "Coefficient of Variation of *In-Situ* Tests in Sand," *Proceedings,* Symposium on Probabilistic Characterization of Soil Properties, Atlanta, pp. 119–139.

Broms, B. B. (1965). "Design of Laterally Loaded Piles," *Journal of the Soil Mechanics and Foundations Division,* American Society of Civil Engineers, Vol. 91, No. SM3, pp. 79–99.

Coyle, H. M., and Castello, R. R. (1981). "New Design Correlations for Piles in Sand," *Journal of the Geotechnical Engineering Division,* American Society of Civil Engineers, Vol. 107, No. GT7, pp. 965–986.

Das, B. M. (1984). "Model Uplift Tests on Pile Groups in Sand," *Transportation Research Record No. 998,* National Academy of Sciences, Washington, D.C., pp. 25–28.

Das, B. M., and Azim, M. F. (1985). "Uplift Capacity of Rigid Pile Groups in Clay," *Soils and Foundations,* Vol. 25, No. 4, pp. 56–60.

Das, B. M., and Seeley, G. R. (1975). "Uplift Capacity of Buried Model Piles in Sand," *Journal of the Geotechnical Engineering Division,* American Society of Civil Engineers, Vol. 101, No. GT10, pp. 1091–1094.

Das B. M., and Seeley, G. R. (1982). "Uplift Capacity of Pipe Piles in Saturated Clay," *Soils and Foundations,* The Japanese Society of Soil Mechanics and Foundation Engineering, Vol. 22, No. 1, pp. 91–94.

Davisson, M. T., and Gill, H. L. (1963). "Laterally Loaded Piles in a Layered Soil System," *Journal of the Soil Mechanics and Foundations Division,* American Society of Civil Engineers, Vol. 89, No. SM3, pp. 63–94.

Feld, J. (1943). "Friction Pile Foundations," *Discussion, Transactions,* American Society of Civil Engineers, Vol. 108.

Gates, M. (1957). "Empirical Formula for Predicting Pile Bearing Capacity," *Civil Engineering,* American Society of Civil Engineers, Vol. 27, No. 3, pp. 65–66.

Goodman, R. E. (1980). *Introduction to Rock Mechanics,* Wiley, New York.

International Conference of Building Officials (1982). "Uniform Building Code," Whittier, Calif.

Janbu, N. (1953). *An Energy Analysis of Pile Driving with the Use of Dimensionless Parameters,* Norwegian Geotechnical Institute, Oslo, Publication No. 3.

Janbu, N. (1976). "Static Bearing Capacity of Friction Piles," *Proceedings*, Sixth European Conference on Soil Mechanics and Foundation Engineering, Vol. 1.2, pp. 479–482.

Kishida, H., and Meyerhof, G. G. (1965). "Bearing Capacity of Pile Groups Under Eccentric Loads in Sand," *Proceedings*, Sixth International Conference on Soil Mechanics and Foundation Engineering, Montreal, Vol. 2, pp. 270–274.

Liu, J. L., Yuan, Z. L., and Zhang, K. P. (1985). "Cap-Pile-Soil Interaction of Bored Pile Groups," *Proceedings,* Eleventh International Conference on Soil Mechanics and Foundation Engineering, San Francisco, Vol. 3, pp. 1433–1436.

Matlock, H., and Reese, L. C. (1960). "Generalized Solution for Laterally Loaded Piles," *Journal of the Soil Mechanics and Foundations Division,* American Society of Civil Engineers, Vol. 86, No. SM5, Part I, pp. 63–91.

McClelland, B. (1974). "Design of Deep Penetration Piles for Ocean Structures," *Journal of the Geotechnical Engineering Division,* American Society of Civil Engineers, Vol. 100, No. GT7, pp. 709–747.

Meyerhof, G. G. (1961). "Compaction of Sands and Bearing Capacity of Piles," *Transactions*, American Society of Civil Engineers, Vol. 126. Part 1, pp. 1292–1323.

Meyerhof, G. G. (1976). "Bearing Capacity and Settlement of Pile Foundations," *Journal of the Geotechnical Engineering Division,* American Society of Civil Engineers, Vol. 102, No. GT3, pp. 197–228.

Meyerhof, G. G. (1995). "Behavior of Pile Foundations Under Special Loading Conditions: 1994 R. M. Hardy Keynote Address," *Canadian Geotechnical Journal,* Vol. 32, No. 2, pp. 204–222.

Meyerhof, G. G., and Adams, J. I. (1968). "The Ultimate Uplift Capacity of Foundations," *Canadian Geotechnical Journal,* Vol. 5, No. 4, pp. 225–244.

Michigan State Highway Commission (1965). *A Performance Investigation of Pile Driving Hammers and Piles,* Lansing, 338 pp.

Olson, R. E., and Flaate, K. S. (1967). "Pile Driving Formulas for Friction Piles in Sand," *Journal of the Soil Mechanics and Foundations Division,* American Society of Civil Engineers, Vol. 93, No. SM6, pp. 279–296.

Seiler, J. F., and Keeney, W. D. (1944). "The Efficiency of Piles in Groups," *Wood Preserving News*, Vol. 22, No. 11 (November).

Sharma, H. D., and Joshi, R. C. (1988). "Drilled Pile Behavior in Granular Deposit," *Canadian Geotechnical Journal,* Vol. 25, No. 2, pp. 222–232.

Terzaghi, K., and Peck, R. B. (1967). *Soil Mechanics in Engineering Practice,* 2nd ed., Wiley, New York.

Vesic, A. S. (1961). "Bending of Beams Resting on Isotropic Elastic Solids," *Journal of the Engineering Mechanics Division,* American Society of Civil Engineers, Vol. 87, No. EM2, pp. 35–53.

Vesic, A. S. (1969). "Experiments with Instrumented Pile Groups in Sand," American Society for Testing and Materials; Special Technical Publication, No. 444, pp. 177–222.

Vesic, A. S. (1970). "Tests on Instrumented Piles — Ogeechee River Site," *Journal of the Soil Mechanics and Foundations Division*, American Society of Civil Engineers, Vol. 96, No. SM2, pp. 561–584.

Vesic, A. S. (1977). *Design of Pile Foundations*, National Cooperative Highway Research Program Synthesis of Practice No. 42, Transportation Research Board, Washington, D.C.

Vijayvergiya, V. N., and Focht, J. A. Jr. (1972). *A New Way to Predict Capacity of Piles in Clay,* Offshore Technology Conference Paper 1718, Fourth Offshore Technology Conference, Houston.

Wong, K. S., and Teh, C. I. (1995). "Negative Skin Friction on Piles in Layered Soil Deposit,"

Journal of Geotechnical and Geoenvironmental Engineering, American Society of Civil Engineers, Vol. 121, No. 6, pp. 457–465.

Woo, S. M., and Juang, C. H. (1995). "Analysis of Pile Test Results," in *Developments in Deep Foundations and Ground Improvement Schemes,* Eds. A. S. Balasubramaniam et al., A. A. Balkema, Rotterdam, pp. 69–88.

Woodward, R. J., Gardner, W. S., and Greer, D. M. (1972). *Drilled Pier Foundations*, McGraw-Hill, New York.

CHAPTER TEN

DRILLED-SHAFT AND CAISSON FOUNDATIONS

10.1 INTRODUCTION

The terms *caisson, pier, drilled shaft,* and *drilled pier* are often used interchangeably in foundation engineering; all refer to a *cast-in-place pile generally having a diameter of about 2.5 ft* (≈750 mm) or more, with or without steel reinforcement and with or without an enlarged bottom. Sometimes the diameter can be as small as 1 ft (≈305 mm).

To avoid confusion, we use the term *drilled shaft* for a hole drilled or excavated to the bottom of a structure's foundation and then filled with concrete. Depending on the soil conditions, casings or *laggings* (boards or sheet piles) may be used to prevent the soil around the hole from caving in during construction. The diameter of the shaft is usually large enough for a person to enter for inspection.

The use of drilled-shaft foundations has several advantages:

1. A single drilled shaft may be used instead of a group of piles and the pile cap.
2. Constructing drilled shafts in deposits of dense sand and gravel is easier than driving piles.
3. Drilled shafts may be constructed before completion of grading operations.
4. When piles are driven by a hammer, the ground vibration may cause damage to nearby structures, which the use of drilled shafts avoids.
5. Piles driven into clay soils may produce ground heaving and cause previously driven piles to move laterally, which does not occur during construction of drilled shafts.
6. There is no hammer noise during the construction of drilled shafts, as there is during pile driving.
7. Because the base of a drilled shaft can be enlarged, it provides great resistance to the uplifting load.
8. The surface over which the base of the drilled shaft is constructed can be visually inspected.

9. Construction of drilled shafts generally utilizes mobile equipment, which, under proper soil conditions, may prove to be more economical than methods of constructing pile foundations.
10. Drilled shafts have high resistance to lateral loads.

There are also several drawbacks to the use of drilled-shaft construction. The concreting operation may be delayed by bad weather and always needs close supervision. Also, as in the case of braced cuts, deep excavations for drilled shafts may induce substantial ground loss and damage to nearby structures.

The term *caisson* refers to a substructure element used at wet construction sites, such as rivers, lakes, and docks. For the construction of caissons, a hollow shaft or a box is sunk into position to rest on firm ground. The lower part of the shaft or the box is provided with a cutting edge to help it penetrate the soft soil layers below the water level and come to rest on a load-bearing stratum. The material inside the shaft or box is dredged through the openings at the top, and then concrete is poured in. Bridge abutments, quay walls, and structures for shore protection can be built over caissons.

DRILLED SHAFTS

10.2   TYPES OF DRILLED SHAFTS

Drilled shafts are classified according to the ways in which they are designed to transfer the structural load to the substratum. Figure 10.1a shows a drilled *straight shaft*. It extends through the upper layer(s) of poor soil, and its tip rests on a strong

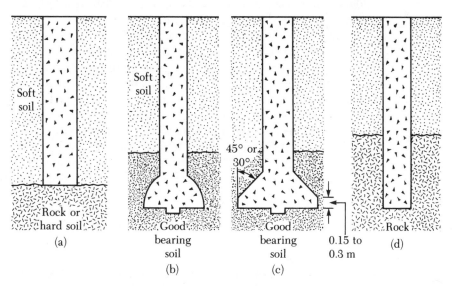

▼ **FIGURE 10.1**   Types of drilled shaft: (a) straight-shaft; (b) and (c) belled shaft; (d) straight-shaft socketed into rock

load-bearing soil layer or rock. The shaft can be cased with steel shell or pipe when required (as in the case of cased, cast-in-place concrete piles; Figure 9.4). For such shafts, the resistance to the applied load may develop from end bearing and also from side friction at the shaft perimeter and soil interface.

A *belled shaft* (Figure 10.1b and c) consists of a straight shaft with a bell at the bottom, which rests on good bearing soil. The bell can be constructed in the shape of a dome (Figure 10.1b), or it can be angled (Figure 10.1c). For angled bells, the underreaming tools commercially available can make 30° to 45° angles with the vertical. For the majority of drilled shafts constructed in the United States, the entire load-carrying capacity is assigned to the end bearing only. However, under certain circumstances, the end-bearing capacity and the side friction are taken into account. In Europe, both the side frictional resistance and the end-bearing capacity are always taken into account.

Straight shafts can also be extended into an underlying rock layer (Figure 10.1d). In the calculation of the load-bearing capacity of such shafts, the end bearing and the shear stress developed along the shaft perimeter and rock interface can be taken into account.

10.3   CONSTRUCTION PROCEDURES

One of the oldest methods of construction of drilled shafts is the *Chicago* method (Figure 10.2a). In this method, circular holes with diameters of 3.5 ft (1.1 m) or more are excavated by hand for depths of 2–6 ft (0.6–1.8 m) at a time. The sides of the excavated hole are then lined with vertical boards, referred to as *laggings*. They are held tightly in place by two circular steel rings. After placement of the rings, the excavation is continued for another 2–6 ft (0.6–1.8 m). When the desired depth of excavation is reached, the bell is excavated. Following the completion of the excavation, the hole is filled with concrete.

In the *Gow* method of construction (Figure 10.2b), the hole is excavated by hand. Telescopic metal shells are used to maintain the shaft. The shells can be removed one section at a time as concreting progresses. The minimum diameter of a Gow drilled shaft is about 4 ft (1.22 m). Any given section of the shell is about 2 in. (50 mm) less in diameter than the section immediately above it. Shafts as deep as 100 ft (30 m) have been installed by this method.

Most shaft excavations are now done mechanically rather than by hand. Open helix augers (flight augers) are common excavation tools. These augers have cutting edges or cutting teeth. Those with cutting edges are used mostly for drilling in soft, homogeneous soil; those with cutting teeth are for drilling in hard soil and hard pan. The auger is attached to a square shaft referred to as the *Kelly* and pushed into the soil and rotated. When the flights are filled with soil, the auger is raised above the ground surface, and the soil is dumped into a pile by rotating the auger at high speed. These augers are available in various diameters; sometimes they may be as large as 10 ft (3 m) or more.

When the excavation is extended to the level of the load-bearing stratum, the auger is replaced by underreaming tools to shape the bell, if required. An underreamer essentially consists of a cylinder with two cutting blades hinged to the

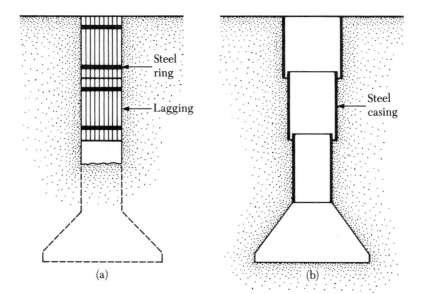

▼ **FIGURE 10.2**   (a) Chicago method of drilled-shaft construction; (b) Gow
method of drilled-shaft construction

top of the cylinder (Figure 10.3). When the underreamer is lowered into the hole,
the cutting blades stay folded inside the cylinder. When the bottom of the hole is
reached, the blades are spread outward, and the underreamer is rotated. The loose
soil falls inside the cylinder, which is raised periodically and emptied until the bell
is completed. Most underreamers can cut bells with diameters as large as three
times the diameter of the shaft.

Another common drilling device is the *bucket-type drill.* It is essentially a bucket
with an opening and cutting edges at the bottom. The bucket is attached to the
Kelly and rotated. The loose soil is collected in the bucket, which is periodically
raised and emptied. Holes as large as 16–18 ft (5–5.5 m) in diameter can be drilled
with this type of equipment.

When rock is encountered during drilling, *core barrels* with *tungsten carbide
teeth* attached to the bottom of the barrels are used. *Shot barrels* are also used for

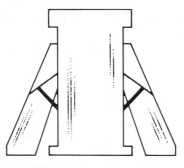

▼ **FIGURE 10.3**   Underreamer

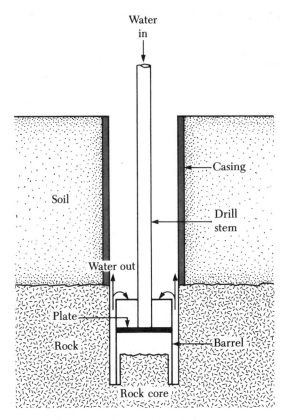

▼ **FIGURE 10.4** Schematic diagram of shot barrel

drilling into very hard rock. The principle of rock coring by a shot barrel is shown in Figure 10.4. The drill stem is attached to the shot barrel's plate. The barrel has some feeder slots through which chilled steel shots are supplied to the bottom of the bore hole. The steel shots cut the rock when the barrel is rotated. Water is supplied to the drill hole through the drill stem. Fine rock and steel particles (produced by the grinding of the steel shots) are washed upward, and they settle on the upper portion of the barrel.

The *Benoto machine* is another type of drilling equipment that is generally used when drilling conditions are difficult and many boulders are in the soil. It essentially consists of a steel tube that can be oscillated and pushed into the soil. A tool usually referred to as the *hammer grab,* which is fitted with cutting blades and jaws, is used to break up the soil and rock inside the tube and remove them.

Use of Casings and Drilling Mud

When holes are driven in soft clays, the soil tends to squeeze in and close the hole. In such situations, casings may be used to keep the hole open and may have to be driven before excavation begins. Holes made in gravelly and sandy soils also tend to cave in. Excavation of drilled-shaft holes in these soils can be continued either by casing as the hole progresses or by using *drilling mud*. As pointed out in Chapter 2, drilling mud is also used during field exploration.

Inspection of the Bottom of the Hole

In many instances, the bottom of the hole must be inspected to ensure that the load-bearing stratum is what was anticipated and that the bell is properly done. For these reasons, an inspector must descend to the bottom of the hole. Several safety precautions must be observed during this procedure:

1. If a casing is not already in the hole, one should be lowered by crane into it to prevent the hole and the bell from collapsing.
2. The hole should be tested for the presence of poisonous or explosive gases, which can be done by using a miner's safety lamp.
3. The inspector should wear a safety harness.
4. The inspector should also carry a safety lamp and an air tank.

10.4   OTHER DESIGN CONSIDERATIONS

For the design of ordinary drilled shafts without casings, a minimum amount of vertical steel reinforcement is always desirable. Minimum reinforcement is 1% of the gross cross-sectional area of the shaft. In California, a reinforcing cage having a length of about 12 ft (3.65 m) is used in the top part of the shaft, and no reinforcement is provided at the bottom. This procedure helps in the construction process because the cage is placed after most of the concreting is complete.

For drilled shafts with nominal reinforcement, most building codes suggest using a design concrete strength, f_c, on the order of $f'_c/4$. Thus the minimum shaft diameter becomes

$$f_c = 0.25f'_c = \frac{Q_w}{A_{gs}} = \frac{Q_w}{\frac{\pi}{4}D_s^2}$$

or

$$D_s = \sqrt{\frac{Q_w}{\left(\frac{\pi}{4}\right)(0.25)f'_c}} = 2.257\sqrt{\frac{Q_w}{f'_c}} \qquad (10.1)$$

where D_s = diameter of the shaft
f'_c = 28-day concrete strength
Q_w = working load of the drilled shaft
A_{gs} = gross cross-sectional area of the shaft

Depending on the loading conditions, the reinforcement percentage may sometimes be too high. In that case, use of a *single rolled-steel section* at the center of the pier (Figure 10.5b) may be considered. In that case,

$$Q_w = (A_{gs} - A_s)f_c + A_s f_s \qquad (10.2)$$

where A_s = area of the steel section
         f_s = allowable strength of steel $\approx 0.5\sigma_{yield}$

When a permanent steel casing is used for construction instead of a central rolled-steel section (Figure 10.5a), Eq. (10.2) may be used. However, f_s for steel should be on the order of $0.4f_s$.

If drilled shafts are likely to be subjected to tensile loads, reinforcement should be continued for the entire length of the shaft.

Concrete Mix Design

The concrete mix design for drilled shafts is not much different from that for any other concrete structure. When a reinforcing cage is used, consideration should be given to the ability of the concrete to flow through the reinforcement. In most cases, a concrete slump of about 6 in. (150 mm) is considered satisfactory. Also, the maximum size of the aggregate should be limited to about 0.75 in. (20 mm).

10.5 LOAD TRANSFER MECHANISM

The load transfer mechanism from drilled shafts to soil is similar to that of piles as described in Section 9.5. Figure 10.6 shows the load test results of a drilled shaft in a clay soil in Houston, Texas (Reese, Touma, and O'Neill, 1976). This drilled

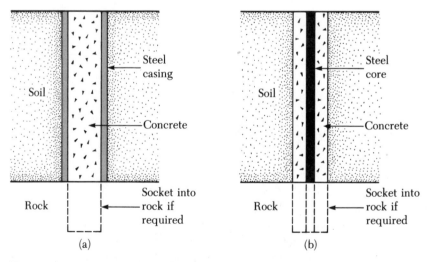

▼ **FIGURE 10.5**   Drilled shafts with (a) steel casing and (b) a central steel core

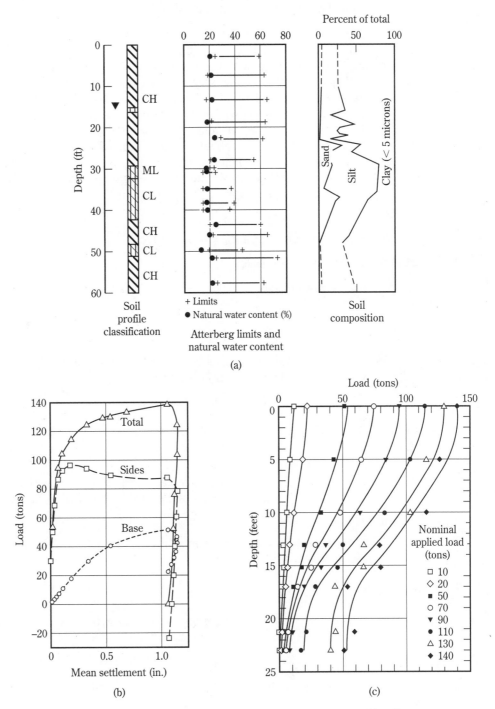

▼ FIGURE 10.6   Load test results for a drilled shaft in Houston, Texas: (a) soil profile, (b) load-displacement curves, (c) load-distribution curves at various stages of loading (after Reese, Touma, and O'Neill, 1976)

shaft had a diameter of 2.5 ft (0.76 m) and a depth of penetration of 23.1 ft (7.04 m). The soil profile at the site is shown in Figure 10.6a. Figure 10.6b shows the load–settlement curves. It may be seen that the total load carried by the drilled shaft was 140 tons (1246 kN). The load carried by side resistance was about 90 tons (801 kN) and the rest was carried by point bearing. It is also interesting to note that, with a downward movement of about 0.25 in. (6.35 mm), full side resistance was mobilized. However, about 1 in. (25.4 mm) of downward movement was required for mobilization of full point resistance. This is similar to that observed in the case of piles. Figure 10.6c shows the load-distribution curves for different stages of the loading, which are similar to those shown in Figure 9.10.

10.6 ESTIMATION OF LOAD-BEARING CAPACITY—GENERAL

The ultimate load-bearing capacity of a drilled shaft (Figure 10.7) is

$$Q_u = Q_p + Q_s \tag{10.3}$$

where Q_u = ultimate load
Q_p = ultimate load-carrying capacity at the base
Q_s = frictional (skin) resistance

The equation for the ultimate base load is similar to that for shallow foundations:

$$Q_p = A_p(cN_c^* + q'N_q^* + 0.3\gamma D_b N_\gamma^*) \tag{10.4}$$

where N_c^*, N_q^*, N_γ^* = the bearing capacity factors
q' = vertical effective stress at the level of the bottom of the pier
D_b = diameter of the base (see Figure 10.7a and b)
A_p = area of the base = $\pi/4 D_b^2$

In most cases, the last term (containing N_γ^*) is neglected except for relatively short drilled shafts, so

$$Q_p = A_p(cN_c^* + q'N_q^*) \tag{10.5}$$

The net load-carrying capacity at the base (that is, the gross load minus the weight of the pier) may be approximated as

$$\boxed{Q_{p(\text{net})} = A_p(cN_c^* + q'N_q^* - q') = A_p[cN_c^* + q'(N_q^* - 1)]} \tag{10.6}$$

The expression for the frictional, or skin, resistance, Q_s, is similar to that for piles:

$$Q_s = \int_0^{L_1} pf\, dz \tag{10.7}$$

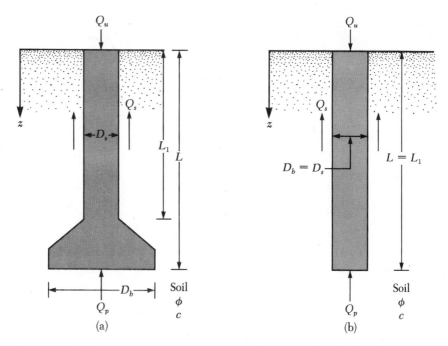

▼ **FIGURE 10.7**   Ultimate bearing capacity of drilled shafts: (a) with bell; (b) straight shaft

where p = shaft perimeter = πD_s
f = unit frictional (or skin) resistance

The following two sections describe the procedures for obtaining the ultimate and allowable load-bearing capacities of drilled shafts in sand and clay.

10.7   DRILLED SHAFTS IN SAND—LOAD-BEARING CAPACITY

For drilled shafts in sand, $c = 0$ and, hence Eq. (10.6) simplifies to

$$Q_{p(\text{net})} = A_p q' (N_q^* - 1) \tag{10.8}$$

Determination of N_q^* is always a problem for deep foundations, as in the case of piles. Note, however, that all shafts are *drilled*, unlike the majority of piles, which are *driven*. For similar initial soil conditions, the actual value of N_q^* may be substantially lower for objects drilled and placed *in situ* compared to that for objects that are driven. Vesic (1967) compared the theoretical results obtained by several investigators relating to the variation of N_q^* with soil friction angle. These investigators include DeBeer, Meyerhof, Hansen, Vesic, and Terzaghi. The values of N_q^* given by Vesic (1963) are approximately the lower bound, and hence are used in this text (see Figure 10.8). We also use Eq. (9.20) to calculate the ultimate point load, Q_p. Thus

$$Q_{p(\text{net})} = A_p(\sigma_o' N_\sigma^* - q')$$

where $\sigma_o' = [(1 + 2K_o)/3]q'$

or

$$Q_{p(\text{net})} = A_p\left[\frac{(1 + 2K_o)}{3} N_\sigma^* - 1\right]q' \tag{10.9}$$

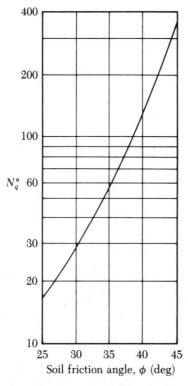

▼ **FIGURE 10.8**   Vesic's bearing capacity factor, N_q^*, for deep foundations

Table D.6 (Appendix D) gives the values of N_σ^* for various magnitudes of I_{rr} and soil friction angles. For ease of calculation, those N_σ^* values are plotted in Figure 10.9.

The frictional resistance at ultimate load, Q_s, developed in a drilled shaft may be calculated from the relation given in Eq. (10.7), in which

$$p = \text{shaft perimeter} = \pi D_s$$

$$f = \text{unit frictional (or skin) resistance} = K\sigma_v' \tan \delta \qquad (10.10)$$

where K = earth pressure coefficient $\approx K_o = 1 - \sin \phi$
σ_v' = effective vertical stress at any depth z

Thus

$$Q_s = \int_0^{L_1} pf\, dz = \pi D_s (1 - \sin \phi) \int_0^{L_1} \sigma_v' \tan \delta\, dz \qquad (10.11)$$

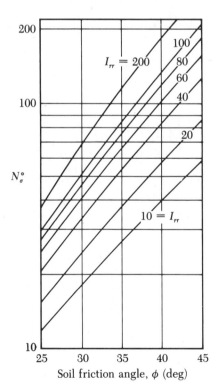

▼ **FIGURE 10.9**   Plot of Vesic's bearing capacity factor, N_σ^* (see Table D.6; Appendix D)

The value of σ_v' will increase to a depth of about $15D_s$ and will remain constant thereafter, as shown in Figure 9.18.

An appropriate factor of safety should be applied to the ultimate load to obtain the net allowable load, or

$$Q_{all(net)} = \frac{Q_{p(net)} + Q_s}{FS} \tag{10.12}$$

A reliable estimate of the soil friction angle, ϕ, must be made to obtain the net base resistance, $Q_{p(net)}$. Figure 10.10 shows a conservative correlation between the soil friction angle and the corresponding corrected standard penetration resistance numbers in granular soils. However, these friction angles are valid only for low confining pressures. At higher confining pressures, which occur in the case of deep foundations, ϕ can decrease substantially for medium to dense sands. This decrease affects the value of N_q^* or N_σ^* (and I_{rr}) to be used for estimating $Q_{p(net)}$. For example, Vesic (1977) showed that, for Chattahoochee River sand at a relative density of about 80%, the triaxial angle of friction is about 45° at a confining pressure of 10 lb/in² (70 kN/m²). However, at a confining pressure of 1500 lb/in² (10.35 MN/m²), the friction angle is about 32.5°, which will ultimately result in a tenfold

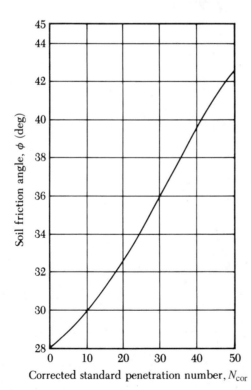

▼ **FIGURE 10.10**   Correlation of corrected standard penetration number with the soil friction angle

decrease of N_q^* or N_σ^*. Thus, for general working conditions of drilled shafts, the estimated friction angle determined from Figure 10.10 should be reduced by about 10%–15%. In general, the existing experimental values show the following range of N_q^* for standard drilled shafts (or cast-in-place piles).

Sand type	Relative density of sand	Range of N_q^*
Loose	40 or less	10–20
Medium	40–60	25–40
Dense	60–80	30–50
Very dense	>80	75–90

Load-Bearing Capacity Based on Settlement

Based on the performance of bored piles in sand with an average diameter of 2.5 ft (750 mm), Touma and Reese (1974) suggested the following procedure for calculating the allowable load-carrying capacity. It is also applicable to drilled shafts in sand.

For $L > 10D_b$ and a *base movement* of 1 in. (25.4 mm), the allowable net point load,

$$Q_{p\text{-all(net)}} = \frac{0.508A_p}{D_b} q_p \qquad (10.13)$$

where $Q_{p\text{-all(net)}}$ is in kN, A_p is in m², D_b is in m, and q_p is the unit point resistance in kN/m²

In English units,

$$Q_{p\text{-all(net)}} = \frac{A_p}{0.6D_b} q_p \qquad (10.14)$$

where $Q_{p\text{-all(net)}}$ is in lb, A_p is in ft², D_b is in ft, and q_p is in lb/ft²

The values of q_p as recommended by Touma and Reese are

Sand type	q_p (kN/m²)	q_p (lb/ft²)
Loose	0	0
Medium	1530	32,000
Very dense	3830	80,000

For sands of intermediate densities, linear interpolation can be used. The shaft friction resistance can be calculated as

$$
Q_s = \int_0^{L_1} (0.7)\, p\sigma_v' \tan\phi\, dz = 0.7(\pi D_s) \int_0^{L_1} \sigma_v' \tan\phi\, dz
$$
$$
= 2.2 D_s \int_0^{L_1} \sigma_v' \tan\phi\, dz
$$

(10.15)

where ϕ = soil friction angle
   σ_v' = vertical effective stress at a depth z

For the definition of L_1, see Figure 10.7. Thus

$$
Q_{\text{all(net)}} = Q_{p\text{-all(net)}} + \frac{Q_s}{FS} \qquad \text{(for a base movement of 1 in.)}
$$

(10.16)

where FS = factor of safety (≈ 2)

Based on a database of 41 loading tests, Reese and O'Neill (1989) also proposed a method to calculate the load-bearing capacity of drilled shafts that is based on settlement. The method is applicable to the following ranges:

1. Shaft diameter: D_s = 1.7 ft to 3.93 ft (0.52 m to 1.2 m)
2. Bell depth: L = 15.4 ft to 100 ft (4.7 m to 30.5 m)
3. Field standard penetration resistance: N_F = 5 to 60
4. Concrete slump = 4 in. to 9 in. (100 mm to 225 mm)

Reese and O'Neill's procedure, with reference to Figure 10.11, gives

$$
Q_{u(\text{net})} = \sum_{i=1}^{N} f_i p\, \Delta L_i + q_p A_p
$$

(10.17)

where f_i = ultimate unit shearing resistance in layer i
   p = perimeter of the shaft = πD_s
   q_p = unit point resistance
   A_p = area of the base = $(\pi/4)D_b^2$

Following are the relationships for determining $Q_{u(\text{net})}$ in granular soils. Based on Eq. 10.17

$$
f_i = \beta \sigma_{vzi}' \le 4\ \text{kip/ft}^2
$$

(10.18)

where σ_{vzi}' = vertical effective stress at the middle of layer i
   $\beta = 1.5 - 0.135 z_i^{0.5}$ $(0.25 \le \beta \le 1.2)$ (10.19)
   z_i = depth to the middle of layer i (ft)

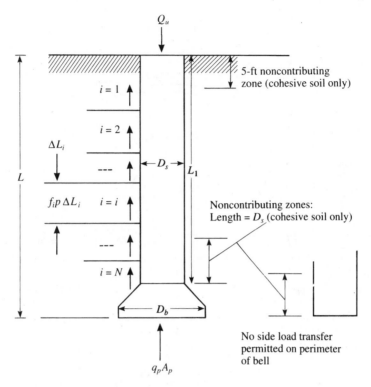

The point bearing capacity is

$$q_p \ (\text{kip/ft}^2) = 1.2 N_F \leq 90 \ \text{kip/ft}^2 \qquad (\text{for } D_b < 50 \ \text{in.}) \qquad (10.20)$$

where N_F = mean *uncorrected* standard penetration number within a distance of $2D_b$ below the base of the drilled shaft

If D_b is equal to or greater than 50 in., excessive settlement may occur. In that case, q_p may be replaced by q_{pr}, or

$$q_{pr} = \frac{50}{D_b \ (\text{in.})} q_p \qquad (\text{for } D_b \geq 50 \ \text{in.}) \qquad (10.21)$$

Figures 10.12 and 10.13 may now be used to calculate the allowable load $Q_{\text{all(net)}}$ based on the desired level of settlement. Example 10.2 shows the method of calculating the net allowable load.

▼ **EXAMPLE 10.1** _____

A soil profile is shown in Figure 10.14 (p. 692). A point bearing drilled shaft with a bell is to be placed in the dense sand and gravel layer. The working load, Q_w, is 2000 kN.

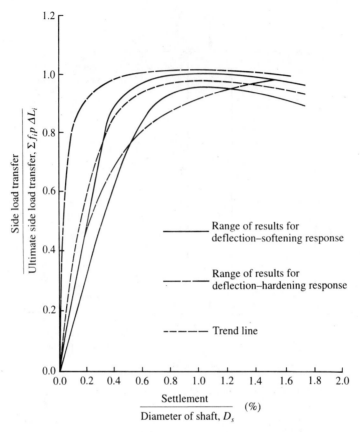

▼ **FIGURE 10.12**   Normalized side load transfer vs. settlement for cohesionless soil (after Reese and O'Neill, 1989)

a.   Determine the shaft diameter for $f'_c = 21,000$ kN/m².
b.   Use Eq. (10.8) and a factor of safety of 4 to determine the bell diameter,
D_b. Ignore the frictional resistance of the shaft.
c.   Use Eq. (10.13) and obtain D_b for a settlement of 25.4 mm. Ignore the
frictional resistance of the shaft. Use $q_p = 3000$ kN/m².

Solution

Part a: Determination of the Shaft Diameter, D_s
From Eq. (10.1),

$$D_s = 2.257 \sqrt{\frac{Q_w}{f'_c}}$$

For $Q_w = 2000$ kN and $f'_c = 21,000$ kN/m²,

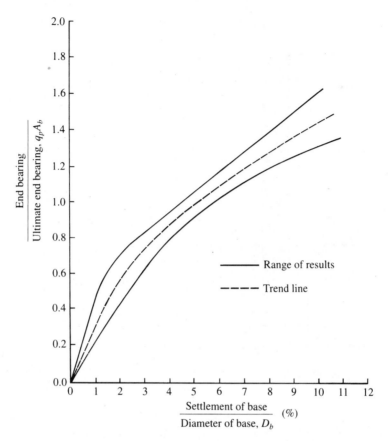

FIGURE 10.13 Normalized base load transfer vs. settlement for cohesionless soil (after Reese and O'Neill, 1989)

$$D_s = 2.257 \sqrt{\frac{2000}{21,000}} = 0.697 \text{ m}$$

Use $D_s = \mathbf{1}$ **m**

Part b: Determination of the Bell Diameter Using Eq. (10.8)

$$Q_{p(\text{net})} = A_p q' (N_q^* - 1)$$

For $N_{\text{cor}} = 40$, Figure 10.10 indicates that $\phi \approx 39.5°$. To be conservative, use a reduction of about 10%, or $\phi = 35.6$. From Figure 10.8, $N_q^* \approx 60$, so

$$q' = 6(16.2) + 2(19.2) = 135.6 \text{ kN/m}^2$$

$$Q_{p(\text{net})} = (Q_w)(FS) = (2000)(4) = 8000 \text{ kN}$$

$$8000 = (A_p)(135.6)(60 - 1)$$

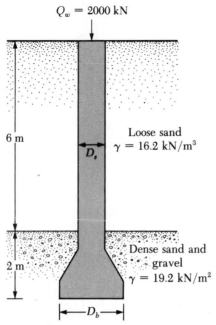

Average corrected standard penetration number $= 40 = N_{cor}$

▼ FIGURE 10.14

$$A_p = 1.0 \text{ m}^2$$

$$D_b = \sqrt{\frac{1.0}{\frac{\pi}{4}}} = \mathbf{1.13\ m}$$

Part c: Determination of Bell Diameter Using Eq. (10.13)

$$Q_{p\text{-all(net)}} = \frac{0.508A_p}{D_b} q_p$$

Because the limit of settlement is 25.4 mm,

$$Q_w = Q_{p\text{-all(net)}}$$

Thus

$$Q_{p\text{-all(net)}} = Q_w = 2000 = \frac{0.508A_p}{D_b} q_p = \frac{(0.508)\,(\pi/4)\,(D_b^2)q_p}{D_b}$$

$$= 0.399D_b q_p$$

or

$$D_b = \frac{2000}{(0.399)\,(3000)} = \mathbf{1.67\,m}$$

Note: The value of D_b determined in Part b corresponds to an allowable bearing capacity that is based on the ultimate bearing capacity. Settlement has not been taken into consideration at all, and the ultimate bearing capacity of drilled shafts may occur at a settlement exceeding 10–15% of the bell diameter. The bell diameter in Part c corresponds to a settlement of 25.4 mm.   ▲

▼ **EXAMPLE 10.2** _____

A drilled shaft is shown in Figure 10.15. The uncorrected average standard penetration number within a distance of $2D_b$ below the base of the shaft is about 30. Determine:

  a. The ultimate load-carrying capacity
  b. The load-carrying capacity for a settlement of 0.5 in. Use Reese and O'Neill's method.

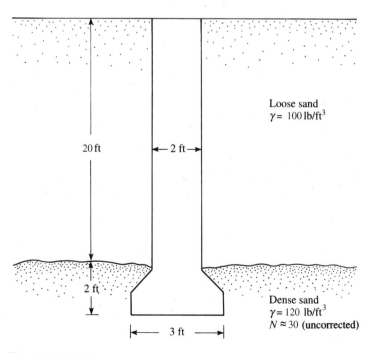

▼ **FIGURE 10.15**

Solution

Part a

From Eqs. (10.18) and (10.19),

$$f_i = \beta \sigma'_{vzi}$$

$$\beta = 1.5 - 0.135z_i^{0.5}$$

For this problem, $z_i = 20/2 = 10$ ft, so

$$\beta = 1.5 - (0.135)(10)^{0.5} = 1.07$$

$$\sigma'_{vzi} = \gamma z_i = (100)(10) = 1000 \text{ lb/ft}^2$$

Thus

$$f_i = (1000)(1.07) = 1070 \text{ lb/ft}^2$$

$$\Sigma f_i p \, \Delta L_i = (1070)(\pi \times 2)(20) = 134{,}460 \text{ lb} = 134.46 \text{ kip}$$

From Eq. (10.20)

$$q_p = 1.2N_F = (1.2)(30) = 36 \text{ kip/ft}^2$$

$$q_p A_p = (36)\left[\frac{\pi}{4}(3)^2\right] = 254.47 \text{ kip}$$

Hence

$$Q_{u(\text{net})} = q_p A_p + \Sigma f_i p \, \Delta L_i = 254.47 + 134.46 = \textbf{388.9 kip}$$

Part b

$$\frac{\text{Allowable settlement}}{D_s} = \frac{0.5}{(2)(12)} = 0.021 = 2.1\%$$

The trend line shown in Figure 10.12 indicates that, for a normalized settlement of 2.1%, the normalized side load is about 0.9. Thus side load transfer is (0.9) $(134.46) \approx 121$ kip. Similarly,

$$\frac{\text{Allowable settlement}}{D_b} = \frac{0.5}{(3)(12)} = 0.014 = 1.4\%$$

The trend line shown in Figure 10.13 indicates that, for a normalized settlement of 1.4%, the normalized base load is 0.312. So the base load is $(0.312)(254.47) = 79.4$ kip. Hence the total load is

$$Q = 121 + 79.4 \approx \textbf{200 kip} \qquad \blacktriangle$$

10.8 DRILLED SHAFTS IN CLAY— LOAD-BEARING CAPACITY

From Eq. (10.6), for saturated clays with $\phi = 0$, $N_q^* = 1$; hence the net base resistance becomes

$$Q_{p(\text{net})} = A_p c_u N_c^*$$ (10.22)

where c_u = undrained cohesion

The bearing capacity factor N_c^* is usually taken to be 9. Figure 9.58 indicates that, when the L/D_b ratio is 4 or more, $N_c^* = 9$, which is the condition for most drilled shafts. Experiments by Whitaker and Cooke (1966) showed that, for belled shafts, the full value of $N_c^* = 9$ is realized with a base movement of about 10%–15% of D_b. Similarly, for straight shafts ($D_b = D_s$), the full value of $N_c^* = 9$ is obtained with a base movement of about 20% of D_b.

The expression for the skin resistance of drilled shafts in clay is similar to Eq. (9.48), or

$$Q_s = \sum_{L=0}^{L=L_1} \alpha^* c_u p \, \Delta L$$ (10.23)

where p = perimeter of the shaft cross section

The value of α^* that can be used in Eq. (10.23) has not yet been fully established. However, the field test results available at this time indicate that α^* may vary between 1.0 to 0.3. Figure 10.16 shows the variation of α^* with depth (α_z^*) at various stages of loading for the case of the drilled shaft discussed in Figure 10.6. The values of α_z^* were derived from Figure 10.6c. At ultimate load, the peak value of α_z^* is about 0.7 with an average of $\alpha^* \approx 0.5$.

Kulhawy and Jackson (1989) reported the field test results of 106 straight drilled shafts — 65 in uplift and 41 in compression. The magnitudes of α^* obtained from these tests are shown in Figure 10.17. The best correlation obtained from these results is

$$\alpha^* = 0.21 + 0.25\left(\frac{p_a}{c_u}\right) \leq 1$$ (10.24)

where p_a = atmospheric pressure = 1.058 ton/ft² (101.3 kN/m²)

So, conservatively, we may assume that

$$\alpha^* = 0.4$$ (10.25)

Reese and O'Neill (1989) suggested the following procedure to estimate the

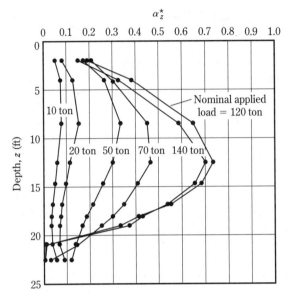

▼ **FIGURE 10.16** Variation of α_z^* with depth for the drilled shaft load test shown in Figure 10.6 (after Reese, Touma, and O'Neill, 1976)

ultimate and allowable (based on settlement) bearing capacities for drilled shafts in clay. According to this procedure, we can use Eq. (10.17) for net ultimate load, or

$$Q_{u(\text{net})} = \sum_{i=1}^{n} f_i p \, \Delta L_i + q_p A_p$$

The unit skin friction resistance can be given as

$$f_i = \alpha_i^* c_{u(i)} \tag{10.26}$$

The following values are recommended for α_i^*:

$\alpha_i^* = 0$ for the top 5 ft (1.5 m) and bottom 1 diameter, D_s, of the drilled shaft. (*Note:* If $D_b > D_s$, then $\alpha^* = 0$ for 1 diameter above the top of the bell and for the peripheral area of the bell itself.)

$\alpha_i^* = 0.55$ elsewhere

and

$$q_p = 6c_{ub} \left(1 + 0.2 \frac{L}{D_b} \right) \le 9c_{ub} \le 80 \, \text{kip/ft}^2 \, (3.83 \, \text{MN/m}^2) \tag{10.27}$$

where c_{ub} = average undrained cohesion within $2D_b$ below the base

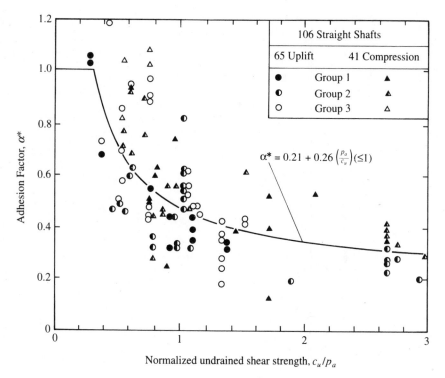

FIGURE 10.17   Variation of α^* with c_u/p_a (after Kulhawy and Jackson, 1989)

If D_b is large, excessive settlement will occur at the ultimate load per unit area, q_p, as given by Eq. (10.27). Thus, for $D_b > 75$ in. (1.91 m), q_p may be replaced by q_{pr}, or

$$q_{pr} = F_r q_p \tag{10.28}$$

where

$$F_r = \frac{2.5}{\psi_1 D_b \text{ (in.)} + \psi_2} \leq 1 \tag{10.29}$$

$$\psi_1 = 0.0071 + 0.0021\left(\frac{L}{D_b}\right) \leq 0.015 \tag{10.30}$$

$$\psi_2 = 0.45(c_{ub})^{0.5} \qquad (0.5 \leq \psi_2 \leq 1.5) \tag{10.31}$$

$$\underset{\uparrow}{} $$
$$\text{kip}/\text{ft}^2$$

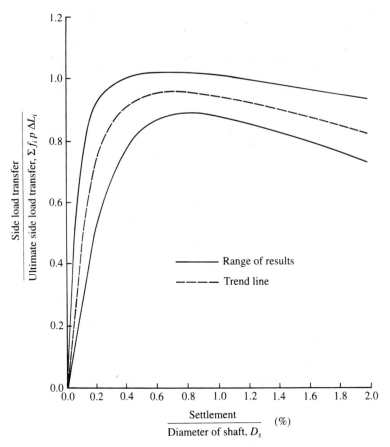

▼ **FIGURE 10.18**   Normalized side load transfer vs. settlement for cohesive soil (after Reese and O'Neill, 1989)

Figures 10.18 and 10.19 may now be used to evaluate the allowable load-bearing capacity based on settlement. (Note that the ultimate bearing capacity in Figure 10.19 is q_p, not q_{pr}.) To do so

1. Select a value of settlement, s.
2. Calculate $\sum_{i=1}^{N} f_i p \, \Delta L_i$ and $q_p A_p$.
3. Using Figures 10.18 and 10.19 and the calculated values in Step 2, determine the *side load* and the *end bearing load*.
4. The sum of the side load and the end bearing load gives the total allowable load.

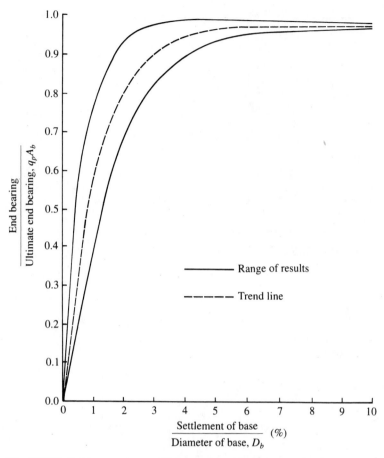

y-axis: $\dfrac{\text{End bearing}}{\text{Ultimate end bearing, } q_p A_b}$

x-axis: $\dfrac{\text{Settlement of base}}{\text{Diameter of base, } D_b}$ (%)

Legend:
—————— Range of results
— — — — Trend line

▼ **FIGURE 10.19**   Normalized base load transfer vs. settlement for cohesive
soil (after Reese and O'Neill, 1989)

▼ **EXAMPLE 10.3**

Figure 10.20 shows a drilled shaft without a bell. Here, $L_1 = 27$ ft, $L_2 = 8.5$ ft,
$D_s = 3.3$ ft, $c_{u(1)} = 1000$ lb/ft², and $c_{u(2)} = 2175$ lb/ft². Determine:

a.   The net ultimate point bearing capacity
b.   The ultimate skin resistance
c.   The working load, Q_w ($FS = 3$)

Use Eqs. (10.22), (10.23), and (10.25).

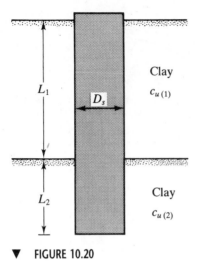

▼ FIGURE 10.20

Solution

Part a From Eq. (10.22),

$$Q_{p(\text{net})} = A_p c_u N_c^* = A_p c_{u(2)} N_c^* = \left[\left(\frac{\pi}{4}\right)(3.3)^2\right](2175)(9)$$
$$= 167,425 \text{ lb} \approx \mathbf{167.4 \text{ kip}}$$

Part b From Eq. (10.23),

$$Q_s = \Sigma\, \alpha^* c_u p\, \Delta L$$

From Eq. (10.25),

$$\alpha^* = 0.4$$
$$p = \pi D_s = (3.14)(3.3) = 10.37 \text{ ft}$$
$$Q_s = (0.4)(10.37)[(1000 \times 27) + (2175 \times 8.5)]$$
$$= 188,682 \text{ lb} \approx \mathbf{188.7 \text{ kip}}$$

Part c

$$Q_w = \frac{Q_{p(\text{net})} + Q_s}{FS} = \frac{167.4 + 188.7}{3} = \mathbf{118.7 \text{ kip}}$$

▲

▼ **EXAMPLE 10.4**

A drilled shaft in a cohesive soil is shown in Figure 10.21. Use Reese and O'Neill's method to determine:

a. The ultimate load-carrying capacity (Eqs. 10.26 through 10.31)
b. The load-carrying capacity for an allowable settlement of 0.5 in.

Solution

Part a

From Eq. (10.26),

$$f_i = \alpha_i^* c_{u(i)}$$

From Figure 10.21,

$$\Delta L_1 = 12 - 5 = 7 \text{ ft}$$

$$\Delta L_2 = (20 - 12) - D_s = (20 - 12) - 2.5 = 5.5 \text{ ft}$$

$$c_{u(1)} = 800 \text{ lb/ft}^2$$

$$c_{u(2)} = 1200 \text{ lb/ft}^2$$

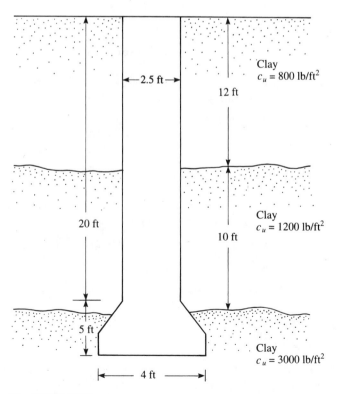

▼ **FIGURE 10.21**

Hence

$$\Sigma f_i p \, \Delta L_i = \Sigma \alpha_i^* c_{u(i)} \, \Delta L_i$$
$$= (0.55)(800)(\pi \times 2.5)(7) + (0.55)(1200)(\pi \times 2.5)(5.5)$$
$$= 52{,}700 \text{ lb} = 52.7 \text{ kip}$$

Again, from Eq. (10.27),

$$q_p = 6c_{ub}\left(1 + 0.2\frac{L}{D_b}\right) = (6)(3000)\left[1 + 0.2\left(\frac{20+5}{4}\right)\right] = 40{,}500 \text{ lb/ft}^2$$
$$= 40.5 \text{ kip/ft}^2$$

Check:

$$q_p = 9c_{ub} = (9)(3000) = 27{,}000 \text{ lb/ft}^2 = 27 \text{ kip/ft}^2 < 40.5 \text{ kip/ft}^2$$

So, use $q_p = 27 \text{ kip/ft}^2$.

$$q_p A_p = q_p\left(\frac{\pi}{4}D_b^2\right) = (27)\left[\left(\frac{\pi}{4}\right)(4)^2\right] \approx 339.3 \text{ kip}$$

Hence

$$Q_u = \Sigma \alpha_i^* c_{u(i)} p \, \Delta L_i + q_p A_p = 52.7 + 339.3 = \mathbf{392 \ kip}$$

Part b

$$\frac{\text{Allowable settlement}}{D_s} = \frac{0.5}{(2.5)(12)} = 0.167 = 1.67\%$$

The trend line shown in Figure 10.18 indicates that, for a normalized settlement of 1.67%, the normalized side load is about 0.89. Thus the side load is

$$(0.89)(\Sigma f_i p \, \Delta L_i) = (0.89)(52.7) = 46.9 \text{ kip}$$

Again,

$$\frac{\text{Allowable settlement}}{D_b} = \frac{0.5}{(4)(12)} = 0.0104 = 1.04\%$$

The trend line shown in Figure 10.19 indicates that, for a normalized settlement of 1.04%, the normalized end bearing is about 0.57, so

$$\text{Base load} = (0.57)(q_p A_p) = (0.57)(339.3) = 193.4 \text{ kip}$$

Thus the total load is

$$Q = 46.9 + 193.4 = \mathbf{240.3 \ kip} \qquad \blacktriangle$$

10.9 SETTLEMENT OF DRILLED SHAFTS AT WORKING LOAD

The settlement of drilled shafts at working load is calculated in a manner similar to the one outlined in Section 9.17. In many cases, the load carried by shaft resistance is small compared to the load carried at the base. In such cases, the contribution of s_3 may be ignored. Note that in Eqs. (9.63) and (9.64) the term D should be replaced by D_b for drilled shafts.

▼ **EXAMPLE 10.5**

Refer to Example 10.3. Estimate the elastic settlement at working loads (that is, $Q_w = 118.7$ kip). Use Eqs. (9.62), (9.64), and (9.65). Given: $\xi = 0.65$, $E_p = 3 \times 10^6$ lb/in², $E_s = 2000$ lb/in², $\mu_s = 0.3$, and $Q_{wp} = 24.35$ kip.

Solution   From Eq. (9.62),

$$s_1 = \frac{(Q_{wp} + \xi Q_{ws})L}{A_p E_p}$$

$$Q_{ws} = 118.7 - 24.35 = 94.35 \text{ kip}$$

$$s_1 = \frac{[24.35 + (0.65 \times 94.35)](35.5)}{\left(\frac{\pi}{4} \times 3.3^2\right)\left(\frac{3 \times 10^6 \times 144}{1000}\right)} = 0.000823 \text{ ft} = 0.0099 \text{ in.}$$

From Eq. (9.64),

$$s_2 = \frac{Q_{wp} C_p}{D_b q_p}$$

From Table 9.7, for stiff clay, $C_p \approx 0.04$,

$$q_p = c_{u(b)} N_c^* = (2.175 \text{ kip/ft}^2)(9) = 19.575 \text{ kip/ft}^2$$

Hence

$$s_2 = \frac{(24.35)(0.04)}{(3.3)(19.575)} = 0.015 \text{ ft} = 0.18 \text{ in.}$$

Again, from Eqs. (9.65) and (9.66),

$$s_3 = \left(\frac{Q_{ws}}{pL}\right)\left(\frac{D_s}{E_s}\right)(1 - \mu_s^2)I_{ws}$$

$$I_{ws} = 2 + 0.35\sqrt{\frac{L}{D_s}} = 2 + 0.35\sqrt{\frac{35}{3.3}} = 3.15$$

So

$$s_3 = \left[\frac{94.35}{(\pi \times 3.3)(35.5)}\right]\left(\frac{3.3}{\frac{2000 \times 144}{1000}}\right)(1 - 0.3^2)(3.15) = 0.0084 \text{ ft} = 0.1 \text{ in.}$$

Total settlement:

$$s = s_1 + s_2 + s_3 = 0.0099 + 0.18 + 0.1 \approx \textbf{0.29 in.}$$ ▲

10.10   UPLIFT CAPACITY OF DRILLED SHAFTS

Sometimes drilled shafts must resist uplifting loads. Field observations of drilled-shaft uplift capacity are relatively scarce. The procedure for determining the ultimate uplifting load for drilled shafts without bells is similar to that for piles described in Chapter 9 (Section 9.18) and will not be repeated here. When a short drilled shaft with a bell is subjected to an uplifting load, the nature of the failure surface in the soil will be like that shown in Figure 10.22. The net ultimate uplift capacity, T_{un}, is

$$T_{un} = T_{ug} - W \qquad (10.32)$$

where T_{ug} = gross ultimate uplift capacity
        W = effective weight of the drilled shaft

The magnitude of T_{un} for drilled shafts in sand can be estimated by the procedure outlined by Meyerhof and Adams (1968) and Das and Seely (1975):

$$T_{un} = B_q A_p \gamma L \qquad (10.33)$$

where B_q = breakout factor
        $A_p = (\pi/4)D_b^2$
        γ = unit weight of soil above the bell (*Note:* If the soil is submerged, the effective unit weight should be used.)

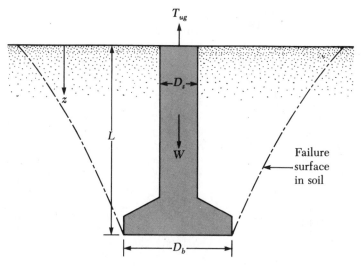

▼ **FIGURE 10.22**   Nature of failure surface in soil caused by uplifting force on drilled shaft with bell

The breakout factor may be expressed as

$$B_q = 2\frac{L}{D_b}K_u' \tan \phi \left(m \frac{L}{D_b} + 1 \right) + 1$$

(10.34)

where K_u' = nominal uplift coefficient
   ϕ = soil friction angle
   m = shape factor coefficient

The value of K_u' may be taken as 0.9 for all values of ϕ from 30–45°. Meyerhof and Adams (1968) gave the variation of m as

Soil friction angle, ϕ (deg)	m
30	0.15
35	0.25
40	0.35
45	0.50

Experiments have shown that the value of B_q increases with the L/D_b ratio to a critical value, $(L/D_b)_{cr}$, and remains constant thereafter. The critical embedment ratio, $(L/D_b)_{cr}$, increases with the soil friction angle. The approximate ranges are

Soil friction angle, ϕ (deg)	$(L/D_b)_{cr}$
30	4
35	5
40	7
45	9

Hence drilled shafts with $L/D_b \leq (L/D_b)_{cr}$ are *shallow foundations,* and shafts with $L/D_b > (L/D_b)_{cr}$ are *deep foundations* with regard to the uplift. The failure surface in soil at ultimate load as shown in Figure 10.22 is for shallow foundations. For deep foundations, local shear failure takes place, and the failure surface in soil *does not extend up to the ground surface.* Based on the preceding considerations, the variation of B_q with L/D_b is shown in Figure 10.23.

Following is a step-by-step procedure for the calculation of the net ultimate uplift capacity of drilled shafts with bells in sand:

1. Determine L, D_b, and L/D_b.
2. Estimate $(L/D_b)_{cr}$ and hence L_{cr}.
3. If $(L/D_b) \leq (L/D_b)_{cr}$, obtain B_q from Figure 10.23. Now,

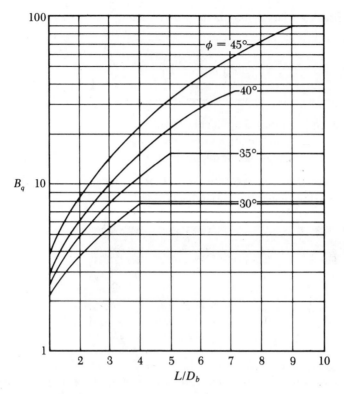

▼ **FIGURE 10.23**   Variation of the breakout factor, B_q, with L/D_b and soil friction angle

$$T_{ug} = B_q A_p \gamma L + W$$

4. If $(L/D_b) > (L/D_b)_{cr}$,

$$T_{ug} = B_q A_p \gamma L + W + \int_0^{L-L_{cr}} (\pi D_s)\, \sigma_v'\, K_u'\, \tan \delta \, dz$$ (10.35)

The last term of Eq. (10.35) is for the frictional resistance developed along the soil–shaft interface from $z = 0$ to $z = L - L_{cr}$ and is similar to Eqs. (9.77) and (9.78). The term σ_v' is the effective stress at any depth z, and K_u and δ are taken from Figure 9.36b and 9.36c, respectively.

The net ultimate uplift capacity of drilled shafts with bell in clay can be estimated according to the procedure outlined by Das (1980):

$$T_{un} = (c_u B_c + \gamma L) A_p$$

(10.36)

where c_u = undrained cohesion
$\quad\quad B_c$ = breakout factor
$\quad\quad \gamma$ = unit weight of clay soil above the bell

As in the case of B_q, the value of B_c increases with the embedment ratio to a critical value of $L/D_b = (L/D_b)_{cr}$ and remains constant thereafter. Beyond the critical depth, $B_c \approx 9$. The critical embedment ratio is related to the undrained cohesion by

$$\left(\frac{L}{D_b}\right)_{cr} = 0.107 c_u + 2.5 \leq 7$$

(10.37)

where c_u is in kN/m²

In English units,

$$\left(\frac{L}{D_b}\right)_{cr} = 0.738 c_u + 2.5 \leq 7$$

(10.38)

where c_u is in lb/in²

Following is a step-by-step procedure for determining the net ultimate uplift capacity of drilled shafts with bell in clay:

1. Determine c_u, L, D_b, and L/D_b.
2. Obtain $(L/D_b)_{cr}$ from Eq. (10.37) or Eq. (10.38) and obtain L_{cr}.
3. If $L/D < (L/D_b)_{cr}$, obtain the value of B_c from Figure 10.24.
4. Use Eq. (10.36) to obtain T_{un}.
5. If $L/D > (L/D_b)_{cr}$, $B_c = 9$. The magnitude of T_{un} may then be obtained from

$$T_{un} = (9c_u + \gamma L) A_p + \Sigma (\pi D_s)(L - L_{cr}) \alpha' c_u$$

(10.39)

The last term of Eq. (10.39) is the skin resistance obtained from the adhesion along the soil–shaft interface and is similar to Eq. (9.70). The magnitude of α' can be obtained from Eqs. (9.71), (9.72), (9.73), and (9.74).

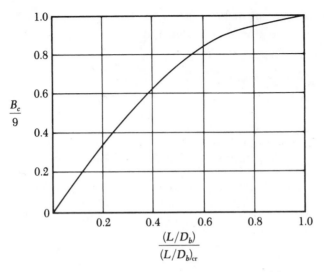

▼ **FIGURE 10.24**   Nondimensional plot of the breakout factor, B_c

▼ **EXAMPLE 10.6**

Refer to Figure 10.22. A drilled shaft with bell has a shaft diameter of 0.76 m, a bell diameter of 1.85 m, and a length of 9.5 m. The bell is supported by a dense sand ($z \geq 9.5$ m) layer. However, a fine, loose sand layer exists above the bell ($z = 0\text{--}9.5$ m). For this sand, $\gamma = 16.4$ kN/m^3, $\phi = 32°$, and the approximate relative density is 30%. The entire structure is located above the water table. Determine the net allowable uplift capacity of the drilled shaft with a factor of safety of 3.

Solution   We begin with $L = 9.5$ m, $D_b = 1.85$ m, and $L/D_b = 9.5/1.85 = 5.14$. For $\phi = 30°$, $(L/D_b)_{cr} = 4$; and for $\phi = 35°$, $(L/D_b)_{cr} = 5$. By interpretation, $(L/D_b)_{cr} \approx 4.2$ for $\phi = 32°$. So $L_{cr} = (4.2)(D_b) = 7.77$. Because $L/D_b = 5.13 > (L/D_b)_{cr} = 4.2$, it is a deep foundation.

According to Eq. (10.34),

$$B_q = 2\left(\frac{L}{D_b}\right)_{cr} K_u' \tan \phi \left[m\left(\frac{L}{D_b}\right)_{cr} + 1 \right] + 1$$

Note that $(L/D_b)_{cr}$ rather than L/D_b was used in the preceding equation because it is a deep foundation. For $\phi = 32°$, $m \approx 0.17$. Hence

$$B_q = (2)(4.2)(0.9)(\tan 32°)[(0.17)(4.2) + 1] + 1 = 9.09$$

From Eq. 10.35,

$$T_{un} = T_{ug} - W = B_q A_p \gamma L + \int_0^{L-L_{cr}} (\pi D_s)(\sigma_v' K_u' \tan \delta)\, dz$$

$$= B_q A_p \gamma L + \frac{\pi}{2} \gamma D_s K_u' \tan \delta (L - L_{cr})^2$$

$$A_p = \left(\frac{\pi}{4}\right)(D_b)^2 = \left(\frac{\pi}{4}\right)(1.85)^2 = 2.687 \text{ m}^2$$

$$L - L_{cr} = 9.5 - 7.77 = 1.73 \text{ m}$$

Also, from Figure 9.36b and 9.36c, for $\phi = 32°$ and relative density = 30%, $K'_u = 1.5$ and $\delta/\phi \approx 0.73$. Hence

$$T_{un} = (9.09)(2.687)(16.4)(9.5) + \left(\frac{\pi}{2}\right)(16.4)(0.76)(1.5)$$

$$\times [\tan(0.73 \times 32)](1.73)^2$$

$$= 3805.4 + 37.96 = 3843.36 \approx 3843$$

So, the net allowable capacity $= 3843/FS = 3843/3 = \textbf{1281 kN.}$ ▲

▼ **EXAMPLE 10.7** _____

Consider the drilled shaft described in Example 10.6. If the soil above the bell is clay with an average value of the undrained shear strength of 95 kN/m², calculate the net ultimate uplift capacity. For clay, $\gamma = 17.9$ kN/m³.

Solution   From Eq. (10.37).

$$\left(\frac{L}{D_b}\right)_{cr} = 0.107c_u + 2.5 = (0.107)(95) + 2.5 = 12.67$$

This quantity is more than 7, so use $(L/D_b)_{cr} = 7$. Hence, $L_{cr} = (7)(1.85) = 12.95$ m. $L_{cr} = 12.95$ is greater than $L = 9.5$ m, so this drilled shaft is a shallow foundation for uplift consideration. For shallow foundations [Eq. (10.36)],

$$T_{un} = (c_u B_c + \gamma L)A_p$$

The magnitude of the breakout factor, B_c, is determined from Figure 10.24:

$$\frac{\left(\frac{L}{D_b}\right)}{\left(\frac{L}{D_b}\right)_{cr}} = \frac{\left(\frac{9.5}{1.85}\right)}{7} = 0.734$$

So, $B_c/9 = 0.92$, or $B_c = 8.28$, and $A_p = (\pi/4)D_b^2 = (\pi/4)(1.85)^2 = 2.688$ m². Thus

$$T_{un} = [(95)(8.28) + (17.9)(9.5)]2.688 = \textbf{2571.5 kN.}$$ ▲

10.11 LATERAL LOAD-CARRYING CAPACITY

The lateral load-carrying capacity of drilled shafts can be analyzed in a manner similar to that presented in Section 9.19 for piles. That method of analysis will not be repeated here.

▼ **EXAMPLE 10.8**

Figure 10.25 shows a drilled shaft in a sand. Given: $L = 6$ m; $D_s = 800$ mm; average horizontal soil modulus $E_s = 35 \times 10^3$ kN/m²; $E_p = 20.7 \times 10^6$ kN/m². Estimate the ultimate lateral load, $Q_{u(g)}$, applied at the ground surface. Use Meyerhof's method given in Section 9.19. Use Eq. (9.107) for check.

Solution   From Eq. (9.105), relative stiffness of the shaft,

$$K_r = \frac{E_p I_p}{E_s L^4}$$

$$I_p = \frac{\pi}{64} D_s^4 = \frac{\pi}{64} \left(\frac{800}{1000}\right)^4 = 0.02 \text{ m}^4$$

$$K_r = \frac{(20.7 \times 10^6)(0.02)}{(35 \times 10^3)(6)^4} = 0.009$$

Since K_r is less than 0.01, this is a flexible drilled shaft. From Eq. (9.110),

$$\frac{L_e}{L} = 1.65 K_r^{0.12}$$

$$L_e = (1.65)(0.009)^{0.12}(6) = 5.63 \text{ m}$$

Thus, from Eq. (9.106)

$$Q_{u(g)} = 0.12\gamma D_s L_e^2 K_{br} \le 0.4 p_l D L_e$$

$$\frac{L_e}{D_s} = \frac{5.63}{0.8} = 7.04$$

From Figure 9.45, for $L_e/D_s = 7.04$ and $\phi = 35°$, the value of $K_{br} \approx 10$, so

$$Q_{u(g)} = (0.12)(17.8)(0.8)(5.63)^2(10) = 541.6 \text{ kN}$$

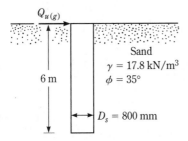

$Q_{u(g)}$

Sand
$\gamma = 17.8$ kN/m³
$\phi = 35°$

6 m

$D_s = 800$ mm

▼ **FIGURE 10.25**

Check:

$$Q_{u(g)} = 0.4p_lD_sL_e = (0.4)(40\ N_q\ \tan\ \phi)D_sL_e$$

↑
Eq. (9.108)

For $\phi = 35°$, $N_q = 33.3$ (Table 3.4),

$$Q_{u(g)} = (0.4)(40)(33.3)(\tan 35)(0.8)(5.63) = 1680.3\ kN$$

So,

$$Q_{u(g)} = \textbf{541.6 kN} \qquad\qquad ▲$$

10.12 DRILLED SHAFTS EXTENDING INTO ROCK

In Section 10.1, we noted that drilled shafts can be extended into rock. This section describes the principles of analysis of the load-bearing capacity of such drilled shafts based on the procedure developed by Reese and O'Neill (1988, 1989). Figure 10.26 shows a drilled shaft whose depth of embedment in rock is equal to L. In the design process recommended below, it is assumed that *there is either side resistance between*

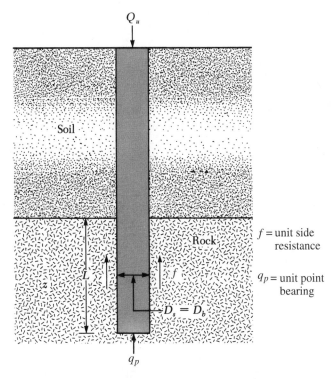

▼ **FIGURE 10.26** Drilled shaft socketed into rock

the shaft and the rock or point resistance at the bottom, but not both. Following is a step-by-step procedure for estimating the ultimate bearing capacity:

1. Calculate the ultimate unit side resistance as

$$f \text{ (lb/in}^2) = 2.5q_u^{0.5} \le 0.15q_u \tag{10.40}$$

   where q_u = unconfined compression strength of a rock core of NW size or larger, or of the drilled shaft concrete, whichever is smaller (in lb/in²)

2. Calculate the ultimate capacity based on side resistance only, or

$$Q_u = \pi D_s L f \tag{10.41}$$

3. Calculate the settlement (*s*) of the shaft at the top of the rock socket, or

$$s = s_e + s_b \tag{10.42}$$

   where s_e = elastic compression of the drilled shaft within the socket assuming no side resistance
   s_b = settlement of the base

   However,

$$s_e = \frac{Q_u L}{A_c E_c} \tag{10.43}$$

   and

$$s_b = \frac{Q_u I_f}{D_s E_{\text{mass}}} \tag{10.44}$$

   where Q_u = ultimate load obtained from Eq. (10.41). (This assumes that the contribution of the overburden to the side shear is negligible.)

   A_c = cross-sectional area of the drilled shaft in the socket $= \dfrac{\pi}{4} D_s^2$
   $$\tag{10.45}$$

   E_c = Young's modulus of the concrete and reinforcing steel in the shaft
   E_{mass} = Young's modulus of the rock mass into which the socket is drilled
   I_f = elastic influence coefficient (Figure 10.27)

The magnitude of E_{mass} can be determined from the average plot shown in Figure 10.28. In this figure, E_{core} is the Young's modulus of intact specimens of rock cores of NW size or larger. However, unless the socket is very long (O'Neill, 1997),

$$s \approx s_b = \frac{Q_u I_f}{D_s E_{\text{mass}}} \tag{10.46}$$

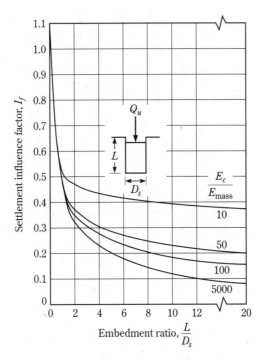

▼ **FIGURE 10.27**   Variation of I_f (after Reese and O'Neill, 1989)

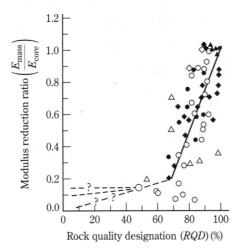

▼ **FIGURE 10.28**   Plot of E_{mass}/E_{core} vs. RQD (after Reese and O'Neill, 1989)

4. If s is less than 0.4 in. (10.2 mm), then the ultimate load-carrying capacity is that calculated by Eq. (10.41). If $s \geq 0.4$ in. (10.2 mm), then go to Step 5.

5. If $s \geq 0.4$ in., there may be rapid, progressive side shear failure in the rock socket resulting in a complete loss of side resistance. In that case, the ultimate capacity is equal to the point resistance, or

$$Q_u = 3A_p \left[\frac{3 + \dfrac{c_s}{D_s}}{10\left(1 + 300\dfrac{\delta}{c_s}\right)^{0.5}} \right] q_u \qquad (10.47)$$

where c_s = spacing of discontinuities (same unit as D_s)
δ = thickness of individual discontinuity (same unit as D_s)
q_u = unconfined compression strength of the rock beneath the base of the socket or the drilled shaft concrete, whichever is smaller

Note that Eq. (10.47) applies for horizontally stratified discontinuities with $c_s > 12$ in. and $\delta < 0.2$ in.

▼ EXAMPLE 10.9

Consider the case of a drilled shaft extending into rock as shown in Figure 10.29. Given: $L = 15$ ft; $D_s = 3$ ft; q_u (rock) = 10,500 lb/in²; q_u (concrete) = 3000 lb/in²; $E_c = 3 \times 10^6$ lb/in²; RQD (rock) = 80%; E_{core} (rock) = 0.36×10^6 lb/in²; $c_s = 18$ in.; $\delta = 0.15$ in. Estimate the allowable load-bearing capacity of the drilled shaft. Use a factor of safety $(FS) = 3$.

Solution

Step 1.   From Eq. (10.40),

$$f \text{ (lb/in}^2) = 2.5\, q_u^{0.5} \leq 0.15 q_u$$

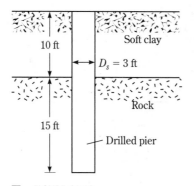

▼ **FIGURE 10.29**

Since q_u (concrete) $< q_u$ (rock), use q_u (concrete) in Eq. (10.40). Hence

$$f = 2.5(3000)^{0.5} = 136.9 \text{ lb/in}^2$$

Check:

$$f = 0.15q_u = (0.15)(3000) = 450 \text{ lb/in}^2 > 136.9 \text{ lb/in}^2$$

So, use $f = 136.9 \text{ lb/in}^2$.

Step 2. From Eq. (10.41),

$$Q_u = \pi D_s Lf = [(\pi)(3 \times 12)(15 \times 12)(136.9)]\frac{1}{1000} = 2787 \text{ kip}$$

Step 3. From Eqs. (10.42), (10.43), and (10.44),

$$s = \frac{Q_u L}{A_c E_c} + \frac{Q_u I_f}{D_s E_{\text{mass}}}$$

For $RQD \approx 80\%$, from Figure 10.28, the value of $E_{\text{mass}}/E_{\text{core}} \approx 0.5$, thus

$$E_{\text{mass}} = 0.5E_{\text{core}} = (0.5)(0.36 \times 10^6) = 0.18 \times 10^6 \text{ lb/in}^2$$

$$\frac{E_c}{E_{\text{mass}}} = \frac{3 \times 10^6}{0.18 \times 10^6} \approx 16.7$$

$$\frac{L}{D_s} = \frac{15}{3} = 5$$

From Figure 10.27, for $E_c/E_{\text{mass}} = 16.7$ and $L/D_s = 5$, the magnitude of I_f is about 0.35. Hence

$$s = \frac{(2787 \times 10^3 \text{ lb})(15 \times 12 \text{ in.})}{\frac{\pi}{4}(3 \times 12 \text{ in.})^2 (3 \times 10^6 \text{ lb/in}^2)} + \frac{(2787 \times 10^3 \text{ lb})(0.35)}{(3 \times 12 \text{ in.})(0.18 \times 10^6 \text{ lb/in}^2)}$$

$$= 0.315 \text{ in.} < 0.4 \text{ in.}$$

Hence $Q_u = 2787 \text{ kip}$

$$Q_{\text{all}} = \frac{Q_u}{FS} = \frac{2787}{3} = \textbf{929 kip}$$

▲

CAISSONS

10.13 TYPES OF CAISSONS

Caissons are divided into three major types: (1) open caissons, (2) box caissons (or closed caissons), and (3) pneumatic caissons.

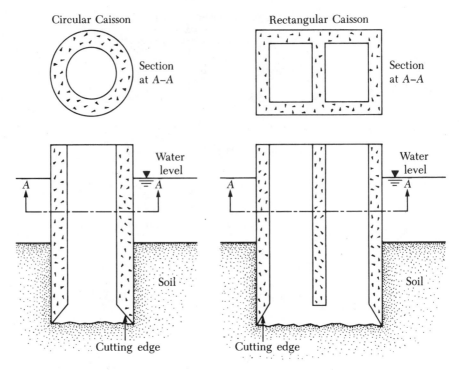

▼   **FIGURE 10.30**   Open caisson

Open caissons (Figure 10.30) are concrete shafts that remain open at the top and bottom during construction. The bottom of the caisson has a cutting edge. The caisson is sunk into place, and soil from the inside of the shaft is removed by grab buckets until the bearing stratum is reached. The shafts may be circular, square, rectangular, or oval. Once the bearing stratum is reached, concrete is poured into the shaft (under water) to form a seal at its bottom. When the concrete seal hardens, the water inside the caisson shaft is pumped out. Concrete is then poured into the shaft to fill it. Open caissons can be extended to great depths, and the cost of construction is relatively low. However, one of their major disadvantages is the lack of quality control over the concrete poured into the shaft for the seal. Also, the bottom of the caisson cannot be thoroughly cleaned out. An alternative method of open-caisson construction is to drive some sheet piles to form an enclosed area, which is filled with sand and is generally referred to as a *sand island*. The caisson is then sunk through the sand to the desired bearing stratum. This procedure is somewhat analogous to sinking a caisson when the ground surface is above the water table.

Box caissons (Figure 10.31) are caissons with closed bottoms. They are constructed on land and then transported to the construction site. They are gradually sunk at the site by filling the inside with sand, ballast, water, or concrete. The cost

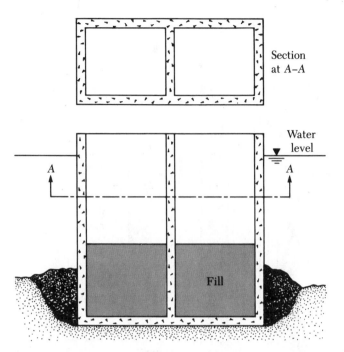

Section
at A–A

Water
level

A   A

Fill

▼ **FIGURE 10.31** Box caisson

for this type of construction is low. The bearing surface must be level, and if it is not, it must be leveled by excavation.

Pneumatic caissons (Figure 10.32) are generally used for depths of about 50–130 ft (15–40 m). This type of caisson is required when an excavation cannot be kept open because the soil flows into the excavated area faster than it can be removed. A pneumatic caisson has a work chamber at the bottom that is at least 10 ft (≈ 3 m) high. In this chamber, the workers excavate the soil and place the concrete. The air pressure in the chamber is kept high enough to prevent water and soil from entering. Workers usually do not encounter severe discomfort when the chamber pressure is raised to about 15 lb/in^2 (≈100 kN/m^2) above atmospheric pressure. Beyond this pressure, decompression periods are required when the workers leave the chamber. When chamber pressures of about 44 lb/in^2 (≈300 kN/m^2) above atmospheric pressure are required, workers should not be kept inside the chamber for more than 1$\frac{1}{2}$–2 hours at a time. Workers enter and leave the chamber through a steel shaft by means of a ladder. This shaft is also used for the removal of excavated soil and the placement of concrete. For large caisson construction, more than one shaft may be necessary; an airlock is provided for each one. Pneumatic caissons gradually sink as excavation proceeds. When the bearing stratum is reached, the work chamber is filled with concrete. Calculation of the load-bearing capacity of caissons is similar to that for drilled shafts. Therefore, it will not be further discussed in this section.

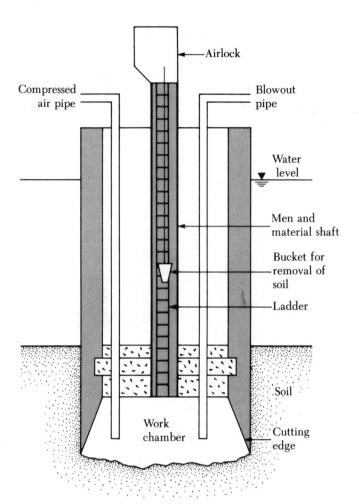

▼   **FIGURE 10.32**   Pneumatic caisson

10.14 THICKNESS OF CONCRETE SEAL IN OPEN CAISSONS

In Section 10.13, we mentioned that, before dewatering the caisson, a concrete seal is placed at the bottom of the shaft (Figure 10.33) and allowed to cure for some time. The concrete seal should be thick enough to withstand an upward hydrostatic force from its bottom after dewatering is complete and before concrete fills the shaft. Based on the theory of elasticity, the thickness, t, according to Teng (1962) is

$$t = 1.18R_i \sqrt{\frac{q}{f_c}} \quad \text{(circular caisson)}$$

(10.48)

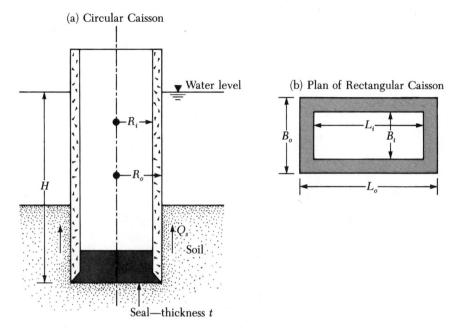

(a) Circular Caisson

Water level

R_i

R_o

H

Q_s

Soil

Seal—thickness t

(b) Plan of Rectangular Caisson

B_o

L_i

B_i

L_o

▼ **FIGURE 10.33** Calculation of the thickness of seal for an open caisson

and

$$t = 0.866 B_i \sqrt{\frac{q}{f_c\left[1 + 1.61\left(\dfrac{L_i}{B_i}\right)\right]}} \qquad \text{(rectangular caisson)} \qquad (10.49)$$

where R_i = inside radius of a circular caisson

           q = unit bearing pressure at the base of the caisson

           f_c = allowable concrete flexural stress (≈ 0.1–0.2 of f_c',

                 where f_c' is the 28-day compressive strength of concrete)

   B_i, L_i = inside width and length, respectively, of rectangular caisson

According to Figure 10.33, the value of q in Eqs. (10.48) and (10.49) can be approximated as

$$q \approx H\gamma_w - t\gamma_c \qquad (10.50)$$

where γ_c = unit weight of concrete

The thickness of the seal calculated by Eqs. (10.48) and (10.49) will be sufficient to protect it from cracking immediately after dewatering. However, two other conditions should also be checked for safety.

1. **Check for Perimeter Shear at Contact Face of Seal and Shaft**
    According to Figure 10.33, the net upward hydrostatic force from the bottom
    of the seal is $A_i H \gamma_w - A_i t \gamma_c$ (where $A_i = \pi R_i^2$ for circular caissons and $A_i =$
    $L_i B_i$ for rectangular caissons). So the perimeter shear developed is

$$v \approx \frac{A_i H \gamma_w - A_i t \gamma_c}{p_i t} \tag{10.51}$$

where p_i = inside perimeter of the caisson

Note that

$$p_i = 2 \pi R_i \qquad \text{(for circular caissons)} \tag{10.52}$$

and that

$$p_i = 2(L_i + B_i) \qquad \text{(for rectangular caissons)} \tag{10.53}$$

The perimeter shear given by Eq. (10.51) should be less than the permissible
shear stress, v_u, or

$$v \, (\text{MN/m}^2) \leq v_u \, (\text{MN/m}^2) = 0.17 \phi \sqrt{f_c'} \, (\text{MN/m}^2) \tag{10.54}$$

where $\phi = 0.85$

In English units,

$$v \, (\text{lb/in}^2) \leq v_u \, (\text{lb/in}^2) = 2 \phi \sqrt{f_c'} \, (\text{lb/in}^2) \tag{10.55}$$

where $\phi = 0.85$

2. **Check for Buoyancy**
    If the shaft is completely dewatered, the buoyant upward force, F_u, is

$$F_u = (\pi R_o^2) H \gamma_w \qquad \text{(for circular caissons)} \tag{10.56}$$

and

$$F_u = (B_o L_o) H \gamma_w \qquad \text{(for rectangular caissons)} \tag{10.57}$$

The downward force, F_d, is caused by the weight of the caisson and the
seal and by the skin friction at the caisson–soil interface, or

$$F_d = W_c + W_s + Q_s \tag{10.58}$$

where W_c = weight of caisson
W_s = weight of seal
Q_s = skin friction

If $F_d > F_u$, the caisson is safe from buoyancy. However, if $F_d < F_u$, dewatering
the shaft completely will be unsafe. For that reason, the thickness of the

seal should be increased by Δt [over the thickness calculated by using Eq. (10.48) or (10.49)], or

$$\Delta t = \frac{F_u - F_d}{A_i \gamma_c} \tag{10.59}$$

▼ **EXAMPLE 10.10**

An open caisson (circular) is shown in Figure 10.34. Determine the thickness of the seal that will enable complete dewatering.

Solution   From Eq. (10.48),

$$t = 1.18 R_i \sqrt{\frac{q}{f_c}}$$

For $R_i = 7.5$ ft,

$$q \approx (45)(62.4) - t\gamma_c$$

With $\gamma_c = 150$ lb/ft³, $q = 2808 - 150t$ and

$$f_c = 0.1 f_c' = 0.1 \times 3 \times 10^3 \, \text{lb/in}^2 = 0.3 \times 10^3 \, \text{lb/in}^2$$

So

$$t = (1.18)(7.5) \sqrt{\frac{2808 - 150t}{300 \times 144}}$$

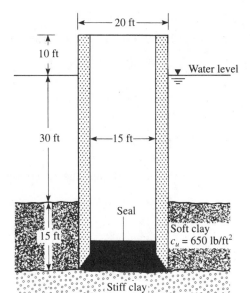

▼ **FIGURE 10.34**

or

$$t^2 + 0.07t - 5.09 = 0$$

$$t = 2.2 \text{ ft}$$

Use $t \approx 2.5$ ft.

Check for Perimeter Shear

According to Eq. (10.51),

$$v = \frac{\pi R_i^2 H \gamma_w - \pi R_i^2 t \gamma_c}{2\pi R_i t} = \frac{(\pi)(7.5)^2[(45)(62.4) - (2.5)(150)]}{(2)(\pi)(7.5)(2.5)} \approx 3650 \text{ lb/ft}^2$$

$$= 25.35 \text{ lb/in}^2$$

The allowable shear stress is

$$v_u = 2\phi\sqrt{f_c} = (2)(0.85)\sqrt{300} = 29.4 \text{ lb/in}^2$$

$$v = 25.35 \text{ lb/in}^2 < v_u = \textbf{29.4 lb/in}^2\textbf{—OK}$$

Check Against Buoyancy

The buoyant upward force is

$$F_u = \pi R_o^2 H \gamma_w$$

For $R_o = 10$ ft,

$$F_u = \frac{(\pi)(10)^2(45)(62.4)}{1000} = 882.2 \text{ kip}$$

The downward force, $F_d = W_c + W_s + Q_s$ and

$$W_c = \pi(R_o^2 - R_i^2)(\gamma_c)(55) = \pi(10^2 - 7.5^2)(150)(55) = 1,133,919 \text{ lb} \approx 1134 \text{ kip}$$

$$W_s = (\pi R_i^2)t\gamma_c = (\pi)(7.5)^2(1)(150) = 26,507 \text{ lb} = 26.5 \text{ kip}$$

Assume that $Q_s \approx 0$. So

$$F_d = 1134 + 26.5 = 1160.5 \text{ kip}$$

Because $F_u < F_d$, it is safe. For design, **assume that t = 2.5 ft.**   ▲

PROBLEMS

10.1   A drilled shaft is shown in Figure P10.1. Determine the net allowable point bearing capacity Given:

$D_b = 6$ ft $\gamma_c = 100 \text{ lb/ft}^3$

$D_s = 3.5$ ft $\gamma_s = 112 \text{ lb/ft}^3$

$L_1 = 18$ ft $\phi = 35°$

$L_2 = 10$ ft $c_u = 720 \text{ lb/ft}^2$

Factor of safety = 4

Reduce the friction angle, ϕ, by 10%. Use Figure 10.8.

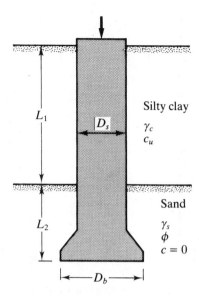

▼ FIGURE P10.1

10.2 Redo Problem 10.1 by Vesic's method [Eq. (10.9)]. Given: $I_{rr} = 100$.

10.3 For the drilled shaft described in Problem 10.1, what skin resistance would develop in the top 18 ft, which are in clay? Use Eq. (10.25).

10.4 Redo Problem 10.1 with the following:

$D_b = 1.75\,\text{ft}$ $\gamma_c = 17.8\,\text{kN/m}^3$

$D_s = 1\,\text{m}$ $\gamma_s = 18.2\,\text{kN/m}^3$

$L_1 = 4\,\text{m}$ $\phi = 32°$

$L_2 = 2.5\,\text{m}$ $c_u = 32\,\text{kN/m}^2$

Factor of safety = 4

10.5 Solve Problem 10.4 by Vesic's method [Eq. (10.9)]. Given: $I_{rr} = 80$.

10.6 For the drilled shaft described in Problem 10.4, what skin resistance would develop in the top 4 meters (that is, the portion in clay soil)?

10.7 Figure P10.7 shows a drilled shaft without a bell. Determine:
   a. The net ultimate point bearing capacity
   b. The ultimate skin resistance
   c. The working load, Q_w (factor of safety = 3)

   Given: $L_1 = 6\,\text{m}$ $c_{u(1)} = 45\,\text{kN/m}^2$

         $L_2 = 5\,\text{m}$ $c_{u(2)} = 74\,\text{kN/m}^2$

         $D_s = 1.5\,\text{m}$

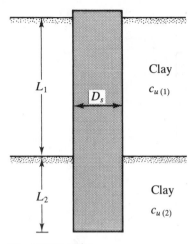

▼   **FIGURE P10.7**

10.8   For the drilled shaft described in Problem 10.7, estimate the total elastic settlement at working load. Use Eqs. (9.62), (9.64), and (9.65). Assume that $E_p = 20 \times 10^6$ kN/ m², $C_p = 0.03$, $\xi = 0.65$, $\mu_s = 0.3$, $E_s = 12000$ kN/m², and $Q_{ws} = 0.5Q_s$.

10.9   Repeat Problem 10.7 with the following:

$L_1 = 25$ ft $c_{u(1)} = 1200$ lb/ft²

$L_2 = 10$ ft $c_{u(2)} = 2000$ lb/ft²

$D_s = 3.5$ ft

10.10   For the drilled shaft described in Problem 10.9, estimate the total elastic settlement at working load. Use Eqs. (9.62), (9.64), and (9.65). Assume that $E_p = 3 \times 10^6$ lb/ in², $C_p = 0.03$, $\xi = 0.65$, $\mu_s = 0.3$, $E_s = 2000$ lb/in², and $Q_{ws} = 0.5Q_s$.

10.11   Refer to the soil profile shown in Figure P10.11. A drilled shaft with a bell is to be

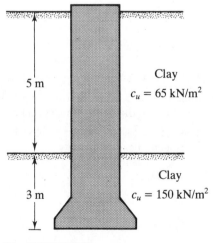

▼   **FIGURE P10.11**

constructed. Given: $f'_c = 21,000$ kN/m². The drilled shaft has to support a working load, $Q_w = 900$ kN (factor of safety = 3).

a. Determine the minimum diameter of the shaft required [use Eq. (10.1)]. Make a reasonable assumption for the diameter to be used.

b. With the shaft diameter determined in part (a), determine the diameter of the bell needed. Skin friction and point bearing capacity are to be considered.

10.12 A drilled shaft in a medium sand is shown in Figure P10.12. Using the method proposed by Touma and Reese, determine the following:

a. The net allowable point resistance for a base movement of 25.4 mm

b. The shaft frictional resistance

c. The total load that can be carried by the drilled shaft for a total base movement of 25.4 mm

Given: $L = 12$ m $\gamma = 18$ kN/m³

   $L_1 = 11$ m $\phi = 38°$

   $D_s = 1$ m $D_r = 65\%$ (medium sand)

   $D_b = 2$ m

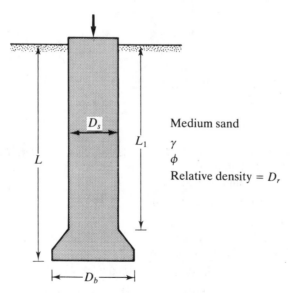

Medium sand

γ

ϕ

Relative density = D_r

▼ **FIGURE P10.12**

10.13 Refer to Figure P10.12, for which $L = 25$ ft, $L_1 = 20$ ft, $D_s = 3.5$ ft, $D_b = 5$ ft, $\gamma = 110$ lb/ft³, and $\phi = 35°$. The average uncorrected standard penetration number within $2D_b$ below the drilled shaft is 29. Determine:

a. The ultimate load-carrying capacity

b. The load-carrying capacity for a settlement of 1 in.

Use Reese and O'Neill's method [Eqs. (10.17)–(10.21) and Figures 10.12 and 10.13].

10.14   For the drilled shaft described in Problem 10.7, determine
   a.   The ultimate load-carrying capacity
   b.   The load-carrying capacity for a settlement of 12.7 mm
   Use the procedure outlined by Reese and O'Neill [Eqs. (10.26)–(10.31) and Figures 10.18 and 10.19].

10.15   Assume the drilled shaft shown in Figure P10.15 to be a point bearing shaft with a working load of 650 kip. Calculate the drilled shaft settlement from Eqs. (9.62) and (9.63) for $E_p = 3 \times 10^6$ lb/in², $\mu_s = 0.35$, and $E_s = 6000$ lb/in².

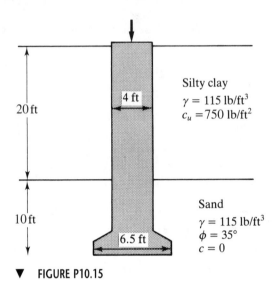

Silty clay
$\gamma = 115$ lb/ft³
$c_u = 750$ lb/ft²

4 ft

20 ft

10 ft

6.5 ft

Sand
$\gamma = 115$ lb/ft³
$\phi = 35°$
$c = 0$

▼   **FIGURE P10.15**

10.16   Refer to Figure 10.22. For the drilled shaft, $D_s = 3$ ft, $D_b = 5$ ft, and $L = 25$ ft. The drilled shaft is in a homogeneous sand with $\phi = 35°$ and $\gamma = 115$ lb/ft³. Determine the net ultimate uplift capacity, T_{un}.

10.17   Repeat Problem 10.16 with $L = 18$ ft.

10.18   For the drilled shaft in Figure 10.22, $D_s = 3$ ft, $D_b = 5$ ft, and $L = 20$ ft. The drilled shaft is in clay with $c_u = 1480$ lb/ft² and a unit weight of $\gamma = 115$ lb/ft³. Estimate the net ultimate uplift capacty.

10.19   Repeat Problem 10.18 for $c_u = 700$ lb/ft².

10.20   Figure P10.20 shows a drilled shaft extending to rock. Given:

$q_{u(concrete)} = 24,000$ kN/m² $E_{(concrete)} = 22$ GN/m²

$q_{u(rock)} = 52,100$ kN/m² $E_{core(rock)} = 12.1$ GN/m²

$RQD_{(rock)} = 75\%$

Spacing of discontinuity in rock = 550 mm

Thickness of individual discontinuity in rock = 2.5 mm

Estimate the allowable load-bearing capacity of the drilled shaft. Use $FS = 4$.

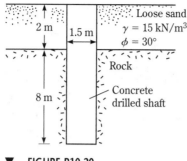

▼   **FIGURE P10.20**

REFERENCES

Das, B. M. (1980). "A Procedure for Estimation of Ultimate Uplift Capacity of Foundations in Clay," *Soils and Foundations,* The Japanese Society of Soil Mechanics and Foundation Engineering, Vol. 20, No. 1, pp. 77–82.

Das, B. M., and Seeley, G. R. (1975). "Breakout Resistance of Shallow Vertical Anchors," *Journal of the Geotechnical Engineering Division,* Americal Society of Civil Engineers, Vol. 101, No. GT9, pp. 999–1003.

Kulhawy, F. H., and Jackson, C. S. (1989). "Some Observations on Undrained Side Resistance of Drilled Shafts," *Proceedings,* Foundation Engineering: Current Principles and Practices, American Society of Civil Engineers, Vol. 2, pp. 1011–1025.

Ladanyi, B. (1977). "Discussion on Friction and Endbearing Tests on Bedrock for High Capacity Socket Design," *Canadian Geotechnical Journal,* Vol. 14, No. 1, pp. 153–156.

Meyerhof, G. G., and Adams, J. I. (1968). "The Ultimate Uplift Capacity of Foundations," *Canadian Geotechnical Journal,* Vol. 5, No. 4, pp. 225–244.

O'Neill, M. W. (1997). Personal communication.

Reese, L. C., and O'Neill, M. W. (1988). *Drilled Shafts: Construction and Design,* FHWA, Publication No. HI-88-042.

Reese, L. C., and O'Neill, M. W. (1989). "New Design Method for Drilled Shafts from Common Soil and Rock Tests," *Proceedings,* Foundation Engineering: Current Principles and Practices, American Society of Civil Engineers, Vol. 2, pp. 1026–1039.

Reese, L. C., Touma, F. T., and O'Neill, M. W. (1976). "Behavior of Drilled Piers Under Axial Loading," *Journal of Geotechnical Engineering Division,* American Society of Civil Engineers, Vol. 102, No. GT5, pp. 493–510.

Teng, W. C. (1962). *Foundation Design,* Prentice-Hall, Englewood Cliffs, N.J.

Touma, F. T., and Reese, L. C. (1974). "Behavior of Bored Piles in Sand," *Journal of the Geotechnical Engineering Division,* American Society of Civil Engineers, Vol. 100, No. GT7, pp. 749–761.

Vesic, A. S. (1963). "Bearing Capacity of Deep Foundations in Sand," *Highway Research Record,* no. 39, Highway Research Board, National Academy of Science, Washington, D. C., pp. 112–153.

Vesic, A. S. (1967). "Ultimate Load and Settlement of Deep Foundations in Sand," *Proceedings,* Symposium on Bearing Capacity and Settlement of Foundations, Duke University, Durham, N.C., p. 53.

Vesic, A. S. (1977). "Design of Pile Foundations," *NCHRP No. 42,* Transportation Research Board, National Research Council, Washington, D.C.

Whitaker, T., and Cooke, R. W. (1966). "An Investigation of the Shaft and Base Resistance of Large Bored Piles in London Clay," *Proceedings,* Conference on Large Bored Piles, Institute of Civil Engineers, London, pp. 7–49.

CHAPTER ELEVEN

FOUNDATIONS ON DIFFICULT SOILS

11.1 INTRODUCTION

In many areas of the United States and other parts of the world, certain soils make construction of foundations extremely difficult. For example, expansive or collapsible soils may cause high differential movements in structures by excessive heave or settlement. Similar problems can also arise when foundations are constructed over sanitary landfills. Foundation engineers must be able to identify difficult soils when they are encountered in the field. Although not all the problems caused by all soils can be solved, preventive measures can be taken to reduce the possibility of damage to structures built on them. This chapter outlines the fundamental properties of three major soil conditions — collapsible soils, expansive soils, and sanitary landfills — and methods of careful foundation construction.

COLLAPSIBLE SOIL

11.2 DEFINITION AND TYPES OF COLLAPSIBLE SOIL

Collapsible soils, which are sometimes referred to as *metastable soils,* are unsaturated soils that undergo a large volume change upon saturation. This volume change may or may not be the result of the application of additional load. The behavior of collapsing soils under load is best explained by the typical void ratio–pressure plot (e against $\log p$) for a collapsing soil, as shown in Figure 11.1. Branch ab is determined from the consolidation test on a specimen at its natural moisture content. At a pressure level of p_w, the equilibrium void ratio is e_1. However, if water is introduced into the specimen for saturation, the soil structure will collapse. After saturation, the equilibrium void ratio at the same pressure level p_w is e_2; cd is the branch of e–$\log p$ curve under additional load after saturation. Foundations that are constructed on such soils

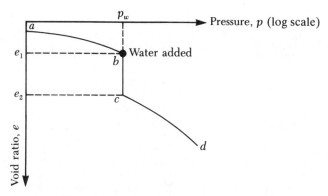

▼ **FIGURE 11.1** Nature of variation of void ratio with pressure for a collapsing soil

may undergo large and sudden settlement if and when the soil under them becomes saturated with an unanticipated supply of moisture. This moisture may come from several sources, such as (a) broken water pipelines, (b) leaky sewers, (c) drainage from reservoirs and swimming pools, (d) slow increase of groundwater, and so on. This type of settlement generally causes considerable structural damage. Hence identification of collapsing soils during field exploration is crucial.

The majority of naturally occurring collapsing soils are *aeolian* — that is, wind-deposited sand and/or silts, such as loess, aeolic beaches, and volcanic dust deposits. These deposits have high void ratios and low unit weights and are cohesionless or only slightly cohesive. *Loess* deposits have silt-sized particles. The cohesion in loess may be the result of the presence of clay coatings around the silt-size particles, which holds them in a rather stable condition in an unsaturated state. The cohesion may also be caused by the presence of chemical precipitates leached by rainwater. When the soil becomes saturated, the clay binders lose their strength and hence undergo a structural collapse. In the United States, large parts of the Midwest and arid West have such types of deposit. *Loess* deposits are also found over 15%–20% of Europe and over large parts of China.

Many collapsing soils may be residual soils that are products of weathering of parent rocks. The weathering process produces soils with a large range of particle-size distribution. Soluble and colloidal materials are leached out by weathering, resulting in large void ratios and thus unstable structures. Many parts of South Africa and Zimbabwe have residual soils that are decomposed granites. Sometimes collapsing soil deposits may be left by flash floods and mud flows. These deposits dry out and are poorly consolidated. An excellent review of collapsing soils is that of Clemence and Finbarr (1981).

11.3 PHYSICAL PARAMETERS FOR IDENTIFICATION

Several investigators have proposed various methods to evaluate the physical parameters of collapsing soils for identification. Some of these methods are discussed briefly in Table 11.1.

▼ **TABLE 11.1** Reported Criteria for Identification of Collapsing Soil[a]

Investigator	Year	Criteria
Denisov	1951	Coefficient of subsidence: $$K = \frac{\text{void ratio at liquid limit}}{\text{natural void ratio}}$$ $K = 0.5\text{-}0.75$: highly collapsible $K = 1.0$: noncollapsible loam $K = 1.5\text{-}2.0$: noncollapsible soils
Clevenger	1958	If dry unit weight is less than 80 lb/ft³ (≈ 12.6 kN/m³), settlement will be large; if dry unit weight is greater than 90 lb/ft³ (≈ 14.1 kN/m³), settlement will be small.
Priklonski	1952	$$K_D = \frac{\text{natural moisture content} - \text{plastic limit}}{\text{plasticity index}}$$ $K_D < 0$: highly collapsible soils $K_D > 0.5$: noncollapsible soils $K_D > 1.0$: swelling soils
Gibbs	1961	Collapse ratio, $R = \dfrac{\text{saturation moisture content}}{\text{liquid limit}}$ This was put into graph form.
Soviet Building Code	1962	$$L = \frac{e_o - e_L}{1 + e_o}$$ where e_o = natural void ratio and e_L = void ratio at liquid limit. For natural degree of saturation less than 60%, if $L > -0.1$, it is a collapsing soil.
Feda	1964	$$K_L = \frac{w_o}{S_r} - \frac{PL}{PI}$$ where w_o = natural water content, S_r = natural degree of saturation, PL = plastic limit, and PI = plasticity index. For $S_r < 100\%$, if $K_L > 0.85$, it is a subsident soil.
Benites	1968	A dispersion test in which 2 g of soil are dropped into 12 ml of distilled water and specimen is timed until dispersed; dispersion times of 20 to 30 s were obtained for collapsing Arizona soils.
Handy	1973	Iowa loess with clay (<0.002 mm) contents: <16%: high probability of collapse 16-24%: probability of collapse 24-32%: less than 50% probability of collapse >32%: usually safe from collapse

[a] Modified after Lutenegger and Saber (1988)

Jennings and Knight (1975) suggested a procedure to describe the *collapse potential* of a soil. It can be determined by taking an undisturbed soil specimen at natural moisture content in a consolidation ring. Step loads are applied to the specimen up to a pressure level of 29 lb/in² (≈ 200 kN/m²). (In Figure 11.1, this is p_w.) At this pressure ($p_w = 29$ lb/in²), the specimen is flooded for saturation and left for 24 hours. This test provides the void ratios (e_1 and e_2) before and after flooding. The collapse potential, C_p, may now be calculated as

▼ **TABLE 11.2** Relation of Collapse Potential to the
Severity of Foundation Problems

C_p (%)	Severity of problem
0–1	No problem
1–5	Moderate trouble
5–10	Trouble
10–20	Severe trouble
20	Very severe trouble

[a] After Clemence and Finbarr (1981)

$$C_p = \Delta\varepsilon = \frac{e_1 - e_2}{1 + e_o} \tag{11.1}$$

where e_o = natural void ratio of the soil
$\Delta\varepsilon$ = vertical strain

The severity of foundation problems associated with a collapsible soil have been
correlated with the collapse potential, C_p, by Jennings and Knight (1975). They were
summarized by Clemence and Finbarr (1981) and are given in Table 11.2.

Holtz and Hilf (1961) suggested that a loessial soil that has a void ratio large
enough to allow its moisture content to exceed its liquid limit upon saturation is
susceptible to collapse. So, for collapse

$$w_{(saturated)} \geq LL \tag{11.2}$$

However, for saturated soils

$$e_o = wG_s \tag{11.3}$$

where LL = liquid limit
G_s = specific gravity of soil solids

Combining Eqs. (11.2) and (11.3), for collapsing soils, yields

$$e_o \geq (LL)(G_s) \tag{11.4}$$

The natural dry unit weight, γ_d, of the soil for collapse is

$$\gamma_d \leq \frac{G_s\gamma_w}{1 + e_o} = \frac{G_s\gamma_w}{1 + (LL)(G_s)} \tag{11.5}$$

For an average value of $G_s = 2.65$, the limiting values of γ_d for various liquid limits
may now be calculated from Eq. (11.5):

Liquid limit (%)	Limiting values of γ_d	
	(lb/ft³)	(kN/m³)
10	130.8	20.56
15	118.3	18.60
20	108.1	16.99
25	99.5	15.64
30	92.1	14.48
35	85.8	13.49
40	80.3	12.62
45	75.4	11.86

Figure 11.2 shows a plot of the preceding limiting dry unit weights against the corresponding liquid limits. For any soil, if the natural dry unit weight falls below the limiting line, the soil is likely to collapse.

Care should be taken to obtain undisturbed samples for determining the collapse potentials and dry unit weights, preferably block samples cut by hand. The reason is that samples obtained by thin wall tubes may undergo some compression during the sampling process. However, if this procedure is used, the boreholes should be made *without water*.

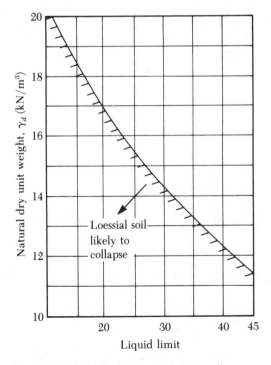

▼ **FIGURE 11.2**   Loessial soil likely to collapse

11.4 PROCEDURE FOR CALCULATING COLLAPSE SETTLEMENT

Jennings and Knight (1975) proposed the following laboratory procedure to determine the collapse settlement of structures upon saturation of soil:

1. Obtain *two* undisturbed soil specimens for tests in a standard consolidation test apparatus (oedometer).
2. Place the two specimens under 0.15 lb/in² (1 kN/m²) pressure for 24 hours.
3. After 24 hours, saturate one specimen by flooding. Keep the other specimen at natural moisture content.
4. After 24 hours of flooding, resume the consolidation test for both specimens by doubling the load (same procedure as the standard consolidation test) to the desired pressure level.
5. Plot the e–log p graphs for both specimens (Figure 11.3a and b).
6. Calculate the *in situ* effective pressure, p_o. Draw a vertical line corresponding to the pressure p_o.
7. From the e–log p curve of the soaked specimen, determine the preconsolidation pressure, p_c. If $p_c/p_o = 0.8$–1.5, the soil is normally consolidated; however, if $p_c/p_o > 1.5$, it is preconsolidated.
8. Determine e'_o, corresponding to p_o from the e–log p curve of the soaked specimen. (This procedure for normally consolidated and overconsolidated soils is shown in Figure 11.3a and b, respectively.)
9. Through point (p_o, e'_o) draw a curve that is similar to the e–log p curve obtained from the specimen tested at natural moisture content.
10. Determine the incremental pressure, Δp, on the soil caused by the construction of the foundation. Draw a vertical line corresponding to the pressure of $p_o + \Delta p$ in the e–log p curve.
11. Now, determine Δe_1 and Δe_2. The settlement of soil without change in the natural moisture content is

$$S_1 = \frac{\Delta e_1}{1 + e'_o} (H) \tag{11.6}$$

Also, the settlement caused by collapse in the soil structure is

$$S_2 = \frac{\Delta e_2}{1 + e'_o} (H) \tag{11.7}$$

where H = thickness of soil susceptible to collapse

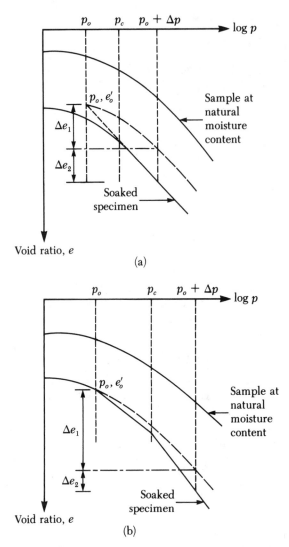

▼ **FIGURE 11.3** Settlement calculation from double oedometer test: (a) normally consolidated soil; (b) overconsolidated soil

11.5 FOUNDATION DESIGN IN SOILS NOT SUSCEPTIBLE TO WETTING

For actual foundation design purposes, some standard field load tests may also be conducted. Figure 11.4 shows the results of some field load tests in loess deposits in Nebraska and Iowa. Note that the load–settlement relationships are essentially linear up to a certain critical pressure, p_{cr}, at which there is a breakdown of the soil structure and hence a large settlement. Sudden breakdown of soil structure is more common with soils having a high natural moisture content than with normally dry soils.

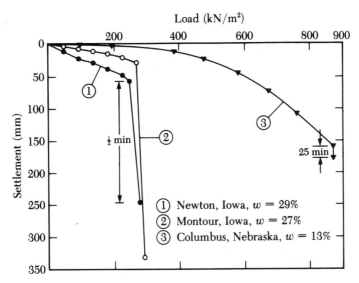

▼ **FIGURE 11.4** Results of standard load test on loess deposits in Iowa and Nebraska (adapted from *Foundation Engineering,* Second Edition, by R. B. Peck, W. E. Hanson, and T. H. Thornburn. Copyright 1974 by John Wiley and Sons. Reprinted by permission.)

If enough precautions are taken in the field to prevent moisture from increasing under structures, spread foundations and raft foundations may be built on potentially collapsible soils. However, the foundations must be proportioned so that the critical stresses (Figure 11.4) in the field are never exceeded. A factor of safety of about 2.5 to 3 should be used to calculate the allowable soil pressure, or

$$p_{all} = \frac{p_{cr}}{FS} \tag{11.8}$$

where p_{all} = allowable soil pressure
        FS = factor of safety (about 2.5 to 3)

The differential and total settlements of these foundations should be similar to those of foundations designed for sandy soils.

Continuous foundations may be safer than isolated foundations over collapsible soils in that they can effectively minimize differential settlement. Figure 11.5 shows a typical procedure for construction of continuous foundations. This procedure uses footing beams and longitudinal load-bearing beams.

In the construction of heavy structures, such as grain elevators, over collapsible soils, settlements up to about 1 ft (≈0.3 m) are sometimes allowed (Peck, Hanson, and Thornburn, 1974). In this case, tilting of the foundation is not likely to occur because there is no eccentric loading. The total expected settlement for such structures can be estimated from standard consolidation tests on samples of field moisture content. Without eccentric loading, the foundations will exhibit uniform settlement over loessial deposits; however, if the soil is of residual or colluvial nature, settlement

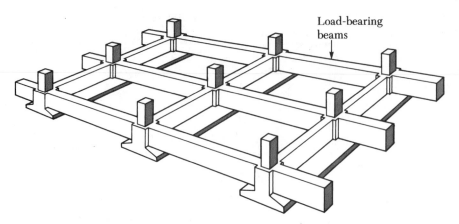

▼ **FIGURE 11.5**   Continuous foundation with load-bearing beams (after Clemence and Finbarr, 1981)

may not be uniform. The reason is the nonuniformity generally encountered in residual soils.

Extreme caution must be used in building heavy structures over collapsible soils. If large settlements are expected, drilled-shaft and pile foundations should be considered. These types of foundation can transfer the load to a stronger load-bearing stratum.

11.6 FOUNDATION DESIGN IN SOILS SUSCEPTIBLE TO WETTING

If the upper layer of soil is likely to get wet and collapse at some time after construction of the foundation, several design techniques to avoid foundation failure may be considered.

1. If the expected depth of wetting is about 5 to 6.5 ft (≈1.5 to 2 m) from the ground surface, the soil may be moistened and recompacted by heavy rollers. Spread footings and rafts may be constructed over the compacted soil. An alternative to recompaction by heavy rollers is *heavy tamping,* which is sometimes referred to as *dynamic compaction* (see Chapter 12). It consists primarily of dropping a heavy weight repeatedly on the ground. The height of the hammer drop can vary from 25 to 100 ft (≈8 to 30 m). The stress waves generated by the hammer drop help in the densification of the soil.

2. If conditions are favorable, foundation trenches can be flooded with solutions of sodium silicate and calcium chloride to stabilize the soil chemically. The soil will behave like a soft sandstone and resist collapse upon saturation. This method is successful only if the solutions can penetrate to the desired depth; thus it is most applicable to fine sand deposits. Silicates are rather costly and are not generally used. However, in some parts of Denver, silicates have been used very successfully.

The injection of a sodium silicate solution for stabilization of collapsible soil deposits has been used extensively in the former Soviet Union and Bulgaria (Houston and Houston, 1989). This process is used for dry collapsible soils and for wet collapsible soils that are likely to compress under the added weight of the structure to be built and consists of three steps:

Step 1. Injection of carbon dioxide for removal of any water present and preliminary activation of soil

Step 2. Injection of sodium silicate grout

Step 3. Injection of carbon dioxide for neutralization of alkali

3. When the soil layer is susceptible to wetting to a depth of about 10 m, several techniques may be used to cause collapse of the soil *before* foundation construction. Two of these are *vibroflotation* and *ponding* (also called *flooding*). Vibroflotation is used successfully in free-draining soil (see Chapter 12). The procedure of ponding — by constructing low dikes — is utilized at sites that have no impervious layers. However, even after saturation and collapse of the soil by ponding, some additional settlement of the soil may occur after foundation construction. Additional settlement may also be caused by incomplete saturation of the soil at the time of construction. Ponding may be used successfully in the construction of earth dams.

4. If precollapsing of soil is not practical, foundations may be extended beyond the zone of possible wetting, which may require drilled shafts and piles. The design of drilled shafts and piles must take into consideration the effect of negative skin friction resulting from the collapse of the soil structure and the associated settlement of the zone of subsequent wetting.

In some cases, a *rock-column type of foundation* (*vibroreplacement*) may also be considered. Rock columns are built with large boulders that penetrate the potentially collapsible soil layer. They act as piles in transferring the load to a more stable soil layer.

11.7   CASE HISTORIES OF STABILIZATION OF COLLAPSIBLE SOIL

Use of Dynamic Compaction

Lutenegger (1986) reported the use of dynamic compaction to stabilize a thick layer of friable loess before construction of a foundation in Russe, Bulgaria. During field exploration, the water table was not encountered to a depth of 33 ft (10 m), and the natural moisture content was below the plastic limit. Initial density measurements made on undisturbed soil specimens indicated that the moisture content at saturation would exceed the liquid limit, a property usually encountered in collapsible loess.

For dynamic compaction of the soil, the upper 5.6 ft (1.7 m) of crust material was excavated. A circular concrete weight of 15 ton ($\approx$133 kN) was used as a hammer. At each grid point, compaction was achieved by dropping the hammer 7 to 12 times through a vertical distance of 8.2 ft (2.5 m).

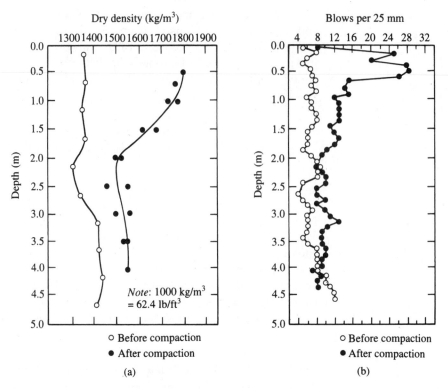

▼ **FIGURE 11.6**   (a) Dry density before and after compaction; (b) penetration resistance before and after compaction (after Lutenegger, 1986)

Figure 11.6a shows the dry density of the soil before and after compaction. Figure 11.6b shows the increase in field standard penetration resistance before and after compaction. The increase in dry density of the soil and standard penetration resistance shows that dynamic compaction can be used effectively to stabilize collapsible soil.

Chemical Stabilization

Semkin et al. (1986) reported on chemical stabilization of a loessial soil deposit with a carbon dioxide, sodium silicate, and carbon dioxide injection scheme (also see Houston and Houston, 1989). The site is that of the Interregional Center in Tashkent, which consists of two buildings — one three stories high with strip foundations and the other one story high with spread foundations. The loessial soil deposit at the site was about 115 ft (35 m) thick, and the water table was at a depth of about 60 ft (18 m). The natural moisture content and porosity of the soil above the water table was 10%–25% and 0.48, respectively.

The Interregional Center was constructed in 1973. Unanticipated leakage from conduits in the center caused differential settlement to occur in 1974. Without shutting down the center, chemical stabilization was used effectively, and the differential settlement was stopped.

EXPANSIVE SOILS

11.8   EXPANSIVE SOILS—GENERAL

Many plastic clays swell considerably when water is added to them and then shrink with the loss of water. Foundations constructed on these clays are subjected to large uplifting forces caused by the swelling. These forces will induce heaving, cracking, and breakup of both building foundations and slab-on-grade members. Expansive clays cover large parts of the United States, South America, Africa, Australia, and India. In the United States, these clays are predominant in Texas, Oklahoma, and the upper Missouri Valley. In general, potentially expansive clays have liquid limits and plasticity indices greater than about 40 and 15, respectively.

As noted, an increase in moisture content causes clay to swell. The depth in a soil to which periodic changes of moisture occur is usually referred to as the *active zone*. The depth of the active zone varies, depending on location. Some typical active-zone depths in American cities are given in Table 11.3. In some clays and clay shales in the western United States, the depth of the active zone can be as much as 50 ft ($\approx$15 m). The active-zone depth can be easily determined by plotting the liquidity index against the depth of the soil profile over several seasons. Figure 11.7 shows such a plot for the Beaumont formation in the Houston area.

An example of the effect of seasonal change in the active zone related to shrinking and swelling of an expansive soil deposit is shown in Figure 11.8. It is a typical record of vertical ground movement at an open-field test plot in Regina, Saskatchewan (Canada), for the depths below the ground surface indicated. The seasonal ground movement virtually ceases at a depth of about 10–12 ft (3–4 m).

▼   **TABLE 11.3** Typical Active-Zone Depths in Some
U.S. Cities[a]

City	Depth of active zone	
	(ft)	(m)
Houston	5 to 10	1.5 to 3
Dallas	7 to 15	2.1 to 4.6
San Antonio	10 to 20	3 to 9
Denver	10 to 15	3 to 4.6
[a] After O'Neill and Poormoayed (1980)		

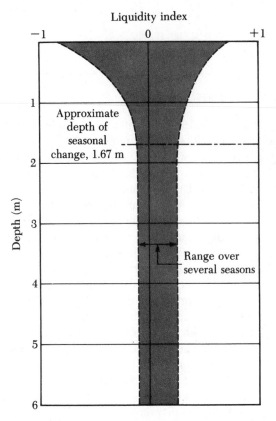

Liquidity index

FIGURE 11.7 Active zone in Houston area — Beaumont formation (after O'Neill and Poormoayed, 1980)

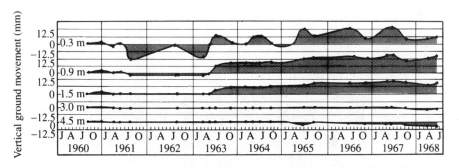

FIGURE 11.8 Vertical ground movements for an open-field test plot at Regina, Saskatchewan, as measured by Hamilton 1968 (after Sattler and Fredlund, 1991)

11.9 LABORATORY MEASUREMENT OF SWELL

To study the magnitude of possible swell in a clay, simple laboratory oedometer tests can be conducted on undisturbed specimens. Two common tests are the unrestrained swell test and swelling pressure test.

In the *unrestrained swell test,* the specimen is placed in an oedometer under a small surcharge of about 1 lb/in² (6.9 kN/m²). Water is then added to the specimen, and the expansion of the volume of the specimen (that is, height; the area of cross section is constant) is measured until equilibrium is reached. The percent of free swell may be expressed as a ratio:

$$s_{w(\text{free})}\ (\%) = \frac{\Delta H}{H}\ (100)$$

(11.9)

where $s_{w(\text{free})}$ = free swell, as a percent
ΔH = height of swell due to saturation
H = original height of the specimen

Vijayvergiya and Ghazzaly (1973) analyzed various soil test results obtained in this manner and prepared a correlation chart of the free swell, liquid limit, and natural moisture content, as shown in Figure 11.9. O'Neill and Poormoayed (1980) developed a relationship for calculating the free surface swell from this chart:

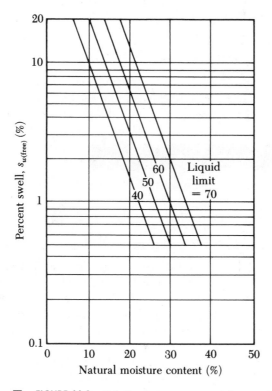

▼ **FIGURE 11.9**   Relation between percent free swell, liquid limit, and natural moisture content (after Vijayvergiya and Ghazzaly, 1973)

$$\Delta S_F = 0.0033 Z s_{w\text{(free)}}$$

(11.10)

where ΔS_F = free surface swell
Z = depth of active zone
$s_{w\text{(free)}}$ = free swell, as a percent (Figure 11.9)

More recently, Sivapullaiah et al. (1987) suggested a new test method for obtaining a *modified free swell index* for clays, which appears to give a better indication for the swelling potential of clayey soils. This test begins with an oven-dried soil with a mass of about 10 g. The soil mass is well pulverized and transferred into a 100-ml graduated jar containing distilled water. After 24 hours, the swollen sediment volume is measured. The *modified free swell index* is then calculated:

$$\text{Modified free swell index} = \frac{V - V_s}{V_s}$$

(11.11)

where V = soil volume after swelling

V_s = volume of soil solid = $\dfrac{W_s}{G_s \gamma_w}$

W_s = weight of oven-dried soil
G_s = specific gravity of soil solids
γ_w = unit weight of water

Based on the modified free swell index, the swelling potential of a soil may be qualitatively classified as follows:

Modified free swell index	Swelling potential
<2.5	Negligible
2.5 to 10	Moderate
10 to 20	High
>20	Very high

Sikh (1993) reported the results of several free swell tests on undisturbed soil specimens obtained from Southern California. The tests were conducted by subjecting the soil specimens to the *actual effective overburden pressure*. The results of these tests are given in Figure 11.10. The upper-bound curve indicates that, for an effective over-burden pressure of about 1.4 kip/ft^2 or greater, the vertical free swell [Eq. (11.9)] generally decreases to less than 1%.

The *swelling pressure test* can be conducted by taking a specimen in a consolidation ring and applying a pressure equal to the effective overburden pressure, p_o, plus the approximate anticipated surcharge caused by the foundation, p_s. Water is then added to the specimen. As the specimen starts to swell, pressure is applied in

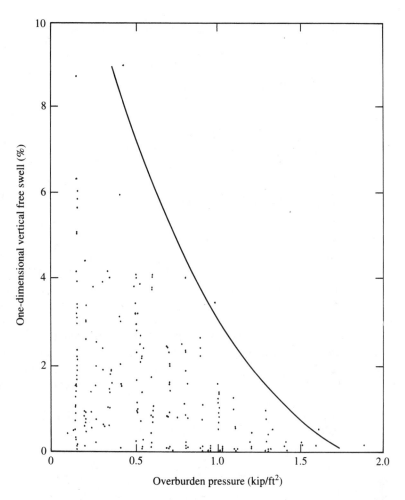

FIGURE 11.10   One-dimensional vertical free swell of some southern California
soils (after Sikh, 1993)

small increments to prevent swelling. It is continued until full swelling pressure is
developed on the specimen. At that time, the total pressure is

$$p_T = p_o + p_s + p_1 \tag{11.12}$$

where p_T = total pressure to prevent swelling, or zero swell pressure
           p_1 = additional pressure added to prevent swelling after addition of water

Figure 11.11 shows the variation of the percentage of swell against pressure during
a swelling pressure test. For more information on this type of test, see Sridharan
et al. (1986).

   A p_T of about 0.4–0.65 kip/ft² (20–30 kN/m²) is considered to be low, and a p_T
of 30–40 kip/ft² (1500–2000 kN/m²) is considered to be high. After zero swell

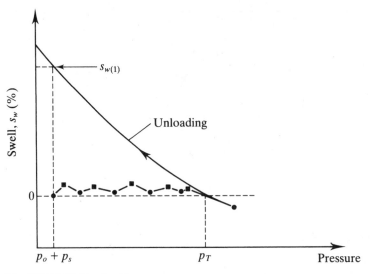

▼ FIGURE 11.11   Swelling pressure test

pressure is attained, the soil specimen can be unloaded in steps to the level of the overburden pressure, p_o. This unloading process will cause the specimen to swell. The equilibrium swell for each pressure level is also recorded. The variation of the swell, in percent, s_w (%), and the applied pressure on the specimen will be like that shown in Figure 11.11.

The *swelling pressure test* can be used to determine the surface heave, ΔS, for a foundation (O'Neill and Poormoayed, 1980) as

$$\Delta S = \sum_{i=1}^{n} [s_{w(1)} \, (\%)] \, (H_i) \, (0.01) \tag{11.13}$$

where $s_{w(1)}$ (%) = swell, in percent, for layer i under a pressure of $p_o + p_s$
       (see Figure 11.11)
   ΔH_i = thickness of layer i

▼ EXAMPLE 11.1

A soil profile has an active zone of expansive soil of 2 m. The liquid limit and the average natural moisture content during the construction season are 60% and 30%, respectively. Determine the free surface swell.

Solution   From Figure 11.9 for $LL = 60\%$ and $w = 30\%$, $s_{w(\text{free})} = 1\%$. From Eq. (11.10),

$$\Delta S_F = 0.0033 Z s_{w(\text{free})}$$

Hence

$$\Delta S_F = 0.0033 (2) (1) (1000) = \textbf{6.6 mm} \qquad \blacktriangle$$

▼ EXAMPLE 11.2

An expansive soil profile has an active-zone thickness of 5.2 m. A shallow foundation is to be constructed 1.2 m below the ground surface. A swelling pressure test provided the following data:

Depth below ground surface (m)	Swell under overburden and estimated foundation sur-charge pressure, $s_{w(1)}$ (%)
1.2	3.0
2.2	2.0
3.2	1.2
4.2	0.55
5.2	0.0

    a.   Estimate the total possible swell under the foundation.
    b.   If the allowable total swell is 15 mm, what would be the necessary undercut?

Solution

Part a:

Figure 11.12 shows the plot of depth versus $s_{w(1)}$ (%). The area of this diagram will be the total swell. Thus

$$\Delta S = \frac{1}{100} \left[\left(\frac{1}{2}\right)(0.55 + 0)(1) + \left(\frac{1}{2}\right)(0.55 + 1.2)(1) \right.$$

$$\left. + \left(\frac{1}{2}\right)(1.2 + 2)(1) + \left(\frac{1}{2}\right)(2 + 3)(1) \right]$$

$$= 0.0525 \text{ m} = \textbf{52.5 m}$$

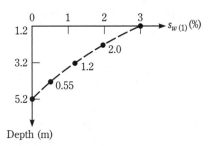

Depth (m)

▼ FIGURE 11.12

Part b:

Total swell at various depths can be calculated as follows:

Depth (m)	Total swell, ΔS (m)
5.2	0
4.2	$0 + \frac{1}{2}(0.55 + 0)(1)(1/100) = 0.00275$
3.2	$0.00275 + \frac{1}{2}(1.2 + 0.55)(1)(1/100) = 0.0115$
2.2	$0.0115 + \frac{1}{2}(2 + 1.2)(1)(1/100) = 0.0275$
1.2	$0.0275 + \frac{1}{2}(2 + 3)(1)(1/100) = 0.0525$

The plot of ΔS versus depth is shown in Figure 11.13. From this figure, the depth of undercut is $2.91 - 1.2 = $ **1.71 m below the bottom of the foundation.**

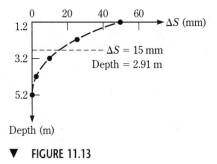

▼   **FIGURE 11.13**   ▲

11.10 CLASSIFICATION OF EXPANSIVE SOIL BASED ON INDEX TESTS

Classification systems for expansive soils are based on the problems they create in the construction of foundations (potential swell). Most of the classifications contained in the literature are summarized in Figure 11.14 and Table 11.4. However, the classification system developed by U.S. Army Waterways Experiment Station (Snethen et al., 1977) is the one most widely used in the United States. It has also been summarized by O'Neill and Poormoayed (1980); see Table 11.5 (p. 749).

11.11 FOUNDATION CONSIDERATIONS FOR EXPANSIVE SOILS

If a soil has a low swell potential, standard construction practices may be followed. However, if the soil possesses a marginal or high swell potential, precautions need to be taken, which may entail

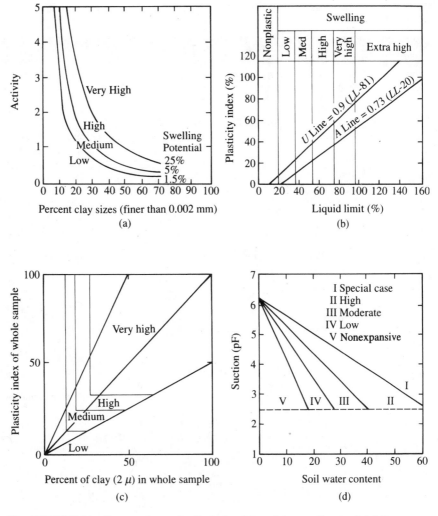

▼ **FIGURE 11.14** Commonly used criteria for determining swell potential (after Abduljauwad and Al-Sulaimani, 1993)

1. Replacing the expansive soil under the foundation

2. Changing the nature of the expansive soil by compaction control, prewetting, installation of moisture barriers, and/or chemical stabilization

3. Strengthening the structures to withstand heave, constructing structures that are flexible enough to withstand the differential soil heave without failure, or constructing isolated deep foundations below the depth of the active zone

One particular method may not be sufficient in all situations. Combining several techniques may be necessary, and local construction experience should always be

▼ **TABLE 11.4** Summary of Some Criteria for Identifying Swell Potential (after Abduljauwad and Al-Sulaimani, 1993)

Reference	Criteria	Remarks
Holtz (1959)	$CC > 28$, $PI > 35$, and $SL < 11$ (very high) $20 \leq CC \leq 31$, $25 \leq PI \leq 41$, and $7 \leq SL \leq 12$ (high) $13 \leq CC \leq 23$, $15 \leq PI \leq 28$, and $10 \leq SL \leq 16$ (medium) $CC \leq 15$, $PI \leq 18$, and $SL \geq 15$ (low)	Based on CC, PI, and SL
Seed et al. (1962)	See Figure 11.14a	Based on oedometer test using compacted specimen, percentage of clay < 2 μm and activity
Altmeyer (1955)	$LS < 5$, $SL > 12$, and $PS < 0.5$ (noncritical) $5 \leq LS \leq 8$, $10 \leq SL \leq 12$, and $0.5 \leq PS \leq 1.5$ (marginal) $LS > 8$, $SL < 10$, and $PS > 1.5$ (critical)	Based on LS, SL, and PS Remolded sample ($\rho_{d(\max)}$ and w_{opt}) Soaked under 6.9 kPa surcharge
Dakshanamanthy and Raman (1973)	See Figure 11.14b	Based on plasticity chart
Raman (1967)	$PI > 32$ and $SI > 40$ (very high) $23 \leq PI \leq 32$ and $30 \leq SI \leq 40$ (high) $12 \leq PI \leq 23$ and $15 \leq SI \leq 30$ (medium) $PI < 12$ and $SI < 15$ (low)	Based on PI and SI
Sowers and Sowers (1970)	$SL < 10$ and $PI > 30$ (high) $10 \leq SL \leq 12$ and $15 \leq PI \leq 30$ (moderate) $SL > 12$ and $PI < 15$ (low)	Little swell will occur when w_0 results in LI of 0.25
Van Der Merwe (1964)	See Figure 11.14c	Based on PI, percentage of clay <2 μm, and activity
Uniform Building Code, 1968	$EI > 130$ (very high) and $91 \leq EI \leq 130$ (high) $51 \leq EI \leq 90$ (medium) and $21 \leq EI \leq 50$ (low) $0 \leq EI \leq 20$ (very low)	Based on oedometer test on compacted specimen with degree of saturation close to 50% and a surcharge of 6.9 kPa
Snethen (1984)	$LL > 60$, $PI > 35$, $\tau_{nat} > 4$, and $SP > 1.5$ (high) $30 \leq LL \leq 60$, $25 \leq PI \leq 35$, $1.5 \leq \tau_{nat} \leq 4$, and $0.5 \leq SP \leq 1.5$ (medium) $LL < 30$, $PI < 25$, $\tau_{nat} < 1.5$, and $SP < 0.5$ (low)	PS is representative for field condition, can be used without τ_{nat}, but accuracy will be reduced
Chen (1988)	$PI \geq 35$ (very high) and $20 \leq PI \leq 55$ (high) $10 \leq PI \leq 35$ (medium) and $PI \leq 15$ (low)	Based on PI
McKeen (1992)	Figure 11.14d	Based on measurements of soft water content, suction, and volume change on drying
Vijayvergiya and Ghazzaly (1973)	$\log SP = (1/12)(0.44LL - w_o + 5.5)$	Empirical equations
Nayak and Christensen (1974)	$SP = (0.00229PI)(1.45C)/w_0 + 6.38$	Empirical equations
Weston (1980)	$SP = 0.00411(LL_w)^{4.17}q^{-3.86}w_0^{-2.33}$	Empirical equations

Note: C = clay, %
CC = colloidal content, %
EI = Expansion index = $100 \times$ percent swell $\times$ fraction passing no. 4 sieve
LI = liquidity index, %
LL = liquid limit, %
LL_w = weighted liquid limit, %
LS = linear shrinkage, %
PI = plasticity index, %
PS = probable swell, %

q = surcharge
SI = shrinkage index = $LL - SL$, %
SL = shrinkage limit, %
SP = swell potential, %
w_0 = natural soil moisture
w_{opt} = optimum moisture content, %
τ_{nat} = natural soil suction in tsf
$\rho_{d(\max)}$ = max dry density

▼ **TABLE 11.5** Expansive Soil Classification System[a]

Liquid limit	Plasticity index	Potential swell (%)	Potential swell classification
<50	<25	<0.5	Low
50–60	25–35	0.5–1.5	Marginal
>60	>35	>1.5	High
Potential swell = vertical swell under a pressure equal to overburden pressure			
[a] Compiled from O'Neill and Poormoayed (1980)			

considered. Following are details of some of the commonly used techniques of dealing with expansive soils.

Replacement of Expansive Soil

When shallow, moderately expansive soils are present at the surface, they can be removed and replaced by less expansive soils and then compacted properly.

Changing the Nature of Expansive Soil

1. *Compaction*: Heave of expansive soils decreases substantially when the soil is compacted to a lower unit weight on the high side of the optimum moisture content (possibly 3–4% above the optimum moisture content). Even under such conditions, a slab-on-ground type of construction should not be considered when the total probable heave is expected to be about 1.5 in. (38 mm) or more. Figure 11.15 shows the recommended limits of soil compaction in the field for reduction of heave. Note that the recommended dry unit weights are based on climatic ratings. According to U.S. Weather Bureau data, a climatic rating of 15 represents an extremely unfavorable climatic condition; a rating of 45 represents a favorable climatic condition. The isobars of climatic rating for the continental United States are shown in Figure 11.15b.

2. *Prewetting*: One technique for increasing the moisture content of the soil is by ponding and hence achieving most of the heave before construction. However, this technique may be time-consuming because the seepage of water through highly plastic clays is slow. After ponding, 4–5% of hydrated lime may be added to the top layer of the soil to make it less plastic and more workable (Gromko, 1974).

3. *Installation of moisture barriers*: The long-term effect of the differential heave can be reduced by controlling the moisture variation in the soil. It is achieved by providing vertical moisture barriers about 5 ft ($\approx$1.5 m) deep around the perimeter of slabs for the slab-on-grade type of construction. These moisture

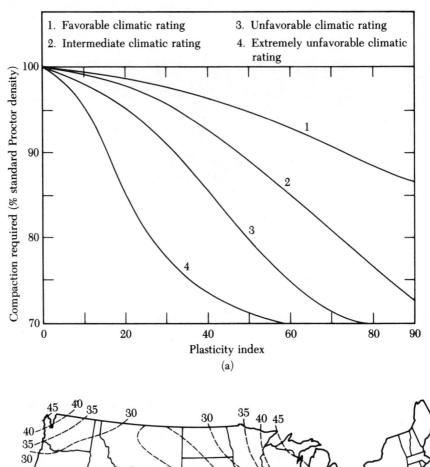

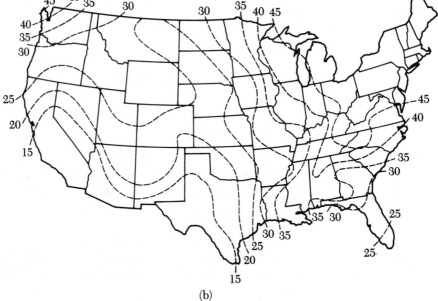

▼ **FIGURE 11.15** (a) Soil compaction requirement based on climatic rating; (b) equivalent climatic rating of the United States (after Gromko, 1974)

barriers may be constructed in trenches filled with gravel, lean concrete, or impervious membranes.

4. *Stabilization of soil*: Chemical stabilization with the aid of lime and cement has often proved useful. A mix containing about 5% lime is sufficient in most cases. Lime or cement and water are mixed with the top layer of soil and compacted. The addition of lime or cement will decrease the liquid limit, the plasticity index, and the swell characteristics of the soil. This type of stabilization work can be done to a depth of 3–5 ft (≈1–1.5 m). Hydrated high-calcium lime and dolomite lime are generally used for lime stabilization.

Another method of stabilization of expansive soil is the *pressure injection* of lime slurry or lime–fly ash slurry into the soil, usually to a depth of 12–16 ft (4–5 m) and occasionally deeper to cover the active zone. Further details of the pressure injection technique are presented in Chapter 12. Depending on the soil conditions at a site, single or multiple injections can be planned, as shown in Figure 11.16. Figure 11.17 shows the slurry pressure injection work for a building pad. The stakes marked are the planned injection points. Figure 11.18 shows lime–fly ash stabilization by pressure injection of the bank of a canal that had experienced sloughs and slides.

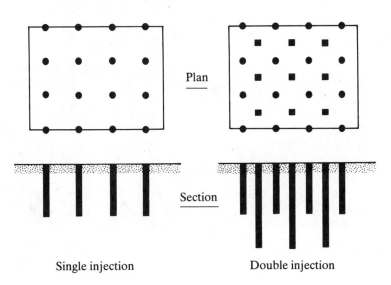

▼ **FIGURE 11.16** Multiple lime slurry injection planning for a building pad

▼ **FIGURE 11.17**   Pressure injection of lime slurry for a building pad (Courtesy of GKN Hayward Baker, Inc., Woodbine Division, Ft. Worth, Texas)

▼ **FIGURE 11.18**   Slope stabilization of a canal bank by pressure injection of lime–fly ash slurry (Courtesy of GKN Hayward Baker, Inc., Woodbine Division, Ft. Worth, Texas)

11.12   CONSTRUCTION ON EXPANSIVE SOILS

Care must be exercised in choosing the type of foundation to be used on expansive soils. Table 11.6 shows some recommended construction procedures based on the total predicted heave, ΔS, and the length-to-height ratio of the wall panels.

▼ **TABLE 11.6** Construction Procedures for Expansive Clay Soils[a]

Total predicted heave (mm)		Recommended construction	Method	Remarks
L/H = 1.25	L/H = 2.5			
0 to 6.35	12.7	No precaution		
6.35 to 12.7	12.7 to 50.8	Rigid building tolerating movement (steel reinforcement as necessary)	*Foundations*: Pads Strip footings Raft (waffle)	Footings should be small and deep, consistent with the soil-bearing capacity. Rafts should resist bending.
			Floor slabs: Waffle Tile	Slabs should be designed to resist bending and should be independent of grade beams.
			Walls:	Walls on a raft should be as flexible as the raft. No rigid connections vertically. Brick works should be strengthened with tie bars or bands.
12.7 to 50.8	50.8 to 101.6	Building damping movement	*Joints*: Clear Flexible	Contacts between structural units should be avoided; or flexible, waterproof material may be inserted in the joints.
			Walls: Flexible Unit construction Steel frame	Walls or rectangular building units should heave as a unit.
			Foundations: Three point Cellular Jacks	Cellular foundations allow slight soil expansion to reduce swelling pressure. Adjustable jacks can be inconvenient to owners. Three-point loading allows motion without duress.
>50.8	>101.6	Building independent of movement	*Foundation drilled shaft*: Straight shaft Bell bottom	Smallest-diameter and widely spaced piers compatible with load should be placed. Clearance should be allowed under grade beams.
			Suspended floor:	Floor should be suspended on grade beams 305 to 460 mm above the soil.

[a] After Gromko, 1974

For example, Table 11.6 proposes the use of waffle slabs as an alternative in designing rigid buildings capable of tolerating movement. Figure 11.19 shows a schematic diagram of a waffle slab. In this type of construction, the ribs hold the structural load. The waffle voids allow the expansion of soil.

Table 11.6 also suggests the use of drilled shaft foundation with a suspended floor slab for the construction of structures independent of movement. Figure 11.20a shows a schematic diagram of such an arrangement. The bottom of the shafts should be placed below the active zone of the expansive soil. For the design of the shafts, the uplifting force, U, may be estimated (Figure 11.20b) from the equation

$$U = \pi D_s Z p_T \tan \phi_{ps} \tag{11.14}$$

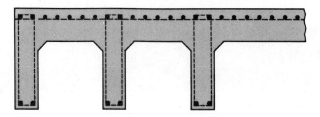

▼ **FIGURE 11.19**   Waffle slab

where D_s = diameter of the shaft
         Z = depth of the active zone
         ϕ_{ps} = effective angle of plinth–soil friction
         p_T = pressure for zero horizontal swell (see Figure 11.11; $p_T = p_o + p_s + p_1$)

In most cases, the value of ϕ_{ps} varies between 10° and 20°. An average value of the zero horizontal swell pressure must be determined in the laboratory. In the absence of laboratory results, $p_T \tan \phi_{ps}$ may be considered equal to the undrained shear strength of clay, c_u, in the active zone.

The belled portion of the drilled shaft will act as an anchor to resist the uplifting force. Ignoring the weight of the drilled shaft

$$Q_{\text{net}} = U - D \tag{11.15}$$

where Q_{net} = net uplift load
         D = dead load

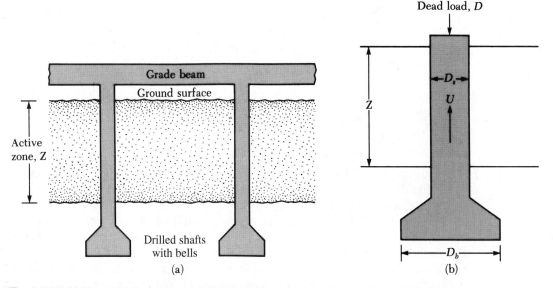

▼ **FIGURE 11.20**   (a) Construction of drilled shafts with bells and grade beam; (b) definition of parameters in Eq. (11.14)

Now

$$Q_{net} \approx \frac{c_u N_c}{FS}\left(\frac{\pi}{4}\right)(D_b^2 - D_s^2) \tag{11.16}$$

where c_u = undrained cohesion of the clay in which the bell is located

Combining Eqs. (11.15) and (11.16) gives

$$U - D = \frac{c_u N_c}{FS}\left(\frac{\pi}{4}\right)(D_b^2 - D_s^2) \tag{11.17}$$

where N_c = bearing capacity factor
FS = factor of safety
D_b = diameter of the bell of the drilled shaft

Conservatively, N_c (Tables 3.4 and 3.5) is

$$N_c \approx N_{c(strip)}F_{cs} = N_{c(strip)}\left(1 + \frac{N_q B}{N_c L}\right) \approx 5.14\left(1 + \frac{1}{5.14}\right) = 6.14$$

An example of a drilled-shaft design is given in Example 11.3.

▼ EXAMPLE 11.3

Figure 11.21 shows a drilled shaft with a bell. The depth of the active zone is 5 m. The zero swell pressure of the swelling clay (p_T) is 450 kN/m². For the drilled shaft, the dead load (D) is 600 kN and the live load is 300 kN. Assume $\phi_{ps} = 12°$.

a. Determine the diameter of the bell, D_b.
b. Check the bearing capacity of the drilled shaft assuming zero uplift force.

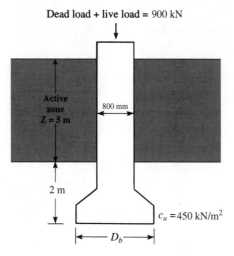

Dead load + live load = 900 kN

Active zone
Z = 5 m

800 mm

2 m

$c_u = 450$ kN/m²

D_b

▼ FIGURE 11.21

Solution

Part a: Determining the Bell Diameter, D_b

The uplift force, Eq. (11.14), is

$$U = \pi D_s Z p_T \tan \phi_{ps}$$

Given: $Z = 5$ m and $p_T = 450$ kN/m^2. Then

$$U = \pi (0.8)(5)(450)\tan 12° = 1202 \text{ kN}$$

Assume the dead load and live load to be zero, and FS in Eq. (11.17) to be 1.25. So, from Eq. (11.17),

$$U = \frac{c_u N_c}{FS}\left(\frac{\pi}{4}\right)(D_b^2 - D_s^2)$$

$$1202 = \frac{(450)(6.14)}{1.25}\left(\frac{\pi}{4}\right)(D_b^2 - 0.8^2); \qquad D_b = \mathbf{1.15\,m}$$

The factor of safety against uplift with the dead load also should be checked. A factor of safety of at least 2 is desirable. So, from Eq. (11.17)

$$FS = \frac{c_u N_c \left(\frac{\pi}{4}\right)(D_b^2 - D_s^2)}{U - D}$$

$$= \frac{(450)(6.14)\left(\frac{\pi}{4}\right)(1.15^2 - 0.8^2)}{1202 - 600} = \mathbf{2.46 > 2 \text{—OK}}$$

Part b: Check for Bearing Capacity

Assume that $U = 0$. Then

$$\text{Dead load + live load} = 600 + 300 = 900 \text{ kN}$$

$$\text{Downward load per unit area} = \frac{900}{\left(\frac{\pi}{4}\right)(D_b^2)} = \frac{900}{\left(\frac{\pi}{4}\right)(1.15)^2} = 866.5 \text{ kN/m}^2$$

Net bearing capacity of the soil under the bell $= q_{u(\text{net})} = c_u N_c$
$$= (450)(6.14) = 2763 \text{ kN/m}^2$$

Hence the factor of safety against bearing capacity failure is

$$FS = \frac{2763}{866.5} = \mathbf{3.19 > 3 \text{—OK}}$$

▲

SANITARY LANDFILLS

11.13 SANITARY LANDFILLS—GENERAL

Sanitary landfills provide a way to dispose of refuse on land without endangering public health. Sanitary landfills are used in almost all countries, with varying degrees of success. The refuse disposed of in sanitary landfills may contain organic, wood, paper, and fibrous wastes or demolition wastes such as bricks and stones. The refuse is dumped and compacted at frequent intervals and then covered with a layer of soil, as shown in Figure 11.22. In the compacted state, the average unit weight of the refuse may vary between 32–64 lb/ft³ (5–10 kN/m³.) A typical city in the United States, with a population of one million, generates about 135×10^6 ft³ ($\approx 3.8 \times 10^6$ m³) of compacted landfill material per year.

As property values continue to increase in densely populated areas, constructing structures over sanitary landfills becomes more and more tempting. In some instances, a visual site inspection may not be enough to detect an old sanitary landfill. However, construction of foundations over sanitary landfills is generally problematic because of poisonous gases (e.g., methane), excessive settlement, and low inherent bearing capacity.

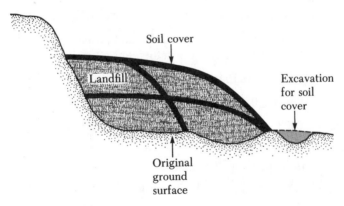

▼ **FIGURE 11.22**   Schematic diagram of sanitary landfill in progress

11.14 SETTLEMENT OF SANITARY LANDFILLS

Sanitary landfills undergo large continuous settlements over a long period of time. Yen and Scanlon (1975) documented the settlement of several landfill sites in California. The settlement rate after completion of the landfill (Figure 11.23) may be expressed as

$$m = \frac{\Delta H_f}{\Delta t} \qquad (11.18)$$

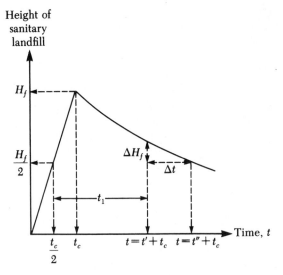

▼ **FIGURE 11.23**   Settlement of sanitary landfills

where m = settlement rate

H_f = maximum height of the sanitary landfill

Based on several field observations, Yen and Scanlon (1975) determined the following empirical correlations for the settlement rate:

$$m = 0.0268 - 0.0116 \log t_1 \qquad \text{(for fill heights ranging from 12–24 m)} \qquad (11.19)$$

$$m = 0.038 - 0.0155 \log t_1 \qquad \text{(for fill heights ranging from 24–30 m)} \qquad (11.20)$$

$$m = 0.0433 - 0.0183 \log t_1 \qquad \text{(for fill heights ranging from 30 m)} \qquad (11.21)$$

where m is in m/mo.

t_1 is the median fill age, in months

The median fill age may be defined from Figure 11.23 as follows:

$$t_1 = t - \frac{t_c}{2} \qquad\qquad\qquad (11.22)$$

where t = time from the beginning of landfill

t_c = time for completion of the landfill

Equations (11.19), (11.20), and (11.21) were based on field data from landfills for which t_c varied from 70 to 82 months. To get an idea of the approximate length of time required for a sanitary landfill to undergo complete settlement, consider Eq. (11.19). For a fill 12 m high and t_c = 72 months,

$$m = 0.0268 - 0.0116 \log t_1$$

$$\log t_1 = \frac{0.0268 - m}{0.0116}$$

If $m = 0$ (zero settlement rate), log $t_1 = 2.31$, or $t_1 \approx 200$ months. Thus settlement will continue for $t_1 - t_c/2 = 200 - 36 = 164$ months (≈ 14 years) after completion of the fill — a fairly long time. This calculation emphasizes the need to pay close attention to the settlement of foundations constructed on sanitary landfills.

A comparison of Eqs. (11.19) to (11.21) for rates of settlement shows that the value of m increases with the height of the fill. However, for fill heights greater than about 30 m, the rate of settlement should not be much different from that obtained from Eq. (11.21). The reason is that decomposition of organic matter close to the surface is mainly the result of an anaerobic environment. For deeper fills, the decomposition is slower. Hence, for fill heights greater than about 30 m, the rate of settlement does not exceed those for fills that are about 30 m in height.

Sowers (1973) also proposed a relation for calculation of the settlement of a sanitary landfill:

$$\Delta H = \frac{\alpha H_f}{1 + e} \log \left(\frac{t''}{t'} \right) \tag{11.23}$$

where H_f = height of the fill
   e = void ratio
   α = a coefficient for settlement
   t'', t' = times (see Figure 11.23)
   ΔH = settlement between times t' and t''

The coefficients α fall between

$\alpha = 0.09e$ (for conditions favorable to decomposition) $\tag{11.24}$

and

$\alpha = 0.03e$ (for conditions unfavorable to decomposition) $\tag{11.25}$

Equation (11.23) is similar to the equation for secondary consolidation settlement.

PROBLEMS

11.1 For loessial soil, given $G_s = 2.74$. Plot a graph of γ_d(kN/m³) versus liquid limit to identify the zone in which the soil is likely to collapse on saturation. If a soil has a liquid limit of 27, $G_s = 2.74$, and $\gamma_d = 14.5$ kN/m³, is collapse likely to occur?

11.2 A collapsible soil layer in the field has a thickness of 3 m. The average effective overburden pressure on the soil layer is 62 kN/m². An undisturbed specimen of this soil was subjected to a double oedometer test. The preconsolidation pressure of the specimen as determined from the soaked specimen was 84 kN/m². Is the soil in the field normally consolidated or preconsolidated?

11.3 An expansive soil has an active-zone thickness of 10 ft. The natural moisture content of the soil is 20% and its liquid limit is 50. Calculate the free surface swell of the expansive soil upon saturation.

11.4   The following are the results of a modified free swell index test:

Mass of dry soil = 10 grams
$G_s = 2.71$

Volume of swollen sediment after 24 hours in water = 26.2 cm^3

a. Determine the modified free swell index.
b. Describe the swell potential.

11.5   An expansive soil profile has an active-zone thickness of 12 ft. A shallow foundation is to be constructed at a depth of 4 ft below the ground surface. Based on a swelling pressure test, the following are given:

Depth from ground surface (ft)	Swell under overburden and estimated foundation surcharge pressure, $s_{w(1)}$ (%)
4	4.75
6	2.75
8	1.5
10	0.6
12	0.0

Estimate the total possible swell under the foundation.

11.6   Refer to Problem 11.5. If the allowable total swell is 1 in., what would be the necessary undercut?

11.7   Repeat Problem 11.5 with the following: active zone thickness = 6 m; depth of shallow foundation = 1.5 m

Depth from ground surface (m)	Swell under overburden and estimated foundation surcharge pressure, $s_{w(1)}$ (%)
1.5	5.5
2.0	3.1
3.0	1.5
4.0	0.75
5.0	0.4
6.0	0.0

11.8   Refer to Problem 11.7. If the allowable total swell is 30 mm, what would be the necessary undercut?

11.9   Refer to Figure 11.20b. For the drilled shaft with bell, given:

Thickness of active zone, Z = 30 ft
Dead load = 300 kips
Live load = 60 kips
Diameter of the shaft, D_s = 3.5 ft

Zero swell pressure for the clay in the active zone = 6 ton/ft^2
Average angle of plinth–soil friction, ϕ_{ps} = 15°
Average undrained cohesion of the clay around the bell = 3020 lb/ft^2

Determine the diameter of the bell, D_b. A factor of safety of 2 against uplift is required with the assumption that dead load plus live load is equal to zero.

11.10   Refer to Problem 11.9. If an additional requirement is that the factor of safety against uplift is at least 3 with the dead load on (live load = 0), what should be the diameter of the bell?

REFERENCES

Abduljauwad, S. N., and Al-Sulaimani, G. J. (1993). "Determination of Swell Potential of Al-Qatif Clay," *Geotechnical Testing Journal,* American Society for Testing and Materials, Vol. 16, No. 4, pp. 469–484.

Altmeyer, W. T. (1955). "Discussion of Engineering Properties of Expansive Clays," *Journal of the Soil Mechanics and Foundations Division,* American Society of Civil Engineers, Vol. 81, No. SM2, pp. 17–19.

Benites, L. A. (1968). "Geotechnical Properties of the Soils Affected by Piping Near the Benson Area, Cochise County, Arizona," M. S. Thesis, University of Arizona, Tucson.

Chen, F. H. (1988). *Foundations on Expansive Soils,* Elsevier, Amsterdam.

Clemence, S. P., and Finbarr, A. O. (1981). "Design Considerations for Collapsible Soils," *Journal of the Geotechnical Engineering Division,* American Society of Civil Engineers, Vol. 107, No. GT3, pp. 305–317.

Clevenger, W. (1958). "Experience with Loess as Foundation Material," *Transactions,* American Society of Civil Engineers, Vol. 123, pp. 151–170.

Dakshanamanthy, V., and Raman, V. (1973). "A Simple Method of Identifying an Expansive Soil," *Soils and Foundations,* Vol. 13, No. 1, pp. 97–104.

Denisov, N. Y. (1951). *The Engineering Properties of Loess and Loess Loams,* Gosstroiizdat, Moscow.

Feda, J. (1964). "Colloidal Activity, Shrinking and Swelling of Some Clays," *Proceedings,* Soil Mechanics Seminar, Loda, Illinois, pp. 531–546.

Gibbs, H. J. (1961). "Properties Which Divide Loose and Dense Uncemented Soils," *Earth Laboratory Report EM-658,* Bureau of Reclamation, U.S. Department of the Interior, Washington, D.C.

Gromko, G. J. (1974). "Review of Expansive Soils," *Journal of the Geotechnical Engineering Division,* American Society of Civil Engineers, Vol. 100, No. GT6, pp. 667–687.

Hamilton, J. J. (1968). "Effect of Natural and Man-Made Environments on the Performance of Shallow Foundations," *Proceedings,* Twenty-First Annual Canadian Soil Mechanics Conference, Winnipeg, Manitoba.

Handy, R. L. (1973). "Collapsible Loess in Iowa," *Proceedings,* Soil Science Society of America, Vol. 37, pp. 281–284.

Holtz, W. G. (1959). "Expansive Clays — Properties and Problems," *Journal of the Colorado School of Mines,* Vol. 54, No. 4, pp. 89–125.

Holtz, W. G., and Hilf, J. W. (1961). "Settlement of Soil Foundations Due to Saturation," *Proceedings,* Fifth International Conference on Soil Mechanics and Foundation Engineering, Paris, Vol. 1, 1961, pp. 673–679.

Houston, W. N., and Houston, S. L. (1989). "State-of-the-Practice Mitigation Measures for Collapsible Soil Sites," *Proceedings,* Foundation Engineering: Current Principles and Practices, American Society of Civil Engineers, Vol. 1, pp. 161–175.

Jennings, J. E., and Knight, K. (1975). "A Guide to Construction on or with Materials Exhibiting Additional Settlements Due to 'Collapse' of Grain Structure," *Proceedings,* Sixth Regional Conference for Africa on Soil Mechanics and Foundation Engineering, Johannesburg, pp. 99–105.

Lutenegger, A. J. (1986). "Dynamic Compaction in Friable Loess," *Journal of Geotechnical Engineering,* American Society of Civil Engineers, Vol. 112, No. GT6, pp. 663–667.

Lutenegger, A. J., and Saber, R. T. (1988). "Determination of Collapse Potential of Soils," *Geotechnical Testing Journal,* American Society for Testing and Materials, Vol. 11, No. 3, pp. 173–178.

McKeen, R. G. (1992). "A Model for Predicting Expansive Soil Behavior," *Proceedings,* Seventh International Conference on Expansive Soils, Dallas, Vol. 1, pp. 1–6.

Nayak, N. V., and Christensen, R. W. (1974). "Swell Characteristics of Compacted Expansive Soils," *Clay and Clay Minerals,* Vol. 19, pp. 251–261.

O'Neill, M. W., and Poormoayed, N. (1980). "Methodology for Foundations on Expansive Clays," *Journal of the Geotechnical Engineering Division,* American Society of Civil Engineers, Vol. 106, No. GT12, p. 1345–1367.

Peck, R. B., Hanson, W. E., and Thornburn, T. B. (1974). *Foundation Engineering,* Wiley, New York.

Priklonski, V. A. (1952). *Gruntovedenia-Vtoraid Chast,* Gosgeolzdat, Moscow.

Raman, V. (1967). "Identification of Expansive Soils from the Plasticity Index and the Shrinkage Index Data," *The Indian Engineer,* Vol. 11, No. 1, pp. 17–22.

Sattler, P. J., and Fredlund, D. G. (1991). "Modelling Vertical Ground Movements Using Surface Climate Flux," *Proceedings,* Geotechnical Engineering Congress, American Society of Civil Engineers, Vol. II, pp. 1292–1306.

Seed, H. B., Woodward, R. J., Jr., and Lundgren, R. (1962). "Prediction of Swelling Potential for Compacted Clays," *Journal of the Soil Mechanics and Foundations Division,* American Society of Civil Engineers, Vol. 88, No. SM3, pp. 53–87.

Semkin, V. V., Ermoshin, V. M., and Okishev, N. D. (1986). "Chemical Stabilization of Loess Soils in Uzbekistan," *Soil Mechanics and Foundation Engineering* (trans. from Russian), Vol. 23, No. 5, pp. 196–199.

Sikh, T. S. (1993). "Swell Potential Versus Overburden Pressure," *Geotechnial Testing Journal,* American Society for Testing and Materials, Vol. 16, No. 3, pp. 393–396.

Sivapullaiah, P. V., Sitharam, T. G., and Rao, K. S. S. (1987). "Modified Free Swell Index for Clay," *Geotechnical Testing Journal,* American Society for Testing and Materials, Vol. 11, No. 2, pp. 80–85.

Snethen, D. R. (1984). "Evaluation of Expedient Methods for Identification and Classification of Potentially Expansive Soils," *Proceedings,* Fifth International Conference on Expansive Soils, Adelaide, Australia, pp. 22–26.

Snethen, D. R., Johnson, L. D., and Patrick, D. M. (1977). "An Evaluation of Expedient Methodology for Identification of Potentially Expansive Soils," *Report No. FHWA-RD-77-94,* U.S. Army Engineers Waterways Experiment Station, Vicksburg, Miss.

Sowers, G. F. (1973). "Settlement of Waste Disposal Fills," *Proceedings,* Eighth International Conference on Soil Mechanics and Foundation Engineering, Moscow, pp. 207–210.

Sowers, G. B., and Sowers, G. F. (1970). *Introductory Soil Mechanics and Foundations,* 3rd ed. Macmillan, New York.

Sridharan, A., Rao, A. S., and Sivapullaiah, P. V. (1986), "Swelling Pressure of Clays," *Geotechnical Testing Journal,* American Society for Testing and Materials, Vol. 9, No. 1, pp. 24–33.

Uniform Building Code (1968). *UBC Standard No. 29-2.*

Van Der Merwe, D. H. (1964), "The Prediction of Heave from the Plasticity Index and

Percentage Clay Fraction of Soils," *Civil Engineer in South Africa,* Vol. 6, No. 6, pp. 103–106.

Vijayvergiya, V. N., and Ghazzaly, O. I. (1973). "Prediction of Swelling Potential of Natural Clays," *Proceedings,* Third International Research and Engineering Conference on Expansive Clays, pp. 227–234.

Weston, D. J. (1980). "Expansive Roadbed Treatment for Southern Africa," *Proceedings,* Fourth International Conference on Expansive Soils, Vol. 1, pp. 339–360.

Yen, B. C., and Scanlon, B. (1975). "Sanitary Landfill Settlement Rates," *Journal of the Geotechnical Engineering Division,* American Society of Civil Engineers, Vol. 101, No. GT5, pp. 475–487.

CHAPTER TWELVE

SOIL IMPROVEMENT AND GROUND MODIFICATION

12.1 INTRODUCTION

The existing soil at a construction site may not always be totally suitable for supporting structures such as building, bridges, highways, and dams. For example, in granular soil deposits the *in situ* soil may be very loose and indicate a large elastic settlement. In such a case, the soil needs to be densified to increase its unit weight and thus the shear strength.

Sometimes the top layers of soil are undesirable and must be removed and replaced with better soil on which the structural foundation can be built. The soil used as fill should be well compacted to sustain the desired structural load. Compacted fills may also be required in low-lying areas to raise the ground elevation for foundation construction.

Soft saturated clay layers are often encountered at shallow depths below foundation(s). Depending on the structural load and the depth of the clay layer(s), unusually large consolidation settlement may occur. Special soil-improvement techniques are required to minimize settlement.

In Chapter 11 we mentioned that the properties of expansive soils could be altered substantially by adding stabilizing agents such as lime. Improving *in situ* soils by using additives is usually referred to as *stabilization.*

Various techniques for improving soil are used to

1. Reduce the settlement of structures
2. Improve the shear strength of soil and thus increase the bearing capacity of shallow foundations
3. Increase the factor of safety against possible slope failure of embankments and earth dams
4. Reduce the shrinkage and swelling of soils

This chapter discusses some of the general principles of soil improvement such as compaction, vibroflotation, precompression, sand drains, wick drains, and stabilization by admixtures, as well as the use of stone columns and sand compaction piles in weak clay for foundation construction.

12.2 COMPACTION—GENERAL PRINCIPLES

If a small amount of water is added to a soil that is then compacted, the soil will have a certain unit weight. If the moisture content of the same soil is gradually increased and the energy of compaction is the same, the dry unit weight of the soil will gradually increase. The reason is that water acts as a lubricant between the soil particles, and under compaction it helps rearrange the solid particles into a denser state. The increase in dry unit weight with increase of moisture content for a soil will reach a limiting value beyond which further addition of water to the soil will result in a *reduction* of dry unit weight. The moisture content at which the *maximum dry unit weight* is obtained is referred to as the *optimum moisture content*.

The standard laboratory tests used for evaluation of maximum dry unit weights and optimum moisture contents for various soils are

a. Standard Proctor test (ASTM designation D-698), and
b. Modified Proctor test (ASTM designation D-1557)

The soil is compacted in a mold in several layers by a hammer. The moisture content of the soil, w, is changed, and the dry unit weight, γ_d, of compaction for each test is determined. The maximum dry unit weight of compaction and the corresponding optimum moisture content are determined by plotting a graph of γ_d against w (%). The standard specifications for the two types of Proctor test are given in Tables 12.1 and 12.2.

▼ **TABLE 12.1** Specifications for Standard Proctor Test (Based on ASTM Designation 698-91)

Item	Method A	Method B	Method C
Diameter of mold	4 in. (101.6 mm)	4 in. (101.6 mm)	6 in. (152.4 mm)
Volume of mold	0.0333 ft^3 (944 cm^3)	0.0333 ft^3 (944 cm^3)	0.075 ft^3 (2124 cm^3)
Weight of hammer	5.5 lb (2.5 kg)	5.5 lb (2.5 kg)	5.5 lb (2.5 kg)
Height of hammer drop	12 in. (304.8 mm)	12 in. (304.8 mm)	12 in. (304.8 mm)
Number of hammer blows per layer of soil	25	25	56
Number of layers of compaction	3	3	3
Energy of compaction	12,400 ft·lb/ft^3 (600 kN·m/m^3)	12,400 ft·lb/ft^3 (600 kN·m/m^3)	12,400 ft·lb/ft^3 (600 kN·m/m^3)
Soil to be used	Portion passing no. 4 (4.57 mm) sieve. May be used if 20% *or less* by weight of material is retained on no. 4 sieve.	Portion passing $\frac{3}{8}$-in. (9.5 mm) sieve. May be used if soil retained on no. 4 sieve *is more* than 20%, and 20% *or less* by weight is retained on $\frac{3}{8}$-in. (9.5 mm) sieve	Portion passing $\frac{3}{4}$-in. (19.0 mm) sieve. May be used if *more than* 20% by weight of material is retained on $\frac{3}{8}$-in. (9.5 mm) sieve, and *less than* 30% by weight is retained on $\frac{3}{4}$-in. (19.00 mm) sieve.

▼ **TABLE 12.2** Specifications for Modified Proctor Test (Based on ASTM Designation 1557-91)

Item	Method A	Method B	Method C
Diameter of mold	4 in. (101.6 mm)	4 in. (101.6 mm)	6 in. (152.4 mm)
Volume of mold	0.0333 ft^3 (944 cm^3)	0.0333 ft^3 (944 cm^3)	0.075 ft^3 (2124 cm^3)
Weight of hammer	10 lb (4.54 kg)	10 lb (4.54 kg)	10 lb (4.54 kg)
Height of hammer drop	18 in. (457.2 mm)	18 in. (457.2 mm)	18 in. (457.2 mm)
Number of hammer blows per layer of soil	25	25	56
Number of layers of compaction	5	5	5
Energy of compaction	56,000 ft·lb/ft^3 (2700 kN·m/m^3)	56,000 ft·lb/ft^3 (2700 kN·m/m^3)	56,000 ft·lb/ft^3 (2700 kN·m/m^3)
Soil to be used	Portion passing no. 4 (4.57 mm) sieve. May be used if 20% *or less* by weight of material is retained on no. 4 sieve.	Portion passing ⅜-in. (9.5 mm) sieve. May be used if soil retained on no. 4 sieve *is more* than 20%, and 20% *or less* by weight is retained on ⅜-in. (9.5 mm) sieve	Portion passing ¾-in. (19.0 mm) sieve. May be used if *more than* 20% by weight of material is retained on ⅜-in. (9.5 mm) sieve, and *less than* 30% by weight is retained on ¾-in. (19.00 mm) sieve.

Figure 12.1 shows the plot of γ_d against w (%) for a clayey silt obtained from standard and modified Proctor tests (method A). The following conclusions may be drawn:

1. The maximum dry unit weight and the optimum moisture content depend on the degree of compaction.

2. The higher the energy of compaction, the higher is the maximum dry unit weight.

3. The higher the energy of compaction, the lower is the optimum moisture content.

4. No portion of the compaction curve can lie to the right of the zero-air-void line. The zero-air-void dry unit weight, γ_{zav}, at a given moisture content is the theoretical maximum value of γ_d, which means that all the void spaces of the compacted soil are filled with water, or

$$\gamma_{zav} = \frac{\gamma_w}{\dfrac{1}{G_s} + w} \tag{12.1}$$

where γ_w = unit weight of water
  G_s = specific gravity of the soil solids
  w = moisture content

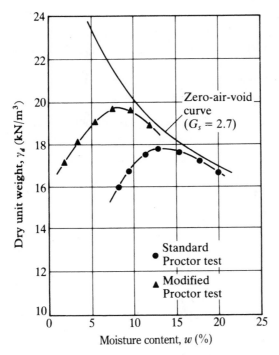

▼ FIGURE 12.1 Standard and modified Proctor compaction curves for a clayey silt (method A)

5. The maximum dry unit weight of compaction and the corresponding optimum moisture content will vary from soil to soil.

Using the results of the laboratory compaction (γ_d against w), specifications may be written for the compaction of a given soil in the field. In most cases, the contractor is required to achieve a relative compaction of 90% or more on the basis of a specific laboratory test (either the standard or the modified Proctor compaction test). Relative compaction, RC, is defined as

$$RC = \frac{\gamma_{d(\text{field})}}{\gamma_{d(\text{max})}} \tag{12.2}$$

Chapter 1 introduced the concept of relative density (for compaction of granular soils). Relative density was defined as

$$D_r = \left[\frac{\gamma_d - \gamma_{d(\text{min})}}{\gamma_{d(\text{max})} - \gamma_{d(\text{min})}} \right] \frac{\gamma_{d(\text{max})}}{\gamma_d}$$

where D_r = relative density
          γ_d = dry unit weight of compaction in the field
          $\gamma_{d(max)}$ = maximum dry unit weight of compaction as determined
                  in the laboratory
          $\gamma_{d(min)}$ = minimum dry unit weight of compaction
                  as determined in the laboratory

For granular soils in the field, the degree of compaction obtained is often measured in terms of relative density. Comparing the expressions for relative density and relative compaction reveals that

$$RC = \frac{A}{1 - D_r(1 - A)}$$

(12.3)

where $A = \dfrac{\gamma_{d(min)}}{\gamma_{d(max)}}$

Lee and Singh (1971) reviewed 47 different soils, and, based on their review, presented the correlation:

$$D_r(\%) = \frac{(RC - 80)}{0.2}$$

(12.4)

12.3 ONE-POINT METHOD OF OBTAINING $\gamma_{d(max)}$

The state highway department of Ohio has developed a family of standard curves for various soil types, as shown in Figure 12.2. Note that they are plots of *wet unit weight*, γ, against moisture content, w (%). These curves are used to obtain $\gamma_{d(max)}$ in the field. This technique, referred to as the *one-point method*, serves as a rapid means of field compaction control. This method first involves a standard Proctor test (method A) with the soil in use and a determination of the moist unit weight of compaction and the corresponding moisture content. Then a plot of the values of γ and w identifies the compaction curve number (Figure 12.2) corresponding to the test results. Using this curve number with Table 12.3 gives the maximum dry unit weight and the corresponding optimum moisture content.

The one-point method appears to be simple and easy to use. However, that may not always be the case. Researchers have determined that not all soils yield the bell-shaped compaction curves shown in Figure 12.2. Lee and Suedkamp (1972) performed 700 compaction tests on 35 soil samples on soil portions passing a no. 4 sieve (method A). Their results show that, depending on the property of the soil, the plot of γ_d against w (%) may exhibit one of four different shapes. These shapes are shown in Figure 12.3 (p. 771) and are marked Types I, II, III, and IV. Type I is

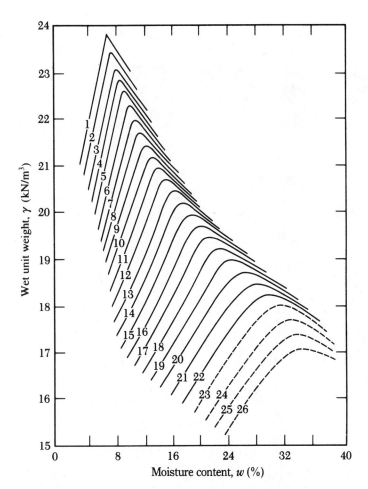

FIGURE 12.2 Ohio compaction curves (from "Factors That Influence Field Compaction of Soils," by A. W. Johnson and J. R. Sallberg, *Bulletin No. 262*, Highway Research Board, 1960. Reprinted by permission)

a standard bell-shaped curve. Type II is a curve showing one and one-half peaks. Type III is a double-peak curve. Type IV is an oddly shaped curve that shows no distinct optimum moisture content. Lee and Suedkamp (1972) then developed the following guidelines to help predict the nature of compaction curves that may be obtained from various soils:

Liquid limit of soil	Nature of compaction curve to be expected
30 to 70	Type I
Less than 30	Types II and III
Greater than 70	Types III and IV

▼ **TABLE 12.3** Maximum Dry Unit Weight and Optimum Moisture Content for the Compaction Curves in Figure 12.2[a]

Curve number	Maximum dry unit weight		Optimum moisture content (%)
	lb/ft³	kN/m³	
1	142.8	22.29	6.6
2	139.1	21.87	7.2
3	136.3	21.43	7.9
4	134.1	21.08	8.5
5	132.0	20.75	9.0
6	129.3	20.33	9.7
7	126.6	19.90	10.5
8	124.2	19.53	11.2
9	121.7	19.13	11.9
10	119.3	18.76	12.7
11	117.0	18.39	13.5
12	114.6	18.02	14.6
13	112.0	17.61	15.8
14	109.6	17.23	16.9
15	107.1	16.84	18.1
16	104.7	16.46	19.2
17	102.4	16.10	20.3
18	99.9	15.71	21.5
19	97.4	15.31	22.7
20	94.6	14.87	24.4
21	92.1	14.48	25.8
22	89.9	14.13	27.4
23	87.5	13.76	29.5
24	85.0	13.36	30.5
25	83.0	13.05	31.5
26	81.1	12.75	32.5

[a] After Johnson and Sallberg (1960)

▼ **EXAMPLE 12.1**

A soil was compacted in the field by the standard Proctor test procedure using the material passing a no. 4 sieve. The weight of compacted wet soil in the mold = 17.2 N, and the moisture content = 14%. Use Figure 12.2 to determine $\gamma_{d(max)}$ and the optimum moisture content.

Solution The moist unit weight of compaction is

$$\gamma = \frac{17.2 \text{ N}}{\text{volume of Proctor mold}} = \frac{17.2}{0.944 \times 10^{-3}} = 18.22 \text{ kN/m}^3$$

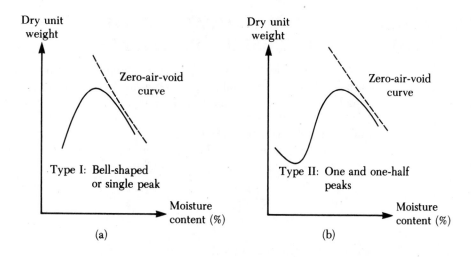

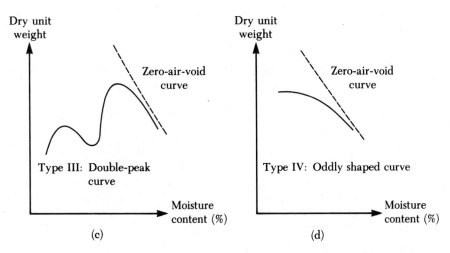

▼   **FIGURE 12.3**   Various types of compaction curve

According to Figure 12.2 (with $\gamma = 18.22$ kN/m³ and $w = 14\%$), the soil appears to fall between curves 15 and 16. According to Table 12.3, $\gamma_{d(\max)}$ is between 16.84 kN/m³ and 16.46 kN/m³, and the optimum moisture content is between 18.1 and 19.2%. So, for this soil,

$$\gamma_{d(\max)} \approx \mathbf{16.6\ kN/m^3}$$

and

optimum moisture content $\approx$ **18.5%**   ▲

12.4 CORRECTION FOR COMPACTION OF SOILS WITH OVERSIZED PARTICLES

As shown in Tables 12.1 and 12.2, depending on the method used, certain oversized particles (such as material retained on a no. 4 sieve or $\frac{3}{4}$-in. sieve) may need to be removed from the soil to conduct laboratory compaction tests. ASTM test designation D-4718-87 provides a method to make corrections for maximum dry unit weight and optimum moisture content in the presence of oversized particles. This is helpful in writing specifications for field compaction. According to this method, the corrected maximum dry unit weight, $\gamma_{d(\max)\text{-}C}$, may be calculated as

$$\gamma_{d(\max)\text{-}C} = \frac{100\gamma_w}{\dfrac{P_C}{G_m} + \dfrac{\gamma_w(100 - P_C)}{\gamma_{d(\max)\text{-}F}}} \tag{12.5}$$

where γ_w = unit weight of water
P_C = percentage of oversized particles by weight
G_m = bulk specific gravity of the oversized particles
$\gamma_{d(\max)\text{-}F}$ = maximum dry unit weight of the finer fraction used in the laboratory compaction

The corrected optimum moisture content, $w_{\text{optimum-}C}$ (%), can be expressed as

$$w_{\text{optimum-}C}\,(\%) = w_{\text{optimum-}F}\,(100 - P_C) + w_C P_C \tag{12.6}$$

where $w_{\text{optimum-}F}$ = optimum moisture content of the finer fraction determined in the laboratory
w_C = saturated surface dry moisture content of the oversized particles (fraction)

The preceding equations for corrected maximum dry unit weight and optimum moisture content are valid when the oversized particles are about 30% or less (by weight) of the total soil sample.

12.5 FIELD COMPACTION

Ordinary compaction in the field is done by rollers. Of the several types of roller used, the most common are

1. Smooth wheel rollers (or smooth drum rollers)
2. Pneumatic rubber-tired rollers
3. Sheepsfoot rollers
4. Vibratory rollers

Figure 12.4 shows a *smooth wheel roller* that can also create vertical vibration during compaction. Smooth wheel rollers are suitable for proof-rolling subgrades and for finishing the construction of fills with sandy or clayey soils. They provide 100% coverage under the wheels, and the contact pressure can be as high as 45–60 lb/in^2 ($\approx$300–400 kN/m^2). However, they do not produce uniform unit weight of compaction when used on thick layers.

▼ **FIGURE 12.4** Vibratory smooth wheel rollers (courtesy of Tampo Manufacturing Co., Inc., San Antonio, Texas)

Pneumatic rubber-tired rollers (Figure 12.5) are better in many respects than smooth wheel rollers. These rollers, which may weigh as much as 450 kip (2000 kN), consist of a heavily loaded wagon with several rows of tires. These tires are closely spaced — four to six in a row. The contact pressure under the tires may range up to 85–100 lb/in^2 ($\approx$600–700 kN/m^2), and they produce about 70%–80% coverage. Pneumatic rollers, which can be used for sandy and clayey soil compaction, produce a combination of pressure and kneading action.

Sheepsfoot rollers (Figure 12.6) consist basically of drums with large numbers of projections. The area of each of the projections may be 4–14 in^2 (25–90 cm^2). These rollers are *most effective in compacting cohesive soils*. The contact pressure under the projections may range from 215–1100 lb/in^2 ($\approx$1500–7500 kN/m^2). During compaction in the field, the initial passes compact the lower portion of a lift. Later, the middle and top of the lift are compacted.

Vibratory rollers are efficient in compacting granular soils. Vibrators can be attached to smooth wheel, pneumatic rubber-tired, or sheepsfoot rollers to send vibrations into the soil being compacted. Figures 12.4 and 12.6 show vibratory smooth wheel rollers and a vibratory sheepsfoot roller.

In general, compaction in the field depends on several factors, such as the type of compactor, soil type, moisture content, lift thickness, towing speed of the compactor, and the number of roller passes. Figure 12.7 shows the variation of the dry unit weight of a heavy clay with the number of passes of pneumatic-tired rollers. Table 12.4 gives the details of the variables for the three curves shown in Figure 12.7.

▼ **FIGURE 12.5**   Pneumatic rubber-tired roller (courtesy of Tampo Manufacturing Co., Inc., San Antonio, Texas)

▼ **FIGURE 12.6**   Vibratory sheepsfoot roller (courtesy of Tampo Manufacturing Co., Inc., San Antonio, Texas)

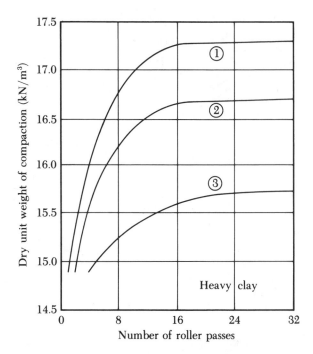

▼ **FIGURE 12.7**   Relation between dry unit weight of compaction for the upper 150 mm of soil and the number of passes of pneumatic-tired roller (from "Factors That Influence Field Compaction of Soils," by A. W. Johnson and J. R. Sallberg, *Bulletin No. 262,* Highway Research Board, 1960. Reprinted by permission)

▼ **TABLE 12.4**   Details of the Variables for the Three Curves Shown in Figure 12.7[a]

Curve no.	1	2	3
Moist content as rolled (%)	19.0	20.0	24.0
Optimum moisture content — standard Proctor test (%)	22.8	22.8	22.8
Roller rating (kN)	416.0	416.0	120.0
Wheel load (kN)	99.6	49.8	13.3
Tire pressure (kN/m²)	966.0	621.0	248.4
Loose lift thickness (mm)	305.0	305.0	229.0
[a] After Johnson and Sallberg (1960)			

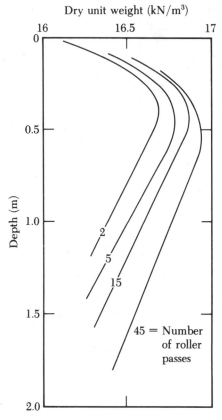

Dry unit weight (kN/m³)

▼   **FIGURE 12.8**   Vibratory compaction of a sand — variation of dry unit weight with depth and number of roller passes; lift thickness = 2.44 m (after D'Appolonia et al., 1969)

Figure 12.8 shows the variation of the unit weight of compaction with depth for a poorly graded dune sand compacted by a vibratory drum roller. Vibration was produced by mounting an eccentric weight on a single rotating shaft within the drum cylinder. The weight of the roller used for this compaction was 55.7 kN and the drum diameter was 1.19 m. The lifts were kept at 2.44 m. Note that, at any depth, the dry unit weight of compaction increases with the number of roller passes. However, the rate of increase of unit weight gradually decreases after about fifteen passes. Note also the variation of dry unit weight with depth by number of roller passes. The dry unit weight and hence the relative density, D_r, reach maximum values at a depth of about 0.5 m and then gradually decrease as the depth decreases. The reason is the lack of confining pressure toward the surface. Once the depth against relative density (or dry unit weight) relation for a soil for a given number

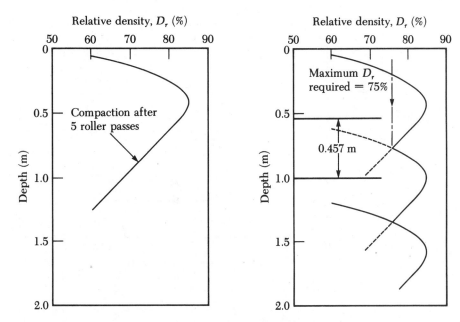

▼   **FIGURE 12.9**   A method for estimating compaction lift thickness. Minimum relative density required is 75% after five roller passes (after D'Appolonia et al., 1969)

of roller passes is determined, estimating the approximate thickness of each lift is
easy. This procedure is shown in Figure 12.9.

12.6   COMPACTION CONTROL FOR CLAY HYDRAULIC BARRIERS

Compacted clays are commonly used as hydraulic barriers in cores of earth dams,
liners and covers of landfills, and liners of surface impoundments. Since the primary
purpose of a barrier is to minimize flow, the hydraulic conductivity, k, is the controlling factor. In many cases, it is desired that the hydraulic conductivity be less than
10^{-7} cm/s. This can be achieved by controlling the minimum degree of saturation
during compaction. The above fact can be explained by referring to the compaction
characteristics of three soils described in Table 12.5 (Othman and Luetrich, 1994).

▼   **TABLE 12.5**   Characteristics of Soils Reported in Figures 12.10, 12.11, and 12.12

Soil	Classification	Liquid limit	Plasticity index	Percent finer than no. 200 sieve (0.075 mm)
Wisconsin A	CL	34	16	85
Wisconsin B	CL	42	19	99
Wisconsin C	CH	84	60	71

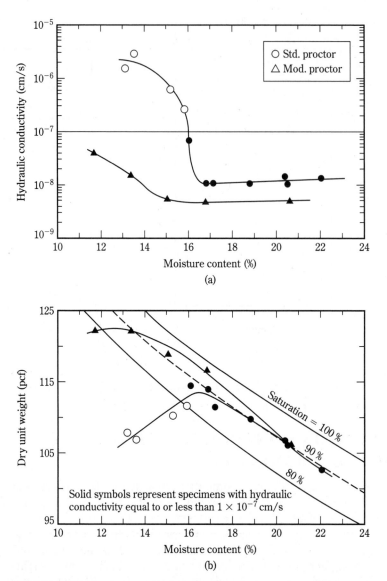

▼ **FIGURE 12.10**   Standard and modified proctor test results and hydraulic conductivity of Wisconsin A soil (after Othman and Luettich, 1994)

Figures 12.10, 12.11, and 12.12 show the standard and modified Proctor test results and the hydraulic conductivities of compacted specimens. Note that the solid symbols represent specimens with hydraulic conductivities of 10^{-7} cm/s or less. As can be seen from these figures, the data points plot generally parallel to the line of full saturation. Figure 12.13 (p. 781) shows the effect of the degree of saturation during compaction on the hydraulic conductivity of the three soils. It is evident from this figure that, if it

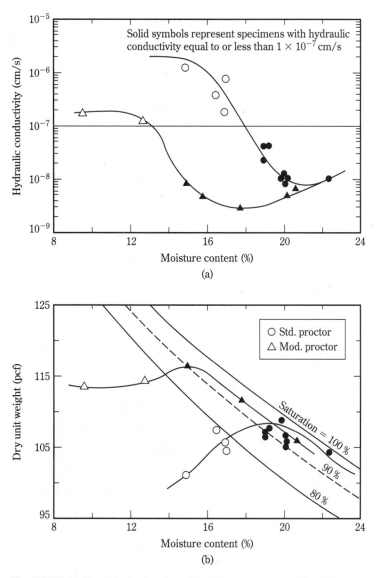

FIGURE 12.11 Standard and modified Proctor test results and hydraulic conductivity of Wisconsin B soil (after Othman and Luettich, 1994)

is desired that the maximum hydraulic conductivity should be 10^{-7} cm/s, then all soils should be compacted at a minimum degree of saturation of 88%.

In field compaction at a given site, soils of various composition may be encountered. Small changes in the content of fines will change the magnitude of hydraulic conductivity. Hence, considering various soils to be encountered at a given site, a

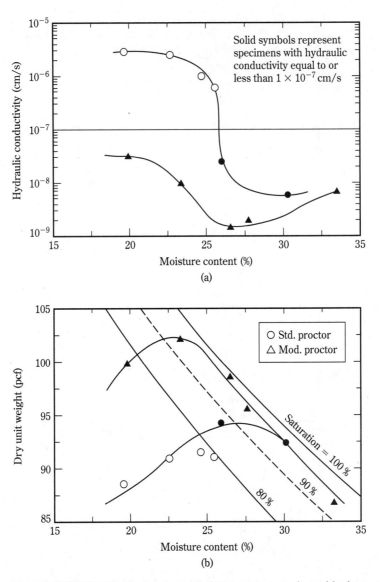

▼ **FIGURE 12.12**   Standard and modified Proctor test results and hydraulic conductivity of Wisconsin C soil (after Othman and Luettich, 1994)

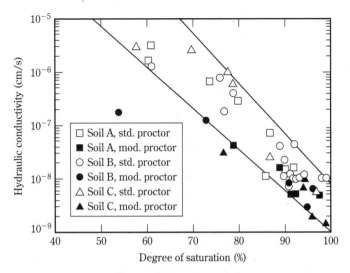

▼ **FIGURE 12.13**   Effect of degree of saturation on hydraulic conductivity of Wisconsin A, B, and C soils (after
Othman and Luettich, 1994)

minimum degree of saturation criterion for compaction can be developed to construct
hydraulic barriers.

12.7 VIBROFLOTATION

Vibroflotation is a technique developed in Germany in the 1930s for *in situ* densification of thick layers of loose granular soil deposits. Vibroflotation was first used in
the United States about ten years later. The process involves the use of a *vibroflot*
(called the *vibrating unit*), as shown in Figure 12.14, which is about 6 ft (2 m) in
length. This vibrating unit has an eccentric weight inside it and can develop a
centrifugal force. The weight enables the vibrating unit to vibrate horizontally. There
are openings at the bottom and top of the vibrating unit for water jets. The vibrating
unit is attached to a follow-up pipe. Figure 12.14 shows the vibroflotation equipment
necessary for compaction in the field.

The entire compaction process can be divided into four stages (Figure 12.15):

Stage 1. The jet at the bottom of the vibroflot is turned on, and the vibroflot
is lowered into the ground.

Stage 2. The water jet creates a quick condition in the soil, which allows the
vibrating unit to sink.

Stage 3. Granular material is poured into the top of the hole. The water from
the lower jet is transferred to the jet at the top of the vibrating unit.
This water carries the granular material down the hole.

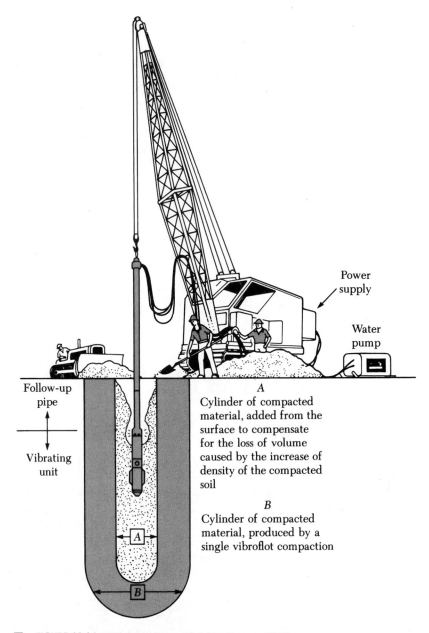

Follow-up pipe

Vibrating unit

A
Cylinder of compacted material, added from the surface to compensate for the loss of volume caused by the increase of density of the compacted soil

B
Cylinder of compacted material, produced by a single vibroflot compaction

Power supply

Water pump

▼   **FIGURE 12.14**   Vibroflotation unit (after Brown, 1977)

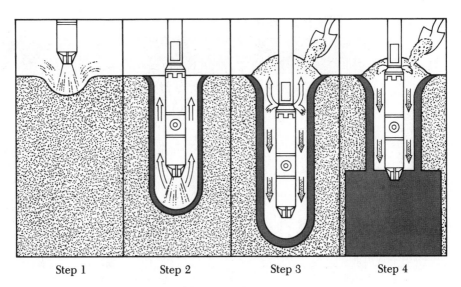

▼ **FIGURE 12.15**   Compaction by the vibroflotation process (after Brown, 1977)

Stage 4.   The vibrating unit is gradually raised in about 1-ft (0.3 m) lifts and
held vibrating for about 30 seconds at a time. This process compacts
the soil to the desired unit weight.

Table 12.6 gives the details of various types of vibroflot unit used in the United
States. The 30-HP electric units have been used since the latter part of the 1940s.
The 100-HP units were introduced in the early 1970s. The zone of compaction around
a single probe will vary according to the type of vibroflot used. The cylindrical zone
of compaction will have a radius of about 6 ft (2 m) for a 30-HP unit. This radius
may extend to about 10 ft (3 m) for a 100-HP unit. Compaction by vibroflotation
involves various probe spacings, depending on the zone of compaction (see Figure
12.16). Mitchell (1970) and Brown (1977) reported several successful cases of foundation design using vibroflotation.

The capacity of successful densification of *in situ* soil depends on several factors,
the most important of which is the grain-size distribution of the soil and also the
nature of backfill used to fill the holes during the withdrawal period of the vibroflot.
The range of the grain-size distribution of *in situ* soil marked Zone 1 in Figure 12.17
is most suitable for compaction by vibroflotation. Soils that contain excessive amounts
of fine sand and silt-size particles are difficult to compact; for them, considerable
effort is needed to reach proper relative density of compaction. Zone 2 in Figure
12.17 is the approximate lower limit of grain-size distribution for compaction by
vibroflotation. Soil deposits whose grain-size distribution falls in Zone 3 contain
appreciable amounts of gravel. For these soils, the rate of probe penetration may
be rather slow, and so compaction by vibroflotation might prove to be uneconomical
in the long run.

▼ **TABLE 12.6** Types of Vibrating Units[a]

	100-HP electric and hydraulic motors	30-HP electric motors
(a) Vibrating tip		
Length	7 ft (2.1 m)	6.11 ft (1.86 m)
Diameter	16 in. (406.4 mm)	15 in. (381 mm)
Weight	4000 lb (17.8 kN)	4000 lb (17.8 kN)
Maximum movement when free	0.49 in. (12.45 mm)	0.3 in. (7.62 mm)
Centrifugal force	18 ton (160 kN)	10 ton (89 kN)
(b) Eccentric		
Weight	260 lb (1.16 kN)	170 lb (0.76 kN)
Offset	1.5 in. (38.1 mm)	1.25 in. (31.75 mm)
Length	24 in. (610 mm)	15.25 in. (387 mm)
Speed	1800 rpm	1800 rpm
(c) Pump		
Operating flow rate	0–400 gal/min (0–1.6 m³/min)	0–150 gal/min (0–6 m³/min)
Pressure	100–150 lb/in² (690–1035 kN/m²)	100–150 lb/in² (690–1035 kN/m²)
(d) Lower follow-up pipe and extensions		
Diameter	12 in. (305 mm)	12 in. (305 mm)
Weight	250 lb/ft (3.65 kN/m)	250 lb/ft (3.65 kN/m)

[a] After Brown (1977)

The grain-size distribution of the backfill material is one of the factors that control the rate of densification. Brown (1977) defined a quantity called *suitability number, S_N,* for rating a backfill material:

$$S_N = 1.7 \sqrt{\frac{3}{(D_{50})^2} + \frac{1}{(D_{20})^2} + \frac{1}{(D_{10})^2}} \qquad (12.7)$$

where D_{50}, D_{20}, and D_{10} are the diameters (in mm) through which 50%, 20%, and

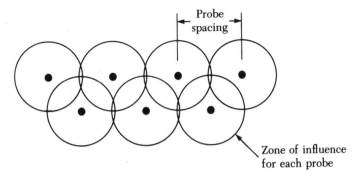

▼ **FIGURE 12.16**   Nature of probe spacing for vibroflotation

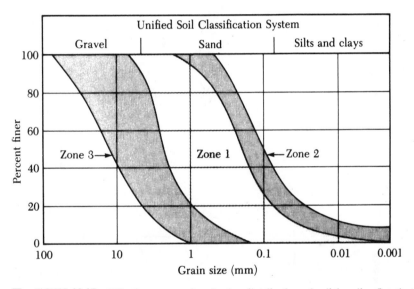

▼ **FIGURE 12.17**   Effective range of grain-size distribution of soil for vibroflotation

10%, respectively, of the material is passing. The smaller the value of S_N, the more desirable is the backfill material. Following is a backfill rating system as proposed by Brown (1977):

Range of S_N	Rating as backfill
0–10	Excellent
10–20	Good
20–30	Fair
30–50	Poor
>50	Unsuitable

An excellent case study that evaluated the benefits of vibroflotation was presented by Basore and Boitano (1969). Densification of granular subsoil was necessary for construction of a three-story office building at the Treasure Island Naval Station in San Francisco, California. The top 30 ft ($\approx$9 m) of soil at the site was loose to medium-dense sand fill that had to be compacted. Figure 12.18 shows the layout of the vibroflotation compaction points and the location of the test borings. Sixteen compaction points were arranged in groups of four, with 4-ft, 5-ft, 6-ft, 7-ft, and 8-ft spacing. Prior to compaction, standard penetration tests were conducted at the centers of groups of three compaction points. After completion of compaction by vibroflotation, the variation of the standard penetration resistance with depth was determined at the same points.

Figure 12.19 shows the variation of the standard penetration resistance, N_F (blows/ft), with depth before and after compaction. The 4-ft spacing produced the

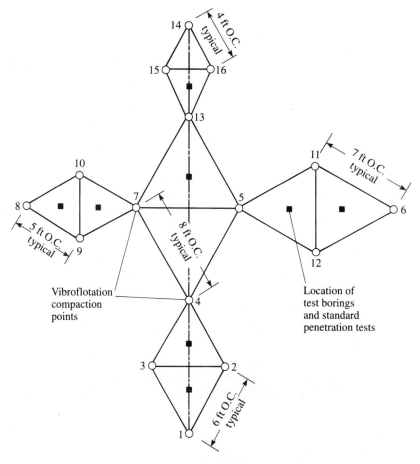

▼   **FIGURE 12.18**   Layout of vibroflotation compaction points and test borings (after Basore and Boitano, 1969)

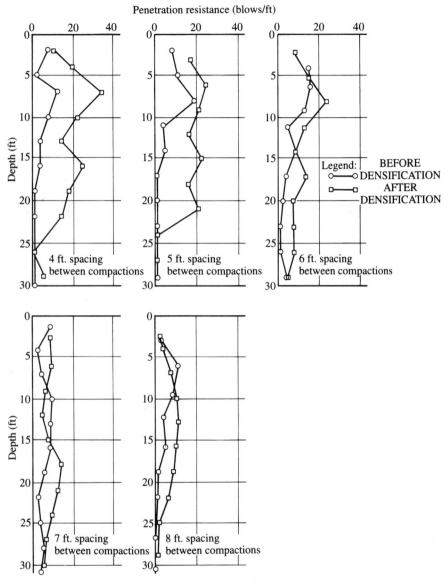

▼ FIGURE 12.19   Variation of standard penetration resistance before and after compaction (after Basore and Boitano, 1969)

greatest increase in density of the sand, whereas the 8-ft spacing had practically no effect. Figure 12.20 shows the relationship between the standard penetration resistance before and after compaction, spacing between compaction points, and the depth below the ground surface. For any given spacing between compaction points, the standard penetration resistance decreased with an increase in depth.

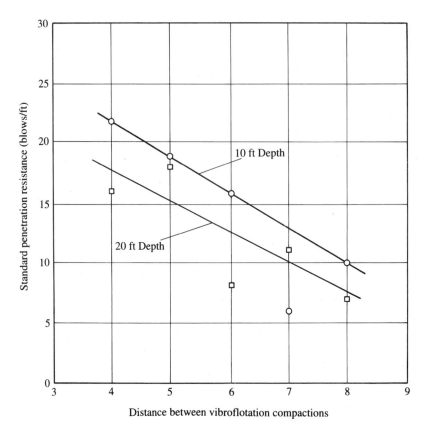

▼ **FIGURE 12.20**   Variation of field standard penetration resistance after compaction with spacing and depth (after Basore and Boitano, 1969)

The increase in the standard penetration resistance at any depth indicates the increase in the relative density of compaction, D_r, of sand. Figure 12.21 shows the variation of D_r before and after compaction for depths up to 30 ft.

During the past 30 to 35 years, the vibroflotation technique has been used successfully on large projects to compact granular subsoils, thereby controlling structural settlement.

12.8 PRECOMPRESSION—GENERAL CONSIDERATIONS

When highly compressible, normally consolidated clayey soil layers lie at a limited depth and large consolidation settlements are expected as the result of the construction of large buildings, highway embankments, or earth dams, precompression of soil may be used to minimize postconstruction settlement. The principles of precompression are best explained by reference to Figure 12.22 (p. 790). Here, the proposed structural load per unit area is $\Delta p_{(p)}$ and the thickness of the clay layer

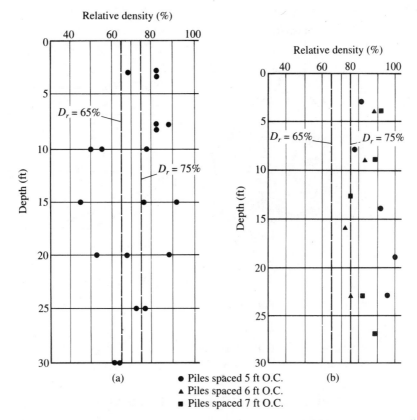

▼ **FIGURE 12.21** Variation of relative density (a) before compaction; (b) after compaction (after Basore and Boitano, 1969)

undergoing consolidation is H_c. The maximum primary consolidation settlement caused by the structural load, $S_{(p)}$, then is

$$S_{(p)} = \frac{C_c H_c}{1 + e_o} \log \frac{p_o + \Delta p_{(p)}}{p_o} \tag{12.8}$$

The settlement–time relationship under the structural load will be like that shown in Figure 12.22b. However, if a surcharge of $\Delta p_{(p)} + \Delta p_{(f)}$ is placed on the ground, the primary consolidation settlement $S_{(p+f)}$ will be

$$S_{(p+f)} = \frac{C_c H_c}{1 + e_o} \log \frac{p_o + [\Delta p_{(p)} + \Delta p_{(f)}]}{p_o} \tag{12.9}$$

The settlement–time relationship under a surcharge of $\Delta p_{(p)} + \Delta p_{(f)}$ is also shown in Figure 12.22b. Note that a total settlement of $S_{(p)}$ would occur at time t_2, which is much shorter than t_1. So, if a temporary total surcharge of $\Delta p_{(f)} + \Delta p_{(p)}$ is applied on the ground surface for time t_2, the settlement will equal $S_{(p)}$. At that time, if the surcharge is removed and a structure with a permanent load per unit area of $\Delta p_{(p)}$

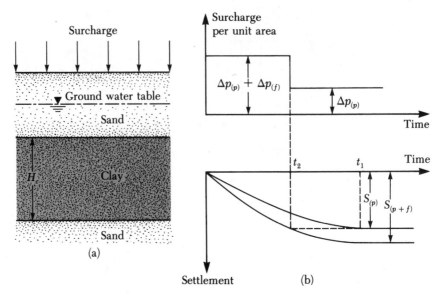

FIGURE 12.22   Principles of precompression

is built, no appreciable settlement will occur. The procedure just described is *precompression*. The total surcharge $\Delta p_{(p)} + \Delta p_{(f)}$ can be applied by means of temporary fills.

Derivation of Equations to Obtain $\Delta p_{(f)}$ and t_2

Figure 12.22b shows that, under a surcharge of $\Delta p_{(p)} + \Delta p_{(f)}$, the degree of consolidation at time t_2 after load application is

$$U = \frac{S_{(p)}}{S_{(p+f)}} \tag{12.10}$$

Substitution of Eqs. (12.8) and (12.9) into Eq. (12.10) yields

$$U = \frac{\log\left[\dfrac{p_o + \Delta p_{(p)}}{p_o}\right]}{\log\left[\dfrac{p_o + \Delta p_{(p)} + \Delta p_{(f)}}{p_o}\right]} = \frac{\log\left[1 + \dfrac{\Delta p_{(p)}}{p_o}\right]}{\log\left\{1 + \dfrac{\Delta p_{(p)}}{p_o}\left[1 + \dfrac{\Delta p_{(f)}}{\Delta p_{(p)}}\right]\right\}} \tag{12.11}$$

Figure 12.23 gives magnitudes of U for various combinations of $\Delta p_{(p)}/p_o$ and $\Delta p_{(f)}/\Delta p_{(p)}$. The degree of consolidation referred to in Eq. (12.11) is actually the average degree of consolidation at time t_2, as shown in Figure 12.22b. However, if the average degree of consolidation is used to determine time t_2, some construction problems might occur. The reason is that, after the removal of the surcharge and placement of the structural load, the portion of clay close to the drainage surface

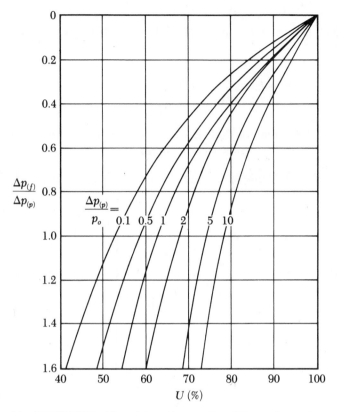

▼   **FIGURE 12.23**   Plot of $\Delta p_{(f)}/\Delta p_{(p)}$ against U for various values
of $\Delta p_{(p)}/p_o$ — Eq. (12.11)

will continue to swell, and the soil close to the midplane will continue to settle
(Figure 12.24). If some cases, net continuous settlement might result. A conservative

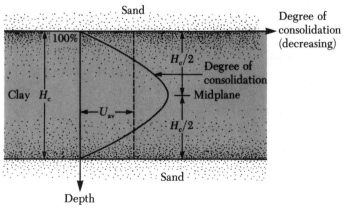

▼   **FIGURE 12.24**

approach may solve this problem — that is, assume that U in Eq. (12.11) is the midplane degree of consolidation (Johnson, 1970a). Now, from Eq. (1.76),

$$U = f(T_v) \tag{1.76}$$

where T_v = time factor = $C_v t_2 / H^2$
 C_v = coefficient of consolidation
 t_2 = time
 H = maximum drainage path ($= H_c/2$ for two-way drainage
 and $= H_c$ for one-way drainage)

The variation of U (midplane degree of consolidation) with T_v is given in Figure 12.25.

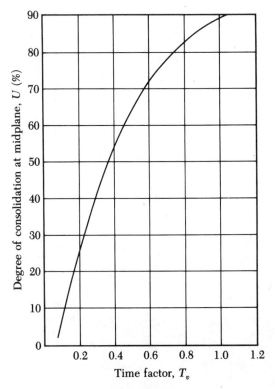

▼ **FIGURE 12.25**   Plot of midplane degree of consolidation against T_v

Procedure for Obtaining Precompression Parameters

Two problems may be encountered by engineers during precompression work in the field:

1. The value of $\Delta p_{(f)}$ is known, but t_2 must be obtained. In such a case, obtain p_o, $\Delta p_{(p)}$, and solve for U using Eq. (12.11) or Figure 12.23. For this value of U, obtain T_v from Figure 12.25. Then

$$t_2 = \frac{T_v H^2}{C_v} \tag{12.12}$$

2. For a specified value of t_2, $\Delta p_{(f)}$ must be obtained. In such a case, calculate T_v. Then refer to Figure 12.25 to obtain the midplane degree of consolidation, U. With the estimated value of U, go to Figure 12.23 to get the required $\Delta p_{(f)}/\Delta p_{(p)}$ and then calculate $\Delta p_{(f)}$.

Examples of Precompression and General Comments

Johnson (1970a) presented an excellent review of the use of precompression for improving foundation soils for several projects, including the Morganza Floodway Control Structure near Baton Rouge, Louisiana; the Old River Low-Sill Control Structure near Natchez, Mississippi; and the Old River Overbank Control Structure, Port Elizabeth Marine Terminal, New York. Figure 12.26 shows the subsoil conditions encountered near the Port Elizabeth Marine Terminal before the construction of warehouse buildings No. 131 and 132 located in the Jersey Meadows just west of New York City. Details of the precompression of the subsoil before the construction of the buildings are shown in Figure 12.27. Also shown is the theoretical variation of the settlement with time.

In most cases, the predicted settlement exceeded the actual consolidation settlement. The reason is that several variables are involved in proper precompression design and performance. The information obtained from only a handful of borings is used in the calculation of both the surcharge load and the time necessary for removal of the surcharge. Hence precise numbers for precompression design may be difficult to obtain. Settlement observations should be continued during the period of surcharge application because they may dictate design changes.

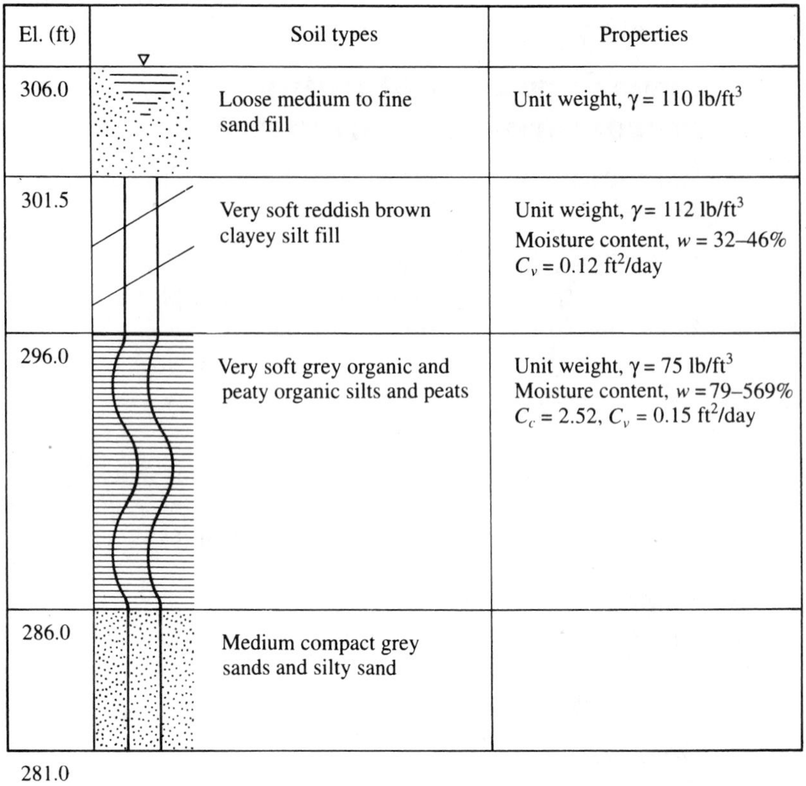

El. (ft)
Soil types
Properties
306.0
Loose medium to fine sand fill
Unit weight, γ = 110 lb/ft³
301.5
Very soft reddish brown clayey silt fill
Unit weight, γ = 112 lb/ft³
Moisture content, w = 32–46%
Cv = 0.12 ft²/day
296.0
Very soft grey organic and peaty organic silts and peats
Unit weight, γ = 75 lb/ft³
Moisture content, w = 79–569%
Cc = 2.52, Cv = 0.15 ft²/day
286.0
Medium compact grey sands and silty sand
281.0

▼ **FIGURE 12.26** Subsoil condition at Port Elizabeth Marine Terminal (after Johnson, 1970a)

Fill el. (ft)
335
325
315
305
Surcharge fill
Settlement (ft)
0
1
2
3
4
5
o Plate 281
Δ Plate 236
∇ Plate 232
□ Plate 241
—— Theoretical
N D J F M A M J J A S O N D J F M A M J J A
1961
1962
1963
Time (months)

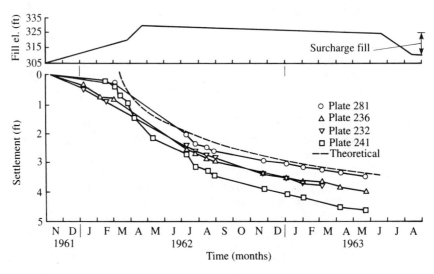

▼ **FIGURE 12.27** Precompression for support of warehouses No. 131 and 132, Port Elizabeth Marine Terminal (after Johnson, 1970a)

▼   **EXAMPLE 12.2**

Refer to Figure 12.22. During the construction of a highway bridge, the average permanent load on the clay layer is expected to increase by about 115 kN/m². The average effective overburden pressure at the middle of the clay layer is 210 kN/m². Here, $H_c = 6$ m, $C_c = 0.28$, $e_o = 0.9$, and $C_v = 0.36$ m²/mo. The clay is normally consolidated. Determine:

a.   The total primary consolidation settlement of the bridge without precompression

b.   The surcharge, $\Delta p_{(f)}$, needed to eliminate by precompression the entire primary consolidation settlement in 9 mo.

Solution

Part a

The total primary consolidation settlement may be calculated from Eq. (12.8):

$$S_{(p)} = \frac{C_c H_c}{1 + e_o} \log \left[\frac{p_o + \Delta p_{(p)}}{p_o} \right] = \frac{(0.28)(6)}{1 + 0.9} \log \left[\frac{210 + 115}{210} \right]$$

$$= 0.1677 \text{ m} = \mathbf{167.7 \text{ mm}}$$

Part b

$$T_v = \frac{C_v t_2}{H^2}$$

$C_v = 0.36$ m²/mo.

$H = 3$ m (two-way drainage)

$t_2 = 9$ mo.

Hence

$$T_v = \frac{(0.36)(9)}{3^2} = 0.36$$

According to Figure 12.25, for $T_v = 0.36$ the value of U is 47%. Now

$$\Delta p_{(p)} = 115 \text{ kN/m}^2$$

$$p_o = 210 \text{ kN/m}^2$$

So

$$\frac{\Delta p_{(p)}}{p_o} = \frac{115}{210} = 0.548$$

According to Figure 12.23, for $U = 47\%$ and $\Delta p_{(p)}/p_o = 0.548$, $\Delta p_{(f)}/\Delta p_{(p)} \approx 1.8$, so

$$\Delta p_{(f)} = (1.8)(115) = \mathbf{207 \text{ kN/m}^2} \qquad ▲$$

12.9 SAND DRAINS

The use of sand drains is another way to accelerate the consolidation settlement of soft, normally consolidated clay layers and achieve precompression before the construction of a desired foundation. Sand drains are constructed by drilling holes through the clay layer(s) in the field at regular intervals. The holes are then backfilled with sand. This can be achieved by several means, such as (a) rotary drilling and then backfilling with sand; (b) drilling by continuous flight auger with hollow stem and backfilling with sand (through the hollow stem); and (c) driving hollow steel piles. The soil inside the pile is then jetted out after which backfilling with sand is done. Figure 12.28 shows the schematic diagram of sand drains. After backfilling the drill holes with sand, a surcharge is applied at the ground surface. This surcharge will increase the pore water pressure in the clay. The excess pore water pressure in the clay will be dissipated by drainage — both vertically and radially to the sand drains — which accelerates settlement of the clay layer. In Figure 12.28a note that the radius of the sand drains is r_w. Figure 12.28b shows the plan of the layout of the sand drains. The effective zone from which the radial drainage will be directed toward a given sand drain is approximately cylindrical, with a diameter of d_e.

To determine the surcharge that needs to be applied at the ground surface and the length of time that it has to be maintained, refer to Figure 12.22 and use the corresponding equation, Eq. (12.11):

$$U_{v,r} = \frac{\log\left[1 + \dfrac{\Delta p_{(p)}}{p_o}\right]}{\log\left\{1 + \dfrac{\Delta p_{(p)}}{p_o}\left[1 + \dfrac{\Delta p_{(f)}}{\Delta p_{(p)}}\right]\right\}} \tag{12.13}$$

The notations $\Delta p_{(p)}$, p_o, and $\Delta p_{(f)}$ are the same as those in Eq. (12.11). However, unlike Eq. (12.11), the left-hand side of Eq. (12.13) is the *average degree* of consolidation instead of the degree of consolidation at midplane. Both *radial* and *vertical* drainage contribute to the average degree of consolidation. If $U_{v,r}$ can be determined for any time t_2 (see Figure 12.22b), the total surcharge $\Delta p_{(f)} + \Delta p_{(p)}$ may be obtained easily from Figure 12.23. The procedure for determination of the average degree of consolidation ($U_{v,r}$) follows:

For a given surcharge and duration, t_2, the average degree of consolidation due to drainage in the vertical and radial directions is

$$U_{v,r} = 1 - (1 - U_r)(1 - U_v) \tag{12.14}$$

where U_r = average degree of consolidation with radial drainage only
U_v = average degree of consolidation with vertical drainage only

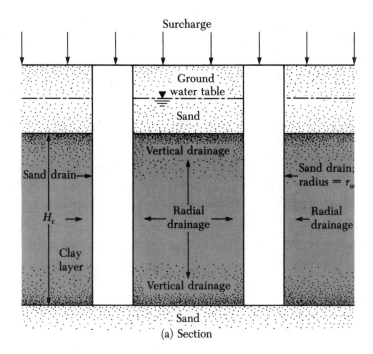

(a) Section

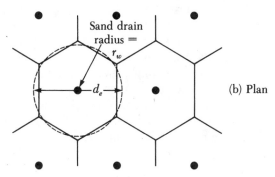

(b) Plan

▼ **FIGURE 12.28** Sand drains

The successful use of sand drains has been described in detail by Johnson (1970b). As for precompression, constant field settlement observations may be necessary during the period of surcharge application.

Average Degree of Consolidation Due to Radial Drainage Only

Figure 12.29 shows the schematic diagram of a sand drain. In this figure, r_w = the radius of the sand drain, the $r_e = d_e/2$ = radius of the effective zone of drainage. It is also important to realize that, during the installation of sand drains, a certain zone

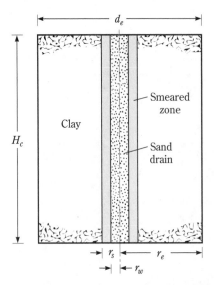

▼ **FIGURE 12.29**   Schematic diagram of a sand drain

of clay surrounding them is smeared, thereby changing the hydraulic conductivity of the clay. In Figure 12.29, r_s is the radial distance from the center of the sand drain to the farthest point of the smeared zone. Now, for the average degree of consolidation relationship, we will use the *theory of equal strain*. Two cases may arise that relate to the nature of the application of surcharge, and they are shown in Figure 12.30 (refer to the notations shown in Figure 12.22). The two cases are (a) the entire surcharge applied instantaneously (Figure 12.30a), and (b) the surcharge applied in the form of a ramp load (Figure 12.30b). When the entire surcharge is applied instantaneously (Barron, 1948)

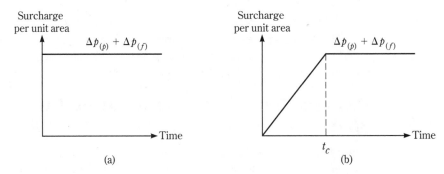

▼ **FIGURE 12.30**   Nature of surcharge application

$$U_r = 1 - \exp\left(\frac{-8T_r}{m}\right)$$

(12.15)

where

$$m = \frac{n^2}{n^2 - S^2} \ln\left(\frac{n}{S}\right) - \frac{3}{4} + \frac{S^2}{4n^2} + \frac{k_h}{k_s}\left(\frac{n^2 - S^2}{n^2}\right) \ln S$$

(12.16)

$$n = \frac{d_e}{2r_w} = \frac{r_e}{r_w}$$

(12.17)

$$S = \frac{r_s}{r_w}$$

(12.18)

k_h = hydraulic conductivity of clay in the horizontal direction in the unsmeared zone

k_s = horizontal hydraulic conductivity in the smeared zone

T_r = nondimensional time factor for radial drainage only = $\dfrac{C_{vr}t_2}{d_e^2}$ (12.19)

C_{vr} = coefficient of consolidation for radial drainage

$$= \frac{k_h}{\left[\dfrac{\Delta e}{\Delta p\,(1 + e_{av})}\right]\gamma_w}$$

(12.20)

For a *no-smear case*, $r_s = r_w$ and $k_h = k_s$, so $S = 1$ and Eq. (12.16) becomes

$$m = \left(\frac{n^2}{n^2 - 1}\right) \ln(n) - \frac{3n^2 - 1}{4n^2}$$

(12.21)

Table 12.7 gives the values of U_r for various values of T_r and n.

If the surcharge is applied in the form of a *ramp* and *there is no smear*, then (Olson, 1977)

$$U_r = \frac{T_r - \frac{1}{A}[1 - \exp(-AT_r)]}{T_{rc}} \quad \text{(for } T_r \le T_{rc})$$

(12.22)

and

$$U_r = 1 - \frac{1}{AT_{rc}}[\exp(AT_{rc}) - 1]\exp(-AT_{rc}) \quad \text{(for } T_r \ge T_{rc})$$

(12.23)

▼ TABLE 12.7 Variation of U_r for Various Values of T_r and n — No-Smear Case [Eqs. (12.15) and (12.21)]

Degree of consolidation U_r (%)	Time factor T_r for value of $n(= r_e/r_w)$				
	5	10	15	20	25
0	0	0	0	0	0
1	0.0012	0.0020	0.0025	0.0028	0.0031
2	0.0024	0.0040	0.0050	0.0057	0.0063
3	0.0036	0.0060	0.0075	0.0086	0.0094
4	0.0048	0.0081	0.0101	0.0115	0.0126
5	0.0060	0.0101	0.0126	0.0145	0.0159
6	0.0072	0.0122	0.0153	0.0174	0.0191
7	0.0085	0.0143	0.0179	0.0205	0.0225
8	0.0098	0.0165	0.0206	0.0235	0.0258
9	0.0110	0.0186	0.0232	0.0266	0.0292
10	0.0123	0.0208	0.0260	0.0297	0.0326
11	0.0136	0.0230	0.0287	0.0328	0.0360
12	0.0150	0.0252	0.0315	0.0360	0.0395
13	0.0163	0.0275	0.0343	0.0392	0.0431
14	0.0177	0.0298	0.0372	0.0425	0.0467
15	0.0190	0.0321	0.0401	0.0458	0.0503
16	0.0204	0.0344	0.0430	0.0491	0.0539
17	0.0218	0.0368	0.0459	0.0525	0.0576
18	0.0232	0.0392	0.0489	0.0559	0.0614
19	0.0247	0.0416	0.0519	0.0594	0.0652
20	0.0261	0.0440	0.0550	0.0629	0.0690
21	0.0276	0.0465	0.0581	0.0664	0.0729
22	0.0291	0.0490	0.0612	0.0700	0.0769
23	0.0306	0.0516	0.0644	0.0736	0.0808
24	0.0321	0.0541	0.0676	0.0773	0.0849
25	0.0337	0.0568	0.0709	0.0811	0.0890
26	0.0353	0.0594	0.0742	0.0848	0.0931
27	0.0368	0.0621	0.0776	0.0887	0.0973
28	0.0385	0.0648	0.0810	0.0926	0.1016
29	0.0401	0.0676	0.0844	0.0965	0.1059
30	0.0418	0.0704	0.0879	0.1005	0.1103
31	0.0434	0.0732	0.0914	0.1045	0.1148
32	0.0452	0.0761	0.0950	0.1087	0.1193
33	0.0469	0.0790	0.0987	0.1128	0.1239
34	0.0486	0.0820	0.1024	0.1171	0.1285
35	0.0504	0.0850	0.1062	0.1214	0.1332
36	0.0522	0.0881	0.1100	0.1257	0.1380
37	0.0541	0.0912	0.1139	0.1302	0.1429
38	0.0560	0.0943	0.1178	0.1347	0.1479
39	0.0579	0.0975	0.1218	0.1393	0.1529
40	0.0598	0.1008	0.1259	0.1439	0.1580
41	0.0618	0.1041	0.1300	0.1487	0.1632
42	0.0638	0.1075	0.1342	0.1535	0.1685
43	0.0658	0.1109	0.1385	0.1584	0.1739

(Continued)

Degree of consolidation U_r (%)	Time factor T_r for value of $n(= r_e/r_w)$				
	5	10	15	20	25
44	0.0679	0.1144	0.1429	0.1634	0.1793
45	0.0700	0.1180	0.1473	0.1684	0.1849
46	0.0721	0.1216	0.1518	0.1736	0.1906
47	0.0743	0.1253	0.1564	0.1789	0.1964
48	0.0766	0.1290	0.1611	0.1842	0.2023
49	0.0788	0.1329	0.1659	0.1897	0.2083
50	0.0811	0.1368	0.1708	0.1953	0.2144
51	0.0835	0.1407	0.1758	0.2020	0.2206
52	0.0859	0.1448	0.1809	0.2068	0.2270
53	0.0884	0.1490	0.1860	0.2127	0.2335
54	0.0909	0.1532	0.1913	0.2188	0.2402
55	0.0935	0.1575	0.1968	0.2250	0.2470
56	0.0961	0.1620	0.2023	0.2313	0.2539
57	0.0988	0.1665	0.2080	0.2378	0.2610
58	0.1016	0.1712	0.2138	0.2444	0.2683
59	0.1044	0.1759	0.2197	0.2512	0.2758
60	0.1073	0.1808	0.2258	0.2582	0.2834
61	0.1102	0.1858	0.2320	0.2653	0.2912
62	0.1133	0.1909	0.2384	0.2726	0.2993
63	0.1164	0.1962	0.2450	0.2801	0.3075
64	0.1196	0.2016	0.2517	0.2878	0.3160
65	0.1229	0.2071	0.2587	0.2958	0.3247
66	0.1263	0.2128	0.2658	0.3039	0.3337
67	0.1298	0.2187	0.2732	0.3124	0.3429
68	0.1334	0.2248	0.2808	0.3210	0.3524
69	0.1371	0.2311	0.2886	0.3300	0.3623
70	0.1409	0.2375	0.2967	0.3392	0.3724
71	0.1449	0.2442	0.3050	0.3488	0.3829
72	0.1490	0.2512	0.3134	0.3586	0.3937
73	0.1533	0.2583	0.3226	0.3689	0.4050
74	0.1577	0.2658	0.3319	0.3795	0.4167
75	0.1623	0.2735	0.3416	0.3906	0.4288
76	0.1671	0.2816	0.3517	0.4021	0.4414
77	0.1720	0.2900	0.3621	0.4141	0.4546
78	0.1773	0.2988	0.3731	0.4266	0.4683
79	0.1827	0.3079	0.3846	0.4397	0.4827
80	0.1884	0.3175	0.3966	0.4534	0.4978
81	0.1944	0.3277	0.4090	0.4679	0.5137
82	0.2007	0.3383	0.4225	0.4831	0.5304
83	0.2074	0.3496	0.4366	0.4992	0.5481
84	0.2146	0.3616	0.4516	0.5163	0.5668
85	0.2221	0.3743	0.4675	0.5345	0.5868
86	0.2302	0.3879	0.4845	0.5539	0.6081
87	0.2388	0.4025	0.5027	0.5748	0.6311
88	0.2482	0.4183	0.5225	0.5974	0.6558

(Continued)

▼ **TABLE 12.7** (Continued)

Degree of consolidation U_r (%)	Time factor T_r for value of $n(= r_e/r_w)$				
	5	10	15	20	25
89	0.2584	0.4355	0.5439	0.6219	0.6827
90	0.2696	0.4543	0.5674	0.6487	0.7122
91	0.2819	0.4751	0.5933	0.6784	0.7448
92	0.2957	0.4983	0.6224	0.7116	0.7812
93	0.3113	0.5247	0.6553	0.7492	0.8225
94	0.3293	0.5551	0.6932	0.7927	0.8702
95	0.3507	0.5910	0.7382	0.8440	0.9266
96	0.3768	0.6351	0.7932	0.9069	0.9956
97	0.4105	0.6918	0.8640	0.9879	1.0846
98	0.4580	0.7718	0.9640	1.1022	1.2100
99	0.5391	0.9086	1.1347	1.2974	1.4244

where

$$T_{rc} = \frac{C_{vr}t_c}{d_e^2} \quad \text{(see Figure 12.30b for definition of } t_c) \tag{12.24}$$

$$A = \frac{2}{m} \tag{12.25}$$

Average Degree of Consolidation Due to Vertical Drainage Only

Referring to Figure 12.30a, for instantaneous surcharge application, the average degree of consolidation due to vertical drainage only may be obtained from Eq. (1.77) and (1.78):

$$T_v = \frac{\pi}{4}\left[\frac{U_v(\%)}{100}\right]^2 \quad \text{(for } U_v = 0\text{–}60\%) \tag{1.77}$$

and

$$T_v = 1.781 - 0.933 \log(100 - U_v(\%)) \quad \text{(for } U_v > 60\%) \tag{1.78}$$

where U_v = average degree of consolidation due to vertical drainage only

$$T_v = \frac{C_v t_2}{H^2} \tag{1.72}$$

C_v = coefficient of consolidation for vertical drainage

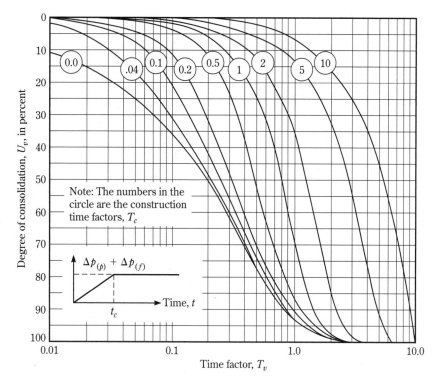

▼ **FIGURE 12.31** Variation of U_v with T_v and T_c (after Olson, 1977)

For the case of ramp loading as shown in Figure 12.30b, the variation of $U_v(\%)$ with T_v and T_c (Olson, 1977) is given in Figure 12.31. Note that

$$T_c = \frac{C_v t_c}{H^2} \tag{12.26}$$

where H = length of maximum vertical drainage path

12.10   AN EXAMPLE OF A SAND DRAIN APPLICATION

Aboshi and Monden (1963) provided details on the field performance of 2700 sand drains used to construct the Toya Quay Wall on reclaimed land in Japan in a study that was summarized by Johnson (1970b). The location of the project site is shown in the insert of Figure 12.32. The soil at the site consisted of 30-m-thick soft, normally consolidated clayey silt. The following data are for the *in situ* soil and the sand drains.

▶ *In situ* soil: Liquid limit (LL) = 110
Plastic limit (PL) = 48
Natural moisture content, w = 74%–65%

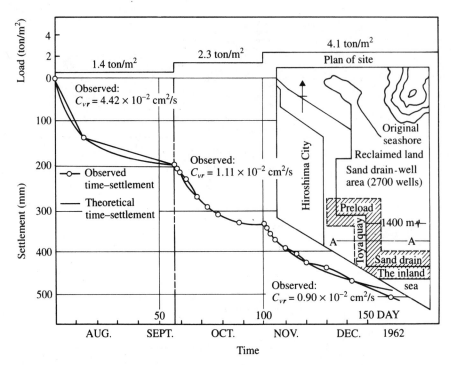

▼ **FIGURE 12.32**   Comparison of observed and theoretical settlements due to
sand drawn only for Toya Quay Wall Construction, Japan (after
Johnson, 1970b)

▶ Sand drains: Total number used = 2700
Length = 15 m
d_e = 3.15 m
r_w = 0.225 m
$\left.\begin{array}{l} \dfrac{C_{vr}}{C_v} = 2.7 \\[4pt] \overline{C_v} = 1.7 \end{array}\right\}$ (from triaxial test)
(from consolidometer)

The top portion of Figure 12.32 shows the variation of the surcharge load application
with time. The bottom portion shows the observed and theoretical variation of
settlement due to *sand drains only*. The agreement appears to be excellent.

▼ **EXAMPLE 12.3** _____

Redo Example 12.2 with the addition of some sand drains. Assume that $r_w = 0.1$ m,
$d_e = 3$ m, $C_v = C_{vr}$, and the surcharge is applied instantaneously (Figure 12.30a).
Also assume that this is a no-smear case.

Solution

Part a

The total primary consolidation settlement will be 167.7 mm as before.

Part b

From Example 12.2, $T_v = 0.36$. Using Eq. (1.77)

$$T_v = \frac{\pi}{4} \left[\frac{U_v(\%)}{100} \right]^2$$

or

$$U_v = \sqrt{\frac{4T_v}{\pi}} \times 100 = \sqrt{\frac{(4)(0.36)}{\pi}} \times 100 = 67.7\%$$

$$n = \frac{d_e}{2r_w} = \frac{3}{2 \times 0.1} = 15$$

Again,

$$T_r = \frac{C_{vr}t_2}{d_e^2} = \frac{(0.36)(9)}{(3)^2} = 0.36$$

From Table 12.7 for $n = 15$ and $T_r = 0.36$, the value of U_r is about 77%. Hence

$$U_{v,r} = 1 - (1 - U_v)(1 - U_r) = 1 - (1 - 0.67)(1 - 0.77)$$

$$= 0.924 = 92.4\%$$

Now, from Figure 12.23, for $\Delta p_{(p)}/p_o = 0.548$ and $U_{v,r} = 92.4\%$ the value of $\Delta p_{(f)}/\Delta p_{(p)} \approx 0.12$. Hence

$$\Delta p_{(f)} = (115)(0.12) = \mathbf{13.8 \ kN/m^2}$$ ▲

▼   **EXAMPLE 12.4** _____

Refer to Figure 12.28, which is a sand drain project. This clay is normally consolidated. Given:

$$\text{Clay: } H_c = 15 \text{ ft (two-way drainage)}$$

$$C_c = 0.31$$

$$e_o = 1.1$$

Effective overburden pressure at the middle of the clay layer

$$= 1000 \text{ lb/ft}^2$$

$$C_v = 0.115 \text{ ft}^2/\text{day}$$

$$\text{Sand drain: } r_w = 0.3 \text{ ft}$$

$$d_e = 6 \text{ ft}$$

$$C_v = C_{vr}$$

A surcharge is applied as shown in Figure 12.33. Assume this to be a no-smear case. Calculate the degree of consolidation 30 days after the beginning of the surcharge application. Also determine the consolidation settlement at that time due to the surcharge.

Solution   From Eq. (12.26),

$$T_c = \frac{C_v t_c}{H^2} = \frac{(0.115 \ \text{ft}^2/\text{day}) \ (60)}{\left(\dfrac{15}{2}\right)^2} = 0.123$$

$$T_v = \frac{C_v t_2}{H^2} = \frac{(0.115) \ (30)}{\left(\dfrac{15}{2}\right)^2} = 0.061$$

Using Figure 12.31 for $T_c = 0.123$ and $T_v = 0.061$, $U_v \approx 9\%$. For the sand drain,

$$n = \frac{d_e}{2r_w} = \frac{6}{(2) \ (0.3)} = 10$$

From Eq. (12.24),

$$T_{rc} = \frac{C_{vr} t_c}{d_e^2} = \frac{(0.115) \ (60)}{(6)^2} = 0.192$$

$$T_r = \frac{C_{vr} t_2}{d_e^2} = \frac{(0.115) \ (30)}{(6)^2} = 0.096$$

Again, from Eq. (12.22),

Surcharge (lb/ft²)

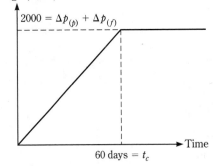

▼ FIGURE 12.33

$$U_r = \frac{T_r - \frac{1}{A}[1 - \exp(-AT_r)]}{T_{rc}}$$

$$m = \frac{n^2}{n^2 - 1}\ln(n) - \frac{3n^2 - 1}{4n^2} = \frac{10^2}{10^2 - 1}\ln(10) - \frac{3(10)^2 - 1}{4(10)^2} = 1.578$$

$$A = \frac{2}{m} = \frac{2}{1.578} = 1.267$$

$$U_r = \frac{0.096 - \frac{1}{1.267}[1 - \exp(-1.267 \times 0.096)]}{0.192} = 0.03 = 3\%$$

From Eq. (12.14),

$$U_{v,r} = 1 - (1 - U_r)(1 - U_v) = 1 - (1 - 0.03)(1 - 0.09) = 0.117 = \mathbf{11.7\%}$$

Total primary settlement is

$$S_{(p)} = \frac{C_c H_c}{1 + e_o}\log\left[\frac{p_o + \Delta p_{(p)} + \Delta p_{(f)}}{p_o}\right] = \frac{(0.31)(15)}{1 + 1.1}\log\left(\frac{1000 + 2000}{1000}\right) = 1.056 \text{ ft}$$

Settlement after 30 days is

$$S_{(p)}\, U_{v,r} = (1.056)(0.117)(12) = \mathbf{1.48 \text{ in.}} \qquad\qquad \blacktriangle$$

12.11 PREFABRICATED VERTICAL DRAINS (PVDs)

The PVDs, also referred to as *wick or strip drains*, were originally developed as a substitute for the commonly used sand drain. With the advent of materials science, these drains are manufactured from synthetic polymers such as polypropylene and high-density polyethylene. PVDs are normally manufactured with a corrugated or channeled synthetic core enclosed by a geotextile filter, as shown schematically in Figure 12.34. Installation rates reported in the literature are on the order of 0.1 to 0.3 m/s, excluding equipment mobilization and setup time. The PVDs have been used extensively in the past for expedient consolidation of low permeability soils under surface surcharge. The main advantage of PVDs over sand drains is that they do not require drilling and, thus, installation is much faster. Design curves for PVDs are given in Appendix E.

12.12 LIME STABILIZATION

As mentioned in Section 12.1, admixtures are occasionally used to stabilize soils in the field — particularly fine-grained soils. The most common admixtures are lime, cement, and lime–fly ash. The main purposes of soil stabilization are to (a) modify the soil, (b) expedite construction, and (c) improve the strength and durability of the soil.

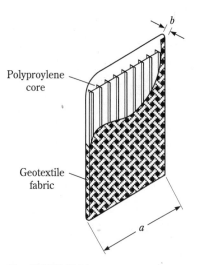

▼   **FIGURE 12.34**   Prefabricated vertical drain (PVD)

The types of *lime* commonly used for stabilization of fine-grained soils are hydrated high-calcium lime [$Ca(OH)_2$], calcitic quick lime (CaO), monohydrated dolomitic lime [$Ca(OH)_2 \cdot MgO$], and dolomitic quick lime. The quantity of lime used for stabilization of most soils usually is in the range of 5%–10%. When lime is added to clayey soils, several chemical reactions occur, *cation exchange* and *flocculation–agglomeration*, and they are also *pozzolanic*. In the cation exchange and flocculation–agglomeration reactions, the *monovalent* cations generally associated with clays are replaced by the *divalent* calcium ions. Based on their affinity for exchange, the cations can be arranged in a series:

$$Al^{3+} > Ca^{2+} > Mg^{2+} > NH_4^+ > K^+ > Na^+ > Li^+$$

Any cation can replace the ions to its right. For example, calcium ions can replace potassium and sodium ions from a clay. Flocculation and agglomeration produce a change in the texture of clay soils. The clay particles tend to clump together to form larger particles. These reactions tend to (a) decrease the liquid limit, (b) increase the plastic limit, (c) decrease the plasticity index, (d) increase the shrinkage limit, (e) increase the workability, and (f) improve the strength and deformation properties of a soil.

Pozzolanic reaction between soil and lime involves a reaction between lime and the silica and alumina of soil to form cementing material. For example,

$$Ca(OH)_2 + SiO_2 \rightarrow CSH$$
$$\uparrow$$
$$\text{Clay silica}$$

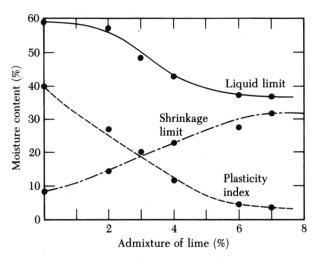

▼ **FIGURE 12.35** Variation of liquid limit, plasticity index, and shrinkage of a clay with lime additive

where C = CaO
       S = SiO$_2$
       H = H$_2$O

The pozzolanic reaction may continue for a long period of time.

Figure 12.35 shows the variation of the liquid limit, the plasticity index, and the shrinkage limit of a clay with percentage of lime admixture. The first 2%–3% lime (on the dry weight basis) substantially influences the workability and the property (such as plasticity) of the soil. The addition of lime to clayey soils affects their compaction characteristics.

Figure 12.36a shows the results of standard Proctor tests for Vicksburg clay without any additives and also with 4% high-calcium hydrated lime additive (uncured). Note that the addition of lime helps reduce the maximum compacted dry unit weight and increase the optimum moisture content. Figure 12.36b also shows the change of the unconfined compressive strength, q_u, of uncured Vicksburg clay with the percentage of high-calcium hydrated lime. The value of q_u with 6% lime is about six times that obtained with no additive. Note that the specimens prepared for the determination of q_u were all at a moisture content of 29%–29.5%. This is shown as the molding moisture content in Figure 12.36a.

Arman and Munfakh (1972) evaluated the lime stabilization of organic clays found in Louisiana. Figure 12.37a shows the change of plasticity with Ca(OH)$_2$ content for an organic soil with 22% organic material. The curing time for these soil specimens was 48 hours. The effects of lime are generally similar to those shown in Figure 12.36. The change of the unconfined compression strength of the same soil with lime additives is shown in Figure 12.37b. Based on their study, Arman and Munfakh concluded that (a) the presence of organic matter does not block the

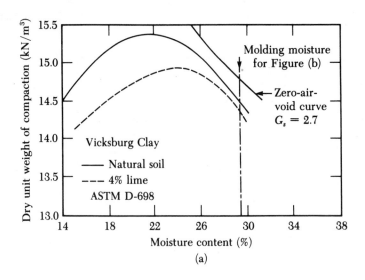

(a)

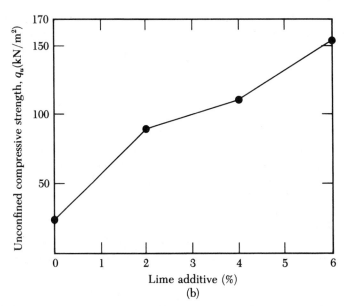

(b)

▼ **FIGURE 12.36** Lime stabilization of Vicksburg clay (% less than 2μ size = 46; liquid limit = 59; plasticity index = 30; A-7-6(20), pH = 6): (a) laboratory dry unit weight against moisture content curve; (b) change of unconfined compression strength with percent of lime. *Note:* The specimens for (b) were molded at a moisture content shown in (a) (from "Stability Properties of Uncured Lime-Treated Fine-Grained Soil," by C. H. Neubauer and M. R. Thompson. In *Highway Research Record No. 381*, Highway Research Board, 1972, pp. 20–26. Reprinted by permission)

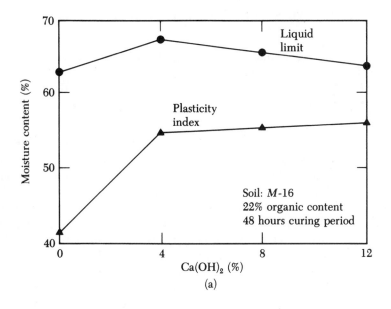

(a)

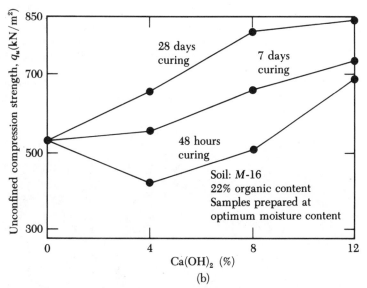

(b)

▼ **FIGURE 12.37** Lime stabilization of organic soil: (a) variation of liquid limit and plasticity index with 48 hours curing; (b) variation of unconfined compression strength (from "Lime Stabilization of Organic Soil," by A. Arman and G. A. Munfakh. In *Highway Research Record No. 381*, Highway Research Board, 1972, pp. 37–45. Reprinted by permission)

pozzolanic reaction that helps change the fundamental soil properties and make the soil more workable and (b) about 2% of lime is sufficient to satisfy the base exchange capacity of organic matters.

Lime stabilization in the field can be done in three ways:

1. The *in situ* material and/or the borrowed material can be mixed with the proper amount of lime at the site and then compacted after the addition of moisture.
2. The soil can be mixed with the proper amount of lime and water at a plant and then hauled back to the site for compaction.
3. Lime slurry can be pressure injected into the soil to a depth of 12 to 16 ft (4–5 m). Figure 12.38 shows a vehicle used for pressure injection of lime slurry. The slurry-injection mechanical unit is mounted to the injection vehicle. A common injection unit is a hydraulic-lift mast with cross beams that contain the injection rods. The injection rods are pushed into the ground by the action of the lift mast beams. The slurry is generally mixed in a batching tank about 10 ft (3 m) in diameter and 36 ft (12 m) long and is pumped at high pressure to the injection rods. Figure 12.39 is a photograph of the lime slurry pressure-injection process. The ratio typically specified for preparation of lime slurry is 2.5 lb of dry lime to a gallon of water. For more information on this technique, see Blacklock and Pengelly (1988).

Because the addition of hydrated lime to soft clayey soils immediately increases the plastic limit, thus changing the soil from plastic to solid and making it appear

▼ **FIGURE 12.38**   Equipment for pressure injection of lime slurry (courtesy of GKN Hayward Baker, Inc., Woodbine Division, Ft. Worth, Texas)

▼ **FIGURE 12.39**   Pressure injection of lime slurry (courtesy of GKN Hayward
Baker, Inc., Woodbine Division, Ft. Worth, Texas)

to "dry up," limited amounts of it can be thrown on muddy and troublesome construction sites. This action improves trafficability and may save money and time. Quick
limes have also been successfully used in drill holes having diameters of 4 in. to
6 in. (100 mm to 150 mm) for stabilization of subgrades and slopes. For this type
of work, holes are drilled in a grid pattern and then filled with quick lime.

12.13 CEMENT STABILIZATION

Cement is increasingly used as a stabilizing material for soil, particularly for the
construction of highways and earth dams. The first controlled soil–cement construction in the United States was carried out near Johnsonville, South Carolina, in 1935.
Cement can be used to stabilize sandy and clayey soils. As in the case of lime, cement

▼ **TABLE 12.8** Cement Requirement by Volume for Effective Stabilization of Various Soils[a]

Soil type		Percent cement by volume
AASHTO classification	Unified classification	
A-2 and A-3	GP, SP, and SW	6–10
A-4 and A-5	CL, ML, and MH	8–12
A-6 and A-7	CL, CH	10–14
[a] After Mitchell and Freitag (1959)		

helps decrease the liquid limit and increase the plasticity index and workability of clayey soils. For clayey soils, cement stabilization is effective when the liquid limit is less than 45–50 and the plasticity index is less than about 25. The optimum requirements of cement by volume for effective stabilization of various types of soil are given in Table 12.8.

Like lime, cement helps increase the strength of soils, and strength increases with curing time. Table 12.9 presents some typical values of the unconfined compressive strength of various types of untreated soil and soil–cement mixture made with approximately 10% cement by weight.

Granular soils and clayey soils with low plasticity obviously are most suitable for cement stabilization. Calcium clays are more easily stabilized by the addition of

▼ **TABLE 12.9** Typical Compressive Strengths of Soils and Soil–Cement Mixtures[a]

Material	Unconfined compressive strength range	
	lb/in²	(kN/m²)[b]
Untreated soil:		
Clay, peat	Less than 50	Less than 350
Well-compacted sandy clay	10–40	70–280
Well-compacted gravel, sand, and clay mixtures	40–100	280–700
Soil–cement (10% cement by weight):		
Clay, organic soils	Less than 50	Less than 350
Silts, silty clays, very poorly graded sands, slightly organic soils	50–150	350–1050
Silty clays, sandy clays, very poorly graded sands, and gravels	100–250	700–1730
Silty sands, sandy clays, sands, and gravels	250–500	1730–3460
Well-graded sand–clay or gravel–sand–clay mixtures and sands and gravels	500–1500	3460–10,350
[a] After Mitchell and Freitag (1959)		
[b] Rounded off		

cement, whereas sodium and hydrogen clays, which are expansive in nature, respond better to lime stabilization. For these reasons, proper care should be given in the selection of the stabilizing material.

For field compaction, the proper amount of cement can be mixed with soil either at the site or at a mixing plant and then carried to the site. The soil is compacted to the required unit weight with a predetermined amount of water.

Similar to lime injection, cement slurry made of Portland Cement and water (water–cement ratio = 0.5 : 5) can be used for pressure grouting of poor soils under foundations of buildings and other structures. Grouting decreases the hydraulic conductivity of soils and increases the strength and the load-bearing capacity. For design of low-frequency machine foundations subjected to vibrating forces, stiffening the foundation soil by grouting and thereby increasing the resonant frequency is sometimes necessary.

12.14 FLY ASH STABILIZATION

Fly ash is a by-product of the pulverized coal combustion process usually associated with electric power generating plants. It is a fine-grained dust and is primarily composed of silica, alumina, and various oxides and alkalies. It is pozzolanic in nature and can react with hydrated lime to produce cementitious products. For that reason, lime–fly ash mixtures can be used for stabilization of highway bases and subbases. Effective mixes can be prepared with 10%–35% fly ash and 2%–10% lime. Soil–lime–fly ash mixes are compacted under controlled conditions with proper amounts of moisture to obtain stabilized soil layers.

A certain type of fly ash is obtained from the burning of coal primarily from the western United States, and it is referred to as "Type C" fly ash. It contains a fairly large proportion (up to about 25%) of free lime that, with the addition of water, will react with other fly ash compounds to form cementitious products. Its use may eliminate the need to add manufactured lime.

12.15 STONE COLUMNS

A method now being used to increase the load-bearing capacity of shallow foundations on soft clay layers is the construction of stone columns. Construction of a stone column generally consists of water-jetting a vibroflot (Section 12.7) into the soft clay layer to make a circular hole that extends through the clay to firmer soil. The hole is then filled with an imported gravel. The gravel in the hole is gradually compacted as the vibrator is withdrawn. The gravel used for the stone column has sizes ranging from 0.25–1.5 in. (6–40 mm). Stone columns usually have diameters of 1.6–2.5 ft (0.5–0.75 m) and are spaced at about 5–10 ft (1.5–3 m) center-to-center.

After the construction of stone columns, a fill material should always be placed over the ground surface and compacted before construction of the foundation. The stone columns tend to reduce the settlement of foundations at allowable loads. Several case histories of construction projects using stone columns are presented by Hughes and Withers (1974), Hughes et al. (1975), Mitchell and Huber (1985), and others.

At this time there is no standard way to estimate the settlement of foundations constructed over stone columns. However, based on the recommendation of Greenwood and Thompson (1984) and on observations of the author, a tentative chart for estimating settlement is given in Figure 12.40. To utilize Figure 12.40, use the following procedure:

1. Determine the cross-sectional area of the stone column, A_S.
2. Determine the average area of the column foundation, A_F.
3. Calculate the ratio of A_F/A_S.
4. Estimate the undrained shear strength of the clay, c_u, and the probable settlement of a column foundation assuming that it was constructed without the stone columns, S_F.
5. With known values of A_F/A_S and c_u, determine the ratio of S_F/S_S (S_S = probable settlement of the foundation constructed over stone columns) from Figure 12.40b.
6. With known values of S_F and S_F/S_S, calculate S_S.

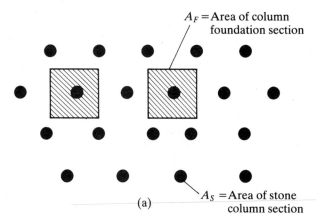

(a)

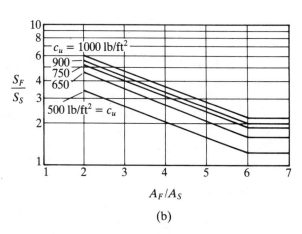

(b)

▼ **FIGURE 12.40**   Settlement of foundation built on stone columns

Hughes et al. (1975) provided an approximate relationship for the allowable bearing capacity (q_{all}) of stone columns, which can be given as

$$q_{all} = \frac{\tan^2\left(45 + \dfrac{\phi}{2}\right)}{FS}(4c_u + \sigma_r') \tag{12.27}$$

where FS = factor of safety (≈ 1.5 to 2)

         c_u = undrained shear strength of the clay

         σ_r' = effective radial stress as measured by a pressuremeter ($\approx 2c_u$)

Stone columns work more effectively when used for stabilizing a large area where the undrained shear strength of the subsoil is in the range of 200–1000 lb/ft² (10–15 kN/m²) than does improving the bearing capacity of structural foundations (Bachus and Barksdale, 1989). Subsoils weaker than that may not provide sufficient lateral support for the stone columns. For large-site improvement, stone columns are most effective to a depth of 20–30 ft (6–10 m). However, stone columns have been constructed to a depth of 100 ft (31 m). Bachus and Barksdale provided the following general guidelines for the design of stone columns for stabilizing large areas:

Figure 12.41a shows the plan view of several stone columns, and Figure 12.41b depicts the unit cell idealization of a stone column. The area replacement ratio, a_s,

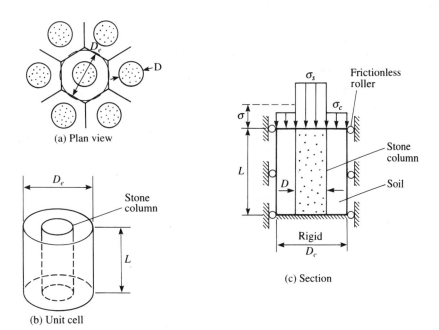

(a) Plan view

(b) Unit cell

(c) Section

▼ **FIGURE 12.41**   Unit cell idealization of stone column (after Bachus and Barksdale, 1989)

for the stone columns may be expressed as

$$a_s = \frac{A_s}{A} \tag{12.28}$$

where A_s = area of the stone column
A = total area within the unit cell

For an *equilateral triangular pattern* of stone columns,

$$a_s = 0.907 \left(\frac{D}{s}\right)^2 \tag{12.29}$$

where D = diameter of the stone column
s = spacing between the stone columns

When a uniform stress by means of a fill operation is applied to an area with stone columns to induce consolidation, a stress concentration occurs due to the change in the stiffness between the stone columns and the surrounding soil (Figure 12.41c). The stress concentration factor, n', is defined as

$$n' = \frac{\sigma_s}{\sigma_c} \tag{12.30}$$

where σ_s = stress in the stone column
σ_c = stress in the subgrade soil

The relationships for σ_s and σ_c are

$$\sigma_s = \sigma \left[\frac{n'}{1 + (n' - 1)a_s}\right] = \mu_s \sigma \tag{12.31}$$

$$\sigma_c = \sigma \left[\frac{1}{1 + (n' - 1)a_s}\right] = \mu_c \sigma \tag{12.32}$$

where σ = average vertical stress
μ_s, μ_c = stress concentration factors

The variation of μ_c and a_s and n' is shown in Figure 12.42. The improvement of the soil owing to the stone columns may be expressed as

$$\frac{S_t}{S} = \mu_c \tag{12.33}$$

where S_t = settlement of the treated soil
S = total settlement of the untreated soil

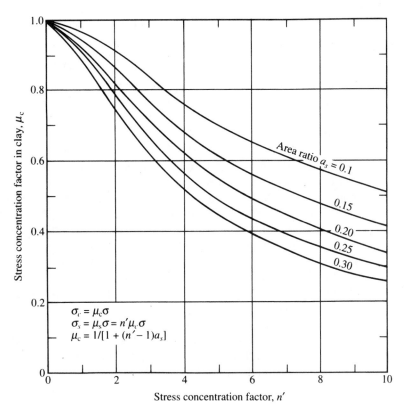

▼ **FIGURE 12.42**   Variation of μ_c with a_s and n' (after Bachus and Barksdale, 1989)

12.16 SAND COMPACTION PILES

Sand compaction piles are similar to stone columns, and they can be used in marginal sites to improve stability, control liquefaction, and reduce settlement of various structures. These piles can significantly accelerate the pore water pressure-dissipation process and hence the time for consolidation when built in soft clay.

Sand piles were first constructed in Japan between 1930 and 1950 (Ichimoto, 1981). Large-diameter compacted sand columns were constructed in 1955 using the Compozer technique (Aboshi et al., 1979). The Vibro-Compozer method of sand pile construction was developed by Murayama in Japan in 1958 (Murayama, 1962).

Sand compaction piles are constructed by driving a hollow mandrel with its bottom closed during driving. On partial withdrawal, the bottom doors open. Sand is poured from the top of the mandrel and compacted in steps by applying air pressure as the mandrel is withdrawn. The piles are usually 1.5–2.5 ft (0.46–0.76 m) in diameter and are placed at about 5 ft to 10 ft (1.5–3 m) center-to-center. The pattern of layout of sand compaction piles is the same as for stone columns. Following is an overview of a successful sand compaction pile project in Korea (Shin, Shin, and Das, 1992).

Sand compaction piles were first used in Korea in 1984 for the construction of the Kwang Yang Steel Mill complex. The project site, Kwang-Yang, is located on the South Sea about 187 miles (300 km) south of Seoul, where a delta is formed at the convergence of the Sum Jin and Su Oh Rivers (Figure 12.43a). This site was selected for the steel mill complex because it is directly connected to the sea, which facilitates the import of raw material and export of finished products.

Figure 12.43b shows the general nature of the *in situ* soil along with a backfilled sand of about 16.5 ft (5 m). Site improvement techniques used in this project included sand compaction piles as well as some sand drains with preloading. The sand compaction piles and the sand drains had diameters of 2.3 ft (0.7 m) and 1.3 ft (0.4 m), respectively, with an average length of about 82 ft (25 m) each. The center-to-center spacing of the sand drains and sand compaction piles ranged from 5.8 ft (1.75 m) to 8.2 ft (2.5 m). The reclaimed area was about 15,602,000 ft² (1,450,000 m²). The volume of sand used for construction of sand piles and preloading, including the backfill sand layer, was about 15.2×10^6 yd³ (11,600,000 m³). The improved site presently supports a stockyard of heavy material, a slab yard, oil tanks, embankments for roads and railways, and steel mill factories.

Figure 12.44 shows the staged preloading along with the variation of pore water pressure and consolidation settlement at the stockyard site with time. The strength of the subsoil was sufficiently improved and, after removal of the preload, all construction work progressed smoothly.

12.17 DYNAMIC COMPACTION

Dynamic compaction is a technique that is beginning to gain popularity in the United States for densification of granular soil deposits. This process primarily involves dropping a heavy weight repeatedly on the ground at regular intervals. The weight of the hammer used varies from 8 to 35 metric tons, and the height of the hammer drop varies between 25 and 100 ft (≈ 7.5 and 30.5 m). The stress waves generated by the hammer drops help in the densification. The degree of compaction achieved depends on the

a. Weight of the hammer
b. Height of hammer drop
c. Spacing of the locations at which the hammer is dropped

Leonards et al. (1980) suggested that the significant depth of influence for compaction is approximately

$$DI \simeq \tfrac{1}{2}\sqrt{W_H h} \tag{12.34}$$

where DI = significant depth of densification (m)
   W_H = dropping weight (metric ton)
   h = height of drop (m)

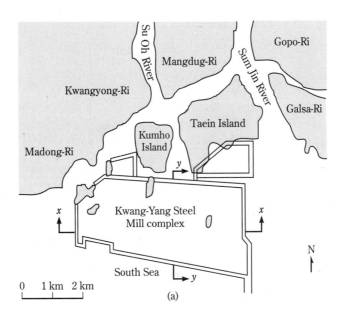

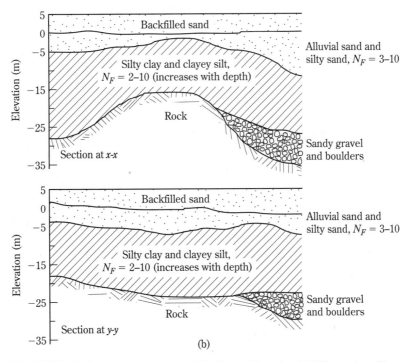

▼ **FIGURE 12.43** (a) Site location of the Kwang-Yang Steel Mill complex; (b) general nature of the soil profile. *Note*: N_F = field standard penetration number (after Shin, Shin, and Das, 1992)

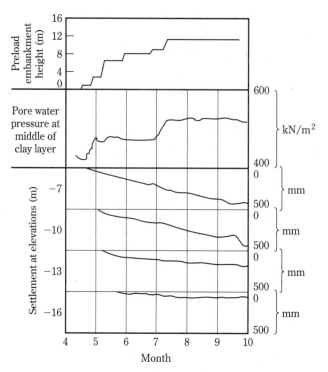

▼ **FIGURE 12.44**   Pore water pressure and settlement measurement with staged preloading at the stockyard site (after Shin, Shin, and Das, 1992)

In English units, Eq. (12.34) becomes

$$DI = 0.61\sqrt{W_H h} \tag{12.35}$$

where DI and h are in ft and W_H is in kip

Partos et al. (1989) provided several case histories of site improvement using dynamic compaction. Figure 12.45 shows the effect of dynamic compaction in improving the standard penetration resistance at the construction site of an office building in the Riverview Executive Park (Trenton, New Jersey). For this dynamic compaction,

Weight of the hammer, $W_H = 18.5$ ton (163 kN)

Height of drop, $h = 85$ ft (26 m)

Spacing between hammer drops = 10.6 ft (3.3 m)

Number of drops at each location = 7

More recently, Poran and Rodriguez (1992) suggested a rational method for conducting dynamic compaction for granular soils in the field. According to this method, for a hammer of width D having a weight W_H and a drop h, the approximate shape of the densified area will be of the type shown in Figure 12.46 (that is, a

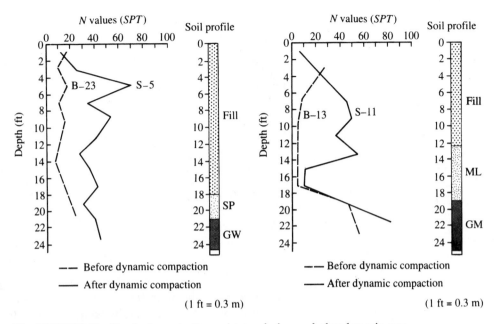

▼ **FIGURE 12.45** Standard penetration resistance before and after dynamic compaction, Riverview Executive Park, Trenton, New Jersey (after Partos et al., 1989)

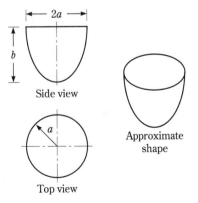

▼ **FIGURE 12.46** Approximate shape of the densified area due to dynamic compaction (after Poran and Rodriguez, 1992)

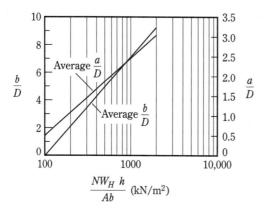

▼ **FIGURE 12.47**   Plot of a/D and b/D vs. NW_Hh/Ab
(after Poran and Rodriguez, 1992)

semiprolate spheroid). Note that in this figure $b = DI$. Figure 12.47 gives the design chart for a/D and b/D versus NW_Hh/Ab (D = width of the hammer if not circular in cross section; A = area of cross section of the hammer; N = number of required hammer drops). Following are the steps in using this method:

1. Determine the required significant depth of densification, $DI(=b)$.
2. Determine the hammer weight (W_H), height of drop (h), dimensions of the cross section, and, thus, the area A and the width D.
3. Determine $DI/D = b/D$.
4. Use Figure 12.47 and determine the magnitude of NW_Hh/Ab for the value of b/D obtained in Step 3.
5. Since the magnitudes of W_H, h, A, and b are known (or assumed) from Step 2, the number of hammer drops can be estimated from the value of NW_Hh/Ab obtained from Step 4.
6. With known values of NW_Hh/Ab, go to Figure 12.47 and determine a/D and thus a.
7. The grid spacing, S_g, for dynamic compaction may now be assumed to be equal to or somewhat less than a (Figure 12.48).

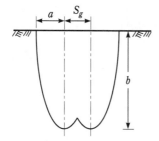

▼ **FIGURE 12.48**   Approximate grid spacing
for dynamic compaction

PROBLEMS

12.1 Make the necessary calculations and prepare the zero-air-void unit-weight curves (in kN/m^3) as related to a Proctor compaction test for G_s = 2.6, 2.65, 2.7, and 2.75.

12.2 A sandy soil has a maximum dry unit weight of 112 lb/ft^3 and a dry unit weight of compaction in the field of 100 lb/ft^3. Estimate the following:
   a. The relative compaction in the field
   b. The relative density in the field
   c. The minimum dry unit weight of the soil

12.3 According to the Ohio one-point method, a soil will have a dry unit weight of 102 lb/ft^3 at a moisture content of 19%. Estimate the dry unit weight of the soil when compacted to a moisture content of 16.5%.

12.4 The following are given for a natural soil deposit:
   Moist unit weight, γ = 102 lb/ft^3
   Moisture content, w = 16%
   G_s = 2.71

This soil is to be excavated and transported to a construction site for use in a compacted fill. If the specification calls for the soil to be compacted to a minimum dry unit weight of 105 lb/ft^3 at the same moisture content of 16%, how many cubic yards of soil from the excavation site are needed to produce 10,000 cubic yards of compacted fill? How many twenty-ton truckloads are needed to transport the excavated soil?

12.5 A proposed embankment fill required 8000 m^3 of compacted soil. The void ratio of the compacted fill is specified to be 0.65. Four available borrow pits are shown below along with the void ratios of the soil and the cost per cubic meter for moving the soil to the proposed construction site.

Borrow pit	Void ratio	Cost ($/m³)
A	0.9	8
B	1.1	5
C	0.95	6
D	0.75	11

Make the necessary calculations to select the pit from which the soil should be brought to minimize the cost. Assume G_s to be the same for all borrow-pit soil.

12.6 For a vibroflotation work, the backfill to be used has the following characteristics:

D_{50} = 2 mm

D_{20} = 0.7 mm

D_{10} = 0.65 mm

Determine the suitability number of the backfill. How would you rate the material?

12.7   Repeat Problem 12.6 with the following:

$D_{50} = 3.2$ mm

$D_{20} = 0.91$ mm

$D_{10} = 0.72$ mm

12.8   Refer to Figure 12.22. For a large fill operation, the average permanent load $[\Delta p_{(p)}]$ on the clay layer will increase by about 75 kN/m². The average effective overburden pressure on the clay layer before the fill operation is 110 kN/m². For the clay layer, which is normally consolidated and drained at top and bottom, given: $H_c = 8$ m, $C_c = 0.27$, $e_o = 1.02$, $C_v = 0.52$ m²/month. Determine the following:

   a.   The primary consolidation settlement of the clay layer caused by the addition of the permanent load $\Delta p_{(p)}$

   b.   The time required for 90% of primary consolidation settlement under the additional permanent load only

   c.   The temporary surcharge, $\Delta p_{(f)}$, that will be required to eliminate the entire primary consolidation settlement in 12 months by the precompression technique

12.9   Repeat Problem 12.8 with the following: $\Delta p_{(p)} = 1200$ lb/ft², average effective overburden pressure on the clay layer = 1000 lb/ft², $H_c = 15$ ft, $C_c = 0.3$, $e_o = 1.0$, $C_v = 1.5 \times 10^{-2}$ in²/min.

12.10   The diagram of a sand drain project is shown in Figures 12.28 and 12.29. Given: $r_w = 0.25$ m, $r_s = 0.35$ m, $d_e = 4.5$ m, $C_v = C_{vr} = 0.3$ m²/month, $k_h/k_s = 2$, $H = 9$ m. Determine:

   a.   The degree of consolidation for the clay layer caused only by the sand drains after six months of surcharge application

   b.   The degree of consolidation for the clay layer that is caused by the combination of vertical drainage (drained on top and bottom) and radial drainage after six months of the application of surcharge. Assume that the surcharge is applied instantaneously.

12.11   A 10-ft-thick clay layer is drained at the top and bottom. Its characteristics are $C_{vr} = C_v$ (for vertical drainage) = 0.042 ft²/day, $r_w = 8$ in., and $d_e = 6$ ft. Estimate the degree of consolidation of the clay layer caused by the combination of vertical and radial drainage at $t = 0.2$, 0.4, 0.8, and 1 yr. Assume that the surcharge is applied instantaneously, and there is no smear.

12.12   For a sand drain project (Figure 12.28), the following are given:

> Clay: Normally consolidated
> $H_c = 20$ ft (one-way drainage)
> $C_c = 0.28$
> $e_o = 0.9$
> $C_v = 0.21$ ft²/day
> Effective overburden pressure at the middle of clay layer = 2000 lb/ft²
> Sand drain: $r_w = 0.25$ ft
> $r_w = r_s$
> $d_e = 7.5$ ft
> $C_v = C_{vr}$

A surcharge is applied as shown in Figure P12.12. Calculate the degree of consolidation and the consolidation settlement 50 days after the beginning of the surcharge application.

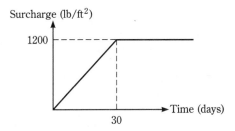

Surcharge (lb/ft^2)

1200

30

Time (days)

▼   **FIGURE P.12.12**

REFERENCES

Aboshi, H., Ichimoto, E., and Harada, K. (1979). "The Compozer—A Method to Improve Characteristics of Soft Clay by Inclusion of Large Diameter Sand Column," *Proceedings, International Conference on Soil Reinforcement, Reinforced Earth and Other Techniques,* Vol. 1, Paris, pp. 211–216.

Aboshi, H., and Monden, H. (1963). "Determination of the Horizontal Coefficient of Consolidation of an Alluvial Clay," *Proceedings,* Fourth Australia–New Zealand Conference on Soil Mechanics and Foundation Engineering, pp. 159–164.

American Society for Testing and Materials (1997). *Annual Book of Standards,* Vol. 04.08, West Conshohocken, Pennsylvania.

Arman, A., and Munfakh, G. A. (1972). "Lime Stabilization of Organic Soils," *Highway Research Record, No. 381,* National Academy of Sciences, pp. 37–45.

Bachus, R. C., and Barksdale, R. D. (1989). "Design Methodology for Foundations on Stone Columns," *Proceedings,* Foundation Engineering: Current Principles and Practices, American Society of Civil Engineers, Vol. 1, pp. 244–257.

Barron, R. A. (1948). "Consolidation of Fine-Grained Soils by Drain Wells," *Transactions,* American Society of Civil Engineers, Vol. 113, pp. 718–754.

Basore, C. E., and Boitano, J. D. (1969). "Sand Densification by Piles and Vibroflotation," *Journal of the Soil Mechanics and Foundations Division,* American Society of Civil Engineers, Vol. 95, No. SM6, pp. 1303–1323.

Blacklock, J. R., and Pengelly, A. D. (1988). "Soil Treatment for Foundations on Expansive Clay," *Special Topics in Foundations,* GSP No. 16 (ed. B. M. Das), American Society of Civil Engineers, pp. 73–92.

Brown, R. E. (1977). "Vibroflotation Compaction of Cohesionless Soils," *Journal of the Geotechnical Engineering Division,* American Society of Civil Engineers, Vol. 103, No. GT12, pp. 1437–1451.

D'Appolonia, D. J., Whitman, R. V., and D'Appolonia, E. (1969). "Sand Compaction with Vibratory Rollers," *Journal of the Soil Mechanics and Foundations Division,* American Society of Civil Engineers, Vol. 95, No. SM1, pp. 263–284.

Greenwood, D. A., and Thompson, G. H. (1984). *Ground Stabilization: Deep Compaction and Grouting,* ICE Works Construction Guides, Thomas Telford Ltd., London.

Hughes, J. M. O., and Withers, N. J. (1974). "Reinforcing of Soft Cohesive Soil with Stone Columns," *Ground Engineering,* Vol. 7, pp. 42–49.

Hughes, J. M. O., Withers, N. J., and Greenwood, D. A. (1975). "A Field Trial of Reinforcing Effects of Stone Columns in Soil," *Geotechnique,* Vol. 25, No. 1, pp. 31–34.

Ichimoto, A. (1981). "Construction and Design of Sand Compaction Piles," *Soil Improvement, General Civil Engineering Laboratory* (in Japanese), Vol. 5, pp. 37–45.

Johnson, S. J. (1970a). "Precompression for Improving Foundation Soils," *Journal of the Soil Mechanics and Foundations Division,* American Society of Civil Engineers, Vol. 96, No. SM1, pp. 114–144.

Johnson, S. J. (1970b). "Foundation Precompression with Vertical Sand Drains," *Journal of the Soil Mechanics and Foundations Division,* American Society of Civil Engineers, Vol. 96, No. SM1, pp. 145–175.

Johnson, A. W., and Sallberg, J. R. (1960). "Factors That Influence Field Compaction of Soils," *Bulletin No. 272,* Highway Research Board, National Academy of Sciences, Washington, D.C.

Lee, K. L., and Singh, A. (1971). "Relative Density and Relative Compaction," *Journal of the Soil Mechanics and Foundations Division,* American Society of Civil Engineers, Vol. 97, No. SM7, pp. 1049–1052.

Lee, P. Y., and Suedkamp, R. J. (1972). "Characteristics of Irregularly Shaped Compaction Curves of Soils," *Highway Research Record No. 381,* National Academy of Sciences, Washington, D.C., pp. 1–9.

Leonards, G. A., Cutter, W. A., and Holtz, R. D. (1980). "Dynamic Compaction of Granular Soils," *Journal of Geotechnical Engineering Division,* ASCE, Vol. 96, No. GT1, pp. 73–110.

Mitchell, J. K. (1970). "In-Place Treatment of Foundation Soils," *Journal of the Soil Mechanics and Foundations Division,* American Society of Civil Engineers, Vol. 96, No. SM1, pp. 73–110.

Mitchell, J. K., and Freitag, D. R. (1959). "A Review and Evaluation of Soil-Cement Pavements," *Journal of the Soil Mechanics and Foundations Division,* American Society of Civil Engineers, Vol. 85, No. SM6, pp. 49–73.

Mitchell, J. K., and Huber, T. R. (1985). "Performance of a Stone Column Foundation," *Journal of Geotechnical Engineering,* American Society of Civil Engineers, Vol. 111, No. GT2, pp. 205–223.

Murayama, S. (1962). "An Analysis of Vibro-Compozer Method on Cohesive Soils," *Construction in Mechanization* (in Japanese), No. 150, pp. 10–15.

Neubauer, C. H., Jr., and Thompson, M. R. (1972). "Stability Properties of Uncured Lime-Treated Fine-Grained Soils," *Highway Research Record No. 381,* National Academy of Sciences, pp. 20–26.

Olson, R. E. (1977). "Consolidation Under Time-Dependent Loading," *Journal of Geotechnical Engineering Division,* ASCE, Vol. 102, No. GT1, pp. 55–60.

Othman, M. A., and Luettich, S. M. (1994). "Compaction Control Criteria for Clay Hydraulic Barriers," *Transportation Research Record,* No. 1462, National Research Council, Washington, D.C. pp. 28–35.

Partos, A., Welsh, J. P., Kazaniwsky, P. W., and Sander, E. (1989). "Case Histories of Shallow Foundation on Improved Soil," *Proceedings,* Foundation Engineering: Current Principles and Practices, American Society of Civil Engineers, Vol. 1, pp. 313–327.

Poran, C. J., and Rodriguez, J. A. (1992). "Design of Dynamic Compaction," *Canadian Geotechnical Journal,* Vol. 2, No. 5, pp. 796–802.

Shin, E. C., Shin, B. W., and Das, B. M. (1992). "Site Improvement for a Steel Mill Complex," *Proceedings,* Specialty Conference on Grouting, Soil Improvement, and Geosynthetics, ASCE, Vol. 2, pp. 816–828.

APPENDIX A

CONVERSION FACTORS

A.1 CONVERSION FACTORS FROM ENGLISH TO SI UNITS

Length:	1 ft	= 0.3048 m
	1 ft	= 30.48 cm
	1 ft	= 304.8 mm
	1 in.	= 0.0254 m
	1 in.	= 2.54 cm
	1 in.	= 25.4 mm
Area:	1 ft^2	= 929.03 $\times$ 10^{-4} m^2
	1 ft^2	= 929.03 cm^2
	1 ft^2	= 929.03 $\times$ 10^2 mm^2
	1 in^2	= 6.452 $\times$ 10^{-4} m^2
	1 in^2	= 6.452 cm^2
	1 in^2	= 645.16 mm^2
Volume:	1 ft^3	= 28.317 $\times$ 10^{-3} m^3
	1 ft^3	= 28.317 cm^3
	1 in^3	= 16.387 $\times$ 10^{-6} m^3
	1 in^3	= 16.387 cm^3
Section modulus:	1 in^3	= 0.16387 $\times$ 10^5 mm^3
	1 in^3	= 0.16387 $\times$ 10^{-4} m^3
Hydraulic conductivity:	1 ft/min	= 0.3048 m/min
	1 ft/min	= 30.48 cm/min
	1 ft/min	= 304.8 mm/min
	1 ft/sec	= 0.3048 m/s
	1 ft/sec	= 304.8 mm/s
	1 in./min	= 0.0254 m/min
	1 in./min	= 2.54 cm/min
	1 in./min	= 25.4 mm/min

Coefficient of consolidation:	1 in²/sec	= 6.452 cm²/s
	1 in²/sec	= 20.346×10^3 m²/yr
	1 ft²/sec	= 929.03 cm²/s

Force:	1 lb	= 4.448 N
	1 lb	= 4.448×10^{-3} kN
	1 lb	= 0.4536 kgf
	1 kip	= 4.448 kN
	1 U.S. ton	= 8.896 kN
	1 lb	= 0.4536×10^{-3} metric ton
	1 lb/ft	= 14.593 N/m

Stress:	1 lb/ft²	= 47.88 N/m²
	1 lb/ft²	= 0.04788 kN/m²
	1 U.S. ton/ft²	= 95.76 kN/m²
	1 kip/ft²	= 47.88 kN/m²
	1 lb/in²	= 6.895 kN/m²

Unit weight:	1 lb/ft³	= 0.1572 kN/m³
	1 lb/in³	= 271.43 kN/m³

Moment:	1 lb-ft	= 1.3558 N · m
	1 lb-in.	= 0.11298 N · m

Energy:	1 ft-lb	= 1.3558 J

Moment of inertia:	1 in⁴	= 0.4162×10^6 mm⁴
	1 in⁴	= 0.4162×10^{-6} m⁴

A.2   CONVERSION FACTORS FROM SI TO ENGLISH UNITS

Length:	1 m	= 3.281 ft
	1 cm	= 3.281×10^{-2} ft
	1 mm	= 3.281×10^{-3} ft
	1 m	= 39.37 in.
	1 cm	= 0.3937 in.
	1 mm	= 0.03937 in.

Area:	1 m²	= 10.764 ft²
	1 cm²	= 10.764×10^{-4} ft²
	1 mm²	= 10.764×10^{-6} ft²
	1 m²	= 1550 in²
	1 cm²	= 0.155 in²
	1 mm²	= 0.155×10^{-2} in²

Volume:	1 m^3	$= 35.32$ ft^3
	1 cm^3	$= 35.32 \times 10^{-4}$ ft^3
	1 m^3	$= 61,023.4$ in^3
	1 cm^3	$= 0.061023$ in^3
Section modulus:	1 mm^3	$= 6.102 \times 10^{-5}$ in^3
	1 m^3	$= 6.102 \times 10^4$ in^3
Hydraulic conductivity:	1 m/min	$= 3.281$ ft/min
	1 cm/min	$= 0.03281$ ft/min
	1 mm/min	$= 0.003281$ ft/min
	1 m/s	$= 3.281$ ft/sec
	1 mm/s	$= 0.03281$ ft/sec
	1 m/min	$= 39.37$ in./min
	1 cm/min	$= 0.3937$ in./min
	1 mm/min	$= 0.03937$ in./min
Coefficient of consolidation:	1 cm^2/s	$= 0.155$ in^2/sec
	1 m^2/yr	$= 4.915 \times 10^{-5}$ in^2/sec
	1 cm^2/s	$= 1.0764 \times 10^{-3}$ ft^2/sec
Force:	1 N	$= 0.2248$ lb
	1 kN	$= 224.8$ lb
	1 kgf	$= 2.2046$ lb
	1 kN	$= 0.2248$ kip
	1 kN	$= 0.1124$ U.S. ton
	1 metric ton	$= 2204.6$ lb
	1 N/m	$= 0.0685$ lb/ft
Stress:	1 N/m^2	$= 20.885 \times 10^{-3}$ lb/ft^2
	1 kN/m^2	$= 20.885$ lb/ft^2
	1 kN/m^2	$= 0.01044$ U.S. ton/ft^2
	1 kN/m^2	$= 20.885 \times 10^{-3}$ kip/ft^2
	1 kN/m^2	$= 0.145$ lb/in^2
Unit weight:	1 kN/m^3	$= 6.361$ lb/ft^3
	1 kN/m^3	$= 0.003682$ lb/in^3
Moment:	1 N · m	$= 0.7375$ lb-ft
	1 N · m	$= 8.851$ lb-in.
Energy:	1 J	$= 0.7375$ ft-lb
Moment of inertia:	1 mm^4	$= 2.402 \times 10^{-6}$ in^4
	1 m^4	$= 2.402 \times 10^6$ in^4

APPENDIX B

BEARING CAPACITY OF SHALLOW FOUNDATIONS

▼ **TABLE B.1** Meyerhof's Bearing Capacity Factor, N_γ

$$N_\gamma = (N_q - 1)\tan(1.4\phi)$$

Eq. (3.26)

ϕ	N_γ	ϕ	N_γ
0	0.00	27	9.46
1	0.002	28	11.19
2	0.01	29	13.24
3	0.02	30	15.67
4	0.04	31	18.56
5	0.07	32	22.02
6	0.11	33	26.17
7	0.15	34	31.15
8	0.21	35	37.15
9	0.28	36	44.43
10	0.37	37	53.27
11	0.47	38	64.07
12	0.60	39	77.33
13	0.74	40	93.69
14	0.92	41	113.99
15	1.13	42	139.32
16	1.38	43	171.14
17	1.66	44	211.41
18	2.00	45	262.74
19	2.40	46	328.73
20	2.87	47	414.32
21	3.42	48	526.44
22	4.07	49	674.91
23	4.82	50	873.84
24	5.72	51	1143.93
25	6.77	52	1516.05
26	8.00	53	2037.26

▼   **TABLE B.2** Hansen's Bearing Capacity Factor, N_γ

$$N_\gamma = 1.5(N_q - 1)\tan\phi$$

Eq. (3.26)

ϕ (deg)	N_γ
0	0.00
1	0.00
2	0.01
3	0.02
4	0.05
5	0.07
6	0.11
7	0.16
8	0.22
9	0.30
10	0.39
11	0.50
12	0.63
13	0.78
14	0.97
15	1.18
16	1.43
17	1.73
18	2.08
19	2.48
20	2.95
21	3.50
22	4.13
23	4.88
24	5.75
25	6.76
26	7.94
27	9.32
28	10.94
29	12.84
30	15.07
31	17.69
32	20.79
33	24.44
34	28.77
35	33.92
36	40.05
37	47.38
38	56.17
39	66.75
40	79.54

(*continued*)

▼   **TABLE B.2**   (Continued)

ϕ (deg)	N_γ
41	95.05
42	113.95
43	137.10
44	165.58
45	200.81
46	244.64
47	299.52
48	368.66
49	456.40
50	568.56

▼   **TABLE B.3**   Lundgren and Mortensen's (1953) Bearing Capacity Factor, N_γ

These values of N_γ were obtained from the theory of plasticity using numerical methods.

ϕ (deg)	N_γ
0	0
5	0.17
10	0.46
15	1.4
25	6.92
30	15.32
35	35.19
40	86.46
45	215.0

▼ **TABLE B.4** Shape, Depth, and Inclination Factors Recommended in Other Texts and References

Factor	Relationship	Source
Shape[a]	*For* $\phi = 0$: $F_{cs} = 1 + 0.2\left(\dfrac{B}{L}\right)$ $F_{qs} = 1$ $F_{\gamma s} = 1$ *For* $\phi \geq 10°$: $F_{cs} = 1 + 0.2\left(\dfrac{B}{L}\right)\tan^2\left(45 + \dfrac{\phi}{2}\right)$ $F_{qs} = F_{\gamma s}$ $= 1 + 0.1\left(\dfrac{B}{L}\right)\tan^2\left(45 + \dfrac{\phi}{2}\right)$	Meyerhof (1963)
Depth	*For* $\phi = 0$: $F_{cd} = 1 + 0.2\left(\dfrac{D_f}{B}\right)$ $F_{qd} = F_{\gamma d} = 1$ *For* $\phi \geq 10°$: $F_{cd} = 1 + 0.2\left(\dfrac{D_f}{B}\right)\tan\left(45 + \dfrac{\phi}{2}\right)$ $F_{qd} = F_{\gamma d}$ $= 1 + 0.1\left(\dfrac{D_f}{B}\right)\tan\left(45 + \dfrac{\phi}{2}\right)$	Meyerhof (1963)
Inclination[a]	$F_{ci} = F_{qi} - \dfrac{(1 - F_{qi})}{(N_q - 1)}$ $F_{qi} = \left[1 - \dfrac{(0.5)(Q_u)\sin\beta}{Q_u\cos\beta + BLc\cot\phi}\right]^5$ $F_{\gamma i} = \left[1 - \dfrac{(0.7)(Q_u)\sin\beta}{Q_u\cos\beta + BLc\cot\phi}\right]^5$	Hansen (1970)

[a] L = length ($\geq B$)

APPENDIX C

SHEET PILE SECTIONS

▼ **TABLE C.1** Properties of Some Sheet Pile Sections (Produced by Bethlehem Steel Corporation)

Section designation	Sketch of section	Section modulus		Moment of inertia	
		in³/ft of wall	m³/m of wall	in⁴/ft of wall	m⁴/m of wall
PZ-40		60.7	326.4×10^{-5}	490.8	670.5×10^{-6}
PZ-35		48.5	260.5×10^{-5}	361.2	493.4×10^{-6}
PZ-27		30.2	162.3×10^{-5}	184.2	251.5×10^{-6}

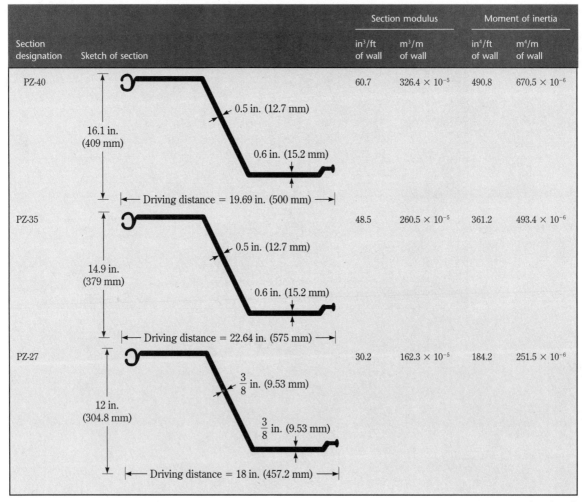

PZ-40
16.1 in. (409 mm)
0.5 in. (12.7 mm)
0.6 in. (15.2 mm)
Driving distance = 19.69 in. (500 mm)

PZ-35
14.9 in. (379 mm)
0.5 in. (12.7 mm)
0.6 in. (15.2 mm)
Driving distance = 22.64 in. (575 mm)

PZ-27
12 in. (304.8 mm)
$\frac{3}{8}$ in. (9.53 mm)
$\frac{3}{8}$ in. (9.53 mm)
Driving distance = 18 in. (457.2 mm)

(continued)

▼ **TABLE C.1** (Continued)

Section designation	Sketch of section	Section modulus		Moment of inertia	
		in³/ft of wall	m³/m of wall	in⁴/ft of wall	m⁴/m of wall
PZ-22		18.1	97×10^{-5}	84.4	115.2×10^{-6}
PSA-31		2.01	10.8×10^{-5}	3.23	4.41×10^{-6}
PSA-23		2.4	12.8×10^{-5}	4.13	5.63×10^{-6}

PZ-22 sketch labels: $\frac{3}{8}$ in. (9.53 mm); $\frac{3}{8}$ in. (9.53 mm); 9 in. (228.6 mm); Driving distance = 22 in. (558.8 mm)

PSA-31 sketch labels: $\frac{1}{2}$ in. (12.7 mm); Driving distance = 19.7 in. (500 mm)

PSA-23 sketch labels: $\frac{3}{8}$ in. (9.53 mm); Driving distance = 16 in. (406.4 mm)

APPENDIX D

PILE FOUNDATIONS

▼ **TABLE D.1a** Common H-Pile Sections Used in the United States (English Units)

Designation size (in.) × weight (lb/ft)	Depth d_1 (in.)	Section area (in²)	Flange and web thickness w (in.)	Flange width d_2 (in.)	Moment of inertia (in⁴)	
					I_{xx}	I_{yy}
HP 8 × 36	8.02	10.6	0.445	8.155	119	40.3
HP 10 × 57	9.99	16.8	0.565	10.225	294	101
× 42	9.70	12.4	0.420	10.075	210	71.7
HP 12 × 84	12.28	24.6	0.685	12.295	650	213
× 74	12.13	21.8	0.610	12.215	570	186
× 63	11.94	18.4	0.515	12.125	472	153
× 53	11.78	15.5	0.435	12.045	394	127
HP 13 × 100	13.15	29.4	0.766	13.21	886	294
× 87	12.95	25.5	0.665	13.11	755	250
× 73	12.74	21.6	0.565	13.01	630	207
× 60	12.54	17.5	0.460	12.90	503	165
HP 14 × 117	14.21	34.4	0.805	14.89	1220	443
× 102	14.01	30.0	0.705	14.78	1050	380
× 89	13.84	26.1	0.615	14.70	904	326
× 73	13.61	21.4	0.505	14.59	729	262

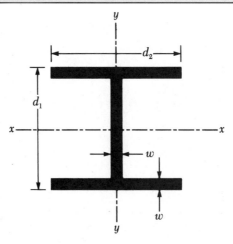

▼ **TABLE D.1b** Common H-Pile Sections Used in the United States (SI Units)

Designation, size (mm) × weight (kN/m)	Depth d_1 (mm)	Section area ($m^2 \times 10^{-3}$)	Flange and web thickness w (mm)	Flange width d_2 (mm)	Moment of inertia ($m^4 \times 10^{-6}$)	
					I_{xx}	I_{yy}
HP 200 × 0.52	204	6.84	11.3	207	49.4	16.8
HP 250 × 0.834	254	10.8	14.4	260	123	42
× 0.608	246	8.0	10.6	256	87.5	24
HP 310 × 1.226	312	15.9	17.5	312	271	89
× 1.079	308	14.1	15.49	310	237	77.5
× 0.912	303	11.9	13.1	308	197	63.7
× 0.775	299	10.0	11.05	306	164	62.9
HP 330 × 1.462	334	19.0	19.45	335	370	123
× 1.264	329	16.5	16.9	333	314	104
× 1.069	324	13.9	14.5	330	263	86
× 0.873	319	11.3	11.7	328	210	69
HP 360 × 1.707	361	22.2	20.45	378	508	184
× 1.491	356	19.4	17.91	376	437	158
× 1.295	351	16.8	15.62	373	374	136
× 1.060	346	13.8	12.82	371	303	109

▼ **TABLE D.2a** Selected Pipe Pile Sections (English Units)

Outside diameter (in.)	Wall thickness (in.)	Area of steel (in²)
$8\frac{5}{8}$	0.125	3.34
	0.188	4.98
	0.219	5.78
	0.312	8.17
10	0.188	5.81
	0.219	6.75
	0.250	7.66
12	0.188	6.96
	0.219	8.11
	0.250	9.25
16	0.188	9.34
	0.219	10.86
	0.250	12.37
18	0.219	12.23
	0.250	13.94
	0.312	17.34
20	0.219	13.62
	0.250	15.51
	0.312	19.30
24	0.250	18.7
	0.312	23.2
	0.375	27.8
	0.500	36.9

▼ **TABLE D.2b** Selected Pipe Pile Sections (SI Units)

Outside diameter (mm)	Wall thickness (mm)	Area of steel (cm²)
219	3.17	21.5
	4.78	32.1
	5.56	37.3
	7.92	52.7
254	4.78	37.5
	5.56	43.6
	6.35	49.4
305	4.78	44.9
	5.56	52.3
	6.35	59.7
406	4.78	60.3
	5.56	70.1
	6.35	79.8
457	5.56	80
	6.35	90
	7.92	112
508	5.56	88
	6.35	100
	7.92	125
610	6.35	121
	7.92	150
	9.53	179
	12.70	238

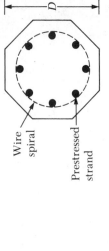

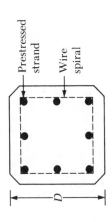

▼ **TABLE D.3a** Typical Prestressed Concrete Pile in Use (English Units)

Pile shape[a]	D (in.)	Area of cross section (in²)	Perimeter (in.)	Number of strands		Minimum effective prestress force (kip)	Section modulus (in³)	Design bearing capacity (kip)	
				$\frac{1}{2}$-in. diameter	$\frac{7}{16}$-in. diameter			Concrete strength	
								5000 psi	6000 psi
S	10	100	40	4	4	70	167	125	175
O	10	83	33	4	4	58	109	104	125
S	12	144	48	5	6	101	288	180	216
O	12	119	40	4	5	83	189	149	178
S	14	196	56	6	8	137	457	245	295
O	14	162	46	5	7	113	300	203	243
S	16	256	64	8	11	179	683	320	385
O	16	212	53	7	9	148	448	265	318
S	18	324	72	10	13	227	972	405	486
O	18	268	60	8	11	188	638	336	402
S	20	400	80	12	16	280	1333	500	600
O	20	331	66	10	14	234	876	414	503
S	22	484	88	15	20	339	1775	605	727
O	22	401	73	12	16	281	1166	502	602
S	24	576	96	18	23	403	2304	710	851
O	24	477	80	15	19	334	2123	596	716

[a] S = square section; O = octagonal section

▶ **TABLE D.3b** Typical Prestressed Concrete Pile in Use (SI Units)

Pile shape[a]	D (mm)	Area of cross section (cm²)	Perimeter (mm)	Number of strands		Minimum effective prestress force (kN)	Section modulus (m³ × 10⁻³)	Design bearing capacity (kN)	
				12.7-mm diameter	11.1-mm diameter			Concrete strength (MN/m²)	
								34.5	41.4
S	254	645	1016	4	4	312	2.737	556	778
O	254	536	838	4	4	258	1.786	462	555
S	305	929	1219	5	6	449	4.719	801	962
O	305	768	1016	4	5	369	3.097	662	795
S	356	1265	1422	6	8	610	7.489	1091	1310
O	356	1045	1168	5	7	503	4.916	901	1082
S	406	1652	1626	8	11	796	11.192	1425	1710
O	406	1368	1346	7	9	658	7.341	1180	1416
S	457	2090	1829	10	13	1010	15.928	1803	2163
O	457	1729	1524	8	11	836	10.455	1491	1790
S	508	2581	2032	12	16	1245	21.844	2226	2672
O	508	2136	1677	10	14	1032	14.355	1842	2239
S	559	3123	2235	15	20	1508	29.087	2694	3232
O	559	2587	1854	12	16	1250	19.107	2231	2678
S	610	3658	2438	18	23	1793	37.756	3155	3786
O	610	3078	2032	15	19	1486	34.794	2655	3186

[a] S = square section; O = octagonal section

▼ **TABLE D.4** Partial List of Typical Air and Steam Hammers

Maker of hammer[a]	Model no.	Type of hammer	Rated energy		Blows per minute	Ram weight	
			kip-ft	kN·m		kip	kN
V	3100	Single acting	300	406.8	58	100	448.8
V	540	Single acting	200	271.2	48	40.9	181.9
V	060	Single acting	180	244.1	62	60	266.9
MKT	OS-60	Single acting	180	244.1	55	60	266.9
V	040	Single acting	120	162.7	60	40	177.9
V	400C	Differential	113.5	153.9	100	40	177.9
R	8/0	Single acting	81.25	110.2	35	25	111.2
MKT	S-20	Single acting	60	81.4	60	20	89
R	5/0	Single acting	56.9	77.2	44	17.5	77.8
V	200-C	Differential	50.2	68.1	98	20	89
R	150-C	Differential	48.75	66.1	95–105	15	66.7
MKT	S-14	Single acting	37.5	50.9	60	14	62.3
V	140C	Differential	36	48.8	103	14	62.3
V	08	Single acting	26	35.3	50	8	35.6
MKT	S-8	Single acting	26	35.3	55	8	35.6
MKT	11B3	Double acting	19.2	26.1	95	5	22.2
MKT	C-5	Double acting	16.0	21.7	110	5	22.2
V	30-C	Double acting	7.3	9.9	133	3	13.3

[a] V—Vulcan Iron Works, Florida
MKT—McKiernan-Terry, New Jersey
R—Raymond International, Inc., Texas

▼ **TABLE D.5** Partial List of Typical Diesel Hammers

Maker of hammer[a]	Model no.	Rated energy		Blows per minute	Piston weight	
		kip-ft	kN·m		kN	kip
K	K150	280	379.7	45–60	147.2	33.1
M	MB70	141–63.4	191.2–86	38–60	70.5	15.84
K	K-60	105.6	143.2	42–60	58.7	13.2
K	K-45	91.1	123.5	39–60	44.0	9.9
M	M-43	84–37.8	113.9–51.3	40–60	42.1	9.46
K	K-35	70.8	96	39–60	34.3	7.7
MKT	DE70B	63–42	85.4–57	40–50	31.1	7.0
K	K-25	50.7	68.8	39–60	24.5	5.51
V	N-46	32.55	44.1	50–60	17.6	3.96
L	520	26.3	35.7	80–84	22.6	5.07
M	M-14S	26–11.88	35.3–16.1	42–60	13.2	2.97
V	N-33	24.6	33.4	50–60	13.3	3.0
L	440	18.2	24.7	86–90	17.8	4.0
MKT	DE20	18.0–12.0	24.4–16.3	40–50	8.9	2.0
MKT	DE-10	8.8	11.9	40–50	4.9	1.1
L	180	8.1	11.0	90–95	7.7	1.73

[a] V—Vulcan Iron Works, Florida
M—Mitsubishi International Corporation
MKT—McKiernan-Terry, New Jersey
L—Link Belt, Cedar Rapids, Iowa
K—Kobe Diesel

TABLE D.6 Bearing Capacity Factors (N_c^*) and (N_σ^*) Based on the Theory of Expansion of Cavities

ϕ	I_{rr}									
	10	20	40	60	80	100	200	300	400	500
0	6.97	7.90	8.82	9.36	9.75	10.04	10.97	11.51	11.89	12.19
	1.00	1.00	1.00	1.00	1.00	1.00	1.00	1.00	1.00	1.00
1	7.34	8.37	9.42	10.04	10.49	10.83	11.92	12.57	13.03	13.39
	1.13	1.15	1.16	1.18	1.18	1.19	1.21	1.22	1.23	1.23
2	7.72	8.87	10.06	10.77	11.28	11.69	12.96	13.73	14.28	14.71
	1.27	1.31	1.35	1.38	1.39	1.41	1.45	1.48	1.50	1.51
3	8.12	9.40	10.74	11.55	12.14	12.61	14.10	15.00	15.66	16.18
	1.43	1.49	1.56	1.61	1.64	1.66	1.74	1.79	1.82	1.85
4	8.54	9.96	11.47	12.40	13.07	13.61	15.34	16.40	17.18	17.80
	1.60	1.70	1.80	1.87	1.91	1.95	2.07	2.15	2.20	2.24
5	8.99	10.56	12.25	13.30	14.07	14.69	16.69	17.94	18.86	19.59
	1.79	1.92	2.07	2.16	2.23	2.28	2.46	2.57	2.65	2.71
6	9.45	11.19	13.08	14.26	15.14	15.85	18.17	19.62	20.70	21.56
	1.99	2.18	2.37	2.50	2.59	2.67	2.91	3.06	3.18	3.27
7	9.94	11.85	13.96	15.30	16.30	17.10	19.77	21.46	22.71	23.73
	2.22	2.46	2.71	2.88	3.00	3.10	3.43	3.63	3.79	3.91
8	10.45	12.55	14.90	16.41	17.54	18.45	21.51	23.46	24.93	26.11
	2.47	2.76	3.09	3.31	3.46	3.59	4.02	4.30	4.50	4.67
9	10.99	13.29	15.91	17.59	18.87	19.90	23.39	25.64	27.35	28.73
	2.74	3.11	3.52	3.79	3.99	4.15	4.70	5.06	5.33	5.55
10	11.55	14.08	16.97	18.86	20.29	21.46	25.43	28.02	29.99	31.59
	3.04	3.48	3.99	4.32	4.58	4.78	5.48	5.94	6.29	6.57
11	12.14	14.90	18.10	20.20	21.81	23.13	27.64	30.61	32.87	34.73
	3.36	3.90	4.52	4.93	5.24	5.50	6.37	6.95	7.39	7.75
12	12.76	15.77	19.30	21.64	23.44	24.92	30.03	33.41	36.02	38.16
	3.71	4.35	5.10	5.60	5.98	6.30	7.38	8.10	8.66	9.11
13	13.41	16.69	20.57	23.17	25.18	26.84	32.60	36.46	39.44	41.89
	4.09	4.85	5.75	6.35	6.81	7.20	8.53	9.42	10.10	10.67
14	14.08	17.65	21.92	24.80	27.04	28.89	35.38	39.75	43.15	45.96
	4.51	5.40	6.47	7.18	7.74	8.20	9.82	10.91	11.76	12.46
15	14.79	18.66	23.35	26.53	29.02	31.08	38.37	43.32	47.18	50.39
	4.96	6.00	7.26	8.11	8.78	9.33	11.28	12.61	13.64	14.50
16	15.53	19.73	24.86	28.37	31.13	33.43	41.58	47.17	51.55	55.20
	5.45	6.66	8.13	9.14	9.93	10.58	12.92	14.53	15.78	16.83

(continued)

I_r

ϕ	10	20	40	60	80	100	200	300	400	500
17	16.30	20.85	26.46	30.33	33.37	35.92	45.04	51.32	56.27	60.42
	5.98	7.37	9.09	10.27	11.20	11.98	14.77	16.99	18.20	19.47
18	17.11	22.03	28.15	32.40	35.76	38.59	48.74	55.80	61.38	66.07
	6.56	8.16	10.15	11.53	12.62	13.54	16.84	19.13	20.94	22.47
19	17.95	23.26	29.93	34.59	38.30	41.42	52.71	60.61	66.89	72.18
	7.18	9.01	11.31	12.91	14.19	15.26	19.15	21.87	24.03	25.85
20	18.83	24.56	31.81	36.92	40.99	44.43	56.97	65.79	72.82	78.78
	7.85	9.94	12.58	14.44	15.92	17.17	21.73	24.94	27.51	29.67
21	19.75	25.92	33.80	39.38	43.85	47.64	61.51	71.34	79.22	85.90
	8.58	10.95	13.97	16.12	17.83	19.29	24.61	28.39	31.41	33.97
22	20.71	27.35	35.89	41.98	46.88	51.04	66.37	77.30	86.09	93.57
	9.37	12.05	15.50	17.96	19.94	21.62	27.82	32.23	35.78	38.81
23	21.71	28.84	38.09	44.73	50.08	54.66	71.56	83.68	93.47	101.83
	10.21	13.24	17.17	19.99	22.26	24.20	31.37	36.52	40.68	44.22
24	22.75	30.41	40.41	47.63	53.48	58.49	77.09	90.51	101.39	110.70
	11.13	14.54	18.99	22.21	24.81	27.04	35.32	41.30	46.14	50.29
25	23.84	32.05	42.85	50.69	57.07	62.54	82.98	97.81	109.88	120.23
	12.12	15.95	20.98	24.64	27.61	30.16	39.70	46.61	52.24	57.06
26	24.98	33.77	45.42	53.93	60.87	66.84	89.25	105.61	118.96	130.44
	13.18	17.47	23.15	27.30	30.69	33.60	44.53	52.51	59.02	64.62
27	26.16	35.57	48.13	57.34	64.88	71.39	95.02	113.92	128.67	141.39
	14.33	19.12	25.52	30.21	34.06	37.37	49.88	59.05	66.56	73.04
28	27.40	37.45	50.96	60.93	69.12	76.20	103.01	122.79	139.04	153.10
	15.57	20.91	28.10	33.40	37.75	41.51	55.77	66.29	74.93	82.40
29	28.69	39.42	53.95	64.71	73.58	81.28	110.54	132.23	150.11	165.61
	16.90	22.85	30.90	36.87	41.79	46.05	62.27	74.30	84.21	92.80
30	30.03	41.49	57.08	68.69	78.30	86.64	118.53	142.27	161.91	178.98
	18.24	24.95	33.95	40.66	46.21	51.02	69.43	83.14	94.48	104.33
31	31.43	43.64	60.37	72.88	83.27	92.31	126.99	152.95	174.49	193.23
	19.88	27.22	37.27	44.79	51.03	56.46	77.31	92.90	105.84	117.11
32	32.89	45.90	63.82	77.29	88.50	98.28	135.96	164.29	187.87	208.43
	21.55	29.68	40.88	49.30	56.30	62.41	85.96	103.66	118.39	131.24
33	34.41	48.26	67.44	81.92	94.01	104.58	145.46	176.33	202.09	224.62
	23.34	32.34	44.80	54.20	62.05	68.92	95.46	115.51	132.24	146.87

(continued)

I_{rr}

φ	10	20	40	60	80	100	200	300	400	500
34	35.99	50.72	71.24	86.80	99.82	111.22	155.51	189.11	217.21	241.84
	25.28	35.21	49.05	59.54	68.33	76.02	105.90	128.55	147.51	164.12
35	37.65	53.30	75.22	91.91	105.92	118.22	166.14	202.64	233.27	260.15
	27.36	38.32	53.67	65.36	75.17	83.78	117.33	142.89	164.33	183.16
36	39.37	55.99	79.39	97.29	112.34	125.59	177.38	216.98	250.30	279.60
	29.60	41.68	58.68	71.69	82.62	92.24	129.87	158.65	182.85	204.14
37	41.17	58.81	83.77	102.94	119.10	133.34	189.25	232.17	268.36	300.26
	32.02	45.31	64.13	78.57	90.75	101.48	143.61	175.95	203.23	227.26
38	43.04	61.75	88.36	108.86	126.20	141.50	201.78	248.23	287.50	322.17
	34.63	49.24	70.03	86.05	99.60	111.56	158.65	194.94	225.62	252.71
39	44.99	64.83	93.17	115.09	133.66	150.09	215.01	265.23	307.78	345.41
	37.44	53.50	76.45	94.20	109.24	122.54	175.11	215.78	250.23	280.71
40	47.03	68.04	98.21	121.62	141.51	159.13	228.97	283.19	329.24	370.04
	40.47	58.10	83.40	103.05	119.74	134.52	193.13	238.62	277.26	311.50
41	49.16	71.41	103.49	128.48	149.75	168.63	243.69	302.17	351.95	396.12
	43.74	63.07	90.96	112.68	131.18	147.59	212.84	263.67	306.94	345.34
42	51.38	74.92	109.02	135.68	158.41	178.62	259.22	322.22	375.97	423.74
	47.27	68.46	99.16	123.16	143.64	161.83	234.40	291.13	339.52	382.53
43	53.70	78.60	114.82	143.23	167.51	189.13	275.59	343.40	401.36	452.96
	51.08	74.30	108.08	134.56	157.21	177.36	257.99	321.22	375.28	423.39
44	56.13	82.45	120.91	151.16	177.07	200.17	292.85	365.75	428.21	483.88
	55.20	80.62	117.76	146.97	172.00	194.31	283.80	354.20	414.51	468.28
45	58.66	86.48	127.28	159.48	187.12	211.79	311.04	389.35	456.57	516.58
	59.66	87.48	128.28	160.48	188.12	212.79	312.03	390.35	457.57	517.58
46	61.30	90.70	133.97	168.22	197.67	224.00	330.20	414.26	486.54	551.16
	64.48	94.92	139.73	175.20	205.70	232.96	342.94	429.98	504.82	571.74
47	64.07	95.12	140.99	177.40	208.77	236.85	350.41	440.54	518.20	587.72
	69.71	103.00	152.19	191.24	224.88	254.99	376.77	473.42	556.70	631.25
48	66.97	99.75	148.35	187.04	220.43	250.36	371.70	468.28	551.64	626.36
	75.38	111.78	165.76	208.73	245.81	279.06	413.82	521.08	613.65	696.64
49	70.01	104.60	156.09	197.17	232.70	264.58	394.15	497.56	586.96	667.21
	81.54	121.33	180.56	227.82	268.69	305.37	454.42	573.38	676.22	768.53
50	73.19	109.70	164.21	207.83	245.60	279.55	417.82	528.46	624.28	710.39
	88.23	131.73	196.70	248.68	293.70	334.15	498.94	630.80	744.99	847.61

From "Design of Pile Foundations," by A. S. Vesic, in NCHRP *Synthesis of Highway Practice 42*, Transportation Research Board, 1977. Reprinted by permission. *Note:* Upper number N_c^*, lower number N_σ^*.

APPENDIX E

DESIGN CURVES FOR PREFABRICATED VERTICAL DRAINS (PVDs)

The relationships for the average degree of consolidation due to radial drainage into sand drains were given in Eqs. (12.15) through (12.20). These are for equal strain cases. Yeung (1997) used these relationships to develop design curves for PVDs. The theoretical developments used by Yeung are given below.

Figure E.1 shows the layout of a square-grid pattern of prefabricated vertical drains (also see Figure 12.34 for the definition of a and b). The equivalent diameter of a PVD can be given as

$$d_w = \frac{2(a + b)}{\pi} \tag{E.1}$$

Now, Eq. (12.15) can be rewritten as

$$U_r = 1 - \exp\left(-\frac{8C_{vr}t}{d_w^2} \frac{d_w^2}{d_e^2 m}\right) = 1 - \exp\left(-\frac{8T_r'}{\alpha'}\right) \tag{E.2}$$

where d_e = diameter of the effective zone of drainage = $2r_e$

$$T_r' = \frac{C_{vr}t}{d_w^2} \tag{E.3}$$

$$\alpha' = n^2 m = \frac{n^4}{n^2 - S^2} \ln\left(\frac{n}{S}\right) - \left(\frac{3n^2 - S^2}{4}\right) + \frac{k_h}{k_s}(n^2 - S^2)\ln S \tag{E.4}$$

$$n = \frac{d_e}{d_w} \tag{E.5}$$

From Eq. (E.2),

$$T_r' = -\frac{\alpha'}{8} \ln(1 - U_r)$$

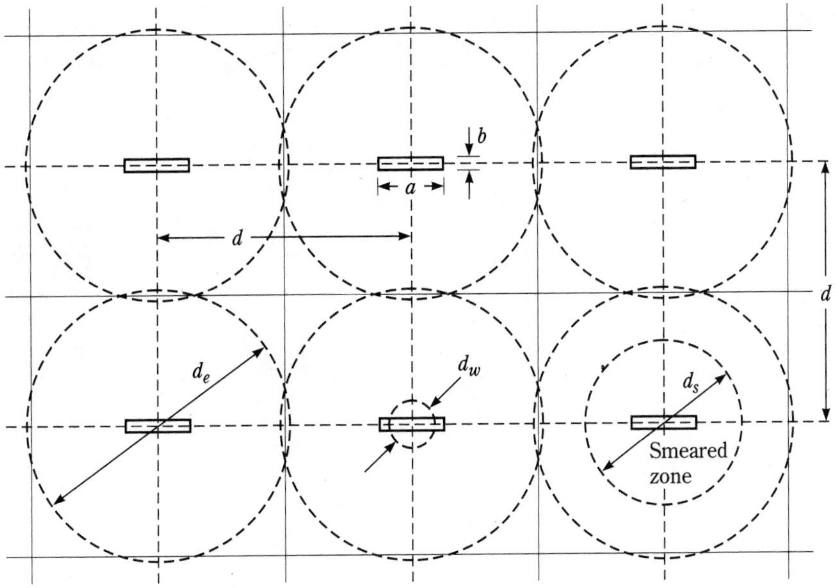

▼ **FIGURE E.1**   Layout of square-grid pattern of prefabricated vertical drains (after Yeung, 1997)

or

$$(T'_r)_1 = \frac{T'_r}{\alpha'} = -\frac{\ln(1 - U_r)}{8} \tag{E.6}$$

Figure E.2 shows the plot of U_r versus $(T'_r)_1$. Also, Figure E.3 shows the plot of n versus α from Eq. (E.4).

Following is a step-by-step procedure for the design of prefabricated vertical drains:

1. Determine time t_2 available for the consolidation process.
2. Determine U_v at time t_2 due to vertical drainage [Eq. (12.14)]. Thus, from Eq. (12.14),

$$U_r = \frac{1 - U_{v,r}}{1 - U_v} \tag{E.7}$$

3. For the PVD to be used, calculate d_w from Eq. (E.1).
4. Determine $(T'_r)_1$ from Eqs. (E.6) and (E.7).
5. Determine T'_r from Eq. (E.3).
6. Determine $\alpha' = \dfrac{T'_r}{(T'_r)_1}$.

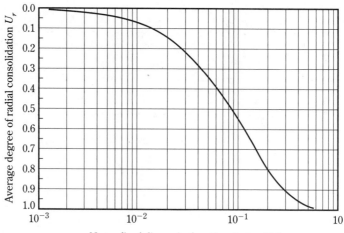

▼ **FIGURE E.2**   Plot of U_r vs. $(T_r')_1$ (after Yeung, 1977)

7. Using Figure E.3 and α', determined from Step 6, determine n.
8. From Eq. (E.5),

$$d_e = \underset{\underset{\text{Step 7}}{\uparrow}}{n} \; \underset{\underset{\text{Step 3}}{\uparrow}}{d_w}$$

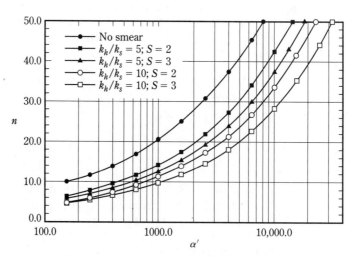

▼ **FIGURE E.3**   Relationships between α' and n (after Yeung, 1997)

9. Choose the drain spacing:

$$d = \frac{d_e}{1.05} \quad \text{(for triangular pattern)}$$

$$d = \frac{d_e}{1.128} \quad \text{(for square pattern)}$$

REFERENCES

Yeung, A. T. (1997). "Design Curves for Prefabricated Vertical Drains," *Journal of Geotechnical and Geoenvironmental Engineering*, Vol. 123, No. 8, pp. 755–759.

ANSWERS TO SELECTED PROBLEMS

Chapter 1

1.1
  a. 0.76
  b. 0.43
  c. 14.93 kN/m³
  d. 17.17 kN/m³
  e. 53%

1.3
  a. 0.45
  b. 69.5%
  c. 17.58 kN/m³
  d. 14.53 kN/m³

1.5
  a. 129.2 lb/ft³
  b. 7.2 lb/ft³
  c. 124.6 lb/ft³

1.7

Soil	Classification
A	A-7-6(9)
B	A-6(5)
C	A-3(0)
D	A-4(0)
E	A-2-6(1)
F	A-7-6(19)

1.9 0.117 cm/sec
1.11 1.9×10^{-6} cm/sec
1.13 -161.2 lb/ft²
1.15
  a. 379 kN/m²
  b. 549 kN/m²

1.17 56 mm
1.19
  a. 0.377
  b. 0.736
1.21 39.06 days
1.23 38°
1.25 387.8 kN/m²
1.27 $c_{cu} = 0$
  $\phi_{cu} = 25°$
  $c = 0$
  $\phi = 34°$
1.29 14.7 kN/m²

Chapter 2

2.1
  a. 13.78%
  b. 1.907 in.
2.3 50.4 kN/m²
2.5 30°
2.7 ϕ (av) = 35°
2.9 16.8 ft
2.11
  a. 51.4 kN/m²
  b. 39.7 kN/m²
  c. 40.6 kN/m²
2.13 4.27
2.15
  a. 30 kN/m²
  b. 1.84

2.17 **a.** 0.65
        b. 1.37
        c. 2131 kN/m^2

2.19 0.00448 ft/min

2.21 Z_1 = 2.6 m
        Z_2 = 2.56 m
        v_1 = 492 m/sec
        v_2 = 1390 m/sec
        v_3 = 3390 m/sec

Chapter 3

3.1 **a.** 5195 lb/ft^2
        b. 372.8 kN/m^2
        c. 280 kN/m^2

3.3 **a.** 5879 lb/ft^2
        b. 372.8 kN/m^2
        c. 368.8 kN/m^2

3.5 3721 kN

3.7 **a.** 40°
        b. 4946.3 kip

3.9 707.3 kN

3.11 6.75 ft

3.13 30.1 kip

3.15 495.2 kN

3.17 **a.** 145.3 kN/m^2

b.

b (m)	q_u (kN/m^2)
0	279.5
1	374
2	435.9
3	476
4	476
5	476
6	476

3.19 700.5 kN/m^2

Chapter 4

4.1 1191 lb/ft^2

4.3 619 lb/ft^2

4.5

z (ft)	Δp (lb/ft^2)
0	2500
5	2125
10	1375
15	875
20	600

4.7 54.85 kN/m^2

4.9 22 kN/m^2

4.11 60.7 mm

4.13 0.54 in.

4.15 246 mm

4.17 13.73 mm

4.19 54.6 mm

4.21 **a.**

Depth (ft)	N_{cor}
5	17
10	13
15	13
20	8
25	12

b. 4.67 kip/ft^2

4.23

z (m)	X_o (m)
0.4	0.59
0.8	0.72
1.2	0.95
1.6	1.26
2.0	1.45

4.25

z (m)	$T_{(N)}$ (kN/m)
0.4	62.3
0.8	70.5
1.2	76.4
1.6	78.6
2.0	79.1

4.27 a.

Layer	t (in.)
1	0.022
2	0.024
3	0.0242
4	1.0253
5	0.026

Use $t = 0.026$ for all layers.

b.

Layer	$2L_o$ (ft)
1	14
2	16
3	21.2
4	25.6
5	28.8

Chapter 5

5.1 a. 771 kN/m²
b. 16,321 lb/ft²
5.3 186.7 kN/m²
5.5 3.39 m
5.7 0.193

5.9

Point	q (kN/m²)
A	36.81
B	31.86
C	26.91
D	25.19
E	30.14
F	35.09

5.11 14.9 lb/in³
5.13 18 kN/m³

Chapter 6

6.1 $P_o = 3888$ lb/ft
$\bar{z} = 4$ ft
6.3 $P_o = 4839.6$ lb/ft
$\bar{z} = 3.65$ ft
6.5 $P_o = 143.2$ kN/m
$\bar{z} = 1.5$ m

6.7 45.64 kN/m
6.9 $P_a = 118.6$ kN/m
$\bar{z} = 1.67$ m
6.11 5598 lb/ft
6.13 $P_{ae} = 103.3$ kN/m
$\bar{z} = 1.87$ m

6.15

n_a	P_a (kN/m)
0.3	47.6
0.4	53.2
0.5	59.4

6.17 a. $P_p = 37,440$ lb/ft
b. $\bar{z} = 7.44$ ft
6.19 71.21 kip/ft

Chapter 7

Problem	$FS_{(overturning)}$	$FS_{(sliding)}$	$FS_{(bearing)}$
7.1	3.41	1.5	5.48
7.3	3.82	1.66	3.78
7.5	—	1.35	—
7.7	6.2	2.35	—

7.9

z (m)	σ_v (kN/m²)
1	46.5
2	54.4
3	66.1
4	79.6
5	95.0
6	110.5

7.11 a. 0.201 in.
b. 43.52 ft
7.13 a. 0.161 in.
b. 40.33 ft
7.15 a. 0.1 in.
b. 29.75 ft
7.17 $FS_{(overturning)} = 3.43$
$FS_{(sliding)} = 1.35$
$FS_{(bearing)} = 9.79$

Chapter 8

8.1 **a.** 23.52 ft
  b. 53.6 ft
  c. 100.6 kip-ft/ft

8.3 **a.** 10.43 m
  b. 22.56 m
  c. 1480.9 kN-m/m

8.5 $D = 3.18$ m
  $M_{max} = 59.8$ kN-m/m

8.7 **a.** 7 m
  b. 16.8 m
  c. 367.04 kN-m/m

8.9 $D = 1.1$ m
  $M_{max} = 32.18$ kN-m/m

8.11 **a.** 940 kN-m/m
  b. *PZ-35*

8.13 *PZ-27*

8.15 **a.** 2.47 m
  b. 116 kN/m
  c. 406.9 kN-m/m

8.17 **a.** 1.15 ft
  b. 3319 lb/ft

8.19 22.61 kip

8.21 $u = 65$ m

B (m)	P (kN)
0.3	15.4
0.6	25.5
0.9	34.1

8.23

Level	Strut load (kN)
A	131.4
B	69.3
C	178.8

8.25

Level	Strut load (kN)
A	148.5
B	78.4
C	202

8.27 **a.** $c_{av} = 19.53$ kN/m^2
  $\gamma_{av} = 17.94$ kN/m^3
  b. 65.4 kN/m^3

8.29

Level	Strut load (kip)
A	78.19
B	93.29
C	43.88

8.31

Level	Strut load (kip)
A	78.19
B	89.56
C	83.53

8.33 1.78

Chapter 9

9.1 **a.** 336 kN
  b. 2362 kN
  c. 1496 kN

9.3 605 kN

9.5 333 kip

9.7 514 kN

9.9 108 kip

9.11 110.6 kip

9.13 10.42 mm

9.15 116 kip

9.17 721.8 kN

9.19 32.5 kN

9.21 1924 kip

9.23 25.3 kN

9.25 56.4 kip

9.27 **a.** 87.96%
  b. 76.06%

9.29 364.7 kip

9.31 217.7 mm

Chapter 10

10.1 744.9 kip

10.3 57 kip

10.5 1933.8 kN

10.7 **a.** 1177 kN
  b. 1205.8 kN
  c. 794.3 kN

10.9 **a.** 173.18 kip
         b. 219.9 kip
         c. 131 kip
10.11 **a.** 0.5 m
         b. 1.5 m
10.13 **a.** 942 kip
         b. 579.8 kip
10.15 1.34 in.
10.17 386 kip
10.19 150 kip

Chapter 11

11.1

LL	γ_d (kN/m³) below which collapse will occur
20	17.36
25	15.95
30	14.75
35	13.72
40	12.82

11.3 1.19 in.
11.5 1.73 in.
11.7 63.5 mm
11.9 12.57 ft

Chapter 12

12.1

w (%)	γ_{zav} (kN/m³)			
	$G_s = 2.6$	$G_s = 2.65$	$G_s = 2.7$	$G_s = 2.75$
5	22.57	22.95	23.34	23.72
10	20.24	20.55	20.86	21.16
15	18.35	18.60	18.85	19.1
20	16.78	16.99	17.20	17.4

12.3 99.9 lb/ft³
12.5 *B*
12.7 3.15—excellent
12.9 **a.** 0.77 ft
         b. 10.6 months
         c. 252 lb/ft²

12.11

Time (year)	$U_{v,r}$
0.2	0.72
0.4	0.91
0.8	0.99
1	0.997

INDEX